THE MOLECULAR AND CELLULAR BIOLOGY OF THE YEAST *SACCHAROMYCES*

GENOME DYNAMICS, PROTEIN SYNTHESIS, AND ENERGETICS

COLD SPRING HARBOR
MONOGRAPH SERIES

THE MOLECULAR AND CELLULAR BIOLOGY OF THE YEAST *SACCHAROMYCES*

GENOME DYNAMICS, PROTEIN SYNTHESIS, AND ENERGETICS

Edited by

James R. Broach
Princeton University

John R. Pringle
University of Michigan

Elizabeth W. Jones
Carnegie-Mellon University

Cold Spring Harbor Laboratory Press
1991

THE MOLECULAR AND CELLULAR BIOLOGY OF THE YEAST *SACCHAROMYCES*
GENOME DYNAMICS, PROTEIN SYNTHESIS, AND ENERGETICS
Monograph 21
Copyright 1991 by Cold Spring Harbor Laboratory Press
All rights reserved
Printed in the United States of America
Book design by Emily Harste

Library of Congress Cataloging-in-Publication Data

The Molecular and Cellular Biology of the Yeast Saccharomyces / edited by
    James R. Broach, John Pringle, Elizabeth Jones.
    p. cm. -- (Cold Spring Harbor monograph series; 21)
    Includes bibliographical references and index.
    ISBN 0-87969-355-X (cloth)
        0-87969-306-0 (pbk.)
    1. Saccharomyces--Cytology   2. Fungal Molecular Biology
I. Broach, James R.   II. Pringle, John, 1943- .   III. Jones, Elizabeth, 1939- .
    IV. Series.
QK623.S23M6   1991   90-43967
589.2'33048—dc20   CIP

All Cold Spring Harbor Laboratory Press publications may be ordered directly from Cold
Spring Harbor Laboratory Press, 10 Skyline Drive, Plainview, New York 11803. (Phone: 1-
800-843-4388 in Continental U.S. and Canada.) All other locations: (516) 349-1930. FAX:
(516) 349-1946.

Contents

Preface

Ten years ago, two of us (J.R.B. and E.W.J.) along with Jeff Strathern presided over the production of the two-volume monograph *The Molecular Biology of the Yeast* Saccharomyces, an opus conceived during the heady days following the development of yeast transformation. Not accidently, this monograph coincided with a transition in the use of yeast as an experimental organism. Yeast transformation offered the prospect of executing precise genetic surgery on the organism, raising us from the role of observers to that of creators. All of us in the yeast field anticipated that this technical breakthrough, coupled with the elegant genetics and ease of manipulation of yeast, presaged an explosive growth in yeast molecular biology. At that time, we perceived the need for a resource that would not only provide a comprehensive review of the biology of *Saccharomyces*, but also highlight the issues and topics to which this new technology was likely to be applied. From the comments we received on the monograph, we appear to have been successful in achieving this goal.

Ten years later, we find that the explosive growth in yeast molecular biology has exceeded what even the most ardent proselytizers of the art would have dreamed. In part, this success followed from our monopoly on performing reverse genetics on a eukaryote organism, an advantage that was anticipated ten years ago. However, what was not anticipated was the extent to which basic biological processes have been obsessively conserved throughout the eukaryotic world. Each fundamental process—cell cycle progression and its control, protein secretion and targeting, transcription and its regulation, mRNA processing, DNA replication, cell-cell signaling, and biogenesis of cytoskeletal architecture—is accomplished in all eukaryotes by essentially identical cellular machinery composed of essentially identical protein components. This conservation of function has catapulted the yeasts *Saccharomyces cerevisiae* and *Schizosaccharomyces pombe* from parochial backwaters to the forefront of experimental molecular biology. Most experimental biologists now fully appreciate that what is true for a yeast is true for an elephant, but one can get the answer a lot faster from a yeast.

The unfettered growth of yeast molecular biology during the last ten years has highlighted the need for an updated version of *The Molecular Biology of the Yeast* Saccharomyces. We hope that this volume, in conjunction with the forthcoming two volumes, fulfills that need. Our goal in producing this monograph series is to address the needs of three groups of investigators. For those of us in the field, we attempted to provide a comprehensive compendium of the advances in molecular biology that have occurred over the last ten years. These volumes are designed to fill in gaps in our knowledge and to serve as an updated reference source to resolve those minor factual questions that arise almost daily. In addition, we sought to provide those investigators new to the field a reference source that would rapidly bring them up to speed in all aspects of yeast molecular biology. In this context, these volumes should be viewed as supplements to, rather than replacement for, the previous two-volume series. The previous monograph is still a unique and valuable resource; with the publication of these volumes, we simply have closed the gap between the state of the field ten years ago and where it stands now. Finally, we hope that these volumes serve as a source of background and context for those investigators outside the field who feel that advances in the molecular biology of yeasts are germane to their own research

Our survey of yeast molecular and cellular biology is presented in three volumes (this 50% increase over the space the field required ten years ago reflects the growth of a healthy enterprise). This volume focuses on the genome organization of the yeast *Saccharomyces* as well as energy metabolism and protein translation. The second volume emphasizes gene expression, detailing transcriptional mechanisms, RNA processing, and the various regulatory circuits governing yeast. The third volume addresses the rapidly expanding field of yeast cell biology, emphasizing organelle biogenesis as well as the cell cycle and its regulation.

We have included three chapters in this series that review certain aspects of the molecular biology of *S. pombe*: mitosis, the control of the cell cycle, and mating-type regulation. We did not intend to provide a comprehensive review of *S. pombe* as an experimental system. Rather, we sought to include those results and observations gleaned from *S. pombe* that have served to inform and elaborate related endeavors using *S. cerevisiae* as the experimental system.

Our perception in participating in the practice of yeast molecular biology and in compiling this monograph is that the enormous potential of *Saccharomyces* as an experimental system evident a decade ago has been realized beyond our expectations. We now approach a new stage, as the yeast field reaches the maturity of middle age. We hope that these volumes help us to take stock of what we have accomplished and to appreciate the questions still to be resolved. Whatever success we have in achieving this goal can be attributed to the outstanding efforts of the authors of these chapters, to whom we are deeply indebted. Their chapters document that despite the emerging maturity of yeast molecular biology, the excitement rampant ten years ago still permeates the field today.

As editors, we have taken the liberty of essaying an experiment in protein nomenclature in these volumes. We have adopted the protocol first used in the yeast secretion field of designating the protein product of a gene as the gene name in Roman typeface, with only the initial letter capitalized and followed by a "p," i.e., the product of the *SEC4* gene is Sec4p. We have rigorously applied this rule throughout these volumes as a means of fully examining the usefulness of this system. This unrelenting application has led to some awkward constructions, such as referring to the *ARS*-binding factor 1 as Abflp. However, we felt that uniformity of treatment took precedence over other considerations, at least for these volumes. We are not by any means advocating a unilateral imposition of a new nomenclature on the yeast field. Rather, we felt that the feasibility of this system was worth exploring in depth.

We wish to thank Nancy Ford, Dorothy Brown, and Joan Ebert for their roles in editing the manuscripts and for making this publication a reality. We also thank our many colleagues who have contributed to these volumes and to whom all credit for a successful monograph belongs.

J.R. Broach
J.R. Pringle
E.W. Jones

1

Genome Structure and Organization in *Saccharomyces cerevisiae*

Maynard V. Olson
Department of Genetics and
Howard Hughes Medical Institute
Washington University School of Medicine
St. Louis, Missouri 63110

I. INTRODUCTION

A relatively simple picture is emerging of the structure of the yeast genome. With the important exception of mating-type interconversion, the genome seems to be quite stable: Other DNA rearrangements appear to be infrequent and of peripheral significance to the yeast life cycle. Genes are tightly packed, with little "spacer" DNA and strikingly few pseudogenes. There is no extensive clustering of functionally related or coordinately regulated genes. Genetic distance is largely proportional to physical distance, at least when averaged over tens or hundreds of kilobase pairs. Finally, the *cis*-acting sequences required for chromosome function—telomeres, centromeres, and replication origins—are only a few hundred base pairs long and function in almost any context.

Volume I. *The Molecular and Cellular Biology of the Yeast* Saccharomyces: *Genome Dynamics, Protein Synthesis, and Energetics.* Copyright 1991 Cold Spring Harbor Laboratory Press 0-87969-355-X/91 $3 + 00

Table 1 Basic features of the yeast genome

Genome size	12.5 Mb + rDNA (rDNA typically 1-2 Mb, but highly variable)
Single-copy DNA[a]	12.0 Mb
Length of genetic map	>4300 cM
Number of chromosomes	16
Organelle DNA	mitochondrial, 75 kb (see Chapter 7)
Extrachromosomal elements	2-micron DNA (see Chapter 6) double-stranded RNA (see Chapter 5)

[a] The amount of single-copy DNA has been calculated by subtracting the amount of DNA known to be associated with *TEL* sequences, subtelomeric repeats, and retrotransposon families from the size of the non-rDNA component of the genome. No correction was made for duplicated genes; there appear to be relatively few cases of gene duplication in which the different copies are not readily distinguished at the DNA level (see Section VI.D).

This overview has emerged after 15 years of recombinant DNA work on yeast. Recombinant DNA techniques have favored a divide and conquer approach to the yeast genome; hence, the present perspective may overemphasize the trees at the expense of the forest. Nevertheless, the view of the yeast chromosomes as a series of nearly independent functional units (like "beads on a string") has been a productive working model. Several recent reviews have covered aspects of the structural and functional organization of the yeast genome. Two articles by Newlon (1988, 1989) provide particularly broad coverage.

II. THE YEAST GENOME: VITAL STATISTICS

The basic features of the yeast genome are summarized in Table 1. The best estimate of the genome size comes from physical mapping of the chromosomes with restriction endonucleases that cleave yeast DNA infrequently (Link and Olson 1991); this estimate is in good agreement with earlier values derived from the renaturation kinetics of yeast DNA (Lauer et al. 1977). The fraction of the genome that is composed of "single-copy" DNA has been estimated by subtracting the amount of DNA associated with the known families of repeated sequences. The length of the genetic map is a minimal value calculated by adding the distances between pairs of adjacent linked markers (Mortimer et al. 1989). The number of chromosomes has been most reliably determined by the use of pulsed-field gel electrophoresis (PFGE) to establish an "electrophoretic karyotype" (Carle and Olson 1985, 1987; Link and Olson 1991); a value of 16 is also supported by genetic data (Mortimer et al. 1989), as well as by both fluorescence and electron microscopy (Kuroiwa et al. 1984; Dresser and Giroux 1988).

Genetic elements other than the nuclear chromosomes are discussed elsewhere in this volume (Wickner; Broach and Volkert; Pon and Schatz). Although mitochondrial DNA is essential for respiratory metabolism, it is not required for viability. The other extrachromosomal elements listed are also dispensable. Indeed, their known genetic functions are all directly related to their own propagation or survival.

III. THE GENETIC MAP

The current genetic map for *Saccharomyces cerevisiae* is presented in the Appendix. Exhaustive documentation of this map has been published in a series of reviews by R. Mortimer and his colleagues. A comprehensive publication that superseded early versions appeared in 1980; updates were published in 1985 and 1989 (see Mortimer and Schild 1980, 1985; Mortimer et al. 1989).

The current map contains more than 1000 markers and spans 4390 cM. This estimate of the length of the genetic map is a lower limit for two reasons. First, for a majority of the 32 chromosome arms, neither the physical distance nor the genetic distance from the most distal genetic marker to the telomere is known. Second, internal gaps occur in the meiotic linkage maps of chromosomes IX, X, and XVI. Nevertheless, the recent history of genetic mapping in yeast suggests that there is only a low probability that a new marker will expand the existing map. During the interval between the 1980 and 1989 versions of the map, the estimated minimal length of the map actually decreased from 4600 cM to 4295 cM as the number of markers increased from 317 to 769. Although some new markers defined short extensions of particular chromosome arms, improved estimates of internal genetic distances led to the overall decrease in genetic length.

On the whole, genetic distance in yeast appears to be remarkably proportional to physical distance, with a global average of 3 kb/cM. The only dramatic exception to this proportionality is the nearly complete absence of meiotic recombination within the large block of ribosomal DNA (rDNA) on chromosome XII (Petes 1979b). This genetic observation has a cytological correlate: In preparations of meiotic chromosomes, the nucleolus remains associated with chromosome XII during pachytene, and the synaptonemal complex appears to be discontinuous in this region of the chromosome (Byers and Goetsch 1975; Dresser and Giroux 1988). There also appears to be a moderate suppression of recombination in the regions immediately surrounding centromeres. The best-documented examples are on chromosomes I (Kaback et al. 1989) and III (Lambie and Roeder 1986). In these cases, the region over which recombination was suppressed was short (10–20 kb) and only two- to fourfold in magnitude. Furthermore, normal levels of recombination were restored when novel sequences were integrated near the chromosome III centromere (Symington and Petes 1988). This observation suggests that local hot spots can counterbalance the centromeric suppression of recombination.

Naturally occurring hot spots for meiotic recombination have also been described. However, even the most dramatic cases, such as the *CDC24-PYK1* interval on chromosome I (Coleman et al. 1986), only increase the probability that a crossover will occur within a 10-kb interval by a factor of 4 relative to the genome-wide average. Genetic recombination is discussed in more detail by Petes et al. (this volume).

A longer-range anomaly in the frequency of genetic recombination appears to correlate with chromosome size (Kaback et al. 1989). The smallest chromosomes display lower ratios of physical to genetic distance than does the genome as a whole. This phenomenon may simply reflect the operation of an "obligatory crossover" mechanism; i.e., in addition to a background level of random crossovers occurring at a frequency consistent with a 3 kb/cM ratio between physical and genetic distance, there may be a mechanism that ensures the occurrence of at least one crossover between each pair of homologs. Such a mechanism would cause the ratio of physical to genetic distance to be significantly lower on a very small chromosome than it is in the genome at large.

IV. THE PHYSICAL MAP

A. Electrophoretic Karyotype

Early efforts to characterize the DNA molecules that corresponded to specific yeast chromosomes were frustrated by the low resolutions of the available separation methods. Estimates of the size of the yeast genome from the reassociation kinetics of yeast DNA and of the number of chromosomes from genetic studies had suggested that the smallest yeast chromosomes were likely to be only a few hundred kilobase pairs in size. Several physical studies using electron microscopy, velocity sedimentation, and viscoelastic relaxation provided tantalizing evidence for the existence of such molecules in yeast lysates. However, until 1984, the only unequivocal success in correlating a specific chromosomal DNA molecule with its genetic counterpart involved a circular derivative of chromosome III (Strathern et al. 1979).

This situation changed rapidly with the development of PFGE. The very first publication on this method by Schwartz et al. (1983) demon-

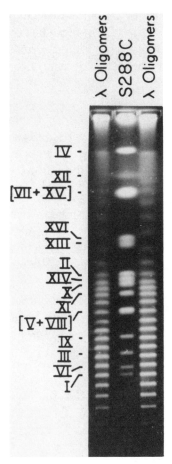

Figure 1 Separation of the yeast chromosome by PFGE. The chromosomes were prepared from AB972 (Olson et al. 1986), a yeast strain that is isogenic with S288C (Mortimer and Johnston 1986). The conditions of the electrophoresis are described by Link and Olson (1991).

strated that a few discrete bands could be obtained when yeast lysates were electrophoresed under conditions that allowed the separation of the chromosomal DNA molecules of the bacteriophages T4 (170 kb) and G (700 kb). The identification of specific bands by DNA-DNA hybridization with specific probes followed quickly (Schwartz and Cantor 1984; Carle and Olson 1984), and a complete electrophoretic karyotype was published in 1985 (Carle and Olson 1985).

A typical modern separation of the yeast chromosomal DNA molecules is shown in Figure 1. These data were obtained for a strain isogenic with the widely used yeast strain S288C (Mortimer and

Johnston 1986). In this genetic background, the 16 chromosomal DNA molecules can be resolved into 14 bands; two pairs of chromosomes (V and VIII and VII and XV) comigrate. Yeast strains have been described in which these comigrating doublets can be resolved (Carle and Olson 1985), and a strain has even been constructed in which all of the chromosomes are readily separable (Vollrath et al. 1988).

As evident from Figure 1, the electrophoretic behavior of chromosome XII, the largest yeast chromosome, is anomalous on typical pulsed-field gels. It now appears that all reported anomalies in the behavior of chromosome XII are due to the effects of one or more of the following phenomena (Olson 1989).

1. On gels such as the one shown in Figure 1, the relationship between size and mobility is double-valued for DNA molecules appreciably larger than chromosome IV (1.6 Mb). Thus, chromosome XII, which is more than 2 Mb in most S288C-related strains, migrates ahead of chromosome IV.
2. DNA molecules more than 2 Mb are subject to irreversible entrapment in the gel matrix unless the electrophoresis is carried out at uncommonly low-voltage gradients. Loss of some molecules to entrap-

Table 2 Sizes of the yeast chromosomes

Ordered by chromosome number		Ordered by chromosome size	
chromosome	size	chromosome	size
I	240 kb	I	240 kb
II	840	VI	280
III	350	III	350
IV	1,640	IX	440
V	590	V	590
VI	280	VIII	590
VII	1,120	XI	680
VIII	590	X	755
IX	440	XIV	810
X	755	II	840
XI	680	XIII	950
XII	1,095 + rDNA	XVI	980
XIII	950	VII	1,120
XIV	810	XV	1,130
XV	1,130	IV	1,640
XVI	980	XII	1,095 + rDNA
Total	12,490 + rDNA	Total	12,490 + rDNA

ment accounts for the lower intensity of the chromosome XII band compared to the chromosome IV band in Figure 1.

3. The size of chromosome XII varies from strain to strain and even from isolate to isolate, depending on the number of copies of the rDNA repeat that are present. In most cases, the size of chromosome XII is between 2 Mb and 3 Mb, reflecting the presence of 100–200 copies of the rDNA repeat.

The best current size estimates of the S288C chromosomes are presented in Table 2. The main practical application of PFGE has been to the mapping of cloned yeast genes. Cloned fragments of yeast DNA can be assigned to a chromosome simply by gel-transfer hybridization (see, e.g., Johnson et al. 1987). PFGE has also simplified the analysis of chromosomal rearrangements such as translocations and large deletions. Even aneuploidy is readily detected by careful assessment of the relative intensities of the bands on PFGE. These capabilities have been of practical value in the analysis of industrial yeasts such as brewing strains (Bilinski and Casey 1989). PFGE is also a useful tool in yeast taxonomy (de Jonge et al. 1986; Johnston and Mortimer 1986).

B. Restriction Maps

The long-term goal for the physical analysis of the yeast genome is the determination of the complete nucleotide sequence. As discussed below, progress is being made toward this objective both by the piecemeal analyses of individual genes and through more global approaches. The electrophoretic karyotype provides an initial low-resolution map of the genome. Global mapping at the next higher level of resolution has employed the restriction enzymes SfiI (GGCCNNNN'NGGCC) and NotI (GC'GGCCGC). Taken together, SfiI and NotI provide a continuous map of the genome with 110-kb resolution (Link and Olson 1991), as shown in Figure 2. The SfiI/NotI map underlies the best current estimates of the sizes of the yeast chromosomes and of the genome as a whole (Table 2).

Several projects are attempting to map long regions of yeast DNA at higher resolution. All of these efforts depend on "contig" mapping (i.e., the analysis of overlapping sets of clones); as a byproduct, they yield systematic sets of clones for the mapped regions. Contig maps have been reported for part of chromosome I (Kaback et al. 1989) and nearly all of chromosome III (Newlon et al. 1986; Yoshikawa and Isono 1990); extensive unpublished data exist for chromosome II (H. Feldmann, pers. comm.).

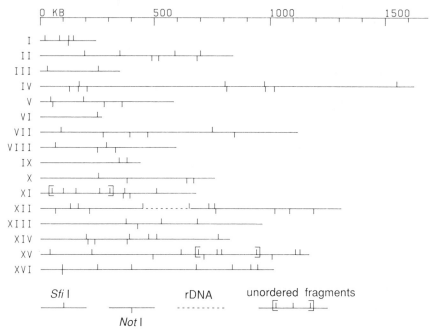

Figure 2 *Sfi*I/*Not*I restriction map of the 16 yeast chromosomes. The mapping was carried out on AB972 (Olson et al. 1986), a yeast strain that is isogenic with S288C (Mortimer and Johnston 1986). The dashed line in the middle of chromosome XII represents the position of the rDNA. The rDNA, which contains one *Sfi*I site per 9-kb repeat, is not drawn to scale; the actual length is highly variable but is commonly 1–2 Mb (see Section VI.A). The map is based on that of Link and Olson (1991).

A global contig-mapping effort is also in progress (Olson et al. 1986). In this project, contigs are built on the basis of "fingerprinting" data on individual λ and cosmid clones. Clone fingerprints are obtained by measuring the sizes of the fragments produced when the clones are digested with a combination of *Eco*RI and *Hin*dIII. The mapping landmarks are referred to as RH sites because no distinction is made between the two enzymes in the initial mapping. Nearly complete RH maps have been constructed for the six smallest chromosomes (I, VI, III, IX, V, and VIII). Of the 2.49 Mb of DNA estimated to be present in these chromosomes, 2.41 Mb (97%) is accounted for in eight contigs. Most of the missing DNA is immediately proximal to the chromosome ends, but single internal gaps also occur on chromosomes VIII and IX.

For the genome as a whole, 12.4 Mb of the 12.5 Mb of non-rDNA is currently represented in 113 contigs, over 90% of which have been assigned to particular intervals on the *Sfi*I/*Not*I map. The high fraction of DNA accounted for in contigs suggests that most of the gaps between contigs are small. In projects of this type, the degree of completion becomes difficult to estimate when it approaches the 100% range, because there can be a few percent error in each of the numbers being compared. A practical benefit of the global contig mapping is that it already allows most cloned fragments of yeast DNA to be assigned to physical map positions by DNA-DNA hybridization to grids of overlapping λ and cosmid clones. Approximately 1000 clones are required to cover the portion of the genome that already exists in contigs.

C. DNA Sequences

Essentially all of the yeast DNA sequences reported to date have been acquired in conjunction with the analysis of individual genes. GENBANK provides the most extensive compilation of these data. GENBANK sequences derived from *S. cerevisiae* are assigned locus names with the prefix YSC. YSC sequences of nuclear genes are placed in the PLANT.DAT file; sequences corresponding to mitochondrial genes are placed in ORGANELLE.DAT file. In Release 65.0 of GENBANK (September 1990), there are 910 YSC entries in PLANT.DAT. The aggregate length of these sequences is 1.62 Mb. There are some duplicate entries and a small contribution from sources other than nuclear genes (e.g., the 2-micron plasmid and the double-stranded RNAs of the killer system). Nevertheless, because most of the data represent single entries for nuclear genes, a reasonable estimate of the fraction of the yeast genome that has been sequenced is 10–15%.

Systematic sequencing of the yeast genome has yet to have a significant impact. Indeed, the longest single tract of sequence in the current GENBANK release is 11.2 kb. Projects directed at the sequencing of all or part of chromosomes III (Thierry et al. 1990; S. Oliver, pers. comm.), VI (E. Soeda, pers. comm.), and IX (B. Barrell, pers. comm.) are under way. However, given that the total amount of DNA in these three chromosomes is only 1.1 Mb, it will be some time before the contributions from systematic sequencing overtake those from the piecemeal analysis of individual genes.

V. *CIS*-ACTING SEQUENCES REQUIRED FOR CHROMOSOME FUNCTION

The 1980s saw dramatic progress in the characterization of the *cis*-acting DNA sequences required for yeast chromosome function. Three classes of sequences—*ARS*s (autonomous replication sequences), *CEN* sequences (centromeres), and *TEL* sequences (telomeres)—appear to account for the basic mitotic and meiotic behavior of yeast chromosomes. *ARS*s, *CEN* sequences, and *TEL* sequences are all short: Full functionality is typically displayed by fragments of only 100–200 bp. Artificial chromosomes with properties remarkably similar to those of the natural yeast chromosomes can be constructed simply by incorporating an *ARS*, a *CEN*, and two *TEL* sequences into a linear DNA molecule. Artificial chromosomes smaller than approximately 50 kb have poor mitotic stability (Murray and Szostak 1983), but their stability increases with length (Hieter et al. 1985a; Murray et al. 1986). At 150 kb, artificial chromosomes that derive most of their length from tandem repetitions of the genome of bacteriophage λ replicate and segregate successfully in 99.9% of cell divisions. This success in constructing model artificial chromosomes has led to the development of yeast artificial chromosome (YAC) vectors for cloning large segments of exogenous DNA (Burke et al. 1987).

A. *ARS*s

Of the three classes of *cis*-acting sequences required for chromosome function, *ARS*s were the first to be described (Hsiao and Carbon 1979; Stinchcomb et al. 1979; Struhl et al. 1979). For nearly a decade, *ARS*s had only the phenomenological definition of conferring high-frequency transformation and the capacity for extrachromosomal replication on small circular plasmids. More recently, direct evidence has been obtained that at least some *ARS*s function in their normal contexts as origins of chromosomal DNA replication (Brewer and Fangman 1987; Huberman et al. 1988). Sequences with relatively strong *ARS* activity, as judged from a plasmid-transformation assay, are found at an average spacing of approximately 30–40 kb in the yeast genome (Beach et al. 1980; Chan and Tye 1980). A detailed study of a 200-kb region of chromosome III revealed 12 *ARS*s; time of replication studies of this region suggest that only a small subset of these *ARS*s is normally functional (Reynolds et al. 1989). *ARS*s are discussed in detail by Campbell and Newlon (this volume).

B. *CEN* Sequences

The first *CEN* sequence described was *CEN3*, isolated from the centromeric region of chromosome III (Clarke and Carbon 1980). The presence of *CEN3* on a small *ARS*-containing plasmid resulted in a low copy number, an improved mitotic stability, and a predominance of 2:0

segregation of the plasmid during meiosis. The basis for the low copy number following transformation appears to involve selection against cells containing an excessive number of centromeres, rather than a direct role of *CEN* sequences in copy-number regulation (Futcher and Carbon 1986).

CEN sequences are necessary and sufficient for the orderly mitotic and meiotic segregation of chromosomes. Deletion of the *CEN* sequence from a natural chromosome by integrative transformation leads to a random pattern of mitotic segregation (Clarke and Carbon 1983). Conversely, the presence of a *CEN* sequence confers nearly normal levels of mitotic stability and an appropriate pattern of meiotic segregation on an artificial chromosome. Full *CEN* function has been displayed by a fragment as short as 125 bp when placed in an apparently neutral context (Cottarel et al. 1989).

The sequences of 12 of the 16 yeast *CEN*s have been published previously (Hieter et al. 1985b; Huberman et al. 1986). Comparison of these sequences reveals a highly conserved motif whose length varies from 111 bp to 119 bp. The motif can be subdivided into three adjacent centromere DNA elements (CDEs): CDEI (8 bp: PuTCACPuTG, Pu = G or A), CDEII (78–86 bp: >90% A/T base pairs, but highly variable in actual sequence), and CDEIII (25 bp: TGTxTxTGyyTTCCGAA$yyyyy$AAA, x = T or A, y = any base). At the level of the whole motif, the degree of sequence conservation is too small to allow significant DNA-DNA hybridization between pairs of *CEN*s; however, an oligodeoxynucleotide homologous to the first 17 nucleotides of CDEIII displays sufficient cross-hybridization between different *CEN* sequences to be useful in localizing these sequences within individual clones (Hieter et al. 1985b).

Extensive structure-function studies have been carried out on *CEN* sequences (for review, see Fitzgerald-Hayes 1987; Newlon 1988; Clarke 1990). In general, these studies confirm that the conserved features in the *CEN* sequence motif are functionally important and that sequences outside the CDEI-CDEIII interval normally have little effect. However, *CEN* function is disrupted when high levels of transcription are directed into the *CEN* DNA. This phenomenon has allowed the construction of conditional centromeres, in which *CEN* activity is under the indirect control of a galactose-inducible promoter (Hill and Bloom 1987, 1989).

Current research on *CEN* sequences focuses on *CEN*-binding proteins (Bloom and Yeh 1989). Cytological evidence suggests that only one microtubule attaches to each yeast chromatid at mitosis (Peterson and Ris 1976). A plausible hypothesis is that protein-DNA and protein-protein interactions lead to the assembly of a single microtubule-binding site on each *CEN* sequence. In some respects, the set of yeast *CEN* sequences may resemble a set of coordinately regulated promoters. In both cases, the same set of proteins must assemble on several different DNA sequences and then interact with one another in highly specific ways. As is the case with coordinately regulated promoters, *CEN* sequences share only short, highly conserved sequences that are subject to relatively imprecise spacing requirements. Furthermore, it is difficult, but not impossible, to eliminate *CEN* function with point mutations, as is also the case for promoters. One point mutation that completely eliminates mitotic *CEN* function is the change of the conserved core of CDEIII from TTCCGAA to TTCTGAA (McGrew et al. 1986). This mutation also eliminates the marked nuclease resistance displayed by *CEN* sequences in chromatin preparations (Saunders et al. 1988). It is therefore likely that the association of a large complex of proteins with a *CEN* sequence is dependent on the successful binding of a single protein to the CDEIII core.

C. *TEL* Sequences

Advances in yeast molecular genetics led directly to the discoveries of *ARS*s and *CEN* sequences. At the time of these discoveries, information about the replication origins and centromeres of other eukaryotic organisms was too sparse to have any direct influence on the yeast studies. In contrast, the discovery of yeast *TEL* sequences grew directly out of prior work on chromosome ends in the macronuclei of protozoans such as *Tetrahymena* and *Oxytricha*. Telomere structure and function have been reviewed by Walmsley (1987), Zakian (1989), Blackburn (1990), and Greider (1990).

During macronuclear development in *Tetrahymena* and *Oxytricha*, the fragmentation of the micronuclear (i.e., germ line) genome leads to the de novo formation of large numbers of linear chromosomes. In particular, the single micronuclear rRNA gene, which is present at an internal site on a large micronuclear chromosome, is amplified to high copy number on a 21-kb linear replicon (for review, see Blackburn and Karrer 1986). These molecules terminate in long tracts of a simple repeat that has the sequence end-$(5'\text{-CCCCAA-}3')_n$-rDNA. The value of n is typically 50–70 but is heterogeneous within a population of rDNA molecules and changes systematically during prolonged periods of vegetative growth (Larson et al. 1987). Although *TEL* sequences are predominantly double-stranded, the actual terminus appears to have a single-strand 3' overhang of two repeat units (Henderson and Blackburn 1989). This single-strand tail on the G-rich strand may have an unusual secondary structure, as short synthetic DNAs of the same sequence engage in extensive intra- and interstrand hydrogen bonding between guanines (Sundquist and Klug 1989).

In an experiment that paved the way for the analysis of yeast telomeres, Szostak and Blackburn (1982) demonstrated that the *Tetrahymena* rDNA termini will stabilize the ends of linear DNA molecules transformed into yeast. Normally, linearized YRp plasmids (i.e., plasmids that contain an *ARS* and a selectable genetic marker) can only survive following transformation if they have either circularized or integrated into a yeast chromosome. However, these molecules persist as linear plasmids if short restriction fragments bearing the natural termini of the *Tetrahymena* rDNA molecules are ligated onto their ends. Szostak and Blackburn (1982) employed a variation on this experiment to clone a natural yeast telomere by functional selection. When a heterogeneous population of yeast restriction fragments was ligated onto one end of a linearized YRp plasmid whose other end was capped with a *Tetrahymena* telomere, only those molecules that had acquired a yeast telomere survived as linear plasmids following transformation.

Similar experiments have subsequently led to the cloning of telomeres from higher organisms, which also stabilize the ends of linear replicons in yeast (Brown 1989; Cheng et al. 1989; Cross et al. 1989; Riethman et al. 1989). Interspecies incompatibilities do occur; e.g., the *S. cerevisiae* telomeres function poorly when transformed into *Schizosaccharomyces pombe* (Hahnenberger et al. 1989), but they appear to be the exception rather than the rule. In this respect, *TEL* sequences differ from *CEN* sequences and *ARS*s. Numerous efforts to clone the centromeres and replication origins of higher organisms by functional selection in yeast have failed.

The primary structure of the yeast *TEL* sequences is similar, but not identical, to that of their protozoan homologs. The yeast motif is end-$(5'\text{-}C_{1-3}A\text{-}3')_n...$, with the most common repeat unit being the dinucleotide CA (Shampay et al. 1984). Typical values of n are 100–200, but variation occurs from cell to cell, even at a particular chromosome end (Shampay and Blackburn 1988). There is also genetic variation among different laboratory yeast strains in the average length of *TEL* sequences (Walmsley and Petes 1985). Subtelomeric repeat sequences (discussed in Section VI.B) are found immediately proximal to the *TEL* sequences.

Genetic studies suggest that telomere length reflects a steady-state balance between the degradation and the elongation of *TEL* sequences. Mutations defective in both processes have been isolated. These mutations characteristically only display their phenotypes after many generations of cell growth (Carson and Hartwell 1985; Lustig and Petes 1986;

Lundblad and Szostak 1989). This observation suggests that most *TEL* DNA is replicated via conventional mechanisms that conserve its length. When the specialized processes responsible for the fine tuning of telomere length fail to function, only small changes in length occur during each cell generation. Telomere replication is discussed further by Campbell and Newlon (this volume).

VI. REPEATED SEQUENCES

A. rDNA Cluster

The most abundant repeated sequence in the yeast genome is the rDNA. As is generally the case in eukaryotic cells, a single RNA polymerase I transcription unit produces a precursor RNA that is processed into 18S, 5.8S, and 25S RNAs. A more unusual feature of the *Saccharomyces* rDNA cluster is that the RNA polymerase I transcription unit is interspersed with a gene for 5S RNA; this gene is transcribed by RNA polymerase III, and the direction of transcription is opposite to that of the RNA polymerase I transcription unit. The two transcription units, together with two short nontranscribed spacers, make up a 9.1-kb repeat unit. Structural and functional features of the repeat are reviewed by Planta and Raue (1988) and Warner (1989); a complete, composite sequence of the repeat can be constructed from several piecemeal sources cited in these reviews. In a pioneering application of restriction-fragment-length polymorphisms, Petes demonstrated that the rDNA cluster maps to a single site on the right arm of chromosome XII (Petes 1979a,b).

Early evidence for a simple tandem arrangement of the rDNA repeats included R-loop studies with the electron microscope, in which as many as 26 repeat units were visualized on a single DNA molecule (Kaback and Davidson 1980). The repeats appeared to be identical and to be organized as a simple head-to-tail array. More recent analyses of restriction digests on PFGE suggest that the entire rDNA cluster has a similar arrangement with no intervening blocks of non-rDNA: Cleavage of high-molecular-weight yeast DNA with a variety of restriction enzymes that do not cleave the rDNA repeat (e.g., *Sal*I, *Bam*HI, and *Pst*I) liberates the rDNA in a single block. The apparent size of this block varies dramatically among different yeast strains and even among different isolates of the same strain (G. Carle and M. Olson, unpubl.). Typical block sizes are 1–2 Mb, which corresponds to 100–200 copies of the repeat.

These data, which suggest that the rDNA is a dynamic structure whose size varies stochastically during cell growth, are consistent with earlier studies of the behavior of foreign DNA introduced into the rDNA repeat by integrative transformation (Petes 1980; Szostak and Wu 1980).

Extrachromosomal circular forms of rDNA may play a role in these dynamic processes. Small circular forms, which contain one or a few repeat units, are readily observed in most strains. The most abundant of these episomes is the 3-micron circle, which contains a single repeat unit (Meyerink et al. 1979; Clark-Walker and Azad 1980; Larionov et al. 1980). However, very large circles would not have been detected in these studies. It is even possible that large changes in the length of the chromosomal rDNA cluster simply reflect the excision and reintegration of rDNA circles. As the rDNA repeat contains an *ARS*, circular forms could persist for many generations. However, unequal exchanges between sister chromatids provide an alternative mechanism for changes in the amount of rDNA during mitosis (Szostak and Wu 1980; for review, see Petes and Hill 1988; Petes et al., this volume).

B. Subtelomeric Repeats

Two classes of subtelomeric repeats have been described in yeast, X and Y'(Szostak and Blackburn 1982; Chan and Tye 1983a,b). Repeated sequences located immediately adjacent to the *TEL* sequences at chromosome ends appear to be a common feature of eukaryotic genomes. The X and Y' sequences of *Saccharomyces* are the best-studied examples, but similar repeats have also been described in organisms as diverse as *Plasmodium falciparum* (Corcoran et al. 1988) and humans (de Lange et al. 1990). There is no evidence that these sequences play any role in chromosome function. However, only relatively crude tests for functionality have been performed. The most commonly employed YAC vectors produce artificial chromosomes that lack subtelomeric repeats (Burke et al. 1987). Although the resultant YACs display qualitatively normal mitotic and meiotic behavior (Brownstein et al. 1989; Green and Olson 1990), YACs differ from the natural chromosomes in too many respects to provide a good test system. Telomeres are readily introduced into the natural yeast chromosomes at new sites by integrative transformation, but the most widely used methods leave a Y' sequence adjacent to the newly formed end (Surosky and Tye 1985; Vollrath et al. 1988). However, integrative transformation has been used to construct a nearly full-length derivative of chromosome III that has only *TEL* sequences at both ends. The meiotic behavior of this construct has not been studied, but its mitotic stability is nearly identical to that of the wild-type chromosome (Murray et al. 1986).

All of the well-characterized telomeres of yeast have an X sequence, a Y' sequence, or both immediately adjacent to their terminal *TEL* sequences (Link and Olson 1991). The only reported exception is the right arm of chromosome VI in strain S288C, which appears not to hybridize to DNA-DNA hybridization probes specific for either the X or the Y'sequences (Link and Olson 1991). However, as discussed below, the X family of repeats is heterogeneous, and it is not possible to score the presence or absence of X-related sequences reliably on the basis of any single DNA-DNA hybridization assay. Most telomeres that have one or more Y' sequences adjacent to their terminal *TEL* sequences also have an X sequence in a more centromere-proximal position (Louis and Haber 1990a; Link and Olson 1991).

The Y' repeats have been more extensively studied than the X family. In the early literature, two segments of the Y' repeat were commonly referred to as Y (Szostak and Blackburn 1982) and 131 sequences (Chan and Tye 1983a). However, as the two segments appear always to occur together, the single name Y' has been adopted to designate the entire repeat. Y' sequences occur in 6.7-kb and 5.2-kb variants, which differ by the presence or absence of an internal 1.5-kb sequence (Chan and Tye 1983a; Louis and Haber 1990a). Sequence data have been reported only for a 1-kb region within Y' that is present in both the 6.7-kb and 5.2-kb variants. Within this 1-kb region, there are 12 repetitions of a 36-bp repeat (Horowitz and Haber 1984), a sequence organization reminiscent of the satellite DNA of higher organisms.

Y' sequences are present at 17 of the chromosome ends in the common laboratory strain S288C (Link and Olson 1991). There is no obvious pattern to their occurrence: Several chromosomes lack Y' sequences at both ends, whereas others have Y' sequences at either or both ends. Furthermore, the number and distribution of Y' sequences show extreme variation from strain to strain (Zakian and Blanton 1988; Jaeger and Philippsen 1989; Louis and Haber 1990a; Link and Olson 1991), and they are undetectable by DNA-DNA hybridization in some *Saccharomyces* species (Jaeger and Philippsen 1989). Y' sequences can be repeated as many as four times at a particular telomere (Chan and Tye 1983a). Given that Y' sequences display *ARS* activity (Chan and Tye 1983b) and are tandemly repeated at some sites, it is not surprising that low levels of circular forms of Y' are detectable (Horowitz and Haber 1985). Excision and reintegration of Y' circles may account for some of the strain-to-strain variation in the distribution of Y' sequences, but other mechanisms involving mitotic and meiotic recombination between non-homologous chromosomes also appear to operate at a high frequency (Horowitz et al. 1984; Louis and Haber 1990b).

The X family of repeats is less well defined than the Y' family of repeats. Like Y' sequences, many X sequences contain an *ARS* (Chan and Tye 1983a,b). Detailed restriction maps of at least eight distinct X sequences, all selected from the subset of Xs that are centromere-proximal to Y′ sequences, have been reported (Chan and Tye 1983b; Walmsley et al. 1984); short tracts of DNA sequences were also determined for two of the cases. A detailed map and extensive sequence data are also available for a single X sequence that is directly adjacent to the terminal *TEL* sequences on the left arm of chromosome III (Button and Astell 1986). These studies suggest that the centromere-distal ends of the various X sequences are 80–90% homologous; in all cases, these distal ends are marked by tracts of a few hundred base pairs of typical *TEL* sequences, which serve either as the true telomere or as a spacer between the X sequence and a distal Y′ sequence. In some instances, the partly conserved tract of X-related sequence extends for only a few hundred base pairs centromere-proximal to the *TEL* sequences. In other cases, the tract appears to extend for 3–4 kb. In these cases, the proximal boundary has only been defined by a transition in the copy number of the DNA, as judged from DNA-DNA hybridization experiments (Chan and Tye 1983b; Button and Astell 1986). Distal to the boundary, the copy number is ten or more, and proximal to the boundary, it is one or two. Cases in which the DNA centromere proximal to an X sequence is not strictly single copy may reflect the duplication of short genomic segments at two or more different telomeres (see Section VI.D).

C. Retrotransposons

Most or all of the dispersed, repetitive DNA in yeast is associated with a few families of retrotransposons. Retrotransposons have the same sequence organization as retroviruses and, like retroviruses, transpose by way of RNA intermediates. These elements are discussed in detail by Boeke and Sandmeyer (this volume). In this section, the emphasis is on their distribution in the genome.

The complete retrotransposons are 5–6 kb long. They are bracketed by long terminal repeats (LTRs), which are 300–400 bp long. The LTRs are important in their own right as repetitive sequences because they occur frequently as isolated elements. Indeed, these "solo" LTRs are the most abundant dispersed repetitive sequences in yeast. Solo LTRs in the genome are presumed to have arisen by a two-step process: transposition of a complete element to a new site via an RNA intermediate and subsequent homologous recombination between the LTRs of the complete element, a process that leaves behind a solo LTR. This two-step process is well established in the case of genetically selected transposition events (see Roeder et al. 1980). Circular forms of retrotransposons have been

Table 3 Properties of the known yeast retrotransposons

Family name		Length		Approximate copy numbers[a]	
whole element	LTR	whole element	LTR	whole element	LTR
Ty-1	δ	5.9 kb	330–338 bp	30	100
Ty-2	δ	5.9 kb	330–338 bp	12	100
Ty-3	σ	5.4 kb	340–341 bp	2	30
Ty-4	τ	6.3 kb	370 bp	2	25

[a] When available, estimates are for the S288C genetic background (Mortimer and Johnston 1986). The Ty-1 and Ty-2 estimates are from Curcio et al. (1990). The estimates for δ elements, from Cameron et al. (1979), are particularly rough in view of this element's abundance, heterogeneity, and tendency to cluster. The Ty-3 estimate is from Sandmeyer et al. (1988) and that for σ is from Brodeur et al. (1983). The estimate for Ty-4 is from Stucka et al. (1989) and that for τ is from Chisholm et al. (1984).

detected at low copy number in yeast cells, but neither their exact structures nor their functional significance is known (Ballario et al. 1983).

Table 3 lists the names of the four known families of yeast retrotransposons and their associated LTRs; it also provides a general guide to their abundances in the genome. However, the copy numbers of these elements show great strain-to-strain variation (Eibel et al. 1981). Indeed, the presence or absence of particular elements at particular sites accounts for many of the restriction-fragment-length polymorphisms observed among different laboratory yeast strains (Cameron et al. 1979; Olson et al. 1979; Sandmeyer and Olson 1982). Despite considerable intrafamily variation, the boundaries between the different families of yeast retrotransposons and LTRs are well defined. Ty1 and Ty2 are the most closely related of the distinct elements. Typical Ty1 and Ty2 elements are nearly identical for approximately half of their overall lengths; two 1–2-kb blocks of dissimilar sequences account for the other half of their lengths (Kingsman et al. 1981; Warmington et al. 1985). At the level of DNA sequence, Ty3 and Ty4 have no significant similarity with each other or with Ty1 or Ty2 (Hansen et al. 1988; Stucka et al. 1989). However, the genetic organization of all four families of Ty elements is strikingly similar, and the encoded proteins appear to be structurally and functionally homologous (Clare and Farabaugh 1985; Fulton et al. 1985; Warmington et al. 1985; Hansen et al. 1988; Stucka et al. 1989). The three classes of LTRs—δ, σ, and τ—are all essentially dissimilar in sequence; vestigial homologies have been hypothesized, but they are undetectable by DNA-DNA hybridization or standard dot-matrix sequence comparisons (Genbauffe et al. 1984).

Considerable sequence heterogeneity is found within particular families of retrotransposons and LTRs. The Ty1- and Ty2-associated δ sequences have been grouped into two subtypes. The sequences of different δ elements of the same subtype generally differ by only a few percent, whereas typical members of one subtype share only approximately 85% sequence similarity with typical members of the other subtype (Roeder et al. 1980; Roeder and Fink 1983). There is no correlation between the class of Ty element (Ty1 or Ty2) and the subtype to which the associated δ sequences belong. However, the two δ sequences associated with a particular Ty1 or Ty2 are always of the same subtype; indeed, they are often identical and never diverge in sequence by more than 2% (Roeder and Fink 1983). This phenomenon, at least in part, presumably reflects the mechanism of transposition, in which the two LTRs of a newly inserted retrotransposon are generated from an intermediate that preserves only slightly more than one net LTR sequence (Boeke et al. 1985). There is greater variation among solo δ sequences than among the Ty-associated elements; e.g., among one set of five solo δ elements near the *SUP4* gene on chromosome X, the pairwise sequence similarities varied from 69% to 97% (Rothstein et al. 1987). Presumably, the average length of time that solo δ sequences have been at their present locations is longer than the corresponding time for Ty-associated elements. The presence of numerous nucleotide substitutions in intact solo δ elements and the frequent occurrence of truncated elements both suggest that there is little or no selection for the maintenance of these elements. There is less information about the τ family; however, the τ family also appears to be extremely heterogeneous. For example, two solo τ elements from chromosomes II and III share only 62% sequence similarity (Stucka et al. 1989). In contrast, the σ family appears to be remarkably homogeneous. In one set of ten σ elements, which included both solo and Ty3-associated elements, the most divergent pair shared 97% sequence similarity (Sandmeyer et al. 1988).

Heterogeneity is also found within the families of complete retrotransposons, but it appears to be much less than the heterogeneity among LTRs. For example, two completely sequenced Ty1 elements shared 99% sequence similarity, with most of the 64 nucleotide differences confined to one 240-bp region (Boeke et al. 1988).

The retrotransposons and solo LTRs are widely dispersed in the yeast genome (Klein and Petes 1984; Fink et al. 1986; Sandmeyer et al. 1988). Clusters of elements do occur, and a given cluster often contains members of two or more families. Indeed, the elements themselves appear to be good targets for the insertion of other elements. For example, a segment of DNA has been described in which a σ element appears to have been interrupted by insertion of a δ element, which was, in turn, interrupted by insertion of a τ element (Chisholm et al. 1984). Complex clusters involving multiple families of complete retrotransposons and solo LTRs have been characterized on chromosomes III and V (Warmington et al. 1985; Lochmueller et al. 1989). The detailed structures of these clusters are invariably highly polymorphic in different yeast strains. Although they are striking features, clusters of repetitive elements appear to account for only a minor proportion of yeast's dispersed repetitive DNA.

The most dramatic nonrandomness in the distribution of retrotransposons and solo LTRs involves their association with tRNA genes. This association is most spectacular in the case of the σ/Ty3 family. These elements are found *only* in the 5′-flanking regions of tRNA genes. Sequence data on 13 solo σ elements and two complete Ty3 elements revealed that all 15 elements began between positions −16 and −19 relative to the 5′ end of a tRNA-coding sequence (Sandmeyer et al. 1988; Chalker and Sandmeyer 1990). The 15 elements are adjacent to genes encoding 11 different tRNAs. Yeast tRNA genes share little or no sequence similarity at the site where Ty3 insertions occur, but the insertion site is close or identical to the site of transcription initiation by RNA polymerase III. Presumably, the transposition machinery associated with Ty3 recognizes some feature of the RNA polymerase III transcription complex.

The association of the other families of retrotransposons with the 5′-flanking regions of tRNA genes is less specific, but still real (Eigel and Feldmann 1982; Gafner et al. 1983). For example, in one study of eight cosmids selected because each contained a single tRNA^{Glu} gene, approximately 300 kb of yeast DNA was surveyed (Hauber et al. 1988). This DNA contained a total of 13 tRNA genes, present as isolated, widely spaced transcription units. Eleven of the tRNA genes had one or more retrotransposon or LTR in its immediate 5′-flanking region, and no such elements were detected at any other sites in the 300 kb. Solo LTR sequences from the δ, σ, and τ families were all well represented, many of them truncated at the end distal to the tRNA gene. Only one complete retrotransposon was present. Unlike the situation with σ, the spacing between the coding region of a tRNA gene and associated δ or τ sequences is quite variable; distances of 100–500 bp are typical.

An intriguing possibility is that additional families of LTRs or even of complete retrotransposons remain to be discovered. The currently known elements have all been discovered initially because of their repetitive nature. Future discoveries are most likely to involve elements present at low copy numbers. The 5′-flanking regions of tRNA genes are obvious places to search for new classes of elements. Indeed, Melnick and Sherman (1990) have reported one good candidate on the basis of DNA sequence data. A 300-bp region in the 5'-flanking region of a tRNA^{Glu} gene on chromosome X shows the same level of sequence similarity with σ, δ, and τ that these element show with each other. However, a DNA probe prepared from this region hybridizes only to itself when used to probe restriction digests of genomic yeast DNA.

In evaluating the significance of the association between retrotransposons and tRNA genes, it is important to distinguish between the preferred insertion sites for these elements and their preferred genomic locations. In the case of Ty3, the insertion specificity appears to be so great that the distinction is moot. However, Ty1 can insert into a wide variety of targets; there is some preference for the 5'-noncoding regions of genes, but insertions into coding regions are also common (Giroux et al. 1988; Natsoulis et al. 1989). Although the preferred insertion sites of transposons are determined by their transposition mechanisms, the distribution of genomic locations at which the elements are found in typical yeast strains is probably determined primarily by selection. It may be that elements accumulate in the 5'-noncoding regions of tRNA genes because these regions are among the most selectively neutral regions of the genome. The primary promoter elements associated with RNA polymerase III transcription are internal to the coding regions of tRNA genes, and there is no evidence that the expression of these genes responds to the presence of nearby enhancers; consequently, there are probably few functional constraints on their 5'-flanking sequences.

There is little doubt that retrotransposons have played a major role in shaping the modern yeast genome. Recombination between elements is a common cause of translocations, inversions, and large deletions in yeast (Roeder et al. 1980; Liebman et al. 1981; Downs et al. 1985; Rothstein et al. 1987). Nevertheless, these events appear to have been rare during yeast evolution in comparison to transposition. Although there is extraordinary polymorphism in the distribution of retrotransposons and LTRs in the genomes of laboratory yeast strains, polymorphisms involving large translocations, inversions, and deletions have not been reported. Presumably, most such large-scale events are highly deleterious in natural yeast populations.

D. Duplicated Genes and Gene Clusters

Gene duplication is common in yeast. Indeed, the yeast genome is replete with examples of different stages of gene duplication and divergence. The simplest case is the tandem repetition of identical copies of a gene, as described for the rDNA (see above). A more common arrangement is for the multiple copies to be dispersed in the genome. In some instances, all the copies encode precisely the same product. More frequently, slight differences occur between the products of different family members, but there is no evidence of functional divergence. Particularly when there are only a few copies of a gene, it is common for strains to survive and even to grow well under laboratory conditions when all but one are deleted. Finally, the real frontier in the study of duplicated genes involves families of closely related genes whose products have overlapping, but subtly distinct, functions.

The *CUP1* locus, at which there are local tandem repetitions of a gene whose copy number is readily altered by selection, is a particularly simple model for gene duplication in yeast (for review, see Butt and Ecker 1987). Yeast strains exist that have anywhere from one to more than ten tandemly repeated copies of the *CUP1* gene, which encodes a protein that sequesters copper ions. There is a direct correlation between the number of tandem repeats and the phenotypic level of copper resistance. In most strains that contain multiple copies of *CUP1*, the repeat unit is 2.0 kb; it codes for two transcripts, but only the one encoding copper chelatin is implicated in copper resistance (Karin et al. 1984). Selection for increased copper resistance readily yields strains with increased copy number, as long as there is more than one copy in the starting strain. Indeed, as is the case with the rDNA locus, unselected variations in copy number occur at a high frequency during nonselective mitotic growth. However, there has been no success in selecting for an initial duplication event in a strain that has a single copy of *CUP1* (Fogel et al. 1983).

Other than the rDNA and *CUP1*, tandem repeats of identical or nearly identical genes have not been reported in yeast. The apparent scarcity of these structures is not surprising. The high level of homologous recombination in yeast is known to make tandem duplications unstable once formed, and the *CUP1* model suggests that they arise infrequently to begin with.

Given the scarcity of tandem repeats and the abundance of dispersed duplicated genes, mechanisms that lead to duplication and dispersal in a single step deserve consideration. Fink (1987) has proposed that most yeast genes have been through cycles of transcription, splicing, reverse transcription, and reintegration of cDNAs by homologous recombination. This hypothesis provides a possible explanation for the paucity of introns in yeast genes. It also would explain the tendency of those introns that do exist to be at the 5' ends of genes (i.e., in positions where they could be removed only by the reintegration of full-length reverse transcripts). Nonhomologous integration of reverse transcripts at distant genomic sites would provide a one-step mechanism of gene duplication and dispersal. However, such a process would be expected to produce a high frequency of nonfunctional genes, particularly with defects at their 5' ends. Even the integration of full-length reverse transcripts would normally create genes that lack essential regulatory sequences. The only described case of a pseudogene that appears to be truncated at its 5' end is a cognate of *CDC4* on chromosome V (Yochem and Byers 1987). This gene has high sequence similarity with the 3' half of the *CDC4*-coding region, although the similarity is disrupted by an in-frame stop codon; the pseudogene does not contain sequences similar to the 5' half of the *CDC4*-coding region. Other examples of nonfunctional genes, most notably *suc2* (Carlson et al. 1981) and *ho* (Jensen et al. 1983), appear simply to represent spontaneous mutations in genes that are functional in most "wild-type" strains. Indeed, these mutations may have been hand selected on the basis of colony morphology or other nonspecific phenotypes during the development of laboratory yeasts.

The paucity of pseudogenes extends to tRNA genes, which pose particular challenges to ideas about the evolution of multigene families. The extensively studied tRNA^{Tyr} gene family appears to be typical (Goodman et al. 1977; Olson et al. 1979; for a review of yeast tRNA genes, see Guthrie and Abelson 1982; for descriptions of other tRNA gene families, see Olson et al. 1981; Hauber et al. 1988). There are eight identical genes that code for the single tRNA^{Tyr} species present in yeast cytoplasm. The eight genes are dispersed over five chromosomes, with no cases of tightly linked pairs. Only the 100 bp that spans the region encoding the mature tRNA and the short intron is conserved among the loci. All eight genes are demonstrably expressed, although there is some evidence for locus-to-locus variation in the level of expression. Only after five of the genes have been deleted is there any detectable effect on cell growth (D.T. Burke and M.V. Olson, unpubl.). The dispersal mechanism for these genes is not known, but a process mediated by reverse transcriptase is plausible. The preservation of the intron does not obviate this hypothesis because splicing is a late step in tRNA maturation and splicing may even be coupled to transport out of the nucleus (De Robertis and Olson 1979). The major nuclear transcript of the tRNA^{Tyr} genes contains the intervening sequence but has mature ends; intriguingly, it is coterminal with the DNA sequence that is conserved among the eight loci. Concerted evolution of the family is almost certainly mediated by interlocus gene conversion events; interlocus gene conversion between dispersed tRNA genes is well documented in *S. pombe* (Amstutz et al. 1985; Szankasi et al. 1986).

The dilemma of how selection operates to maintain eight functional tRNA^{Tyr} genes, which appear to represent excess coding capacity, is general to discussions of all multigene families. Undoubtedly, laboratory growth conditions fail to mimic the selective processes that occur in natural populations. However, in an organism with a small genome and a high rate of homologous recombination, an additional possibility is that there may be selection against redundant genes that have been inactivated by mutation. Such genes may increase the mutational load on the species by providing a reservoir of mutations that can spread to active genes by ectopic gene conversion, which occurs at a remarkably high frequency in yeast (for review, see Petes and Hill 1988; Petes et al., this volume).

Two families of duplicated genes that illustrate the complexity of yeast genealogy are the *SUC* and *MAL* gene families, which encode enzymes necessary for the utilization of sucrose and maltose. Modern laboratory yeasts contain variable numbers of *SUC* and *MAL* loci, primarily at telomeric locations. The *SUC* genes encode invertase, the only specialized enzyme required for sucrose utilization (Carlson et al. 1985). Strains that contain one or more functional *SUC* genes can ferment sucrose, whereas those that have no functional *SUC* gene cannot. In crosses between particular fermenting and nonfermenting strains, it is commonplace for multiple functional *SUC* genes to segregate (Mortimer and Hawthorne 1969). These genes have been mapped to six different loci, all of which are near the ends of chromosomes. The only locus common to all laboratory yeasts is *SUC2*, which maps near the left end of chromosome IX. Although some strains with no history of genetic selection carry mutant *suc2* alleles, the gene is always physically present (Carlson et al. 1981, 1985). *SUC2* is more than 30 kb from the telomere and is embedded in ordinary DNA sequences, whereas the five other *SUC* loci are all immediately adjacent to telomeres and embedded in subtelomeric repeats. A plausible evolutionary scenario is that a copy of the *SUC2* gene migrated to a telomeric location in a rare event and was then dispersed to other telomeres by the same processes that result in a high degree of polymorphism in subtelomeric regions (see Section VI.B). The *MAL* gene family offers a number of parallels with the *SUC* genes, but data are more limited. In particular, comparisons between the structures of particular chromosome ends with and without associated *MAL* genes are unavailable (Charron et al. 1989).

The complexity of the *SUC* and *MAL* gene families in modern laboratory yeasts may be an artifact of domestication. Yeast, like maize, is an organism with no well-defined wild type. Contemporary strains were developed by the uncontrolled hybridization of diverse yeasts, many of which had been propagated under uncontrolled culture conditions for hundreds or thousands of years. By requiring interfertility, early yeast breeders excluded yeasts with highly divergent karyotypes from the gene pool of modern laboratory strains. However, the chromosome ends would not have been heavily constrained by this process.

In contrast to the distinctive cases discussed above, the most common form of gene duplication in yeast is illustrated by the histone genes. For example, histones H2A and H2B are expressed from two unlinked loci, each of which contains H2A- and H2B-coding regions separated by approximately 700 bp (Hereford et al. 1979). Little sequence similarity occurs between the loci except in the coding regions. The two H2B-coding regions differ at 49 of 390 nucleotide positions (13%), most of which are in the third positions of codons (Wallis et al. 1980). The predicted protein sequences differ at only 4 of 130 amino acid positions (3%), and all 4 differences are conservative substitutions. Null mutations at one of the loci produce a slight slow-germination phenotype, whereas those at the other locus cause no detectable phenotypic change (Rykowski et al. 1981). The double mutant, of course, is inviable. Similar results apply to the two H2A genes (Kolodrubetz et al. 1982).

In view of the extensive divergence between the two loci, the duplication of the H2A and H2B genes is clearly ancient. Nevertheless, there has been no convincing functional divergence between the two loci. Strikingly similar findings apply to many other genes that are present in two or three unlinked copies. Well-studied examples include the genes coding for glyceraldehyde-3-phosphate dehydrogenase (McAllister and Holland 1985), the MFα mating pheromone (Kurjan 1985), and a number of the genes encoding ribosomal proteins (Abovich and Rosbash 1984; for review, see Warner 1989; Woolford and Warner, this volume). In all of these cases, the selective advantage of the duplications appears simply to involve increased rates of protein synthesis, rather than functional specialization. However, gene duplication is only one mechanism by which biosynthetic capacity has been fine-tuned during evolution. There is no obvious logic to the pattern of genes in which duplication has occurred as opposed to those whose level of expression has been modulated by other mechanisms.

The most complex gene families in yeast are those in which the members have developed varying degrees of specialization. Although gene disruption is a powerful tool for analyzing these families, its application is not always straightforward. Single null mutations often cause no detectable phenotypic change, whereas the phenotypes of strains carrying more than one mutation may depend on the precise subsets of family members that are still functional. An example is provided by the genes encoding the family of hsp70 proteins (for review, see Craig 1989 and these volumes). These proteins affect the response of yeast cells to heat shock, as well as to other metabolic stresses and are also required for cell growth under normal physiological conditions. In general, the function of the hsp70 proteins appears to be to facilitate the intracellular transport and assembly into multimeric complexes of a wide variety of proteins.

Nine hsp70 genes have been described. Nucleotide sequence similarities in the coding regions range from 50% to 96%. Subfamilies can be defined on the basis of sequence comparisons. Although such groupings do correlate with function, the complexity of individual subfamilies is remarkable. For example, all pairs of genes in the subfamily *SSA1*, *SSA2*, *SSA3*, and *SSA4* share at least 85% sequence similarity. Nevertheless, each of these genes shows a distinct pattern of regulation. Although it is difficult to prove that there is a selective advantage to this regulatory diversity, the differences would appear to be physiologically relevant. For example, only *SSA3* is induced by starvation, and only *SSA2* is expressed at a high level at all temperatures. Collectively, the *SSA* subfamily is essential for growth under all conditions. To a first approximation, this requirement can be met by adequate levels of any one of the four gene products, but there are clear differences in the heat tolerance of strains that are dependent on one or another subfamily member for meeting the growth requirement. Other subfamilies of hsp70 genes perform functions that cannot be complemented by any of the *SSA* genes.

There are many other examples of gene families in yeast whose members have diverged in their properties. For example, the two closest homologs of the mammalian *ras* oncogene, *RAS1* and *RAS2*, behave much like the *SSA* subfamily of hsp70 genes. Substantial expression of either gene is sufficient and necessary for cell growth, but secondary phenotypes depend on which gene provides the Ras function (Tatchell et al. 1985). In particular, *ras2* mutants have a variety of phenotypes that differ from those of *ras1* mutants (e. g., they grow poorly on nonfermentable carbon sources, accumulate abnormally high levels of storage carbohydrates, and sporulate in rich media). It is presently unclear whether the differing phenotypes of *ras1* and *ras2* mutants are simply due to differing levels of Ras protein present under particular growth conditions or if functional differences between Ras1p and Ras2p also contribute. Yeast also contains many other more-or-less divergent *ras* homologs; these have apparently been adapted to a wide variety of different functions (Bourne et al. 1990; Johnson and Pringle 1990). An example of a pair of closely related genes in which functional specialization of the proteins appears to have occurred is provided by the two unlinked hexokinase genes, *HXK1* and *HXK2*. These genes encode enzymes that are 78% identical at the amino acid level. The genes are dispensable for growth on glucose, presumably because yeast also has a glucose-specific kinase; however, at least one of the genes must be expressed to support growth on fructose. Surprisingly, expression of *HXK2*, but not *HXK1*, is required for normal catabolite repression (Ma and Botstein 1986). The mechanism by which the *HXK2* gene product exerts this effect is not known.

In contrast to the wealth of information about duplicated genes, little is known about the duplication of larger segments of the yeast genome. A nearly perfect duplication of 20 kb at the right end of chromosome I and one of the ends of chromosome VIII has been reported (Steensma et al. 1989). The chromosome I segment ends at the telomere, whereas the chromosome VIII segment is offset from the telomere by 10 kb. The duplicated segment contains an alkaline phosphatase gene, which is present on both chromosomes. The restriction maps of the two segments match exactly, except for a Ty1 insertion present on chromosome VIII but not on chromosome I. A more distant evolutionary homology has been proposed between a region on chromosome X and a region on chromosome V. The 7.5-kb *COR* region, which has been completely sequenced, contains the *CYC1*, *OSM1*, and *RAD7* genes on chromosome X. These genes may be homologs of the *CYC7*, *ANP1*, and *RAD23* genes of the *ARC* region on chromosome V (Melnick and Sherman 1990). The homology between *CYC1* and *CYC7*, both of which code for cytochrome *c* isozymes that appear to be functionally interchangeable, is undisputed.

VII. GENE ORGANIZATION

A. Local Arrangement of Genes

The most detailed knowledge of local gene organization in yeast comes from surveys of the transcripts produced from regions that are long enough to include several genes (for review, see Coleman et al. 1986). In all cases examined, most of the DNA has been transcribed. The transcripts originate, seemingly at random, from both strands. In most instances, transcripts from adjacent genes do not overlap. For example, in a 10-kb region surrounding the *HIS3* gene, five or six nonoverlapping transcripts that account for 85% of the DNA were detected (Struhl and Davis 1981; Struhl 1985). Similarly, six nonoverlapping transcripts that account for 75% of the DNA were found in the 7.5-kb *COR* region on chromosome X (Barry et al. 1987; Melnick and Sherman 1990). Finally, ten nonoverlapping transcripts, including a Ty1 and two tRNA genes, were detected in a 23-kb interval surrounding *ADE1* on chromosome I; 80% of the DNA was transcribed (Steensma et al. 1987). A survey of nearly all of chromosome III detected 156 poly(A)[+] transcripts, whose average size was 1.9 kb, in 335 kb of DNA (Yoshikawa and Isono 1990); thus, there appeared to be approximately one transcript per 2.1 kb of DNA, and a high fraction of the DNA appeared to be transcribed. In summary, it appears that approximately 80% of yeast DNA is transcribed and that there is one gene for every 1.5–2 kb of DNA. This figure corresponds to an estimate of 6000–8000 genes for the whole genome.

More complex local arrangements of genes have also been described. For example, the *SPO12* and *SPO16* genes are in a tail-to-tail arrangement with only 103 bp between the termination codons of their coding regions (Malavasic and Elder 1990). A major *SPO16* transcript continues beyond the 3′ end of the *SPO16*-coding region and is complementary to nearly the entire *SPO12* transcript. However, mutational analysis indicates that this antisense RNA has no effect on *SPO12* expression. Overlapping transcripts have also been detected at the *HAP3* (Hahn et al. 1988), *PET122* (Ohmen et al. 1990), *CDC9/CDC36* (Barker et al. 1985), and *UBA1* (McGrath et al. 1991) loci.

In general, yeast genes tend to be independent regulatory units. The major exceptions involve coregulated pairs of divergently transcribed genes. The well-studied case of the *GAL1/GAL7/GAL10* gene cluster on chromosome II illustrates both the rule and a typical exception to it. The three genes are all induced by galactose in a process that requires the product of the *GAL4* gene (for review, see Johnston 1987; Johnston and Carlson, these volumes). *GAL1* and *GAL10* are divergently transcribed, with 680 bp separating their translational start codons. The two genes share a centrally located regulatory region that contains two *GAL4*-binding sites; both sites appear to facilitate transcription of both genes (West et al. 1984). *GAL7*, which is immediately downstream from *GAL1*, has an entirely separate regulatory region with its own *GAL4*-binding sites (Tajima et al. 1986). Coregulated divergent promoters, such as the *GAL1/GAL10* promoter, are quite common in yeast (e.g., the histone H2A/H2B genes [Hereford et al. 1979], the histone H3/H4 genes [Smith and Murray 1983], and the *MATa1/MATa2* and *MATα1/MATα2* genes [Astell et al. 1981]). However, there is no indication that the mechanism of regulation associated with these promoters is any different from that associated with isolated transcription units. Divergent promoters may simply represent the economical utilization of upstream activation sequences, whose orientational independence allows their shared use by genes that are clustered in a head-to-head arrangement.

True regional transcriptional control does occur in yeast, but it appears to be rare. The clearest described case involves the unexpressed mating-type loci *HMR* and *HML*. In this case, the presence of "silencer" sequences a few kilobase pairs from divergently transcribed pairs of genes results in transcriptional inactivation (Brand et al. 1985). Silencer function appears to involve the nonspecific exclusion of transcription factors from the affected region (Schnell and Rine 1986).

B. Distribution of Essential Genes

A long-standing issue in yeast genetics concerns the number of essential genes. A minimal definition of an essential gene is one that is required for growth on rich medium. Simple extrapolations from the results of classical mutational analyses would suggest the presence of only a few hundred essential genes in yeast (Kaback et al. 1984; Riles and Olson 1988; Harris and Pringle 1991). However, these estimates are known to be low because of the limitations of available methods for recovering conditional lethal mutants. For example, detailed molecular analyses of chromosome I have revealed that the majority of the essential genes are identified rarely, if ever, in classical genetic studies that depend on temperature-sensitive mutations (Diehl and Pringle 1991 and pers. comm.; A. Barton and D. Kaback, pers. comm.).

Gene disruption in diploid strains, followed by sporulation to test the viability of haploid segregants, provides the least biased technique for defining essential genes. In one sizeable study that applied this approach to the genome as a whole, 216 insertions that had been constructed to be as randomly distributed as possible were analyzed; only 12% of the insertions produced recessive lethal mutations (Goebl and Petes 1986). If one assumes that 20% of the yeast genome is nonessential intergenic DNA, that all insertions into an essential gene disrupt its essential function, and that there are 6000–8000 (see above), this result extrapolates to a total of approximately 900–1200 essential genes.

Given the implication that 85% of yeast genes are nonessential, another striking feature of this study was the paucity of detectable nonlethal phenotypes. Viable haploid segregants bearing the insertion mutations were tested for auxotrophy, sensitivity to radiation, sensitivity to a chemical mutagen, ability to mate, heat sensitivity, and cold sensitivity. Only two mutants (1%) displayed any of these phenotypes. Nondescript phenotypes such as slow growth were more common, but fully 70% of the insertions displayed no detectable phenotype. Similarly, approximately 65% of the 27 genes disrupted to date in the chromosome I studies have yielded no readily detectable phenotypes (Diehl and Pringle 1991; B. Diehl et al.; A. Barton and D. Kaback; both pers. comm.).

These data suggest that only a small fraction of yeast genes can be analyzed genetically on the basis of simple loss-of-function mutations. Many additional genes are undoubtedly accessible using genetic techniques that allow mutants to be recognized by virtue of their contributions to composite phenotypes (see, e.g., Basson et al. 1987; Bender and Pringle 1989, 1991). However, these techniques do not lend themselves to broad surveys for essential genes.

Many genes that play critical roles in yeast biology, whether or not they meet the minimalist definition of being essential, are unlikely to be discovered until large-scale sequence analysis of the yeast genome progresses. In some instances, sequence data will allow genes to be grouped into families for which it will be possible to demonstrate collective functions by the construction of multiple mutants. In other cases, it will be possible to infer functions by sequence similarity with genes in other organisms for which biochemical data are available. However, the prevalence of slow-growth phenotypes in current gene-disruption studies is a warning that many genes in yeast—perhaps a significant fraction of the total—have only subtle functions or are functionally redundant with other genes. This circumstance presumably reflects the high premium that selection under natural conditions places on the fine tuning of the yeast life cycle. However, as important as fine tuning may be in nature, the myriad of mechanisms by which it is achieved may prove difficult to elucidate in the laboratory.

REFERENCES

Abovich, N. and M. Rosbash. 1984. Two genes for ribosomal protein 51 of *Saccharomyces cerevisiae* complement and contribute to the ribosomes. *Mol. Cell. Biol.* **4:** 1871.

Amstutz, H., P. Munz, W.-D. Heyer, U. Leupold, and J. Kohli. 1985. Concerted evolution of tRNA genes: Intergenic conversion among three unlinked serine tRNA genes in *S. pombe. Cell* **40:** 879.

Astell, C.R., L. Ahlstrom-Jonasson, and M. Smith. 1981. The sequence of the DNAs coding for the mating-type loci of *Saccharomyces cerevisiae. Cell* **27:** 15.

Ballario, P., P. Filetici, N. Junakovic, and F. Pedone. 1983. Ty1 extrachromosomal circular copies in *Saccharomyces cerevisiae. FEBS Lett.* **155:** 225.

Barker, D.G., J.H.M. White, and L.H. Johnston. 1985. The nucleotide sequence of the DNA ligase gene (*CDC9*) from *Saccharomyces cerevisiae*: A gene which is cell-cycle regulated and induced in response to DNA damage. *Nucleic Acids Res.* **13:** 8323.

Barry, K., J.I. Stiles, D.F. Pietras, L. Melnick, and F. Sherman. 1987. Physical analysis of the COR region: A cluster of six genes in *Saccharomyces cerevisiae. Mol. Cell. Biol.* **7:** 632.

Basson, M.E., R.L. Moore, J. O'Rear, and J. Rine. 1987. Identifying mutations in duplicated functions in *Saccharomyces cerevisiae*: Recessive mutations in HMG-CoA reductase genes. *Genetics* **117:** 645.

Beach, D., M. Piper, and S. Shall. 1980. Isolation of chromosomal origins of replication in yeast. *Nature* **284:** 185.

Bender, A. and J.R. Pringle. 1989. Multicopy suppression of the *cdc24* budding defect in yeast by *CDC42* and three newly identified genes including the *ras*-related gene *RSR1. Proc. Natl. Acad. Sci.* **86:** 9976.

———. 1991. Use of a screen for synthetic lethal and multicopy suppressee mutants to identify two new genes involved in morphogenesis in *Saccharomyces cerevisiae. Mol. Cell. Biol.* **11:** (in press).

Bilinski, C.A. and G.P. Casey. 1989. Developments in sporulation and breeding of brewer's yeast. *Yeast* **5**: 429.

Blackburn, E.H. 1990. Telomeres: Structure and synthesis. *J. Biol. Chem.* **265**: 59

Blackburn, E.H. and K.M. Karrer. 1986. Genomic reorganization in ciliated protozoans. *Annu. Rev. Genet.* **20**: 501.

Bloom, K. and E. Yeh. 1989. Centromeres and telomeres: Structural elements of eukaryotic chromosomes. *Curr. Opin. Cell Biol.* **1**: 526.

Boeke, J.D., D. Eichinger, D. Castrillon, and G.R. Fink. 1988. The *Saccharomyces cerevisiae* genome contains functional and nonfunctional copies of transposon Ty1. *Mol. Cell. Biol.* **8**: 1432.

Boeke, J.D., D.J. Garfinkel, C.A. Styles, and G.R. Fink. 1985. Ty elements transpose through an RNA intermediate. *Cell* **40**: 491.

Bourne, H.R., D.A. Sanders, and F. McCormick. 1990. The GTPase superfamily: A conserved switch for diverse cell functions. *Nature* **348**: 125.

Brand, A.H., L. Breeden, J. Abraham, R. Sternglanz, and K. Nasmyth. 1985. Characterization of a "silencer" in yeast: A DNA sequence with properties opposite to those of a transcriptional enhancer. *Cell* **41**: 41.

Brewer, B.J. and W.L. Fangman. 1987. The localization of replication origins on *ARS* plasmids in *S. cerevisiae. Cell* **51**: 463.

Brodeur, G.M., S.B. Sandmeyer, and M.V. Olson. 1983. Consistent association between *sigma* elements and tRNA genes in yeast. *Proc Natl. Acad. Sci.* **80**: 3292.

Brown, W.R.A. 1989. Molecular cloning of human telomeres in yeast. *Nature* **338**: 774.

Brownstein, B.H., G.A. Silverman, R.D. Little, D.T. Burke, S.J. Korsmeyer, D. Schlessinger, and M.V. Olson. 1989. Isolation of single-copy human genes from a library of yeast artificial chromosome clones. *Science* **244**: 1348.

Burke, D.T., G.F. Carle, and M.V. Olson. 1987. Cloning of large segments of exogenous DNA into yeast by means of artificial chromosome vectors. *Science* **236**: 806.

Butt, T.R. and D.J. Ecker. 1987. Yeast metallothionein and applications in biotechnology. *Microbiol. Rev.* **51**: 351.

Button, L.L. and C.R. Astell. 1986. The *Saccharomyces cerevisiae* chromosome III left telomere has a type X, but not a type Y ', *ARS* region. *Mol. Cell. Biol.* **6**: 1352.

Byers, B. and L. Goetsch. 1975. Electron microscopic observations on the meiotic karyotype of diploid and tetraploid *Saccharomyces cerevisiae. Proc. Natl. Acad. Sci.* **72**: 5056.

Cameron, J.R., E.Y. Loh, and R.W. Davis. 1979. Evidence for transposition of dispersed repetitive DNA families in yeast. *Cell* **16**: 739.

Carle, G. and M.V. Olson. 1984. Separation of chromosomal DNA molecules from yeast by orthogonal-field-alternation gel electrophoresis. *Nucleic Acids Res.* **12**: 5647.

Carle, G. and M.V. Olson. 1985. An electrophoretic karyotype for yeast. *Proc. Natl. Acad. Sci.* **82**: 3756.

———. 1987. Orthogonal-field-alternation gel electrophoresis. *Methods Enzymol.* **155**: 468.

Carlson, M., J.L. Celenza, and F.J. Eng. 1985. Evolution of the dispersed *SUC* gene family of *Saccharomyces* by rearrangements of chromosome telomeres. *Mol. Cell. Biol.* **5**: 2894.

Carlson, M., B.C. Osmond, and D. Botstein. 1981. Genetic evidence for a silent *SUC* gene in yeast. *Genetics* **98**: 41.

Carson, M.J. and L. Hartwell. 1985. *CDC17*: An essential gene that prevents telomere elongation in yeast. *Cell* **42**: 249.

Chalker, D.L. and S.B. Sandmeyer. 1990. Transfer RNA genes are genomic targets for *de novo* transposition of the yeast retrotransposon Ty3. *Genetics* **126**: 837.

Chan, C.S.M. and B.-K. Tye. 1980. Autonomously replicating sequences in *Saccharomyces cerevisiae*. *Proc. Natl. Acad. Sci.* **77**: 6329.

———. 1983a. Organization of DNA sequences and replication origins at yeast telomeres. *Cell* **33**: 563.

———. 1983b. A family of *Saccharomyces cerevisiae* repetitive autonomously replicating sequences that have very similar genomic environments. *J. Mol. Biol.* **168**: 505.

Charron, M.J., E. Read, S.R. Haut, and C.A. Michels. 1989. Molecular evolution of the telomere-associated *MAL* loci of *Saccharomyces*. *Genetics* **122**: 307.

Cheng, J.-F., C.L. Smith, and C.R. Cantor. 1989. Isolation and characterization of a human telomere. *Nucleic Acids Res.* **17**: 6109.

Chisholm, G.E., F.S. Genbauffe, and T.G. Cooper. 1984. *tau*, a repeated DNA sequence in yeast. *Proc. Natl. Acad. Sci.* **81**: 2965.

Clare, J. and P. Farabaugh. 1985. Nucleotide sequence of a yeast Ty element: Evidence for an unusual mechanism of gene expression. *Proc. Natl. Acad. Sci.* **82**: 2829.

Clark-Walker, G.D. and A.A. Azad. 1980. Hybridizable sequences between cytoplasmic ribosomal RNAs and 3 micron circular DNAs of *Saccharomyces cerevisiae* and *Torulopsis glabrata*. *Nucleic Acids Res.* **8**: 1009.

Clarke, L. 1990. Centromeres of budding and fission yeasts. *Trends Genet.* **6**: 150.

Clarke, L. and J. Carbon. 1980. Isolation of a yeast centromere and construction of functional small circular chromosomes. *Nature* **287**: 504.

———. 1983. Genomic substitution of centromeres in *Saccharomyces cerevisiae*. *Nature* **305**: 23.

Coleman, K.G., H. Yde Steensma, D. B. Kaback, and J.R. Pringle. 1986. Molecular cloning of chromosome I DNA from *Saccharomyces cerevisiae*: Isolation and characterization of the *CDC24* gene and adjacent regions of the chromosome. *Mol. Cell. Biol.* **6**: 4516.

Corcoran, L.M., J.K. Thompson, D. Walliker, and D.J. Kemp. 1988. Homologous recombination within subtelomeric repeat sequences generates chromosome size polymorphisms in *P. falciparum*. *Cell* **53**: 807.

Cottarel, G., J.H. Shero, P. Hieter, and J.H. Hegemann. 1989. A 125-base-pair *CEN6* DNA fragment is sufficient for complete meiotic and mitotic centromere functions in *Saccharomyces cerevisiae*. *Mol. Cell. Biol.* **9**: 3342.

Craig, E.A. 1989. Essential roles of 70kDa heat inducible proteins. *BioEssays* **11**: 48.

Cross, S.H., R.C. Allshire, S.J. McKay, N.I. McGill, and H.J. Cooke. 1989. Cloning of human telomeres by complementation in yeast. *Nature* **338**: 771.

Curcio, M.J., A.-M. Hedge, J.D. Boeke, and D.J. Garfinkel. 1990. Ty RNA levels determine the spectrum of retrotransposition events that activate gene expression in *Saccharomyces cerevisiae*. *Mol. Gen. Genet.* **220**: 213.

De Jonge, P., F.C.M. De Jongh, R. Meijers, H.Y. Steensma, and W.A. Scheffers. 1986. Orthogonal-field-alternation gel electrophoresis banding patterns of DNA from yeasts. *Yeast* **2**: 193.

de Lange, T., L. Shiue, R.M. Myers, D.R. Cox, S.L. Naylor, A.M. Killery, and H.E. Varmus. 1990. Structure and variability of human chromosome ends. *Mol. Cell. Biol.* **10**: 518.

De Robertis, E.M. and M.V. Olson. 1979. Transcription and processing of cloned yeast tyrosine tRNA genes microinjected into frog oocytes. *Nature* **278**: 137.

Diehl, B.E. and J.R. Pringle. 1991. Molecular analysis of *Saccharomyces cerevisiae* chromosome I: Identification of additional transcribed regions and demonstration that some encode essential functions. *Genetics* **127**: 837.

Downs, K.M., G. Brennan, and S.W. Liebman. 1985. Deletions extending from a single

Ty1 element in *Saccharomyces cerevisiae. Mol. Cell. Biol.* **5:** 3451.

Dresser, M.E. and C.N. Giroux. 1988. Meiotic chromosome behavior in spread preparations of yeast. *J. Cell Biol.* **106:** 567.

Eibel, H., J. Gafner, A. Stotz, and P. Philippsen. 1981. Characterization of the yeast mobile element Ty1. *Cold Spring Harbor Symp. Quant. Biol.* **45:** 609.

Eigel, A. and H. Feldmann. 1982. Ty1 and delta elements occur adjacent to several tRNA genes in yeast. *EMBO J.* **1:** 1245.

Fink, G.R. 1987. Pseudogenes in yeast? *Cell* **49:** 5.

Fink, G.R., J.D. Boeke, and D.J. Garfinkel. 1986. The mechanism and consequences of retrotransposition. *Trends Genet.* **2:** 118.

Fitzgerald-Hayes, M. 1987. Yeast centromeres. *Yeast* **3:** 187.

Fogel, S., J.W. Welch, G. Cathala, and M. Karin. 1983. Gene amplification in yeast: *CUP1* copy number regulates copper resistance. *Curr. Genet.* **7:** 347.

Fulton, A.M., J. Mellor, M.J. Dobson, J. Chester, J.R. Warmington, K.J. Indge, S.G. Oliver, P. de la Paz, W. Wilson, A.J. Kingsman, and S.M. Kingsman. 1985. Variants within the yeast Ty sequence family encode a class of structurally conserved proteins. *Nucleic Acids Res.* **13:** 4097.

Futcher, B. and J. Carbon. 1986. Toxic effects of excess cloned centromeres. *Mol. Cell. Biol.* **6:** 2213.

Gafner, J., E.M. De Robertis, and P. Philippsen. 1983. Delta sequences in the 5′ non-coding region of yeast tRNA genes. *EMBO J.* **2:** 583.

Genbauffe, F.S., G.E. Chisholm, and T.G. Cooper. 1984. Tau, Sigma and Delta. A family of repeated elements in yeast. *J. Biol. Chem.* **259:** 10518.

Giroux, C.N., J.R.A. Mis, M.K. Pierce, S.E. Kohalmi, and B.A. Kunz. 1988. DNA sequence analysis of spontaneous mutations in the *SUP4-o* gene of *Saccharomyces cerevisiae. Mol. Cell. Biol.* **8:** 978.

Goebl, M.G. and T.D. Petes. 1986. Most of the yeast genomic sequences are not essential for cell growth and division. *Cell* **46:** 983.

Goodman, N.M., M.V. Olson, and B.D. Hall. 1977. Nucleotide sequence of a mutant eukaryotic gene: The yeast tyrosine-inserting ochre suppressor *SUP4-o. Proc. Natl. Acad. Sci.* **74:** 5453.

Green, E.D. and M.V. Olson. 1990. Chromosomal region of the cystic fibrosis gene in yeast artificial chromosomes: A model for human genome mapping. *Science* **250:** 94.

Greider, C.W. 1990. Telomeres, telomerase and senescence. *BioEssays* **12:** 363.

Guthrie, C. and J. Abelson. 1982. Organization and expression of tRNA genes in *Saccharomyces cerevisiae.* In *The molecular biology of the yeast* Saccharomyces: *Metabolism and gene expression* (ed. J.N. Strathern et al.), p. 487. Cold Spring Harbor Laboratory, Cold Spring Harbor, New York.

Hahn, S., J. Pinkham, R. Wei, R. Miller, and L. Guarente. 1988. The *HAP3* regulatory locus of *Saccharomyces cerevisiae* encodes divergent overlapping transcripts. *Mol. Cell. Biol.* **8:** 655.

Hahnenberger, K.M., M.P. Baum, C.M. Polizzi, J. Carbon, and L. Clarke. 1989. Construction of functional artificial minichromosomes in the fission yeast *Schizosaccharomyces pombe. Proc. Natl. Acad. Sci.* **86:** 577.

Hansen, L.J., D.L. Chalker, and S.B. Sandmeyer. 1988. Ty3, a yeast retrotransposon associated with tRNA genes, has homology to animal retroviruses. *Mol. Cell. Biol.* **8:** 5245.

Harris, S.D. and J.R. Pringle. 1991. Genetic analysis of *Saccharomyces cerevisiae* chromosome I: On the role of mutagen specificity in delimiting the set of genes identifiable using temperature-sensitive-lethal mutations. *Genetics* **127:** 279.

Hauber, J., R. Stucka, R. Krieg, and H. Feldmann. 1988. Analysis of yeast chromosomal regions carrying members of the glutamate tRNA gene family: Various transposable elements are associated with them. *Nucleic Acids Res.* **16:** 10623.

Henderson, E.R. and E.N. Blackburn. 1989. An overhanging 3' terminus is a conserved feature of telomeres. *Mol. Cell. Biol.* **9:** 345.

Hereford, L., K. Fahrner, J. Woolford, Jr., M. Rosbash, and D.B. Kaback. 1979. Isolation of yeast histone genes *H2A* and *H2B*. *Cell* **18:** 1261.

Hieter, P., C. Mann, M. Snyder and R.W. Davis. 1985a. Mitotic stability of yeast chromosomes: A colony color assay that measures nondisjunction and chromosome loss. *Cell* **40:** 381.

Hieter, P., D. Pridmore, J.H. Hegemann, M. Thomas, R.W. Davis, and P. Philippsen. 1985b. Functional selection and analysis of yeast centromeric DNA. *Cell* **42:** 913.

Hill, A. and K. Bloom. 1987. Genetic manipulation of centromere function. *Mol. Cell. Biol.* **7:** 2397.

————. 1989. Acquisition and processing of a conditional dicentric chromosome in *Saccharomyces cerevisiae*. *Mol. Cell. Biol.* **9:** 1368.

Horowitz, H. and J.E. Haber. 1984. Subtelomeric regions of yeast chromosomes contain a 36 base-pair tandemly repeated sequence. *Nucleic Acids Res.* **12:** 7105.

————. 1985. Identification of autonomously replicating circular subtelomeric Y' elements in *Saccharomyces cerevisiae*. *Mol. Cell. Biol.* **5:** 2369.

Horowitz, N., P. Thorburn, and J.E. Haber. 1984. Rearrangements of highly polymorphic regions near telomeres of *Saccharomyces cerevisiae*. *Mol. Cell. Biol.* **4:** 2509.

Hsiao, C.-L. and J. Carbon. 1979. High-frequency transformation of yeast by plasmids containing the cloned yeast *ARG4* gene. *Proc. Natl. Acad. Sci.* **76:** 3829.

Huberman, J.A., J. Zhu, L.R. Davis, and C.S. Newlon. 1988. Close association of a DNA replication origin and an *ARS* element on chromosome III of the yeast, *Saccharomyces cerevisiae*. *Nucleic Acids Res.* **16:** 6373.

Huberman, J.A., R.D. Pridmore, D. Jaeger, B. Zonneveld, and P. Philippsen. 1986. Centromeric DNA from *Saccharomyces uvarum* is functional in *Saccharomyces cerevisiae*. *Chromosoma* **94:** 162.

Jaeger, D. and P. Philippsen. 1989. Many yeast chromosomes lack the telomere-specific Y' sequence. *Mol. Cell. Biol.* **9:** 5754.

Jensen, R., G.F. Sprague, Jr., and I. Herskowitz. 1983. Regulation of yeast mating-type interconversion: Feedback control of *HO* gene expression by the mating-type locus. *Proc. Natl. Acad. Sci.* **80:** 3035.

Johnson, D.I. and J.R. Pringle. 1990. Molecular characterization of *CDC42*, a *Saccharomyces cerevisiae* gene involved in the development of cell polarity. *J. Cell Biol.* **111:** 143.

Johnson, D.I., C.W. Jacobs, J.R. Pringle, L.C. Robinson, G.F. Carle, and M.V. Olson. 1987. Mapping of the *Saccharomyces cerevisiae CDC3*, *CDC25*, and *CDC42* genes to chromosome XII by chromosome blotting and tetrad analysis. *Yeast* **3:** 243.

Johnston, J.R. and R.K. Mortimer. 1986. Electrophoretic karyotyping of laboratory and commercial strains of *Saccharomyces* and other yeasts. *Int. J. Syst. Bacteriol.* **36:** 569.

Johnston, M. 1987. A model fungal gene regulatory mechanism: The *GAL* genes of *Saccharomyces cerevisiae*. *Microbiol. Rev.* **51:** 458.

Kaback, D.B. and N. Davidson. 1980. Organization of the ribosomal RNA gene cluster in the yeast *Saccharomyces cerevisiae*. *J. Mol. Biol.* **138:** 745.

Kaback, D.B., N.Y. Steensma, and P. De Jonge. 1989. Enhanced meiotic recombination on the smallest chromosome of *Saccharomyces cerevisiae*. *Proc. Natl. Acad. Sci.* **86:** 3694.

Kaback, D.B., P.W. Oeller, H.Y. Steensma, J. Hirschman, D. Ruezinsky, K.G. Coleman, and J.R. Pringle. 1984. Temperature-sensitive lethal mutations on yeast chromosome I appear to define only a small number of genes. *Genetics* **108:** 67.

Karin, M., R. Najarian, A. Haslinger, P. Valenzuela, J. Welch, and S. Fogel. 1984. Primary structure and transcription of an amplified genetic locus: The *CUP1* locus of yeast. *Proc. Natl. Acad. Sci.* **81:** 337.

Kingsman, A.J., R.L. Gimlich, L. Clarke, A.C. Chinault, and J. Carbon. 1981. Sequence variation in dispersed repetitive sequences in *Saccharomyces cerevisiae. J. Mol. Biol.* **145:** 619.

Klein, H.L. and T.D. Petes. 1984. Genetic mapping of Ty elements in *Saccharomyces cerevisiae. Mol. Cell. Biol.* **4:** 329.

Kolodrubetz, D., M.C. Rykowski, and M. Grunstein. 1982. Histone H2A subtypes associate interchangeably *in vivo* with histone H2B subtypes. *Proc. Natl. Acad. Sci.* **79:** 7814.

Kurjan, J. 1985. α-Factor structural gene mutations in *Saccharomyces cerevisiae*: Effects on α-factor production and mating. *Mol. Cell. Biol.* **5:** 787.

Kuroiwa, T., H. Kojima, I. Miyakawa, and N. Sando. 1984. Meiotic karyotype of the yeast *Saccharomyces cerevisiae. Exp. Cell Res.* **153:** 259.

Lambie, E.J. and G.S. Roeder. 1986. Repression of meiotic crossing over by a centromere (*CEN3*) in *Saccharomyces cerevisiae. Genetics* **114:** 769.

Larionov, V.L., A.V. Grishin, and M.N. Smirnov. 1980. 3 μm DNA—An extrachromosomal ribosomal DNA in the yeast *Saccharomyces cerevisiae. Gene* **12:** 41.

Larson D.D., E.A. Spangler, and E.H. Blackburn. 1987. Dynamics of telomere length variation in *Tetrahymena thermophila. Cell* **50:** 477.

Lauer, G.D., T.M. Roberts, and L.C. Klotz. 1977. Determination of the nuclear DNA content of *Saccharomyces cerevisiae* and implications for the organization of DNA in yeast chromosomes. *J. Mol. Biol.* **114:** 507.

Liebman, S., P. Shalit, and S. Picologlou. 1981. Ty elements are involved in the formation of deletions in *DEL1* strains of *Saccharomyces cerevisiae. Cell* **26:** 401.

Link, A.J. and M.V. Olson. 1991. Physical map of the *Saccharomyces cerevisiae* genome at 110-kb resolution. *Genetics* (in press).

Lochmueller, H., R. Stucka, and H. Feldmann. 1989. A hot-spot for transposition of various Ty elements on chromosome V in *Saccharomyces cerevisiae. Curr. Genet.* **16:** 247.

Louis, E.J. and J.E. Haber. 1990a. The subtelomeric Y′ repeat family in *Saccharomyces cerevisiae*: An experimental system for repeated sequence evolution. *Genetics* **124:** 533.

―――. 1990b. Mitotic recombination among subtelomeric Y′ repeats in *Saccharomyces cerevisiae. Genetics* **124:** 547.

Lundblad, V. and J.W. Szostak. 1989. A mutant with a defect in telomere elongation leads to senescence in yeast. *Cell* **57:** 633.

Lustig, A.J. and T.D. Petes. 1986. Identification of yeast mutants with altered telomere structure. *Proc. Natl. Acad. Sci.* **83:** 1398.

Ma, H. and D. Botstein. 1986. Effects of null mutations in the hexokinase genes of *Saccharomyces cerevisiae* on catabolite repression. *Mol. Cell. Biol.* **6:** 4046.

Malavasic, M.J. and R.T. Elder. 1990. Complementary transcripts from two genes necessary for normal meiosis in the yeast *Saccharomyces cerevisiae. Mol. Cell. Biol.* **10:** 2809.

McAlister, L. and M.J. Holland. 1985. Isolation and characterization of yeast strains carrying mutations in the glyceraldehyde-3-phosphate dehydrogenase genes. *J. Biol. Chem.* **260:** 15013.

McGrath, J.P., S. Jentsch, and A. Varshavsky. 1991. *UBA1:* An essential yeast gene encoding ubiquitin-activating enzyme. *EMBO J.* **10:** 227.

McGrew, J., B. Diehl, and M. Fitzgerald-Hayes. 1986. Single base-pair mutations in centromere element III cause aberrant chromosome segregation in *Saccharomyces cerevisiae. Mol. Cell. Biol.* **6:** 530.

Melnick, L. and F. Sherman. 1990. Nucleotide sequence of the *COR* region: A cluster of six genes in the yeast *Saccharomyces cerevisiae. Gene* **87:** 157.

Meyerink, J.A., J. Klootwijk, R.J. Planta, A. van der Ende, and E.F.J. van Bruggen. 1979. Extrachromosomal circular ribosomal DNA in the yeast *Saccharomyces carlsbergensis. Nucleic Acids Res.* **7:** 69.

Mortimer, R.K. and D.C. Hawthorne. 1969. Yeast genetics. In *The yeasts* (ed. A.H. Rose and J.S. Harrison), vol. I, p. 385. Academic Press, New York.

Mortimer, R.K. and J.R. Johnston. 1986. Genealogy of principal strains of the yeast genetic stock center. *Genetics* **113:** 35.

Mortimer, R.K. and D. Schild. 1980. Genetic map of *Saccharomyces cerevisiae. Microbiol. Rev.* **44:** 519.

————. 1985. Genetic map of *Saccharomyces cerevisiae*, edition 9. *Microbiol. Rev.* **49:** 181.

Mortimer, R.K., D. Schild, C.R. Contopoulou, and J.A. Kans. 1989. Genetic map of *Saccharomyces cerevisiae*, edition 10. *Yeast* **5:** 321.

Murray, A.W. and J.W. Szostak. 1983. Construction of artificial chromosomes in yeast. *Nature* **305:** 189.

Murray, A.W., N.P. Schultes, and J.W. Szostak. 1986. Chromosome length controls mitotic chromosome segregation in yeast. *Cell* **5:** 529.

Natsoulis, G., W. Thomas, M.C. Roghmann, F. Winston, and J.D. Boeke. 1989. Ty1 transposition in *Saccharomyces cerevisiae* is nonrandom. *Genetics* **123:** 269.

Newlon, C.S. 1988. Yeast chromosome replication and segregation. *Microbiol. Rev.* **52:** 568.

————. 1989. Deoxyribonucleic acid organization and replication. In *The yeasts* (ed. A.H. Rose and J.S. Harrison), vol. III, p. 57. Academic Press, San Diego.

Newlon, C.S., R.P. Green, K.J. Hardeman, K.E. Kim, L.R. Lipchitz, T.G. Palzkill, S. Synn, and S.T. Woody. 1986. Structure and organization of yeast chromosome III. In *Yeast cell biology* (ed. J.B. Hicks), p. 211. A.R. Liss, New York.

Ohmen, J.D., K.A. Burke, and J.E. McEwen. 1990. Divergent overlapping transcripts at the *PET122* locus in *Saccharomyces cerevisiae. Mol. Cell. Biol.* **10:** 3027.

Olson, M.V. 1989. Pulsed-field gel electrophoresis. In *Genetic engineering* (ed. J.K. Setlow), p. 183. Plenum Press, New York.

Olson, M.V., K. Loughney, and B.D. Hall. 1979. Identification of the yeast DNA sequences that correspond to specific tyrosine-inserting nonsense suppressor loci. *J. Mol. Biol.* **132:** 387.

Olson, M.V., G.S. Page, A. Sentenac, P.W. Piper, M. Worthington, R.B. Weiss, and B.D. Hall. 1981. Only one of two closely related yeast suppressor tRNA genes contains an intervening sequence. *Nature* **291:** 464.

Olson, M.V., J.E. Dutchik, M.Y. Graham, G.M. Brodeur, C. Helms, M. Frank, M. MacCollin, R. Scheinman, and T. Frank. 1986. Random-clone strategy for genomic restriction mapping in yeast. *Proc. Natl. Acad. Sci.* **83:** 7826.

Peterson, J.B. and H. Ris. 1976. Electron-microscopic study of the spindle and chromosome movement in the yeast *Saccharomyces cerevisiae. J. Cell Sci.* **22:** 219.

Petes, T.D. 1979a. Meiotic mapping of yeast ribosomal deoxyribonucleic acid on chromosome XII. *J. Bacteriol.* **138:** 185.

————. 1979b. Yeast ribosomal DNA genes are located on chromosome XII. *Proc. Natl.*

Acad. Sci. **76:** 410.

————. 1980. Unequal meiotic recombination within tandem arrays of yeast ribosomal DNA genes. *Cell* **19:** 765.

Petes, T.D. and C.W. Hill. 1988. Recombination between repeated genes in microorganisms. *Annu. Rev. Genet.* **22:** 147.

Planta, R.J. and H.A. Raue. 1988. Control of ribosome biogenesis in yeast. *Trends Genet.* **4:** 64.

Reynolds, A.E., R.M. McCarroll, C.S. Newlon, and W.L. Fangman. 1989. Time of replication of *ARS* elements along yeast chromosome III. *Mol. Cell. Biol.* **9:** 4488.

Riethman, H.C., R.K. Moyzis, J. Meyne, D.T. Burke, and M.V. Olson. 1989. Cloning human telomeric DNA fragments into *Saccharomyces cerevisiae* using a yeast-artificial-chromosome vector. *Proc. Natl. Acad. Sci.* **86:** 6240.

Riles, L. and M.V. Olson. 1988. Nonsense mutations in essential genes of *Saccharomyces cerevisiae. Genetics* **118:** 601.

Roeder, G.S. and G.R. Fink. 1983. Transposable elements in yeast. In *Mobile genetic elements* (ed. J.A. Shapiro), p. 299. Academic Press, New York.

Roeder, G.S., P.J. Farabaugh, D.T. Chaleff, and G.R. Fink. 1980. The origins of gene instability in yeast. *Science* **209:** 1375.

Rothstein, R., C. Helms, and N. Rosenberg. 1987. Concerted deletions and inversions are caused by mitotic recombination between delta sequences in *Saccharomyces cerevisiae. Mol. Cell. Biol.* **7:** 1198.

Rykowski, M.C., J.W. Wallis, J. Choe, and M. Grunstein. 1981. Histone H2B subtypes are dispensable during the yeast cell cycle. *Cell* **25:** 477.

Sandmeyer, S.B. and M.V. Olson. 1982. Insertion of a repetitive element at the same position in the 5′-flanking regions of two dissimilar yeast tRNA genes. *Proc. Natl. Acad. Sci.* **79:** 7674.

Sandmeyer, S.B., V.W. Bilanchone, D.J. Clark, P. Morcos, G.F. Carle, and G.M. Brodeur. 1988. Sigma elements are position-specific for many different yeast tRNA genes. *Nucleic Acids Res.* **16:** 1499.

Saunders, M., M. Fitzgerald-Hayes, and K. Bloom. 1988. Chromatin structure of altered yeast centromeres. *Proc. Natl. Acad. Sci.* **85:** 175.

Schnell, R. and J. Rine. 1986. A position effect on the expression of a tRNA gene mediated by the *SIR* genes in *Saccharomyces cerevisiae. Mol. Cell. Biol.* **6:** 494.

Schwartz, D.C. and C.R. Cantor. 1984. Separation of yeast chromosome-sized DNAs by pulsed field gradient gel electrophoresis. *Cell* **37:** 67.

Schwartz, D.C., W. Saffran, J. Welsh, R. Haas, M. Goldenberg, and C.R. Cantor. 1983. New techniques for purifying large DNAs and studying their properties and packaging. *Cold Spring Harbor Symp. Quant. Biol.* **47:** 189.

Shampay, J. and E.H. Blackburn. 1988. Generation of telomere-length heterogeneity in *Saccharomyces cerevisiae. Proc. Natl. Acad. Sci.* **85:** 534.

Shampay, J., J.W. Szostak, and E.H. Blackburn. 1984. DNA sequences of telomeres maintained in yeast. *Nature* **310:** 154.

Smith, M.M. and K. Murray. 1983. Yeast H3 and H4 histone messenger RNAs are transcribed from two non-allelic gene sets. *J. Mol. Biol.* **169:** 641.

Steensma, H.Y., J.C Crowley, and D.B. Kaback. 1987. Molecular cloning of chromosome I DNA from *Saccharomyces cerevisiae*: Isolation and analysis of the *CEN1-ADE1-CDC15* region. *Mol. Cell. Biol.* **7:** 410.

Steensma, H.Y., P. de Jonge, A. Kaptein, and D.B. Kaback. 1989. Molecular cloning of chromosome I DNA from *Saccharomyces cerevisiae*: Localization of a repeated sequence containing an acid phosphatase gene near a telomere of chromosome I and chromosome VIII. *Curr. Genet.* **16:** 131.

Stinchcomb, D.T., K. Struhl, and R.W. Davis. 1979. Isolation and characterisation of a yeast chromosomal replicator. *Nature* **282:** 39.

Strathern, J.N., C.S. Newlon, I. Herskowitz, and J.B. Hicks. 1979. Isolation of a circular derivative of yeast chromosome III: Implications for the mechanism of mating type interconversion. *Cell* **18:** 309.

Struhl, K. 1985. Nucleotide sequence and transcriptional mapping of the yeast *pet56-his3-ded1* region. *Nucleic Acids Res.* **13:** 8587.

Struhl, K. and R.W. Davis. 1981. Transcription of the *his3* gene region in *Saccharomyces cerevisiae. J. Mol. Biol.* **152:** 535.

Struhl, K., D.T. Stinchcomb, S. Scherer, and R.W. Davis. 1979. High-frequency transformation of yeast: Autonomous replication of hybrid DNA molecules. *Proc. Natl. Acad. Sci.* **76:** 1035.

Stucka, R., H. Lochmueller, and H. Feldmann. 1989. Ty4, a novel low-copy number element in *Saccharomyces cerevisiae*: One copy is located in a cluster of Ty elements and tRNA genes. *Nucleic Acids Res.* **17:** 4993.

Sundquist, W.I. and A. Klug. 1989. Telomeric DNA dimerizes by formation of guanine tetrads between hairpin loops. *Nature* **342:** 825.

Surosky, R.T. and B.-K. Tye. 1985. Construction of telocentric chromosomes in *Saccharomyces cerevisiae. Proc. Natl. Acad. Sci.* **82:** 2106.

Symington, L.S. and T.D. Petes. 1988. Meiotic recombination within the centromere of a yeast chromosome. *Cell* **52:** 237.

Szankasi, P., C. Gysler, U. Zehntner, U. Leupold, J. Kohli, and P. Munz. 1986. Mitotic recombination between dispersed but related tRNA genes of *Schizosaccharomyces pombe* generates a reciprocal translocation. *Mol. Gen. Genet.* **202:** 394.

Szostak, J.W. and E.H. Blackburn. 1982. Cloning yeast telomeres on linear plasmid vectors. *Cell* **29:** 245.

Szostak, J.W. and R. Wu. 1980. Unequal crossing over in the ribosomal DNA of *Saccharomyces cerevisiae. Nature* **284:** 426.

Tajima, M., Y. Nogi, and T. Fukasawa. 1986. Duplicate upstream activating sequences in the promoter region of the *Saccharomyces cerevisiae GAL7* gene. *Mol. Cell. Biol.* **6:** 246.

Tatchell, K., L.C. Robinson, and M. Breitenbach. 1985. *RAS2* of *Saccharomyces cerevisiae* is required for gluconeogenic growth and proper response to nutrient limitation. *Proc. Natl. Acad. Sci.* **82:** 3785.

Thierry, A., C. Fairhead, and B. Dujon. 1990. The complete sequence of the 8.2 kb segment left of *MAT* on chromosome III reveals five ORFs, including a gene for a yeast ribokinase. *Yeast* **6:** 521.

Vollrath, D., R.W. Davis, C. Connelly, and P. Hieter. 1988. Physical mapping of large DNA by chromosome fragmentation. *Proc. Natl. Acad. Sci.* **85:** 6027.

Wallis, J.W., L. Hereford, and M. Grunstein. 1980. Histone H2B genes of yeast encode two different proteins. *Cell* **22:** 799.

Walmsley, R.M. 1987. Yeast telomeres: The end of the chromosome story? *Yeast* **3:** 139.

Walmsley, R.M. and T.D. Petes. 1985. Genetic control of chromosome length in yeast. *Proc. Natl. Acad. Sci.* **82:** 506.

Walmsley, R.W., C.S.M. Chan, B.-K. Tye, and T.D. Petes. 1984. Unusual DNA sequences associated with the ends of yeast chromosomes. *Nature* **310:** 157.

Warmington, J.R., R.W. Waring, C.S. Newlon, K.J. Indge, and S.G. Oliver. 1985. Nucleotide sequence characterization of Ty 1-17, a class II transposon from yeast. *Nucleic Acids Res.* **13:** 6679.

Warner, J.R. 1989. Synthesis of ribosomes in *Saccharomyces cerevisiae. Microbiol. Rev.* **53:** 256.

West, R.W., R.R. Yocum, and M. Ptashne. 1984. *Saccharomyces cerevisiae GAL1-GAL10* divergent promoter region: Location and function of the upstream activating sequence UAS_G. *Mol. Cell. Biol.* **4:** 2467.

Yochem, J. and B. Byers. 1987. Structural comparison of the yeast cell division cycle gene *CDC4* and a related pseudogene. *J. Mol. Biol.* **195:** 233.

Yoshikawa, A. and K. Isono. 1990. Chromosome III of *Saccharomyces cerevisiae*: An ordered clone bank, a detailed restriction map and analysis of transcripts suggest the presence of 160 genes. *Yeast* **6:** 401.

Zakian, V. 1989. Structure and function of telomeres. *Annu. Rev. Genet.* **23:** 579.

Zakian, V.A. and H.M. Blanton. 1988. Distribution of telomere-associated sequences on natural chromosomes in *Saccharomyces cerevisiae. Mol. Cell. Biol.* **8:** 2257.

2

Chromosomal DNA Replication

Judith L. Campbell
Divisions of Chemistry and Biology
California Institute of Technology
Pasadena, California 91125

Carol S. Newlon
Department of Microbiology and Molecular Genetics
UMDNJ-New Jersey Medical School
Newark, New Jersey 07103

I. **Introduction**
II. **Replication of Chromosomal DNA**
    A. Timing and Duration of S Phase
    B. Chromosomal DNA Replication
        1. Replicon Spacing and Fork Movement
        2. Identification of Replication Origins
        3. Replication of Ribosomal DNA
        4. Replication of Chromosome III
        5. Replication Termini
        6. Telomere Replication
    C. Temporal Order of Replication
III. **Autonomously Replicating Sequence Elements**
    A. Assays
    B. DNA Sequence Analysis
        1. *ARS* Consensus Sequence
        2. 3′-flanking Sequence
    C. *ARS*-binding Proteins
        1. *ARS*-binding Factor 1 (Abf1p)
        2. *ARS*-binding Factors II and III
        3. Chromatin Proteins and Nuclear Scaffold Interactions
IV. **Replication Apparatus**
    A. DNA Polymerases
        1. DNA Polymerase α
        2. Requirements and Inhibitors
        3. Mechanism
        4. Genetics
        5. Structure/Function Studies of Yeast DNA Polymerase α
        6. Amino Terminus
        7. Carboxyl Terminus
        8. Reverse Genetics and Subunits of Yeast DNA Polymerase α
        9. Primase Genes

Volume I. The Molecular and Cellular Biology of the Yeast Saccharomyces: *Genome Dynamics, Protein Synthesis, and Energetics.* Copyright 1991 Cold Spring Harbor Laboratory Press 0-87969-355-X/91 $3 + 00   **41**

I. INTRODUCTION

A consistent theme in the first edition of this series (Strathern et al. 1981, 1982) was the potential of yeast as a model eukaryote. During the past few years, much of the promise of yeast as a model for eukaryotic DNA replication has been realized. The long-standing question of whether there are specific origins of replication in chromosomal DNA has been answered by the demonstration that replication initiates at or close to several *a*utonomously *r*eplicating *s*equence (*ARS*) elements in ribosomal DNA (rDNA) and on chromosome III. From the analysis of large segments of chromosomes, a picture of the spatial and temporal activation of replicons is beginning to emerge. The analysis of *ARS* elements has progressed to the point that nucleotide sequences required for origin function can be at least partially specified. Progress has also been made in identifying and characterizing proteins that function in DNA replication, particularly DNA polymerases and topoisomerases. In virtually every respect examined, yeast replication proteins are homologous to replication proteins from other eukaryotes, further validating yeast as a model system.

The features of yeast that make it particularly attractive for the study of DNA replication include its small genome size and the unparalleled power of its genetic system. The small size of yeast chromosomes, which are nevertheless typical of the larger chromosomes of other eukaryotes in organization and replication pattern, has facilitated cloning and mapping large segments of chromosomes. Cloning of chromosomal elements such as centromeres and telomeres was made easier by their relatively higher concentration in yeast chromosomal DNA. The characteristics of the *Saccharomyces* transformation system, in which nonreplicative plasmids integrate only by homologous recombination, have facilitated the genetic analysis and systematic mutagenesis of genes encoding replication functions. In addition, the relative inefficiency of transformation by integrative plasmids made efficient transformation by replicative plasmids stand out and quickly focused attention on *ARS* elements, *cis*-acting sequences that allow the extrachromosomal maintenance of plasmids in yeast, as potential replication origins.

In this paper, we focus on recent progress in understanding the organization and activation of chromosomal replicons and the proteins that participate in DNA replication. Other recent reviews focus on DNA polymerases (Burgers 1989), topoisomerases (Wang 1985, 1987; Sternglanz 1989), *ARS* elements (Kearsey 1986; Campbell 1988), and replication origins (Umek et al. 1989), or provide comprehensive coverage of yeast DNA replication (Campbell 1986; Newlon 1988, 1989).

II. REPLICATION OF CHROMOSOMAL DNA

A. Timing and Duration of S Phase

The replication of nuclear DNA species is restricted to the S phase of the *Saccharomyces cerevisiae* cell cycle (Zakian et al. 1979; Brewer et al. 1980). The length of the S phase and its timing relative to other cell-cycle events have been measured in several ways (Williamson 1965; Barford and Hall 1976; Slater et al. 1977; Johnston et al. 1980; Rivin and Fangman 1980a; Brewer et al. 1984). In cells growing in rich medium at the highest growth rates (90–120-min doubling times), the S phase occupies 25–50% of the cell cycle. The differences in the relative lengths of the S phase must result from either strain differences or variation in growth conditions, because studies on a set of five isogenic strains have demonstrated that neither ploidy nor mating type affect the duration of cell-cycle phases (Brewer et al. 1984). Daughter cells, which are generally smaller than mother cells at the time of cell division, characteristically have longer cell cycles than mother cells in the same population. Their longer cell cycles result from an expansion of the G_1 phase, during which daughter cells increase in size (Pringle and Hartwell 1981; Brewer et al. 1984).

When growth rate is decreased by growth on a poor carbon source, the length of the S and G_2 phases remain relatively constant and the length of the G_1 phase expands (Barford and Hall 1976; Slater et al. 1977). However, when growth rate is decreased by growth on a poor nitrogen source, either all phases of the cell cycle expand by the same factor (Rivin and Fangman 1980a) or both G_1 and S phases expand, with no effect on G_2 (Johnston et al. 1980). Whether the different cell-cycle responses are related to strain differences or to specific growth conditions is not clear. Although it seems that all phases of the cell cycle can be expanded independently, no easily discerned rules predict the pattern or degree of expansion when growth conditions are changed.

Bud emergence and bud growth are easily scored and are often used to monitor progress through the cell cycle. Analysis of cell-cycle mutants has revealed that the events controlling bud emergence and the events controlling DNA replication are on independent pathways (Pringle and Hartwell 1981; Adams et al. 1990). Thus, the relationship between the time of bud emergence and the beginning of S phase need not be fixed and, in fact, varies from strain to strain, with bud emergence occurring before or at the beginning of S phase in some strains (Williamson 1965; Barford and Hall 1976; Brewer et al. 1984) and as late as halfway through the S phase in other strains (Brewer et al. 1984; Rivin and Fangman 1980a).

B. Chromosomal DNA Replication

1. Replicon Spacing and Fork Movement

Each of the 16 yeast chromosomes contains multiple replication origins that occur at an average spacing of 36 kb, with a wide variation in individual interorigin distances (for discussion, see Newlon and Burke 1980; Newlon 1988). Replication forks move bidirectionally from most origins (Petes and Williamson 1975; Rivin and Fangman 1980b). The observation that most segments of replicating DNA contain several active replicons of similar extents of replication has led to the conclusion that adjacent replicons are activated at similar times and that they are therefore regulated coordinately (Newlon and Burke 1980; Rivin and Fangman 1980b). However, 20–30% of adjacent origins may differ in time of initiation by one fourth of S phase, and at least some origins appear to initiate replication in the second half of S phase (Rivin and Fangman 1980b).

2. Identification of Replication Origins

Yeast DNA replication initiates at specific sites, which correspond to previously identified *ARS* elements. *ARS* elements were described in 1979 as sequences that promote extrachromosomal replication of plasmids in yeast cells, and they were recognized by their ability to cause high-frequency transformation (Hsiao and Carbon 1979; Kingsman et al. 1979; Stinchcomb et al. 1979; Struhl et al. 1979). Because bacterial replication origins have similar properties, it was suggested that *ARS* elements are chromosomal replication origins. Consistent with this proposal, the spacing between *ARS* elements is similar to the spacing between origins measured from electron microscopy data. From the fraction of chromosomal DNA restriction fragments that carry *ARS* elements, it was estimated that there is one *ARS* per 32 kb (Chan and Tye 1980) or one *ARS* per 40 kb (Beach et al. 1980).

Two distinct origin-mapping techniques (Brewer and Fangman 1987; Huberman et al. 1987; Nawotka and Huberman 1988) have recently been used to document that *ARS* elements are origins of replication. Although they differ in detail, both methods rely on unique properties of restriction fragments derived from replicating DNA that can be detected by two-dimensional gel electrophoresis (see Fig. 1). The Brewer and Fangman technique takes advantage of unique migration properties of DNA fragments containing replication forks during second-dimension electrophoresis in high-percentage-agarose gels in the presence of ethidium bromide and under relatively high-voltage conditions. The Huberman method relies on alkaline gel conditions to separate nascent daughter strands from parental strands in the second dimension. In each case, restriction fragments of interest are identified by Southern blotting and hybridization with a radioactively labeled probe. The methods are independent and complementary and can be used to map replication origins, termination sites, and direction of fork movement.

These methods were first applied to multicopy plasmids, the endogenous 2-micron plasmid (Brewer and Fangman 1987; Huberman et al. 1987), and a plasmid containing *ARS1* from chromosome IV (Brewer and Fangman 1987). In both cases, replication was shown to initiate at or near the *ARS* element and to terminate approximately half way around the plasmid from the *ARS* element. In the case of the 2-micron plasmid, there was no evidence of the second origin predicted from earlier electron microscopy studies (Newlon et al. 1981), suggesting either that the second origin was an artifact of the preparation method or that it functions only under some conditions. Thus, in the two cases examined, *ARS* elements serve as replication origins on plasmids. This conclusion raises the question of whether *ARS* elements are chromosomal replication

A

SIMPLE Y   BUBBLE   DOUBLE Y   BUBBLE → Y   Y → DOUBLE Y

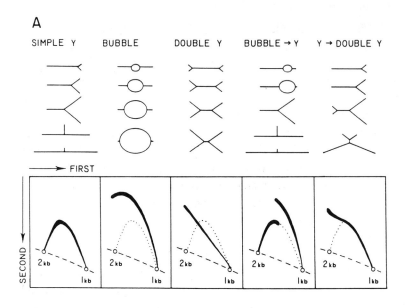

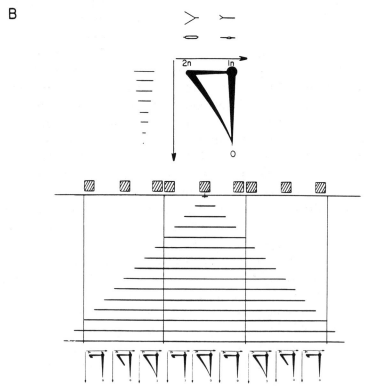

B

Figure 1 (See facing page for legend.)

origins. If so, then are all *ARS* elements origins and do all origins have *ARS* activity on plasmids? Studies of the tandem array of rRNA cistrons and a 200-kb region of chromosome III clearly show that some *ARS* elements serve as chromosomal replication origins, although the correlation is not simply one to one.

3. Replication of Ribosomal DNA

One of the first analyses of replication of a specific region of chromosomal DNA focused on the rRNA cistrons that map to chromosome XII (Petes 1979). This region of chromosome XII consists of approximately 120 tandem copies of a 9.1-kb repeat that contains two transcription units: one for the 5S rRNA and one that encodes the 37S RNA that is processed into the 25S, 18S, and 5.8S rRNAs (Udem and Warner 1972; Bell et al. 1977; Cramer et al. 1977; Petes et al. 1978; Philippsen et al. 1978). The transcription units are in opposite orientations and are separated from each other by *non*transcribed *s*pacer regions (NTS1 and NTS2). The 9.1-kb repeat contains an *ARS* element (Szostak and Wu 1979; Kouprina and Larionov 1983) that has been mapped to a 570-bp region of NTS2 (Skryabin et al. 1984).

Early analysis of nascent DNA from this region using alkaline sucrose gradients suggested that the average rDNA replicon was five repeats in length (Walmsley et al. 1984b), and in their study of replicating rDNA chromatin, Saffer and Miller (1986) were able to visualize replicons one to three repeats in length. Their analysis of small replica-

---

Figure 1 Replication origin mapping techniques. (*A*) Brewer and Fangman neutral-neutral technique: The first-dimension electrophoresis separates primarily on the basis of size. The second dimension separates on the basis of shape. (*Upper* panels) Five replication patterns for an arbitrary 1-kb DNA fragment; (*lower* panels) expected migration in two-dimensional gels of the replication intermediates shown above. Dashed lines mark the locations of linear molecules of various sizes and dotted lines show the location of simple Y replication intermediates (*A*) for reference (from Brewer et al. 1988). (*B*) Huberman neutral alkaline technique: The first-dimension electrophoresis separates primarily on the basis of size, and in the alkaline second dimension, molecules are denatured and nascent single strands are separated from full-length parental strands. The parental strands form a horizontal line between a mass of 2n and 1n. (*Upper* panel) Nascent strands fall along a diagonal arc. (*Lower* panel) Line at the top of the panel depicts three adjacent DNA fragments with an origin in the middle of the middle fragment. Probes from different parts of the fragments (indicated by the hatched boxes) detect different extents of the nascent strand diagonal, with the probe closest to the origin detecting the shortest nascent strands. (Supplied by J. Hamlin.)

tion bubbles mapped the replication origin to NTS2, coincident with the *ARS* element. Application of two-dimensional gel mapping techniques to this region has confirmed that replication usually initiates within NTS2 at a position that is within experimental error of the mapped *ARS* element (Linskens and Huberman 1988). In addition, although replication is initially bidirectional from the origin, the fork that moves in the direction opposite to transcription of the 37S gene stops at the 3′ end of the transcript, and only the other fork, moving in the same direction as transcription of the 37S gene, continues replication (Brewer and Fangman 1988; Linskens and Huberman 1988). Therefore, rDNA replication is asymmetric. In addition, replication initiates only once in three to ten repeats, and therefore most of the rDNA *ARS* elements are not active as origins in any given S phase.

4. Replication of Chromosome III

The other chromosomal region whose replication has been analyzed extensively is an approximately 200-kb contiguous region of chromosome III that includes sequences from near the left telomere to the *MAT* locus on the right arm halfway between the centromere and telomere. Newlon and collaborators have cloned and restriction-mapped this region and have identified and mapped 14 *ARS* elements. Figure 2 summarizes the physical map of this region, with the location of genes and *ARS* elements indicated (Newlon et al. 1986; Button and Astell 1988; Reynolds et al. 1989). Also shown is a summary of the replicon mapping of this region carried out in the laboratories of Newlon and Huberman (see below).

The *ARS* elements in Figure 2 are named according to a new convention in which the first digit indicates the chromosome on which the *ARS* element is located and the second two digits provide a unique identifier for the *ARS*. Since most or all of the *ARS* elements in this region have been identified, they are numbered from left to right. Table 1 lists the new designations for these *ARS* elements as well as the old designations that were based on the restriction fragments in which they resided.

Only four of the nine *ARS* elements tested to date show evidence of origin activity in vivo. Replication initiates at or near *ARS305* (Huberman et al. 1988), *ARS306* (J. Zhu, L.R. Davis, J.A. Huberman, and C.S. Newlon, in prep.), and *ARS307* and *ARS309* (S.A. Greenfeder and C.S. Newlon, in prep.). Regions of replication termination have been identified between *ARS305* and *ARS306* (J. Zhu, L.R. Davis, C.S. Newlon, and J.A. Huberman, in prep.) and between *ARS307* and *ARS309* (S.A. Greenfeder and C.S. Newlon, in prep.). Whether each of these *ARS* elements is active as an origin in every S phase is not yet resolved. Two-

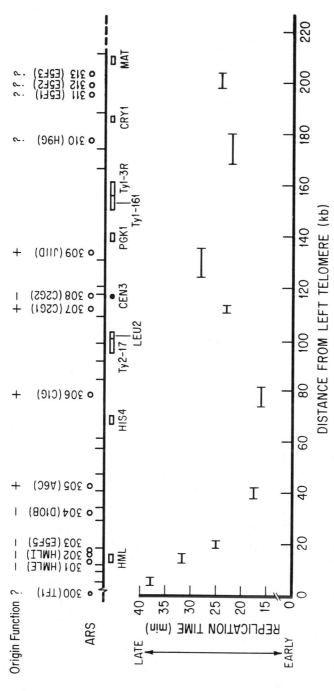

Figure 2 Replication of chromosome III. (*Heavy line*) Cloned region of chromosome III; (*vertical lines*) *Bam*HI sites; (*open circles*) positions of *ARS* elements above the restriction map; (*zigzag line*) left telomere on the restriction map; (+) *ARS* elements shown to have chromosomal origin function; (−) *ARS* elements without detectable chromosomal origin function; (?) *ARS* elements not yet tested (see text for references). Replication time data are redrawn from Reynolds et al. (1989); times are in minutes following release from the *cdc7* block.

Table 1 Designations of chromosome III *ARS* elements

New name	Previous name	References
ARS300	TF1-*ARS*	1
ARS301	*HML E ARS*	2
ARS302	*HML I ARS*, E5F4	2, 3
ARS303	E5F5	3
ARS304	D1OB	3
ARS305	A6C	3↑5
ARS306	C1G	5
ARS307	C2G1	3–6
ARS308	C2G2	3
ARS309	J11D	3–5
ARS310	H9G	3–5
ARS311	E5F1	3
ARS312	E5F2	3
ARS313	E5F3	3, 5

References to the identification and characterization of chromosome III *ARS* elements: (1) Button and Astell 1988; (2) Feldman et al. 1984; (3) Newlon et al. 1986; (4) Palzkill et al. 1986; (5) Reynolds et al. 1989; (6) Palzkill and Newlon 1988.

dimensional gel patterns obtained with the Brewer and Fangman (1987) technique show patterns consistent with initiation occurring at these *ARS* elements in a fraction of the population, with the *ARS*-containing fragments being replicated by forks from external origins in the remainder of the population. However, in the case of *ARS305*, analysis of the direction of fork movement through the fragment demonstrated that the patterns indicative of forks from external origins arose from breakage of replication bubbles (Linskens and Huberman 1990). Further experiments will be required to determine the frequency of origin usage.

Of the five *ARS* elements that have no detectable origin activity in vivo, two (*ARS308*, which is associated with *CEN3*, and *ARS304*) show weak *ARS* function in plasmid stability assays (J.V. Van Houten and C.S. Newlon, unpubl.) and could simply function at a level below detection in vivo. The other three *ARS* elements, which are located in the vicinity of the *HML* locus (*ARS301*, *ARS302*, and *ARS303*), show strong *ARS* function in plasmid stability assays (S. Synn and C.S. Newlon, unpubl.). *ARS301* and *ARS302* have been analyzed on plasmids and have been shown to act as origins in that context. The reason for their apparent lack of chromosomal origin function is under investigation. One possibility is that they are prevented from functioning by the mechanism that turns off expression of the *HML* locus; however, this possibility has been eliminated by demonstrating that they are still not active in mutant strains (*sir4*) that express *HML*, although *sir1*, *sir2*, and *sir3* strains have not been examined. Another possibility is that their failure to function is the result of their proximity to the telomere of the chromosome. This possibility has been tested in two sets of experiments and found to be unlikely. First, moving *ARS301* to the vicinity of *MAT* does not permit it to function as an origin. Second, neither *ARS302* nor *ARS303* functions as an origin in a circular derivative of chromosome III that lacks telomeres (D.D. Dubey et al., in prep.).

In summary, the left 57 kb of chromosome III appears to be part of a single replicon whose origin is coincident with *ARS305*. The failure to detect active replication origins in other locations, the direction of replication fork movement through the region (J. Huberman, pers. comm.), and the timing of replication (see Fig. 2) (Reynolds et al. 1989) all support this conclusion. The next 160 kb of DNA contains at least three origins, but replicon boundaries have not yet been precisely mapped. The spacing between active origins in rDNA and in chromosome III is in good agreement with the estimates of average origin spacing obtained from electron microscopy and DNA autoradiography.

Replication origins that do not have *ARS* function have not been detected. Overlapping restriction fragments covering about 130 kb of chromosome III sequences have been examined, and the only origins detected are coincident with the four *ARS* elements indicated in Figure 2. Moreover, analysis of *ARS* deletions in a 66-kb ring chromosome III containing sequences between Ty2-17 and Ty1-161 (Fig. 2) demonstrates that the two strong *ARS* elements in the chromosome are essential for its mitotic stability (A. Dershowitz and C.S. Newlon, in prep.). Deletion of either *ARS307* or *ARS309* caused the chromosome to be lost at four times the normal rate (10^{-3} per division), whereas deletion of both *ARS* elements caused the chromosome to be lost at a rate greater than 0.5 per division. Therefore, no cryptic sequences within this ring chromosome are able to compensate for the loss of *ARS* function.

The effect of *ARS* deletions on the stability of the 66-kb ring chromosome raises the issue of whether all *ARS* elements are necessary for normal chromosome stability. If the only function of *ARS* elements is to act as replication origins, then the amount of DNA that can be replicated from a single origin and the efficiency of initiation at the origin would set an upper limit on the spacing between *ARS* elements. The increased loss rate caused by *ARS* deletions in the 66-kb ring chromosome could result either from the length of DNA that must be replicated from the remaining *ARS* element or from its failure to initiate replication in some fraction of the population. From estimates of fork rate (2.4–6.3 kb per min at 30°C) and the length of S phase (25–40 min at 30°C), 120–500 kb of DNA could be replicated from a single origin that initiated early in S phase (Johnston and Williamson 1978; Rivin and Fangman 1980b). Replication origins are therefore considerably closer together than would be required for complete replication of DNA, and replication of the 66-kb ring chromosome from a single origin should be possible. In support of these observations, deletion of several *ARS* elements, singly and in combination, from the full-length chromosome III had no effect on its rate of loss (Newlon et al. 1986; A. Dershowitz and C.S. Newlon, in prep.). Because several of these deletions leave a longer region without an *ARS* than the deletions on the 66-kb ring chromosome, the increased loss rate of the ring chromosome is likely to result from failure of the remaining *ARS* to initiate replication. Further *ARS* deletion analysis, in combination with analysis of origin function, should help to resolve the issue of how many origins are required for chromosome replication.

The issue of whether there are differences in origin usage in premeiotic S phase compared to mitotic S phase is largely unexplored. The regulation of S phase appears to be different in yeast mitosis and meiosis. Temperature-sensitive cell-cycle mutants that arrest just prior to S phase in mitosis arrest following premeiotic S phase (Schild and Byers 1978; Shuster and Byers 1989). However, replicon spacing in premeiotic S phase is similar to that in mitotic S phase (Johnston et al. 1982). To test directly whether the DNA replication origins used during premeiotic S phase are the same as those used during mitosis, the pattern of origin usage on chromosome III is being examined (I. Collins and C.S. Newlon, unpubl.). In the region analyzed, the pattern of origin usage is the same in premeiotic S phase as in mitotic S phase: *ARS305*, *ARS306*, and *ARS307* act as origins, whereas *ARS304* and *ARS308* do not function as origins at detectable levels. No origins have been found in DNA fragments that do not have *ARS* function.

5. Replication Termini

Specific *cis*-acting sequences that result in replication termination have been identified in the *Escherichia coli* chromosome, the *Bacillus subtilis* chromosome, and the R6K plasmid (for review, see Kuempel et al. 1989). These terminators act in a polar manner, inhibiting replication forks from one direction but not the other. In these circular prokaryotic chromosomes, at least two copies of the terminator are present, and they are oriented so that replication forks can pass through the first terminator encountered but are blocked by the second or third. The terminators therefore define the boundaries of a region in which replication forks meet and serve to trap forks in the region. However, several well-studied replicons—simian virus 40 (SV40) and bacteriophage λ—do not contain replication termini (Lai and Nathans 1975; Valenzuela et al. 1976).

Specific replication termini do not appear to be present in yeast chromosomes, except perhaps in the rDNA cluster. Replication forks meet approximately 180° from the origin of replication in an *ARS1*-containing plasmid (Brewer and Fangman 1987) and in the 2-micron plasmid (Brewer and Fangman 1987; Huberman et al. 1987), as would be expected if the forks traverse the molecule at equal rates. In addition, the replication forks that move toward each other from *ARS305* and *ARS306* and from *ARS307* and *ARS309* on chromosome III meet and terminate over a 4–8-kb region about halfway between the *ARS* elements (J. Zhu, L.R. Davis, J.A. Huberman, and C.S. Newlon; S.A. Greenfeder and C.S. Newlon; both in prep.). Finally, since *ARS* elements used as origins can be deleted from chromosome III without affecting its stability (Newlon et al. 1986; A. Dershowitz and C.S. Newlon, in prep.), and since replication initiates at only a fraction of potential origins in the rDNA region of chromosome XII, fixed termini of replication, capable of arresting forks in a nonpolar manner, clearly do not lie between every pair of origins.

The finding that replication forks moving in the direction opposite to that of transcription are arrested at the 3′ end of the 37S rRNA transcription unit raises the possibility that a *cis*-acting replication terminus may lie at this point (Brewer and Fangman 1988; Linskens and Huberman 1988). This putative replication terminus would have to act in a polar manner because replication forks moving in the direction of transcription are not arrested. Alternatively, the barrier to fork movement could be transcription itself, a possibility discussed by Brewer (1988).

6. Telomere Replication

Telomeres are the specialized structures at the very ends of chromosomes. Because DNA polymerases synthesize DNA only in the 5′ to 3′ direction by adding deoxynucleotides to a preexisting primer, one strand of the DNA duplex at each chromosome end could not be replicated completely if the chromosome end were simply the blunt end of a DNA molecule. Yeast telomeres consist of a few hundred base pairs of an irregular sequence, one strand of which can be represented as $(C_{1-3}A)_n$ (Shampay et al. 1984; Walmsley et al. 1984a). The length of the telomere tract is heterogeneous and varies from cell to cell within a clonal population. The repetitive Y′ and X elements found internal to many telomeres (Chan and Tye 1983b) are not required for telomere function (Murray and Szostak 1986; Zakian and Blanton 1988; Jäger and Philippsen 1989). For a review of telomere structure and function, see Zakian (1989).

Several different mechanisms have evolved for telomere replication.

These include protein primers, hairpin loops, addition of telomeric repeats by a telomerase, and, possibly, recombination or polymerase slippage. The first two mechanisms, well documented in a variety of prokaryotic and eukaryotic viruses and organellar DNAs, are unlikely to function on yeast chromosomes because they cannot easily account for the heterogeneity in telomere length (for review, see Walmsley 1987).

Elegant work by Blackburn and colleagues has demonstrated that in *Tetrahymena*, telomeres are added to macronuclear chromosomal DNA molecules by an enzyme that they have named telomerase. This enzyme is a ribonucleoprotein that recognizes the G-rich strand of the hexanucleotide terminal repeat ($C_4A_2 \cdot T_2G_4$) and adds additional repeats. The RNA component of the enzyme contains a sequence complementary to the T_2G_4 repeat and probably serves as a template for the repeat (Greider and Blackburn 1985, 1987, 1989; Yu et al. 1990).

Yeast most likely uses a telomerase-based mechanism for replicating its telomeres. The evidence for this is as follows. Yeast and *Tetrahymena* telomeres are quite similar. *Tetrahymena* telomerase recognizes the yeast telomere sequence (Greider and Blackburn 1985, 1987). Yeast cells are capable of adding their own telomeres onto linear plasmids carrying *Tetrahymena* telomeres (Shampay et al. 1984; Walmsley et al. 1984a). In addition, the recent characterization of a telomerase from human cells argues for a conserved role for telomerase in telomere maintenance (Morin 1989). Finally, the sequence of *EST1*, mutations in which cause gradual loss of telomeres, suggests that it likely encodes a yeast telomerase (Lundblad and Blackburn 1990). Although the irregularity of the yeast telomere repeat is difficult to reconcile with an RNA-templated mechanism, a telomerase most likely plays a central role in telomere maintenance.

A recombinational mechanism for the replication of telomeres provides an adjunct to the telomerase paradigm (Van der Ploeg et al. 1984; Walmsley et al. 1984a). Consistent with this model, recent work has demonstrated the transfer of DNA sequences from one telomere to another by a *RAD52*-independent recombination event during de novo telomere formation in yeast (Pluta and Zakian 1989; Wang and Zakian 1990).

Telomere replication has also been approached genetically. Mutations in five genes yield altered telomere length. Mutations in *TEL1*, *TEL2*, and *EST1* result in shortened telomeres. Whereas strains carrying *tel1* and *tel2* mutations have no other obvious phenotypes (Lustig and Petes 1986), *est1* mutants die after many generations, apparently because of their gradual loss of telomeric sequences and the destabilization of chromosomes (Lundblad and Szostak 1989). In contrast, strains carrying a temperature-sensitive *cdc17* mutation have longer than normal telomeres when grown at semipermissive temperatures (Carson and Hartwell 1985). Since *CDC17* encodes the catalytic subunit of yeast DNA polymerase α (Carson 1987), defects in DNA polymerase α cause changes in telomere metabolism, but whether the role of polymerase is direct or indirect has not been determined. One possible mechanism is that the polymerase "slips" in the simple repeat sequences.

A fifth protein implicated in telomere maintenance is a telomere-binding activity identified on the basis of its binding to the $C_{1-3}A$ tracts of yeast telomeres (Berman et al. 1986). The binding activity corresponds to the Rap1p/TUF/GRF1 transcriptional repressor-activator protein (Longtine et al. 1989). Recent analysis of strains carrying temperature-sensitive *rap1* alleles and strains overproducing Rap1p implicate Rap1p in telomere metabolism. Some temperature-sensitive *rap1* alleles cause a decrease in telomere length at semipermissive temperatures (Lustig et al. 1990), and overproduction of Rap1p causes an increase in telomere length (Conrad et al. 1990).

The expression of all mutations that affect telomere length shows a long phenotypic lag, suggesting that telomere length changes only slightly with each replication cycle. The rather long length of the telomeric tract (300–700 bp) makes it quite possible that telomeres may be replicated by the normal replication machinery most of the time, with periodic lengthening by either telomerase or recombination occurring only once in 10–30 replication cycles.

C. Temporal Order of Replication

Direct measurements of the time of replication of specific yeast DNA sequences have been made on synchronized cultures, using either radioactive isotope labeling techniques or density labeling to assess the extent of DNA replication (Zakian et al. 1979; Brewer et al. 1980; Fangman et al. 1983; McCarroll and Fangman 1988; Reynolds et al. 1989). These experiments demonstrate a temporal program of replication in yeast. In one study (McCarroll and Fangman 1988), 34 chromosomal restriction fragments between 2 and 6 kb in length were found to replicate at specific and reproducible times during S phase. Moreover, the length of S phase defined by the time interval between the beginning of replication of the earliest sequence and the end of replication of the latest sequence was the same as the length of S phase measured by incorporation of radioactive precursors.

In the same study, the time of replication of a number of centromeres and telomeres was determined (McCarroll and Fangman 1988). Centromeres act in *cis* to direct the segregation of chromosomes by functioning as the assembly site for the kinetochore, which serves as the binding site for microtubules of the spindle apparatus. Because sister chromatids in larger eukaryotes remain associated at their kinetochores until anaphase of mitosis, several investigators have suggested that this association is caused by the failure of centromeres to replicate until anaphase (Tschumper and Carbon 1983; Murray and Szostak 1985). This proposal was tested by determining the time of replication of 9 of the 16 yeast centromeres. In contrast to the prediction, all replicated during the first third of S phase. Fragments containing *CEN4* and *CEN6* were further tested for evidence of very short unreplicated regions, which would not have been obvious from density measurements. No evidence for such short unreplicated regions was found. Therefore, incomplete replication of centromeric DNA cannot account for the association of sister chromatids in mitosis in yeast.

In contrast to centromeres, telomeres replicate late in S phase. These results suggest that each yeast chromosome arm must have at least one region where a transition from early to late replication occurs and raise the question of whether the transition is sharp or gradual and whether there are late initiating origins in the late-replicating regions.

To examine the pattern of replication of a single chromosome in more detail, Reynolds et al. (1989) studied the 200-kb cloned region of chromosome III. In the region between the left telomere and *MAT* was a short region of late replication near the telomere and a large, early-replicating internal region (see Fig. 2). Most of the fragments analyzed contain *ARS* elements. The *ARS* elements known to function as origins all replicate early, during the first third of S phase. The *ARS* elements in the late-replicating region do not appear to function as origins in vivo (D.D. Dubey et al., in prep.). The transition from the early- to the late-replicating region is gradual, and the simplest explanation of the data is that the timing transition results from the movement of a single fork from an early-replicating origin to the end of the chromosome. The average rate at which this fork would move, 1.3 kb per minute, is slower by almost a factor of 3 than the fork rate previously measured by Rivin and Fangman (1980b) under similar culture conditions. Whether the lower than expected fork rate is the result of pauses in fork movement, caused perhaps by specific DNA sequences, proteins bound to DNA, chromatin structure, or active transcription units, has not been determined.

These results raise several questions. First, do all active origins initiate replication early in S phase? Current data suggest that most origins do indeed replicate early: In addition to the four origins on chromosome III, most 2-micron plasmid molecules replicate within the first third of S phase (Zakian et al. 1979) and *ARS1* replicates early both in its chromosomal location (Fangman et al. 1983; McCarroll and Fangman 1988; Reynolds et al. 1989) and on a plasmid (Fangman et al. 1983). Although rDNA replicates throughout S phase (Brewer et al. 1980), all initiations could occur early. The length of time required for replication could reflect the time required for forks to pass through rDNA repeats in which replication did not initiate (Saffer and Miller 1986; Brewer and Fangman 1988; Linskens and Huberman 1988). Similarly, the later replication times of *ARS307* and *ARS309* relative to *ARS305* and *ARS310* could reflect either their later times of initiation or their less efficient use as origins. In the second case, their apparent replication time would reflect the weighted average of two distinct times of replication, the first reflecting the time of initiation and the second reflecting the time that a fork from an adjacent origin replicates the fragment when replication failed to initiate at the *ARS* element.

Two lines of evidence suggest that, in contrast to the majority of origins, a few origins initiate replication late in S phase. First, a late-replicating region at the distal end of the right arm of chromosome V contains a strong *ARS* element, *ARS501* (previously called *ARSZR1*), that replicates during mid-S phase (Ferguson and Fangman 1988). *ARS501* is active as an origin: Flanking sequences on each side show a gradient of progressively later replication times, and two-dimensional gel patterns are diagnostic of an active replication origin (B. Ferguson and W. Fangman, pers. comm.). Second, there is an active origin in the telomere-adjacent Y′ element (B. Brewer and W. Fangman, pers. comm.). This repetitive sequence contains an *ARS* element (Chan and Tye 1983a,b) and replicates late (McCarroll and Fangman 1988).

A second question concerning the temporal order of replication is, What controls the timing of replication initiation? Do *cis*-acting timing sequences interact with initiation proteins or does the structure or location of a chromosomal segment within the nucleus determine its time of replication? As a way of asking whether timing information is contained within *ARS* elements, the times of replication of *ARS1* and *ARS501* have been measured both in their normal chromosomal contexts and on plasmids (Fangman et al. 1983; Ferguson and Fangman 1988). *ARS1* replicates early in both contexts, but the timing of *ARS501* replication changes with its context. In contrast to its late replication in its chromosomal context (see above), *ARS501* replicates as early as *ARS1* when it is present with 14 kb of flanking sequence in a centromere-containing plasmid. Moreover, when *ARS1* is inserted into the late-replicating chromosome V region, it replicates late. Thus, the determinants for replication timing are separable from the determinants of *ARS* activity.

Recent results suggest that the presence of telomeric $C_{1-3}A$ sequences nearby may influence time of replication (B. Ferguson and W.L. Fangman, pers. comm.). When the early-replicating *ARS501* plasmid was converted to a linear plasmid by the addition of telomeric sequences, the linear plasmid replicated late, at the normal time for *ARS501*. Moreover, addition of a block of telomeric repeats to the early-replicating circular plasmid shifted its replication to a time that was intermediate between its original time of replication and the time of replication of the linear plasmid. Finally, a fragmentation of chromosome V that placed its telomere adjacent to a region that replicates early in the full-length chromosome caused the region to replicate late. How these telomeric repeats act to influence replication time remains to be determined.

A third question concerns the boundaries between early- and late-replicating regions. Is the gradual transition seen on the left arm of chromosome III typical of all such boundaries or do replication fork barriers or termini of replication lie between some early- and late-replicating regions? To answer this question, several more replication timing transitions need to be examined in detail.

Finally, are all late-replicating regions of yeast chromosomes adjacent to telomeres? The answer to this question is not yet clear, but a possible internal late-replicating region has been identified on chromosome XIV (B. Ferguson and W. Fangman, pers. comm.). A fragment containing an *ARS* element, the *KEX2* locus, and a *Not*I site 200 kb from the end of chromosome XIV replicates late. This observation suggests either that chromosome XIV contains an internal late-replicating region or that the transition from early to late replication is a long distance from the telomere. In addition, two late-replicating sequences identified in a survey of the time at which restriction fragments replicate (McCarroll and Fangman 1988) have not yet been mapped.

III. AUTONOMOUSLY REPLICATING SEQUENCE ELEMENTS

Although at least a subset of *ARS* elements function as replication origins, they may also have other roles that may or may not be related to their function as origins. For example, nuclear-scaffold-binding sites have been found near a number of *ARS* elements (Amati and Gasser 1988). In addition, Laskey and colleagues have shown that DNA replication occurs at discrete sites in *Xenopus* nuclei (Mills et al. 1989). *ARS* elements may be required for recognition of equivalent sites in yeast. A detailed analysis of *ARS* elements has been undertaken to define the sequences and the proteins that interact with them to mediate these roles.

A. Assays

ARS elements were first recognized by their ability to promote high-frequency transformation in *S. cerevisiae* (Hsiao and Carbon 1979; Kingsman et al. 1979; Stinchcomb et al. 1979; Struhl et al. 1979). In contrast to plasmids that transform at low efficiency as a result of integrating into chromosomes by homologous recombination, *ARS*-containing plasmids are maintained extrachromosomally. They are unstable and are typically lost at a rate of 0.2–0.3 losses per division during mitotic growth (Dani and Zakian 1983; Celniker et al. 1984). However, although fewer than half the cells in a population growing under selection for the plasmid contain plasmid, the average plasmid copy number is high, in the range of 20–50 copies per plasmid-containing cell (Fitzgerald-Hayes et al. 1982; Hyman et al. 1982; Zakian and Kupfer 1982). Both the instability and the high copy number of *ARS*-containing plasmids result from their failure to partition efficiently (Zakian and Kupfer 1982; Murray and Szostak 1983). In 30–60% of cell divisions, all plasmid molecules segregate together, most often to the mother cell. Stable plasmids, which are maintained at one or two copies per cell, are produced by adding a centromere to *ARS*-containing plasmids.

The high-frequency transformation assay has remained the standard for the identification of *ARS* elements. However, transformation efficiencies are not related to the efficiency of *ARS* function (Celniker et al. 1984; Strich et al. 1986; Palzkill and Newlon 1988; Williams et al. 1988). Quantitative measures of *ARS* function can be obtained either from plasmid loss rates under nonselective conditions (Dani and Zakian 1983; Futcher and Cox 1984; Maine et al. 1984; Hieter et al. 1985; Koshland et al. 1985), from culture growth rates under selection for the plasmid (Srienc et al. 1985), or from the fraction of cells that contain plasmid when selection is in force (Stinchcomb et al. 1979; Hsiao and Carbon 1981; Hyman et al. 1982). The last measure, called the "mitotic stability," is a function of plasmid loss rates and the number of times a cell that has lost plasmid is able to divide. Murray and Szostak (1983) derived equations that relate mitotic stability to plasmid loss rate. It should be emphasized that none of these assays measure DNA replication specifically.

Several vectors have been designed for quantitative stability assays. These vectors allow ready identification of cells or colonies that contain plasmid. One plasmid carries the *E. coli lacZ* gene, whose product can be detected in single cells through the use of a fluorescent substrate and fluorescence-activated cell sorting (Srienc et al. 1983, 1985). Others take advantage of the observation that blockage of purine nucleotide biosynthesis at either of two steps results in accumulation of a pigment that colors cells and colonies red. The plasmids carry genes that either suppress (Hieter et al. 1985) or cause (Koshland et al. 1985) the formation of pigment, allowing the presence or absence of plasmid to be inferred from colony color. In addition, the red pigment can be detected by fluorescence-activated cell sorting (Bruschi and Chuba 1988).

The use of plasmids for the measurement of *ARS* efficiency is limited by the inherent instability of small plasmids, by the requirement that plasmid-containing cells be able to form colonies under selective conditions, and by the effect of sequence context on the apparent efficiency of *ARS* function (for discussion, see Newlon 1988). For these reasons, plasmid stability assays are useful for measuring *ARS* efficiency only in a limited range and only when the plasmid context of the *ARS* is kept constant. Despite these reservations about the plasmid stability assay, it is the best functional *ARS* assay currently available, and it has been useful in dissecting DNA sequences required for *ARS* function.

B. DNA Sequence Analysis

Intensive analysis of six *ARS* elements has led to a picture of DNA sequence requirements that can be summarized as follows. An A+T-rich consensus sequence 12 bp in length and a variable amount of sequence 3′ to the T-rich strand of the consensus sequence are absolutely required for *ARS* function.

1. ARS *Consensus Sequence*

With the exception of the *ARS* elements embedded in X and Y′ elements at telomeres (Chan and Tye 1980, 1983a,b), yeast *ARS* elements do not have sufficient homology to cross-hybridize. However, early DNA sequence comparisons of *ARS*-containing fragments led to two generalizations that have withstood the test of time. First, all DNA fragments with *ARS* function in yeast contain one or more copies of a consensus sequence 5′-(A/T)TTTAT(A/G)TTT(A/T)-3′(Stinchcomb et al. 1981; Broach et al. 1983). This sequence has been called the *ARS* consensus, the core consensus, and domain A. Second, short DNA fragments with *ARS* activity have a higher A+T content (73–82%) than chromosomal DNA (60%) (Broach et al. 1983).

The construction and analysis of the effects of point mutations, small deletions, and small substitutions within the consensus sequence have demonstrated that it is essential for *ARS* function. In *ARS1* (Celniker et al. 1984; Srienc et al. 1985), the *HO ARS* (Kearsey 1983, 1984), the histone H4 *ARS* (Bouton and Smith 1986; Bouton et al. 1987), the *HMR*

E ARS (Brand et al. 1985, 1987), *ARS307* (Palzkill and Newlon 1988), and *ARS121* (Walker et al. 1990), the sequences required for *ARS* function include the core consensus sequence, two to three nucleotides 5′ to the T-rich strand of the consensus sequence, and a variable number of nucleotides 3′ to the T-rich strand of the consensus sequence.

Numerous *ARS* elements do not contain an exact match to the core consensus sequence, although all but *ARS121* have one or more elements with 10 of 11 matches to the consensus. In fact, nearly all *ARS* elements contain multiple near matches to the consensus sequence, and it is often not clear which, if any, of the matches is essential for function. In addition, many *ARS* elements contain near matches that overlap the exact match to the consensus (Palzkill et al. 1986; Palzkill and Newlon 1988).

A recent systematic mutagenesis of the *ARS307* consensus sequence (Van Houten and Newlon 1990) and recent quantitative analysis of the consequences of *HO ARS* mutations and newly created *ARS* elements in M13 vector sequences (Kipling and Kearsey 1990) have helped to clarify the requirement for specific nucleotides in the consensus sequence. These results suggest that the consensus sequence should be modified as follows:

```
                            T
 A          T A        A    C
   TTTA              TTT
 T          C G        T
                            G

 1   2345   6 7   8910 11  12
```

The modifications of the original consensus sequence include (1) the substitution of a pyrimidine for the T at position 6 because substitution of a C at position 6 in *ARS307* leaves wild-type function and (2) the inclusion of a twelfth position because substitution of an A at position 12 in the *HO ARS* and *ARS307* reduces function. The Ts at positions 3, 8, 9, and 10, the A at position 5, and the purine at position 7 are essential for *ARS307* activity; however, other active *ARS* elements have a consensus sequence that deviates at one of these positions. For example, the single consensus sequence in the tRNA[Glu] *ARS* has a G at position 9 (Feldmann et al. 1981), and the essential consensus sequence in *ARS121* has a T at position 5 as well as a G at position 2 and an A at position 12 (Walker et al. 1990). Therefore, the DNA context in which the consensus sequence resides is likely to be important for *ARS* function. At the remaining positions, one or more single base-pair substitutions reduce or eliminate *ARS307* function. Thus, every nucleotide in the consensus sequence appears to play a critical role in efficient *ARS* activity, a situation reminiscent of bacterial promoter elements (for review, see McClure 1985), which are known to bind RNA polymerase, and the DNA replication initiator protein-binding sites in *oriC* (for review, see Baker et al. 1988) and SV40 (for review, see Stillman 1989). By analogy, it seems likely that the *ARS* core consensus sequence is a binding site for an initiator protein, although the protein has not yet been identified.

2. 3′-flanking Sequence

A region of somewhat variable length located 3′ to the T-rich strand of the core consensus sequence is also required for *ARS* function. Progressive deletions into this region reduce and eventually abolish *ARS* function (Celniker et al. 1984; Kearsey 1984; Srienc et al. 1985; Bouton and Smith 1986; Strich et al. 1986; Palzkill and Newlon 1988) and large insertions in the region also abolish function (Palzkill et al. 1986; Diffley and Stillman 1988b). Moreover, numerous DNA segments containing *ARS* consensus sequences, both in chromosomal DNA (Broach et al. 1983; Palzkill et al. 1986; Bouton et al. 1987) and in synthetic oligonucleotides (Palzkill and Newlon 1988), do not have *ARS* activity alone. Thus, sequences in addition to the consensus sequence are required for *ARS* function. However, in contrast to the core consensus sequence, small substitutions, insertions, and deletions in the 3′-flanking region have small effects on *ARS* function (Bouton and Smith 1986; Strich et al. 1986; Diffley and Stillman 1988b; Palzkill and Newlon 1988). The function of the 3′-flanking region depends on its placement relative to the consensus sequence. Precise inversion of the essential 14-bp core element relative to the required flanking region of the H4 *ARS* abolished function (Holmes and Smith 1989).

The apparent size of the 3′-flanking region is dependent on the assay used. In the cases where plasmid stabilities have been measured, effects of deletions become apparent approximately 90 bp 3′ to the consensus sequence (Snyder et al. 1986; Palzkill and Newlon 1988). *ARS* function, as measured by high-frequency transformation, is not lost until deletions encroach to within 15–40 bp of the consensus sequence (Celniker et al. 1984; Kearsey 1984; Bouton and Smith 1986; Palzkill and Newlon 1988). In the case of *ARS1*, a 19-bp fragment has *ARS* activity in one context (Srienc et al. 1985) but not in a different context (Celniker et al. 1984).

In contrast to the requirement for 3′-flanking sequences, 5′-flanking sequences contribute variably to plasmid stability. In *ARS121*, *ARS*-binding factor 1 (Abf1p)-binding sites in the 5′-flanking region make a substantial contribution to plasmid stability (Walker et al. 1990), whereas in *ARS1*, the effect of 5′ deletions is modest (Celniker et al. 1984; Srienc et al. 1985; Strich et al. 1986). In *ARS307* and the H4 *ARS*, there is no effect of deleting 5'-flanking sequences (Palzkill and Newlon 1988; Holmes and Smith 1989). In the case of *ARS1*, the 5'-flanking sequences contain the promoter for the *GAL3* gene (Bajwa et al. 1988). The possible effects of the binding of transcription factors on *ARS* activity have not been investigated.

The function of the 3'-flanking region (also called domain B) is not yet clear. Models to explain the function of this domain fall into two groups: those that involve the primary DNA sequence of the domain and those that involve the structural properties of its DNA sequence or chromatin.

In their analysis of *ARS307*, Palzkill and Newlon (1988) found that the flanking region contained several degenerate matches (82% homology) to the core consensus sequence, and sequential deletion of these matches resulted in progressively reduced *ARS* activity. Consistent with the hypothesis that core consensus sequence elements in the 3'-flanking region are necessary for *ARS* function, most if not all *ARS* elements have several near matches in their flanking region. These near matches to the core consensus sequence would be predicted to bind the hypothetical initiator protein. Moreover, Palzkill and Newlon were able to construct synthetic *ARS* elements containing two, three, or four copies of a core consensus sequence. This hypothesis is attractive because both *oriC* and the SV40 origin contain multiple initiator protein-binding sites. However, the spacing of dnaA-binding sites at *oriC* is critical for origin function, whereas the spacing between degenerate matches in *ARS* elements appears less important. In addition, the recent findings that the degenerate matches to the consensus sequence in the 3'-flanking region of the H4 *ARS* and *ARS121* could be eliminated by directed mutagenesis without major effects on *ARS* function suggest either that this model is not correct or that *ARS121* and the H4 *ARS* are in a different class than *ARS307* (Holmes and Smith 1989; Walker et al. 1990).

Another model for the function of the 3'-flanking region invokes the binding of protein(s) in this region to either facilitate *ARS* function(s) or protect the *ARS* element from interference. *ARS* elements are closely apposed to the 3' and/or 5' ends of yeast transcription units (for review, see Newlon 1989). Because experiments have shown that a plasmid carrying a construct in which a strong promoter can be induced to transcribe toward *ARS1* was destabilized by inducing transcription (Snyder et al. 1988), sequences that interfere with transcription may be an important component of *ARS* elements. Similarly, binding of proteins in the promoters of flanking genes could have a positive or negative effect on *ARS* activity. In this model, the protein-binding sites might be expected to be different in different *ARS* elements, and therefore no highly conserved sequence would be expected.

Three other hypotheses assume that the physical properties of the flanking sequence rather than the specific DNA sequence elements are important for *ARS* function. The first hypothesis, spawned in part by analogy with the bacteriophage λ (Zahn and Blattner 1985) and SV40 (Ryder et al. 1986) replication origins, proposed that bent DNA plays a role in origin function (Snyder et al. 1986; Williams et al. 1988). *ARS1* contains a region of bent DNA in its 3′-flanking region, about 80 bp from the consensus sequence (Anderson 1986; Snyder et al. 1986). Deletions of *ARS1* that remove this bent DNA have a small but significant effect on plasmid stability (Snyder et al. 1986), although these deletions also remove the Abf1p-binding site (see below). Insertion of synthetic bent DNA into an *ARS1* deletion lacking most 3′-flanking DNA increases plasmid stability (Williams et al. 1988). Some other *ARS* elements contain tracts of A and T that computer analysis predicts would form a bend, but these have not been directly tested (Eckdahl and Anderson 1987). However, *ARS307* does not contain bent DNA (Palzkill 1988), demonstrating that bent DNA is not an essential component of *ARS* elements.

A second hypothesis proposes that the 3′-flanking region contributes to *ARS* function by facilitating DNA unwinding (Broach et al. 1983; Umek and Kowalski 1988, 1990; Umek et al. 1989). Umek and Kowalski (1987, 1988, 1990) analyzed the unwinding potential of yeast *ARS* elements by measuring the sensitivity to mung bean nuclease of *ARS* elements in supercoiled plasmids and showed that the nuclease sensitivity is the result of thermodynamically stable DNA unwinding. The region of easily unwound DNA, which they have named the DUE (*DNA u*nwinding *e*lement), maps to the 3′-flanking region of *ARS* elements. Their observations that H4 *ARS* deletions that partially remove the DUE can be suppressed by growth at elevated temperatures and that an easily unwound pBR322 sequence could restore *ARS* function to an Ars⁻ deletion when inserted next to the consensus sequence provide support for the hypothesis.

A third hypothesis postulates that the chromatin structure of the 3′-flanking region is the feature important for *ARS* activity (Bouton and Smith 1986; Holmes and Smith 1989). Support for this model has recently been provided by studies of in vitro SV40 replication which suggest that the role of origin auxiliary sequences is to bind factors that protect the origin from encroachment by nucleosomes (Cheng and Kelly 1989). Inappropriate positionings of nucleosomes at *ARS1* do, in fact, restrict *ARS* function (Simpson 1990).

These hypotheses are not all mutually exclusive and cannot be distinguished on the basis of present data. DNA sequences that result in DNA bending are not clearly distinct from sequences that confer a low energy for unwinding or from sequences that affect chromatin structure. The ARS consensus sequence has tracts of A that might confer bending and the sequence is sufficiently A+T-rich that it might be easily unwound (although A+T content is not an infallible predictor of the energy of unwinding; see Umek and Kowalski 1987). The easily unwound pBR322 sequence that rescued ARS function contains near matches to the core consensus sequence, further confounding the issue. Isolation and characterization of a consensus sequence binding protein should help to clarify this issue.

Whatever the function of the 3'-flanking sequences, recent work suggests that ARS function can be created from redundant modules that alone do not have measurable ARS activity. First, Kipling and Kearsey (1990), in a search for revertants of HO ars mutations that differed from the wild-type sequence by one to three single-base changes in and around the core consensus sequence, recovered three new ARS elements that were created de novo by mutation of A+T-rich M13 vector sequences. In one case, a single nucleotide change created a better match to the core consensus sequence, and in two other cases, small duplications of vector sequences created additional matches to the consensus sequence. This observation lends strength to the proposal that multiple matches to the core consensus sequence are important for ARS function. Second, Zweifel and Fangman (1990) have studied a 64-bp fragment of mitochondrial DNA that is 100% A+T and contains numerous 9 of 11 and 10 of 11 matches to the core consensus sequence. No ARS function is exhibited by an insert containing four tandem repeats of this fragment, but eight repeats either in tandem or in two separate modules exhibit ARS function. Third, a mutant of ARS307 that contains two overlapping, individually inactive copies of the ARS consensus sequence has ARS function (Van Houten and Newlon 1990). These results are reminiscent of those obtained by Herr and Clarke (1986) in their studies of the SV40 enhancer, in which they showed that duplication of an enhancer-derived sequence, which itself had no enhancer function, could generate enhancer function.

In summary, sequences required for ARS function on plasmids include a consensus sequence that is likely to bind an initiator protein and flanking sequences that may include additional near matches to the consensus and be involved in protein recruitment or binding, or may facilitate unwinding of the origin. We emphasize though that this simple picture is based on DNA sequence comparisons and a relatively limited mutational analysis. Furthermore, it is based on analysis of *ARS* function on plasmids, which may not reflect origin activity in the chromosome. None of the minimal sequences with *ARS* function on plasmids has been tested for origin function in its chromosomal location. The identification of *ARS*-binding proteins that contribute to, but are not required for, *ARS* function on plasmids (see below) raises the possibility that origin function is more complicated than is now apparent.

C. *ARS*-binding Proteins

ARS function is probably mediated by the interaction of proteins with specific DNA sequences, and a complete understanding of origin function requires a knowledge of both protein factors and the DNA sequences to which they bind. The biochemical identification of proteins that interact with *ARS* elements provides an approach for studying *ARS* function complementary to the molecular genetic studies described above. Several proteins that bind to *ARS* elements have now been identified. Although none of these proteins recognize the core consensus sequence, the evidence summarized below indicates that one or more of these proteins contributes to the function of several *ARS* elements.

1. ARS-*binding Factor 1 (Abf1p)*

At least seven laboratories have independently identified Abf1p. (For consistency within this volume, we refer to ABF1 as Abf1p, for reasons discussed in the Preface.) Interestingly, these laboratories used widely differing sites in the yeast genome to detect DNA binding, including the *HMR E* silencer element or *ARS1* (Shore et al. 1987; Buchman et al. 1988b; Diffley and Stillman 1988b; Sweder et al. 1988); the telomeric X region *ARS*, *ARS120* (Eisenberg et al. 1988), a region between the divergently transcribed genes *YPT1* and *TUB2* (Halfter et al. 1989a); and the genes encoding ribosomal proteins L2A and L2B and the RNA polymerase genes *RPC160* and *RPC40* (Della Seta et al. 1990a,b). In addition to ABF1 and Abf1p, this factor has been called SBF-B (Shore et al. 1987), OBF1 (Eisenberg et al. 1988 and pers. comm.), and BAF1 (Halfter et al. 1989a). The consensus binding sequence for Abf1p, derived from mutational analysis, is $RTCRYN_5ACG$ (Della Seta et al. 1990a).

Abf1p-binding sites have been found at many, but not all, *ARS* elements. The affinity of Abf1p for the various *ARS* elements correlates with the similarity of the binding site to the consensus. The binding site is within 200 bp of the core consensus sequence in *ARS1*, *ARS121*, *ARS2*,

ARS300 (TFI-ARS), and the *HMR E, HMR I*, and *HML I* (*ARS302*) silencers, but it is 700 bp away from the 2-micron *ARS* consensus sequence (Abraham et al. 1984; Shore et al. 1987; Buchman et al. 1988b;
Diffley and Stillman 1988b; Sweder et al. 1988; Biswas and Biswas
1990; Walker et al. 1990). Abf1p does not bind anywhere within several
hundred base pairs of *ARS301* (*HML E*) or the H2B and H4 *ARS* elements (Buchman et al. 1988a).

The regions containing Abf1p-binding sites contribute to, but are not
essential for, *ARS* function. Deletions that remove the Abf1p-binding site
of *ARS1* result in a small but detectable increase in the rates of plasmid
loss (Snyder et al. 1986; Diffley and Stillman 1988b). Mutations in the
Abf1p-binding sites of *ARS121* cause significant reductions of plasmid
stability that are correlated with defects in the in vitro binding of Abf1p
(Walker et al. 1989). In addition, synthetic Abf1p-binding sites in either
orientation and up to 1.2 kb away from the *ARS* consensus sequence are
able to substitute for the entire 5'-flanking domain of *ARS121* (Walker
et al. 1990). The interpretation of experiments on the *ARS* at *HMR E* is
complicated by the apparent presence of a second *ARS* element within a
few hundred base pairs of the *HMR E ARS*. When a 138-bp fragment
containing the *HMR E ARS* was analyzed, a linker mutation within the
Abf1p-binding site reduced plasmid stability, and a linker mutation
within the core consensus sequence abolished *ARS* activity (Brand et al.
1987). However, when a longer *HMR E* fragment was used to assess *ARS*
function, neither a single-base insertion in the Abf1p site that abolished
in vitro binding of Abf1p (Kimmerly et al. 1988), nor point mutations
within the core consensus sequence that abolished function of the *HO
ARS* (Kearsey 1984) and *ARS307* (Van Houten and Newlon 1990) had
any detectable effect on plasmid stability (Kimmerly and Rine 1987). Although it has not been demonstrated directly, the failure by Kimmerly et
al. (1988) to detect the expected mutant phenotype was likely caused by
the presence of the second *ARS* in the fragment they analyzed (Brand et
al. 1987).

The role of Abf1p in stimulating replication has not been resolved. As
noted below, Abf1p is also a transcription factor, and the role of Abf1p
in stimulating DNA replication may be analogous to the role of nuclear
factor I (NF-I) and NF-III in adenovirus replication (Rosenfeld and Kelly
1986; O'Neill and Kelly 1988) and the role of the SV40 enhancer elements in SV40 replication (Cheng and Kelly 1989; Guo et al. 1989). The
role could be indirect, for example, by physically preventing nucleosome
assembly at the *ARS*, and thus keeping the region open for interactions
with other proteins. Alternatively, the role could be more direct, either
causing a conformational change in the DNA that facilitates binding of

an initiator protein, acting as an entry site for other proteins through protein-protein interactions, or producing a transcript necessary for transcriptional activation of the origin.

Abf1p is encoded by a single-copy essential gene (Diffley and Stillman 1989; Halfter et al. 1989b; Rhode et al. 1989; S. Eisenberg, pers. comm.). DNA sequence analysis revealed an open reading frame (ORF) predicted to encode an 81,662-dalton protein with an unusual amino acid composition that may account for its slower than expected migration in SDS gels. From analysis of the predicted amino acid sequence, two different metal-binding, DNA-binding motifs have been proposed that might function in DNA binding (Diffley and Stillman 1989; Halfter et al. 1989b; Rhode et al. 1989). Although additional data will be required to assess which, if either, motif is required for Abf1p function, DNA binding by the protein requires zinc and at least one unmodified cysteine residue (Diffley and Stillman 1989), and mutations that change His-57 and Cys-71 to other amino acids abolish DNA binding (Halfter et al. 1989b). Diffley and Stillman (1989) have also reported that Abf1p shares significant homology with Rap1p protein (*r*epressor/*a*ctivator *p*rotein), also called GRF1 (*g*eneral *r*egulatory *f*actor), a protein with which Abf1p may interact to mediate silencer function (Shore et al. 1987; Shore and Nasmyth 1987; Buchman et al. 1988b).

Abf1p has been purified to homogeneity and has been shown to migrate in SDS gels as two closely spaced bands with mobilities suggesting molecular weights between 120,000 and 135,000 (Diffley and Stillman 1988b; Sweder et al. 1988; Francesconi and Eisenberg 1989; Halfter et al. 1989a). The two molecular-weight species appear to be related on the basis of binding specificity (Sweder et al. 1988) and antigenic reactivity (Rhode et al. 1989). The *E. coli*-produced protein has the mobility of the lower-molecular-weight form (Rhode et al. 1989). Abf1p is a phosphoprotein, and treatment of Abf1p with acid phosphatase yields only the lower-molecular-weight form (P.R. Rhode et al., in prep.). The equilibrium binding constant of Abf1p for the optimal binding site is very high, 5×10^{-10} M (Della Seta et al. 1990a). Abf1p is present at 10^3 to 10^4 molecules per cell (Buchman et al. 1988b; Sweder et al. 1988; Francesconi and Eisenberg 1989). Thus, Abf1p belongs to the class of abundant, high-affinity DNA-binding proteins.

In addition to its apparent function in DNA replication, Abf1p has been clearly implicated in the regulation of gene expression. Its binding site at *YPT1* has been correlated with transcriptional activation (Halfter et al. 1989a), and its binding site at *HMR E* plays a role in the transcriptional repression mediated by the silencer and possibly also in plasmid segregation (Brand et al. 1985, 1987; Kimmerly and Rine 1987; Kimmerly et al. 1988). An Abf1p-binding site also functions as an upstream activating sequence (UAS) in *lacZ* reporter gene constructs (Brand et al. 1987; Buchman and Kornberg 1990; Della Seta et al. 1990b). In addition, a number of independently identified proteins that bind to promoter elements at sites that appear to have UAS function (GF1, SUF, TAF, and Y protein) have been shown to bind to DNA sequences conforming to the Abf1p consensus and may be identical to Abf1p. GF1 binds upstream of several nuclear genes encoding proteins that are imported into mitochondria (Dorsman et al. 1988, 1989). A factor that binds upstream of *COX6* has been shown to be antigenically related to Abf1p (P.R. Rhode et al., in prep.). SUF and TAF bind to the promoters of genes encoding ribosomal proteins (Vignais et al. 1987; Hamil et al. 1988), and Y protein binds to the promoter of the *p*hosphoglycerate *k*inase gene, *PGK1* (Stanway et al. 1987). Five functional Abf1p-binding sites have been identified in the promoter of the *ABF1* gene itself, suggesting that it may be autoregulated (Halfter et al. 1989b; P.R. Rhode and J.L. Campbell, unpubl.). A search of *S. cerevisiae* DNA sequence databases revealed at least 80 genes with an Abf1p-binding site in their promoter regions (Della Seta et al. 1990b). Many of the genes are regulated by the carbon source used for growth (for discussion, see Rhode et al. 1989; Della Seta et al. 1990a).

If all of these binding activities are mediated by Abf1p, then Afb1p may serve to coordinate transcription, replication, and translation in some general sense. However, the possibility that as yet unidentified proteins, with binding specificities similar to that of Abf1p, mediate some of these functions has not yet been ruled out.

2. ARS-*binding Factors II and III*

ABFII binds all DNAs with similar affinities. However, at *ARS1*, a specific DNase I protection pattern is observed that is not present when pBR322 DNA/ABFII complexes are digested with DNase I (Diffley and Stillman 1988a). ABFII resembles the mitochondrial protein HM in its ability to supercoil DNA in the presence of a topoisomerase and in molecular weight; gene disruptions show that it is not involved in DNA replication (Caron et al. 1979; Diffley and Stillman 1988b; J.F.X. Diffley, pers. comm.). ABFIII binds distal to Abf1p at *ARS1*, but deletion of the binding site has no effect on *ARS* activity (Sweder et al. 1988).

3. *Chromatin Proteins and Nuclear Scaffold Interactions*

Another approach to examination of protein-DNA interactions at *ARS* elements is to examine chromatin structure and binding to the nuclear scaffold. In *ARS1*-containing plasmids, both the core consensus sequence and the flanking region are hypersensitive to nuclease, suggesting that the region is nucleosome-free (Thoma et al. 1984; Long et al. 1985). However, as revealed by in vivo footprinting, the region is associated with protein, with one protected area over the core consensus sequence and other protected regions 3′ and 5′ to the consensus (Lohr and Torchia 1988). The proteins responsible for the observed protection have not been purified. On the basis of nuclease sensitivity, a model for nucleosome positioning at *ARS1* has been proposed. Deletions in domain C that move the core consensus into the nucleosome proposed to reside there decrease the copy number of *ARS1* plasmids (Simpson 1990).

Gasser and colleagues have shown that *ARS*-containing DNAs bind tightly to nuclear scaffold preparations in vitro. On this basis, they have proposed that *ARS* elements contain scaffold attachment regions (SARs) and that *ARS* elements are associated with the scaffold in vivo (Amati and Gasser 1988; Hoffman et al. 1989). In *ARS1*, the SAR appears to include two attachment sites that flank the core consensus region. The Abf1p-binding site does not appear to mediate scaffold attachment because the binding site at *ARS1* is not within the SAR and other Abf1p-binding DNA fragments do not bind the scaffold (Amati and Gasser 1988). In a detailed study of the *HML* silencers, which are scaffold-associated, an activity that purified with scaffolds and was capable of directing the formation of specific DNA loops between the silencer elements and between the silencer elements and the promoter was shown to be Rap1p (Hoffman et al. 1989). These results suggest that *ARS* elements may contain sequences that serve a structural role or a nuclear localization role and that proteins that bind to *ARS* elements may serve to stabilize higher-order DNA structures in the nucleus. It will be of interest to assess the role of these sequence elements in origin function.

IV. REPLICATION APPARATUS

Countering the many advantages of the yeast system has been the difficulty of devising appropriate in vitro replication systems that allow efficient initiation of DNA replication at chromosomal origins of replication (Jazwinski and Edelman 1979, 1984; Kojo et al. 1981; Sugino et al. 1981; Celniker and Campbell 1982; Jazwinski et al. 1983; Kuo et al. 1983; Tsuchiya and Fukui 1983; Eberly et al. 1989). This initially slowed the pace of identifying the yeast analogs of proteins known to be essential for replication in other systems. The challenge of identifying bona fide replication proteins without an in vitro system has been met by combining genetics and biochemistry in novel ways, such as by the use of "reverse genetics." The term reverse genetics implies that a purified protein is used to generate appropriate reagents for cloning the gene encoding the protein. In vitro mutagenesis of the gene followed by homologous gene replacement and in vivo analysis of the phenotype of mutants can reveal the biological function of the protein. Furthermore, such methods provide a powerful tool for systematically correlating effects of changes in structure on catalytic function in vitro and biological function in vivo.

In addition to reverse genetics, the SV40 in vitro replication system has filled a major gap left by the absence of a yeast in vitro replication system. Namely, it has served as a reliable basis upon which to establish priorities as to which proteins to study. As the crude SV40 system has been resolved into component proteins of known function (for review, see Stillman 1989), structural and functional yeast analogs have been rapidly identified, e.g., yeast DNA polymerase δ, yPCNA, and most recently, yRP-A (see below). The degree of conservation between the replication apparatus in yeast and that in metazoans was unanticipated, even by those who were the most optimistic about yeast as a model system at the writing of the previous edition of this monograph (Strathern et al. 1981, 1982).

To date, two categories of enzyme, the DNA polymerases and topoisomerases, have been the focus of biochemistry and reverse genetics in yeast. Results obtained in yeast have provided clarification of both the cellular and enzymatic functions of the polymerases and topoisomerases in all eukaryotes, an understanding enhanced by the genetic analysis available in yeast. Description of work on these proteins therefore naturally receives the greatest emphasis in the following discussion. Work on other replication proteins—single-stranded DNA-binding proteins, helicases, and RNaseHs—is at an earlier stage, but the progress in identifying these proteins in yeast and the genes encoding them is also described in this section.

A. DNA Polymerases

Since the discovery of calf thymus DNA polymerase α in 1960, numerous additional DNA polymerases, called β, γ, δ_1, and δ_2, have been described in eukaryotic cells. Because of the small amounts of these proteins present in cells and the inability to control proteolysis during purification, it was difficult to resolve whether some of these polymerases were simply forms of a single polymerase or whether they were the products of different genes. In addition, some ambiguity has persisted in assigning these polymerases specific roles in replication, repair, and recombination, due in part to the lack of appropriate mutant cell lines. A major goal of reverse genetics in yeast has been to identify the yeast analogs of eukaryotic DNA polymerases and to define the division of labor among the various cellular DNA polymerases genetically.

To date, four DNA-dependent DNA polymerases have been purified from yeast, and the existence of one additional DNA polymerase has been inferred from genetic and DNA sequence data. The first DNA polymerase identified in yeast was associated exclusively with mitochondrial fractions (Wintersberger 1966). The mitochondrial enzyme was subsequently purified and shown to be the product of a nuclear gene (Wintersberger and Wintersberger 1970b). Mutations in this gene block mitochondrial DNA replication (Genga et al. 1986) and have been used to clone the gene, *MIP1* (Foury 1990). In 1970, the first two nuclear DNA polymerases, DNA polymerase I(A) and II(B) were described (Wintersberger and Wintersberger 1970a; E. Wintersberger 1974). They were extensively characterized and shown to be biochemically and immunologically distinct (Chang 1977; Plevani et al. 1980). The fourth DNA-dependent DNA polymerase in yeast, DNA polymerase III, was not reported until 1988 (Bauer et al. 1988). A fifth DNA polymerase may be encoded by the *REV3* (*rev*ersionless) gene, as deduced from DNA sequence, but has not yet been identified biochemically (Morrison et al. 1989). Yeast is apparently missing DNA polymerase β (U. Wintersberger 1974), a polymerase found in higher cells that is related to terminal transferase in amino acid sequence. Finally, at least one single-copy RNA-dependent DNA-polymerizing activity has been reported in yeast, SRT(160). It appears to be different from any of the DNA polymerases and from the reverse transcriptases associated with the repetitive Ty elements (Wintersberger et al. 1988).

Because of the rapid flux in our understanding of yeast and mammalian polymerases over the past year, there is currently a significant problem with nomenclature. A new general nomenclature derived largely from the original Greek letter designations of eukaryotic DNA polymerases will be used in this paper (Burgers et al. 1990). On the basis of similarity in biochemical properties to mammalian polymerases, DNA polymerase I(A) will be called yeast DNA polymerase α, DNA polymerase II(B) will be called DNA polymerase ε, and DNA polymerase III will be called DNA polymerase δ. Table 2 summarizes the distinguishing characteristics of the α, δ, and ε polymerases.

1. DNA Polymerase α

On the basis of subunit structure, tight association with a DNA primase, sensitivity to inhibitors, biochemical properties, and amino acid se-

Table 2 DNA polymerase

Characteristic	DNA polymerase		
	α	δ	ε
Catalytic subunits (kD)	167 (*POL1, CDC17*)	124 (*POL3, CDC2*)	170
Accessory subunits (kD)	70 58 (*PRI2*) primase 48 (*PRI1*) primase	50	?
Exonuclease	no	intrinsic	intrinsic
I_{50} aphidicolin	1 μg/ml	0.6 μg/ml	1 μg/ml
I_{50} BuPdGTP	1 μg/ml	>100 μg/ml	>100 μg/ml
Processivity	100	<100	>1000
PCNA	independent	dependent	stimulated
Major function	replication	replication	(?)

quence, DNA polymerase I(A) is the counterpart of mammalian DNA polymerase α (for review, see Campbell 1986; Burgers 1989). Immunoaffinity chromatography reveals a core enzyme assembly that contains the four-subunit structure conserved in DNA polymerase α from all sources: a 180-, a "70"-, a 58- and a 48-kD polypeptide (Plevani et al. 1984, 1985). The 180-kD subunit contains the DNA polymerase active site, since this polypeptide can be renatured in SDS gels and be shown to catalyze DNA polymerization (Badaracco et al. 1983a, 1985; Johnson et al. 1985). The subunit appears on SDS gels as a heterogeneous set of bands ranging in size from 140 kD to 180 kD, which may arise due to proteolysis or which may be modified forms of the 180-kD species. The extent and nature of posttranslational modifications of yeast DNA polymerase α have not been studied to date. Peptides of 110 kD and even 70 kD appear to retain catalytic activity (Badaracco et al. 1983a; Johnson et al. 1985). The other subunits are described below.

Earlier studies of yeast DNA polymerase α using conventional purification and traditional nuclease substrates suggested that none of the subunits contain a 3′ to 5′ exonuclease. An extremely sensitive assay has recently been used to verify that no 3′ to 5′ exonuclease is tightly associated with affinity-purified yeast DNA polymerase α (Kunkel et al. 1989). Like DNA polymerase III of *E. coli*, yeast DNA polymerase α may have a less tightly associated subunit that contains the 3′ to 5′ exonuclease. Immunoaffinity-purified human and calf thymus DNA polymerase α also appear to lack 3′ to 5′ exonuclease. *Drosophila* DNA polymerase α, which has only been purified by conventional chromatography, has been reported to have 3' to 5' exonuclease activity, although whether the nuclease activity is intrinsic to one of the four DNA polymerase α core subunits has not been rigorously demonstrated (Cotterill et al. 1987b; see also Kaguni et al. 1983).

2. Requirements and Inhibitors

Yeast DNA polymerase α is sensitive to aphidicolin and butylphenyl dGTP (BuPdGTP) but insensitive to dideoxy TTP (ddTTP), as is mammalian DNA polymerase α. The K_m for dNTPs is similar for the yeast and mammalian enzymes. The preferred template is activated salmon sperm DNA. Templates with low ratios of primer to template, i.e., long regions of single-stranded DNA, are used inefficiently by the four-subunit core polymerase, indicating that it is not an inherently processive enzyme. As with calf thymus DNA polymerase α, processivity is very sensitive to changes in Mg^{++} ion concentration and pH (Sabatino et al. 1988; Burgers 1989).

3. Mechanism

Few kinetic studies have been carried out using yeast DNA polymerase α, and one must rely on extensive characterization of the KB cell DNA polymerase to infer that the basic mechanism is an ordered binding of primer template and nucleotide. However, studies of products of polymerization aimed at measuring the fidelity of the polymerase have nevertheless provided detailed insight into the allowed interactions between the polymerase and its primer-template. Nucleotide selection, the relative affinity of polymerase for a mismatched terminus, the ability to extend from a mismatched primer terminus, and the ability to remove damage will all contribute to the fidelity of a DNA polymerase, and information about all of these properties of the polymerase can be inferred by careful analysis of polymerization errors.

The pattern of mutations produced by highly purified yeast DNA polymerase α and ideas as to how they may occur have been reported by Kunkel et al. (1989). An in vivo marker rescue assay was used to measure errors arising during copying of phage M13 templates by yeast DNA polymerase α. Yeast DNA polymerase α commits substitution, frameshift, and deletion errors with fairly high frequencies: 1 in 12,000 and 1 in 17,000 nucleotides incorporated for substitutions and frameshifts, respectively. Substitutions were most often due to simple misincorporation. Frameshifts occurred most frequently at reiterated bases, as expected, for instance, from the Streisinger slippage model (Streisinger et al. 1967). Seventeen percent of all mutations were deletions that occurred between directly repeated sequences, suggesting that rearrangement and realignment of primer and template occur easily during synthesis by yeast DNA polymerase α. A small group of more complex mutations were also detected with yeast DNA polymerase α, but whether they arose in vitro or in vivo was not resolved.

The error rates observed with yeast DNA polymerase α, of course, are far higher than the in vivo rate. The discrepancy is not surprising considering that yeast DNA polymerase α has no intrinsic 3′ to 5′ exonuclease. In addition, at least one other DNA polymerase and several other accessory proteins are involved in DNA replication in vivo. These factors increase the affinity of core enzyme for paired primers (Pritchard et al. 1983) and affect processivity, both of which would tend to reduce the in vivo error rate.

4. Genetics

POL1, the gene encoding the catalytic subunit of yeast DNA polymerase α, has been cloned (Johnson et al. 1985; Lucchini et al. 1985). The gene maps to chromosome XIV, 4 cM from *TOP2* (Budd and Campbell 1987; Lucchini et al. 1988), and is both single copy and essential. The lethal effect of gene disruptions showed that no other DNA polymerase in the cell could compensate for loss of yeast DNA polymerase α (Johnson et al. 1985; Lucchini et al. 1985) and unequivocally documented for the first time that yeast DNA polymerase α is essential for DNA replication.

To investigate the basis for the lethality of *pol1* deletions and to define the in vivo roles of yeast DNA polymerase α further, conditional lethal mutants were derived by in vitro mutagenesis. Budd and Campbell (1987) and Budd et al. (1989b) isolated 12 temperature-sensitive *pol1* alleles, of which seven (*pol1-11* through *pol1-17*) have been characterized. Lucchini et al. (1988) have isolated and characterized one temperature-sensitive mutation, *pol1-1*. The phenotypes of the various *pol1* mutants obtained by in vitro mutagenesis and those of *pol1* alleles recovered in screens for cell-cycle mutants and recombination mutants provide a good example of why it is useful to be able to isolate many different mutant alleles of a gene (see Table 3).

Most important in defining the actual role of yeast DNA polymerase α in replication is the tightest allele, *pol1-17*. All other alleles show extensive residual DNA synthesis at restrictive temperatures. *pol1-17* is a classic "quick-stop" replication mutant. Upon shift of a *pol1-17* culture to the nonpermissive temperature, nuclear DNA synthesis stops immedi-

Table 3 Phenotypes of different *poll* alleles

Allele	Mutation	Phenotype	References
poll-1	aa 493 Gly→Arg in P region	ts; slow-stop mutant after shift to npt; on-going DNA synthesis is completed; cells divide once before arresting at second cycle; terminal phenotype: large bud containing an undivided nucleus between the mother and daughter cell; protein may be stable in a replication complex.	1, 2
poll-12 and *poll-13*	aa 814 Asp→Asn	ts; quick stop	9
poll-17	aa 1004 Thr→Ileu located 4 aa carboxy-terminal to YGDTDS (2nd change at aa 1050, Ala→Val)	ts; quick-stop mutant of nuclear, not mtDNA synthesis; first division arrest in G_1 or S; terminal phenotype: large bud with undivided nucleus between mother and daughter cell; defect at elongation; necessary for premeiotic DNA synthesis and commitment to recombination; not necessary for repair of X-ray damage, not UV-sensitive at 30°C.	3–4
poll-14		ts; quick stop; "haploid" cells have >2 N DNA; replication and mitosis may be uncoupled	4
poll-15	aa 882 Thr→Ileu	ts; quick stop	9
cdc17-1	aa 904 of *POL1* active site Gly→Asp	ts; extensive DNA replication at npt, hyperrecombination, elongated telomeres	2, 5–7
cdc17-2	aa 634 Gly→Asp	ts; extensive DNA replication at npt, hyperrecombination, elongated telomeres	7
hpr3	aa 493 Gly→Glu	ts; intrachromosomal gene conversion, reciprocal recombination, and mitotic recombination between homologous chromosomes elevated, sensitive to MMS, UV survival similar to wt	7–8

See facing page for footnote to Table 3.

ately, whereas mitochondrial DNA synthesis continues. After 3 hours at the nonpermissive temperature, more than 90% of the cells are arrested, with large buds and a nucleus stuck between mother and daughter, a classic Cdc⁻ phenotype. Cells show first division arrest. Only those cells that have completed DNA replication at the time of shift up divide, whereas those in G_1 or S phase arrest without dividing (Budd et al. 1989b). The quick-stop phenotype of *poll-17* mutants suggests a defect in the elongation stage of DNA replication (Budd et al. 1989b). It also suggests that denaturation of yeast DNA polymerase α interferes with any further synthesis by any other DNA polymerase involved in replication as well.

5. Structure/Function Studies of Yeast DNA Polymerase α

DNA sequence analysis of *POL1* (Pizzagalli et al. 1988; Wong et al. 1988), which contains the sequence of the gene identified in Johnson et

Abbreviations: (npt) nonpermissive temperature; (ts) temperature sensitive; (mt) mitochondrial; (MMS) methylmethanesulfonate; (wt) wild type; (aa) amino acid.

References: (1) Lucchini et al. 1988; (2) Pizzagalli et al. 1988; (3) Budd and Campbell 1987; (4) Budd et al. 1989b; (5) Carson 1987; (6) Carson and Hartwell 1985; (7) Lucchini et al. 1990; (8) Aguilera and Klein 1988; (9) K.C. Sitney et al., unpubl.

A number of different mutant alleles of *POL1* have been isolated, analysis of which has clarified the various roles in DNA metabolism played by yeast DNA polymerase α. These alleles have been obtained either following direct mutagenesis of the cloned *POL1* gene (Budd and Campbell 1987; Lucchini et al. 1988; Budd et al. 1989b) or through screens designed to identify genes involved in the cell division cycle (see Hartwell et al., these volumes) or in recombination. The properties of a number of these alleles are summarized above in the table.

Consistent with the expectation that yeast DNA polymerase α is essential for duplication of chromosomal DNA, most of the alleles exhibit a cell division cycle arrest as large budded cells. This corresponds to the position of arrest of cells whose DNA is underreplicated or damaged (Weinert and Hartwell 1988). Some of the alleles of *poll* (e.g., *poll-17* and *cdc17-1*) exhibit quick stop in DNA synthesis and first cycle arrest, consistent with inactivation of the mutant enzyme on a shift to the nonpermissive temperature, whereas others (e.g., *poll-1*) arrest only after multiple cycles of growth, suggesting that these alleles are temperature sensitive for assembly, rather than activity.

Diploids homozygous for the *poll-17* allele are sporulation-deficient, show less than 5% of the wild-type level of meiotic recombination, and fail to complete premeiotic DNA synthesis. This indicates that yeast DNA polymerase α is required for premeiotic DNA synthesis and cannot be substituted by other DNA polymerases. In contrast, both the rate and extent of repair of single- and double-stranded DNA lesions induced by X-rays are normal in *poll-17* strains (Budd et al. 1989b). Thus, either other polymerases can compensate for the absence of yeast DNA polymerase α in the process of repair of DNA damage or yeast DNA polymerase α is not involved in this type of repair.

A number of other phenotypes have been noted to be attendant on specific mutation of *POL1*. These include enhanced rates of mitotic recombination and gene conversion, increased spontaneous mutation rates, sensitivity to alkalating agents, increased telomere length, and polyploidy. The recombination effects are likely to be an indirect consequence of the accumulation of recombinogenic damage in strains with a mutant polymerase, a situation previously noted for *E. coli* DNA replication mutants (Konrad 1977). Other phenotypes may address the various properties of the polymerase α. However, the polyploidy observed in the *poll-14* strain is difficult to explain. This phenotype may indicate a coupling between replication and mitosis that can be disrupted by mutation of yeast DNA polymerase α. This observation has a correlate in mammalian cells, in which drugs that inhibit DNA polymerase α yield a >4 N DNA content (Johnson et al. 1986).

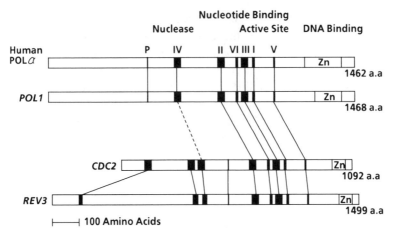

Figure 3 Conserved regions in various DNA polymerases.

al. (1985), combined with genetic analysis has led to important new insights into the structure of DNA polymerases. The predicted yeast DNA polymerase α contains 1496 amino acids. Comparison of the primary sequence with that derived from cDNA clones of human DNA polymerase α, three polymerases from herpesvirus and related viruses, adenovirus polymerase, four bacteriophage polymerases, the S1 mitochondrial DNA polymerase, a DNA polymerase encoded by a yeast linear plasmid, and two other nuclear yeast DNA polymerases (*CDC2* and *REV3*, see below) reveals six regions of extensive similarity (regions I–VI) in a 400–500-amino-acid stretch in the mid- to carboxy-terminal portion of the protein (amino acids 609–1085 in human DNA polymerase α; amino acids 610–1079 in yeast DNA polymerase α). These regions have an invariant amino- to carboxy-terminal order and similar, although not identical, spacing in all of the polymerases (see Fig. 3) (Jung et al. 1987; Wang et al. 1989). Although many of these polymerases are required for replication, one of them, yeast Rev3p, is not essential for growth and therefore is not required for DNA replication (Morrison et al. 1989).

Although regions I–VI are common to these polymerases, the sequencing of the second essential cellular DNA polymerase, yeast DNA polymerase δ encoded by *CDC2* (*POL3*), highlighted the fact that the polymerases do separate into two distinct families distinguished both by differences within regions I–VI and by the presence of several additional regions conserved in only one of the families (Boulet et al. 1989). The α polymerases constitute one class, represented by yeast DNA polymerase α and human DNA polymerase α. They can be aligned over their entire sequence, although the greatest similarities are in the regions already mentioned (Wong et al. 1989). The second class, which will be referred to as the δ class, includes yeast DNA polymerase δ and the polymerases from herpes simplex virus (HSV) and related viruses. Several regions, first pointed out as being characteristic of the herpesvirus and related polymerases, are found in yeast DNA polymerase δ and the yeast Rev3p, but not in DNA polymerase α. These regions and mutations affecting them are further discussed in the context of *CDC2* and *REV3* below.

Single-base-pair changes in region II, III, or V of the HSV DNA polymerase confer altered sensitivity to multiple inhibitors, which include nucleoside analogs used as antiviral drugs, the pyrophosphate analog phosphonoacetic acid, and/or aphidicolin, a competitor of dCTP. The apparent lack of specificity for any particular chemical moiety in the substrates prevents assignment of specific binding functions to specific amino acids (Gibbs et al. 1985, 1988). The location of the mutations does suggest, however, that amino acids spanning regions II, III, and V fold into a substrate-binding pocket in the polymerase active site. There is no homology with the GXXXXGKS motif conserved in other nucleotide-binding proteins such as helicases, ATPases, and kinases (Möller and Amons 1985). The most highly conserved motif, in region II, is the "SLYPSI" run of amino acids. The lysine conserved in region III is of interest in view of studies that have shown that the active site of *Drosophila* DNA polymerase is cross-linked by pyridoxal phosphate (Diffley 1988).

Region I is the most highly conserved domain (90% overall) and contains the motif YGDTDS. Although none of the HSV mutations mapped to this region, mutations in five out of six of the amino acids of the conserved motif inactivate the φ29 polymerase, and the *pol1-17* mutation, Thr-1004→Ile, also maps in region I (Bernad et al. 1989; Ma 1989). Thus, the region may be active in catalysis.

The finding that all characterized temperature-sensitive *pol1* mutations, except *pol1-1* and *cdc17-2*, map to region I, II, or III (Lucchini et al. 1988; Budd et al. 1989b) further suggests that this part of the protein may be critical for protein folding or stability of substrate-binding sites. Although all of the temperature-sensitive mutations fall in the same region as the drug-resistant mutations of HSV, none of the temperature-sensitive mutations confer on yeast resistance to aphidicolin, acyclovir, or arabinosyl thymidine (Ma 1989). This was anticipated because the other essential yeast DNA polymerases are also sensitive to these drugs (see below). The temperature-sensitive mutant polymerases have not been tested for drug resistance in vitro, because the mutant proteins are inactive in lysates. Isolation of yeast mutants resistant to substrate analogs will require either site-directed mutagenesis or use of an inhibitor that affects yeast DNA polymerase α more strongly than yeast DNA polymerase δ, such as butylphenyl triphosphates (Khan et al. 1984; Bauer et al. 1988; Burgers and Bauer 1988; Sitney et al. 1989).

Region IV, which is significantly amino-terminal to the II-III-I-V region, may constitute an entirely separate enzymatic domain from the polymerase domain. The sequence of region IV is similar to the active site of the *E. coli* DNA polymerase I (pol I) 3′ to 5′ exonuclease, which also resides in an "independent" domain amino-terminal to the polymerase active site. The ε subunit of *E. coli* DNA polymerase III, which encodes a 3′ to 5′ exonuclease, also belongs to this family (Bernad et al. 1987, 1989; Leavitt and Ito 1989). Three subregions, called Box 1, Box 2, and Box 3, appear to contain the residues critical for exonuclease activity. For example, crystallography shows two metals in the exonuclease-active site of *E. coli* pol I. One of them, site A, is coordinated to the carboxylate groups of Glu-355 and Asp-357 (Box 1 in Bernad et al. 1989), with the dNMP as an additional ligand. The second metal, site B, is coordinated to the dNMP and Asp-424 (Box 3) (Joyce and Steitz 1987; Freemont et al. 1988). Mutation of Asp-424 to alanine or the double alteration of Glu-355 to alanine and Asp-357 to alanine in *E. coli* pol I and the equivalent mutations in φ29 polymerase inactivate the nuclease without affecting the level of polymerase activity (Bernad et al. 1987, 1989; Freemont et al. 1988). Mutations of bacteriophage T4 that cause an antimutator phenotype also map in the region coinciding with region IV (amino acids 202 and 213), whereas mutations causing a mutator phenotype affect either amino acid 189, 191, or 342 (L. Reha-Krantz, pers. comm.). Yeast DNA polymerase α and human DNA polymerase α, which appear to lack exonuclease, differ substantially from the other DNA polymerases in region IV, although both retain some similarity. The presence of recognizable region IV in human DNA polymerase α and yeast DNA polymerase α may reflect vestigial nuclease that has diverged and become inactive. It is even possible that inactivation is due to the replacement of a single amino acid. The aspartate in Box 1, which is essential for nuclease activity in *E. coli* pol I and φ29 polymerase, is conserved in all DNA polymerases with active nucleases but is not conserved in the human DNA polymerase α or yeast DNA polymerase α. On the other hand, the absence of nuclease in DNA polymerase α may be due to a more substantial evolutionary difference, since a third region, called Box 3, has also been proposed as being important for nuclease activity. This region is not as clearly conserved in the DNA polymerase α family. Box 3 contains the Tyr-497 and Asp-501 residues thought to be essential in the *E. coli* pol I exonuclease (Freemont et al. 1988) and the Asp-324 residue implicated in T4 exonuclease (Reha-Krantz 1988 and pers. comm.). The DNA polymerase δ family, including the HSV-type polymerases, contains Box 3 (see *CDC2* sequence below).

6. Amino Terminus

The function of the amino terminus of yeast DNA polymerase α is unclear. The amino-terminal 155 amino acids are not essential, because a truncated protein, initiated at an internal ATG and expressed under the control of the *GAL10* promoter, fully complements a *pol1-17* mutation and retains all of the in vitro properties of the intact polymerase (C. Gordon and J. L. Campbell, unpubl.). Further into the gene, the *pol1-1* and *hpr3* mutations, which result from changes at amino acid 493, cause a temperature-sensitive phenotype (Lucchini et al. 1988, 1990). This has led to the proposal that amino acids 493–497 are a specific locus, the P region, for interaction between the catalytic polymerase subunit and one of the primase subunits. However, it is difficult to distinguish a specific effect from general weakening of the protein-protein interaction due to denaturation of the polymerase (Pakula and Sauer 1989).

7. Carboxyl Terminus

The carboxyl terminus has been shown to contain two putative zinc fingers about 250 amino acids downstream from the region I YGDTDS motif and has therefore been proposed to be important in binding to DNA. A deletion removing 219 amino acids of the carboxyl terminus of yeast DNA polymerase α is lethal to cells (Johnson et al. 1985). All polymerases of the α and δ classes contain at least one zinc finger in the carboxyl terminus, and the T4amB22 mutant, which alters a zinc finger, is deficient in DNA binding (L. Reha-Krantz, pers. comm.).

In summary, like many multifunctional enzymes, DNA polymerase α appears to be composed of a set of modules assembled in linear array from the amino to the carboxyl terminus. The degree to which these domains are structurally and functionally independent remains to be seen, however. For instance, the conservation of region IV in DNA polymerase α in the apparent absence of conservation of nuclease activity suggests that either interaction with other parts of the polymerase or other proteins of the replication complex constrain the sequence in this domain to some extent.

8. Reverse Genetics and Subunits of Yeast DNA Polymerase α

The demonstration that *Drosophila* DNA polymerase α could synthesize DNA de novo on single-stranded homopolymer and phage DNA templates suggested that DNA polymerase α contained an intrinsic RNA-synthesizing, or primase, activity. Soon after, it was shown that a poly(dT) replicating activity also copurified with yeast DNA polymerase α (Plevani et al. 1984, 1985; Singh and Dumas 1984; Singh et al. 1986) and subsequently that this was a primase activity, which can be selectively separated from the polymerase-primase complex (Plevani et al. 1984, 1985; Brooks and Dumas 1989). The active primase fraction contains the 48-kD and 58-kD subunits, but not the 180-kD catalytic polypeptide or the 86-kD subunit. Several lines of evidence suggest that both p48 and p58 are necessary for primer synthesis. Polyclonal antibodies specific for either p48 or p58 inhibit primase activity (Lucchini et al. 1987). A monoclonal antibody, 21A6, that inhibits the primase but not polymerase immunoprecipitates both the 58-kD and 48-kD subunits but not the higher-molecular-mass subunits in the polymerase-primase complex (Brooks and Dumas 1989). The native molecular mass of the primase is 110 kD, consistent with the primase being either a homodimer of one of these polypeptides or a heterodimer of the two subunits (Plevani et al. 1985). Primase from *Drosophila* has been shown to synthesize primers that are not complementary to the DNA template (Cotterill et al. 1987a). Yeast primase activity is observed on phage DNA in the absence of a full complement of nucleotides and may represent a similar reaction (Brooks and Dumas 1989).

The reactions catalyzed by the polymerase-primase complex and the purified primase have been compared. Both synthesize short oligoribonucleotides, 11–12 nucleotides in length, in a processive fashion on both natural and synthetic templates. In addition, both the polymerase-primase complex and the purified primase extend the unit length primers into multiples of unit length, the purified primase differing only in that it is unable to extend the oligoribonucleotides beyond dimeric length, whereas the polymerase-primase is capable of multiple extensions. The enzyme seems to dissociate from the template after the synthesis of one primer and rebind for further extension (Singh et al. 1986; Brooks and Dumas 1989). The primase appears to be an RNA polymerizing activity, and not a DNA polymerizing activity (Brooks and Dumas 1989). Purified mouse primase and polymerase-primase complex have mechanisms apparently identical to that of the yeast enzymes (Suzuki et al. 1989).

Despite years of study, a fundamental gap remains in our knowledge of the mechanism regulating the transition from RNA to DNA synthesis in the polymerase-primase complex. The fact that DNA polymerase appears to have a higher affinity for primer than does primase may aid the transition from RNA to DNA synthesis (Singh et al. 1986; Brooks and

Dumas 1989; Suzuki et al. 1989). It is not known whether polymerase-primase dissociates after primer synthesis and then rebinds for chain extension.

An affinity-labeling technique originally developed for identifying the polypeptide carrying the active site of RNA polymerase has been used to investigate the primase active site (Riva et al. 1987; Foiani et al. 1989b). In polymerase-primase and in purified primase, both p58 and p48, but none of the other subunits, were labeled by this method. Thus, both subunits appear to participate in forming the active site. Although the active site may consist of both subunits, UV cross-linking experiments show that only p48 binds ATP, suggesting that the ribonucleotide-binding site is localized to a single subunit. Interestingly, when the primase reaction was allowed to proceed in the presence of the active-site affinity label, template, and ribonucleoside triphosphate before stable fixation by a reducing agent, the 180-kD polymerase catalytic subunit was labeled in addition to the primase subunits. Furthermore, when priming was coupled to DNA synthesis, by inclusion of deoxynucleotides in addition to ribonucleotides, only p180 was labeled. These observations suggest that the enzyme rearranges during the transition from priming to chain extension, translocating the primer from a site on the primase to the primer site on the polymerase subunit. A similar model has been proposed for initiation and elongation by RNA polymerase (Riva et al. 1987).

9. Primase Genes

Lucchini et al. (1985) separated the 180-, 70-, 58-, and 48-kD subunits of yeast DNA polymerase α on SDS gels and prepared polyclonal antisera to the individual polypeptides. None of the specific antibodies cross-reacted with any other subunit. The genes encoding p48 and p58 have been cloned (Lucchini et al. 1987; Foiani et al. 1989a). *PRI1*, the p48 gene, encodes a 409-amino-acid protein of 46,210 kD. Gene disruptions show that *PRI1* is an essential single-copy gene on chromosome IX. The DNA sequence reveals a protein with five regions (labeled I–IV from amino terminus to carboxyl terminus) of significant similarity to the mouse primase subunit of similar size, as deduced from the mouse cDNA sequence (Plevani et al. 1987; Prussak et al., 1989). Regions I–IV fall in the amino-terminal portion of the enzyme and can be aligned with a shift of only two amino acids; thus, both sequence and spacing are conserved. The most highly conserved region, region IV (yeast amino acids 103–182), contains 48 of 72 amino acid identities and both sequences reveal the potential metal-binding/DNA-binding motif $CX_2CX_{35}CX_5C$.

Residues 103–115 are rich in acidic amino acids. The carboxy-terminal portion of p48 diverges between yeast and mouse, as it does between mouse and other mammals, as demonstrated by the fact that a 3'probe from the mouse primase fails to detect an RNA in human and hamster cells, whereas a 5'probe does. There is only one stretch of conserved sequence, region V, in the carboxy-terminal domain (20 of 34 identical amino acids) (Prussak et al. 1989). These similarities are not surprising in view of the virtually complete conservation of mechanism between the primases from various sources (for review, see Campbell 1986; Brooks and Dumas 1989; Burgers 1989; Suzuki et al. 1989).

The gene encoding the 58-kD primase subunit, *PRI2*, is also a single-copy essential gene (Foiani et al. 1989a). *PRI2* is predicted to encode a 528-amino-acid protein of 62,262 daltons. The gene shares no homology with prokaryotic DNA primases. When the mouse p58 is cloned, it can be expected to show informative similarities.

10. A Second Primase

Several groups have attempted to purify primase directly from cell extracts, rather than from the purified polymerase-primase complex. A protein that stimulates yeast DNA polymerase α or *E. coli* DNA pol I activity on unprimed templates such as poly(dT) or single-stranded phage DNA has been identified and purified (Jazwinski and Edelman 1985; Wilson and Sugino 1985; Biswas et al. 1987). This protein is not antigenically related to the p48 or p58 primase subunits (Brooks and Dumas 1989), nor is it a known RNA or DNA polymerase. The biological role of this activity is currently unknown.

11. The "70"-kD Subunit

The intact 70-kD protein migrates with an apparent molecular mass of 86 kD on SDS gels but is easily proteolyzed to the 70-kD species previously reported in yeast DNA polymerase α preparations (D. Hinkle, pers. comm.). The 70-kD subunit has also now been cloned and sequenced. The DNA sequence predicts a 78.7-kD protein. The gene is located on chromosome 11 and thus is not allelic to *POL1* (chromosome XlV) or any other known replication gene. The gene is essential for growth (D. Hinkle, pers. comm.). Thus, all the subunits of yeast DNA polymerase α are genetically and structurally independent.

12. Processivity Factors

DNA polymerases do not usually have intrinsic abilities to copy long stretches of DNA without dissociating and rebinding. Most require additional subunits that increase the affinity of the polymerase for the primer terminus. Such proteins can be identified by assay for factors that affect the ability of polymerases to copy DNA with a low ratio of primer to template. Two mechanisms allow stimulation: (1) reduction of K_m for the preformed primer, an event reflecting the incorporation of the first nucleotide, or (2) increase in processivity, which reflects increase in affinity for the primer terminus after insertion of the first nucleotide and during each subsequent polymerization step. Although the core, four-subunit DNA polymerase α purified by immunoaffinity chromatography does not efficiently use templates with a low ratio of primer to template, a high-molecular-mass DNA polymerase α complex has been isolated from monkey and human cells that uses poly(dA)·oligo(dT), 20:1, 25 times more efficiently than the core enzyme. A complex of two proteins, called C1C2, is responsible for this stimulation (Lamothe et al. 1981; Vishwanatha et al. 1986). C1 and C2 are called primer-recognition proteins because they decrease the K_m for primers and thus stimulate the rate of incorporation of the first dNTPs. They do not increase processivity (Pritchard and DePamphilis 1983; Pritchard et al. 1983). The C1 protein is a tetramer of four identical subunits of 40 kD and the C2 polypeptide is a monomeric protein of 52 kD (Vishwanatha et al. 1986).

A protein with similar chromatographic properties (elution from phosphocellulose at >0.5 M salt and lack of binding to DEAE) that stimulates yeast DNA polymerase α was purified from yeast many years ago. It is called protein C and its estimated size is 37 kD (Chang et al. 1979). In addition to increasing the activity of yeast DNA polymerase α (but not yeast DNA polymerase ϵ) on long stretches of single-stranded DNA, protein C binds to single-stranded and double-stranded DNAs. Maximum yeast DNA polymerase α stimulation is observed at one protein C molecule per 100 nucleotides. Although data supporting a relationship of protein C with the C1C2 complex are tenuous, further characterization of protein C and protein C mutants would be interesting. Another possibility is that protein C is SSB1 (single-stranded DNA-binding protein), a 34-kD protein, although protein C appears to be much less abundant than SSB1 (Jong et al. 1985).

13. DNA Polymerase δ

DNA polymerase δ is the designation given to a class of DNA polymerases first described in rabbit bone marrow that, unlike DNA polymerase α, copurify with a 3′ to 5′ exonuclease (Byrnes et al. 1976; Goscin and Byrnes 1982). Despite many similarities, the various enzymes that have been collectively called DNA polymerase δ represent two different polymerases, DNA polymerase δ and DNA polymerase ε, distinguishable by subunit structure, template/primer preference, and response to some inhibitors. In yeast, the two enzymes are products of different genes (Sitney et al. 1989). In mammalian cells, where the genes have not yet been identified, the simplest way to distinguish them is on the basis of their inherent processivity and response to PCNA (proliferating cell nuclear antigen; a processivity factor, described in more detail below, required for SV40 DNA replication in vitro). DNA polymerase δ is a moderately processive enzyme that becomes highly processive in the presence of PCNA, whereas DNA polymerase ε is processive even in the absence of PCNA, although PCNA stimulates further.

Yeast DNA polymerase δ has the subunit structure and biochemical properties (Bauer et al. 1988; Burgers and Bauer 1988) of the PCNA-dependent DNA polymerase δ from calf thymus (Lee et al. 1984) and human placenta (Lee and Toomey 1987) and is therefore yeast DNA polymerase δ. Highly purified yeast DNA polymerase δ preparations contain two polypeptides that migrate as 120-kD and 55-kD polypeptides on SDS gels (Bauer and Burgers 1988; Burgers and Bauer 1988; Downey et al. 1988). The enzyme is sensitive to aphidicolin but resistant to 100 μM BuPdGTP and ddTTP. Yeast DNA polymerase δ has been reported to be more sensitive to aphidicolin than yeast DNA polymerase α (I_{50}, yeast DNA polymerase α = 4 μg/ml, yeast DNA polymerase δ = 0.6 μg/ml). However, yeast DNA polymerase α preparations themselves differ widely in their sensitivity to aphidicolin (Plevani et al. 1985). Yeast DNA polymerase δ copurifies with a tightly associated 3′ to 5′ exonuclease. It is a moderately processive enzyme, and its processivity is stimulated by calf thymus PCNA (Burgers 1988).

14. Genetic analysis

The identification of the yeast gene encoding yeast DNA polymerase δ established simultaneously not only that yeast DNA polymerase δ is distinct from DNA polymerase α, but also that it is required for DNA replication in addition to yeast DNA polymerase α. The identification of the gene for yeast DNA polymerase δ came not from reverse genetics, but from the demonstration that yeast DNA polymerase δ was deficient in extracts of a previously described replication mutant, cdc2 (Boulet et al. 1989; Sitney et al. 1989), following the demonstration that the DNA sequence of the CDC2 gene, cloned by complementation of a cdc2 mutation, predicted a protein with sequence similarity to DNA polymerases (Boulet et al. 1988). Fractionation of extracts of CDC2 poll-17 and cdc2-1 poll-17 double-mutant cells demonstrated that yeast DNA polymerase δ was deficient in *cdc2-1* cells and that the *pol1-17* mutation had no effect on yeast DNA polymerase δ (Sitney et al. 1989). Yeast DNA polymerase ε was present in normal amounts in the *cdc2-1 pol1-17* double mutant, and thus yeast DNA polymerase ε, which has many properties in common with yeast DNA polymerase δ, is encoded neither by *CDC2* nor by *POL1* (Boulet et al. 1989; Sitney et al. 1989). It was also shown that a different purification procedure than that originally described for yeast DNA polymerase δ yielded both yeast DNA polymerase δ and yeast DNA polymerase α in normal amounts from wild-type cells (Sitney et al. 1989). The demonstration that yeast DNA polymerase δ is encoded by an essential gene is especially interesting in view of recent data directly implicating mammalian DNA polymerase δ in replication in permeabilized cells and in the SV40 in vitro DNA replication system (Dresler and Frattini 1986; Decker et al. 1987; Dresler and Kimbro 1987; Hammond et al. 1987; Wold et al. 1989), and the indirect argument for a requirement for DNA polymerase δ based on the requirement for PCNA in SV40 DNA replication (Prelich et al. 1987a,b).

The 3′ to 5′ exonuclease that normally copurifies with yeast DNA polymerase δ is missing in the *cdc2* mutants (Boulet et al. 1989; Sitney et al 1989). This result, taken together with the conservation of the putative exonuclease domain in the *CDC2* gene, strongly suggests that the nuclease is contained in the 125-kD subunit. It has not, however, been formally shown that the 55-kD subunit is not the nuclease (for discussion, see Boulet et al. 1989).

The phenotype of existing *cdc2* mutants is complex and does not demonstrate unambiguously that yeast DNA polymerase δ is required continuously during DNA synthesis, as has been shown for DNA polymerase α. Hartwell has shown that Cdc2p is essential for DNA synthesis and that a *cdc2* mutant cannot complete its function during arrest of cells by hydroxyurea, a drug that inhibits DNA replication (Hartwell 1976). *cdc2* mutants have a "first cycle arrest" phenotype insofar as they do not divide after shift to the nonpermissive temperature. Although chromosomal replication is incomplete, 70% of a genome equivalent of DNA is synthesized after shift of a synchronous culture, arrested in G_1 phase, to the restrictive temperature (Conrad and Newlon 1983). These observations are consistent either with a defect in assembly of the replication complex and hence failure to initiate at a subset of replicons or with a partial defect in elongation. The currently favored idea is that both yeast DNA polymerase δ and yeast DNA polymerase α are required for elongation, based on the apparent requirement for DNA polymerase δ in the SV40 in vitro replication system.

Although substantial DNA synthesis occurs in vivo in *cdc2* mutants, the same mutants are defective in DNA synthesis in permeabilized yeast cells. When yeast cells are permeabilized with Brij in the presence of sucrose, they can continue propagation of replication forks initiated before treatment, but they cannot initiate new rounds of replication (Kuo et al. 1983). DNA synthesis in the Brij-sucrose-treated *cdc2* mutants ts328, ts370, and ts346 is normal at 23°C but completely defective at 37°C. Because synthesis at 23°C is normal, the number of replication forks is the same in the *cdc2* mutant and in wild type. Thus, the defect at 37°C is most consistent with a defect in elongation in *cdc2* mutants (Kuo et al. 1983). The extensive in vivo synthesis observed in the same mutants may be due to partial function of the mutant yeast DNA polymerase δ in vivo or to synthesis by yeast DNA polymerase α or another polymerase in the absence of yeast DNA polymerase δ.

One additional allele of *CDC2*, *hpr6*, has recently been uncovered in the search for strains with a hyperrecombination phenotype. In addition to the hyperrecombination phenotype, *hpr6* mutants are similar to *hpr3* mutants in their sensitivity to methylmethane sulfonate (MMS) and resistance to UV-induced lesions at semipermissive temperatures (Aguilera and Klein 1988).

15. Structure/Function

The *CDC2* sequence (Boulet et al. 1989) preserves the order and spacing of regions IV-II-VI-III-I-V found in many polymerases (see Fig. 3), and like the others, it contains potential DNA-binding zinc fingers in the carboxy-terminal portion of the protein. However, the *CDC2* gene clearly belongs to a family different from the DNA polymerase α family. The yeast *POL1* sequence is much more closely related to that derived from the human DNA polymerase α cDNA than it is to *CDC2*. For instance, regions I, II, III, and IV in yeast DNA polymerase α show more extensive homology with human DNA polymerase α than with yeast DNA polymerase δ. In addition, yeast DNA polymerase α and yeast DNA polymerase δ diverge completely outside of the core. When the sequence for mammalian DNA polymerase δ becomes available, it might be expected to show more similarity to the *CDC2* sequence than to the human DNA polymerase α sequence.

More dramatic differences between the DNA polymerase α and DNA polymerase δ families are seen in the amino-terminal regions and region IV, the putative exonuclease-active site. Cdc2p preserves all of the amino acids in Box 1 and Box 2 of region IV that are important for *E. coli* pol I 3' to 5' exonuclease function, in keeping with the fact that yeast DNA polymerase δ appears to have an intrinsic exonuclease activity (Bernad et al. 1989). A third conserved nuclease region, Box 3, is also found in Cdc2p. It lies within region A, which was originally defined by several drug-resistant HSV mutants and by homology in the region between all HSV-related polymerases (Gibbs et al. 1988). The divergence of region A in the α polymerases had led to the proposal that this might be a target for specific antivirus drugs; however, its conservation in yeast DNA polymerase δ, an essential cellular enzyme, makes this dubious. Five additional regions similar to those of the HSV-related polymerase are found in *CDC2* but are absent from *POL1* (Kouzarides et al. 1987; Boulet et al. 1989). No function has been correlated with these regions.

16. PCNA: A Polymerase Accessory Protein

In early studies using high-resolution two-dimensional gel electrophoresis to monitor global patterns of protein synthesis, two abundant proteins were identified whose synthesis occurred preferentially during an S phase following serum stimulation of quiescent cells. One of these proteins is an acidic nuclear protein of 36 kD that corresponds to an antigen that reacts with autoantibodies in the sera of certain patients with systemic lupus erythematosus (SLE) and that is only found in proliferating cells (Bravo and Celis 1980; Bravo and Macdonald-Bravo 1984, 1985; Mathews et al. 1984). This protein is now called PCNA. In contrast to the increase of PCNA RNA and protein synthesis observed after serum or growth factor stimulation of arrested cells, the level of PCNA does not fluctuate dramatically during the cell cycle of exponentially growing cells (Wold et al. 1988; Liu et al. 1989; Morris and Mathews 1989). PCNA has, however, been shown to be important for DNA replication (Zuber et al. 1989).

PCNA is an accessory protein of DNA polymerase δ that is required to enable DNA polymerase δ to copy templates with low ratios of primer to template efficiently, such as poly(dA)·oligo(dT) (Tan et al. 1986; Bravo et al. 1987; Prelich et al. 1987a,b). PCNA increases the processivity of DNA polymerase δ from a few nucleotides to several hundred nucleotides and also increases the affinity of DNA polymerase δ for double-stranded DNA, as measured by nitrocellulose-filter-binding assays. The most significant finding is that PCNA is required for SV40 DNA replication in vitro (Bravo et al. 1987; Prelich et al. 1987a,b).

The conservation of PCNA across species suggested that there might be a PCNA analog in yeast (Almendral et al. 1987; Matsumoto et al. 1987; O'Reilly et al. 1989). Burgers (1988) showed that calf thymus PCNA enhanced the activity of yeast DNA polymerase δ in a manner identical to that described for calf thymus DNA polymerase δ. The ability of mammalian PCNA to interact with yeast DNA polymerase δ is remarkable and is the first demonstration of functional conservation between the yeast and mammalian replication complexes, similar to the recently demonstrated interaction between yeast and mammalian components of the transcription apparatus.

Yeast PCNA (yPCNA) was purified by measuring stimulation of calf thymus DNA polymerase δ, which is entirely inactive on poly(dA)·oligo(dT) (Bauer and Burgers 1988). yPCNA is an acidic protein consisting of a trimer or tetramer of 26-kD subunits. It is immunologically distinct from mammalian PCNA. It stimulates yeast DNA polymerase δ, but not yeast DNA polymerase α or yeast DNA polymerase ε. On poly(dA)$_{500}$·oligo(dT)$_{10}$, 40:1, the processivity of yeast DNA polymerase δ is enhanced to several hundred nucleotides by yPCNA; however, there is little stimulation of synthesis by available preparations of yeast DNA polymerase δ on primed phage DNA templates. Thus, in assays to date, the upper limit in estimates of the degree of processivity of yPCNA is set by the length of the poly(dA) and not the inherent processivity of the polymerase-PCNA complex. The gene encoding PCNA, *POL30*, has been cloned and sequenced. Gene disruptions show that it is essential for growth (Bauer and Burgers 1990).

17. "Two Polymerase" Models

Why are two DNA polymerases required for replication? It is possible that one is involved in initiation and one is involved in elongation, i.e., the polymerases function sequentially. Precedent for this kind of replicon organization is well documented for colicin E1 (ColE1) plasmids in *E. coli*, where DNA polymerase I (*E. coli*) is required for initiation of DNA replication, but DNA polymerase III (*E. coli*) takes over for elongation of both the leading and lagging strands (Marians 1984). In the SV40 system, DNA polymerase α appears to initiate synthesis at the SV40 origin of replication in the absence of DNA polymerase δ. Synthesis appears to be continued by a second polymerase (for references, see Stillman 1989; Tsurimoto et al. 1990).

An additional model, which remains to be tested, proposes that two different polymerases function together at the replication fork to carry out concurrent replication of the leading and lagging strands (Sinha et al. 1980; Kornberg 1981). Studies showing a requirement for yeast DNA polymerase α during elongation in vivo (Budd et al. 1989b) and for yeast DNA polymerase δ in permeabilized *cdc2* cells (Kuo et al. 1983) and the studies of Prelich and Stillman (1988), which demonstrate that only lagging-strand synthesis occurs in the absence of PCNA, suggest that it is likely that two DNA polymerases are required for elongation. Further-more, circumstantial evidence for concerted action of yeast DNA polymerase α and yeast DNA polymerase δ at the yeast replication fork indicates that elongation is blocked completely at 36°C by the *poll-17* mutation, which directly affects only yeast DNA polymerase α. Viewed in the context of these results, the enzymatic properties of the two polymerases suggest specifically that DNA polymerase α copies the lag-ging strand and DNA polymerase δ/PCNA copies the leading strand. The association of primase with DNA polymerases α makes DNA poly-merase α ideal for discontinuous synthesis. The processivity of the DNA polymerase δ/PCNA complex, combined with its ability to catalyze strand displacement (Downey et al. 1988), makes it suitable for the leading-strand polymerase. One attractive feature of the concurrent two polymerase model is that affinity of DNA polymerase α for DNA polymerase δ would provide continuous association of the lagging-strand polymerase with the replication fork while allowing for dissociation from one Okazaki piece and recycling to the next primer. The availability of yeast mutations in both of the replicative polymerase genes should allow us to unravel the roles of DNA polymerase α and DNA polymerase δ in all eukaryotes. With the recent demonstration that yeast DNA polymerase ε is essential for viability (Morrison et al. 1990), the question arises as to where a third polymerase might be required.

18. DNA Polymerase ε

Yeast DNA polymerase II has recently been purified to homogeneity (Budd et al. 1989a; Hamatake et al. 1990), and it is found at normal levels in both *poll-17* and *poll-17 cdc2* mutants (Budd et al. 1989a; Sit-ney et al. 1989). It is not inhibited by polyclonal antibodies against yeast DNA polymerase α or yeast DNA polymerase δ (Chang 1977; Winters-berger 1978; Burgers and Bauer 1988). Thus, it is the product of a sepa-rate gene from yeast DNA polymerase α and yeast DNA polymerase δ. As noted above, we will refer to DNA polymerase II as yeast DNA polymerase ε in order to conform to current nomenclature. Subunit struc-ture, biochemical properties, and sensitivity to inhibitors imply that this is the analog of the so-called PCNA-independent DNA polymerase δ in mammalian cells. Like PCNA-independent DNA polymerase δ, this polymerase is a high-molecular-weight, aphidicolin-sensitive DNA polymerase that copurifies with a tightly associated 3′ to 5′ exonuclease. In addition, yeast DNA polymerase ε is highly processive in the absence of cofactors. Under some conditions, but not others, it is stimulated by purified PCNA (Crute et al. 1986; Burgers and Bauer 1988; Focher et al.

1988; Nishida et al. 1988; Syvaoja and Linn 1989; Brown et al. 1990; Hamatake et al. 1990;).

Yeast DNA polymerase ε has four main distinguishing properties. First, it copurifies through all purification steps with a 3′ to 5′ exonuclease. Since the polymerase appears to be homogeneous, it is likely that the nuclease is intrinsic to the catalytic core subunit of yeast DNA polymerase ε. Second, the size of yeast DNA polymerase ε is interesting. The polymerase migrates on SDS gels as a 170-kD protein, and activity gels reveal that this polypeptide is catalytically active. An active species of higher molecular mass corresponding to the reported size of the mammalian enzyme (~200 kD), can also be detected (Budd et al. 1989a and unpubl.; Hamatake et al. 1990). The third important property of yeast DNA polymerase ε is its preference for primer templates with long stretches of single-stranded templates, i.e., low ratios of primer to template, such as poly(dA)·oligo(dT), 40:1. Use of this assay has allowed purification of yeast DNA polymerase ε in amounts enabling the production of antibody and extensive characterization (Budd et al. 1989a; Hamatake et al. 1990). Finally, yeast DNA polymerase ε is comparatively resistant to BuPdGTP, like yeast DNA polymerase δ and unlike yeast DNA polymerase α.

A chromatographically distinct DNA polymerase was found during purification of yeast DNA polymerase ε. In all but chromatographic properties, however, the enzyme is identical to yeast DNA polymerase ε and may simply represent an altered form of yeast DNA polymerase ε (Budd et al. 1989a; Hamatake et al. 1990).

The yeast DNA polymerase ε gene has been cloned and has been shown to be an essential gene (Morrison et al. 1990). One can infer from work with mammalian PCNA-independent DNA polymerase δ, however, that yeast DNA polymerase ε plays a role in DNA repair. Human PCNA-independent DNA polymerase δ has been shown to be able to complement permeabilized human cells depleted of a factor essential for repair of UV-induced damage and was partially purified using that assay (Nishida et al. 1988). Because yeast DNA polymerase ε can be measured specifically in extracts by using its preferred template, poly(dA)· oligo(dT), a screen of existing *rad* mutants might be worthwhile as an approach to finding a mutation affecting yeast DNA polymerase ε.

19. REV3 *DNA Polymerase*

Cloning and sequencing of a gene called *REV3* has unexpectedly uncovered a fourth nuclear DNA polymerase species. The *REV3* gene was first identified by screening for mutants with decreased frequency of UV-induced reversion (Lemontt 1971, 1976; Quah et al. 1980). A number of different genes—*REV1*, *REV6*, *REV7*, *REV3*—having the same anti-mutator phenotype were isolated, but *rev3* mutants have the strongest defect. The *REV3* gene has recently been cloned by complementation of the defect in UV-induced mutagenesis (Morrison et al. 1989). The gene maps to chromosome XVI, 2.8 cM from *CDC60*. DNA sequencing revealed an ORF encoding a protein of 172,956 daltons that has substantial sequence similarity to the other nuclear yeast DNA polymerases (Morrison et al. 1989). *REV3* contains regions I through VI and seems to fall in the δ class of DNA polymerases, having closest homology with EBV DNA polymerase and diverging extensively from the α class outside of regions I–VI. Like the other polymerases, Rev3p contains several potential zinc fingers about 250 amino acids carboxy-terminal to the conserved YGDTDS region I. It is interesting that the *REV1* gene encodes an analog of the *E. coli umuC,D* gene products, which have been proposed to enhance the processivity of DNA polymerases in *E. coli* (Morrison et al. 1989).

A complete deletion of the *REV3* ORF does not cause lethality, suggesting that the activity encoded by *REV3* may be involved only in repair DNA synthesis and not in normal replication. *rev3-1* and *rev3Δ* strains are slightly more sensitive to killing by DNA-damaging agents than is wild type, but they are not as sensitive as mutants deficient in either excision repair or recombination-dependent repair. The *rev3* mutants are also not deficient in either meiotic recombination or gene conversion. Thus, it has been proposed that *REV3* and the other genes with similar mutant phenotypes function in a pathway of mutagenic *trans*-lesion DNA replication.

Biochemical fractionation of extracts of strains carrying the *rev3* deletion show normal levels of yeast DNA polymerase α, yeast DNA polymerase δ, and yeast DNA polymerase ε (including the altered chromatographic form of yeast DNA polymerase ε described by Budd et al. (1989a and unpubl.). In fact, no polymerase has yet been detected that may correspond to *REV3*. One possible explanation is that the enzyme actually encodes a nuclease, despite the apparent conservation of the polymerase-active site. Alternatively, the *REV3* polymerase may be of such low abundance that its activity has not yet been detected.

B. Topoisomerases

During replication of DNA circles or loops, unwinding of the DNA in advance of the replication fork leads to the introduction of positive supercoils, which must be removed by some kind of swivel. Topoisomerases, which catalyze transient-strand breakage and rejoining, are likely candidates for this swivel, and thus one potential role for topoisomerases in replication is maintenance of appropriate torsional stress during DNA synthesis. Indeed, bacterial topoisomerase mutants are defective in DNA replication, and the phenotypes of the topoisomerase mutants suggest that either a type I or a type II topoisomerase can provide the essential function. Topoisomerases may also be essential during termination of replication. For example, in vitro replication of SV40 requires a type II topoisomerase for separation of the multiply intertwined catenated dimers observed during the terminal stages of SV40 replication (Sundin and Varshavsky 1980, 1981). In addition to direct participation of topoisomerases in replication, local topology of DNA most likely plays a major role in protein/DNA interactions during transcription, replication, recombination, and repair, i.e., in virtually all DNA transactions. Protein/DNA interactions at origins of replication have already been shown to be affected by topology in prokaryotic replication systems (see Wang 1985).

Although only two topoisomerases have been detected biochemically in yeast, four distinct topoisomerase genes have been identified. The two biochemically defined topoisomerases, known as topoisomerase I and II, are encoded by *TOP1* and *TOP2*, respectively. They each have well-studied counterparts in other eukaryotes (Wang 1985). Identification of the third, or *TOP3*, topoisomerase, rests solely on predicted amino acid sequence similarity of *TOP3* to the *topA* gene of *E. coli* (Wallis et al. 1989). No corresponding topoisomerase activity has been detected in yeast extracts, nor has a potential counterpart been described in any other eukaryote. Similarly, identification of the fourth putative topoisomerase, the product of the *HPR1* gene, rests on its homology with the *TOP1* gene (Aguilera and Klein 1990).

1. Topoisomerase I

Yeast topoisomerase I is a 90-kD, monomeric protein. It is a type I topoisomerase that relaxes positively and negatively supercoiled DNA at equal rates. ATP and Mg^{++} are not required, but divalent cations stimulate the reaction up to 20-fold (Badaracco et al. 1983b; Goto et al. 1984). The enzyme can also break single-stranded circular DNA at low Mg^{++} and religate upon addition of Mg^{++} (Badaracco et al. 1983b). At low ionic strength and high protein concentration, partially purified topoisomerase I catalyzes catenation of nicked circular DNAs (Badaracco et al. 1983b). At long reaction times, linear DNAs with contour lengths of 5–57 monomeric lengths are observed. The rearrangement to linear multimers appears to be sequence-specific when ϕX174 is the substrate. A majority of the catenated molecules show intermolecular strand interactions, and "χ" structures can also be detected after restriction enzyme digestion, suggesting that catenation proceeds through transient fusion of rings (Badaracco et al. 1983b). Highly purified topoisomerase I does not appear to catalyze production of catenated networks or linear concatemer formation (Goto et al. 1984).

The topoisomerase I mechanism involves two *trans*-esterification reactions (for references, see Wang 1987). First, the enzyme cleaves a phosphodiester bond, simultaneously forming an O^4-phosphotyrosine linkage with the free 3′-phosphate end of the cleaved DNA. This protein/DNA complex is transient and is reversed in the second *trans*-esterification, or rejoining stage of the reaction, releasing the protein for further catalysis. The protein/nicked DNA complex has been shown to be a true intermediate by separation of the complex from free topoisomerase and demonstration of subsequent inter- and intramolecular joining. The initial protein/DNA complex is sometimes called the "cleavable complex," because treatment of the complex with denaturing agents such as SDS or alkali inhibit the rejoining step and lead to accumulation of the nicked DNA/protein intermediate. Certain drugs, such as the plant alkaloid and antitumor agent camptothecin, which likely inhibit the eukaryotic enzyme at the rejoining step, stimulate formation of the cleavable complex (see, e.g., Hsiang et al. 1985; Nitiss and Wang 1988).

The details of the reactions catalyzed by mammalian topoisomerase I and *E. coli* ω protein, or *topA* topoisomerase, differ significantly. First, although both proteins form a transient covalent protein/DNA intermediate, the bacterial enzyme attaches to the 5′ end of the DNA, whereas the eukaryotic enzyme is esterified to the 3′ end. Second, the two stages of the reaction are more tightly coupled for the bacterial enzyme. Unlike the eukaryotic topoisomerase I, the prokaryotic enzyme seems to form a stable bridge between the 5′ and 3′ termini, since the covalent intermediate can be trapped only by denaturing the protein. Finally, positive supercoils are removed by the eukaryotic enzyme but are not removed by the bacterial enzyme unless a single-stranded loop is present in the substrate. This suggests that the bacterial enzyme requires single-stranded DNA but that the eukaryotic enzyme does not.

The gene encoding yeast topoisomerase I has been sequenced and includes an ORF for a protein of 90,020 daltons (Thrash et al. 1985; and see below). Consistent with the mechanistic differences between the bacterial and yeast enzymes, little, if any, recognizable sequence homology exists between yeast topoisomerase I and the *E. coli topA* protein. Genes homologous to *TOP1* have been identified in *Schizosaccharomyces pombe*, vaccinia virus, and humans. Regions of similarity fall in two blocks, one containing yeast amino acids 240–530 and the other containing the carboxy-terminal 70 amino acids of the protein (Lynn et al. 1989). Tyr-727, which is in the conserved carboxy-terminal domain, is the acceptor for the $3'PO_4$ of the DNA chain (Eng et al. 1989), and changing this tyrosine to either phenylalanine or serine inactivates the topoisomerase (Lynn et al. 1989).

Several studies suggest that topoisomerase I is concentrated in the nucleolus in animal cells (Muller et al. 1985). In *Tetrahymena thermophila*, topoisomerase I is associated with specific sites in the ribosomal spacer DNA that are conserved in yeast, *Drosophila*, and several other organisms, and the yeast enzyme has been shown to bind to the *Tetrahymena* rDNA sites (Bonven et al. 1985).

2. Topoisomerase II

Yeast topoisomerase II is the analog of the prototypic type II DNA topoisomerase of all eukaryotes (Goto and Wang 1982). Like bacterial gyrase, topoisomerase II catalyzes relaxation of positively or negatively supercoiled DNA, catenation and decatenation, and knotting and unknotting of DNA rings. Unlike bacterial DNA gyrase and similar to phage T4 gene 52 protein, topoisomerase II does not catalyze supercoiling. Nevertheless, all reactions require ATP and a divalent cation and are inhibited by nonhydrolyzable ATP analogs. The enzyme was first detected as an activity that catalyzed ATP-dependent aggregation of DNA rings, which were later shown to be topologically interlocked networks. The aggregation reaction required a second 30-kD protein that binds both double- and single-stranded DNAs, which may act as a condensing agent in catenation because it is not required for decatenation (Goto and Wang 1982). Relaxed DNA is a better substrate for the aggregation than supercoiled DNA. All of the reactions catalyzed by topoisomerase II show a very sharp optimum salt concentration, 150–200 mM. This may account for the ability to block decatenation of SV40 replicative intermediates by hypertonic shock (Sundin and Varshavsky 1981). Topoisomerase II is found as a component of the nuclear scaffold and mitotic chromosomes in *Drosophila* and metazoan organisms and therefore may serve a structural role in the yeast nucleus in addition to its enzymatic role in topoisomerase reactions (Earnshaw et al. 1985; Gasser and Laemmli 1986). The protein is one of the most abundant chromatin-associated proteins aside from histones.

The reaction mechanism of eukaryotic type II topoisomerases involves the transient formation of a staggered, double-strand break to form an intermediate with a 4-bp single-stranded 5'-phosphoryl end covalently linked to protein through an O_4-phosphotyrosine and a protein-free, base-paired 3'-OH end (Wang 1987). The protein passes the duplex strand through the break and reseals. The bacterial *gyrA* quinolone inhibitors, nalidixic acid and oxolinic acid, which abort the strand passage and rejoining reaction by trapping a tight, possibly covalent, enzyme/DNA complex, called the cleavable complex, inhibit yeast topoisomerase II only at very high concentrations. However, several drugs (mAMSA, VM26, daunomycin, adriamycin) that inhibit the mammalian topoisomerase II by forming a reversible cleavable complex do inhibit yeast topoisomerase II and are cytotoxic, further suggesting conservation in mechanism between yeast topoisomerase II and other eukaryotic type II topoisomerases (Goto and Wang 1982; Wang 1987; Nitiss and Wang 1988).

Despite some differences from gyrase, such as the inability to super-coil DNA, the amino acid sequences derived from the DNA sequence of the human *TOP2, Drosophila TOP2, S. pombe TOP2*, T4 gene 52, and *E. coli* and *B. subtilis gyrA* and *gyrB* genes show extensive similarity (Lynn et al. 1986). Interestingly, the amino terminus of yeast *TOP2* is colinear with and similar to that of *gyrB*, and the carboxy terminus of yeast *TOP2* is similar to that of *gyrA* (22% identity, 14% similarity), suggesting that the yeast gene is the product of a gene fusion. In keeping with this idea, the bacterial gyrase is a tetramer of two each of two nonidentical subunits, A2B2, and the eukaryotic type II DNA topoisomerase is probably a homodimer of 150-kD subunits. As predicted from homology with the *E. coli gyrA* subunit and its known active site, Tyr-783 of yeast topoisomerase II has been shown to be covalently linked to the 5' end of the DNA molecule in the transient intermediate (Worland and Wang 1989).

3. Genetic Analysis of Topoisomerases I and II

The genes encoding topoisomerase I and topoisomerase II have been identified and cloned (DiNardo et al. 1984; Goto and Wang 1984, 1985; Thrash et al. 1985; Voelkel-Meiman et al. 1986). *TOP1* is allelic to the killer factor maintenance gene, *MAK1*, and is nonessential. *top1* null mutants do, however, show up to a 30% reduction in growth rate relative to wild type under certain conditions (Thrash et al. 1985; Kim and Wang 1989).

Despite the mechanistic differences between yeast topoisomerase I and *E. coli* topoisomerase I, biologically, the bacterial and yeast enzymes are to some extent interchangeable. The yeast *TOP1* gene, when expressed in *E. coli*, can complement a temperature-sensitive mutation in the *topA* gene (Bjornsti and Wang 1987). Complementation depends on the ability of the yeast enzyme to maintain a moderate level of negative supercoiling in the DNA, since in the absence of yeast topoisomerase I function, plasmids from the *topA* strain are highly negatively super-coiled. A deletion of 415 bp from the amino terminus of the yeast *TOP1* gene yields a protein with topoisomerase activity in extracts of *E. coli*, but the mutant protein cannot complement the *topA* defect, suggesting that the amino terminus is required for the essential function of topo-isomerase I. Yeast topoisomerase I is sensitive to the drug camptothecin, which interferes with the rejoining reaction, and it may be possible to study the effect of various antitumor drugs on the eukaryotic enzymes expressed in *E. coli*. Expression of *E. coli topA* in yeast has also been studied (Giaever and Wang 1988).

Temperature-sensitive *top2* mutants rapidly lose viability when placed at the restrictive temperature (DiNardo et at. 1984; Holm et al. 1985). In contrast to *cdc* mutants, asynchronous *top2* cultures fail to ar-rest with a unique, characteristic morphology upon temperature shift up, but instead show a mixture of unbudded cells and cells with large buds. However, when a synchronous culture is examined, the cells arrest uniformly with a single, large bud and an undivided nucleus (DiNardo et al. 1984; Holm et al. 1985, 1989). Using temperature-shift experiments, Holm et al. (1985) found that the timing of topoisomerase II function correlated with the time of mitosis. As discussed in detail in the chapter by Hartwell et al. (these volumes), the essential role of topoisomerase II in mitosis is probably in the segregation of newly replicated chromo-somes.

Another essential function that requires either topoisomerase I or topoisomerase II is DNA replication. Careful analysis by alkaline sucrose gradient sedimentation of the size and conformation of the DNA synthesized at various times during a synchronous S phase indicates that the rate of formation of high-molecular-weight DNA is significantly retarded in a *top1* deletion mutant but completely unaffected in a temperature-sensitive *top2* strain at the nonpermissive temperature. New-ly synthesized DNA accumulates in *top1*Δ mutants as strands about 5 kb in length. After a lag period, however, the chromosomes are fully repli-cated. This suggests that topoisomerase I is dispensable for at least the majority of initiations of DNA replication but that it normally partici-pates in the elongation phase of DNA replication, perhaps acting as a swivel. However, topoisomerase I is not absolutely required for elonga-tion, since the delay in extension of nascent chains is temporary, and eventually the chromosomes do reach full length in *top1*Δ mutants (Kim and Wang 1989).

Studies with *top1Δ top2*^{ts} double mutants suggest that topoisomerase II can compensate for the absence of topoisomerase I in DNA replication (Kim and Wang 1989). When an α-factor-arrested culture of a *top1Δ top2*^{ts} mutant is released from the block at the nonpermissive temperature, DNA synthesis at the elevated temperature is only about 20% of that seen at permissive temperature, and the cells remain unbudded (Goto and Wang 1985; Brill et al. 1987; Kim and Wang 1989). Alkaline sucrose gradient analysis of the DNA made under restrictive conditions shows that nascent DNA does appear but that it ceases to grow beyond several thousand nucleotides in length. When cells are shifted to the nonpermissive temperature at any time during S phase, DNA synthesis ceases rapidly and does not recover (Brill et al. 1987). Assuming that topoisomerase II activity is completely eliminated at the restrictive temperature, these observations imply that neither topoisomerase is necessary for initiation and that topoisomerase II alone compensates for loss of topoisomerase I in fork elongation in *top1Δ* mutants. These results in yeast are consistent with the finding that either a type I or type II topoisomerase is required continuously during replication of SV40 in vitro (Yang et al. 1987).

Recent studies have demonstrated a role for topoisomerase I and topoisomerase II in modulating torsional stress associated with transcripti n and in the metabolism of the rDNA repeats on chromosome XII. However, these topics are beyond the scope of this paper and are not discussed here. The interested reader is directed to chapters on transcription (Guarente; Thurieux and Sentenac) and ribosome biogenesis (Woolford and Warner).

4. The TOP3 Topoisomerase

top3-7 was identified as a mutation that caused hyperrecombination between the δ repeats of yeast and was formerly called *edr1-1* (Rothstein 1984). The sequence of the *TOP3* gene predicts a protein of approximately 74 kD with 39.1% identify with *E. coli* ω protein (Wallis et al. 1989). The latter type I topoisomerase is encoded by the *topA* gene, which is thought to be essential for *E. coli* growth. The *E. coli topA* gene cloned under control of the yeast *GAL1* promoter complements the *top3Δ* mutant on galactose medium, strengthening the argument that *TOP3* encodes a topoisomerase (Wallis et al. 1989). Indirect evidence for a topoisomerase activity in *top1Δtop2-4*^{ts} mutants, which are presumably totally lacking topoisomerase I and topoisomerase II, suggests that there is at least one additional topoisomerase in yeast not encoded by *TOP1* or *TOP2* (Brill and Sternglanz 1988). The *TOP3* gene product has not, however, been purified and strictly shown to carry topoisomerase activity. Although *TOP3* almost certainly encodes a topoisomerase of some description, the preliminary nature of our understanding of both the biological and biochemical functions of Top3p makes it difficult to assess the degree of conservation of function implied by the similarity in the *TOP3* protein and ω protein primary sequence (Fink 1989).

C. DNA Helicases

Helicases were originally defined as proteins that unwind duplex DNA and hydrolyze ATP and other nucleotides in a DNA-dependent reaction. RNA helicases have also now been shown to participate in transcription and translation. During DNA replication, DNA helicases are required during both initiation and elongation. During initiation of replication, helicases form stable nuclear protein complexes with the initiator protein and the origin of replication. Such complexes can be isolated in active form and have been shown to serve as substrates for replication by primase and polymerase. At least one and perhaps two helicases are also required for movement of the primosome on the lagging strand during elongation and seem to fulfill two functions: unwinding of the replication fork in front of the moving replication apparatus and translocation of primase to successive priming sites. The prototypic prokaryotic replication helicases are the dnaB protein, the phage T7-encoded gene 4 protein, and phage T4-encoded gene 41 protein. Recently, n′ protein (factor Y) has also been shown to be a helicase. SV40 large T antigen is one of the few eukaryotic replication proteins for which helicase activity has been demonstrated (Stahl et al. 1986). Although all helicases characterized to date move unidirectionally along the strand to which they bind, two categories of helicases are distinguishable: those that translocate in a 5′ to 3′ direction (i.e., in the direction of fork movement when bound to the lagging strand), such as the *dnaB* helicase, gene 4 of T7, and gene 41 of T4, and those that translocate in a 3′ to 5′ direction, such as n′ protein of *E. coli* and SV40 large T antigen.

Helicases have not been systematically sought in yeast thus far because assaying specifically for helicases in extracts is not currently possible. However, several studies have resulted in the identification of at least two helicases. First, ATPase III has been shown to possess helicase activity on φX174 DNA annealed to a short oligonucleotide (Sugino et al. 1986). ATPase III is a 63-kD, DNA-dependent ATPase. ATPase III stimulates yeast DNA polymerase α on both poly(dA)·oligo(dT) and activated calf thymus DNA templates five- to sevenfold, stimulation of polymerase being a property associated with many helicases.

Second, as discussed by Friedberg et al. (this volume), the product of the *RAD3* gene is also a helicase. At the moment, whether the *RAD3* helicase participates in replication is not clear, although extensive in vitro mutagenesis of the *RAD3* gene suggests that the helicase and ATPase can be inactivated without affecting cell viability (Sung et al. 1988). An unexplained finding is that ATPase III is absent in a *rad3* mutant, but does not appear to be the *RAD3* gene product (Sugino et al. 1986). Polyclonal antiserum to *RAD3* does not react with ATPase III, and ATPase III activity is not overproduced in strains that overproduce Rad3p (Sugino et al. 1986). The absence of ATPase III from the particular *rad3* strain used may have been adventitious.

D. Other Replication Proteins

1. SSBs

The term SSB (single-stranded DNA-binding protein) was originally used to describe proteins whose in vivo roles depend on their binding to single-stranded DNA. In prokaryotes, a single SSB is involved in DNA unwinding during replication, recombination, and repair. SSBs have also been isolated from eukaryotic cells, as they were from bacteria, on the basis of differential affinity for double-stranded and single-stranded DNAs. They were postulated to be involved in replication because many seemed to stimulate certain eukaryotic DNA polymerases specifically. However, protein and DNA sequencing of several prototypic eukaryotic SSBs and their cDNAs suggested that it was more likely that they played a role in RNA processing. For example, UP1 and UP2 from calf thymus show extensive homology with heterogeneous nuclear ribonucleoproteins (hnRNPs), as does mouse myeloma helix-destabilizing protein (HDP1) (for reviews, see Kowalczykowski et al. 1981; Willams et al. 1985; Chase and Williams 1986). Although UP1, UP2, and related proteins may be involved in replication as well as in RNA processing, this seems unlikely because the stimulation of polymerase α by these proteins does not depend on specific SSB-polymerase interactions similar to those found in prokaryotic systems. Instead, the stimulation results from coating DNA to mask the inhibitory effect of naked DNA.

Yeast contains multiple species of protein that bind preferentially to single-stranded DNA, and several groups have studied SSBs that stimulate DNA polymerase I (LaBonne and Dumas 1983; Jong et al. 1985; Jong and Campbell 1986). One of these, SSB1, is not essential, but rather is involved in a small nuclear RNP (snRNP) required for rRNA processing (Jong et al. 1987; M.W. Clark et al., in prep.). Therefore, a role for this protein in replication is unlikely.

Two other SSBs have been purified by affinity chromatography from yeast, SSB2, and a mitochondrial SSB (SSBm) (Jong and Campbell 1986). SSB2 is 55 kD and inhibits DNA polymerase I on salmon sperm DNA templates. SSBm is found only in mitochondria, and its relationship to the mitochondrial HM protein is unknown (Caron et al. 1979; Jong and Campbell 1986). All of these SSBs are abundant and elute from single-stranded DNA cellulose between 0.4 and 0.6 M NaCl, a much lower ionic strength than is required to elute prokaryotic replicative SSBs.

Another approach to identifying DNA-binding proteins, and replication proteins, in general, is to look for accessory proteins to the replicative polymerases. Several viral proteins and one nuclear host protein with SSB "activity" stimulate eukaryotic polymerases and are required for viral DNA replication. Both HSV and adenovirus type 2 encode SSBs essential for the replication of the viruses (see Campbell 1986; O'Donnell et al. 1987a,b). In addition, a nuclear protein with the properties of an SSB has been shown to be required for both the initiation and elongation phases of SV40 DNA replication in vitro and also stimulates DNA polymerases (for review, see Stillman 1989). This protein is called SSB, RF-A, or RP-A, and unlike the single-subunit prokaryotic prototypes, it consists of three subunits (Wobbe et al. 1987; Fairman and Stillman 1988; Wold and Kelly 1988). Recently, a yeast RP-A analog has been purified independently by two different approaches (Brill and Stillman 1989; Heyer and Kolodner 1989). Brill and Stillman (1989) identified a yeast protein that would substitute for human RP-A in the formation of an "unwound" complex at an early step in the initiation of SV40 DNA replication. This protein was purified and shown to be similar to its human analog in chromatographic behavior, subunit structure, and DNA-binding activity. The 66-kD subunit turned out to be encoded by the same gene, *RPA1*, as a previously described yeast SSB, purified by Heyer and Kolodner (1989) as a protein that stimulated a strand-transfer protein (Heyer et al. 1990). Rpa1p was identified as a single-subunit protein of 34 kD that, like *E. coli* SSB or phage T4 gene 32 protein, elutes from single-stranded DNA cellulose at 1.5 M salt (Heyer and Kolodner 1989). The protein binds cooperatively to single-stranded DNA and also stimulates strand exchange mediated by the *E. coli* RecA protein. Subsequently, it was shown that the 34-kD protein was encoded by the same gene as the 66-kD subunit of yeast RP-A, suggesting that the smaller protein is a proteolytic product of the 66-kD protein carrying the DNA-binding site (Heyer et al. 1990). The gene, *RPA1*, is essential for yeast growth (Heyer et al. 1990).

An additional SSB was identified by Brown et al. (1990), who purified SFI, one of two factors that were identified on the basis of their ability to stimulate yeast DNA polymerase ε on templates that have long regions of single-stranded DNA (Budd et al. 1989a). This protein was shown to bind single-stranded DNA specifically.

Three additional DNA-binding proteins, DBPI, DBPII, and DBPIII (M_r = 31,000, 14,000, and 20,000, and 22,500, respectively), have been identified on the basis of modulation of topoisomerase reactions (Goto and Wang [1982] referred to in Hamatake et al. 1989). These three proteins and several more species of SSB isolated as proteins that bind single-stranded DNA cellulose, but not double-stranded DNA cellulose, have been characterized with respect to stimulating strand-transfer proteins thought to be involved in mitotic and meiotic recombination (Hamatake et al. 1989; Heyer and Kolodner 1989). $ySSB_{14K}$ and $ySSB_{26K}$, and its breakdown product $ySSB_{20K}$, are basic proteins that stimulate DNA aggregation. This property is probably responsible for the observed enhancement by these SSBs of the activity of strand-transfer proteins. These activities of the SSBs may not be biologically significant, however, since spermidine or histone HI have the same effect. A group of acidic SSBs, $ySSB_{40K}$, $ySSB_{42K}$, and $ySSB_{55K}$, do not stimulate strand transfer. None of these proteins have the molecular weight or subunit structure of yRP-A.

2. Ribonuclease H

Enzymes that degrade the RNA component of RNA-DNA hybrids, designated RNase H activity, may play a role in replication, for instance, by removing RNA primers or in providing specificity to initiation. Several RNase H activities have been described in *S. cerevisiae*. RNase HI is a small, abundant 48-kD basic protein, active at acid and alkaline pH. The products are mostly 5′-NMPs and pNpN, suggesting an exonucleolytic mode of cleavage. RNase H2 is a smaller (21-kD) basic protein that is most active at alkaline pH and that produces primarily oligoribonucleotide products, suggesting an endonucleolytic mechanism (Wyers et al. 1973, 1976a,b). A third RNase H has been observed to copurify with RNA polymerase II (Huet et al. 1976).

A fourth RNase H activity, designated RNase H(70), was discovered during a search among DNA-binding proteins for proteins that specifically stimulated yeast DNA polymerase α (Karwan et al. 1983). Highly purified RNase H(70) retains the ability to stimulate yeast DNA polymerase α on homopolymer primer templates. The enzyme degrades RNA in RNA/DNA hybrids to oligoribonucleotides, suggesting an endonucleolytic mode, and will hydrolyze a phosphodiester bond joining a ribonucleotide and deoxyribonucleotide, as would be required during processing of Okazaki pieces. The enzyme exhibits some specificity, since the protein stimulates yeast DNA polymerase α but not yeast DNA polymerase ε. RNase H activities that specifically stimulate DNA polymerase α have also been purified from other organisms (see, e.g., DiFrancesco and Lehman 1985).

Antiserum to RNase H(70) cross-reacts with both a 70-kD polypeptide and a 160-kD species (Karwan et al. 1986). Antibody prepared by immunopurification of the antiserum using the 160-kD protein as a ligand reacts strongly with the 160-kD protein but only very weakly with the 70-kD protein, and a strain carrying a partial deletion of the gene encoding the 160-kD protein (see below) still produces the 70-kD protein, suggesting that they are probably different gene products (Wintersberger et al. 1988).

Under conditions where the RNase H activity was inhibited, RNase H(70) was shown to possess an RNA-dependent, DNA-synthesizing activity (reverse transcriptase) (see Karwan et al. 1986). Activity gels demonstrated that both the 160-kB and 70-kD proteins have reverse transcriptase activity, and the p160 has been provisionally named SRT(160) (Karwan et al. 1986; Wintersberger et al. 1988). Antisera containing antibodies that recognize epitopes in both the 70-kD and 160-kD polypeptides have been used to screen a λgt11 library, and two sets of nonoverlapping clones have been identified. One set encodes SRT(160). Gene disruptions are not lethal, but the mutant cells have an abnormal morphology and possess at least two times the normal amount of DNA. Interestingly, this is also the phenotype of one mutant *pol1* allele, *pol1-14*. RNase H(70) has been purified from the SRT(160) deletion mutant and has been shown to retain RNase H activity and the ability to stimulate yeast DNA polymerase α. The purified protein stimulates the extent of synthesis but not the length of the product and hence is not a processivity factor.

The association of a reverse transcriptase with the 70-kD and 160-kD polypeptides suggested that one or the other might be encoded by one of the Ty transposable elements, which produce a 65-kD reverse transcriptase with RNase H activity as a cleavage product from larger (100–190-kD) precursor polypeptide (see Boeke and Sandmeyer, this volume). However, the cloned SRT(160) gene does not hybridize to Ty DNA (Karwan et al. 1986; Wintersberger et al. 1988). Sequences related to reverse transcriptases have also been found in yeast mitochondrial DNA. However, strains lacking the mitochondrial genome contain normal amounts of RNase H(70) activity (Karwan et al. 1986).

Recently, two basic polypeptides of 55 kD and 48 kD, each of which has RNase H activity, have been shown to be immunologically related to each other and to share several tryptic peptides and are therefore probably the same protein. The proteins have been designated RNase H(55) (Karwan and Wintersberger 1988). How they are related to the 48-kD protein described by Wyers et al. (1976a,b) is hard to judge, but they are clearly immunologically distinct from RNase H(70) (Karwan and Wintersberger 1988). Using monospecific anti-RNase H(55) antibodies, it has been shown that RNase H(55) is associated with a number of small RNAs, that it cross-reacts with human anti-Sm autoantibodies that inhibit mammalian splicing, and that it contains epitopes related to the human Sm snRNP proteins B/B' (R. Karwan and I. Kindes-Mugge, *Yeast 4* [*Spec. Iss.*]: *490* [1988]). Therefore, RNase H(55) is more likely involved in RNA than in DNA metabolism.

3. Nucleases

Various exonucleases have been identified in yeast, although none of their genes have been cloned. As in prokaryotes, 3' to 5' exonuclease activity is likely to be essential in eukaryotes for high-fidelity DNA replication. Although yeast DNA polymerase α has no intrinsic exonuclease activity, both yeast DNA polymerase ε (Chang 1977; Wintersberger 1978; Budd et al. 1989a) and yeast DNA polymerase δ (Boulet et al. 1989) have associated 3' to 5' exonuclease activities. A number of exonucleases are described in Burgers et al. (1988).

A yeast endonuclease has been purified to near homogeneity, and its gene has been identified through reverse genetics (Burbee et al. 1988). This protein, the *NUC3* gene product, appears to play a role in DNA repair and is not discussed here. Activity gel analysis during the course of the *NUC3* purification revealed six active endonuclease polypeptides in addition to *NUC3* (Burbee et al. 1988). Although some of these may be active proteolytic fragments of *NUC3*, it will be of interest to determine the in vivo roles of other endo- and exonucleases.

V. IDENTIFICATION OF REPLICATION PROTEINS BY IDENTIFICATION OF REPLICATION MUTANTS

A. Plasmid Maintenance Mutants

One strategy for identifying proteins that interact with *ARS* elements is based on experiments designed by Jacob et al. (1964) for detecting mutations in genes affecting F-factor replication. Mutagenized populations are simply screened for strains that fail to support efficient autonomous replication of episomal DNA. In one yeast study using centromere vectors as test plasmids, 43 mutations, falling into 13 complementation groups, were identified (Maine et al. 1984; Gibson et al. 1987). They comprised two classes, the first of which affected all plasmids to the same extent, regardless of which *ARS* or *CEN* was present. This group could arise due to defects in *trans*-acting factors specific for either the *ARS* (replication) or the *CEN* (segregation) or other defects leading to plasmid loss. The second group affected plasmids carrying different *ARS* elements to varying extents, and the pattern was independent of the identity of the *CEN*. This class was therefore considered more likely to identify *ARS*-specific genes, even though the actual pattern of *ARS* stability was the same in all mutants. This class defines five different genes, three of which have been further characterized and designated *MCM1*, *MCM2*, and *MCM3* (*m*inichromosome *m*aintenance).

mcm1 mutants have been characterized in the most detail. The *MCM1* gene has been cloned and has been shown to be an essential gene (Passmore et al. 1988). The product of *MCM1* has recently been identified as the transcription factor PRTF/GRM (*p*heromone *r*eceptor *t*ranscription *f*actor/*g*eneral *r*egulator of *m*ating), which acts synergistically with the α1 protein as a coactivator of α-specific genes and with the α2 protein as corepressor of **a**-specific genes in α cells (Bender and Sprague 1987; Hayes et al. 1988; Jarvis et al. 1988, 1989; Keleher et al. 1988; Passmore et al. 1989; see also Guarente, these volumes).

How do these findings relate to the minichromosome maintenance defect, and why is *MCM1* an essential gene, when the mating-type system is not essential for growth? One possibility is that Mcm1p is required for transcription of a gene essential for replication. Mcm1p bears striking structural and functional similarity to the DNA-binding domain of human serum response factor, SRF (Norman et al. 1988; Passmore et al. 1989). Because of the similarity to SRF, the mammalian activator of growth-regulated genes, the promoters of various yeast *CDC* genes were searched for potential Mcm1p-binding sites. One recognition site was found 142 bp upstream of the transcription start site of the *CDC28* gene, which encodes a protein kinase whose activity is required both for entry into S phase and for mitosis. Indeed, Mcm1p interacts with an oligonucleotide containing this element. There is also an Mcm1p-binding site upstream of *CDC31*, whose product is involved in spindle-body duplication (Passmore et al. 1989). It is not known whether there is an effect on transcription of these genes in *mcm1* mutants.

Another possibility is that Mcm1p is an *ARS*-binding protein. Although binding of Mcm1p to *ARS* elements has not been demonstrated, the involvement of Mcm1p in one of the best-studied cases of combinatorial gene regulation as described above suggests the possibility that Mcm1p may bind to an *ARS* only in conjunction with another protein.

A second *mcm* mutation, *mcm2*, affects the same set of *ARS* elements as *mcm1* at low temperature, but at high temperature, it affects all *ARS* elements tested (Sinha et al. 1986; Gibson et al. 1987). The *MCM2* gene has been cloned and mapped to the left arm of chromosome II between *pet9* and *AMY2*. Linear plasmids, YEp13-based plasmids, and the *TRP1ARS1* circle are also unstable in the *mcm2* strain. During growth in nonselective medium, the copy number of the *TRP1ARS1* circle in the *mcm2* strain was reduced from 5.6 to 1.1 in five generations, after which the plasmid was rapidly lost. Thus, the defect appears even in the absence of a centromere. A 60-kb linear or circular derivative of chromosome III containing three *ARS* elements was stable at 25°C but unstable at 35°C. At 35°C, there was a 50-fold increase in recombination in the *HIS4-LEU2* region. Hyperrecombination and chromosome loss are phenotypes of mutants defective in DNA replication, but no direct effect on replication has yet been reported.

Different alleles at a third gene, *MCM3*, cause different phenotypes. The *mcm3 2B-61* mutation affects all *ARS* elements; addition of a second intragenic mutation to give *mcm3 2B-61R* results in the same phenotype as caused by *mcm1* or *mcm2*. The *mcm3 2B-61* mutant is temperature-sensitive for growth, and mutants arrest at the nonpermissive temperature with a large bud. Interestingly, the *mcm3* mutation is partially dominant in an *mcm3/+* heterozygote but is fully recessive to a *MCM3* gene inserted into the chromosome of a haploid *mcm3* strain. This suggests that gene dosage is not sufficient in heterozygous diploids to activate all the chromosomal *ARS* elements, but is sufficient in the partial diploid (Gibson et al. 1987). *MCM3* has been cloned and maps to chromosome V, between *anp1* and *cup3*. Interestingly, the *MCM2* and *MCM3* genes have recently been shown to be highly homologous to each other (B. Tye, pers. comm.).

Taken together, the phenotypes of the *mcm1*, *mcm2*, and *mcm3* mutants suggest that the *ARS* specificity observed simply reflects the degree of defectiveness of the mutation combined with the inherent strength of the *ARS*. Thus, with the less severe mutations, the weaker *ARS* elements will fall below the threshold required for activation but the stronger ones will still function. If any of the *MCM* gene products is an *ARS*-binding protein, this would mean that affinity for the *MCM* gene product differs in the different *ARS* elements.

A similar strategy has been used to isolate other mutants that have been less well characterized. For example, Larionov et al. (1985) selected a mutant, *smc1* (stability of minichromosomes), that affected the stability of both *ARS1* and 2-micron *ARS* plasmids carrying *CEN3*, *CEN6*, or *CEN11*. Addition of an extra copy of *ARS1* on the test plasmid did not stabilize it. Using a selection for loss of chromosome III in a strain disomic for chromosome III, the *chl2*, *chl3*, *chl4*, and *chl5* mutants were isolated. *ARS1-CEN4* plasmids show 1:0 segregation in *chl2*, *chl4*, and *chl5* mutants, suggesting that they undergo loss at mitosis, due either to a defect in mitosis or to a failure to replicate. In the same study, *cdc6* and *cdc14*, which have defects in S phase and mitosis, respectively, were shown to have an increased frequency of loss of an *ARS1-CEN3* plasmid at the semipermissive temperature (Kouprina et al. 1988).

Kikuchi and Toh-e (1986), looking explicitly for mutations that affected 2-micron plasmid replication, used a test plasmid containing the 2-micron *ori*, IR1 (inverted *r*epeat), and the *STB* (*st*a*b*le maintenance) region. The plasmids differ from those used by the other two groups in that they are high-copy plasmids whose stable maintenance is promoted by the *STB* locus, rather than single-copy plasmids whose stable maintenance results from the segregation function of the centromere. One mutation was identified, *map1* (*ma*intenance of *p*lasmid), that decreases the stability of 2-micron plasmids. YCp50 and YCp19, both of which carry *ARS1* and *CEN4* but no 2-micron sequences, were also unstable in *map1* strains. It is not known if *map1* is allelic to *mcm2*, which is the only *mcm* mutation known to affect the 2-micron plasmid. The fact that *CEN*-stabilized as well as *STB*-stabilized plasmids were affected by the *map1* mutations suggests that Map1p acts through the *ARS*, and further support for this conclusion came from the isolation of plasmid mutants that are stable in *map1* strains. A YEp13 library was transformed into the *map1* strain, and two types of plasmids were isolated that were stable in the mutant. One class had acquired an additional chromosomal *ARS* on the plasmid, called *ARSX*, and the other had acquired an additional 2-micron repeat (IR) in inverted orientation. YEp13 has a mitotic stability of 13% in *map1*, compared to 63% for YEp13*ARSX*. When *ARSX* was inserted into YCp50, the mitotic stability in *map1* was increased from 31% to 90%. Since a plasmid with *ARSX* alone was unstable in the *map1* mutant, the number of *ARS* elements appeared to affect plasmid stability. Consistent with this idea, a series of YCp19 plasmids carrying one, two, or three copies of the 1.4-kb *TRP1ARS1* segment showed mitotic stabilities of 31%, 91%, and 98%, respectively.

An alternative genetic approach to identifying *ARS*-binding proteins has been to look for suppressors of *cis*-acting *ARS* mutations. A major milestone in the understanding of SV40 DNA replication came from the isolation of second-site revertants of point mutations in the 65-bp minimal replication origin region. These second-site revertants all mapped in the large-T-antigen gene, which established that recognition of the origin was at least one of the essential functions of T antigen in the initiation of

SV40 DNA replication (Shortle and Nathans 1979). A similar approach has been taken in yeast by two groups. In the first study (Kearsey and Edwards 1987), mutagenized haploid yeast were transformed with a plasmid that contained a point mutation in the core consensus of the *HO ARS* (position 1026) that led to dramatic reduction of plasmid stability. Several mutants in which the plasmid was more stable were isolated. One, *rar1-1* (regulation of *a*utonomous *r*eplication), that was temperature sensitive for growth was characterized in detail. *RAR1* maps to chromosome XIII, near *rna1*, and therefore does not coincide with any other mapped gene that affects DNA replication. The sequence of *RAR1* was not similar to any other protein in the database.

In a separate study, plasmids carrying three different insertion mutations in the nonessential region of *ARS1*, domain C, were used to obtain *trans*-acting suppressors (Thrash-Bingham and Fangman 1989). The *ARS* mutations appear to affect plasmid replication rather than segregation. However, direct evidence that they interfere with the replication function of *ARS1* is lacking, and mutations in this region on plasmids that do not contain a *CEN* element do not detectably destabilize the plasmid. Nevertheless, at least four mutant complementation groups were found that increased stability of these *ARS1* mutants. One of these, called *amm1* (*a*ltered *m*inichromosome *m*aintenance), was further studied and shown to be allelic with *tup1*, also known as *flk1*, *umr7*, and *cyc9* (for references, see Schultz and Carlson 1987). Null alleles of *AMM1/TUP1* are not lethal and produce the same phenotype as the original *amm1* mutations. The phenotype of *tup1* mutants is very similar to that of *ssn6*. Both mutants fail to grow on glycerol, express glucose-repressible enzymes constitutively, clump, contain elevated levels of iso-2-cytochrome *c*, and exhibit defects involving cell type and sporulation. In addition, *tup1* and *ssn6* mutations suppress the same subset of *snf* mutations (Schultz and Carlson 1987). Schultz and Carlson have proposed that *TUP1* and *SSN6* might encode transcription factors in a common regulatory pathway. Interestingly, *ssn6* mutants have an Mcm$^-$ phenotype (M. Carlson, pers. comm.). Thrash-Bingham and Fangman (1989) have further suggested that transcriptional regulation may have a bearing on *ARS* function, a theory explored in more detail by Veit and Fangman (1988).

B. *cdc* Mutants

The cell division cycle (*cdc*) mutants of *S. cerevisiae* were isolated as temperature-sensitive mutants that arrest at the nonpermissive temperature with a uniform cellular morphology (Hartwell et al. 1970; Pringle and Hartwell 1981; and these volumes). Since failure to function apparently caused a block in progression through the cell cycle at a certain point, the gene products were postulated to be required for specific stages of the cell cycle. To date, however, it is not known how many *cdc* mutations directly affect DNA replication.

cdc mutants that have defects in the elongation stage of DNA replication, *cdc2*, *cdc17*, *cdc9*, *cdc8*, and *cdc21* (the latter two through failure to produce dTTP), share common mutant phenotypes: Cells arrest after a period at the nonpermissive temperature with a large bud and an undivided nucleus, usually located at the isthmus between the two cells. The spindle pole body is duplicated and divided, but the spindle is not elongated, and the chromosomes are therefore presumably not segregated at the nonpermissive temperature (Byers and Goetsch 1974). The mutants show hyperrecombination and increased chromosome loss (Hartwell and Smith 1985). Cells held at the nonpermissive temperature are still sensitive to hydroxyurea, an inhibitor of precursor synthesis, when returned to the permissive temperature (Hartwell 1976). In addition to the mutants named above, the *cdc13* mutant shares these phenotypes and is likely to be affected in replication (T.A. Weinert and L.H. Hartwell, pers. comm.). A feature common to these replication genes is that, where it has been tested, their mRNAs accumulate at a common point in the G_1 phase.

One recent addition to the list of phenotypes of some replication mutants is loss of viability in a *cdcX rad9* double mutant. *RAD9* is a key control gene (a checkpoint gene) that functions in G_2 to arrest the cell cycle before entry into mitosis until damaged DNA is repaired (Weinert and Hartwell 1988; Hartwell and Weinert 1989; Schiestl et al. 1989). Because *rad9* mutants do not show this G_2 arrest, Weinert and Hartwell have suggested that lesions in DNA due to slow or incomplete DNA synthesis in a temperature-sensitive replication mutant should lead to rapid cell death at the restrictive temperature in a *rad9* mutant. As predicted, *cdc9 rad9* double mutants divide and die quickly when shifted to the nonpermissive temperature (Weinert and Hartwell 1988). In contrast, however, the high-temperature arrest of *cdc6*, *cdc7*, and *cdc8* strains is *RAD9*-independent, and the arrest of *cdc17* mutants is *RAD9*-dependent at 34°C, but *RAD9*-independent at 37°C. Thus, the cell-cycle block in replication mutants able to synthesize nearly complete chromosomes appears to be *RAD9*-dependent. However, additional genes may be involved in the arrest of mutants able to replicate only a small fraction of their DNA that are *RAD9*-independent (T.A. Weinert and L.H. Hartwell, pers. comm.).

A number of other *cdc* mutants share many, but not all, of these phenotypes and may be involved in DNA replication. In some cases, where one of the phenotypes differs, it is difficult to decide whether the difference is due to the degree of severity of the mutation in vivo or to a true difference in the event affected. For instance, *cdc5*, *cdc14*, *cdc15*, and *cdc16* mutants have all of the "replication-defective" phenotypes except that they become insensitive to hydroxyurea, or elongate their spindle, or show normal levels of recombination. This would be consistent with either a leaky defect in replication or a defect in mitosis rather than in replication. In these cases, the mutations could slow replication, but ultimately allow its completion. Perhaps the best example of difficulty in interpreting phenotype is the case of *TOP1*, discussed by Kim and Wang (1989).

Reviews the phenotypes of all the *cdc* mutants that appear to affect DNA synthesis, precursor synthesis, or nuclear division are provided in Table 2 of Newlon (1988) and elsewhere in these volumes. Only the genes that have already been cloned and sequenced and are strongly implicated in DNA replication are described here (see Table 4), although others may later turn out to be important.

Reciprocal shift experiments suggest that Cdc6p is required before the hydroxyurea block and therefore that the gene product is involved in initiation of replication. *cdc6* mutants undergo increased chromosome loss and hyperrecombination and also show increased minichromosome loss (Hartwell and Smith 1985; Kouprina et al. 1988), arguing for a fairly direct role for Cdc6p in replication. *cdc6* cells seem to undergo mitotic and premeiotic DNA synthesis at 37°C, but synthesis at the restrictive temperature in both mitosis and meiosis could be due to leakiness of the mutation (Schild and Byers 1978).

The *CDC6* gene has recently been cloned and sequenced (Lisziewicz et al. 1988; Zhou et al. 1989). The sequence predicts an ORF of 58 kD (Zhou et al. 1989) or 48 kD (Lisziewicz et al. 1988). The transcription start site is more consistent with the 58-kD ORF, and the 48-kD ORF can be explained by a difference in the two reported sequences in the 5'-non-coding leader sequence. The predicted Cdc6p sequence has two interesting features. First, near the amino terminus is a consensus type I nucleotide-binding site, indicating that the protein probably utilizes ATP or GTP. The conserved lysine in this site has been changed to glycine, and the resulting mutant allele does not complement the temperature-sensitive phenotype of *cdc6-1* (A.Y.S. Jong et al., pers. comm.). The second interesting feature is that the protein shows 35% identity with the product of *S. pombe cdc10+*, a gene required for "start" in that organism. Preliminary results suggest that the *S. pombe cdc10+* gene does not complement the *cdc6-1* mutation (Zhou et al. 1989). *S. pombe cdc10+* is also

Table 4 Genes relevant to DNA replication in *Saccharomyces cerevisiae*

Gene	Other names	Product	Remarks	References
POL1	*CDC17* *HPR3*	DNA polymerase α, 180-kD subunit	Ts⁻ mutants demonstrate cell division cycle arrest as large budded cell; single copy gene; essential (cf. Table 3)	see Table 3
PRI1		DNA polymerase α, 48-kD subunit DNA primase	single copy, essential gene on chromosome IX; gene has homology with mouse DNA polymerase α primase subunit; protein of 409 amino acids	1–2
PRI2		DNA polymerase α, 58-kD subunit	single copy, essential gene; protein of 528 amino acids	2
POL12		DNA polymerase α, 86-kD subunit	essential gene on chromosome II; predicted 78.7-kD protein	3
POL2		DNA polymerase ε	essential gene	67
POL3	*CDC2* *HPR6*	DNA polymerase δ	Ts⁻ mutants demonstrate cell division cycle arrest as large budded cells; essential gene	4–8
POL30		yPCNA, polymerase δ accessory protein	essential gene	9
REV3		DNA polymerase	mutants demonstrate decreased frequency of induced reversion and are repair deficient; non-essential gene on chromosome XVI	10–13

Gene	Synonym	Product	Phenotype	References
TOP1	*MAK1*	topoisomerase I	mutants exhibit 30% reduction in growth rate under certain conditions	14–16
TOP2		topoisomerase II	asynchronous Ts⁻ mutants arrest as a mixture of unbudded cells and cells with large buds; synchronized cultures arrest with a single large bud and an undivided nucleus; probably functions in mitosis to segregate newly synthesized chromosomes	17–21
TOP3	*EDR1*	topoisomerase III (?)	nonessential gene; mutants have a δ-δ hyperrecombination phenotype and 50% of wild-type growth rates	22
HPR1		topoisomerase IV (?)	nonessential gene; mutants show hyperrecombination phenotype	65
RAD3		helicase	essential gene; ATPase III absent in mutants; mutants are sensitive to UV radiation and bulky adducts and deficient in excision repair (see Friedberg et al., this volume)	23–28
SSB1		RNA-binding protein	nonessential gene; probably involved in ribosomal RNA processing	29–31
RPA1		66-kD subunit of yRPA	essential gene; terminal phenotype of null mutant spores like *cdc4*	32

(Continued on next page.)

Table 4 (Continued.)

Gene	Other names	Product	Remarks	References
SRT		160-kD RNase H, reverse transcriptase	nonessential gene; mutants have abnormal morphology and >2 N DNA	33–34
MCM1	*FUN80*	PRTF/GRM	essential gene; mutants are deficient in minichromosome maintenance	35, 66
MCM2			mutants are deficient in minichromosome maintenance; gene on chromosome II; hyperrecombination and chromosome loss phenotypes	36–37
MCM3			mutants are deficient in minichromosome maintenance; maps to chromosome V; Ts⁻ mutants arrest at nonpermissive temperature with a large bud; partially partially dominant mutations	36
SMC1			necessary for stability of *ARS-CEN* plasmids	38
CHL2			same as *SMC1*	39
CHL4			same as *SMC1*, etc.	39
CHL5			same as *SMC1*, etc.	39
MAP1			necessary for stability of 2-micron plasmids and *ARS-CEN* plasmids	40
RAR1			necessary for *ARS-CEN* plasmid stability; maps to chromosome XIII	41

AMMI	TUP1 FLK1 UMR7 CYC9		altered minichromosome maintenance; isolated as a suppressor of *ARS1* domain C mutations; nonessential gene; mutants do not grow on glycerol, constitutively express glucose-repressible enzymes, clump, have elevated levels of iso-2-cytochrome *c*, and have defects involving cell type and sporulation	42–43
ABF1	see text	ARS-binding factor I	essential gene; overproduction is lethal	44–45
RAP1	GRF1	DNA binding protein, binds to the *HMR E ARS* and to telomeres	essential gene; overproduction is lethal	46–47
CDC6		58 kD protein	probably involved in initiation of replication; mutants display increased chromosome loss, hyper-recombination, and increased minichromosome loss	8, 39, 48–49, 70–71
CDC7		protein kinase, 58 kD	mutants arrest at late G_1/S boundary with dumbbell-shaped terminal phenotype; necessary for initiation of mitotic DNA synthesis and for	8, 50–58, 72

(Continued on next page.)

Table 4 (Continued.)

Gene	Other names	Product	Remarks	References
CDC7 (cont'd.)			commitment to recombination and sporulation; not necessary for premeiotic DNA synthesis, but necessary for meiosis; mutants demonstrate reduced UV-induced mutability	
CDC8		thymidylate kinase	Ts⁻ mutants display "quick-stop" DNA replication phenotype and are defective in UV-induced mutagenesis	59–62, 75,76
CDC9		DNA ligase	Ts⁻ mutants synthesize DNA at 36°C but remain partially sensitive to hydroxyurea; hyperrecombination phenotype	8, 49, 73–74
DBF2		protein kinase, 64.8 kD	Ts⁻ mutants arrest with large bud and one nucleus in each daughter; delay in DNA synthesis; block in nuclear division	64
CDC16		95 kD protein	mutants arrest with a terminal morphology typical of nuclear division defects and display increased chromosome loss; DNA synthesis occurs at nonpermissive temperature in Ts⁻ mutants; several alleles have deficient DNA synthesis in permeabilized cells	8, 63–64,77

Abbreviations: (Ts⁻) temperature sensitive.

References: (1) Lucchini et al. 1987; (2) Foiani et al. 1989a; (3) D. Hinkle, pers. comm.; (4) Boulet et al. 1988; (5) Sitney et al. 1989; (6) Boulet et al. 1989; (7) Conrad and Newlon 1983; (8) Hartwell 1976; (9) Bauer and Burgers 1990; (10) Lemontt 1971; (11) Lemontt 1976; (12) Quah et al. 1980; (13) Morrison et al. 1989; (14) Thrash et al. 1985; (15) Goto and Wang 1985; (16) Kim and Wang 1989; (17) DiNardo et al. 1984; (18) Voelkel-Meiman et al. 1986; (19) Goto and Wang 1984; (20) Holm et al. 1985; (21) Holm et al. 1989; (22) Wallis et al. 1989; (23) Naumovski and Friedberg 1982; (24) Naumovski and Friedberg 1983; (25) Higgins et al. 1983; (26) Sugino et al. 1986; (27) Reynolds and Friedberg 1981; (28) Wilcox and Prakash 1981; (29) Jong et al. 1987; (30) M.W. Clark et al., in prep.; (31) Jong and Campbell 1986; (32) Heyer et al. 1990; (33) Karwan et al. 1986; (34) Wintersberger et al. 1988; (36) Gibson et al. 1987; (37) Sinha et al. 1986; (38) Larionov et al. 1985; (39) Kouprina et al. 1988; (40) Kikuchi and Toh-e 1986; (41) Kearsey and Edwards 1987; (42) Thrash-Bingham and Fangman 1989; (43) see Schultz and Carlson 1987 for references; (44) Rhode et al. 1989; (45) Diffley and Stillman 1989; (46) Shore et al. 1987; (47) Shore and Nasmyth 1987; (48) Culotti and Hartwell 1971; (49) Hartwell and Smith 1985; (50) Byers and Goetsch 1974; (51) Hartwell 1973; (52) Hereford and Hartwell 1974; (53) Schild and Byers 1978; (54) Njagi and Kilbey 1982; (55) Bahman et al. 1988; (56) Y.J. Yoon and J.L. Campbell, in prep.; (57) Jazwinski 1988; (58) Sclafani et al. 1988; (59) Prakash et al. 1979; (60) Sclafani and Fangman 1984; (61) Jong and Campbell 1984; (62) Jong et al. 1984; (63) Kuo et al. 1983; (64) Johnston et al. 1990; (65) Aguilera and Klein 1990; (66) Dubois et al. 1987a,b; (67) Morrison et al. 1990; (68) Aguilera and Klein 1988; (69) Jarvis et al. 1989; (70) Zhou et al. 1989; (71) Lis-ziewicz et al. 1988; (72) Patterson et al. 1986; (73) Game et al. 1979; (74) Barker et al. 1985; (75) Kuo and Campbell 1983; (76) Wintersberger et al. 1974; (77) Icho and Wickner 1987.

117

similar to the *lin12* gene of *Caenorhabditis elegans*, the *Notch* locus of *D. melanogaster*, and *S. cerevisiae SWI6*, but the similarity falls in a unique sequence motif that *CDC6* does not share with any of these genes. The types of proteins that function in DNA replication initiation that have nucleotide-binding sites are the initiators (SV40 T antigen, the dnaA protein of *E. coli*, and the λ O protein), the helicases (T antigen and the dnaB helicase), and also proteins that alter the conformation of other proteins (the dnaK heat-shock protein and the *groE* protein).

cdc7 mutants arrest at the late G_1/S boundary, just prior to the initiation of mitotic DNA synthesis, and show the dumbbell-shaped terminal morphology typically associated with a DNA synthesis defect or defect in nuclear division (Hartwell 1973; Byers and Goetsch 1974). Reciprocal shift experiments show that Cdc7p is required after the α-factor-sensitive step in G_1 (Hereford and Hartwell 1974) and before the hydroxyurea-sensitive step in S phase (Hartwell 1976). *cdc7* mutants do not synthesize enough DNA to be detected as an increase in the amount of DNA in the culture (Hartwell 1976). However, a small amount of DNA synthesis specific to chromosome III and no other chromosome is observed (Reynolds et al. 1989). Protein synthesis required for DNA replication can be completed at the *cdc7* block, because cells incubated at the nonpermissive temperature and shifted to the permissive temperature in the presence of cycloheximide can complete S phase (Hereford and Hartwell 1974). In contrast to the requirement for Cdc7p to initiate mitotic DNA synthesis, premeiotic DNA synthesis appears to be completed in homozygous *cdc7/cdc7* diploids (Schild and Byers 1978). Such diploids are deficient in commitment to recombination and sporulation, however, suggesting that Cdc7p may have different roles in mitosis and meiosis. Another phenotype of the *cdc7* mutants is reduced UV-induced mutability (Njagi and Kilbey 1982). Unlike *cdc6* mutants, *cdc7* mutants do not undergo chromosome loss and do not show hyperrecombination (Hartwell and Smith 1985).

The DNA sequence of *CDC7* predicts a protein with a relative molecular weight of 56,000 or 58,200 (depending on which ATG in the ORF is used) that has homology with catalytic domains characteristic of protein kinases (Patterson et al. 1986). Cdc7p shares homology with the products of *S. pombe cdc2*$^+$, *S. cerevisiae CDC28*, *Aspergillus nidulans Nim1*, and a number of kinase-related transforming proteins. Consistent with this expectation, Cdc7p can phosphorylate exogenous histone H1 and can also autophosphorylate (Bahman et al. 1988; Y.-J. Yoon and J.L. Campbell, in prep.), and Cdc7p purified from yeast is a phosphoprotein. Jazwinski (1988) has reported that a high-molecular-weight complex containing DNA-polymerizing activity prepared from *cdc7-1* cells contains a thermolabile protein kinase activity. The function of Cdc7p in the transition from G_1 to S phase and/or in the initiation of mitotic DNA replication may be phosphorylation of a key replication protein or other protein necessary for entry into S phase.

CDC7 mRNA is present at a constant level of one copy per cell throughout the mitotic cell cycle, whereas the level of its RNA does vary during meiosis, reaching a maximum near the time at which recombination occurs (Sclafani et al. 1988). Furthermore, mitotic cells can divide at least eight times before arresting after the loss of the CDC7 gene, suggesting that Cdc7p is normally present in at least 200-fold excess over what is required for division and hence that its periodic function cannot be attributed to periodic transcription or translation. Recently, Cdc7p has been shown by indirect immunofluorescence and biochemical fractionation to be localized to the nucleus, suggesting that it has a nuclear target (Y.-J. Yoon and J.L. Campbell, in prep.).

The cdc8 mutation enjoys a special place in the history of the yeast DNA replication field, since for a long time it was the only cdc mutant that seemed to block DNA synthesis specifically with a quick-stop phenotype. cdc8 mutants are also defective in UV-induced mutagenesis (Prakash et al. 1979). Cdc8p did not at first appear to be involved in precursor synthesis, since it was required in two quite different types of permeabilized cell systems (Hereford and Hartwell 1971; Kuo and Campbell 1982). In fact, a fraction that complemented the in vitro defect was partially purified, but no enzymatic activity could be associated with it (Kuo and Campbell 1982; Sugino et al. 1983). However, it was later shown that expression of the herpes simplex virus thymidine kinase (HSV TK) gene in yeast complemented all of the cdc8 temperature-sensitive alleles and even a deletion of the CDC8 gene, suggesting that Cdc8p was involved in precursor synthesis (Sclafani and Fangman 1984). Indeed, the CDC8 gene has been shown to encode thymidylate kinase, a second activity associated with the HSV TK enzyme (Jong and Campbell 1984; Sclafani and Fangman 1984; Jong et al. 1985). It has been suggested that Cdc8p plays a role in a multienzyme DNA replication complex. To address this question, suppressors of cdc8 mutations have been isolated. A plasmid that is able to suppress cdc8 mutations when present in high copy number has been shown to carry URA6, which encodes UMP kinase (Choi et al. 1989). Two other dominant suppressors of cdc8 were shown to result from the increased dosage of the partially functional cdc8 allele or the creation of a missense suppressor tRNA (Su et al. 1990). Thus, the suppressors analyzed to date do not provide evidence that Cdc8p interacts with other proteins in a multienzyme complex.

cdc16 mutants arrest with the terminal morphology typical of a defect in nuclear division. The spindle has not elongated, and the chromosomes have not been partitioned. *cdc16* mutants are not sensitive to hydroxyurea when returned to the permissive temperature after incubation at the nonpermissive temperature (36°C), suggesting that they complete DNA synthesis at 36°C (Hartwell 1976). *cdc16* mutants undergo significant chromosome loss but do not show a hyperrecombination phenotype. This phenotype may be more consistent with a defect in chromosome segregation than in DNA synthesis, and *cdc16* mutants have been shown to accumulate microtubule-associated protein (see Icho and Wickner 1987). One puzzling observation, however, is that strains carrying several different *cdc16* alleles are defective in DNA synthesis in cells permeabilized with sucrose and Brij (Kuo et al. 1983). Since DNA synthesis in this system is thought to represent the elongation stage of replication, the defect suggests that Cdc16p may have a role in DNA replication as well. The *CDC16* gene has been cloned and sequenced and encodes a 94,967-dalton protein. The sequence reveals three potential zinc finger motifs and a run of threonines. The zinc finger motifs suggest DNA binding, but further insight into the molecular function of *CDC16* has not emerged from the sequence (Icho and Wickner 1987).

C. Other Mutants

Other attempts to isolate replication mutants, by terminal phenotype and by assay of temperature-sensitive mutants for defects in DNA synthesis in vitro or in vivo, have provided additional *cdc* mutants, but most have not yet been well characterized (Dumas et al. 1982; Johnston and Thomas 1982a,b; Kuo et al. 1983; Eberly et al. 1989; Hasegawa et al. 1989). *DBF2* has been cloned and the sequence predicts that it encodes a protein kinase (Johnston et al. 1990). The precise selections used in bacteria based on bromodeoxyuridine (BrdU) suicide have not been possible in yeast because of the lack of the scavenger pathway for synthesis of dTTP involving thymidine kinase. Expression of HSV TK in yeast, which does allow incorporation of BrdU into DNA in the presence of *tut1* mutations, has not been a good solution because the strains grow poorly and the DNA contains many interruptions when analyzed by alkaline sucrose gradients (Sclafani and Fangman 1986; Budd et al. 1989b).

VI. EXPRESSION OF REPLICATION GENES

In *S. cerevisiae*, as in all eukaryotes, DNA replication occurs during a discrete portion of the cell cycle, called S phase. Although little is known about the controls that directly trigger DNA replication, it does appear to be dependent on a specific sequence of events in G_1. These events have been defined by mutations in a number of *CDC* genes that arrest the cell cycle in G_1 and prevent entry into S phase. The completion of DNA replication is normally required for progression of the cell cycle through mitosis and cell division. Again, little is known about the genetic regulation here, but one gene involved in the arrest of cells with incompletely replicated or damaged DNA is the *RAD9* gene (Weinert and Hartwell 1988). The availability of a large number of cloned replication genes allows us to study the role of the expression of these genes in the periodic nature of DNA synthesis in the cell cycle.

The events in G_1 that are required for entry into S phase center on the point called "Start" at which the cells become committed to progression through the mitotic cycle (see Pringle and Hartwell 1981; Mann et al., these volumes). Start appears to monitor the nutrient environment, attainment of a certain cell size, presence or absence of factors that promote conjugation, and completion of the previous mitosis. In *S. cerevisiae*, several genes have been shown to function sequentially in the events set in motion at Start. For example, the *cdc28*, *cdc4*, and *cdc7* arrest points define a dependent series of events in G_1. *CDC28* is the *S. cerevisiae* homolog of the 34-kD subunit of MPF, a protein kinase that acts both at Start and at the G_2/M boundary to promote entry into the cell cycle or mitosis (Reed et al. 1985; for review, see Dunphy and Newport 1989). *CDC4*, which encodes a protein that shows homology with the β-subunit of the signaling molecule transducin, acts after *CDC28* (Peterson et al. 1984; Yochem and Byers 1987). *CDC7*, which encodes a protein kinase, acts after *CDC4* (Patterson et al. 1986). Whatever the molecular function of these genes, one result of their action, either directly or indirectly, appears to be expression of a number of gene products required for proliferation (for review, see Breeden 1988). It is rather striking when viewed in this context that although most genes in yeast, including *CDC28* and *CDC7*, are constitutively transcribed, many genes whose products function in S phase have been shown to be periodically expressed. Synchronized cell culture analysis has shown that the mRNAs of *POL1*, *PRI1*, *PRI2*, *POL3*, *POL30*, *TOP2*, *CDC21* (*TMP1*), *CDC8*, *RNR2* (ribonucleotide reductase), *CDC9* (DNA ligase), *CDC6*, *HO* (the mating-type interconversion endonuclease), and histones accumulate during late G_1 phase and return to low levels during or after S phase (Storms et al. 1984; Peterson et al. 1985; Elledge and Davis 1987; Johnston et al. 1987; White et al. 1987; Breeden 1988; Budd et al. 1988; Foiani et al. 1989b; Bauer and Burgers 1990; A.Y.S. Jong, pers. comm.). The peak of histone mRNA occurs about 0.1 of a cell cycle after those for the other genes. Several of these genes are also induced in meiosis (Johnston et al. 1986). As yet, it has not been shown whether a transcriptional or posttranscriptional mechanism underlies the observed mRNA accumulation. The enzymatic activities of both DNA polymerase and ribonucleotide reductase have also been shown to increase near the beginning of S phase (Lowden and Vitols 1973; Golombek et al. 1974; Elledge and Davis 1987).

To investigate whether this pattern of expression is dependent on *CDC28*, *CDC4*, or *CDC7* function, the ability of mutants blocked at these points in G_1 to express the replication genes was assessed (Johnston et al. 1987; White et al. 1987). Cells blocked at the *cdc28* step failed to induce accumulation of *POL1*, *PRI1*, *CDC21*, *CDC8*, *CDC9*, *RNR2*, *HO*, and histone mRNAs (Breeden and Nasmyth 1987; Elledge and Davis 1987; Johnston et al. 1987; White et al. 1987). Thus, passage through Start is necessary to induce all of the replication genes, and this transcriptional (or posttranscriptional) event may be one of the results of signals set in motion at Start (see Breeden 1988). In cells blocked subsequent to Start, at the *CDC4* step, all replication genes except histones showed their normal increase in mRNA. Thus, histones are subject to a pathway of regulation different from that of the other genes. Cells blocked at the *CDC7* stage, which falls at the G_1/S transition, showed the same mRNA expression pattern as wild-type cells for all replication genes, including histones. Although these dependency experiments suggest at least two different regulatory circuits for these genes, other data suggest possible additional complexity. In particular, the sequence in the *HO* promoter essential for cell-cycle regulation, $CACGA_4$, does not appear in the promoters of the other genes (Breeden and Nasmyth 1987).

Deletion analyses of the promoter regions of the *TMP1* and *POL1* genes suggest that only about 200 bp upstream of the genes are required to program periodic expression (McIntosh et al. 1986, 1988; C. Gordon and J.L. Campbell, unpubl.). Inspection of the promoter sequences of many of the replication genes has revealed a single conserved sequence, ACGCGT, containing the recognition site for the restriction endonuclease *Mlu*I, sometimes repeated twice about 50–60 bp apart, 100–200 bp upstream of the ATG start site. Preliminary experiments suggest that these sequences are necessary for promoting periodic transcription of *CDC21* (McIntosh and Haynes 1986; McIntosh et al. 1988). Further experiments demonstrate clearly that a single *Mlu*I site is part of a 10-bp, cell-cycle-regulated UAS in the *POL1* promoter (C. Gordon and J.L. Campbell, in prep.).

Whether periodic mRNA regulation contributes to regulation of DNA synthesis is not known. The transient increase in levels of mRNAs encoded by the replication genes at a specific point in G_1 phase appears to be superimposed upon a constant constitutive level of the RNAs present throughout the cell cycle. Furthermore, in experiments that measured the number of divisions possible after loss of plasmids containing many of the *CDC* genes, it was found that up to eight additional cell divisions could occur in the absence of the genes (Byers and Sowder 1980; Sclafani et al. 1988). This suggests either that the cells undergo slow death in the absence of new RNA synthesis from these genes or that sufficient RNA or protein is present to support progression through several cell cycles. In addition, *DCD1* (*d*CMP deaminase) and *DUT1* (*d*UTPase) may not be cell-cycle-regulated (McIntosh and Haynes 1986; McIntosh et al. 1988). The effect of mutations that alter the cell-cycle response of any of these genes is not known, and no *trans*-acting factors involved in the regulation have as yet been identified.

In contrast to the situation with yeast *POL1*, neither the level of the human DNA polymerase α transcript nor the level of DNA polymerase α enzymatic activity fluctuates during the cell cycle in exponentially growing cells. Regulation of levels of polymerase may occur either posttranscriptionally or posttranslationally in these systems. In contrast to results obtained with exponentially growing human cells, however, stimulation of quiescent cells by addition of serum or growth factors does lead to an increase in DNA polymerase α mRNA level (see Marraccino et al. 1987). It is interesting that using a feed/starve synchronization protocol in *S. cerevisiae*, Johnston et al. (1987) showed that the levels of *POL1* mRNA were 100 times the constitutive level, whereas in elutriated cells, which had not been starved, the cell cycle fluctuation in levels was only 6- to 20-fold (Budd et al. 1988). Also puzzling is the fact that in the unrelated yeast *S. pombe*, the *CDC17* ligase gene is not cell-cycle-regulated (White et al. 1986).

Although mRNA levels have been shown to fluctuate, the patterns of specific protein levels as a function of the cell cycle have not been extensively explored. In an approach that measures global patterns of protein synthesis during the cell cycle, Lörincz et al. (1982) examined the synthesis of 200 yeast proteins in synchronized cells by radioactive pulse-labeling at discrete intervals and two-dimensional gel analysis of proteins synthesized during the pulse. These authors found that the rates of labeling of most proteins were constant but that at two points in the cycle, the rate of synthesis of a subset of proteins changed. One group (class I) peaked in late mitosis or early G_1 phase and showed minimal synthesis in S phase, whereas a second group (class II) peaked from late in G_1 to mid-S phase. Among class I proteins were two species that were unstable and disappeared during S phase. Histones H2A, H2B, and H4 were identified among the class II species. Other proteins with known activities have not been identified. The limitation of the technique is its sensitivity; if only moderately abundant or abundant proteins are detected, the most interesting periodically synthesized proteins may not yet have been identified.

Regardless of the role of the regulation of replication genes, as pointed out by Breeden (1988), the dependence of the promoters on Start offers the opportunity of insight into the molecular nature of Start. By determining the *trans*-acting factors and *cis*-acting sequences that mediate the regulation of the replication genes, one may be able to work backward through the signaling pathway to Start.

ACKNOWLEDGMENTS

We thank Walt Fangman and Joel Huberman for comments on the manuscript and all the colleagues who shared their results prior to publication. We are grateful to Dorothy Lloyd for expert preparation of the manuscript. The work was supported by National Institutes of Health grants GM-25508 (to J.L.C.) and GM-35679 (to C.S.N.).

REFERENCES

Abraham, J., K.A. Nasmyth, J.N. Strathern, A.J.S. Klar, and J.B. Hicks. 1984. Regulation of mating type information in yeast. *J. Mol. Biol.* **176:** 307.

Adams, A.E.M., D.I. Johnson, R.M. Longnecker, B.F. Sloat, and J.R. Pringle. 1990. *CDC42* and *CDC43*, two additional genes involved in budding and the establishment of cell polarity in the yeast *Saccharomyces cerevisiae. J. Cell Biol.* **111:** 131.

Aguilera, A. and H.L. Klein. 1988. Genetic control of intrachromosomal recombination in *Saccharomyces cerevisiae*. I. Isolation and genetic characterization of hyper-recombination mutations. *Genetics* **119:** 779.

————. 1990. *HPR1*, a novel yeast gene that prevents intrachromosomal excision recombination, shows carboxy-terminal homology to the *Saccharomyces cerevisiae TOP1* gene. *Mol. Cell. Biol.* **10:** 1439.

Almendral, J.M., D. Huebsch, P.A. Blundell, H. Macdonald-Bravo, and R. Bravo. 1987. Cloning and sequence of the human nuclear protein cyclin: Homology with DNA-binding proteins. *Proc. Natl. Acad. Sci.* **84:** 1575.

Amati, B.D. and S.M. Gasser. 1988. Chromosomal *ARS* and *CEN* elements bind specifically to the yeast nuclear scaffold. *Cell* **54:** 967.

Anderson, J.N. 1986. Detection, sequence patterns and function of unusual DNA structures. *Nucleic Acids Res.* **14:** 8513.

Badaracco, G., L. Capucci, P. Plevani, and L.M.S. Chang. 1983a. Polypeptide structure of DNA polymerase I from *Saccharomyces cerevisiae. J. Biol. Chem.* **258:** 10720.

Badaracco, G., P. Plevani, W.T. Ruyechen, and L.M.S. Chang. 1983b. Purification and characterization of yeast topoisomerase I. *J. Biol. Chem.* **258:** 2022.

Badaracco, G., M. Bianchi, P. Valsanini, G. Magni, and P. Plevani. 1985. Initiation, elongation and pausing of in vitro DNA synthesis catalyzed by immunopurified yeast DNA primase: DNA polymerase complex. *EMBO J.* **4:** 1313.

Bahman, M., V. Buck, A. White, and J. Rosamond. 1988. Characterization of the *CDC7* gene product of *Saccharomyces cerevisiae* as a protein kinase needed for the initiation of mitotic DNA synthesis. *Biochim. Biophys. Acta* **951:** 335.

Bajwa, W., T. Torchia, and J.E. Hopper. 1988. Yeast regulatory gene *GAL3*: Carbon regulation; UAS$_{Gal}$ elements in common with *GAL1, GAL2, GAL7, GAL10, GAL80,* and *MEL1*; and encoded protein strikingly similar to yeast and *Escherichia coli* galactokinases. *Mol. Cell. Biol.* **8:** 3439.

Baker, T.A., L.L. Bertsch, D. Bramhill, K. Sekimizu, E. Wahle, B. Yung, and A. Kornberg. 1988. Enzymatic mechanism of initiation of replication from the origin of the *E. coli* chromosome. *Cancer Cells* **6:** 19.

Barford, J.P. and R.J. Hall. 1976. Estimation of the length of cell cycle phases from asynchronous cultures of *Saccharomyces cerevisiae. Exp. Cell Res.* **102:** 276.

Barker, D.G., A.L. Johnson, and L.H. Johnston. 1985. An improved assay for DNA ligase reveals temperature-sensitive activity in *cdc9* mutants of *Saccharomyces cerevisiae. Mol. Gen. Genet.* **200:** 458.

Bauer, G.A. and P.M.J. Burgers. 1988. The yeast analog of mammalian cyclin/proliferating-cell nuclear antigen interacts with mammalian DNA polymerase δ. *Proc. Natl. Acad. Sci.* **85:** 7506.

―――. 1990. Molecular cloning, structure and expression of the yeast proliferating cell nuclear antigen gene. *Nucleic Acids Res.* **18:** 261.

Bauer, G.A., H.M. Heller, and P.M.J. Burgers. 1988. DNA polymerase III from *Saccharomyces cerevisiae.* I. Purification and characterization. *J. Biol. Chem.* **263:** 917.

Beach, D., M. Piper, and S. Shall. 1980. Isolation of chromosomal origins of replication in yeast. *Nature* **284:** 185.

Bell, G.I., L.J. DeGennaro, D.H. Gelfand, R.J. Bishop, P. Valenzuela, and W.J. Rutter. 1977. Ribosomal RNA genes of *Saccharomyces cerevisiae.* I. Physical map of the repeating unit and location of the regions coding for 5S, 5.8S, 18S, and 25S ribosomal RNAs. *J. Biol. Chem.* **252:** 8118.

Bender, A. and G.F. Sprague. 1987. MATα1 protein, a yeast transcription activator, binds synergistically with a second protein to a set of cell-type specific genes. *Cell* **50:** 681.

Berman, J., C.Y. Tachibana, and B.-K. Tye. 1986. Identification of a telomere-binding activity from yeast. *Proc. Natl. Acad. Sci.* **83:** 3713.

Bernad, A., A. Zaballos, M. Salas, and L. Blanco. 1987. Structural and functional relationships between prokaryotic and eukaryotic DNA polymerases. *EMBO J.* **6:** 4219.

Bernad, A., L. Blanco, J.M. Lázaro, G. Martín, and M. Salas. 1989. A conserved 3′→5′ exonuclease active site in prokaryotic and eukaryotic DNA polymerases. *Cell* **59:** 219.

Biswas, E.E., P.E. Joseph, and S.B. Biswas. 1987. Yeast DNA primase is encoded by a 59-kilodalton polypeptide: Purification and immunochemical characterization. *Biochemistry* **26:** 5377.

Biswas, S.B. and E.E. Biswas. 1990. *ARS* binding factor I of the yeast *Saccharomyces cerevisiae* binds to sequences in telomere and nontelomeric autonomously replicating sequences. *Mol. Cell. Biol.* **10:** 810.

Bjornsti, M.A. and J.C. Wang. 1987. Expression of yeast DNA topoisomerase I can complement a conditional-lethal DNA topoisomerase I mutation in *Escherichia coli. Proc. Natl. Acad. Sci.* **84:** 8971.

Bonven, B.J., E. Gocke, and O. Westergaard. 1985. A high affinity topoisomerase I binding sequence is clustered at DNase I hypersensitive sites in *Tetrahymena* R-chromatin.

Cell **41:** 541.

Boulet, A., M. Simon, and G. Faye. 1988. Isolation and characterization of the cell-division-cycle *CDC2* gene of *Saccharomyces cerevisiae. Yeast* **4:** s119.

Boulet, A., M. Simon, G. Faye, G.A. Bauer, and P.M.J. Burgers. 1989. Structure and function of the *Saccharomyces cerevisiae CDC2* gene encoding the large subunit of DNA polymerase III. *EMBO J.* **8:** 1849.

Bouton, A.H. and M.M. Smith. 1986. Fine-structure analysis of the DNA sequence requirements for autonomous replication of *Saccharomyces cerevisiae* plasmids. *Mol. Cell. Biol.* **6:** 2354.

Bouton, A.H., V.B. Stirling, and M.M. Smith. 1987. Analysis of DNA sequences homologous with the *ARS* core consensus in *Saccharomyces cerevisiae. Yeast* **3:** 107.

Brand, A.H., G. Micklem, and K. Nasmyth. 1987. A yeast silencer contains sequences that can promote autonomous plasmid replication and transcriptional activation. *Cell* **51:** 709.

Brand, A.H., L. Breeden, J. Abraham, R. Sternglanz, and K. Nasmyth. 1985. Characterization of a "silencer" in yeast: A DNA sequence with properties opposite to those of a transcriptional enhancer. *Cell* **41:** 41.

Bravo, R. and J.E. Celis. 1980. A search for differential polypeptide synthesis throughout the cell cycle of HeLa cells. *J. Cell Biol.* **84:** 795.

Bravo, R. and H. Macdonald-Bravo. 1984. Induction of nuclear protein "cyclin" in quiescent mouse 3T3 cells stimulated by serum and growth factors. Correlation with DNA synthesis. *EMBO J.* **3:** 3177.

―――. 1985. Changes in the nuclear distribution of cyclin (PCNA) but not its synthesis depend on DNA replication. *EMBO J.* **4:** 655.

Bravo, R., R. Frank, P.A. Blundell, and H. Macdonald-Bravo. 1987. Cyclin/PCNA is the auxiliary protein of DNA polymerase-δ. *Nature* **326:** 515.

Breeden, L. 1988. Cell cycle-regulated promoters in budding yeast. *Trends Genet.* **4:** 249.

Breeden, L. and K. Nasmyth. 1987. Cell cycle control of the yeast *HO* gene: *cis*- and *trans*-acting regulators. *Cell* **48:** 389.

Brewer, B.J. 1988. When polymerases collide: Replication and the transcriptional organization of the *E. coli* chromosome. *Cell* **53:** 679.

Brewer, B.J. and W.L. Fangman. 1987. The localization of replication origins on *ARS* plasmids in *S. cerevisiae. Cell* **51:** 463.

―――. 1988. A replication fork barrier at the 3′ end of yeast ribosomal RNA genes. *Cell* **55:** 637.

Brewer, B.J., E. Chlebowicz-Sledziewska, and W.L. Fangman. 1984. Cell cycle phases in the unequal mother/daughter cell cycles of *Saccharomyces cerevisiae. Mol. Cell. Biol.* **4:** 2529.

Brewer, B.J., E.P. Sena, and W.L. Fangman. 1988. Analysis of replication intermediates by two-dimensional agarose gel electrophoresis. *Cancer Cells* **6:** 229.

Brewer, B.J., V.A. Zakian, and W.L. Fangman. 1980. Replication and meiotic transmission of yeast ribosomal RNA genes. *Proc. Natl. Acad. Sci.* **77:** 6739.

Brill, S.J. and R. Sternglanz. 1988. Transcription-dependent DNA supercoiling in yeast DNA topoisomerase mutants. *Cell* **54:** 403.

Brill, S.J. and B. Stillman. 1989. Yeast replication factor-A functions in the unwinding of the SV40 origin of DNA replication. *Nature* **342:** 92.

Brill, S.J., S. DiNardo, K. Voelkel-Meiman, and R. Sternglanz. 1987. Need for DNA topoisomerase activity as a swivel for DNA replication and transcription of ribosomal DNA. *Nature* **326:** 414.

Broach, J.R., Y-.Y. Li, J. Feldman, M. Jayaram, J. Abraham, K.A. Nasmyth, and J.B.

Hicks. 1983. Localization and sequence analysis of yeast origins of DNA replication. *Cold Spring Harbor Symp. Quant. Biol.* **47:** 1165.

Brooks, M. and L.B. Dumas. 1989. DNA primase isolated from the yeast DNA primase-DNA polymerase complex. *J. Biol. Chem.* **264:** 3602.

Brown, W.C., J.K. Smiley, and J.L. Campbell. 1990. Yeast DNA polymerase II stimulatory factor I is a functional and structural analog of human replication protein A. *Proc. Natl. Acad. Sci.* **87:** 667.

Bruschi, C.V. and P.J. Chuba. 1988. Nonselective enrichment for yeast adenine mutants by flow cytometry. *Cytometry* **9:** 60.

Buchman, A.R. and R.D. Kornberg. 1990. A yeast *ARS*-binding protein activates transcription synergistically in combination with other weak activating factors. *Mol. Cell. Biol.* **10:** 887.

Buchman, A.R., N.F. Lue, and R.D. Kornberg. 1988a. Connections between transcriptional activators, silencers, and telomeres as revealed by functional analysis of a yeast DNA binding protein. *Mol. Cell. Biol.* **8:** 5086.

Buchman, A.R., W.J. Kimmerly, J. Rine, and R.D. Kornberg. 1988b. Two DNA-binding factors recognize specific sequences at silencers, upstream activating sequences, autonomously replicating sequences, and telomeres in *Saccharomyces cerevisiae. Mol. Cell. Biol.* **8:** 210.

Budd, M. and J.L. Campbell. 1987. Temperature sensitive mutants of yeast DNA polymerase I. *Proc. Natl. Acad. Sci.* **84:** 2838.

Budd, M.E., K. Sitney, and J.L. Campbell. 1989a. Purification of DNA polymerase II, a distinct polymerase, from *Saccharomyces cerevisiae. J. Biol. Chem.* **264:** 6557.

Budd, M.E., K.D. Wittrup, J.E. Bailey, and J.L. Campbell. 1989b. DNA polymerase I is required for DNA replication but not for repair in *Saccharomyces cerevisiae. Mol. Cell. Biol.* **9:** 365.

Budd, M., C. Gordon, K. Sitney, K. Sweder, and J.L. Campbell. 1988. Yeast DNA polymerase and ARS-binding proteins. *Cancer Cells* **6:** 347.

Burbee, D., J.L. Campbell, and F. Heffron. 1988. The purification and characterization of the *NUC1* gene product, a yeast endodeoxyribonuclease required for double-strand break repair. *UCLA Symp. Mol. Cell. Biol. New Ser.* **83:** 223.

Burgers, P.M.J. 1988. Mammalian cyclin/PCNA (DNA polymerase δ auxiliary protein) stimulates processive DNA synthesis by yeast DNA polymerase III. *Nucleic Acids Res.* **16:** 6297.

–––––. 1989. Eukaryotic DNA polymerases α and δ: Conserved properties and interactions from yeast to mammalian cells. *Prog. Nucleic Acid Res. Mol. Biol.* **37:** 235.

Burgers, P.M.J. and G.A. Bauer. 1988. DNA polymerase III from *Saccharomyces cerevisiae*. II. Inhibition studies and comparison with DNA polymerases I and II. *J. Biol. Chem.* **263:** 925.

Burgers, P.M.J., G.A. Bauer, and L. Tam. 1988. Exonuclease V from *Saccharomyces cerevisiae*: A 5′ to 3′-deoxyribonuclease that produces dinucleotides in a sequential fashion. *J. Biol. Chem.* **263:** 8099.

Burgers, P.M.J., R.A. Bambara, J.A. Campbell, L.M.S. Chang, K.M. Downey, U. Hubscher, M.Y.W.T. Lee, S.M. Linn, A.G. So, and S. Spadari. 1990. Revised nomenclature for eukaryotic DNA polymerases. *Eur. J. Biochem.* (in press).

Button, L.L. and C.R. Astell. 1988. DNA fragments isolated from the left end of chromosome III in yeast are repaired to generate functional telomeres. *Genome* **30:** 758.

Byers, B. and L. Goetsch. 1974. Duplication of spindle plaques and integration of the yeast cell cycle. *Cold Spring Harbor Symp. Quant. Biol.* **38:** 123.

Byers, B. and L. Sowder. 1980. Gene expression in the yeast cell cycle. *J. Cell Biol.* **87:** 6a.

Byrnes, J.J., K.M. Downey, V.L. Black, and A.G. So. 1976. A new mammalian DNA polymerase with 3′ to 5′ exonuclease activity: DNA polymerase δ. *Biochemistry* **15:** 2817.

Campbell, J.L. 1986. Eukaryotic DNA replication. *Annu. Rev. Biochem.* **55:** 733.

———. 1988. Eukaryotic DNA replication: Yeast bares its ARSs. *Trends Biochem. Sci.* **13:** 212.

Caron, F., C. Jacq, and J. Rouviére-Yaniv. 1979. Characterization of a histone-like protein extracted from yeast mitochondria. *Proc. Natl. Acad. Sci.* **76:** 4265.

Carson, M.J. 1987. "*CDC17*, the structural gene for DNA polymerase I of yeast: Mitotic hyper-recombination and effects on telomere metabolism." Ph.D. thesis, University of Washington, Seattle.

Carson, M.J. and L. Hartwell. 1985. *CDC17*: An essential gene that prevents telomere elongation in yeast. *Cell* **42:** 249.

Celniker, S.E. and J.L. Campbell. 1982. Yeast DNA replication in vitro: Initiation and elongation events mimic in vivo processes. *Cell* **31:** 201.

Celniker, S.E., K. Sweder, F. Srienc, J.E. Bailey, and J.L. Campbell. 1984. Deletion mutations affecting autonomously replicating sequence *ARS1* of *Saccharomyces cerevisiae*. *Mol. Cell. Biol.* **4:** 2455.

Chan, C.S.M. and B.-K. Tye. 1980. Autonomously replicating sequences in *Saccharomyces cerevisiae*. *Proc. Natl. Acad. Sci.* **77:** 6329.

———. 1983a. A family of *Saccharomyces cerevisiae* autonomously replicating sequences that have a very similar genomic environments. *J. Mol. Biol.* **168:** 563.

———. 1983b. Organization of DNA sequences and replication origins at yeast telomeres. *Cell* **33:** 505.

Chang, L.M.S. 1977. DNA polymerases from baker's yeast. *J. Biol. Chem.* **252:** 1873.

Chang, L.M.S., K. Lurie, and P. Plevani. 1979. A stimulatory factor for yeast DNA polymerase. *Cold Spring Harbor Symp. Quant. Biol.* **43:** 587.

Chase, J.W. and K.R. Williams. 1986. Single-stranded DNA binding proteins required for DNA replication. *Annu. Rev. Biochem.* **55:** 103.

Cheng, L. and T.J. Kelly. 1989. Transcriptional activator nuclear factor I stimulates the replication of SV40 minichromosomes *in vivo* and *in vitro*. *Cell* **59:** 541.

Choi, W.-J., J.L. Campbell, C.-L. Kuo, and A.Y. Jong. 1989. The *Saccharomyces cerevisiae SOC8-1* gene and its relationship to a nucleotide kinase. *J. Biol. Chem.* **264:** 15593.

Conrad, M.N. and C.S. Newlon. 1983. *cdc2* mutants of yeast fail to replicate approximately one-third of their nuclear genome. *Mol. Cell. Biol.* **3:** 1000.

Conrad, M.N., J.H. Wright, A.J. Wolf, and V.A. Zakian. 1990. RAP1 protein interacts with yeast telomeres in vivo: Overproduction alters telomere structure and decreases chromosome stability. *Cell* **63:** 739.

Cotterill, S., G. Chui, and I.R. Lehman. 1987a. DNA polymerase-primase from embryos of *Drosophila melanogaster*. *J. Biol. Chem.* **262:** 16105.

Cotterill, S.M., M.E. Reylard, L.A. Loeb, and I.R. Lehman. 1987b. A cryptic proofreading 3′ to 5′ exonuclease associated with the polymerase subunit of the DNA polymerase-primase from *Drosophila melanogaster*. *Proc. Natl. Acad. Sci.* **84:** 5635.

Cramer, J.H., F.W. Farrelly, J. Barnitz, and R.H. Rownd. 1977. Construction and restriction endonuclease mapping of hybrid plasmids containing *Saccharomyces cerevisiae* ribosomal DNA. *Mol. Gen. Genet.* **151:** 229.

Crute, J.J., A.F. Wahl, and R.A. Bambara. 1986. Purification and characterization of two new high molecular weight forms of DNA polymerase δ. *Biochemistry* **25**: 26.

Culotti, J. and L.H. Hartwell. 1971. Genetic control of the cell division cycle in yeast. *Exp. Cell Res.* **67**: 389.

Dani, G.M. and V.A. Zakian. 1983. Mitotic and meiotic stability of linear plasmids in yeast. *Proc. Natl. Acad. Sci.* **80**: 3406.

Decker, R.S., M. Yamaguchi, R. Possenti, M.K. Bradley, and M.L. DePamphilis. 1987. *In vitro* initiation of DNA replication in simian virus 40 chromosomes. *J. Biol. Chem.* **262**: 10863.

Della Seta, F., I. Treich, J.-M. Buhler, and A. Sentenac. 1990a. ABF1 binding sites in yeast RNA polymerase genes. *J. Biol. Chem.* **265**: 15168.

Della Seta, F., S.-A. Ciafré, C. Marck, B. Santoro, C. Presutti, A. Sentenac, and I. Bozzoni. 1990b. The ABF1 factor is the transcriptional activator of the L2 ribosomal protein genes in *Saccharomyces cerevisiae*. *Mol. Cell. Biol.* **10**: 2437.

Diffley, J.F.X. 1988. Affinity labeling the DNA polymerase a complex. *J. Biol. Chem.* **263**: 14669.

Diffley, J.F.X. and B. Stillman. 1988a. Purification of a yeast protein that binds to origins of DNA replication and a transcriptional silencer. *Proc. Natl. Acad. Sci.* **85**: 2120.

―――. 1988b. Interactions between purified cellular proteins and yeast origins of DNA replication. *Cancer Cells* **6**: 243.

―――. 1989. Similarity between the transcriptional silencer binding proteins ABF1 and RAP1. *Science* **246**: 1034.

DiFrancesco, R.A. and I.R. Lehman. 1985. Interaction of ribonuclease H from *Drosophila melanogaster* embryos with DNA polymerase-primase. *J. Biol. Chem.* **260**: 14764.

DiNardo, S., K. Voelkel, and R. Sternglanz. 1984. DNA topoisomerase II mutant of *Saccharomyces cerevisiae*: Topoisomerase II is required for segregation of daughter molecules at the termination of DNA replication. *Proc. Natl. Acad. Sci.* **81**: 2616.

Dorsman, J.C., W.C. van Heeswijk, and L.A. Grivell. 1988. Identification of two factors which bind to the upstream sequences of a number of nuclear genes coding for mitochondrial proteins and to genetic elements important for cell division in yeast. *Nucleic Acids Res.* **16**: 7287.

Dorsman, J.C., M.M. Doorenbosch, C.T.C. Maurer, J.H. de Winde, W.H. Mager, R.J. Planta, and L.A. Grivell. 1989. An *ARS*/silencer binding factor also activates two ribosomal protein genes in yeast. *Nucleic Acids Res.* **17**: 4917.

Downey, K.M., C.K. Tan, D.M. Andrews, X. Li, and A.G. So. 1988. Proposed roles for DNA polymerases α and δ at the replication fork. *Cancer Cells* **6**: 403.

Dresler, S.L. and M.G. Frattini. 1986. DNA replication and UV-induced repair synthesis in human fibroblasts are much less sensitive than DNA polymerase α to inhibition by butylphenyl deoxy triphosphate. *Fed. Proc.* **46**: 2208.

Dresler, S.L. and K.S. Kimbro. 1987. 2',3'-Dideoxythymidine 5'-triphosphate inhibition of DNA replication and ultraviolet-induced DNA repair synthesis in human cells: Evidence for involvement of DNA polymerase δ. *Biochemistry* **26**: 2664.

Dubois, E., J. Bercy, and F. Messenguy. 1987a. Characterization of two genes *ARGRI* and *ARGRIII* required for specific regulation of arginine metabolism in yeast. *Mol. Gen. Genet.* **207**: 142.

Dubois, E., J. Bercy, F. Descamps, and F. Messenguy. 1987b. Characterization of two new genes essential for vegetative growth in *Saccharomyces cerevisiae*: Nucleotide sequence determination and chromosome mapping. *Gene* **55**: 265.

Dumas, L.B., J.P. Lussky, E.J. McFarland, and J. Shampay. 1982. New temperature-sensitive mutants of *Saccharomyces cerevisiae* affecting DNA replication. *Mol. Gen.*

Genet. **187**: 42.

Dunphy, W.G. and J.W. Newport. 1989. Unraveling of mitotic control mechanisms. *Cell* **55**: 925.

Earnshaw, W.C., B. Halligan, C.A. Cooke, M.M.S. Heck, and L.F. Liu. 1985. Topoisomerase II is a structural component of mitotic chromosome scaffolds. *J. Cell Biol.* **100**: 1706.

Eberly, S.L., A. Sakai, and A. Sugino. 1989. Mapping and characterizing a new DNA replication mutant in *Saccharomyces cerevisiae. Yeast* **5**: 117.

Eckdahl, T.T. and J.N. Anderson. 1987. Computer modeling of DNA structures involved in chromosome maintenance. *Nucleic Acids Res.* **15**: 8531.

Eisenberg, S., C. Civalier, and B.K. Tye. 1988. Specific interaction between a *Saccharomyces cerevisiae* protein and a DNA element associated with certain autonomously replicating sequences. *Proc. Natl. Acad. Sci.* **85**: 743.

Elledge, S.J. and R.W. Davis. 1987. Identification and isolation of the gene encoding the small subunit of ribonucleotide reductase from *Saccharomyces cerevisiae*: DNA damage-inducible gene required for mitotic viability. *Mol. Cell. Biol.* **7**: 2783.

Eng, W.-K., S.D. Pandit, and R. Sternglanz. 1989. Mapping of the active site tyrosine of eukaryotic DNA topoisomerase I. *J. Biol. Chem.* **264**: 13373.

Fairman, M.P. and B. Stillman. 1988. Cellular factors required for multiple stages of SV40 DNA replication *in vitro. EMBO J.* **7**: 1211.

Fangman, W.L., R. Hice, and E. Chlebowicz-Sledziewska. 1983. ARS replication during the yeast S phase. *Cell* **32**: 831.

Feldman, J.B., J.B. Hicks, and J.R. Broach. 1984. Identification of sites required for repression of a silent mating type locus in yeast. *J. Mol. Biol.* **178**: 815.

Feldmann, H., J. Olah, and H. Friedenreich. 1981. Sequence of a DNA fragment containing a chromosomal replicator and a $tRNA_3$ Glu gene. *Nucleic Acids Res.* **9**: 2949.

Ferguson, B. and W.L. Fangman. 1988. Origin activation in a late replicating region on chromosome V. *Yeast* **4**: S126.

Fink, G.R. 1989. A new twist to the topoisomerase I problem. *Cell* **58**: 225.

Fitzgerald-Hayes, M., L. Clarke, and J. Carbon. 1982. Nucleotide sequence comparisons and functional analysis of yeast centromere DNAs. *Cell* **29**: 235.

Focher, F., S. Spadari, B. Ginelli, M. Hottiger, M. Gassman, and U. Hubscher. 1988. Calf thymus DNA polymerase δ: Purification, biochemical and functional properties of the enzyme after its separation from DNA polymerase α, a DNA dependent ATPase and proliferating cell nuclear antigen. *Nucleic Acids Res.* **16**: 6279.

Foiani, M., C. Santocanale, P. Plevani, and G. Lucchini. 1989a. A single essential gene, *PRI2*, encodes the large subunit of DNA primase in *Saccharomyces cerevisiae. Mol. Cell. Biol.* **9**: 3081.

Foiani, M., A.J. Lindner, G.R. Hartmann, G. Lucchini, and P. Plevani. 1989b. Affinity labeling of the active center and ribonucleoside triphosphate binding site of yeast DNA primase. *J. Biol. Chem.* **264**: 2189.

Foury, F. 1990. Cloning and sequencing of the nuclear gene *MIP1* encoding the catalytic subunit of the yeast mitochondrial DNA polymerase. *J. Biol. Chem.* **264**: 20552.

Francesconi, S.C. and S. Eisenberg. 1989. Purification and characterization of OBF1: A *Saccharomyces cerevisiae* protein that binds to autonomously replicating sequences. *Mol. Cell. Biol.* **9**: 2906.

Freemont, P.S., J.M. Friedman, L.S. Beese, M.R. Sanderson, and T.A. Steitz. 1988. Cocrystal structure of an editing complex of Klenow fragment with DNA. *Proc. Natl. Acad. Sci.* **85**: 8924.

Futcher, A.B. and B.S. Cox. 1984. Copy number and stability of 2-μm circle-based artificial plasmids of *Saccharomyces cerevisiae. J. Bacteriol.* **157**: 283.

Game, J.C., L.H. Johnston, and R.C. von Borstel. 1979. Enhanced mitotic recombination in a ligase-defective mutant of the yeast *Saccharomyces cerevisiae. Proc. Natl. Acad. Sci.* **76**: 4589.

Gasser, S.M. and U.K. Laemmli. 1986. The organization of chromatin loops: Characterization of a scaffold attachment site. *EMBO J.* **5**: 511.

Genga, A., L. Bianchi, and F. Foury. 1986. A nuclear mutant of *Saccharomyces cerevisiae* deficient in mitochondrial DNA replication and polymerase activity. *J. Biol. Chem.* **261**: 9328.

Giaever, G.N. and J.C. Wang. 1988. Supercoiling of intracellular DNA can occur in eukaryotic cells. *Cell* **55**: 849.

Gibbs, J.S., H.C. Chiou, K.F. Bastow, Y.-C. Cheng, and D.M. Coen. 1988. Identification of amino acids in herpes simplex virus DNA polymerase involved in substrate and drug recognition. *Proc. Natl. Acad. Sci.* **85**: 6672.

Gibbs, J.S., H.C. Chiou, J.D. Hall, D.W. Mount, M.J. Retondo, S.K. Weller, and D.M. Coen. 1985. Sequence and mapping analyses of the herpes simplex virus DNA polymerase gene predict a C-terminal substrate binding domain. *Proc. Natl. Acad. Sci.* **82**: 7969.

Gibson, S., R. Surosky, P. Sinha, G. Maine, and B.K. Tye. 1987. Complexity of the enzyme system for the initiation of DNA replication in yeast. In *DNA replication and recombination* (ed. R. McMacken and T.J. Kelly), p. 341. A.R. Liss, New York.

Golombek, J., W. Wolf, and E. Wintersberger. 1974. DNA synthesis and DNA-polymerase activity in synchronized yeast cells. *Mol. Gen. Genet.* **132**: 137.

Goscin, L.P. and J.J. Byrnes. 1982. DNA polymerase δ: One polypeptide, two activities. *Biochemistry* **21**: 2513.

Goto, T. and J.C. Wang. 1982. An ATP-dependent type II topoisomerase that catalyzes the catenation, decatenation, unknotting, and relaxation of double-stranded DNA rings. *J. Biol. Chem.* **257**: 5866.

———. 1984. Yeast DNA topoisomerase II is encoded by a single copy, essential gene. *Cell* **36**: 1073.

———. 1985. Cloning of yeast *TOP1*, the gene encoding DNA topoisomerase I, and construction of mutants defective in both DNA topoisomerase I and DNA topoisomerase II. *Proc. Natl. Acad. Sci.* **82**: 7178.

Goto, T., P. Laipis, and J.C. Wang. 1984. The purification and characterization of DNA topoisomerases I and II of the yeast *Saccharomyces cerevisiae. J. Biol. Chem.* **259**: 10422.

Greider, C.W. and E.H. Blackburn. 1985. Identification of a specific telomere terminal transferase activity in *Tetrahymena* extracts. *Cell* **43**: 405.

———. 1987. The telomere terminal transferase of *Tetrahymena* is a ribonucleoprotein enzyme with two kinds of primer specificity. *Cell* **51**: 887.

———. 1989. A telomeric sequence in the RNA of *Tetrahymena* telomerase required for telomere repeat synthesis. *Nature* **337**: 331.

Guo, Z.-S., C. Gutierrez, U. Heine, J.M. Sogo, and M. DePamphilis. 1989. Origin auxiliary sequences can facilitate initiation of simian virus 40 DNA replication *in vitro* as they do *in vivo. Mol. Cell. Biol.* **9**: 3593.

Halfter, N., U. Muller, E.-L. Winnacker, and D. Gallwitz. 1989a. Isolation and DNA-binding characteristics of a protein involved in transcription activation of two divergently transcribed, essential yeast genes. *EMBO J.* **8**: 3029.

Halfter, N., B. Kavety, J. Vandekerckhove, F. Kiefer, and D. Gallwitz. 1989b. Sequence, expression and mutational analysis of BAF1, a transcriptional activator and *ARS1*-

binding protein of the yeast *Saccharomyces cerevisiae. EMBO J.* **8:** 4265.

Hamatake, R.K., C.C. Dykstra, and A. Sugino. 1989. Presynapsis and synapsis of DNA promoted by the STPα and single-stranded DNA-binding proteins from *Saccharomyces cerevisiae. J. Biol. Chem.* **264:** 13336.

Hamatake, R.K., H. Hasegawa, A.B. Clark, K. Bebenek, T.A. Kunkel, and A. Sugino. 1990. Purification and characterization of DNA polymerase II from the yeast *Saccharomyces cerevisiae. J. Biol. Chem.* **265:** 4072.

Hamil, K.G., H.G. Nam, and H.M. Fried. 1988. Constitutive transcription of yeast ribosomal protein gene *TCM1* is promoted by uncommon *cis-* and *trans*-acting elements. *Mol. Cell. Biol.* **8:** 4328.

Hammond, R.A., J.J. Byrnes, and M.R. Miller. 1987. Identification of DNA polymerase δ in CV-1 cells: Studies implicating both DNA polymerase δ and DNA polymerase α in DNA replication. *Biochemistry* **26:** 6817.

Hartwell, L.H. 1973. Three additional genes required for deoxyribonucleic acid synthesis in *Saccharomyces cerevisiae. J. Bacteriol.* **115:** 966.

———. 1976. Sequential function of gene products relative to DNA synthesis in the yeast cell cycle. *J. Mol. Biol.* **104:** 803.

Hartwell, L.H. and D. Smith. 1985. Altered fidelity of mitotic chromosome transmission in cell cycle mutants of *S. cerevisiae. Genetics* **110:** 381.

Hartwell, L.H. and T.A. Weinert. 1989. Checkpoints: Controls that ensure the order of cell cycle events. *Science* **246:** 629.

Hartwell, L.H., J. Culotti, and B. Reid. 1970. Genetic control of the cell division cycle in yeast. I. Detection of mutants. *Proc. Natl. Acad. Sci.* **66:** 352.

Hasegawa, H., A. Sakai, and A. Sugino. 1989. Isolation, DNA sequence and regulation of a new cell division cycle gene from the yeast *Saccharomyces cerevisiae. Yeast* **5:** 509.

Hayes, T.E., P. Sengupta, and B.H. Cochran. 1988. The human *c-fos* serum response factor and the yeast factors GRM/PRTF have related DNA binding specificities. *Genes Dev.* **2:** 1713.

Hereford, L.M. and L.H. Hartwell. 1971. Defective DNA synthesis in permeabilized yeast mutants. *Nature* **234:** 171.

———. 1974. Sequential gene function in the initiation of *Saccharomyces cerevisiae* DNA synthesis. *J. Mol. Biol.* **84:** 445.

Herr, W. and J. Clarke. 1986. The SV40 enhancer is composed of multiple functional elements that can compensate for one another. *Cell* **45:** 461.

Heyer, W.-D. and R.D. Kolodner. 1989. Purification and characterization of a protein from *Saccharomyces cerevisiae* that binds tightly to single-stranded DNA and stimulates a cognate strand exchange protein. *Biochemistry* **28:** 2856.

Heyer, W.D., M. Rao, L.F. Erdile, T.J. Kelly, and R.D. Kolodner. 1990. An essential *Saccharomyces cerevisiae* single-stranded DNA binding protein is homologous to the large subunit of human RP-A. *EMBO J.* **9:** 2321.

Hieter, P., C. Mann, M. Snyder, and R.W. Davis. 1985. Mitotic stability of yeast chromosomes: A colony color assay that measures nondisjunction and chromosome loss. *Cell* **40:** 381.

Higgins, D.R., S. Prakash, P. Reynolds, R. Polakowska, S. Weber, and L. Prakash. 1983. Isolation and characterization of the *RAD3* gene of *S. cerevisiae* and inviability of *rad3* deletion mutants. *Proc. Natl. Acad. Sci.* **80:** 5680.

Hoffman, J.F.-X., T. Laroche, A.H. Brand, and S.M. Gasser. 1989. RAP-I factor is necessary for DNA loop formation *in vitro* at the silent mating type locus *HML. Cell* **57:** 725.

Holm, C., T. Stearns, and D. Botstein. 1989. DNA topoisomerase II must act at mitosis to prevent nondisjunction and chromosome breakage. *Mol. Cell. Biol.* **9**: 159.

Holm, C., T. Goto, J.C. Wang, and D. Botstein. 1985. DNA topoisomerase II is required at the time of mitosis in yeast. *Cell* **41**: 553.

Holmes, S.G. and M.M. Smith. 1989. Interaction of the H4 autonomously replicating sequence core consensus sequence and its 3′-flanking domain. *Mol. Cell. Biol.* **9**: 5464.

Hsiang, Y.-H., R. Hertzberg, S. Hecht, and R.J. Liu. 1985. Camptothecin induces protein-linked DNA breaks via mammalian DNA topoisomerase I. *J. Biol. Chem.* **260**: 14873.

Hsiao, C.-L. and J. Carbon. 1979. High-frequency transformation of yeast by plasmids containing the cloned yeast *ARG4* gene. *Proc. Natl. Acad. Sci.* **76**: 3829.

―――. 1981. Direct selection procedure for the isolation of functional centromeric DNA. *Proc. Natl. Acad. Sci.* **78**: 3760.

Huberman, J.A., J. Zhu, L.R. Davis, and C.S. Newlon. 1988. Close association of a DNA replication origin and an ARS element on chromosome III of the yeast *Saccharomyces cerevisiae*. *Nucleic Acids Res.* **16**: 6373.

Huberman, J.A., L.D. Spotila, K.A. Nawotka, S.M. El-Assouli, and L.R. Davis. 1987. The *in vivo* replication origin of the yeast 2 micron plasmid. *Cell* **51**: 473.

Huet, J., F. Wyers, J.-M. Buhler, A. Sentenac, and P. Fromageot. 1976. Association of RNase H activity with yeast RNA polymerase A. *Nature* **261**: 431.

Hyman, B.C., J.H. Cramer, and R.H. Rownd. 1982. Properties of a *Saccharomyces cerevisiae* mtDNA segment conferring high-frequency yeast transformation. *Proc. Natl. Acad. Sci.* **79**: 1578.

Icho, T. and R.B. Wickner. 1987. Metal-binding, nucleic acid binding finger sequences in the *CDC16* gene of *Saccharomyces cerevisiae*. *Nucleic Acids Res.* **15**: 8439.

Jacob, F., S. Brenner, and F. Cuzin. 1964. On the regulation of DNA replication in bacteria. *Cold Spring Harbor Symp. Quant. Biol.* **28**: 329.

Jäger, D. and P. Philippsen. 1989. Many yeast chromosomes lack the telomere-specific Y′ sequence. *Mol. Cell. Biol.* **9**: 5754.

Jarvis, E.E., K.L. Clark, and G.F. Sprague. 1989. The yeast transcription activator PRTF, a homolog of the mammalian serum response factor, is encoded by the *MCM1* gene. *Genes Dev.* **3**: 936.

Jarvis, E.E., D.C. Hagen, and G.F. Sprague. 1988. Identification of a DNA segment that is necessary and sufficient for α-specific gene control in *Saccharomyces cerevisiae:* Implications for regulation of α-specific and a-specific genes. *Mol. Cell. Biol.* **8**: 309.

Jazwinski, S.M. 1988. *CDC7*-dependent protein kinase activity in yeast replicative-complex preparations. *Proc. Natl. Acad. Sci.* **85**: 2101.

Jazwinski, S.M. and G.M. Edelman. 1979 Replication in vitro of the 2 micron DNA plasmid of yeast. *Proc. Natl. Acad. Sci.* **76**: 1223.

―――. 1984. Evidence for participation of a multiprotein complex in yeast DNA replication in vitro. *J. Biol. Chem.* **259**: 6852.

―――. 1985. A DNA primase from yeast: Purification and partial characterization. *J. Biol. Chem.* **260**: 4995.

Jazwinski, S.M., A. Niedzwiecka, and G.M. Edelman. 1983. In vitro association of a replication complex with a yeast chromosomal replicator. *J. Biol. Chem.* **258**: 2754.

Johnson, L.M., M. Snyder, L.M.S. Chang, R.W. Davis, and J.L. Campbell. 1985. Isolation of the gene encoding yeast DNA polymerase I. *Cell* **43**: 369.

Johnson, R.N., J. Feder, A.B. Hill, S.W. Sherwood, and R.T. Schimke. 1986. Transient inhibition of DNA synthesis results in increased dihydrofolate reductase synthesis and subsequent increased DNA content per cell. *Mol. Cell. Biol.* **6**: 3373.

Johnston, G.C., R.A. Singer, S.O. Sharrow, and M.L. Slater. 1980. Cell division in the yeast *Saccharomyces cerevisiae* growing at different rates. *J. Gen. Microbiol.* **118**: 479.

Johnston, L.H. and A.P. Thomas. 1982a. The isolation of new DNA synthesis mutants in the yeast *Saccharomyces cerevisiae. Mol. Gen. Genet.* **186:** 439.

————. 1982b. A further two mutants defective in initiation of the S phase in the yeast *Saccharomyces cerevisiae. Mol. Gen. Genet.* **186:** 445.

Johnston, L.H. and D.H. Williamson. 1978. An alkaline sucrose gradient analysis of the mechanism of nuclear DNA synthesis in the yeast *Saccharomyces cerevisiae. Mol. Gen. Genet.* **164:** 217.

Johnston, L.H., A.L. Johnson, and D.G. Barker. 1986. The expression in meiosis of genes which are transcribed periodically in the mitotic cell cycle of budding yeast. *Exp. Cell Res.* **165:** 541.

Johnston, L.H., D.H. Williamson, A.L. Johnson, and D.J. Fennell. 1982. On the mechanism of premeiotic DNA synthesis in the yeast *Saccharomyces cerevisiae. Exp. Cell Res.* **141:** 53.

Johnston, L.H., S.L. Eberly, J.W. Chapman, H. Araki, and A. Sugino. 1990. The product of the *Saccharomyces cerevisiae* cell cycle gene *DBF2* has homology with protein kinases and is periodically expressed in the cell cycle. *Mol. Cell. Biol.* **10:** 1358.

Johnston, L.H., J.H.M. White, L. Johnson, G. Lucchini, and P. Plevani. 1987. The yeast DNA polymerase I transcript is regulated in both the mitotic cell cycle and in meiosis and is also induced after DNA damage. *Nucleic Acids Res.* **15:** 5017.

Jong, A.Y.S. and J.L. Campbell. 1984. Characterization of *Saccharomyces cerevisiae* thymidylate kinase, the *CDC8* gene product. *J. Biol. Chem.* **259:** 14394.

————. 1986. Isolation of the gene encoding yeast single-stranded nucleic acid binding protein 1 (SSB-1). *Proc. Natl. Acad. Sci.* **83:** 877.

Jong, A.Y.S., R. Aebersold, and J.L. Campbell. 1985. Multiple species of single-stranded nucleic acid binding proteins in *Saccharomyces cerevisiae. J. Biol. Chem.* **260:** 16367.

Jong, A.Y.S., C.-L. Kuo, and J.L. Campbell. 1984. The *CDC8* gene of yeast encodes thymidylate kinase. *J. Biol. Chem.* **259:** 11052.

Jong, A., M.W. Clark, M. Gilbert, and J.L. Campbell. 1987. Yeast SSB-1 and its relationship to nucleolar RNA binding proteins. *Mol. Cell. Biol.* **7:** 2947.

Joyce, C.M. and T.A. Steitz. 1987. DNA polymerase I: From crystal structure to function via genetics. *Trends Biochem. Sci.* **12:** 288.

Jung, G., M.C. Leavitt, J.C. Hsieh, and J. Ito. 1987. Bacteriophage PRDI DNA polymerase: Evolution of DNA polymerases. *Proc. Natl. Acad. Sci.* **84:** 8287.

Kaguni, L.S., J-.M. Rossignol, R.C. Conaway, and I.R. Lehman. 1983. Isolation of an intact DNA polymerase-primase from embryos of *Drosophila melanogaster. Proc. Natl. Acad. Sci.* **80:** 2221.

Karwan, R. and U. Wintersberger. 1988. In addition to RNase H(70) two other proteins of *Saccharomyces cerevisiae* exhibit ribonuclease H activity. *J. Biol. Chem.* **263:** 14970.

Karwan, R., H. Blutsch, and U. Wintersberger. 1983. Physical association of a DNA polymerase stimulating activity with a ribonuclease H purified from yeast. *Biochemistry* **22:** 5500.

Karwan, R., C. Kühne, and U. Wintersberger. 1986. Ribonuclease H(70) from *Saccharomyces cerevisiae* possesses cryptic reverse transcriptase activity. *Proc. Natl. Acad. Sci.* **83:** 5919.

Kearsey, S. 1983. Analysis of sequences conferring autonomous replication in baker's yeast. *EMBO J.* **2:** 1571.

————. 1984. Structural requirements for the function of a yeast chromosomal replicator. *Cell* **37:** 299.

————. 1986. Replication origins in yeast chromosomes. *Bioessays* **4**: 157.

Kearsey, S.E. and J. Edwards. 1987. Mutations that increase the mitotic stability of mini-chromosomes in yeast: Characterization of *RAR1*. *Mol. Gen. Genet.* **210**: 509.

Keleher, C.A., C. Goutte, and A.D. Johnson. 1988. The yeast cell-type-specific repressor α2 acts cooperatively with a non-cell-type-specific protein. *Cell* **53**: 927.

Khan, N.N., G.E. Wright, L.W. Dudycx, and N.C. Brown. 1984. Butylphenyl dGTP: A selective and potent inhibitor of DNA polymerase α. *Nucleic Acids Res.* **12**: 3695.

Kikuchi, Y. and A. Toh-e. 1986. A nuclear gene of *Saccharomyces cerevisiae* needed for stable maintenance of plasmids. *Mol. Cell. Biol.* **6**: 4053.

Kim, R.A. and J.C. Wang. 1989. The function of DNA topoisomerases as replication swivels in *Saccharomyces cerevisiae*. *J. Mol. Biol.* **208**: 257.

Kimmerly, W.J. and J. Rine. 1987. Replication and segregation of plasmids containing *cis*-acting regulatory sites of silent mating-type genes in *Saccharomyces cerevisiae* are controlled by the *SIR* genes. *Mol. Cell. Biol.* **7**: 4225.

Kimmerly, W., A. Buchman, R. Kornberg, and J. Rine. 1988. Roles of two DNA-binding factors in replication, segregation and transcriptional repression mediated by a yeast silencer. *EMBO J.* **7**: 2241.

Kingsman, A.J., L. Clarke, R.K. Mortimer, and J. Carbon. 1979. Replication in *Saccharomyces cerevisiae* of plasmid pBR313 carrying DNA from the yeast *trp1* region. *Gene* **7**: 141.

Kipling, D. and S. Kearsey. 1990. Reversion of autonomously replicating sequence mutations in *Saccharomyces cerevisiae*: Creation of a eucaryotic replication origin within procaryotic vector DNA. *Mol. Cell. Biol.* **10**: 265.

Kojo, H., B.D. Greenberg, and A. Sugino. 1981. Yeast 2 micron plasmid DNA replication in vitro: Origin and direction. *Proc. Natl. Acad. Sci.* **78**: 7261.

Konrad, E.B. 1977. Method for the isolation of *Escherichia coli* mutants with enhanced recombination between chromosomal duplications. *J. Bacteriol.* **130**: 167.

Kornberg, A. 1981. *DNA replication—Supplement.* Freeman, San Francisco.

Koshland, D., J.C. Kent, and L.H. Hartwell. 1985. Genetic analysis of the mitotic transmission of minichromosomes. *Cell* **40**: 393.

Kouprina, N.Y. and V.L. Larionov. 1983. Study of a rDNA replicator in *Saccharomyces*. *Curr. Genet.* **7**: 433.

Kouprina, N.Y., O.B. Pashina, N.T. Nikolaischwili, N.T. Tsouladze, and V.L. Larionov. 1988. Genetic control of chromosome stability in the yeast *Saccharomyces cerevisiae*. *Yeast* **4**: 257.

Kouzarides, S.T., A.T. Bankier, S.C. Satchwell, K. Weston, P. Tomlinson, and B.G. Barrell. 1987. Sequence and transcription analysis of the human cytomegalovirus DNA polymerase gene. *J. Virol.* **61**: 125.

Kowalczykowski, S.T., D.G. Bear, and P.H. von Hippel. 1981. Single-stranded DNA binding proteins. *The Enzymes* **14A**: 373.

Kuempel, P.L., A.J. Pelletier, and T.M. Hill. 1989. Tus and the terminators: The arrest of replication in prokaryotes. *Cell* **59**: 581.

Kunkel, T.A., R.K. Hamatake, J. Motto-Fox, M.P. Fitzgerald, and A. Sugino. 1989. The fidelity of DNA polymerase I and the DNA polymerase I-DNA primase complex from *Saccharomyces cerevisiae*. *Mol. Cell. Biol.* **9**: 4447.

Kuo, C.-L. and J.L. Campbell. 1982. Purification of the cdc8 protein of *Saccharomyces cerevisiae* by complementation in an aphidicolin-sensitive *in vitro* DNA replication system. *Proc. Natl. Acad. Sci.* **79**: 4243.

————. 1983. Cloning of *Saccharomyces cerevisiae* DNA replication genes: Isolation of the *CDC8* gene and two genes that compensate for the *cdc8-1* mutation. *Mol. Cell.*

Biol. **3**: 1730.

Kuo, C.-L., N.-H. Huang, and J.L. Campbell. 1983. Isolation of yeast DNA replication mutants using permeabilized cells. *Proc. Natl. Acad. Sci.* **80**: 6465.

LaBonne, S.G. and L.B. Dumas. 1983. Isolation of a yeast single-strand deoxyribonucleic acid binding protein that specifically stimulates yeast DNA polymerase I. *Biochemistry* **22**: 3214.

Lai, C.J. and D. Nathans. 1975. Non-specific termination of simian virus 40 DNA replication. *J. Mol. Biol.* **97**: 113.

Lamothe, P., B. Baril, A. Chi, L. Lee, and E. Baril. 1981. Accessory proteins for DNA polymerase α activity with single-strand DNA templates. *Proc. Natl. Acad. Sci.* **78**: 4723.

Larionov, V.L., T.S. Karpova, N.Y. Kouprina, and G.A. Jouravleva. 1985. A mutant of *Saccharomyces cerevisiae* with impaired maintenance of centromeric plasmids. *Curr. Genet.* **10**: 15.

Leavitt, M.C. and J. Ito. 1989. T5 DNA polymerase: Structural-functional relationships to other DNA polymerases. *Proc. Natl. Acad. Sci.* **86**: 4465.

Lee, M.Y.W.T. and N.L. Toomey. 1987. Human placental DNA polymerase δ: Identification of a 170-kilodalton polypeptide by activity staining and immunoblotting. *Biochemistry* **26**: 1076.

Lee, M.Y.W.T., C. Tan, K.M. Downey, and A.G. So. 1984. Further studies on calf thymus DNA polymerase δ purified to homogeneity by a new procedure. *Biochemistry* **23**: 1906.

Lemontt, J.F. 1971. Mutants of yeast defective in mutation induced by ultraviolet light. *Genetics* **68**: 21.

———. 1976. Pathways of ultraviolet mutability in *Saccharomyces cerevisiae*. I. Some properties of double mutants involving *uvs9* and *rev. Mutat. Res.* **13**: 311.

Linskens, M.H.K. and J.A. Huberman. 1988. Organization of replication in ribosomal DNA. *Mol. Cell. Biol.* **8**: 4927.

———. 1990. Ambiguities in results obtained with 2D gel replicon mapping techniques. *Nucleic Acids Res.* **18**: 647.

Lisziewicz, J., A. Godany, D.V. Agoston, and H. Küntzel. 1988. Cloning and characterization of the *Saccharomyces cerevisiae CDC6* gene. *Nucleic Acids Res.* **16**: 11507.

Liu, Y.-C., R.L. Marraccino, P.C. Keng, R.A. Bambara, E.M. Lord, W.-G. Chou, and S.B. Zain. 1989. Requirement for proliferating cell nuclear antigen expression during stages of the Chinese hamster ovary cell cycle. *Biochemistry* **28**: 2967.

Lohr, D. and T. Torchia. 1988. Structure of the chromosomal copy of yeast ARS1. *Biochemistry* **27**: 3961.

Long, C.M., C.M. Brajkovich, and J.F. Scott. 1985. Alternative model for chromatin organization of the *Saccharomyces cerevisiae* chromosomal DNA plasmid TRP1 RI circle (YARpl). *Mol. Cell. Biol.* **5**: 3124.

Longtine, M.S., N.M. Wilson, M.E. Petracek, and J. Berman. 1989. A yeast telomere binding activity binds to two related telomere sequence motifs and is indistinguishable from RAP1. *Curr. Genet.* **16**: 225.

Lörincz, A.T., M.J. Miller, N.-H. Xuong, and E.P. Geiduschek. 1982. Identification of proteins whose synthesis is modulated during the cell cycle of *Saccharomyces cerevisiae*. *Mol. Cell. Biol.* **2**: 1532.

Lowden, M. and E. Vitols. 1973. Ribonucleotide reductase activity during the cell cycle of *Saccharomyces cerevisiae*. *Arch. Biochem. Biophys* **158**: 177.

Lucchini, G., C. Mazza, E. Scacheri, and P. Plevani. 1988. Genetic mapping of the *Saccharomyces cerevisiae* DNA polymerase I gene and characterization of a *poll* temperature-sensitive mutant altered in DNA primase-polymerase complex stability. *Mol. Gen. Genet.* **212:** 459.

Lucchini, G., A. Brandazza, G. Badaracco, M. Bianchi, and P. Plevani. 1985. Identification of the yeast DNA polymerase I gene with antibody probes. *Curr. Genet.* **10:** 245.

Lucchini, G., S. Francesconi, M. Foiani, G. Badaracco, and P. Plevani. 1987. Yeast DNA polymerase-DNA primase complex: Cloning of *PRI1*, a single essential gene related to DNA primase activity. *EMBO J.* **6:** 737.

Lucchini, G., M.M. Falconi, A. Pizzagalli, A. Aguilera, H.L. Klein, and P. Plevani. 1990. Nucleotide sequence and characterization of temperature sensitive *pol1* mutants of *Saccharomyces cerevisiae. Gene* **90:** 99.

Lundblad, V. and E.H. Blackburn. 1990. RNA-dependent polymerase motifs in *EST1*: Tentative identification of a protein component of an essential yeast telomerase. *Cell* **60:** 529.

Lundblad, V. and J.W. Szostak. 1989. A mutant with a defect in telomere elongation leads to senescence in yeast. *Cell* **57:** 633.

Lustig, A.J. and T.D. Petes. 1986. Identification of yeast mutants with altered telomere structure. *Proc. Natl. Acad. Sci.* **83:** 1398.

Lustig, A.J., S. Kurtz, and D. Shore. 1990. Involvement of the silencer and UAS binding protein RAP1 in regulation of telomere length. *Science* **250:** 549.

Lynn, R.M., M.-A. Bjornsti, P.R. Caron, and J.C. Wang. 1989. Peptide sequencing and site-directed mutagenesis identify tyrosine-727 as the active site tyrosine of *Saccharomyces cerevisiae* DNA topoisomerase I. *Proc. Natl. Acad. Sci.* **86:** 3559.

Lynn, R., G. Giaever, S.L. Swanberg, and J.C. Wang. 1986. Tandem regions of yeast DNA topoisomerase II share homology with different subunits of bacterial gyrase. *Science* **233:** 647.

Ma, D. 1989. "Regulatory elements in ColE1 DNA replication in *Escherichia coli.* Mutants of *Saccharomyces cerevisiae* DNA polymerase I resistant to nucleotide analogs: dNTP binding site definition." Ph.D. thesis, California Institute of Technology, Pasadena.

Maine, G.T., P. Sinha, and B.-K. Tye. 1984. Mutants of *S. cerevisiae* defective in the maintenance of minichromosomes. *Genetics* **106:** 365.

Marians, K.J. 1984. Enzymology of DNA replication in prokaryotes. *CRC Crit. Rev. Biochem.* **17:** 153.

Marraccino, R.L., A.F. Wahl, P.C. Keng, E.M. Lord, and R.A. Bambara. 1987. Cell cycle dependent activities of DNA polymerases α and δ in Chinese hamster ovary cells. *Biochemistry* **26:** 7864.

Mathews, M.B., R.M. Bernstein, B.R. Franza, Jr., and J.I. Garrels. 1984. Identity of the proliferating cell nuclear antigen and cyclin. *Nature* **309:** 374.

Matsumoto, K., T. Moriuchi, T. Koji, and P.K. Nakane. 1987. Molecular cloning of cDNA coding for rat proliferating cell nuclear antigen (PCNA)/cyclin. *EMBO J.* **6:** 637.

McCarroll, R.M. and W.L. Fangman. 1988. Time of replication of yeast centromeres and telomeres. *Cell* **54:** 505.

McClure, W.R. 1985. Mechanism and control of transcription initiation in prokaryotes. *Annu. Rev. Biochem.* **54:** 171.

McIntosh, E.M. and R.H. Haynes. 1986. Sequence and expression of the dCMP deaminase gene (*DCD1*) of *Saccharomyces cerevisiae. Mol. Cell. Biol.* **6:** 1711.

McIntosh, E.M., M.H. Gadsden, and R.H. Haynes. 1986. Transcription of genes encoding enzymes involved in DNA synthesis during the cell cycle of *Saccharomyces cerevisiae. Mol. Gen. Genet.* **204:** 363.

McIntosh, E. M., R.W. Ord, and R.K. Storms. 1988. Transcriptional regulation of the cell cycle-dependent thymidylate synthase gene of *Saccharomyces cerevisiae. Mol. Cell. Biol.* **8:** 4616.

Mills, A.D., J.J. Blow, J.G. White, W.B. Amos, D. Wilcock, and R.A. Laskey. 1989. Replication occurs at discrete foci spaced throughout nuclei replicating *in vitro. J. Cell Sci.* **94:** 471.

Möller, W. and R. Amons. 1985. Phosphate-binding sequences in nucleotide-binding proteins. *FEBS Lett.* **186:** 1.

Morin, G.B. 1989. The human telomere terminal transferase enzyme is a ribonucleoprotein that synthesizes TTAGGG repeats. *Cell* **59:** 521.

Morris, G.F. and M.B. Mathews. 1989. Regulation of proliferating cell nuclear antigen during the cell cycle. *J. Biol. Chem.* **264:** 13856.

Morrison, A., H. Araki, A.B. Clark, R.K. Hamatake, and A. Sugino. 1990. A third essential DNA polymerase in *S. cerevisiae. Cell* **62:** 1143.

Morrison, A., R.B. Christensen, J. Alley, A.K. Beck, E.G. Bernstine, J.F. Lemontt, and C.W. Lawrence. 1989. *REV3,* a *Saccaromyces cerevisiae* gene whose function is required for induced mutagenesis, is predicted to encode a nonessential DNA polymerase. *J. Bacteriol.* **171:** 5659.

Muller, M.T., W.P. Pfund, V.B. Mehta, and D.K. Trask. 1985. Eukaryotic type I topoisomerase is enriched in the nucleolus and catalytically active on ribosomal DNA. *EMBO J.* **4:** 1237.

Murray, A. and J. Szostak. 1983. Pedigree analysts of plasmid segregation in yeast. *Cell* **34:** 961.

———. 1985. Chromosome segregation in mitosis and meiosis. *Annu. Rev. Cell Biol.* **1:** 289.

———. 1986. Construction and behavior of circularly permuted and telocentric chromosomes in *Saccharomyces cerevisiae. Mol. Cell. Biol.* **6:** 3166.

Naumovski, L. and E.C. Friedberg. 1982. Molecular cloning of eucaryotic genes required for excision repair of UV-irradiated DNA: Isolation and partial characterization of the *RAD3* gene of *Saccharomyces cerevisiae. J. Bacteriol.* **152:** 323.

———. 1983. A DNA repair gene required for incision of damaged DNA is essential for viability in *Saccharomyces cerevisiae. Proc. Natl. Acad. Sci.* **80:** 4818.

Nawotka, K.A. and J.A. Huberman. 1988. Two-dimensional gel electrophoretic method for mapping DNA replicons. *Mol. Cell. Biol.* **8:** 1408.

Newlon, C.S. 1988. Yeast chromosome replication and segregation. *Microbiol. Rev.* **52:** 568.

———. 1989. DNA organization and replication in yeast. In *The yeasts* (ed. A.H. Rose and J.S. Harrison), vol. 4, p. 57. Academic Press, London.

Newlon, C.S. and W. Burke. 1980. Replication of small chromosomal DNAs in yeast. *ICN-UCLA Symp. Mol. Cell. Biol.* **19:** 399.

Newlon, C.S., R.J. Devenish, P.A. Suci, and C.J. Rofiis. 1981. Replication origins used *in vivo* in yeast. In *The initiation of DNA replication* (ed. D.S. Ray and C.F. Fox), p. 501. Academic Press, New York.

Newlon, C.S., R.P. Green, K.J. Hardeman, K.E. Kim, L.R. Lipchitz, T.G. Palzkill, S. Synn, and S.T. Woody. 1986. Structure and organization of yeast chromosome III. *ICN-UCLA Symp. Mol. Cell. Biol. New Ser.* **33:** 211.

Nishida, C., P. Reinhard, and S. Linn. 1988. DNA repair synthesis in human fibroblasts requires DNA polymerase δ. *J. Biol. Chem.* **263:** 501.

Nitiss, J. and J.C. Wang. 1988. DNA topoisomerase-targeting antitumor drugs can be studied in yeast. *Proc. Natl. Acad. Sci.* **85:** 7501.

Njagi, G.D.E. and B.J. Kilbey. 1982. *cdc7-1* a temperature sensitive cell-cycle mutant which interferes with induced mutagenesis in *Saccharomyces cerevisiae. Mol. Gen. Genet.* **186:** 478.

Norman, C., M. Runswick, R. Pollock, and R. Treisman. 1988. Isolation and properties of cDNA clones encoding SRF, a transcription factor that binds to the c-*fos* serum response element. *Cell* **55:** 989.

O'Donnell, M.E., P. Elias, and I.R. Lehman. 1987a. Processive replication of single-stranded DNA templates by the herpes simplex virus-induced DNA polymerase. *J. Biol. Chem.* **262:** 4252.

O'Donnell, M.E., P. Elias, B.E. Funnell, and I.R. Lehman. 1987b. Interaction between the DNA polymerase and single-stranded DNA-binding protein (infected cell protein 8) of herpes simplex virus 1. *J. Biol. Chem.* **262:** 4260.

O'Neill, E.A. and T.J. Kelly. 1988. Purification and characterization of nuclear factor III (origin recognition protein C), a sequence-specific DNA binding protein required for efficient initiation of adenovirus DNA replication. *J. Biol. Chem.* **263:** 931.

O'Reilly, D.R., A.M. Crawford, and L.K. Miller. 1989. Viral proliferating cell nuclear antigen. *Nature* **337:** 606.

Pakula, A.A. and R.T. Sauer. 1989. Genetic analysis of protein stability and function. *Annu. Rev. Genet.* **23:** 265.

Palzkill, T.D. 1988. "DNA sequence requirements for *ARS* function in *Saccharomyces cerevisiae.*" Ph.D. thesis, University of Iowa, Ames.

Palzkill, T.G. and C.S. Newlon. 1988. A yeast replication origin consists of multiple copies of a small conserved sequence. *Cell* **53:** 441.

Palzkill, T.G., S.G. Oliver, and C.S. Newlon. 1986. DNA sequence analysis of *ARS* elements from chromosome III of *Saccharomyces cerevisiae:* Identification of a new conserved sequence. *Nucleic Acids Res.* **14:** 6247.

Passmore, S., R. Elble, and B.-K. Tye. 1989. A protein involved in minichromosome maintenance in yeast binds a transcriptional enhancer conserved in eukaryotes. *Genes Dev.* **3:** 921.

Passmore, S., G.T. Maine, R. Elble, C. Christ, and B.-K. Tye. 1988. *Saccharomyces cerevisiae* protein involved in plasmid maintenance is necessary for mating of MATα cells. *J. Mol. Biol.* **204:** 593.

Patterson, M., R.A. Sclafani, W.L. Fangman, and J. Rosamond. 1986. Molecular characterization of cell cycle gene *CDC7* from *Saccharomyces cerevisiae. Mol. Cell. Biol.* **6:** 1590.

Peterson, T.A., L. Prakash, S. Prakash, M.A. Osley, and S.I. Reed. 1985. Regulation of *CDC9,* the *Saccharomyces cerevisiae* gene that encodes DNA ligase. *Mol. Cell. Biol.* **5:** 226.

Peterson, T.A., J. Yochem, B. Byers, M.F. Nunn, P.H. Duesberg, R.F. Doolittle, and S.I. Reed. 1984. A relationship between the yeast cell cycle genes *CDC4* and *CDC36* and the *ets* sequence of oncogenic virus E26. *Nature* **309:** 556.

Petes, T.D. 1979. Yeast ribosomal DNA genes are located on chromosome XII. *Proc. Natl. Acad. Sci.* **76:** 410.

Petes, T.D. and D.H. Williamson. 1975. Fiber autoradiography of replicating yeast DNA. *Exp. Cell Res.* **95:** 103.

Petes, T.D., L.M. Hereford, and K.G. Skryabin. 1978. Characterization of two types of yeast ribosomal DNA genes. *J. Bacteriol.* **134:** 295.

Philippsen, P., M. Thomas, R.A. Kramer, and R.W. Davis. 1978. Unique arrangement of coding sequences for 5S, 5.8S, 18S and 28S ribosomal RNA in *Saccharomyces cerevisiae as* determined by R-loop hybridization analysis. *J. Mol. Biol.* **123:** 387.

Pizzagalli, A., P. Valsasnini, P. Plevani, and G. Lucchini. 1988. DNA polymerase I gene of *Saccharomyces cerevisiae*: Nucleotide sequence, mapping of a temperature-sensitive mutation, and protein homology with other DNA polymerases. *Proc. Natl. Acad. Sci.* **85**: 3772.

Plevani, P., G. Badaracco, and L.M.S. Chang. 1980. Purification and characterization of two forms of DNA-dependent ATPase from yeast. *J. Biol. Chem.* **255**: 4957.

Plevani, P., S. Francesconi, and G. Lucchini. 1987. The nucleotide sequence of the *PRI1* gene related to DNA primase in *Saccharomyces cerevisiae*. *Nucleic Acids Res.* **15**: 7975.

Plevani, P., G. Badaracco, C. Augl, and L.M.S. Chang. 1984. DNA polymerase I and DNA primase complex in yeast. *J. Biol. Chem.* **259**: 7532.

Plevani, P., M. Foiani, P. Valsasnini, G. Badaracco, E. Cheriathundam, and L.M.S. Chang. 1985. Polypeptide structure of DNA primase from a yeast DNA polymerase-primase complex. *J. Biol. Chem.* **260**: 7102.

Pluta, A.F. and V.A. Zakian. 1989. Recombination occurs during telomere formation in yeast. *Nature* **337**: 429.

Prakash, L.D., D. Hinkle, and S. Prakash. 1979. Decreased UV mutagenesis in *cdc8*, a DNA replication mutant of *Saccharomyces cerevisiae*. *Mol. Gen. Genet.* **172**: 249.

Prelich, G. and B. Stillman. 1988. Coordinated leading and lagging strand synthesis during SV40 DNA replication *in vitro* requires PCNA. *Cell* **53**: 117.

Prelich, G., M. Kostura, D.R. Marshak, M.B. Mathews, and B. Stillman. 1987a. The cell-cycle regulated proliferating cell nuclear antigen is required for SV40 DNA replication *in vitro*. *Nature* **326**: 471.

Prelich, G., C.-K. Tan, M. Kostura, M.B. Mathews, A.G. So, K.M. Downey, and B. Stillman. 1987b. Functional identity of proliferating cell nuclear antigen and a DNA polymerase-δ auxiliary protein. *Nature* **326**: 517.

Pringle, J.R. and L.H. Hartwell. 1981. *Saccharomyces cerevisiae* cell cycle. In *The molecular biology of the yeast* Saccharomyces: *Life cycle and inheritance* (ed. J.N. Strathern et al.), p. 97. Cold Spring Harbor Laboratory, Cold Spring Harbor, New York.

Pritchard, C.G. and M.L. DePamphilis. 1983. Preparation of DNA α-C1C2 by reconstituting DNA polymerase α with its specific stimulatory cofactors, C1C2. *J. Biol. Chem.* **258**: 9801.

Pritchard, C.G., P.T. Weaver, E.F. Baril, and M.L. DePamphilis. 1983. DNA polymerase α cofactors C1C2 function as primer recognition proteins. *J. Biol. Chem.* **258**: 9810.

Prussak, C.E., M.T. Almazan, and B.Y. Tseng. 1989. Mouse primase p49 subunit molecular cloning indicates conserved and divergent regions. *J. Biol. Chem.* **264**: 4957.

Quah, S.-K., R.C. von Borstel, and P.J. Hastings. 1980. The origin of spontaneous mutation in *Saccharomyces cerevisiae*. *Genetics* **96**: 819.

Reed, S.I., J.A. Hadwiger, and A.T. Lörincz. 1985. Protein kinase activity associated with the product of the yeast cell division cycle gene *CDC28*. *Proc. Natl. Acad. Sci.* **82**: 4055.

Reha-Krantz, L.J. 1988. Amino acid changes coded by bacteriophage T4 DNA polymerase mutator mutants: Relating structure to function. *J. Mol. Biol.* **202**: 711.

Reynolds, A.E., R.M. McCarroll, C.S. Newlon, and W.L. Fangman. 1989. Time of replication of ARS elements along yeast chromosome III. *Mol. Cell. Biol.* **9**: 4488.

Reynolds, R.J. and E.C. Friedberg. 1981. Molecular mechanisms of pyrimidine dimer excision in *Saccharomyces cerevisiae*: Incision of ultraviolet-irradiated deoxyribonucleic acid *in vivo*. *J. Bacteriol.* **146**: 692.

Rhode, P.R., K.S. Sweder, K.F. Oegema, and J.L. Campbell. 1989. The gene encoding

ARS binding factor I is essential for the viability of yeast. *Genes Dev.* **3:** 1926.

Riva, M., A.R. Schaffner, A. Sentenac, G.R. Hartmann, A.A. Mustsev, E.F. Zychikov, and M.A. Grachev. 1987. Active site labeling of the RNA polymerase A, B, and C from yeast. *J. Biol. Chem.* **262:** 14377.

Rivin, C.J. and W.L. Fangman. 1980a. Cell cycle phase expansion in nitrogen-limited cultures of *Saccharomyces cerevisiae. J. Cell Biol.* **85:** 96.

―――. 1980b. Replication fork rate and origin activation during the S phase of *Saccharomyces cerevisiae. J. Cell. Biol.* **85:** 108.

Rosenfeld, P.J. and T.J. Kelly. 1986. Purification of nuclear factor 1 by DNA recognition site chromatography. *J. Biol. Chem.* **261:** 1398.

Rothstein, R. 1984. Double-strand-break repair, gene conversion, and postdivision segregation. *Cold Spring Harbor Symp. Quant. Biol.* **49:** 629.

Ryder, K., S. Silver, A.L. DeLucia, E. Fanning, and P. Tegtmeyer. 1986. An altered DNA conformation in origin region I is a determinant for the binding of SV40 large T antigen. *Cell* **44:** 719.

Sabatino, R.D., T.W. Myers, R.A. Bambera, O. Kwan-Shin, R.L. Marraccino, and P.H. Frickey. 1988. Calf thymus polymerases α and δ are capable of highly processive DNA synthesis. *Biochemistry* **27:** 2998.

Saffer, L.D. and O.L. Miller, Jr. 1986. Electron microscopic study of *Saccharomyces cerevisiae* rDNA chromatin replication. *Mol. Cell. Biol.* **6:** 1148.

Schiestl, R.H., P. Reynolds, S. Prakash, and L. Prakash. 1989. Cloning and sequence analysis of the *Saccharomyces cerevisiae RAD9* gene and further evidence that its product is required for cell cycle arrest induced by DNA damage. *Mol. Cell. Biol.* **9:** 1882.

Schild, D. and B. Byers. 1978. Meiotic effects of DNA defective cell division cycle mutations of *Saccharomyces cerevisiae. Chromosoma* **70:** 109.

Schultz, J. and M. Carlson. 1987. Molecular analysis of *SSN6,* a gene functionally related to the *SNF1* protein kinase of *Saccharomyces cerevisiae. Mol. Cell. Biol.* **7:** 3637.

Sclafani, R.A. and W.L. Fangman. 1984. Yeast gene *CDC8* encodes thymidylate kinase and is complemented by herpes thymidine kinase gene TK. *Proc. Natl. Acad. Sci.* **81:** 5821.

―――. 1986. Thymidine utilization by *tut* mutants and facile cloning of mutant alleles by plasmid conversion in *S. cerevisiae. Genetics* **114:** 753.

Sclafani, R.A., M. Patterson, J. Rosamond, and W.L. Fangman. 1988. Differential regulation of the yeast *CDC7* gene during mitosis and meiosis. *Mol. Cell. Biol.* **8:** 293.

Shampay, J., J.W. Szostak, and E.H. Blackburn. 1984. DNA sequences of telomeres retained in yeast. *Nature* **310:** 154.

Shore, D. and K. Nasmyth. 1987. Purification and cloning of a DNA binding protein from yeast that binds to both silencer and activator elements. *Cell* **51:** 721.

Shore, D., D.J. Stillman, A.H. Brand, and K.A. Nasmyth. 1987. Identification of silencer binding proteins from yeast: Possible roles in SIR control and DNA replication. *EMBO J.* **6:** 461.

Shortle, D. and D. Nathans. 1979. Regulatory mutants of simian virus 40: Constructed mutants with base substitutions at the origin of DNA replication. *J. Mol. Biol.* **131:** 801.

Shuster, E.O. and B. Byers. 1989. Pachytene arrest and other meiotic effects of the Start mutations in *Saccharomyces cerevisiae. Genetics* **123:** 29.

Simpson, R.T. 1990. Nucleosome positioning can affect the function of a *cis*-acting DNA element *in vivo. Nature* **343:** 387.

Singh, H. and L.B. Dumas. 1984. A DNA primase that copurifies with the major DNA polymerase from the yeast *Saccharomyces cerevisiae. J. Biol. Chem.* **259:** 7936.

Singh, H., R.G. Brooke, N.H. Pausch, G.T. Williams, C. Crainor, and L. Dumas. 1986. Yeast DNA primase and DNA polymerase activities. *J. Biol. Chem.* **261:** 8564.

Sinha, N.K., C.F. Morris, and B.M. Alberts. 1980. Efficient *in vitro* replication of double-stranded DNA templates by a purified T4 bacteriophage replication system. *J. Biol. Chem.* **255:** 4290.

Sinha, P., V. Chang, and B.K. Tye. 1986. A mutant that affects the function of autonomously replicating sequences in yeast. *J. Mol. Biol.* **192:** 805.

Sitney, K.C., M.E. Budd, and J.L. Campbell. 1989. DNA polymerase III, a second essential DNA polymerase, is encoded by the *Saccharomyces cerevisiae CDC2* gene. *Cell* **56:** 599.

Skryabin, K.G., M.A. Eldarov, V.L. Larionov, A.A. Bayev, J. Klootwijk, V.C.H.F. de Regt, G.M. Veldman, R.J. Planta, O.I. Georgiev, and A.A. Hadjiolov. 1984. Structure and function of the nontranscribed spacer regions of yeast rDNA. *Nucleic Acids Res.* **12:** 2955.

Slater, M.L., S.O. Sharrow, and J.J. Gart. 1977. Cell cycle of *Saccharomyces cerevisiae* in populations growing at different rates. *Proc. Natl. Acad. Sci.* **74:** 3850.

Snyder, M., A.R. Buchman, and R.W. Davis. 1986. Bent DNA at a yeast autonomously replicating sequence. *Nature* **324:** 87.

Snyder, M., R.J. Sapolsky, and R.W. Davis. 1988. Transcription interferes with elements important for chromosome maintenance in *Saccharomyces cerevisiae. Mol. Cell. Biol.* **8:** 2184.

Srienc, F., J.E. Bailey, and J.L. Campbell. 1985. Effect of *ARS1* mutations on chromosome stability in *Saccharomyces cerevisiae. Mol. Cell. Biol.* **5:** 1676.

Srienc, F., J.L. Campbell, and J. Bailey. 1983. Detection of bacterial beta-galactosidase activity in individual *Saccharomyces cerevisiae* cells by flow cytometry. *Biotechnol. Lett.* **5:** 43.

Stahl, H., P. Dröge, and R. Knippers. 1986. DNA helicase activity of SV40 large tumor antigen. *EMBO J.* **5:** 1939.

Stanway, C., J. Mellor, J.E. Ogden, A.J. Kingsman, and S.M. Kingsman. 1987. The UAS of the yeast *PGK* gene contains functionally distinct domains. *Nucleic Acids Res.* **15:** 6855.

Sternglanz, R. 1989. DNA topoisomerases. *Curr. Opin. Cell Biol.* **1:** 533.

Stillman, B. 1989. Initiation of eukaryotic DNA replication *in vitro. Annu. Rev. Cell Biol.* **5:** 197.

Stinchcomb, D.T., K. Struhl, and R.W. Davis. 1979. Isolation and characterization of a yeast chromosomal replicator. *Nature* **282:** 39.

Stinchcomb, D.T., C. Mann, E. Selker, and R.W. Davis. 1981. DNA sequences that allow the replication and segregation of yeast chromosomes. *ICN-UCLA Symp. Mol. Cell. Biol.* **22:** 473.

Storms, R.K., R.W. Ord, M.T. Greenwood, B. Mirdamadi, F.K. Chu, and M. Belfort. 1984. Cell-cycle-dependent expression of thymidylate synthase in *Saccharomyces cerevisiae. Mol. Cell. Biol.* **4:** 2858.

Strathern, J.N., E.W. Jones, and J.R. Broach, eds. 1981. *The molecular biology of the yeast* Saccharomyces: *Life cycle and inheritance.* Cold Spring Harbor Laboratory, Cold Spring Harbor, New York.

————. 1982. *The molecular biology of the yeast* Saccharomyces: *Metabolism and gene expression.* Cold Spring Harbor Laboratory, Cold Spring Harbor, New York.

Streisinger, G., Y. Okada, J. Emrich, J. Newton, A. Tsugita, E. Terzaghi, and M. Inouye. 1967. Frameshift mutations and the genetic code. *Cold Spring Harbor Symp. Quant. Biol.* **31:** 77.

Strich, R.M., M. Wootner, and J.F. Scott. 1986. Mutations in ARS1 increase the rate of simple loss of plasmids in *Saccharomyces cerevisiae. Yeast* **2:** 169.

Struhl, K., D.T. Stinchcomb, S. Scherer, and R.W. Davis. 1979. High-frequency transformation of yeast: Autonomous replication of hybrid DNA molecules. *Proc. Natl. Acad. Sci.* **76:** 1035.

Su, J.-Y., L. Belmont, and R.A. Sclafani. 1990. Genetic and molecular analysis of the *SOE1* gene: A tRNA$^{Glu}_3$ missense suppressor of yeast *cdc8* mutations. *Genetics* **124:** 523.

Sugino, A., H. Kojo, B.D. Greenberg, P.O. Brown, and K.C. Kim. 1981. In vitro replication of yeast 2μ plasmid DNA. *ICN-UCLA Symp. Mol. Cell. Biol.* **22:** 529.

Sugino, A., B.H. Ryn, T. Sugino, L. Naumovski, and E.C. Friedberg. 1986. A new DNA dependent ATPase which stimulates yeast DNA polymerase I and has DNA unwinding activity. *J. Biol. Chem.* **261:** 11744.

Sugino, A., A. Sakai, F. Wilson-Coleman, J. Arendes, and K.C. Kim. 1983. *In vitro* reconstitution of yeast 2 μm plasmid DNA replication. In *Mechanisms of DNA replication and recombination* (ed. N.R. Cozzarelli), p. 527. A.R. Liss, New York.

Sundin, O. and A. Varshavsky. 1980. Terminal stages of SV40 DNA replication proceed via multiple intertwined catenated dimers. *Cell* **21:** 103.

―――. 1981. Arrest of segregation leads to accumulation of highly intertwined catenated dimers: Dissection of the final stages of SV40 DNA replication. *Cell* **25:** 659.

Sung, P., D. Higgins, L. Prakash, and S. Prakash. 1988. Mutation of lysine-48 to arginine in the yeast *RAD3* protein abolishes its ATPase and DNA helicase activities but not the ability to bind ATP. *EMBO J.* **7:** 3263.

Suzuki, M., T. Enomoto, C. Masutani, F. Hanaoka, M. Yamada, and M. Ui. 1989. DNA primase-DNA polymerase α assembly from mouse FM3A cells: Purification of constituting enzymes, reconstitution, and analysis of RNA priming as coupled to DNA synthesis. *J. Biol. Chem.* **264:** 10065.

Sweder, K.S., P.R. Rhode, and J.L. Campbell. 1988. Purification and characterization of proteins that bind to yeast *ARS*s. *J. Biol. Chem.* **263:** 17270.

Syvaoja, J. and S. Linn. 1989. Characterization of a large form of DNA polymerase δ from HeLa cells that is insensitive to proliferating cell nuclear antigen. *J. Biol. Chem.* **264:** 2489.

Szostak, J.W. and R. Wu. 1979. Insertion of a genetic marker into the ribosomal DNA of yeast. *Plasmid* **2:** 536.

Tan, C.-K., C. Castillo, A.G. So, and K.M. Downey. 1986. An auxiliary protein for DNA polymerase-δ from fetal calf thymus. *J. Biol. Chem.* **261:** 12310.

Thoma, F., L.W. Bergman, and R.T. Simpson. 1984. Nuclease digestion of circular TRP1 ARS1 chromatin reveals positioned nucleosomes separated by nuclease-sensitive regions. *J. Mol. Biol.* **177:** 713.

Thrash, C., A.T. Bankier, B.G. Barrell, and R. Sternglanz. 1985. Cloning, characterization and sequence of the yeast DNA topoisomerase I gene. *Proc. Natl. Acad. Sci.* **82:** 4374.

Thrash-Bingham, C. and W.L. Fangman. 1989. A yeast mutation that stabilizes a plasmid bearing a mutated *ARS1* element. *Mol. Cell. Biol.* **9:** 809.

Tschumper, G. and J. Carbon. 1983. Copy number control by a yeast centromere. *Gene* **23:** 221.

Tsuchiya, E. and S. Fukui. 1983. *In vitro* deoxyribonucleic-acid-replicating system solubilized from isolated nuclei of *Saccharomyces cerevisiae. Agric. Biol. Chem.* **47:** 907.

Tsurimoto, T., T. Melendy, and B. Stillman. 1990. Sequential initiation of lagging and leading strand synthesis by two different polymerase complexes at the SV40 DNA replication origin. *Nature* **346:** 534.

Udem, S.A. and J.R. Warner. 1972. Ribosomal RNA synthesis in *Saccharomyces cerevisiae. J. Mol. Biol.* **65:** 227.

Umek, R.M. and D. Kowalski. 1987. Yeast regulatory sequences preferentially adopt a non-B conformation in supercoiled DNA. *Nucleic Acids Res.* **15:** 4467.

————. 1988. The ease of DNA unwinding as a determinant of initiation at yeast replication origins. *Cell* **52:** 559.

————. 1990. Thermal energy suppresses mutational defects in DNA unwinding at a yeast replication origin. *Proc. Natl. Acad. Sci.* **87:** 2486.

Umek, R.M., M.H.K. Linskens, D. Kowalski, and J.A. Huberman. 1989. New beginnings in studies of eukaryotic DNA replication origins. *Biochim. Biophys. Acta* **1007:** 1.

Valenzuela, M.S., D. Freifelder, and R.B. Inman. 1976. Lack of a unique termination site for the first round of bacteriophage lambda DNA replication. *J. Mol. Biol.* **103:** 569.

Van der Ploeg, L.H.T., A.Y.C. Liu, and P. Borst. 1984. Structure of the growing telomeres of trypanosomes. *Cell* **36:** 459.

Van Houten, J.V. and C.S. Newlon. 1990. Mutational analysis of the consensus sequence of a replication origin from yeast chromosome III. *Mol. Cell. Biol.* **10:** 3917.

Veit, B.E. and W.L. Fangman. 1988. Copy number and partition of the *Saccharomyces cerevisiae* 2 micron plasmid controlled by transcription regulators. *Mol. Cell. Biol.* **8:** 4949.

Vignais, M.-L., L.P. Woudt, G.M. Wassenaar, W.H. Mager, A. Sentenac, and R.J. Planta. 1987. Specific binding of TUF factor to upstream activation sites of yeast ribosomal protein genes. *EMBO J.* **6:** 1451.

Vishwanatha, J.K., S.A. Coughlin, M. Wesolowski-Owens, and E.F. Baril. 1986. A multiprotein form of DNA polymerase α from HeLa cells: Resolution of its associated catalytic activities. *J. Biol. Chem.* **261:** 6619.

Voelkel-Meiman, K., S. DiNardo, and R. Sternglanz. 1986. Molecular cloning and genetic mapping of the DNA topoisomerase II gene of *Saccharomyces cerevisiae. Gene* **42:** 193.

Walker, S.S., S.C. Francesconi, and S. Eisenberg. 1990. A DNA replication enhancer in *Saccharomyces cerevisiae. Proc. Natl. Acad. Sci.* **87:** 4665.

Walker, S.S., S.C. Francesconi, B.-K. Tye, and S. Eisenberg. 1989. The OBF1 protein and its DNA-binding site are important for the function of an autonomously replicating sequence in *Saccharomyces cerevisiae. Mol. Cell. Biol.* **9:** 2914.

Wallis, J.W., G. Chrebet, G. Brodsky, M. Rolfe, and R. Rothstein. 1989. A hyper- recombination mutation in *S. cerevisiae* identifies a novel eukaryotic topoisomerase. *Cell* **58:** 409.

Walmsley, R.W. 1987. Yeast telomeres: The end of the chromosome story? *Yeast* **3:** 139.

Walmsley, R.W., C.S. Chan, B.-K. Tye, and T.D. Petes. 1984a. Unusual DNA sequences associated with the ends of yeast chromosomes. *Nature* **310:** 157.

Walmsley, R.W., L.H. Johnston, D.H. Williamson, and S.G. Oliver. 1984b. Replicon size of yeast ribosomal DNA. *Mol. Gen Genet.* **195:** 260.

Wang, J.C. 1985. DNA topoisomerases. *Annu. Rev. Biochem.* **54:** 665.

————. 1987. Recent studies of DNA topoisomerases. *Biochim. Biophys. Acta* **909:** 1.

Wang, S.-S. and V.A. Zakian. 1990. Telomere-telomere recombination provides an express pathway for telomere acquisition. *Nature* **345:** 456.

Wang, T.S.-F., S.W. Wong, and D. Korn. 1989. Human DNA polymerase α: Predicted functional domains and relationships with viral DNA polymerases. *FASEB J.* **3:** 14.

Weinert, T.A. and L.H. Hartwell. 1988. The *RAD9* gene controls the cell cycle response to DNA damage in *Saccharomyces cerevisiae*. *Science* **241**: 317.

White, J.H.M, D. Barker, P. Nurse, and L.H. Johnston. 1986. Periodic transcription as a means of regulating gene expression during the cell cycle: Contrasting modes of expression of DNA ligase genes in budding and fission yeast. *EMBO J.* **5**: 1705.

White, J.H.M., S.R. Green, D.G. Barker, L.M. Dumas, and L.H. Johnston. 1987. The *CDC8* transcript is cell cycle regulated in yeast and is expressed coordinately with *CDC9* and *CDC21* at a point preceding histone transcription. *Exp. Cell Res.* **171**: 223.

Wilcox, D.R. and L. Prakash. 1981. Incision and post-incision steps of pyrimidine dimer removal in excision defective mutants of *Saccharomyces cerevisiae*. *J. Bacteriol.* **148**: 618.

Williams, J.S., T.T. Eckdahl, and J.N. Anderson. 1988. Bent DNA functions as a replication enhancer in *Saccharomyces cerevisiae*. *Mol. Cell. Biol.* **8**: 2763.

Williams, K.R., K.L. Stone, M.B. LoPresti, B.M. Merrill, and S.R. Planck. 1985. Amino acid sequence of the UP1 calf thymus helix-destabilizing protein and its homology to an analogous protein from mouse myeloma. *Proc. Natl. Acad. Sci.* **82**: 5666.

Williamson, D.H. 1965. The timing of deoxyribonucleic acid synthesis in the cell cycle of *Saccharomyces cerevisiae*. *J. Cell Biol.* **25**: 517.

Wilson, F.E. and A. Sugino. 1985. Purification of a DNA primase activity from the yeast *Saccharomyces cerevisiae*: Primase can be separated from DNA polymerase I. *J. Biol. Chem.* **260**: 8173.

Wintersberger, E. 1966. Occurrence of a DNA-polymerase in isolated yeast mitochondria. *Biochem. Biophys. Res. Commun.* **25**: 1.

⸻. 1974. Deoxyribonucleic acid polymerases from yeast. Further purification and characterization of DNA dependent DNA polymerases A and B. *Eur. J. Biochem.* **50**: 41.

⸻. 1978. Yeast DNA polymerases: Antigenic relationship, use of RNA primer and associated exonuclease activity. *Eur. J. Biochem.* **84**: 167.

Wintersberger, U. 1974. Absence of a low-molecular-weight DNA polymerase from nuclei of the yeast *Saccharomyces cerevisiae*. *Eur. J. Biochem.* **50**: 197.

Wintersberger, U. and E. Wintersberger. 1970a. Studies on deoxyribonucleic acid polymerases from yeast. 1. Partial purification and properties of two DNA polymerases from mitochondria-free cell extracts. *Eur. J. Biochem.* **13**: 11.

⸻. 1970b. Studies on deoxyribonucleic acid polymerases from yeast. 2. Partial purification and characterization of mitochondrial DNA polymerase from wild-type and respiration deficient yeast. *Eur. J. Biochem.* **13**: 20.

Wintersberger, U., J. Hirsch, and A.M. Fink. 1974. Studies on nuclear and mitochondrial DNA-replication in a temperature-sensitive mutant of *Saccharomyces cerevisiae*. *Mol. Gen. Genet.* **131**: 291.

Wintersberger, U., C. Kuhne, and R. Karwan. 1988. Three ribonucleases H and a reverse transcriptase from the yeast, *Saccharomyces cerevisiae*. *Biochim. Biophys. Acta* **951**: 322.

Wobbe, C.R., L. Weissbach, J.A. Borowiec, F.B. Dean, Y. Murakami, P. Bullock, and J. Hurwitz. 1987. Replication of simian virus 40 origin-containing DNA *in vitro* with purified proteins. *Proc. Natl. Acad. Sci.* **84**: 1834.

Wold, M.S. and T. Kelly. 1988. Purification and characterization of replication protein A, a cellular protein required for *in vitro* replication of simian virus 40 DNA. *Proc. Natl. Acad. Sci.* **85**: 2523.

Wold, M.S., D.H. Weinberg, D.M. Virshup, J.J. Li, and T. Kelly. 1989. Identification of cellular proteins required for SV40 DNA replication. *J. Biol. Chem.* **264**: 2801.

Wold, M.S., J.J. Li, D.M. Weinberg, D.M. Virshup, J.L. Sherley, E. Verheyen, and T.

Kelly. 1988. Cellular proteins required for SV40 DNA replication in vitro. *Cancer Cells* **6**: 133.

Wong, S.W., J. Syvaoja, C.-K. Tan, K.M. Downey, A.G. So, S. Linn, and T.S.-F. Wang. 1989. DNA polymerases α and δ are immunologically and structurally distinct. *J. Biol. Chem.* **264**: 5924.

Wong, S.W., A.F. Wahl, P.-M., Yuan, N. Arai, B.E. Pearson, K. Arai, D. Korn, M.W. Hunkapiller, and T.S.-F. Wang. 1988. Human DNA polymerase α gene expression is cell proliferation dependent and its primary structure is similar to both prokaryotic and eukaryotic replicative DNA polymerases. *EMBO J.* **7**: 37.

Worland, S.T. and J.C. Wang. 1989. Inducible overexpression, purification, and active site mapping of DNA topoisomerase II from the yeast *Saccharomyces cerevisiae. J. Biol. Chem.* **264**: 4412.

Wyers, F., A. Sentenac, and P. Fromageot. 1973. Role of DNA-RNA hybrids in eukaryotes. Ribonucleases H in yeast. *Eur. J. Biochem.* **35**: 270.

————. 1976a. Role of DNA-RNA hybrids in eukaryotes: Purification of two ribonucleases H from yeast cells. *Eur. J. Biochem.* **69**: 377.

Wyers, F., J. Huet, A. Sentenac, and P. Fromageot. 1976b. Role of DNA-RNA hybrids in eukaryotes: Characterization of yeast ribonucleases H1 and H2. *Eur. J. Biochem.* **69**: 385.

Yang, L., M.S. Wold, J.J. Li, T.J. Kelly, and L.F. Liu. 1987. Role of DNA topoisomerases in simian virus 40 DNA replication *in vitro. Proc. Natl. Acad. Sci.* **84**: 950.

Yochem, J. and B. Byers. 1987. Structural comparison of the yeast cell division cycle gene *CDC4* and a related pseudogene. *J. Mol. Biol.* **195**: 233.

Yu, G.-L., J.D. Bradley, L.D. Attardi, and E.H. Blackburn. 1990. *In vivo* alteration of telomere sequences and senescence caused by mutated *Tetrahymena* telomerase RNAs. *Nature* **344**: 126.

Zahn, K. and F.R. Blattner. 1985. Sequence-induced DNA curvature at the bacteriophage λ origin of replication. *Nature* **317**: 451.

Zakian, V.A. 1989. Structure and function of telomeres. *Annu. Rev. Genet.* **23**: 579.

Zakian, V.A. and H.M. Blanton. 1988. Distribution of telomere-associated sequences on natural chromosomes in *Saccharomyces cerevisiae. Mol. Cell. Biol.* **8**: 2257.

Zakian, V.A. and D.M. Kupfer. 1982. Replication and segregation of an unstable plasmid in yeast. *Plasmid* **8**: 15.

Zakian, V.A., B.J. Brewer, and W.L. Fangman. 1979. Replication of each copy of the yeast 2 micron DNA plasmid occurs during the S phase. *Cell* **17**: 923.

Zhou, C., S.-H. Huang, and A.Y. Jong. 1989. Molecular cloning of *Saccharomyces cerevisiae CDC6* gene. *J. Biol. Chem.* **264**: 9022.

Zuber, M., E.M. Tan, and M. Ryoji. 1989. Involvement of proliferating cell nuclear antigen (cyclin) in DNA replication in living cells. *Mol. Cell. Biol.* **9**: 57.

Zweifel, S.G. and W.L. Fangman. 1990. Creation of ARS activity in yeast through iteration of non-functional sequences. *Yeast* **6**: 179.

3

Cellular Responses to DNA Damage in Yeast

Errol C. Friedberg,[1] Wolfram Siede,[1] and A. Jane Cooper[1]
Department of Pathology
Stanford University School of Medicine
Stanford, California 94305

I. INTRODUCTION

The general topic of cellular responses to DNA damage in the yeast *Saccharomyces cerevisiae* has been reviewed in several contexts (Lemontt 1980; Haynes and Kunz 1981; Lawrence 1982; Friedberg 1985a,b, 1988; von Borstel and Hastings 1985; Cooper and Kelly 1987; Moustacchi 1987). These reviews have documented the basic genetics and general biology of DNA repair and mutagenesis. In 1991, the essential question is no longer what are the phenomena of DNA repair and mutagenesis in

[1]Present address: Department of Pathology, University of Texas, Southwestern Medical Center at Dallas, Dallas, Texas 75235.

Volume I. *The Molecular and Cellular Biology of the Yeast* Saccharomyces: *Genome Dynamics, Protein Synthesis, and Energetics.* Copyright 1991 Cold Spring Harbor Laboratory Press 0-87969-355-X/91 $3 + 00

yeast, but, rather, how are these elements of DNA metabolism transacted at the molecular level? Although the answers to this question are far from complete, the emphasis of this chapter is on the molecular biology and biochemistry of cellular responses to DNA damage, in keeping with the intended focus of these volumes.

Strictly defined, the term DNA repair refers to cellular events associated with the removal of damaged, inappropriate, or mispaired bases from the genome of living cells (see Friedberg 1985a). In this narrow context, the topic is confined to (1) DNA damage reversal (e.g., monomerization of cyclobutyldipyrimidines [pyrimidine dimers] by DNA photolyase or demethylation of methylated [alkylated] bases by specific DNA methyltransferases); (2) base excision repair, whereby damaged bases (e.g., alkylated) or inappropriate bases (e.g., uracil) are excised by specific classes of repair enzymes designated DNA glycosylases and AP (apurinic/apyrimidinic) endonucleases; (3) nucleotide excision repair, whereby oligonucleotide tracts containing distortive bulky base adducts are excised; and (4) mismatch repair, whereby mismatched bases generated during semiconservative DNA synthesis or during recombination are excised.

This review addresses recent progress in our understanding of most of these repair processes in yeast. An exception is the repair of many types of chemical damage, including alkylation damage and interstrand cross-links in DNA. The general biology and genetics of the repair of chemical damage in yeast have been reviewed elsewhere (Haynes and Kunz 1981; Moustacchi et al. 1983; Brendel and Ruhland 1984; Friedberg 1988). The removal of cross-links from DNA is of particular interest because of the indications that at least one gene (designated *PSO2* [or *SNM1*]) is required specifically for this type of repair (Henriques and Moustacchi 1980; Ruhland et al. 1981; Cassier-Chauvat and Moustacchi 1988; Haase et al. 1989). However, as little is known about the molecular biology or biochemistry of the repair of chemical damage, this topic is not discussed extensively here.

Primarily on the basis of historically established usage, the term DNA repair also embraces other cellular responses to DNA damage, which endow organisms with increased resistance to the potentially lethal effects of genomic injury but do not result in the removal of damage from the genome. These DNA-damage tolerance mechanisms include a variety of recombinational and mutagenic processes that are not necessarily unique to the processing of damaged DNA (see Friedberg 1985a). The molecular mechanisms of these processes in yeast are still unknown. However, because impressive progress is being made, these topics are discussed in this chapter.

II. GENETIC FRAMEWORK

Cells and organisms defective in processing DNA damage typically display increased sensitivity to killing by the agent(s) in question. This phenotype has proved to be extremely useful for the isolation and genetic characterization of mutants. Yeast mutants with hypersensitivity to ultraviolet (UV) or ionizing radiation are designated as *rad* mutants (for *rad*iation-sensitive) (see Haynes and Kunz 1981; Friedberg 1988 and references therein). Many of the corresponding mutations have been grouped into multiple complementation and allelic groups (see Haynes and Kunz 1981; Friedberg 1988). Additionally, mutations from many (but not all) of the established complementation groups have been tested for epistatic interactions by comparing the sensitivity of double mutants to each of the relevant single mutants. Such analyses have revealed three largely nonoverlapping epistasis groups designated *RAD3*, *RAD52*, and *RAD6* (after well-characterized mutations in each). Additional genes have been added to these epistasis groups on the basis of phenotypic similarities to mutant strains already tested (see Friedberg 1988). The current assignment of genes to these epistasis groups is shown in Table 1.

Table 1 Epistasis groups for yeast genes involved in cellular responses to DNA damage

RAD3	RAD52	RAD6
RAD1	RAD24	RAD5 (REV2) (SNM2)
RAD2	RAD50	RAD6
RAD3	RAD51	* RAD8
RAD4	RAD52	* RAD9
RAD7	RAD53	* RAD15
RAD10	RAD54	RAD18
RAD14	RAD55	* RADH
RAD16	RAD56	REV1
RAD23	RAD57	REV3 (PSO1)
RAD24	CDC9	* REV5
CDC8		* REV6
CDC9		REV7
MMS19		* CDC7
PSO2 (SNM1)		CDC8
PSO3		* MMS3
UVS12		* PSO4
		* UMR1-7

The genes are shown as wild-type alleles for simplicity. The assignment of genes to a particular epistasis group is not based on comprehensive analysis of UV or ionizing radiation sensitivity in all cases. In some cases (*), genes have been assigned strictly on the basis of limited phenotypic characterization. Genes shown in parentheses are allelic to those primarily listed. (Adapted from Friedberg 1988.)

Phenotypic characterization of many of these mutants indicates that the *RAD3* epistasis group comprises genes involved in nucleotide excision repair (NER). Mutants in the *RAD52* epistasis group are defective in genetic recombination and in the repair of strand breaks in DNA, and these genes are therefore believed to be required for recombinational DNA repair modes. Many of the genes in the *RAD6* epistasis group are required for spontaneous and/or damage-induced mutagenesis, suggesting that they participate in biochemical events associated with altered replicational fidelity.

Mutants from the three established epistasis groups comprise a useful genetic framework for many of the molecular studies reviewed here; however, this classification has distinct limitations and should not be interpreted as a comprehensive definition of the multiple repair pathways in yeast. Indeed, genes not included in the three major epistasis groups have been identified for specific cellular responses to DNA damage. These include genes for enzymatic photoreactivation, for the base excision of uracil from DNA, and for mismatch repair.

The phenotype of radiation and/or chemical sensitivity has greatly facilitated the molecular cloning of genes by phenotypic complementation. Many of these genes have been characterized in detail, and in some cases, so have the proteins they encode (Table 2). Hence, clues are beginning to emerge about the biochemical and molecular interpretation of the complex genetics suggested by the multiplicity of genes in each of the epistasis groups; however, to date, no biochemical pathways have been deciphered in detail.

III. ENZYMATIC PHOTOREACTIVATION

Yeast is one of the organisms in which DNA photolyase (photoreactivating enzyme) was first discovered in cell-free extracts (see Friedberg 1985a). The early literature contains reports of multiple yeast photoreactivating enzymes with different properties (see Friedberg 1985a). However, in recent years, a single yeast *PHR* gene has been isolated (Yasui and Chevallier 1983; Schild et al. 1984), and its overexpression has facilitated the detailed characterization of what appears to be the major, if not the exclusive, DNA photolyase in *S. cerevisiae*.

A yeast mutant defective in enzymatic photoreactivation (EPR) defined the *PHR1* gene (Resnick 1969; Resnick and Setlow 1972). The isolation of a second mutant suggested the existence of another gene designated *PHR2*, genetically closely linked to *PHR1* (MacQuillan et al. 1981). The *PHR2* gene has not yet been isolated, and its function is un-

Table 2 Cloned genes from *S. cerevisiae* implicated in cellular responses to DNA damage

Gene	Coding region (codons)	Polypeptide size (calculated molecular mass kD)	Biochemical function of gene product	Remarks
RAD1	1100	126.2	?	required for NER, multifunctional
RAD2	1031	117.7	?	required for NER, inducible
RAD3	778	89.7	ATPase/DNA helicase	required for NER, multifunctional
RAD4	754	87.1	?	required for NER, toxic to *E. coli*
RAD7	565	63.7	?	involved in NER
RAD10	210	24.3	?	required for NER
RAD50	1312	157.0	?	required for recombinational repair
RAD51	?	?	?	required for recombinational repair
RAD52	504	56.1	?	required for recombinational repair
RAD54	?	?	?	required for recombinational repair, inducible
RAD55	?	?	?	required for recombinational repair
RAD57	?	?	?	required for recombinational repair
RAD6	172	19.7	ubiquitin-ligase enzyme	multifunctional
RAD9	1309	148.4	?	required for G$_2$ arrest
RAD18	487	55.5	?	multifunctional
RADH	1175	134.2	probable DNA helicase	required for mutagenesis
REV1	985	112.2	?	required for mutagenesis
REV3	1504	173.0	probable DNA polymerase	required for mutagenesis
CDC7	489 or 507	56.0 or 58.2	protein kinase	required for mutagenesis
CDC8	216	24.6	thymidylate kinase	required for mutagenesis
CDC9	755	84.4	DNA ligase	required for mutagenesis
CDC40	?	?	?	required in many DNA transactions
PHR1	565	66.2	DNA photolyase	required for EPR
SNM1 (PSO2)	?	?	?	required for cross-link repair
UNG1	359	40.5	uracil-DNA glycosylase	required for excision of uracil from DNA
NUC1	?	140.0	endonuclease	required for repair of MMS damage[b]
PMS1	904	103.0	?	required for mismatch repair
APN1	367	41.4	AP endonuclease	required for repair of MMS and H$_2$O$_2$ damage

Except where noted, references on the cloning of these genes are found in the text or in Petes et al. (this volume).
[a]Kassir et al. (1985); Kupiec and Simchen (1986b).
[b]Burbee et al. 1988.

151

known. Conceivably, it encodes a distinct DNA photolyase, or possibly it regulates the *PHR1* gene.

The *PHR1* gene was cloned by phenotypic complementation of the *phr1* mutation (Yasui and Chevallier 1983; Schild et al. 1984). The gene contains an open reading frame (ORF) of 565 codons (Sancar 1985a; Yasui and Langeveld 1985), which could encode a polypeptide of about 66.2 kD. Following overexpression of the cloned yeast gene in *Escherichia coli*, a monomeric protein of 60 kD was purified and shown specifically to catalyze the monomerization of pyrimidine dimers in DNA (Sancar et al. 1987).

The *PHR1* gene complements an *E. coli phr* mutant defective in EPR (Sancar 1985b) and vice versa (Langeveld et al. 1985). These observations suggest that the two enzymes share common chromophores, a suggestion borne out by the isolation and characterization of chromophoric elements from both purified enzymes (Sancar et al. 1987). Spectroscopic, fluorescence, and chromatographic studies have demonstrated that one of these chromophores is a 1,5-reduced flavine-adenine dinucleotide (FAD) (Sancar and Sancar 1988); the second chromophore is a pterin (Sancar et al. 1987; Sancar and Sancar 1988).

The amino acid sequences of the yeast and *E. coli PHR* genes show considerable similarity (Sancar 1985a; Yasui and Langeveld 1985). Overall, there is 36.2% identity; however, two short regions near the amino and carboxyl termini of the two polypeptides show even greater sequence conservation. These features suggest that the yeast and *E. coli* enzymes possess conserved structural and functional domains involved in the binding of substrate and/or chromophores.

Such evolutionary conservation notwithstanding, the cloned genes do not cross-hybridize, and neither has yet provided a useful probe for the detection of homologous sequences in higher organisms (see Meechan et al. 1986).

DNA photolyase apparently enhances the efficiency of NER of pyrimidine dimers. Overexpression of the yeast *PHR1* gene increased recovery in the dark from the effects of UV irradiation of yeast mutants defective in the *RAD18* gene (a member of the *RAD6* epistasis group). However, inactivation of the *RAD2* gene (a member of the *RAD3* epistasis group and required for NER) blocked this stimulation of recovery (Sancar and Smith 1989). Similar interactive effects have been observed between *E. coli* DNA photolyase and the UvrABC NER complex from *E. coli* (Sancar et al. 1984). The apparent generality of this phenomenon has led to the suggestion that DNA photolyase may be an accessory protein in the NER pathway (Sancar and Smith 1989). Surprisingly, however, DNA photolyase from yeast failed to enhance NER in *E. coli*, despite the structural and functional similarities between the yeast and *E. coli* photolyases (Baer and Sancar 1989). In fact, the yeast enzyme inhibited NER in *E. coli* by competing for pyrimidine dimer sites (Sancar and Smith 1989).

IV. EXCISION REPAIR

A. Nucleotide Excision Repair

Nucleotide excision defines a DNA repair mode by which damaged bases are removed from the genome as nucleotides rather than free bases (see Friedberg 1985a). This excision phenomenon was one of the earliest modes of DNA repair discovered, and before it was appreciated that the removal of damaged nucleotides can take place by several biochemically distinct mechanisms, nucleotide excision repair (NER) was loosely designated as excision repair, or dark repair, to distinguish it from the light-dependent enzymatic photoreactivation process discussed above.

By analogy with studies in prokaryotes and in higher eukaryotes, NER in yeast most likely involves the removal of damaged bases from DNA as components of oligonucleotide fragments released from the genome following incision by a DNA-damage-specific endonuclease. This endonuclease is presumably targeted to helix-distorting bulky base adducts produced by physical agents, such as UV radiation, and by a variety of chemical agents. Pyrimidine dimers produced by UV radiation constitute a particularly well-characterized and popular model substrate for the study of this repair mode.

The molecular basis for the specific recognition of helix-distorting damage by a putative damage-specific endonuclease in yeast is not understood. In recent years, considerable information has emerged on the biochemistry of NER in *E. coli* (see Weiss and Grossman 1987; Rubin 1988; Sancar and Sancar 1988). The biochemical mechanism defined for *E. coli* may be an informative general paradigm. However, such a working hypothesis must be tempered by the observations that the genetics of NER in yeast is considerably more complex than that in *E. coli*, that there is little similarity between the amino acid sequences of genes required for NER in *E. coli* and *S. cerevisiae*, and that cloned NER genes from these organisms do not cross-complement in mutant strains (see Friedberg 1988).

These reservations notwithstanding, the generality of the genetic complexity of NER in *E. coli*, yeast, and mammalian cells suggests that NER is effected by a multiprotein complex (repairosome) at all levels of biological organization, which is assembled in an ordered fashion at sites of conformational distortion generated by bulky adducts in DNA.

Extensive genetic analysis has highlighted a large number of genes in *S. cerevisiae* involved in NER (see Haynes and Kunz 1981). The products of at least five genes (*RAD1–RAD4* and *RAD10*) appear to be absolutely required for damage-specific incision of UV-irradiated DNA and for excision of pyrimidine dimers (Reynolds and Friedberg 1981; Wilcox and Prakash 1981). Hence, we presume that these gene products are essential components of the putative yeast repairosome. Other genes in the *RAD3* epistasis group (*RAD7, RAD14, RAD16, RAD23*, and *MMS19*) are apparently also involved in these biochemical events; however, mutations in the genes confer a phenotype of deficient rather than defective incision of DNA and excision of dimers (Prakash 1977; Miller et al. 1982).

NER in yeast probably involves even more genes than those listed above. Cox and Parry (1968) attempted to saturate the yeast genome with mutations in all genes involved in resistance to UV radiation. However, they observed that of the 22 genetic complementation groups identified, only 12 were represented by more than a single allele. Hence, on statistical grounds alone, the existing collection of UV-sensitive mutants most likely does not reflect all of the DNA repair genes in *S. cerevisiae*. The recent identification of a hitherto unrecognized yeast gene, using the cloned human NER gene *ERCC3* as a hybridization probe (J.H.J. Hoeijmakers, pers. comm.), provides support for this contention, although whether or not this gene is involved in NER in yeast remains to be demonstrated.

B. Cloned Genes Involved in NER in *S. cerevisiae*

A number of genes in the *RAD3* epistasis group have been isolated by molecular cloning. Detailed characterization of some of them has yielded interesting clues about selected aspects of the molecular biology of NER. Additionally, some of these genes have been tailored for overexpression, and the purification and characterization of individual proteins are beginning to provide the essential framework for elucidating the biochemistry of this process in yeast.

1. RAD3 *Gene*

The *RAD3* gene was cloned by phenotypic complementation of a highly UV-sensitive *rad3* mutant (Naumovski and Friedberg 1982; Higgins et al. 1983b). The *RAD3* ORF comprises 778 codons and is expected to encode a polypeptide approximately 89.7 kD in size (Naumovski et al. 1985; Reynolds et al. 1985b). Detection of a consensus nucleotide-binding domain characteristic of many proteins that catalyze the hydrolysis of ATP or GTP provided an early clue as to a possible biochemical function of the *RAD3* gene product.

Rad3p has been purified to physical homogeneity (Sung et al. 1987b). The purified protein catalyzes the hydrolysis of ATP (or dATP) to ADP (or dADP) and P_i, in the presence of Mg^{++} (or Mn^{++}) and single-stranded DNA. Double-stranded DNA does not support ATP hydrolysis. The ATPase activity has an extremely narrow pH dependence, with a sharp optimum at pH 5.6. Essentially no ATPase activity is detectable at pH values below 5.0 or above 6.0, thereby distinguishing it from other yeast ATPases (Sung et al. 1987a). The K_m for ATP hydrolysis has been variously measured at 50 and 67 μM (Sung et al. 1987b; Harosh et al. 1989).

The hydrolysis of ATP or dATP by purified Rad3p drives a DNA helicase activity with the same narrow pH dependence (Sung et al. 1987a). The Rad3p DNA helicase requires single-stranded regions to initiate unwinding of duplex DNA. Both circular and linear partially duplex molecules are utilized as substrates (Sung et al. 1987a). The enzyme unwinds duplex DNA unidirectionally, with $5' \rightarrow 3'$ polarity relative to the single strand to which it is bound. Duplex regions as long as 850 bp can be unwound in the absence of other proteins (Sung et al. 1987a). Additionally, the enzyme can displace oligonucleotides as short as 11 nucleotides annealed to complementary circular single-stranded molecules and does so in a strictly processive fashion (Harosh et al. 1989). The enzyme can unwind oligonucleotides from gapped regions as small as 21 nucleotides; however, single-stranded gaps 4 nucleotides long are not utilized as substrate, and the enzyme does not catalyze unwinding of duplex nicked DNA (Harosh et al. 1989).

The identification of a specific catalytic activity for Rad3p invites speculation about the role of this activity in vivo. In this regard, it is important to note that by genetic criteria, the *RAD3* gene is multifunctional (see Friedberg 1988). In addition to being required for early events during NER, *RAD3* is an essential gene (Higgins et al. 1983b; Naumovski and Friedberg 1983). Furthermore, certain mutant alleles result in increased spontaneous mutation, increased mitotic recombination, and lethality in the presence of the *rad52* mutation (which confers defective repair of double-strand breaks in DNA [Malone and Hoekstra 1984; Hoekstra and Malone 1987; Montelone et al. 1988]). This combination of characteristics is referred to as the Rem phenotype of *rad3* mutants (Golin and Esposito 1977).

In an effort to dissect the multiple functions of *RAD3*, a series of mutants carrying mutations that generate one or more of the phenotypes described above have been isolated and characterized (Naumovski and Friedberg 1986; Sung et al. 1988a). The protein encoded by one of these mutant alleles, which carries a substitution of arginine for lysine at codon 48, was recently purified (Sung et al. 1988b). This lysine residue is completely conserved in a consensus sequence that defines the nucleotide-binding domain of many ATPases/DNA helicases, and mutations in this codon of the *RAD3* gene render cells extremely sensitive to killing by UV radiation (Sung et al. 1988b). Rad3p^{Arg48} retains the ability to bind ATP, albeit with weaker affinity than the native protein (Sung et al. 1988b). However, the protein is inactivated for both ATPase and DNA helicase activities. These observations suggest that the ATPase/ DNA helicase function of Rad3p is involved in NER. Because the *rad3*^{Arg48} mutant is perfectly viable, however, the ATPase/DNA helicase function of Rad3p is apparently not required for the essential function of the *RAD3* gene.

How might a DNA helicase be involved in NER? One possibility is that such an enzyme might facilitate unwinding of damage-containing oligonucleotides as part of the excision process following damage-specific incision of DNA. However, as indicated above, biochemical studies of the purified protein indicate a requirement for single-stranded regions at least 5 nucleotides long to initiate unwinding of duplex DNA. Conceivably, nicks generated during NER in vivo are converted to single-stranded regions by other components of the multiprotein yeast repairosome. Alternatively, when associated with other proteins, the properties of Rad3p helicase may be modified to accommodate the unwinding of oligonucleotides from nicks.

In *E. coli*, the UvrD protein is required for normal NER (see Weiss and Grossman 1987; Rubin 1988; Sancar and Sancar 1988). This protein is also a DNA helicase (Matson 1986); however, biochemical studies have demonstrated that UvrD protein is not required for postincisional oligonucleotide displacement and repair synthesis in vitro. Rather, together with DNA polymerase I, it plays a role in the displacement and turnover of the UvrABC protein complex that catalyzes damage-specific incision (see Weiss and Grossman 1987; Rubin 1988; Sancar and Sancar 1988). Sung et al. (1988b) have suggested that a similar function might be entertained for Rad3p during NER in yeast.

A third possible role for a DNA helicase in NER is also suggested by observations on the *E. coli* Uvr proteins. Oh and Grossman (1987) reported that the UvrAB protein complex functions as a DNA helicase in vitro. They suggested that this complex binds near sites of bulky damage in DNA and that limited unwinding of DNA may be required to generate a conformation specifically recognized as substrate by the UvrABC endonuclease. Interestingly, both the *E. coli* UvrAB helicase activity and that of Rad3p have the same (5′→3′) polarity, and both are inhibited by the presence of pyrimidine dimers in duplex DNA (Harosh et al. 1989; Oh and Grossman 1989). The hypermutability phenotype associated with certain *rad3* alleles suggests that the Rad3p DNA helicase might also be required for normal replicational fidelity. Indeed, recent studies have shown that a number of *rad3* alleles that result in marked UV sensitivity, but do not confer the complete Rem phenotype (e.g., lethality in the presence of the *rad52* mutation), result in significantly increased rates of spontaneous reverse mutagenesis at the *lys2-1* locus (Song et al. 1990).

As indicated above, Sung et al. (1988b) have shown that the essential function of the *RAD3* gene is apparently served by some attribute of Rad3p other than its DNA helicase activity. The observation that a gene required for specific parameters of DNA metabolism is also an essential gene suggests that the essential function of *RAD3* is related to a vital aspect of DNA biochemistry such as DNA replication. At present, there is no experimental evidence in support of this hypothesis. On the contrary, total DNA synthesis is unaffected by temperature shifts in the single temperature-sensitive *rad3* mutant characterized thus far (Naumovski and Friedberg 1987). However, these studies do not eliminate the possibility of more subtle defects in DNA replication such as defective initiation. Furthermore, studies on a single mutant are not considered to be conclusive, and there is clearly a need to isolate and characterize more mutants of this type.

The amino acid sequence of Rad3p is homologous to that of the product of the human excision repair gene *ERCC2* (C. Weber and L.H. Thompson, pers. comm.). *ERCC2* was isolated by complementation of an NER-defective Chinese hamster cell line with human genomic DNA. Overall, there is 51.7% identity between the two sequences and 72.5% homology when conservative amino acid substitutions are included. Remarkably, every yeast *rad3* mutation thus far sequenced (a total of 15, including 2 *rem* alleles and the single temperature-sensitive allele referred to earlier) is located in a codon that is conserved in the Rad3p and Ercc2 polypeptides. Presumably, mutations in conserved functional domains are more likely to yield mutants with detectable phenotypes.

This homology between yeast and human NER gene products provides persuasive evidence for the utility of *S. cerevisiae* as a model system to dissect the molecular biology and biochemistry of NER in eukaryotes. This structural conservation also suggests that yeast and human NER proteins may be functionally related. Evidence in support of this idea comes from studies with two other homologous NER genes (see below).

2. RAD1, RAD2, RAD4, *and* RAD10 *Genes*

Considerably less is known about the possible roles of the other four *RAD* genes required for damage-specific incision of UV-irradiated DNA. All of these genes have been cloned and sequenced (Higgins et al. 1983a, 1984; Yasui and Chevallier 1983; Naumovski and Friedberg 1984; Yang and Friedberg 1984; Prakash et al. 1985; Reynolds et al. 1985c, 1987; Weiss and Friedberg 1985; Madura and Prakash 1986; Fleer et al. 1987; Gietz and Prakash 1988; Couto and Friedberg 1989), but the nucleotide and amino acid sequences have been uninformative with respect to potential catalytic or structural functions. Like *RAD3*, *RAD1* is a multi-functional gene with a recently defined role in certain recombinational mechanisms (see Petes et al., this volume). Among the five *RAD* genes considered thus far, *RAD2* has the unique property of inducibility follow-ing exposure of cells to a variety of DNA-damaging agents (see below, Section VII).

Translation of the nucleotide sequences of these four *RAD* genes pre-dicts the expression of polypeptides ~126 kD (Rad1p), ~118 kD (Rad2p), ~87 kD (Rad4p), and ~24 kD (Rad10p) in size (see Friedberg 1988). Rad10p has been purified to more than 90% homogeneity and is a polypeptide of the expected molecular weight (Bardwell et al. 1990). The translated *RAD10* sequence shares about 28% identity with that of the product of a second human NER gene designated *ERCC1* (van Duin et al. 1986). Expression of Rad10p in Chinese hamster cell mutants defec-tive in the rodent *ERCC1* homolog results in partial complementation of both UV sensitivity and defective NER (Lambert et al. 1988). This pro-vides a second example of evolutionary conservation of NER proteins in eukaryotes.

Both *RAD10* and *ERCC1* harbor overlapping antisense transcription units in their 3′regions (van Duin et al. 1989). The biological sig-nificance of these antisense transcripts is unclear. It has been suggested that the *RAD10* antisense transcript might hybridize to the 3′end of *RAD10* mRNA, thereby reducing its rate of translation. This suggestion is consistent with the observation that *RAD10* is normally very weakly expressed, like the other *RAD* genes considered thus far (see Friedberg 1988).

3. Other Genes Involved in NER

As indicated above, some mutants in the *RAD3* epistasis group show a deficiency rather than a complete defect in early biochemical events as-sociated with NER. Correspondingly, these mutants confer a phenotype of only modest sensitivity to DNA-damaging agents. Hence, molecular cloning of the wild-type genes by phenotypic complementation has been difficult. One member of this group of genes (*RAD7*) maps genetically very close to a cytochrome oxidase gene (McKnight et al. 1981). Cloning of the latter gene facilitated chromosome walking to *RAD7* (Perozzi and Prakash 1986). The cloned *RAD7* gene has an ORF of 565 codons and could encode a protein 63.7 kD in size. The amino acid sequence of the translated gene provides no indication of the function of Rad7p.

Disruption of *RAD7* confers a phenotype identical to that conferred by point mutants. Hence, the limited UV sensitivity of *rad7* mutants (and presumably of other mutants in this group) and their residual capacity for NER are apparently not the result of simple leakiness. The products of these genes may have regulatory roles in the biochemistry of NER, or, like the *uvrD* and *polA* gene products of *E. coli* discussed earlier, they may be principally involved in postincision events that nevertheless affect the overall kinetics and extent of damage-specific incision. Another possible explanation for the role of these genes in NER derives from recent observations with mammalian cells. Highly UV-sensitive mammalian cell mutants defective in NER are unable to excise either pyrimidine(6-4)pyrimidone photoproducts or cyclobutyl pyrimidine dimers (Cleaver et al. 1987; Thompson et al. 1989). However, some mutants exhibiting a less dramatic UV sensitivity repair (6-4) photoproducts normally but are defective in the repair of pyrimidine dimers (Cleaver et al. 1987; Thompson et al. 1989). These observations suggest that specific components of the mammalian NER complex may determine the relative affinity of the damage-specific endonuclease for different types of distortive damage in DNA. The *RAD7* group of genes may play a similar role in yeast. On the basis of this model, one would predict that the *rad7, rad14, rad16, rad23*, and *mms19* mutants might be defective in NER of pyrimidine dimers, but not of (6-4) photoproducts.

Recent studies have shown that, like mammalian cells (see Hanawalt 1987; Bohr and Wassermann 1988), yeast cells excise pyrimidine dimers from actively transcribing genes at a faster rate than from transcriptionally silent genes. Terleth et al. (1989) showed that the transcriptionally active *MATα* locus was repaired about 2.5 times faster than the silent *HMLα* locus. This preference was lost in a *sir3* mutant, in which both genes are transcriptionally active.

The availability of yeast mutants defective for selected aspects of transcription offers the potential for establishing whether the preferential repair of actively transcribed genes reflects a NER mode specifically associated with transcription or simply reflects accessibility of the large multiprotein repairosome to the open conformation of transcribing DNA.

The former explanation would provide another possible role for the *RAD7* group of genes in NER.

C. Enzymes Involved in Excision Repair

Several yeast enzymes likely to be involved in excision repair (either nucleotide excision or base excision) have been purified and characterized. With the exception of uracil-DNA glycosylase and AP endonuclease, the genes that encode these enzymes are unknown. Hence, functional information potentially derived from the characterization of mutant phenotypes is unavailable as yet.

The removal of 3′-deoxyribose fragments and restoration of normal 3′-OH termini is a likely early step in the repair of certain types of DNA strand breaks induced by free radicals generated by ionizing radiation or oxidizing agents. Using a synthetic DNA substrate containing 3′ phosphoglycoaldehyde esters, Johnson and Demple (1988a) isolated a 40.5-kD yeast protein designated DNA diesterase. The enzyme hydrolyzes a variety of 3′ esters and attacks DNA treated with H_2O_2 or the X-ray-mimetic agent bleomycin (Johnson and Demple 1988b). This DNA diesterase also contains a quantitatively major apurinic/apyrimidinic (AP) endonuclease activity (Johnson and Demple 1988b). The enzyme is not induced by oxidative damage and is not regulated during the cell cycle (Johnson and Demple 1988a). The enzyme can excise a variety of 3′ esters in DNA, including 3′-phosphoglycoaldehyde, 3′-phosphoryl groups, and 3′-α,β-unsaturated aldehydes. The enzyme hydrolyzes DNA 5′ to AP sites, generating 3′-OH and 5′-deoxyribose-phosphate termini. The protein is optimally stimulated by Co^{++}.

The structural gene encoding this AP endonuclease (designated *APN1*, for *AP* endonuclease), has been cloned and sequenced (Popoff et al. 1990). Gene disruption confirms this as the major, if not the exclusive, AP endonuclease in yeast. The predicted amino acid sequence of the *APN1* gene product is homologous to that of the *E. coli nfo* gene product, which is an AP endonuclease called endonuclease IV.

Other yeast AP endonucleases have been purified and characterized to varying degrees (Piñon 1970; Chlebowicz and Jachymczyk 1977; Armel and Wallace 1978, 1984; Bryant and Haynes 1978; Futcher and Morgan 1979; Thielmann and Hess 1981; Akhmedov et al. 1982; Chang et al. 1987). Additionally, a so-called redoxyendonuclease, which attacks osmium tetroxide-treated or UV-irradiated DNA at sites of thymine glycols and other photoproducts, has been purified from yeast (Gossett et al. 1988). Several other enzymes, possibly involved in recombinational repair of DNA, are discussed in the next section.

The many DNA glycosylases isolated from prokaryotes and higher eukaryotes (see Friedberg 1985a) are presumably also present in yeast; however, to date, the only such yeast enzyme characterized extensively is uracil-DNA glycosylase (Crosby et al. 1981). This enzyme catalyzes the exclusive excision of uracil from DNA by cleavage of the *N*-glycosylic bond linking this base to the deoxyribose-phosphate backbone. The isolation of a yeast mutant defective in this enzyme (Burgers and Klein 1986) facilitated the molecular cloning of the *UNG1* gene by complementation of defective enzyme activity (Percival et al. 1989). The cloned gene has an ORF that could encode a protein of about 40.5 kD, a size considerably larger than the 27.8-kD protein purified from yeast cells. This suggests that the primary translation product is modified in vivo and/or that the enzyme is sensitive to proteolyic degradation in vitro (Percival et al. 1989).

The amino acid sequence of the carboxy-terminal two thirds of the yeast protein is homologous to that of the entire *E. coli* protein. Genetic mapping has localized the *UNG1* gene to the left arm of chromosome XIII. Deletion of the *UNG1* gene has no effect on the viability of haploid yeast cells.

V. RECOMBINATIONAL REPAIR

Recombinational repair is believed to be required for the repair of DNA double-strand breaks (Ho 1975; Resnick and Martin 1976). Hence, mutations in the genes involved (*RAD50–RAD57*) confer sensitivity to X-rays and to chemicals that cause strand breaks in DNA (see Haynes and Kunz 1981; Friedberg 1988). Many of these genes also are required in meiotic recombination, most types of gene conversion and mitotic recombination, and the integration of gapped or linearized plasmids (Haynes and Kunz 1981; Orr-Weaver and Szostak 1985; Petes et al., this volume). Additionally, *RAD51*, *RAD52*, and *RAD54* (although not *RAD50*, *RAD55*, and *RAD57*) are involved in recombinational events associated with mating-type switching initiated by double-strand breaks catalyzed by the HO endonuclease (Malone 1983 and pers. comm.).

These findings have provided the premise for a model postulating that the essential recombinogenic substrate in DNA is a double-strand break (Szostak et al. 1983). Because double-strand breaks that result from genomic injury or as DNA repair intermediates are probably also resolved by a recombinational pathway(s), a complete consideration of cellular responses to DNA damage requires discussion of the genetics and enzymology of recombination. However, because this topic is examined in detail by Petes et al. (this volume), this discussion is not recapitulated here. Nevertheless, the reader is encouraged to regard the information presented in Chapter 8 as an integral component of the analysis of the repair of DNA damage in yeast.

VI. MUTAGENESIS

A consideration of the topic of mutagenesis in yeast requires a consideration of the *RAD6* epistasis group, perhaps the most complex and certainly the least understood of the three epistasis groups (Table 1). In part, this complexity reflects the assignment of all genes that confer sensitivity to both UV irradiation and γ-rays to this group. Direct evidence for epistatic interaction is lacking for a number of these genes, and nonepistatic interactions have been shown in some instances (Game and Mortimer 1974; Schiestl et al. 1989). None of the mutants in this epistasis group thus far studied are defective in NER (Prakash 1977; Reynolds and Friedberg 1981); however, there is evidence for defective postreplicative repair in two cases (*rad6* and *rad18*) (DiCaprio and Cox 1981; Prakash 1981).

The *RAD6* group includes most of the mutants defective in damage-induced mutagenesis (for review, see Lemontt 1980; Lawrence 1982). Therefore, this group is often equated with an error-prone pathway of repair, a term that should be avoided, as it implies a specific mutational mechanism analogous to SOS repair in *E. coli*. Furthermore, some of the *RAD6* group of genes clearly do not influence damage-induced mutagenesis (Fabre et al. 1989; Schiestl et al. 1989).

A. Genes in the *RAD6* Epistasis Group

1. RAD6 *Gene*

rad6 mutants are characterized by a pleiotropic phenotype. They are sensitive to a variety of DNA-damaging agents, including UV light, γ-rays, alkylating agents, and cross-linking agents (Cox and Parry 1968; Game and Mortimer 1974; Prakash 1974, 1976; Henriques and Moustacchi 1981; Siede and Brendel 1982). *rad6* mutants are also spontaneous mutators (Hastings et al. 1976); however, they do not show damage-induced mutability, regardless of the type of damage or the type of mutation scored (see Lemontt 1980; Prakash and Prakash 1980; Lawrence 1982). A diploid *rad6-1/rad6-1* strain does not sporulate or carry out meiotic recombination. On the other hand, the *rad6-3/rad6-3* strain is proficient in these functions (Montelone et al. 1981). Both strains show increased spontaneous and damage-induced mitotic recombination (Hunnable and Cox 1971; Kern and Zimmermann 1978). Further evidence in support of the multifunctionality of the *RAD6* gene comes from observations that suppression of the sensitivity of a *rad6* mutant to UV radiation or to trimethoprim does not affect its sensitivity to γ radiation or its defect in mutagenesis (Lawrence and Christensen 1979; Tuite and Cox 1981).

The *RAD6* gene was cloned by complementing the methylmethanesulfonate (MMS) sensitivity of a *rad6* mutant (Prakash et al. 1983; Kupiec and Simchen 1984). Disruption of the chromosomal gene is not lethal, but such mutants are characterized by low plating efficiency and a prolonged S phase (Kupiec and Simchen 1984). *RAD6* is transcriptionally regulated during the cell cycle, and an increase in the steady-state level of *RAD6* mRNA has also been observed early in meiosis (Kupiec and Simchen 1986a).

The *RAD6* ORF comprises 172 codons, and the molecular mass of the translated gene product is about 19.7 kD. The predicted polypeptide contains a highly acidic carboxyl terminus with 13 consecutive aspartic acid residues (Reynolds et al. 1985a). Rad6p has been purified to homogeneity (Morrison et al. 1988). Deletion of the acidic tail yields a mutant protein designated Rad6-149p. This deletion leads to defective sporulation, but UV sensitivity and mutability are unaffected (Morrison et al. 1988).

An entirely different line of investigation revealed an unexpected function for Rad6p. Varshavsky and his colleagues have investigated the role of ubiquitin in yeast (see Finley, these volumes). Ubiquitin is a highly conserved protein of 76 residues, and its conjugation to certain proteins triggers their degradation (Finley and Varshavsky 1985). Reduced ubiquitin content in yeast leads to defective sporulation and sensitivity to heat stress and starvation (Finley et al. 1987). Components of the ubiquitin-ligase system (E1, E2 enzymes) have been purified from yeast extracts (Jentsch et al. 1987). One of five E2 enzymes identified was found to transfer activated ubiquitin to histones H2A and H2B. This protein was partially sequenced, and its identity with Rad6p was established. Additionally, extracts of *E. coli* in which *RAD6* is overexpressed ubiquitinate histone H2B in vitro (Jentsch et al. 1987).

These results have led to the working hypothesis that Rad6p-mediated ubiquitination of chromosomal proteins might be involved in metabolic transactions required for DNA replication and sporulation and DNA repair and mutagenesis. Sung et al. (1988a) reported that Rad6p polyubiquitinates histones H2A and H2B and that the polyacidic tail is essential for this reaction. However, although the *rad6-149* mutant (deleted of the polyacidic tail) is defective in sporulation, its repair characteristics are unaffected. Hence, histone modification may only be essential for sporulation, and other ubiquitinated proteins may mediate the role of *RAD6* in DNA repair and mutagenesis. Alternatively, *RAD6* may play a completely different role in cellular responses to DNA damage.

The *CDC34* gene product has been identified as a second E2 enzyme. It shares predicted amino acid sequence similarity with Rad6p (Goebl et al. 1988). No DNA repair defect has been reported in *cdc34* mutants. The ubiquitin-conjugating system is highly conserved among eukaryotes (Jentsch et al. 1987); hence, studies in yeast might have important implications for repair phenomena in higher eukaryotic cells. A Rad6p-specific antibody has been shown to recognize proteins from *Drosophila* and human cells (Prakash and Morrison 1988).

2. RAD18 *Gene*

Like *rad6* mutants, *rad18* mutants are highly sensitive to UV light, γ-rays, and certain chemicals (see Haynes and Kunz 1981; Lawrence 1982). Unlike *RAD6*, mutations in *RAD18* do not influence UV-induced mutagenesis, sporulation, or meiotic recombination (Resnick 1969; Game and Mortimer 1974; Lawrence and Christensen 1976). However, spontaneous mitotic recombination and mutagenesis are enhanced (Boram and Roman 1976; Quah et al. 1980). X-ray-induced strand breaks are rejoined in *rad18* mutants, but such mutants show abnormalities in reconstituting high-molecular-weight DNA during postirradiation incubation (Mowat et al. 1983). Defective repair of γ-ray-induced base damage has also been observed recently in *rad18* mutants (Eckardt-Schupp et al. 1988).

The *RAD18* gene has been cloned and sequenced (Chanet et al. 1988; Jones et al. 1988; Fabre et al. 1989). Disruption mutations are nonlethal, and the associated phenotypes resemble those of the most UV-sensitive *rad18* point mutations (Jones et al. 1988; Fabre et al. 1989). The *RAD18* ORF can encode a protein of 487 amino acids with a predicted molecular mass of approximately 55.5 kD. The deduced protein sequence has a number of interesting features (Chanet et al. 1988; Jones et al. 1988). Three regions that resemble DNA-binding domains (zinc fingers) have been identified. Additionally, the sequence contains a consensus nucleotide-binding domain. The predicted protein has two acidic regions: one near the carboxyl terminus and another closer to the amino-terminal end, a distribution that resembles that of several known transcriptional activator proteins in yeast. No homology with ubiquitin-conjugating enzymes is evident (Jones et al. 1988). At present, it is not obvious how these structural motifs might contribute to the role of *RAD18* in DNA repair.

3. RAD9 Gene

Certain DNA-damaging agents, particularly those that induce DNA strand breaks, are known to cause G_2 arrest of asynchronously dividing wild-type yeast cultures (Kupiec and Simchen 1985). Weinert and Hartwell (1988) demonstrated that the rad9-1 mutant is defective in G_2 arrest after X-ray treatment. Additionally, extended postirradiation treatment of rad9 cells with the microtubule-inhibiting drug methylbenzimidazole-2-yl-carbamate before plating suppresses the X-ray sensitivity of the mutant. These observations suggest that the X-ray sensitivity of rad9-1 cells is not due to a defect in DNA repair per se but results from the failure of cells to arrest in G_2, the phase in which repair takes place in wild-type cells. Schiestl et al. (1989) have confirmed these results with MMS-treated cells, and defective cell-cycle arrest in a rad9 background was also observed when DNA chain elongation was blocked by shifting the DNA ligase mutant cdc9 to the restrictive temperature.

The RAD9 gene has been cloned and sequenced (Schiestl et al. 1989). The predicted protein consists of 1309 amino acids with a calculated M_r of 148,412. Deletion mutations are nonlethal and produce UV and γ-ray sensitivity comparable to that of the rad9-1 mutant. Deletion of RAD9 had no effect on spontaneous or UV-induced mutagenesis or on recombination (Schiestl et al. 1989). Hence, apparently neither RAD9 nor RAD18 plays a significant role in damage-induced recombination or mutagenesis.

4. REV Genes

Several schemes have been employed to isolate mutants defective in mutagenesis after DNA-damaging treatments. The rev mutants (for defective mutation reversion) were identified by screening for clones showing deficiencies in reversion of various point mutant alleles (Lemontt 1971; Lawrence et al. 1985a,b). The less well characterized umr mutants (for UV mutation resistance) were identified by screening for deficiencies in induction of forward mutations (Lemontt 1977).

These mutants are clearly of interest in elucidating the mechanism(s) of damage-induced mutagenesis in yeast. In general, rev mutations confer only moderate sensitivity to UV light, γ-irradiation, or chemicals (Lemontt 1971; Lawrence et al. 1985a,b). Three mutants (rev3, rev6, and rev7) have a pronounced deficiency in mutation induction by these agents (Lemontt 1971, 1972, 1980; Lawrence et al. 1984, 1985a,b). The rev3 mutant is proficient in postreplication repair (Prakash 1981). Other mutants (rev1, rev2 [rad5], rev4, and rev5) have a more subtle effect, and mutation deficiency depends on the particular marker scored (Lemontt

1972, 1980; Lawrence and Christensen 1978a,b; McKee and Lawrence 1979; Lawrence et al. 1985a,b).

The *REV1* and *REV3* genes have been cloned and sequenced (Larimer et al. 1989; Morrison et al. 1989). A plasmid has also been isolated that complements the *rev2* mutation (Siede and Eckardt-Schupp 1986a). *REV3* encodes a large protein with a predicted molecular mass of approximately 173 kD (Morrison et al. 1989). Its function as a DNA polymerase has been inferred from sequence homology with Epstein-Barr virus DNA polymerase, yeast DNA polymerase I, and human DNA polymerase α (Morrison et al. 1989). However, it is apparently a non-essential DNA polymerase (perhaps exclusively concerned with replication of damaged DNA templates), because deletion mutations are not lethal.

The *REV1* gene encodes a protein of 985 amino acid residues (~112.2 kD). Interestingly, some deletion mutations cause less UV sensitivity than does the missense point mutation *rev1-1*. Total deletion of *REV1* had no effect on viability. A 152-amino-acid domain of the predicted Rev1p shares 25% identity with the *E. coli* UmuC protein (Larimer et al. 1989), which is required for damage-induced mutagenesis in this prokaryote.

B. DNA Damage-induced Mutagenesis

Despite extensive inroads into its genetics and molecular biology, the molecular mechanism(s) of damage-induced mutagenesis has defied biochemical elucidation, even in *E. coli*. The reader with a special interest in mutagenesis in yeast is referred to earlier reviews on this topic (Lemontt 1980; Lawrence 1982). In the present discussion, we confine ourselves to a summary of some of the more challenging observations concerning this aspect of cellular responses to DNA damage. Our remarks are focused mainly on UV-radiation-induced mutagenesis, and we have avoided borrowing potentially confusing and probably mechanistically inappropriate terminology from *E. coli* (e.g., inducible or SOS repair), because such processes have not been specifically demonstrated in yeast.

About 30 mutations are known to affect, directly or indirectly, events that lead to elevated mutation frequencies after exposing cells to DNA-damaging agents (Table 3). Only in several cases (e.g., *rad6, pso2*) is the phenotype of defective mutability associated with a marked sensitivity to killing by radiation and/or chemicals. Most nonmutable mutants are only slightly UV-sensitive. Hence, at least for UV radiation, it is reasonable to conclude that the gene products and processes involved in mutagenesis do not contribute significantly to cellular resistance.

Many of these mutants have locus-specific effects; i.e., defective mutagenesis is not observed for all reversion and forward systems examined (see Lemontt 1980). Neither the type of mutation generated nor the type of DNA damage used accounts for this unexplained specificity (Lemontt 1980; Lawrence et al. 1984).

In contrast to *E. coli*, in which the *recA* gene plays a pivotal role in both recombination and mutagenesis, mutagenesis in yeast is generally independent of recombinational repair, and the quest for the yeast equivalent of *recA* has been fruitless. Mutants known to be severely deficient in recombination (e.g., the *rad52* group and *rad1*) are not defective in damage-induced mutagenesis (Lawrence and Christensen 1976), and the recombinational deficiency of *rad6* mutants is confined to meiosis (Montelone et al. 1981). However, a general deficiency in damage induction of mutation and recombination has been reported recently in a *pso4* mutant (Andrade et al. 1989). Although further characterization of this mutant is clearly in order, this mutation might affect a function common to both processes.

A new gene apparently involved in damage-induced mutagenesis (designated *RADH*) has been isolated recently (Aboussekhra et al. 1989). The gene was originally identified as a suppressor of the radiation sensitivity of a *rad18* deletion mutant. Haploid *radH* mutants are UV-sensitive, but only if irradiated in the G_1 mitotic phase. These mutants are also severely defective in UV-induced mutagenesis. Haploid mutants are resistant to γ radiation and to simple alkylating agents such as MMS. However, homozygous diploid cells are highly sensitive to these agents. The *RADH* gene was cloned by phenotypic complementation. The cloned gene contains an ORF of 1175 codons, which could encode a polypeptide of about 134.2 kD. The predicted amino acid sequence shows homology with several DNA helicases, notably, the *E. coli* Rep and UvrD proteins. Additionally, the amino acid sequence of Radhp contains a leucine zipper motif, suggesting that it interacts with itself and/or other proteins.

Not unexpectedly, proficiency for NER reduces the mutational effect of many DNA-damaging treatments. Higher mutation frequencies, as well as different dose-response curves, are observed in excision-deficient strains relative to excision-proficient strains (see Eckardt and Haynes 1977a). The general conclusion that excision repair is strictly error-free is, however, unwarranted. Functional excision repair alters the spectrum of UV-induced mutations (Ivanov et al. 1986) and plays an essential role in the kinetics of damage-induced mutagenesis. After UV radiation, NER proficiency results mainly in prereplicative fixation of premutagenic lesions, whereas a substantial increase in postreplicative mutation fixation is evident in an excision-deficient background (Eckardt and Haynes

Table 3 Genes involved in DNA-damage-induced mutagenesis in yeast

Gene	Characterization	References
RAD6	required for damage-induced mutagenesis, meiotic recombination, and sporulation; mutants are UV-sensitive; encodes ubiquitin-conjugating enzyme; nonessential gene	6, 9, 10, 13, 21, 22, 27, 28, 33
RAD8	required for mutation induction by UV radiation in diploid cells	10
RAD13	mutants have locus-specific defects in mutation induction by UV radiation in diploid cells	10
RAD15	required for damage-induced mutagenesis, especially after nitrous acid treatment	28
RAD17	required for damage-induced mutagenesis, especially after 4NQO treatment	28
RADH	required for UV mutagenesis; encodes probable DNA helicase	1
PSO2 (SNM1)	required for damage-induced mutagenesis, especially after UV radiation and psoralen treatment; results dependent on pso2 allele used	4, 5, 30
PSO3	required for UV mutagenesis	4
PSO4	required for damage-induced mutagenesis and recombination; mutants are more sensitive in G_2 phase	2
REV1	mutants have locus-specific defects in mutation induction	8, 11, 17, 18, 21
REV2 (RAD5, SNM2)	mutants have locus-specific defects in mutation induction; temperature-sensitive mutant characterized in detail	12, 17, 18, 21, 31, 32

Gene	Description	References
REV3 (PSO1)	required for damage-induced mutagenesis; encodes nonessential DNA polymerase	10, 17, 18, 21, 23
REV4	mutants have locus-specific defects in mutation induction	14
REV5	mutants have locus-specific defects in mutation induction	14
REV6	required for damage-induced mutagenesis; mutants are UV-sensitive	14
REV7	required for damage-induced mutagenesis; mutants are UV-sensitive	15, 16
UMR1↑UMR7	mutants are defective in UV-induced forward mutations (*can1*); only slightly UV sensitive; more general defects in *umr1* and *umr3*; sporulation defects	19
CDC7	required for UV mutagenesis; encodes a protein kinase necessary for initiation of DNA synthesis	3, 25, 26
CDC8	required for UV mutagenesis; encodes thymidylate kinase	7, 29
MMS3	mutants are defective in mutation induction only in diploid cells	20
PRB1	mutants defective in mutation induction; encodes proteinase B	24

References: (1) Aboussekhra et al. 1989; (2) Andrade et al. 1989; (3) Bahman et al. 1988; (4) Cassier et al. 1980; (5) Cassier and Moustacchi 1981; (6) Jentsch et al. 1987; (7) Jong et al. 1984; (8) Larimer et al. 1989; (9) Lawrence et al. 1974; (10) Lawrence and Christensen 1976; (11) Lawrence and Christensen 1978a; (12) Lawrence and Christensen 1978b; (13) Lawrence et al. 1984; (14) Lawrence et al. 1985a; (15) Lawrence et al. 1985b; (16) Lawrence et al. 1985c; (17) Lemontt 1971; (18) Lemontt 1972; (19) Lemontt 1977; (20) Martin et al. 1981; (21) McKee and Lawrence 1979; (22) Montelone et al. 1981; (23) Morrison et al. 1989; (24) E. Moustacchi, pers. comm.; (25) Njagi and Kilbey 1982; (26) Patterson et al. 1986; (27) Prakash 1974; (28) Prakash 1976; (29) Prakash et al. 1979; (30) A. Ruhland et al., unpubl.; (31) Siede and Eckardt-Schupp 1986a; (32) Siede and Eckardt-Schupp 1986b; (33) Sung et al. 1988a.

1977b; James and Kilbey 1977; James et al. 1978; Kilbey et al. 1978; Eckardt et al. 1980; Nasim et al. 1981; Siede and Eckardt 1986b).

The issue of regulated gene functions associated with damage-induced mutagenesis in yeast is still an open question (see Siede and Eckardt 1984). Some evidence suggests the involvement of inducible gene functions during UV mutagenesis. Thus, for example, if a "conditioning" dose of UV radiation is administered to cells prior to a challenging dose, mutability and UV resistance are enhanced (Eckardt et al. 1978). These effects are eliminated in the presence of cycloheximide after the conditioning treatment (Eckardt et al. 1978). However, in recent studies, similar experimental protocols did not enhance mutability in a forward mutation system using the *CAN1* gene (Schenk et al. 1989).

In stationary-phase, repair-competent yeast cells exposed to UV radiation, reverse mutations arise with biphasic, linear-quadratic kinetics relative to the UV dose. Extended incubation in the presence of cycloheximide eliminates the quadratic component (Eckardt et al. 1978). These observations have been interpreted as evidence for both constitutive and induced components of the mutational process (Haynes and Eckardt 1980; Haynes et al. 1985). Induction of gene functions is also suggested by the inhibitory effect of cycloheximide on *REV2*-dependent UV mutagenesis and by the altered kinetics of mutagenesis observed in a thermoconditional *rev2* mutant (Siede et al. 1983a,b; Siede and Eckardt 1986a).

The spectrum of both spontaneous and UV-induced forward mutations in yeast has been documented in several experimental systems (Kunz et al. 1987; Lee et al. 1988). Kunz et al. (1987) reported a predominance of UV-induced transition mutations at sites of adjacent pyrimidines in the *SUP4-o* gene. Because adjacent pyrimidines are required both for pyrimidine dimer and for pyrimidine(6-4)pyrimidone photoproducts, these results suggest that most UV-induced mutations were targeted to sites of DNA damage. The majority of base changes occurred at the 3′ base of potential UV photoproducts, preferentially TC and CC sites, which are the predominant targets for 6-4 photoproducts (Brash and Haseltine 1982). However, pyrimidine dimers most likely account for a substantial fraction of transition mutations at CT and TT sequences. Deletions and double mutations were also recovered. The distribution of UV mutations in the *SUP4-o* gene was not random; several hot spots were identified.

The mutational spectrum has also been measured in the *URA3* gene carried on an integrating plasmid. Using this system, Lee et al. (1988) observed a preference for transitions at adjacent pyrimidines, mainly at TT sequences. However, in contrast to results with the *SUP4-o* gene, the fraction of transversion mutations was substantially higher, and there was no suggestion of hot spots.

A third experimental system has shown that untargeted mutations also occur following UV radiation of yeast. Lawrence and Christensen (1982) irradiated a haploid strain carrying a nonrevertible (deletion) mutation in the *CYC1* gene (*cyc1-363*) and then mated it with an unirradiated heteroallelic partner carrying a point mutation (*cyc1-91*) in the region deleted in *cyc1-363*. Hence, Cyc+ revertants among the diploid progeny must result from untargeted mutations, i.e., mutations outside the *cyc1-91* locus. Using this protocol in an excision-defective background, they observed a mutation frequency approximately 40% of that obtained by irradiation of the *cyc1-91* parent alone. Untargeted mutations were eliminated in a *kar* mutant in which nuclear fusion was prevented.

Studies on the biochemistry of damage-induced mutagenesis are in their infancy. As indicated earlier, the observation that Rad6p is a ubiquitin-conjugating enzyme (Jentsch et al. 1987) suggests potentially interesting regulatory and/or structural roles in damage-induced mutagenesis. The recent identification of the *REV3* gene product as a nonessential DNA polymerase represents another important step toward our understanding of mutational responses in yeast (Morrison et al. 1989). The *CDC7* gene encodes a protein kinase essential for the initiation of DNA synthesis (Patterson et al. 1986; Bahman et al. 1988) and represents another replicative component possibly concerned with mutation induction. It has been reported that overexpression of *CDC7* leads to increased mutagenesis (Patterson et al. 1986). The homology between the amino acid sequence of the *RADH* gene and that of several DNA helicases (Aboussekhra et al. 1989) suggests a third distinct biochemical component required for replication fidelity. Finally, as indicated above, Rev1p has a sequence motif similar to that in the *E. coli* UmuCD proteins (Larimer et al. 1989), and overexpression of the *umuCD*-related *mucAB* genes in yeast results in a moderate increase in spontaneous and damage-induced mutagenesis (Potter et al. 1984).

These fragmentary clues suggest that yeast may assemble a specialized replicative complex that mediates *trans*-lesion DNA replication in a manner similar to that postulated in *E. coli* (Morrison et al. 1989). Clearly, much more needs to be learned about the gene products involved, their regulation, and the interrelations between DNA replication, NER, and heteroduplex repair.

C. Spontaneous Mutagenesis and Mismatch Repair

Spontaneous mutagenesis defines mutagenic processes that arise in the absence of deliberate exposure of cells to DNA damage. Although typically considered in the context of replicative and/or recombinational errors, it should be remembered that mutations can also arise following spontaneous chemical damage (e.g., spontaneous hydrolysis of bases), and the molecular mechanism(s) of these mutations may be indistinguishable from that of damage-induced mutagenesis as discussed above (see Sargentini and Smith 1985).

Perturbations that alter the fidelity of DNA replication are likely to result in a marked mutator phenotype. Such perturbations may involve the replicative machinery directly or may affect replication through less direct mechanisms, for example, by imbalance of dNTP pools. Indeed, a yeast mutant deficient in dCMP deaminase has been shown to be a mutator (Maus et al. 1984). The extensive literature on the recombinogenic and mutagenic effects of nucleotide pool imbalances and their influence on repair-related phenomena will not be considered in this review. The interested reader is referred to the studies of Haynes and his co-workers (Kunz 1982, 1988; Kunz and Haynes 1982; Eckardt et al. 1983).

Replicational fidelity can be enhanced by postreplicative correction of mispaired bases, a process known as mismatch repair, which has been well characterized in prokaryotes (for reviews, see Claverys and Lacks 1986; Radman and Wagner 1986; Modrich 1987). In this section, we summarize the influence of known DNA repair mutations on spontaneous mutability in yeast and then focus on more recent studies on mismatch repair in this organism. We confine our discussion to replicative mismatches.

DNA repair mutations that enhance spontaneous mutability include mutations in *RAD1–RAD3*, *RAD6*, *RAD18*, *REV2*, *RAD51*, *RAD52*, and *UVS2* (Lawrence 1982; Kunz et al. 1989). As indicated earlier, *RAD3* is of particular interest in this context. The *rem* alleles of *RAD3* confer only slightly increased UV sensitivity; however, frequencies of spontaneous mutations and mitotic recombination are considerably enhanced (Malone and Hoekstra 1984; Hoekstra and Malone 1987; Montelone et al. 1988). The combination of *rad3-101* (formerly designated *rem1-1*) and *rad52*, a mutation causing a deficiency in double-strand break repair, is lethal. But lethality of the double mutant can be prevented by an additional defect in NER (Malone and Hoekstra 1984). This has led to the suggestion that *rem* mutants are defective in the repair of replicative mismatches, resulting in double-strand breaks (due to overlapping excision tracts) that require repair by recombination.

Direct visual screening for mutator strains using the *ade* mutant color system resulted in the isolation of the *mut1* and *mut2* mutants (von Borstel et al. 1973; Gottlieb and von Borstel 1976). Antimutator mutants (*ant*) have also been isolated, and *ant2* has been shown to be an allele of

REV3 (Quah et al. 1980). More recently, mutants were isolated that showed increased postmeiotic segregation (*pms*), an expected phenotype of defective postreplicative mismatch correction (Williamson et al. 1985; Williamson and Fogel 1990). A group of strong mutator mutants (*pms1–pms3*) has been studied in detail, and there is now compelling evidence that these mutants are defective in mismatch repair of DNA. The *pms* mutants also show enhanced levels of meiotic recombination between closely linked markers (Williamson et al. 1985; Williamson and Fogel 1990).

Mismatch correction in yeast has been demonstrated in a number of studies. Muster-Nassal and Kolodner (1986) showed repair of mismatches in a cell-free system. A preference for insertion/deletion mismatches and for AC and GT mismatches was found. The in vitro reaction is dependent on Mg^{++} and partially dependent on ATP and dNTPs. Bishop and Kolodner (1986) observed mismatch repair in vivo by using restriction analysis of heteroduplex plasmid DNA containing 8- or 12-bp insertion mismatches or AC or CT single-base-pair mismatches. Insertion mismatches separated by about 1 kb were repaired independently of each other at least 55% of the time, suggesting that repair tracts are often shorter than 1 kb. In the same system, the mutant strain *pms1-1* showed reduced repair efficiency and a significant disparity in favor of removal rather than copying of the inserted bases (Bishop et al. 1987).

Mismatch repair of all eight possible single-base-pair mismatches, single nucleotide loops, and a 38-bp loop was investigated by B. Kramer et al. (1989). Single-base loops were repaired very efficiently (~95%). Single-base-pair mismatches were corrected with frequencies of 75–95%, with the exception of CC mismatches, which were corrected with the same low efficiency (~30–40%) as the 38-bp loop. A general reduction in mismatch repair efficiency was found in all three *pms* mutants, with the exception of single-base loops, which were efficiently corrected in the *pms3* mutant. Postmeiotic segregation frequencies for *arg4* mutants bearing different alleles correlated with the substrate specificity for mismatch correction (B. Kramer et al. 1989). This specificity contrasts somewhat with that observed in the in vitro studies cited above (Muster-Nassal and Kolodner 1986) and with that observed in mammalian cells (Brown and Jiricny 1988). However, it parallels the situation in *E. coli* and *Streptococcus pneumoniae* (Claverys and Lacks 1986; Radman and Wagner 1986).

Bishop et al. (1989) reported a similar study in yeast, using mismatches constructed in a plasmid-borne *lacZ* gene. Although the results were generally in accord with those of B. Kramer et al. (1989), there are some unexplained differences. AA and TT mismatches were found to be repaired as poorly as CC mismatches. Additionally, a 3-bp mismatch was repaired rather poorly. In related studies, Nag et al. (1989) demonstrated inefficient repair of palindromic insertion mismatches.

The *PMS1* gene has been cloned and sequenced (W. Kramer et al. 1989). Mutants bearing deletions in *PMS1* are viable. In diploid strains homozygous for a *pms1* deletion allele, haploid lethal mutations accumulate. This may account (at least in part) for the low spore viability observed. An ORF of 904 codons was identified that could encode a protein of approximately 103 kD. The deduced amino acid sequence has clusters of extensive similarity with the mismatch repair proteins MutL from *Salmonella typhimurium* and HexB from *S. pneumoniae*. Overall, the homology with these two proteins is 32% and 33%, respectively. These studies indicate conservation of mismatch repair genes among prokaryotes and yeast; however, the molecular mechanisms and biochemistry of postreplicative mismatch correction in yeast have not yet been elucidated. Yeast DNA is not methylated to a significant extent (Proffitt et al. 1984); hence, the strand specificity of repair must be determined some other way. Conceivably, strand breaks could provide the basis for strand specificity, analogous to the system in *S. pneumoniae* (Balganesh and Lacks 1985).

VII. DAMAGE-INDUCIBLE GENES

One of the most interesting aspects of the molecular biology of DNA repair is the question of its regulation. In *E. coli*, four major regulatory systems have been identified that control the expression of sets of genes coordinately induced by DNA damage or environmental stress: the SOS response, the adaptive response to alkylation damage, the response to oxidative damage, and the heat-shock response (see Walker 1985). Of these regulatory systems, the SOS response is understood best and includes regulation of the excision repair genes *uvrA* and *uvrB* (see Walker 1985). Studies on damage-inducible genes in yeast are just beginning. However, the available data do not indicate any obvious resemblance to the SOS system, in which genes are negatively regulated by the LexA repressor, and derepression is effected by cleavage of LexA protein following activation of the protease function of RecA protein.

DNA damage inducibilty of genes in *S. cerevisiae* has been demonstrated in several studies. Ruby and Szostak (1985) isolated several inducible genes from a plasmid library that contained random fragments of yeast genomic DNA fused to the *E. coli lacZ* gene; transformants were identified that displayed enhanced expression of β-galactosidase activity after treatment with the UV-mimetic chemical 4-

nitroquinoline-1-oxide (4NQO). These genes are designated *DIN1–DIN6* (for *d*amage *in*ducible). From the efficiency of their screen, these investigators concluded that as many as 50 yeast genes may be inducible by such treatment. McClanahan and McEntee (1984) isolated damage-inducible genes, designated *DDR* (for *D*NA *d*amage *r*esponsive), by differential plaque hybridization, using cDNA made from mRNA isolated from treated and untreated cells.

The nature and function of the majority of these genes are not known. Nevertheless, these studies have generated several interesting general insights. First, multiple and diverse DNA-damaging agents can induce expression of genes, but the spectrum of inducing agents varies considerably for different damage-inducible genes (McClanahan and McEntee 1984; Ruby and Szostak 1985). Second, constitutive expression and inducibility of genes can be influenced by the DNA repair capacity of cells. Thus, for example, induction of the *DDRA2* transcript by 4NQO or nitrosoguanidine is dependent on a functional *RAD3* gene (Maga et al. 1986). Third, despite some overlap, heat-shock treatment induces a set of genes different from that induced by DNA-damaging agents (see McClanahan and McEntee 1986).

Among the *DDR* genes, one was found to be present in multiple copies in the yeast genome and to be homologous to Ty elements (McClanahan and McEntee 1984). Subsequently, Ty transcription was shown to be inducible by UV radiation and other DNA-damaging agents (Rolfe 1985; Rolfe et al. 1986; Bradshaw and McEntee 1989). A corresponding increase in insertion mutations of the *ADH1* promoter mediated by Ty transposition was detected (Morawetz 1987; Bradshaw and McEntee 1989).

Other yeast genes of known function are inducible by DNA-damaging treatments. These include *RNR2*, a gene that encodes the small subunit of ribonucleotide reductase (Elledge and Davis 1987); *CDC8*, which encodes thymidylate kinase (Elledge and Davis 1987); *POL1*, the gene for yeast DNA polymerase I (Johnston et al. 1987); and *CDC9*, which encodes DNA ligase (Peterson et al. 1985). *RNR2* and *CDC8* could play indirect roles in DNA repair by providing precursors for repair synthesis. More direct roles are anticipated for *CDC9* and, possibly, *POL1*. Repair defects have been described in *cdc9* mutants (Johnston 1979); however, studies with the *POL1* gene have not shown its requirement in UV or X-ray repair (Budd et al. 1989).

The *CDC8*, *CDC9*, and *POL1* genes are also cell-cycle-regulated (Peterson et al. 1985; White et al. 1986, 1987; Johnston et al. 1987). For cell-cycle-regulated genes, it is important to establish that enhanced gene expression following exposure to DNA-damaging agents is due to transcriptional induction as a direct response to the damage, rather than partial synchronization of asynchronously dividing cultures by cell-cycle arrest (Kupiec and Simchen 1985). In the case of *CDC9*, enhanced expression following exposure to UV radiation has been demonstrated in noncycling stationary-phase cultures (Johnson et al. 1986) and, hence, clearly cannot be the result simply of synchronization by cell-cycle arrest.

The *UBI4* gene, which encodes polyubiquitin, is required for protein degradation and resistance to starvation conditions (Finley et al. 1987; Finley, these volumes). This gene is induced by 4NQO treatment and also in meiosis (Treger et al. 1988). Proteinase B activity is also induced by UV radiation (Schwencke and Moustacchi 1982). These results suggest that some genes might be induced in response to aberrant proteins generated as the result of DNA- and/or protein-damaging treatments.

Increased activities of catalase and cytochrome P-450 were observed after UV treatment of logarithmically growing diploid yeast cells (Morichetti et al. 1989). These observations provide the first indications for an inducible response to oxidative stress, which might overlap with the network(s) of DNA-damage-inducible genes. Although several yeast genes are induced following treatment with alkylating agents, there is no evidence for an adaptive response to alkylating agents analogous to that in *E. coli* (Maga and McEntee 1985; Polakowska et al. 1986)

Of the many yeast genes known to be involved in DNA repair, two have been shown to be transcriptionally induced by DNA-damaging agents: *RAD2*, a gene necessary for damage-specific incision of DNA during NER (discussed previously in Section IV; Madura and Prakash 1986; Robinson et al. 1986), and *RAD54*, required for double-strand break repair, as well as meiotic and mitotic recombination (see above, Section V; Cole et al. 1987; Petes et al., this volume). The steady-state levels of transcripts encoded by these genes increase as much as sixfold after treatment with a variety of DNA-damaging agents, including some to which the corresponding mutants are not sensitive (e.g., γ-rays in the case of *rad2* mutants and UV radiation in the case of *rad54* mutants).

In both cases, deletion analysis of the promoter regions demonstrated gene regulation by positive control (Cole and Mortimer 1989; Siede et al. 1989). In the case of *RAD2*, four extremely AT-rich tracts are required for constitutive expression of this weakly expressed gene (Siede et al. 1989). Deletion of these elements results in loss of UV inducibility and reduced constitutive expression. AT-rich elements are common to many yeast promoters, and it has been suggested that they may activate transcription by exclusion of nucleosomes (Chen et al. 1987). However, recent in vitro transcription studies have demonstrated sequence-specific binding of *trans*-acting factors to certain AT-rich regions (see Lue et al.

1989), and a protein (called datin) that binds specifically to oligo[d(AT)] tracts has recently been purified (Winter and Varshavsky 1989). This protein requires an uninterrupted stretch of at least nine A or T residues for binding (Winter and Varshavsky 1989). All four AT-rich tracts in the *RAD2* promoter contain such stretches.

Two apparently homologous sequences (with the consensus motif AGGNATTPuAAA), located ~70 and ~140 bp upstream of the translational start site in *RAD2*, are more likely to be *cis*-acting regions for DNA-damage-specific regulation (Siede et al. 1989). Recent gel retardation studies have confirmed this interpretation (W. Siede and E.C. Friedberg, in prep.). Deletion of either of these sequences results in defective or reduced inducibility following UV radiation, and deletion of the more proximal sequence also leads to reduced constitutive expression (Siede et al. 1989).

UV induction of *RAD2* requires a stage of the cell cycle outside the G_1 phase. Thus, induction is reduced dramatically if cells are held in stationary phase or arrested in G_1 by α-factor treatment (Siede et al. 1989). This result contrasts with those reported for *CDC9* and *RAD54* (see above; Johnson et al. 1986; Cole et al. 1987).

Neither *RAD2* nor any of the other *RAD* genes required for damage-specific incision are cell-cycle-regulated (Nagpal et al. 1985; Siede et al. 1989). However, increased *RAD2* transcript levels are observed during meiosis. No *cis*-acting sequences have been identified that ablate this increase (Siede et al. 1989), suggesting that this reflects a general increase in transcriptional activity rather than a specifically regulated phenomenon. Additionally, there is no apparent meiotic phenotype caused by *rad2* disruption mutations (W. Siede and E.C. Friedberg, unpubl.).

The *RAD54* gene is transcriptionally induced by γ-rays or UV radiation and also following the introduction of strand breaks in DNA by overexpression of the *Eco*RI endonuclease (Cole et al. 1987). *RAD54* and *RAD52* (which is not inducible by DNA-damaging treatments) are also induced in meiosis (Cole et al. 1989). Deletion analysis of the *RAD54* promoter defined positive regulation of this gene, and discrete elements for DNA-damage induction and constitutive expression have been identified, some of which are distinct from those in *RAD2* (Cole and Mortimer 1989). AT-rich regions are required for efficient constitutive expression of *RAD54*, and one of these includes a potential binding site for datin protein. The sequences responsible for DNA-damage induction of *RAD54* lie within a 29-bp region, 229–258 bp upstream of the translational start site (Cole and Mortimer 1989).

The biological significance of the induction of *RAD2* and *RAD54* is unknown. Constructs have been made that place these genes under the control of promoters deleted of regions necessary for induction (Cole and Mortimer 1989; Siede et al. 1989). Defective inducibility of *RAD54* did not result in increased sensitivity to γ-rays or MMS in haploid or diploid cells (Cole and Mortimer 1989). Sporulation, recombination, and mating-type switching were also unaffected. Similarly, negative results were found for *RAD2* (Siede et al. 1989).

In summary, by direct screening or by studying the regulation of known genes, a considerable number of yeast genes have been identified as DNA-damage-inducible. Induction patterns or promoter structure have not revealed common regulatory networks or signaling systems, and the biological significance of enhanced expression after DNA damage or in meiosis must be elucidated. Quite clearly, there is still much to be learned about naturally occurring DNA damage, relevant genetic end points, and the physiology of DNA repair in the wild.

VIII. DNA REPAIR IN *SCHIZOSACCHAROMYCES POMBE*

The molecular biology of DNA repair and mutagenesis in the fission yeast *Schizosaccharomyces pombe* is largely unknown; this is unfortunate, because there are persuasive indications that *S. pombe* may be a more relevant biological model for multicellular eukaryotes than *S. cerevisiae*. *S. pombe* is significantly more resistant to UV radiation and γ-rays than is *S. cerevisiae*, especially in the G_2 phase of the cell cycle. Approximately 22 different repair mutants have been characterized in *S. pombe*. As in *S. cerevisiae*, they can be grouped phenotypically as being mainly sensitive to UV radiation, to γ-rays, or to both (for review, see Phipps et al. 1985). The existence of multiple repair pathways is suggested by limited epistasis analysis. The group of highly UV-radiation-sensitive mutants can probably be equated with the NER-defective mutants of *S. cerevisiae*; however, the *S. pombe* mutants are much less sensitive than those from *S. cerevisiae*. Furthermore, none of the *S. pombe* mutants thus far characterized are totally defective in pyrimidine dimer excision. This might reflect leakiness in the existing mutants. Alternatively, there may be several partially redundant excision repair pathways in *S. pombe*.

These differences notwithstanding, McCready et al. (1989) attempted complementation of seven *S. pombe* mutants presumably impaired in excision repair with the *S. cerevisiae RAD1* and *RAD2* genes. Complementation to near wild-type levels of UV resistance was observed following transformation of the *rad13* mutant with *RAD2*. As is the case with *rad2* mutations in *S. cerevisiae*, the *rad13* mutation in *S. pombe* enhances UV mutability, suggesting that excision repair in both organisms is an error-avoiding mechanism.

IX. SUMMARY AND CONCLUSIONS

Yeast cells have evolved multiple mechanisms for the processing of genomic injury. At present, a fairly complete genetic picture of these processing events is in place and suggests profound biochemical complexity. None of the multiple biochemical pathways suggested by the many mutant phenotypes characterized have been deciphered in detail. Nevertheless, progress in the past decade has been impressive. More than 25 genes involved in cellular responses to DNA damage have been isolated by molecular cloning, and discrete biochemical functions for at least a half dozen of these have been defined.

Although the precise functions of many of these genes remain unknown, recent suggestions that some of them may be multifunctional confronts us with new and interesting challenges about the biochemistry of DNA repair. Good examples are provided by the *RAD1*, *RAD3*, and *RAD6* genes. Although unequivocally required for damage-specific incision of DNA during nucleotide excision repair, *RAD1* is also clearly involved in mitotic recombinational pathways. Does this imply that Rad1p has more than one biochemical function or is a single biochemical function common to different pathways for DNA damage processing? The answer to this question awaits the purification and characterization of this and other Rad proteins. In the case of Rad3p, the available evidence suggests that this protein is indeed endowed with more than one biochemical activity, because its ATPase/DNA helicase function is apparently not required for viability.

The complexity of DNA repair processes in yeast is shared by other major transactions of DNA metabolism, notably DNA replication and transcription. Each of these events is characterized by the involvement of multiple gene products that probably assemble into multiprotein complexes at some stage during the transaction in question. As we move into the twenty-first century, the basic tools are at hand to overexpress and purify each of the many components of such putative protein machines and to reconstitute them systematically in vitro. However, this is a formidable task, likely to take many years to complete, and the technological challenge for the future is to devise new experimental strategies that facilitate the isolation of intact functional multiprotein complexes.

ACKNOWLEDGMENTS

We thank numerous colleagues for communicating unpublished results and results in press and members of our laboratory for helpful discussions and detailed review of the manuscript. Studies carried out in the laboratory of E.C.F. were supported by research grant CA-12428 from the U.S. Public Health Service.

REFERENCES

Aboussekhra, A., R. Chanet, Z. Zgaga, C. Cassier-Chauvat, M. Heude, and F. Fabre. 1989. *RADH*, a gene of *Saccharomyces cerevisiae* encoding a putative DNA helicase involved in DNA repair. Characteristics of *radH* mutants and sequence of the gene. *Nucleic Acids Res.* **17:** 7211.

Akhmedov, A.T., O.K. Kaboev, and M.L. Bekker. 1982. Purification and properties of two endonucleases specific for apurinic/apyrimidinic sites in DNA from *Saccharomyces cerevisiae*. *Biochim. Biophys. Acta* **696:** 163.

Andrade, H.H., E.K. Marques, A.C.G. Schenberg, and J.A.P. Henriques. 1989. The *PSO4* gene is responsible for error-prone recombinational repair in *Saccharomyces cerevisiae*. *Mol. Gen. Genet.* **217:** 419.

Armel, P.R. and S.S. Wallace. 1978. Apurinic endonucleases from *Saccharomyces cerevisiae*. *Nucleic Acids Res.* **5:** 3347.

―――. 1984. DNA repair in *Saccharomyces cerevisiae*: Purification and characterization of apurinic endonucleases. *J. Bacteriol.* **160:** 895.

Baer, M. and G.B. Sancar. 1989. Photolyases from *Saccharomyces cerevisiae* and *Escherichia coli* recognize common binding determinants in DNA containing pyrimidine dimers. *Mol. Cell. Biol.* **9:** 4777.

Bahman, M., V. Buck, A. White, and J. Rosamond. 1988. Characterization of the *CDC7* gene product of *Saccharomyces cerevisiae* as a protein needed for the initiation of mitotic DNA synthesis. *Biochim. Biophys. Acta* 951: 335.

Balganesh, T.S. and S.A. Lacks. 1985. Heteroduplex DNA mismatch repair system of *Streptococcus pneumoniae*: Cloning and expression of the *hexA* gene. *J. Bacteriol.* **162:** 979.

Bardwell, L., H. Burtscher, W.A. Weiss, C.N. Nicolet, and E.C. Friedberg. 1990. Characterization of the *RAD10* gene of *Saccharomyces cerevisiae* and purification of Rad10 protein. *Biochemistry* **29:** 3119.

Bishop, D.K. and R.D. Kolodner. 1986. Repair of heteroduplex plasmid DNA after transformation into *Saccharomyces cerevisiae*. *Mol. Cell. Biol.* **6:** 3401.

Bishop, D.K., J. Andersen, and R.D. Kolodner. 1989. Specificity of mismatch repair following transformation of *Saccharomyces cerevisiae* with heteroduplex DNA. *Proc. Natl. Acad. Sci.* **86:** 3713.

Bishop, D.K., M.S. Williamson, S. Fogel, and R.D. Kolodner. 1987. The role of heteroduplex correction in gene conversion in *Saccharomyces cerevisiae*. *Nature* **328:** 362.

Bohr, V.A. and K. Wassermann. 1988. DNA repair at the level of the genes. *Trends Biochem. Sci.* **13:** 429.

Boram, R. and H. Roman. 1976. Recombination in *Saccharomyces cerevisiae*: A DNA repair mutation associated with elevated mitotic gene conversion. *Proc. Natl. Acad. Sci.* **73:** 2828.

Bradshaw, V.A. and K. McEntee. 1989. DNA damage activates transcription and transposition of yeast retrotransposons. *Mol. Gen. Genet.* **218:** 465.

Brash, D.E. and W.A. Haseltine. 1982. UV-induced mutation hotspots occur at DNA damage hotspots. *Nature* **298:** 189.

Brendel, M. and A. Ruhland. 1984. Relationships between functionality and genetic toxicology of selected DNA damaging agents. *Mutat. Res.* **133:** 51.

Brown, T.C. and J. Jiricny. 1988. Different base/base mismatches are corrected with different efficiencies and specificities in monkey kidney cells. *Cell* **54:** 705.

Bryant, D.W. and R.H. Haynes. 1978. A DNA endonuclease isolated from yeast nuclear extract. *Can. J. Biochem.* **56:** 181.

Budd, M.E., K.D. Wittrup, and J.L. Campbell. 1989. DNA polymerase I is required for premeiotic DNA replication and sporulation but not for X-ray repair in *Saccharomyces cerevisiae. Mol. Cell. Biol.* **9:** 365.

Burbee, D., J. Campbell, and F. Heffron. 1988. The purification and characterization of the *NUC1* gene product, a yeast endodeoxyribonuclease required for double strand break repair. In *Mechanisms and consequences of DNA damage processing* (ed. E.C. Friedberg and P.C. Hanawalt), p. 223. A.R. Liss, New York.

Burgers, P.M.J. and M.B. Klein. 1986. Selection by genetic transformation of a *Saccharomyces cerevisiae* mutant defective for the nuclear uracil-DNA-glycosylase. *J. Bacteriol.* **166:** 905.

Cassier, C. and E. Moustacchi. 1981. Mutagenesis induced by mono- and bi-functional alkylating agents in yeast mutants sensitive to photo-addition of furocoumarins (*pso*). *Mutat. Res.* **84:** 37.

Cassier, C., R. Chanet, J.A.P. Henriques, and E. Moustacchi. 1980. The effects of three *pso* genes on induced mutagenesis: A novel class of mutationally defective yeast. *Genetics* **96:** 841.

Cassier-Chauvat, C. and E. Moustacchi. 1988. Allelism between *pso1-1* and *rev3-1* and between *pso2-1* and *snm1-1* mutants in *Saccharomyces cerevisiae. Curr. Genet.* **13:** 37.

Chanet, R., N. Magana-Schwencke, and F. Fabre. 1988. Potential DNA-binding domains in the *RAD18* gene product of *Saccharomyces cerevisiae. Gene* **74:** 543.

Chang, C.-C., Y.W. Kow, and S.S. Wallace. 1987. Apurinic endonucleases from *Saccharomyces cerevisiae* also recognize urea residues in oxidized DNA. *J. Bacteriol.* **169:** 180.

Chen, W., S. Tabor, and K. Struhl. 1987. Distinguishing between mechanisms of eukaryotic transcriptional activation with T7 RNA polymerase. *Cell* **50:** 1047.

Chlebowicz, E. and W.J. Jachymczyk. 1977. Endonuclease for apurinic sites in yeast: Comparison of the enzyme activity in the wild type and in *rad* mutants of *Saccharomyces cerevisiae* to MMS. *Mol. Gen. Genet.* **154:** 221.

Claverys, J.P. and S.A. Lacks. 1986. Heteroduplex deoxyribonucleic acid base mismatch repair in bacteria. *Microbiol. Rev.* **50:** 133.

Cleaver, J.E., F. Cortés, L.H.Lutze, W.F. Morgan, A.N. Player, and D.L. Mitchell. 1987. Unique DNA repair properties of a xeroderma pigmentosum revertant. *Mol. Cell. Biol.* **7:** 3353.

Cole, G.M. and R.K. Mortimer. 1989. Failure to induce a DNA repair gene, *RAD54*, in *Saccharomyces cerevisiae* does not affect DNA repair or recombination pathways. *Mol. Cell. Biol.* **9:** 3314.

Cole, G.M., D. Schild, and R.K. Mortimer. 1989. Two DNA repair and recombination genes in *Saccharomyces cerevisiae, RAD52* and *RAD54*, are induced during meiosis. *Mol. Cell. Biol.* **9:** 3101.

Cole, G.M., D. Schild, S.T. Lovett, and R.K. Mortimer. 1987. Regulation of *RAD54-* and *RAD52-lacZ* gene fusions in *Saccharomyces cerevisiae* in response to DNA damage. *Mol. Cell. Biol.* **7:** 1078.

Cooper, A.J. and S.L. Kelly. 1987. DNA repair and mutagenesis in *Saccharomyces cerevisiae*. In *Enzyme induction, mutagen activation and carcinogen testing in yeast* (ed. A. Wiseman), p. 73. Ellis Horwood, London.

Couto, L.B. and E.C. Friedberg. 1989. Nucleotide sequence of the wild-type *RAD4* gene of *Saccharomyces cerevisiae* and characterization of mutant *rad4* alleles. *J. Bacteriol.* **171:** 1862.

Cox, B.S. and J.M. Parry. 1968. The isolation, genetics and survival characteristics of ultraviolet light-sensitive mutants in yeast. *Mutat. Res.* **6**: 37.

Crosby, B., L. Prakash, H. Davis, and D.C. Hinkle. 1981. Purification and characterization of an uracil-DNA-glycosylase from the yeast *Saccharomyces cerevisiae*. *Nucleic Acids Res.* **9**: 5797.

DiCaprio, L. and B.S. Cox. 1981. DNA synthesis in UV-irradiated yeast. *Mutat. Res.* **82**: 69.

Eckardt, F. and R.H. Haynes. 1977a. Kinetics of mutation induction by ultraviolet light in excision-deficient yeast. *Genetics* **85**: 225.

―――. 1977b. Induction of pure and sectored mutant clones in excision-proficient and deficient strains of yeast. *Mutat. Res.* **43**: 327.

Eckardt, F., B.A. Kunz, and R.H. Haynes. 1983. Variation of mutation and recombination frequencies over a range of thymidylate concentrations in a diploid thymidylate auxotroph. *Curr. Genet.* **7**: 399.

Eckardt, F., E. Moustacchi, and R.H. Haynes. 1978. On the inducibility of error-prone repair in yeast. In *DNA repair mechanisms* (ed. P.C. Hanawalt et al.), p. 421. Academic Press, New York.

Eckardt, F., S.-J. Teh, and R.H. Haynes. 1980. Heteroduplex repair as an intermediate step of UV-mutagenesis in yeast. *Genetics* **95**: 63.

Eckardt-Schupp, F.F. Ahne, E.-M. Geigle, and W. Siede. 1988. Is mismatch repair involved in UV-induced mutagenesis in *Saccharomyces cerevisiae*? In *DNA replication and mutagenesis* (ed. R.E. Moses and W.C. Summers), p. 355. American Society for Microbiology, Washington, D.C.

Elledge, S.J. and R.W. Davis. 1987. Identification and isolation of the gene encoding the small subunit of ribonucleotide reductase from *Saccharomyces cerevisiae*: DNA damage-inducible gene required for mitotic viability. *Mol. Cell. Biol.* **7**: 2783.

Fabre, F., N. Magaña-Schwencke, and R. Chanet. 1989. Isolation of the *RAD18* gene of *Saccharomyces cerevisiae* and construction of *rad18* deletion mutants. *Mol. Gen. Genet.* **215**: 425.

Finley, D. and A. Varshavsky. 1985. The ubiquitin system: Functions and mechanisms. *Trends Biochem. Sci.* **10**: 343.

Finley, D., E. Özkaynak, and A. Varshavsky. 1987. The yeast polyubiquitin gene is essential for resistance to high temperatures, starvation and other stresses. *Cell* **48**: 1035.

Fleer, R., C.M. Nicolet, G.A. Pure, and E.C. Friedberg. 1987. The *RAD4* gene of *Saccharomyces cerevisiae*: Molecular cloning and partial characterization of a gene which is inactivated in *E. coli. Mol. Cell. Biol.* **7**: 1180.

Friedberg, E.C. 1985a. *DNA repair*. W.H. Freeman, New York.

―――. 1985b. Nucleotide excision repair of DNA in eukaryotes: Comparison between human cells and yeast. *Cancer Surv.* **4**: 529.

―――. 1988. Deoxyribonucleic acid repair in the yeast *Saccharomyces cerevisiae*. *Microbiol. Rev.* **52**: 70.

Futcher, A.B. and A.R. Morgan. 1979. A novel assay for endonuclease activity in repair-deficient mutants of *Saccharomyces cerevisiae*. *Can. J. Biochem.* **57**: 932.

Game, J.C. and R.K. Mortimer. 1974. A genetic study of X-ray sensitive mutants in yeast. *Mutat. Res.* **24**: 281.

Gietz, R.D. and S. Prakash. 1988. Cloning and nucleotide sequence analysis of the *Saccharomyces cerevisiae RAD4* gene required for excision repair of UV-damaged DNA. *Gene* **74**: 535.

Goebl, M.G., J. Yochem, S. Jentsch, J.P. McGrath, A. Varshavsky, and B. Byers. 1988. The yeast cell cycle gene *CDC34* encodes a ubiquitin-conjugating enzyme. *Science* **241**: 1331.

Golin, J.E. and M.S. Esposito. 1977. Evidence for joint genic control of spontaneous mutation and genetic recombination during mitosis in *Saccharomyces cerevisiae*. *Mol. Gen. Genet.* **150:** 127.

Gossett, J., K. Lee, R.P. Cunningham, and P.W. Doetsch. 1988. Yeast redoxyendonuclease, a DNA repair enzyme similar to *Escherichia coli* endonuclease III. *Biochemistry* **27:** 2629.

Gottlieb, D.J.C. and J.C. von Borstel. 1976. Mutators in *Saccharomyces cerevisiae, mut1-1, mut1-2,* and *mut2-1. Genetics* **83:** 655.

Haase, E., D. Riehl, M. Mack, and M. Brendel. 1989. Molecular cloning of *SNM1*, a yeast gene responsible for a specific step in the repair of cross-linked DNA. *Mol. Gen. Genet.* **218:** 64.

Hanawalt, P.C. 1987. Preferential DNA repair in expressed genes. *Environ. Health Perspect.* **76:** 9.

Harosh, I., L. Naumovski, and E.C. Friedberg. 1989. Purification and characterization of the Rad3 ATPase/DNA helicase from *Saccharomyces cerevisiae. J. Biol. Chem.* **264:** 20532.

Hastings, P.J., S.K. Quah, and R.C. von Borstel. 1976. Spontaneous mutation by mutagenic repair of spontaneous lesions in DNA. *Nature* **264:** 719.

Haynes, R.H. and F. Eckardt. 1980. Mathematical analysis of mutation induction kinetics. In *Chemical mutagens* (ed. F.J. de Serres and A. Hollaender), p. 271. Plenum Press, New York.

Haynes, R.H. and B.A. Kunz. 1981. DNA repair and mutagenesis in yeast. In *The molecular biology of the yeast* Saccharomyces: *Life cycle and inheritance* (ed. J.N. Strathern et al.), p. 371. Cold Spring Harbor Laboratory, Cold Spring Harbor, New York.

Haynes, R.H., F. Eckardt, and B.A. Kunz. 1985. Analysis of non-linearities in mutation frequency curves. *Mutat. Res.* **150:** 51.

Henriques, J.A.P. and E. Moustacchi. 1980. Isolation and characterization of *pso*-mutants sensitive to photo-addition of psoralen derivatives in *Saccharomyces cerevisiae. Genetics* **95:** 273.

――――. 1981. Interactions between mutations for sensitivity to psoralen photoaddition (*pso*) and radiations (*rad*) in *Saccharomyces cerevisiae. J. Bacteriol.* **148:** 248.

Higgins, D.R., L. Prakash, P. Reynolds, and S. Prakash. 1984. Isolation and characterization of the *RAD2* gene of *Saccharomyces cerevisiae. Gene* **30:** 121.

Higgins, D.R., S. Prakash, P. Reynolds, and L. Prakash. 1983a. Molecular cloning and characterization of the *RAD1* gene of *Saccharomyces cerevisiae. Gene* **26:** 119.

Higgins, D.R., S. Prakash, P. Reynolds, R. Polakowska, S. Weber, and L. Prakash. 1983b. Isolation and characterization of the *RAD3* gene of *Saccharomyces cerevisiae* and inviability of *rad3* deletion mutants. *Proc. Natl. Acad. Sci.* **80:** 5680.

Ho, K.S.Y. 1975. Induction of DNA double-strand breaks by X-rays in a radiosensitive strain of the yeast *Saccharomyces cerevisiae. Mutat. Res.* **30:** 327.

Hoekstra, M.F. and R.E. Malone. 1987. Hyper-mutation caused by the *rem1* mutation in yeast is not dependent on error-prone or excision repair. *Mutat. Res.* **178:** 201.

Hunnable, E.G. and B.S. Cox. 1971. The genetic control of dark recombination in yeast. *Mutat. Res.* **13:** 297.

Ivanov, E.L., S.V. Kovaltzova, G.V. Kassinova, L.M. Gracheva, V.G. Korolev, and I.A. Zakharov. 1986. The *rad2* mutation affects the molecular nature of UV and acridine-mustard-induced mutations in the *ADE2* gene of *Saccharomyces cerevisiae. Mutat. Res.* **160:** 207.

James, A.P. and B.J. Kilbey. 1977. The timing of UV mutagenesis in yeast: A pedigree analysis of induced recessive mutation. *Genetics* **87:** 237.

James, A.P., B.J. Kilbey, and G.J. Prefontaine. 1978. The timing of UV mutagenesis in yeast: Continuing mutation in an excision-defective (*rad1-1*) strain. *Mol. Gen. Genet.* **165:** 207.

Jentsch, S., J.P. McGrath, and A. Varshavsky. 1987. The yeast DNA repair gene *RAD6* encodes a ubiquitin-conjugating enzyme. *Nature* **329:** 131.

Johnson, A.L., D.G. Barker, and L.H. Johnston. 1986. Induction of yeast DNA ligase genes in exponential and stationary phase cultures in response to DNA damaging agents. *Curr. Genet.* **11:** 107.

Johnson, A.W. and B. Demple. 1988a. Yeast DNA 3′-repair diesterase is the major cellular apurinic/apyrimidinic endonuclease: Substrate specificity and kinetics. *J. Biol. Chem.* **263:** 18017.

―――. Yeast DNA diesterase for 3′-fragments of deoxyribose: Purification and physical properties of a repair enzyme for oxidative DNA damage. *J. Biol. Chem.* **263:** 18009.

Johnston, L.H. 1979. The DNA repair capability of *cdc9*, the *Saccharomyces cerevisiae* mutant defective in DNA ligase. *Mol. Gen. Genet.* **170:** 89.

Johnston, L.H., J.H.M. White, A.L. Johnson, G. Lucchini, and P. Plevani. 1987. The yeast DNA polymerase I transcript is regulated in both the mitotic cell cycle and in meiosis and is also induced after DNA damage. *Nucleic Acids Res.* **15:** 5017.

Jones, J.S., S. Weber, and L. Prakash. 1988. The *Saccharomyces cerevisiae RAD18* gene encodes a protein that contains potential zinc finger domains for nucleic acid binding and a putative nucleotide binding sequence. *Nucleic Acids Res.* **16:** 7119.

Jong, A.Y.S., C.-L. Kuo, and J.L. Campbell. 1984. The *CDC8* gene of yeast encodes thymidylate kinase. *J. Biol. Chem.* **259:** 11052.

Kassir, Y., M. Kupiec, A. Shalom, and G. Simchen. 1985. Cloning and mapping of *CDC40*, a *Saccharomyces cerevisiae* gene with a role in DNA repair. *Curr. Genet.* **9:** 253.

Kern, R. and F.K. Zimmermann. 1978. The influence of defects in excision and error prone repair on spontaneous and induced mitotic recombination and mutation in *Saccharomyces cerevisiae*. *Mol. Gen. Genet.* **161:** 81.

Kilbey, B.J., T. Brychcy, and A. Nasim. 1978. Initiation of UV mutagenesis in *Saccharomyces cerevisiae*. *Nature* **274:** 889.

Kramer, B., W. Kramer, M.S. Williamson, and S. Fogel. 1989. Heteroduplex DNA correction in yeast is mismatch-specific and requires functional *PMS* genes. *Mol. Cell. Biol.* **9:** 4432.

Kramer, W., B. Kramer, M.S. Williamson, and S. Fogel. 1989. Cloning and nucleotide sequence of DNA mismatch repair gene *PMS1* from *Saccharomyces cerevisiae*: Homology to procaryotic MutL and HexB. *J. Bacteriol.* **171:** 5339

Kunz, B.A. 1982. Genetic effects of deoxyribonucleotide pool imbalances. *Environ. Mutagen.* **4:** 695.

―――. 1988. Mutagenesis and deoxyribonucleotide pool imbalance. *Mutat. Res.* **200:** 133.

Kunz, B.A. and R.H. Haynes. 1982. DNA repair and the genetic effects of thymidilate stress in yeast. *Mutat. Res.* **93:** 353.

Kunz, B.A., M.K. Pierce, J.R.A. Mis, and C.N. Giroux. 1987. DNA sequence analysis of the mutational specificity of u.v. light in the *SUP4-o* gene of yeast. *Mutagenesis* **2:** 445.

Kunz, B.A., M.G. Peters, S.E. Kohalmi, J.D. Armstrong, M. Glattke, and K. Badiani. 1989. Disruption of the *RAD52* gene alters the spectrum of spontaneous *SUP4-o* mutations in *Saccharomyces cerevisiae*. *Genetics* **122:** 535.

Kupiec, M. and G. Simchen. 1984. Cloning and integrative deletion of the *RAD6* gene of *Saccharomyces cerevisiae. Curr. Genet.* **8:** 559.

――――. 1985. Arrest of the mitotic cell cycle and of meiosis in *Saccharomyces cerevisiae* by MMS. *Mol. Gen. Genet.* **201:** 558.

――――. 1986a. Regulation of the *RAD6* gene of *Saccharomyces cerevisiae* in the mitotic cell cycle and in meiosis. *Mol. Gen. Genet.* **203:** 538.

――――. 1986b. DNA-repair characterization of *cdc40*-1, a cell-cycle mutant of *Saccharomyces cerevisiae. Mutat. Res.* **162:** 33.

Lambert, C., L.B. Couto, W.A. Weiss, R.A. Schultz, L.H. Thompson, and E.C. Friedberg. 1988. A yeast DNA repair gene partially complements defective excision repair in mammalian cells. *EMBO J.* **7:** 3245.

Langeveld, S.A., A. Yasui, and A.P.M. Eker. 1985. Expression of an *Escherichia coli phr* gene in the yeast *Saccharomyces cerevisiae. Mol. Gen. Genet.* **199:** 396.

Larimer, F.W., J.R. Perry, and A.A. Hardigree. 1989. The *REV1* gene of *Saccharomyces cerevisiae*: Isolation, sequence and functional analysis. *J. Bacteriol.* **171:** 230.

Lawrence, C.W. 1982. Mutagenesis in *Saccharomyces cerevisiae. Adv. Genet.* **21:** 173.

Lawrence, C.W. and R.B. Christensen. 1976. UV mutagenesis in radiation-sensitive strains of yeast. *Genetics* **82:** 207.

――――. 1978a. Ultra-violet induced reversion of *cyc1* alleles in radiation sensitive strains of yeast. I. *rev1* mutant strains. *J. Mol. Biol.* **122:** 1.

――――. 1978b. Ultra-violet induced reversion of *cyc1* alleles in radiation sensitive strains of yeast. II. *rev2* mutant strains. *Genetics* **90:** 213.

――――. 1979. Metabolic suppressors of trimethoprim and ultraviolet light sensitivities of *Saccharomyces cerevisiae rad6* mutants. *J. Bacteriol.* **139:** 866.

――――. 1982. The mechanism of untargeted mutagenesis in UV-irradiated yeast. *Mol. Gen. Genet.* **186:** 1.

Lawrence, C.W., G. Das, and R.B. Christensen. 1985a. *REV7*, a new gene concerned with UV mutagenesis in yeast. *Mol. Gen. Genet.* **200:** 80.

Lawrence, C.W., B.R. Krauss, and R.B. Christensen. 1985b. New mutations affecting induced mutagenesis in yeast. *Mutat. Res.* **150:** 211.

Lawrence, C.W., P.E. Nisson, and R.B. Christensen. 1985c. UV and chemical mutagenesis in *rev7* mutants of yeast. *Mol. Gen. Genet.* **200:** 86.

Lawrence, C.W., T. O'Brien, and J. Bond. 1984. UV-induced reversion of *his4* frameshift mutations in *rad6, rev1,* and *rev3* mutants of yeast. *Mol. Gen. Genet.* **195:** 487.

Lawrence, C.W., J.W. Stewart, F. Sherman, and R.B. Christensen. 1974. Specificity and frequency of UV-induced reversion of an iso-1-cytochrome-C ochre mutant in radiation-sensitive strains of yeast. *J. Mol. Biol.* **85:** 137.

Lee, G.S.-F., E.A. Savage, R.G. Ritzel, and R.C. von Borstel. 1988. The base-alteration spectrum of spontaneous and ultraviolet radiation-induced forward mutations in the *URA3* locus of *Saccharomyces cerevisiae. Mol. Gen. Genet.* **214:** 396.

Lemontt, J.F. 1971. Mutants of yeast defective in mutation induction by ultraviolet light. *Genetics* **68:** 21.

――――. 1972. Induction of forward mutations in mutationally defective yeast. *Mol. Gen. Genet.* **119:** 27.

――――. 1977. Pathways of ultraviolet mutability in *Saccharomyces cerevisiae*. III. Genetic analysis and properties of mutants resistant to ultraviolet-induced forward mutation. *Mutat. Res.* **43:** 179.

――――. 1980. Genetic and physiological factors affecting repair and mutagenesis in yeast. In *DNA-repair and mutagenesis in eukaryotes* (ed. W.M. Generoso et al.), p. 85. Plenum Press, New York.

Lue, N.F, A.R. Buchman, and R.D. Kornberg. 1989. Activation of yeast RNA polymerase II transcription by a thymidine-rich upstream element in vitro. *Proc. Natl. Acad. Sci.* **86:** 486.

MacQuillan, A.M., A. Herman, J.S. Coberly, and G. Green. 1981. A second photore-activation-deficient mutation in *Saccharomyces cerevisiae. Photochem. Photobiol.* **34:** 673.

Madura, K. and S. Prakash. 1986. Nucleotide sequence, transcript mapping, and regulation of the *RAD2* gene of *Saccharomyces cerevisiae. J. Bacteriol.* **166:** 914.

Maga, J.A. and K. McEntee. 1985. Response of *S. cerevisiae* to *N*-methyl-*N'*-nitro-*N*-nitrosoguanidine: Mutagenesis, survival and *DDR* gene expression. *Mol. Gen. Genet.* **200:** 313.

Maga, J.A., T.A. McClanahan, and K. McEntee. 1986. Transcriptional regulation of DNA damage responsive (*DDR*) genes in different *rad* mutant strains of *Saccharomyces cerevisiae. Mol. Gen. Genet.* **205:** 276.

Malone, R.E. 1983. Multiple mutant analysis of recombination in yeast. *Mol. Gen. Genet.* **189:** 405.

Malone, R.E. and M.T. Hoekstra. 1984. Relationships between a hyper-rec mutation (*rem1*) and other recombination and repair genes in yeast. *Genetics* **107:** 33.

Martin, P., L. Prakash, and S. Prakash. 1981. a/α-specific effect of the *mms3* mutation on ultraviolet mutagenesis in *Saccharomyces cerevisiae. J. Bacteriol.* **146:** 684.

Matson, S.W. 1986. *Escherichia coli* helicase II (*uvrD* gene product) translocates unidirectionally in a 3' to 5' direction. *J. Biol. Chem.* **261:** 10169.

Maus, K.L., E.M. McIntosh, and R.H. Haynes. 1984. Defective dCMP deaminase confers a mutator phenotype on *Saccharomyces cerevisiae. Environ. Mutagen.* **6:** 415.

McClanahan, T. and K. McEntee. 1984. Specific transcripts are elevated in *Saccharomyces cerevisiae* in response to DNA damage. *Mol. Cell. Biol.* **4:** 2356.

———. 1986. DNA damage and heat shock dually regulate genes in *Saccharomyces cerevisiae. Mol. Cell. Biol.* **6:** 90.

McCready, S.J., H. Burkill, S. Evans, and B.S. Cox. 1989. The *Saccharomyces cerevisiae RAD2* gene complements a *Schizosaccharomyces pombe* repair mutation. *Curr. Genet.* **15:** 27.

McKee, R.H. and C.W. Lawrence. 1979. Genetic analysis of gamma ray mutagenesis in yeast. I. Reversion in radiation sensitive strains. *Genetics* **93:** 361.

McKnight, G.L., T.S. Cardillo, and F. Sherman. 1981. An extensive deletion causing overproduction of yeast iso-2-cytochrome C. *Cell* **25:** 409.

Meechan, P.J., K.M. Milam, and J.E. Cleaver. 1986. Evaluation of homology between cloned *Escherichia coli* and yeast DNA photolyase genes and higher eukaryotic genomes. *Mutat. Res.* **166:** 143.

Miller, R.D., L. Prakash, and S. Prakash. 1982. Defective excision of pyrimidine dimers and interstrand DNA crosslinks in *rad7* and *rad23* mutants of *Saccharomyces cerevisiae. Mol. Gen. Genet.* **188:** 235.

Modrich, P. 1987. DNA mismatch correction. *Annu. Rev. Genet.* **20:** 435.

Montelone, B.A., M.F. Hoekstra, and R.E. Malone. 1988. Spontaneous mitotic recombination in yeast: The hyper-recombinational *rem1* mutations are alleles of the *RAD3* gene. *Genetics* **119:** 289.

Montelone, B.A., S. Prakash, and L. Prakash. 1981. Recombination and mutagenesis in *rad6* mutants of *Saccharomyces cerevisiae*: Evidence for multiple functions of the *RAD6* gene. *Mol. Gen. Genet.* **184:** 410.

Morawetz, C. 1987. Effect of irradiation and mutagenic chemicals on the generation of *ADH2*-constitutive mutants in yeast. Significance for the inducibility of Ty transposition. *Mutat. Res.* **177:** 53.

Morichetti, E., E. Cundari, R. del Carratore, and G. Bronzetti. 1989. Induction of cytochrome P-450 and catalase activity in *Saccharomyces cerevisiae* by UV and X-ray irradiation. Possible role for cytochrome P-450 in cell protection against oxidative damage. *Yeast* **5:** 141.

Morrison, A., E.J. Miller, and L. Prakash. 1988. Domain structure and functional analysis of the carboxyl-terminal polyacidic sequence of the RAD6 protein of *Saccharomyces cerevisiae. Mol. Cell. Biol.* **8:** 1179.

Morrison, A., R.B. Christensen, J. Alley, A.K. Beck, E.G. Bernstine, J.F. Lemontt, and C.W. Lawrence. 1989. *REV3,* a yeast gene whose function is required for induced mutagenesis, is predicted to encode a non-essential DNA polymerase. *J. Bacteriol.* **171:** 5659.

Moustacchi, E. 1987. DNA repair in yeast: Genetic control and biological consequences. *Adv. Radiat. Biol.* **13:** 1.

Moustacchi, E., C. Cassier, R. Chanet, N. Magaña-Schwencke, T. Saeki, and J.A.P. Henriques. 1983. Biological role of photo-induced crosslinks and monoadducts in yeast DNA: Genetic control and steps involved in their repair. In *Cellular responses to DNA damage* (ed. E.C. Friedberg and B.A. Bridges), p. 87. A.R. Liss, New York.

Mowat, M.R.A., W.J. Jachymczyk, P.J. Hastings, and R.C. von Borstel. 1983. Repair of γ-ray induced DNA strand breaks in the radiation-sensitive mutant *rad18-2* of *Saccharomyces cerevisiae. Mol. Gen. Genet.* **189:** 256.

Muster-Nassal, C. and R. Kolodner. 1986. Mismatch correction catalyzed by cell-free extracts of *Saccharomyces cerevisiae. Proc. Natl. Acad. Sci.* **83:** 7618.

Nag, D.K., M.A. White, and T.D. Petes. 1989. Palindromic sequences in heteroduplex DNA inhibit mismatch repair in yeast. *Nature* **340:** 318.

Nagpal, M.L., D.R. Higgins, and S. Prakash. 1985. Expression of the *RAD1* and *RAD3* genes of *Saccharomyces cerevisiae* is not affected by DNA damage or during cell division cycle. *Mol. Gen. Genet.* **199:** 59.

Nasim, A., M.A. Hannan, and E.R. Nestmann. 1981. Pure and mosaic clones—A reflection of differences in mechanisms of mutagenesis by different agents in *Saccharomyces cerevisiae. Can. J. Genet. Cytol.* **23:** 73.

Naumovski, L. and E.C. Friedberg. 1982. Molecular cloning of eucaryotic genes required for excision repair of UV-irradiated DNA: Isolation and partial characterization of the *RAD3* gene of *Saccharomyces cerevisiae. J. Bacteriol.* **152:** 323.

――――. 1983. A DNA repair gene required for the incision of damaged DNA is essential for viability in *Saccharomyces cerevisiae. Proc. Natl. Acad. Sci.* **80:** 4818.

――――. 1984. *Saccharomyces cerevisiae RAD2* gene: Isolation, subcloning and partial characterization. *Mol. Cell. Biol.* **4:** 290.

――――. 1986. Analysis of the essential and excision repair functions of the *RAD3* gene of *Saccharomyces cerevisiae* by mutagenesis. *Mol. Cell. Biol.* **6:** 1218.

――――. 1987. The *RAD3* gene of *Saccharomyces cerevisiae*: Isolation and characterization of a temperature-sensitive mutant in the essential function and of extragenic suppressors of this mutant. *Mol. Gen. Genet.* **209:** 458.

Naumovski, L., G. Chu, P. Berg, and E.C. Friedberg. 1985. *RAD3* gene of *Saccharomyces cerevisiae*: Nucleotide sequence of wild-type and mutant alleles, transcript mapping and aspects of gene regulation. *Mol. Cell. Biol.* **5:** 17.

Njagi, G.D.E. and B.J. Kilbey. 1982. *cdc7-1* a temperature sensitive cell-cycle mutant which interferes with induced mutagenesis in *Saccharomyces cerevisiae. Mol. Gen. Genet.* **186:** 478.

Oh, E.Y. and L. Grossman. 1987. Helicase properties of the *Escherichia coli* Uvr AB protein complex. *Proc. Natl. Acad. Sci.* **84**: 3638.

————. 1989. Characterization of the helicase activity of the *Escherichia coli* UvrAB protein complex. *J. Biol. Chem.* **264**: 1336.

Orr-Weaver, T.L. and J. Szostak. 1985. Fungal recombination. *Microbiol. Rev.* **49**: 33.

Patterson, M., R.A. Sclafani, W.L. Fangman, and J. Rosamond. 1986. Molecular characterization of cell cycle gene *CDC7* from *Saccharomyces cerevisiae. Mol. Cell. Biol.* **6**: 1590.

Percival, K.J., M.B. Klein, and P.M.J. Burgers. 1989. Molecular cloning and primary structure of the uracil-DNA-glycosylase gene from *Saccharomyces cerevisiae. J. Biol. Chem.* **264**: 2593.

Perozzi, G. and S. Prakash. 1986. *RAD7* gene of *Saccharomyces cerevisiae*: Transcripts, nucleotide sequence analysis and functional relationship between the *RAD7* and *RAD23* gene product. *Mol. Cell. Biol.* **6**: 1497.

Peterson, T.A., L. Prakash, S. Prakash, M.A. Osley, and S.I. Reed. 1985. Regulation of *CDC9*, the *Saccharomyces cerevisiae* gene that encodes DNA ligase. *Mol. Cell. Biol.* **5**: 226.

Phipps, J., A. Nasim, and D.R. Miller. 1985. Recovery, repair and mutagenesis in *Schizosaccharomyces pombe. Adv. Genet.* **23**: 1.

Piñon, R. 1970. Characterization of a yeast endonuclease. *Biochemistry* **9**: 6397.

Polakowska, R., G. Perozzi, and L. Prakash. 1986. Alkylation mutagenesis in *Saccharomyces cerevisiae*: Lack of evidence for an adaptive reponse. *Curr. Genet.* **10**: 647.

Popoff, C.A., I.A. Spira, A.W. Johnson, and B. Demple. 1990. Yeast structural gene (*APN1*) for the major apurinic endonuclease: Homology to *Escherichia coli* endonuclease IV. *Proc. Natl. Acad. Sci.* **87**: 4193.

Potter, A.A., E.R. Nestmann, and V.N. Iyer. 1984. Introduction of the plasmid pKM101-associated *muc* genes in *Saccharomyces cerevisiae. Mutat. Res.* **131**: 197.

Prakash, L. 1974. Lack of chemically induced mutation in repair deficient mutants of yeast. *Genetics* **78**: 1101.

————. 1976. Effect of genes controlling radiation sensitivity on chemically induced mutations in *Saccharomyces cerevisae. Genetics* **83**: 285.

————. 1977. Defective thymine dimer excision in radiation-sensitive mutants *rad10* and *rad16* of *Saccharomyces cerevisiae. Mol. Gen. Genet.* **152**: 125.

————. 1981. Characterization of postreplication repair in *Saccharomyces cerevisiae* and effects of *rad6, rad18, rev3* and *rad52* mutations. *Mol. Gen. Genet.* **184**: 471.

Prakash, L. and A. Morrison. 1988. The *RAD6* gene and protein of *Saccharomyces cerevisiae*. In *Mechanisms and consequences of DNA damage processing* (ed. E.C. Friedberg and P.C. Hanawalt), p. 237. A.R. Liss, New York.

Prakash, L. and S. Prakash. 1980. Genetic analysis of error-prone repair systems in *Saccharomyces cerevisiae*. In *DNA repair and mutagenesis in eukaryotes* (ed. W.M. Generoso et al.), p. 141. Plenum Press, New York.

Prakash, L., D. Hinkle, and S. Prakash. 1979. Decreased UV mutagenesis in *cdc8*, a DNA replication mutant of *Saccharomyces cerevisiae. Mol. Gen. Genet.* **172**: 249.

Prakash, L., R. Polakowska, P. Reynolds, and S. Weber. 1983. Molecular cloning and preliminary characterization of the *RAD6* gene of the yeast *Saccharomyces cerevisiae*. In *Cellular response to DNA damage* (ed. E.C. Friedberg and B.A. Bridges), p. 559. A.R. Liss, New York.

Prakash, L., D. Dumais, R. Polakowska, G. Perozzi, and S. Prakash. 1985. Molecular cloning of the *RAD10* gene of *Saccharomyces cerevisiae. Gene* **34**: 55.

Proffitt, J.H., J.R. Davie, D. Swinton, and S. Hattman. 1984. 5-Methylcytosine is not detectable in *Saccharomyces cerevisiae* DNA. *Mol. Cell. Biol.* **4**: 985.

Quah, S.K., R.C. von Borstel, and P.J. Hastings. 1980. The origin of spontaneous mutation in *Saccharomyces cerevisiae*. *Genetics* **96**: 819.

Radman, M. and R. Wagner. 1986. Mismatch repair in *Escherichia coli*. *Annu. Rev. Genet.* **20**: 523.

Resnick, M.A. 1969. Genetic control of radiation sensitivity in *Saccharomyces cerevisiae*. *Genetics* **62**: 519.

Resnick, M.A. and P. Martin. 1976. The repair of double-strand breaks in the nuclear DNA of *Saccharomyces cerevisiae* and its genetic control. *Mol. Gen. Genet.* **143**: 119.

Resnick, M.A. and J.K. Setlow. 1972. Photoreactivation and gene dosage effect in yeast. *J. Bacteriol.* **109**: 1307.

Reynolds, P., L. Prakash, and S. Prakash. 1987. Nucleotide sequence and functional analysis of the *RAD1* gene of *Saccharomyces cerevisiae*. *Mol. Cell. Biol.* **7**: 1012.

Reynolds, P., S. Weber, and L. Prakash. 1985a. *RAD6* gene of *Saccharomyces cerevisiae* encodes a protein containing a tract of 13 consecutive aspartates. *Proc. Natl. Acad. Sci.* **82**: 168.

Reynolds, P., D.R. Higgins, L. Prakash, and S. Prakash. 1985b. The nucleotide sequence of the *RAD3* gene of *Saccharomyces cerevisiae*: A potential adenine nucleotide binding amino acid sequence and a nonessential acidic carboxyl terminal region. *Nucleic Acids Res.* **13**: 2457.

Reynolds, P., L. Prakash, D. Dumais, G. Perozzi, and S. Prakash. 1985c. Nucleotide sequence of the *RAD10* gene of *Saccharomyces cerevisiae*. *EMBO J.* **4**: 3549.

Reynolds, R.J. and E.C. Friedberg. 1981. Molecular mechanism of pyrimidine dimer excision in *Saccharomyces cerevisiae*: Incision of ultraviolet-irradiated deoxyribonucleic acid in vivo. *J. Bacteriol.* **146**: 692.

Robinson, G.W., C.M. Nicolet, D. Kalainov, and E.C. Friedberg. 1986. A yeast excision repair gene is inducible by DNA damaging agents. *Proc. Natl. Acad. Sci.* **83**: 1842.

Rolfe, M. 1985. UV-inducible transcripts in *Saccharomyces cerevisiae*. *Curr. Genet.* **9**: 533.

Rolfe, M., A. Spanos, and G. Banks. 1986. Induction of yeast Ty element transcription by ultraviolet light. *Nature* **319**: 339.

Rubin, J.S. 1988. The molecular genetics of the incision step in the DNA excision repair process. *Int. J. Radiat. Biol.* **54**: 309.

Ruby, S.W. and J.W. Szostak. 1985. Specific *Saccharomyces cerevisiae* genes are expressed in response to DNA-damaging agents. *Mol. Cell. Biol.* **5**: 75.

Ruhland, A., M. Kircher, F. Wilborn, and M. Brendel. 1981. A yeast mutant specifically sensitive to bifunctional alkylation. *Mutat. Res.* **91**: 457.

Sancar, A. and G.B. Sancar. 1988. DNA repair enzymes. *Annu. Rev. Biochem.* **57**: 29.

Sancar, A., K.A. Franklin, and G.B. Sancar. 1984. *Escherichia coli* DNA photolyase stimulates *uvrABC* excision nuclease in vitro. *Proc. Natl. Acad. Sci.* **81**: 7397.

Sancar, G.B. 1985a. Sequence of *Saccharomyces cerevisiae PHR1* gene and homology of the *PHR1* photolyase to *E. coli* photolyase. *Nucleic Acids Res.* **13**: 8231.

―――. 1985b. Expression of a *Saccharomyces cerevisiae* photolyase gene in *Escherichia coli*. *J. Bacteriol.* **161**: 769.

Sancar, G.B. and F.W. Smith. 1989. Interactions between yeast photolyase and nucleotide excision repair proteins in *Saccharomyces* and *E. coli*. *Mol. Cell. Biol.* **9**: 4767.

Sancar, G.B., F.W. Smith, and P.F. Heelis. 1987. Purification of the yeast PHR1 photolyase from an *Escherichia coli* overproducing strain and characterization of the intrinsic chromophores of the enzyme. *J. Biol. Chem.* **262**: 15457.

Sargentini, N.J. and K.C. Smith. 1985. Spontaneous mutagenesis: The roles of DNA repair, replication, and recombination. *Mutat. Res.* **154**: 1.

Schenk, K., F. Zölzer, and J. Kiefer. 1989. Mutation induction in haploid yeast after split-dose radiation exposure. I. Fractionated UV-irradiation. *Radiat. Environ. Biophys.* **28**: 101.

Schiestl, R.H., P. Reynolds, S. Prakash, and L. Prakash. 1989. Cloning and sequence analysis of the *Saccharomyces cerevisiae RAD9* gene and further evidence that its product is required for cell cycle arrest induced by DNA damage. *Mol. Cell. Biol.* **9**: 1882.

Schild, D., J. Johnston, C. Chang, and R.K. Mortimer. 1984. Cloning and mapping of *Saccharomyces cerevisiae* photoreactivation gene *PHR1*. *Mol. Cell. Biol.* **4**: 1864.

Schwencke, J. and E. Moustacchi. 1982. Proteolytic activities in yeast after UV irradiation. I. Variation in proteinase levels in repair proficient *RAD+* strains. *Mol. Gen. Genet.* **185**: 290.

Siede, W. and M. Brendel. 1982. Interactions among genes controlling sensitivity to radiation (*RAD*) and to alkylation by nitrogen mustard (*SNM*) in yeast. *Curr. Genet.* **5**: 33.

Siede, W. and F. Eckardt. 1984. Inducibility of error-prone DNA repair in yeast? *Mutat. Res.* **129**: 3.

———. 1986a. Analysis of mutagenic DNA repair in a thermoconditional mutant of *Saccharomyces cerevisiae*. III. Dose-response pattern of mutation induction in UV-irradiated *rev2*[ts] cells. *Mol. Gen. Genet.* **202**: 68.

———. 1986b. Analysis of mutagenic DNA repair in a thermoconditional mutant of *Saccharomyces cerevisiae*. IV. Influence of DNA replication and excision repair on *REV2* dependent UV-mutagenesis and repair. *Curr. Genet.* **10**: 871.

Siede, W. and F. Eckardt-Schupp. 1986a. DNA repair genes of *Saccharomyces cerevisiae*: Complementing *rad4* and *rev2* mutations by plasmids which cannot be propagated in *Escherichia coli*. *Curr. Genet.* **11**: 205.

———. 1986b. A mismatch repair-based model can explain some features of u.v. mutagenesis in yeast. *Mutagenesis* **1**: 471.

Siede, W., F. Eckardt, and M. Brendel. 1983a. Analysis of mutagenic DNA repair in a thermoconditional repair mutant of *Saccharomyces cerevisiae*. I. Influence of cycloheximide on UV-irradiated stationary phase *rev2*[ts] cells. *Mol. Gen. Genet.* **190**: 406.

———. 1983b. Analysis of mutagenic DNA repair in a thermoconditional mutant of *Saccharomyces cerevisae*. II. Influence of cycloheximide on UV-irradiated exponentially growing *rev2*[ts] cells. *Mol. Gen. Genet.* **190**: 413.

Siede, W., G.W. Robinson, D. Kalainov, T. Malley, and E.C. Friedberg. 1989. Regulation of the *RAD2* gene of *Saccharomyces cerevisiae*. *Mol. Microbiol.* **3**: 1697.

Song, J.M., B.A. Montelone, W. Siede, and E.C. Friedberg. 1990. Effects of multiple yeast *rad3* mutant alleles on mitotic recombination. *J. Bacteriol.* (in press).

Sung, P., S. Prakash, and L. Prakash. 1988a. The RAD6 protein of *Saccharomyces cerevisiae* polyubiquitinates histones and its acidic domain mediates this activity. *Genes Dev.* **2**: 1476.

Sung, P., D. Higgins, L. Prakash, and S. Prakash. 1988b. Mutation of lysine-48 to arginine in the yeast RAD3 protein abolishes its ATPase and DNA helicase activities but not the ability to bind ATP. *EMBO J.* **7**: 3263.

Sung, P., L. Prakash, S.W. Matson, and S. Prakash. 1987a. RAD3 protein of *Saccharomyces cerevisiae* is a DNA helicase. *Proc. Natl. Acad. Sci.* **84**: 8951.

Sung, P., L. Prakash, S. Weber, and S. Prakash. 1987b. The *RAD3* gene of *Saccharomyces cerevisiae* encodes a DNA-dependent ATPase. *Proc. Natl. Acad. Sci.* **84**:

6045.

Szostak, J. W., T.L. Orr-Weaver, R.J. Rothstein, and F.W. Stahl. 1983. The double-strand-break repair model for recombination. *Cell* **33**: 25.

Terleth, C., C.A. van Sluis, and P. van de Putte. 1989. Differential repair of UV damage in *Saccharomyces cerevisiae*. *Nucleic Acids Res.* **17**: 4433.

Thielmann, H.W. and U. Hess. 1981. Apurinic endonuclease from *Saccharomyces cerevisiae*. *Biochem. J.* **195**: 407.

Thompson, L.H., D.L. Mitchell, J.D. Regan, S.D. Bouffler, S.A. Stewart, W.L. Carrier, R.S. Nairn, and R.T. Johnson. 1989. CHO mutant UV61 removes (6-4) photoproducts but not cyclobutane dimers. *Mutagenesis* **4**: 140.

Treger, J.M., K.A. Heichman, and K. McEntee. 1988. Expression of the yeast *UBI4* gene increases in response to DNA-damaging agents and in meiosis. *Mol. Cell. Biol.* **8**: 1132.

Tuite, M.F. and B.S. Cox. 1981. *RAD6+* gene of *Saccharomyces cerevisiae* codes for two mutationally separable deoxyribonucleic acid repair functions. *Mol. Cell. Biol.* **1**: 153.

van Duin, M., J. de Wit, H. Odijk, A. Westerveld, A. Yasui, M.H.M. Koken, J.H.J. Hoeij-makers, and D. Bootsma. 1986. Molecular characterization of the human excision repair gene ERCC-1: cDNA cloning and amino acid homology with the yeast DNA repair gene *RAD10*. *Cell* **44**: 913.

van Duin, M., J van den Tol, J.H.J. Hoeijmakers, D. Bootsma, I.P. Rupp, P. Reynolds, L. Prakash, and S. Prakash. 1989. Conserved pattern of antisense overlapping transcription in the homologous human ERCC-1 and yeast *RAD10* DNA repair gene regions. *Mol. Cell. Biol.* **9**: 1794.

von Borstel, R.C. and P.J. Hastings. 1985. Situation-dependent repair of DNA damage in yeast. In *Basic and applied mutagenesis* (ed. A. Muhammed and R.C. von Borstel), p. 121. Plenum Press, New York.

von Borstel, R.C., S.-K. Quah, C.M. Steinberg, F. Fleury, and D.J.C. Gottlieb. 1973. Mutants of yeast with enhanced spontaneous mutation rates. *Genetics* (suppl.) **73**: 141.

Walker, G.C. 1985. Inducible DNA repair systems. *Annu. Rev. Biochem.* **54**: 425.

Weinert, T.A. and L.H. Hartwell. 1988. The *RAD9* gene controls the cell cycle response to DNA damage in *Saccharomyces cerevisiae*. *Science* **241**: 317.

Weiss, B. and L. Grossman. 1987. Phosphodiesterases involved in DNA repair. *Adv. Enzymol.* **60**: 1.

Weiss, W.A. and E.C. Friedberg. 1985. Molecular cloning and characterization of the yeast *RAD10* gene and expression of RAD10 protein in *E. coli*. *EMBO J.* **4**: 1575.

White, J.H.M., D.G. Barker, P. Nurse, and L.H. Johnston. 1986. Periodic transcription as a means of regulating gene expression during the cell cycle: Contrasting modes of expression of DNA ligase genes in budding and fission yeast. *EMBO J.* **5**: 1705.

White, J.H.M., S.R. Green, D.G. Barker, L.B. Dumas, and L.H. Johnston. 1987. The *CDC8* transcript is cell cycle regulated in yeast and is expressed coordinately with *CDC9* and *CDC21* at a point preceding histone transcription. *Exp. Cell Res.* **171**: 223.

Wilcox, D.R. and L. Prakash. 1981. Incision and postincision step of pyrimidine dimer removal in excision-defective mutants of *Saccharomyces cerevisiae*. *J. Bacteriol.* **148**: 618.

Williamson, M.S. and S. Fogel. 1990. Meiotic gene conversion mutants in *Saccharomyces cerevisiae*. Characterization of *pms2-1*, *pms3-1*, *pms4-1* and double mutants. *Genetics* (in press).

Williamson, M.S., J.C. Game, and S. Fogel. 1985. Meiotic gene conversion mutants in *Saccharomyces cerevisiae*. I. Isolation and characterization of *pms1-1* and *pms1-2*. *Genetics* **110**: 609.

Winter, E. and A. Varshavsky. 1989. A DNA binding protein that recognizes oligo(dA)-oligo(dT) tracts. *EMBO J.* **8:** 1867.

Yang, E. and E. C. Friedberg. 1984. Molecular cloning and nucleotide sequence analysis of the *Saccharomyces cerevisiae RAD1* gene. *Mol. Cell. Biol.* **4:** 2161.

Yasui, A. and M.-R. Chevallier. 1983. Cloning of photoreactivation repair gene and excision repair gene of the yeast *Saccharomyces cerevisiae*. *Curr. Genet.* **7:** 191.

Yasui, A. and S.A. Langeveld. 1985. Homology between the photoreactivation genes of *Saccharomyces cerevisiae* and *Escherichia coli*. *Gene* **36:** 349.

4

Yeast Transposable Elements

Jef D. Boeke
Department of Molecular Biology and Genetics
Johns Hopkins University School of Medicine
Baltimore, Maryland 21205

Suzanne B. Sandmeyer
Department of Microbiology and Molecular Genetics
California College of Medicine, University of California
Irvine, California 92717

I. INTRODUCTION

The transposons of yeast, Ty elements, are members of a widely distributed family of eukaryotic elements called long terminal repeat (LTR)-containing retrotransposons. Retrotransposons are organizationally and functionally similar to animal retroviruses. However, they differ in that they do not encode proteins analogous to the *env* gene products of retroviruses, and they replicate without an obligatory extracellular infection cycle (Fig. 1).

The yeast retrotransposons Ty1, Ty2, Ty3, and Ty4, discussed in this review, can be divided into two classes of elements widely represented among eukaryotic species. Ty1 and Ty2 are related to the *copia* class of plant and animal elements, whereas Ty3 is more closely related to the *gypsy* class of plant and animal elements and viruses. Ty4 has been discovered recently and seems to be more closely related to Ty1 and Ty2

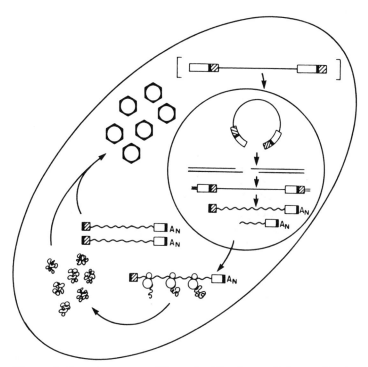

Figure 1 Retrotransposon life cycle. The intracellular replication and integration of a retrotransposon is shown. U3, R, and U5 regions within the LTRs are symbolized by open, solid, and hatched boxes, respectively. Host DNA is indicated by a double line. The RNA transcript is shown as a wavy line. VLP structures are indicated by hexagons. The full-length DNA is shown with the proposed terminal base pairs of the Ty brought together, but a covalent circular form is not implied. Drawing is not to scale.

than to Ty3 (R. Stucka et al., pers. comm.). The greater relatedness of these fungal elements to elements in plants and insects than to each other suggests that they may have moved into ancestors of their current hosts by horizontal transmission.

Yeast transposon biology has developed in exciting ways since the discovery of Ty1 10 years ago. This review summarizes those developments and attempts to highlight the ways in which yeast molecular genetics is now being applied to improve our understanding of retrotransposon transcription, replication, and integration. Further information on yeast Ty elements and retrotransposons from a variety of eukaryotes can be found in a number of books and review articles (Shapiro 1983; Varmus and Swanstrom 1984; Baltimore 1985; Varmus 1985; Fink et al. 1986; Weiner et al. 1986; Boeke 1988, 1989; Berg and

Howe 1989; Boeke and Corces 1989; Doolittle et al. 1989; Kingsman et al. 1989).

II. STRUCTURE OF YEAST TRANSPOSONS TY1–TY4 AND TF1

A. DNA Structure

The yeast transposons are similar in overall structure. Each consists of an internal domain of several kilobase pairs (kbp) flanked by two LTRs of several hundred base pairs (Fig. 2) (Cameron et al. 1979; Clare and Farabaugh 1985; Hauber et al. 1985; Warmington et al. 1985; Hansen et al. 1988). The LTRs are analogous to the LTRs of retroviruses and other retrotransposons. They contain promoter and $3'$-end formation signals for transcription of the genomic RNA. The internal domains of Ty1 and Ty2, at least, contain additional sequences required for transcription activation, and the internal domains of all of the Ty elements encode proteins required for replication and integration. Replication occurs through reverse transcription primed from specific locations within the almost full-length Ty transcript and results in regeneration of a full-length DNA copy of the element (cf. Fig. 1). Genomic insertions of Ty elements are flanked by 5-bp repeats generated from host sequences upon integration.

Isolated LTR sequences, which are also flanked by 5-bp target repeats, occur in the genome and can arise from intact elements by recombination. The solo LTR sequences of the Ty elements are more numerous than those of the complete elements. There are actually very few complete Ty3 and Ty4 elements compared to Ty1 and Ty2 elements in the genomes that have been examined. The LTR sequences of Ty1 and Ty2, Ty3, and Ty4 have been designated with the Greek letters δ, σ, and τ, respectively (Table 1). Genbauffe et al. (1984) noted that multiple short regions of sequence similarity exist among these sequences.

There are some indications that sequences within the different Ty families are not equally conserved. Ty1, the largest family, seems to exhibit the highest degree of variation. Ty2 differs from Ty1 by two large regions spanning about half of the coding region of the elements. The predicted protein sequences of these two elements, with the exception of much of *TYA* and a region of *TYB* of unknown function, are fairly similar (Kingsman et al. 1981; Warmington et al. 1985). There is no evidence for hybrid Ty1 and Ty2 elements despite their apparent relatedness. δ elements within the Ty1 and Ty2 classes vary widely in sequence, and isolated δ elements with up to 30% divergence have been reported (Rothstein et al. 1987). Because only two Ty3 elements have been analyzed, it is not yet possible to draw conclusions concerning overall sequence con-

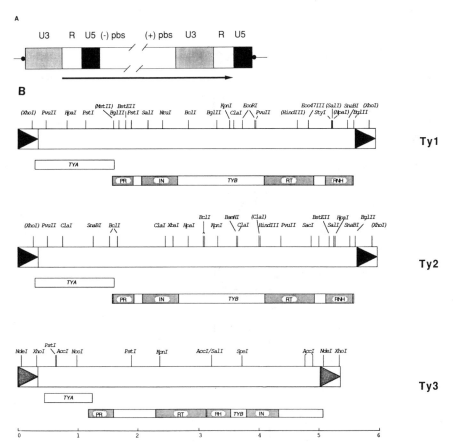

Figure 2 Comparative anatomy of Ty elements. (*A*) Generic retrotransposon structural features referred to in text. Open broken box indicates internal coding region; shaded boxes indicate LTR segments; arrow indicates RNA; dots indicate target site duplications; (–) pbs and (+) pbs are primer binding sites for (–) and (+) strand reverse transcription. (*B*) Ty elements. Ty DNA sequences are indicated by the upper open box. LTR sequences are shown as solid triangles. Restriction sites in parentheses are polymorphic. *TYA* and *TYB* ORFs are shown below the restriction map as open boxes. Domains within the predicted proteins are as follows: (PR) Protease; (IN) integrase; (RT) reverse transcriptase polymerase domain; (RH) RNase H domain. Scale is in kbp.

servation of this element. A number of σ elements have been characterized, however, and it is clear that these are highly conserved, both at the ends of Ty3 elements and as isolated repeats (Hansen et al. 1988; Sandmeyer et al. 1988).

Several factors are likely to contribute to the heterogeneity that is observed within Ty families. In general, the LTR regions are more hetero-

Table 1 Subdivision of Ty element sequence and sequence features of Ty elements

(a) *Subdivision of Ty element sequence*

Ty	Length			Position							Ref.
	Ty (kbp)	LTR (bp)	i.r.[a] (bp)	U3	R	U5	−pbs	+pbs	TYA	TYB	
Ty1-912	5.9	334	2	1–240	241–289	289–334	335–344	5576– 5584	294 (ATG)–1616(TGA) 440 codons	1576(ACA)–5562(TAG) 1328 codons	1
Ty2-117	5.9	332	2	—	—	—	333–342	5623– 5630	292(ATG)–1608(TAA) 438 codons	1562(GCG)–5608(TAG) 1349 codons	2
Ty3-1	5.4	340	8	1–222	223–240	241–340	243–250	4998– 5011	416(ATG)–1288(TGA) 290 codons	1248(ACG)–5060(TGA) 1270 codons	3
Ty4	6.3	371	5	—	—	—	—	—	—	—	4,5

(b) *Sequence features of Ty elements*

Ty (kbp)	LTR (bp)	i.r.[a] 5′-3′	-pbs	+pbs	Ref.
Ty1-912	δ	TG/CA	TGGTAGCGCC	GGGTGGTA	1
Ty2-117	δ	TG/CA	TGGTAGCGCC	GGGTGGTA	2
Ty3-1	σ	TGTTGTAT/ATACAACA	TGGTAGCG(TT)	GAGAGAGAGGAAGA	3
Ty4	τ	TGTTTG/CAACA			4,5

[a] Inverted terminal repeats within each LTR.

References: (1) Clare and Farabaugh 1985; (2) Warmington et al. 1985; (3) Hansen et al. 1988; (4) Genbauffe et al. 1984; (5) Stucka et al. 1989.

199

geneous than the coding regions, probably reflecting the fact that these noncoding regions, especially in isolated LTRs, are under less stringent selection. Because it is possible to isolate yeast strains with no Ty1 elements (M. Ciriacy and P. Philippsen, pers. comm.) or no Ty3 elements (V. Bilanchone and S.B. Sandmeyer, unpubl.), it is also likely that Ty sequences are not required for host viability. (The issue of whether they might be advantageous to populations has not yet been addressed.) Because active high-frequency Ty1 transposition probably causes reduced viability (Boeke et al. 1985), cells in which the dispersion of replication-competent copies of Ty elements is limited by mutation may have a selective advantage. A major source of Ty structural heterogeneity is undoubtedly reverse transcription, a process known to be both error-prone and recombinogenic (Gopinathan et al. 1979; Roberts et al. 1988).

1. Ty1

Ty1 is the most highly repeated Ty element in common laboratory strains of *Saccharomyces cerevisiae*. Typically, about 25–35 copies of Ty1 are present per haploid genome (Cameron et al. 1979; Curcio et al. 1989); however, some wild strains contain few or no copies of Ty1, although they usually contain 10–15 copies of Ty2. Ty1 is 5.9 kbp in length and is flanked by δ elements. δ elements are 334–338 bp in length and terminate in the inverted repeat TG...CA (Farabaugh and Fink 1980; Gafner and Philippsen 1980). Solo δ LTR sequences are present in more than 100 copies in most yeast strains.

The restriction maps of Ty1 elements vary at a number of sites, both within the coding region and within the LTR sequences (Fig. 2 indicates some of the variable sites). The restriction site heterogeneity within the coding regions analyzed consists of base substitutions. Most of these differences are third-base changes that do not affect the encoded protein. One point mutation that has been characterized in detail is the mutation that decreases Ty1-173 transposition efficiency by about 60-fold. This mutation, *tyb-2098*, results in the replacement of a leucine residue with an isoleucine in the p90-TYB protein. It is found in seven of the copies of Ty1 in the strain studied, suggesting that many of the chromosomal copies of Ty1 elements encode proteins that are deficient in catalyzing transposition events (Boeke et al. 1988a,c).

In addition to point mutations, DNA rearrangements, such as deletions, inversions, and possibly duplications, have been found within Ty1 elements (Errede et al. 1986; Boeke 1989). For one case, Errede et al. (1986) hypothesized that an inversion mutation was caused by an aberrant reverse transcription event. In the Ty1 family, these genomic rearrangements are less commonly observed than base substitutions.

2. Ty2

Only six Ty2 elements were present in the yeast strain in which they were originally characterized (Kingsman et al. 1981); 5–15 Ty2 elements are present in the majority of common laboratory strains (Curcio et al. 1989; J.D. Boeke, unpubl.). The two Ty2 elements for which the DNA sequence is known (Ty2-117 and Ty2-917) are extremely similar (Roeder et al. 1985a; Warmington et al. 1985; P.J. Farabaugh, pers. comm.). Restriction mapping of Ty2 elements also fails to reveal extensive structural polymorphism. Undoubtedly, the lower copy number of Ty2 elements in laboratory strains contributes to this lack of polymorphism, but it is our impression, even taking this into account, that the diversity of structure in Ty2 elements is lower than that among Ty1 elements.

3. Ty3

There are one to four Ty3 elements present in typical haploid strains of *S. cerevisiae*. The Ty3 element is 5.4 kbp in length and is composed of an internal domain flanked by direct repeats of the 340-bp σ element (Clark et al. 1988). The σ element, and hence the Ty3 element, terminates in an 8-bp perfect inverted repeat (5´-TGTTGTAT...324 bp...ATACAACA-3´) (del Rey et al. 1982; Sandmeyer and Olson 1982; Clark et al. 1988).

Two Ty3 elements (Ty3-1 and Ty3-2) have been cloned and characterized by restriction mapping and DNA sequence analysis. They occur at position −16 from a tRNA^{Cys} gene and position −17 from a tRNA^{Ile} gene, respectively. Sequence comparison shows that the two elements differ only by a 78-bp duplication and several point mutations. The significance of one of these mutations is discussed below. The four σ elements occurring at the ends of the two Ty3 elements are identical in sequence (Clark et al. 1988; Hansen et al. 1988; Hansen and Sandmeyer 1990).

σ elements, which compose the LTRs of Ty3, are also found as isolated elements throughout the genome (del Rey et al. 1982; Sandmeyer and Olson 1982). Southern blot analysis shows that there are about 30 isolated copies of this sequence present in standard haploid laboratory strains and that they are all associated with tRNA genes (Brodeur et al. 1983). A comparison of different σ elements from one strain of yeast (Sandmeyer et al. 1988) shows that even isolated σ elements are highly conserved, differing typically at no more than a few positions.

4. Ty4

Ty4 is the most recently characterized of the Ty elements. Only one to four copies of this element have been found in the haploid strains analyzed thus far. The characterized Ty4 element is 6.3 kbp in length and contains LTRs of the 371-bp τ element (Stucka et al. 1989). τ also occurs in 15–25 isolated copies in the genome. Both Ty4 and τ have the terminal inverted repeat TGTTG...CAACA and are flanked by 5-bp repeats of the apparent target sequence (Chisholm et al. 1984; Genbauffe et al. 1984).

5. Tf1

Tf1 is a recently discovered LTR-containing retrotransposon from the fission yeast *Schizosaccharomyces pombe*. It is found in about 10–40 copies per haploid genome. Preliminary sequence analysis suggests that it is most closely related to Ty3 and elements of the *gypsy* family of retrotransposons (Levin and Boeke 1990).

B. RNA Structure

The most abundant transcript of Ty elements is an almost full-length, polyadenylated transcript that resembles retroviral genomic RNA in structure (see Elder et al. 1983). The transcription of this RNA by polymerase II begins at a site earlier in the upstream LTR than the site at which it terminates in the 3′ LTR. This generates a terminally redundant RNA molecule. The first and last nucleotides of this transcript define retrovirus-like domains of the LTR called R (the *r*epeated terminus of the transcript), U5 (*u*nique to the 5′ end of the mRNA), and U3 (*u*nique to the 3′ end of the transcript). Messenger and genomic RNA molecules have not been distinguished biochemically, and it is likely that one type of molecule is capable of performing both functions.

1. Ty1

The major Ty1 transcript is 5.7 kbp in length and is one of the most abundant RNA species in the cell. The unusual stability of this molecule probably contributes to its abundance. When temperature-sensitive mutants of RNA polymerase II are shifted to the nonpermissive temperature, Ty1 mRNA is degraded with a half-life of at least 3 hours, which is much greater than that of most cellular mRNAs (M. Nonet and R. Young, pers. comm.). The extraordinary stability of these transcripts may be attributable to protection within virus-like particles (VLPs).

The major 5'end of the Ty1 transcript was mapped by primer extension to 95 bp before the downstream end of the δ element (Ty1-912 nucleotide 240; cf. Fig. 3). The 3'ends of Ty transcripts were mapped by sequence analysis of cDNA clones to positions 38–46 bp before the end of the δ element (Elder et al. 1983). This transcript structure defines the regions analogous to proviral U3-R-U5 regions at nucleotides 1–240, 241–289, and 289–335, respectively.

Subgenomic Ty1 transcripts of 5.0 and 2.2 kb are also present in yeast cells. Northern blot analyses of nested deletions from the upstream δ element by Fulton et al. (1988) are consistent with initiation of the 2.2- and 5.7-kb transcripts in the δ sequence and initiation of the 5.0 kb downstream in the ε sequence. In addition, Winston et al. (1984a) have mapped the 5'end of a set of 5.0-kb transcripts in *spt3* cells to a position approximately 800 bp downstream from the major species (i.e., within the ε region). Elder et al. (1983), however, interpreted R-loop analysis to suggest that the 5.0-kb transcript might initiate within the upstream δ sequence. Whether or not two different 5.0-kb transcripts are synthesized has not been resolved.

In addition to terminal redundancy corresponding to the organization of the retroviral genomic RNA, the Ty1 element transcript has other retrovirus-like features (Varmus and Swanstrom 1984). The minus-strand primer-binding site (pbs) is a 10-bp sequence immediately downstream from the 5'LTR that is perfectly complementary to the 3'end of the initiator methionine tRNA (cf. Table 1; Fig. 3) (Simsek and RajBhandary 1972; Eibel et al. 1981; Cigan and Donahue 1986). Indirect evidence suggests that this tRNA primes minus-strand synthesis, the initial step in reverse transcription. Mutagenesis of the pbs showed that it was essential, and tRNA$^{i\text{-met}}$ is selectively packaged into Ty1-VLPs (K. Chapman et al., unpubl.). Immediately upstream of the 3'LTR is a sequence that is somewhat purine-rich and has been suggested to be the site for priming of plus-strand synthesis. Direct evidence that plus-strand priming requires this sequence is not yet available.

Sequences required for packaging of retroviral genomes are located within the upstream portion of the internal domain. Analysis of helper-dependent mini-Ty elements has shown that all of the sequences required in *cis* for Ty1-H3 element transposition occur from nucleotides 240 to 620 and from 5614 to 5918 of the element (Xu and Boeke 1990a). (Ty1-H3 numbering is the same as that for Ty1-912; Fig. 3.)

2. Ty2

Ty2 also produces an end-to-end RNA of about 5.7 kb. The termini of this transcript are likely to be identical to those of Ty1 transcripts because both of these elements have very similar δ LTRs, and primer extension experiments indicate that the 5' end of the RNA is identical to that of Ty1.

The almost full-length transcript from the Ty2 element is much less abundant than the equivalent RNA from Ty1, even after correction for relative copy number of the two types of elements (Curcio et al. 1989). This discrepancy is probably not due to differential RNA instability, because when the Ty2 sequences are fused to the *GAL1* promoter, the resulting transcripts are as abundant as Ty1 transcripts produced from the same expression vector. This suggests that the transcription of Ty2 may be regulated differently than the transcription of Ty1.

Both pbs sequences are identical in Ty1 and Ty2, and the region thought to be required for packaging of RNA into VLPs is conserved between Ty1 and Ty2 as well. Mini-Ty1 elements are complemented in *trans* by both Ty1 and Ty2 helper elements, suggesting that Ty2 proteins can encapsidate, reverse transcribe, and catalyze the integration of Ty1 RNA (H. Xu and J.D. Boeke, unpubl.).

3. Ty3

Ty3 is transcribed into a major 5.2-kb and minor 3.1-kb RNA species (Clark et al. 1988). The ends of the 3.1-kb RNA have not been mapped, but hybridization with different internal fragments suggests that the 5' end is located within the internal domain and the 3' end is located in the downstream σ element (V. Bilanchone and S.B. Sandmeyer, unpubl.). The 5.2-kb RNA fulfills the structural requisites of an intermediate in

Figure 3 Comparison of Ty1 and Ty2 regulatory sequences. The sequences aligned are as follows: Ty1-912 (Clare and Farabaugh 1985); Ty1-15 (Fulton et al. 1985); Ty1*CYC7-H2* (Errede et al. 1985); Ty2-917 (Roeder et al. 1985a); and Ty2-117 (Warmington et al. 1985). Ty elements on lines 1–3 are of the Ty1 class, and Ty elements on lines 4 and 5 are of the Ty2 class. Regions shown to be involved in transcription regulation are labeled, and the particular Ty elements with which activities were demonstrated are referenced in the text. (Note that the variant of Ty2-917 that binds TyBF has the same TyBF site sequence as Ty2-117.) Numbering is according to Ty1-912. Putative a1-α2 mating-type control sequences are indicated by the thin lines. This is not a complete listing of protein-binding sites within Ty1 and Ty2. Two terminator-like sequences TS-1, and TS-2, conforming to the Zaret and Sherman consensus (1982, 1984) identified by Yu and Elder (1989) in the 3' LTR of Ty1-D15, are indicated by double wavy lines. (TS-1 is a poor match to the consensus in the Ty elements shown here.)

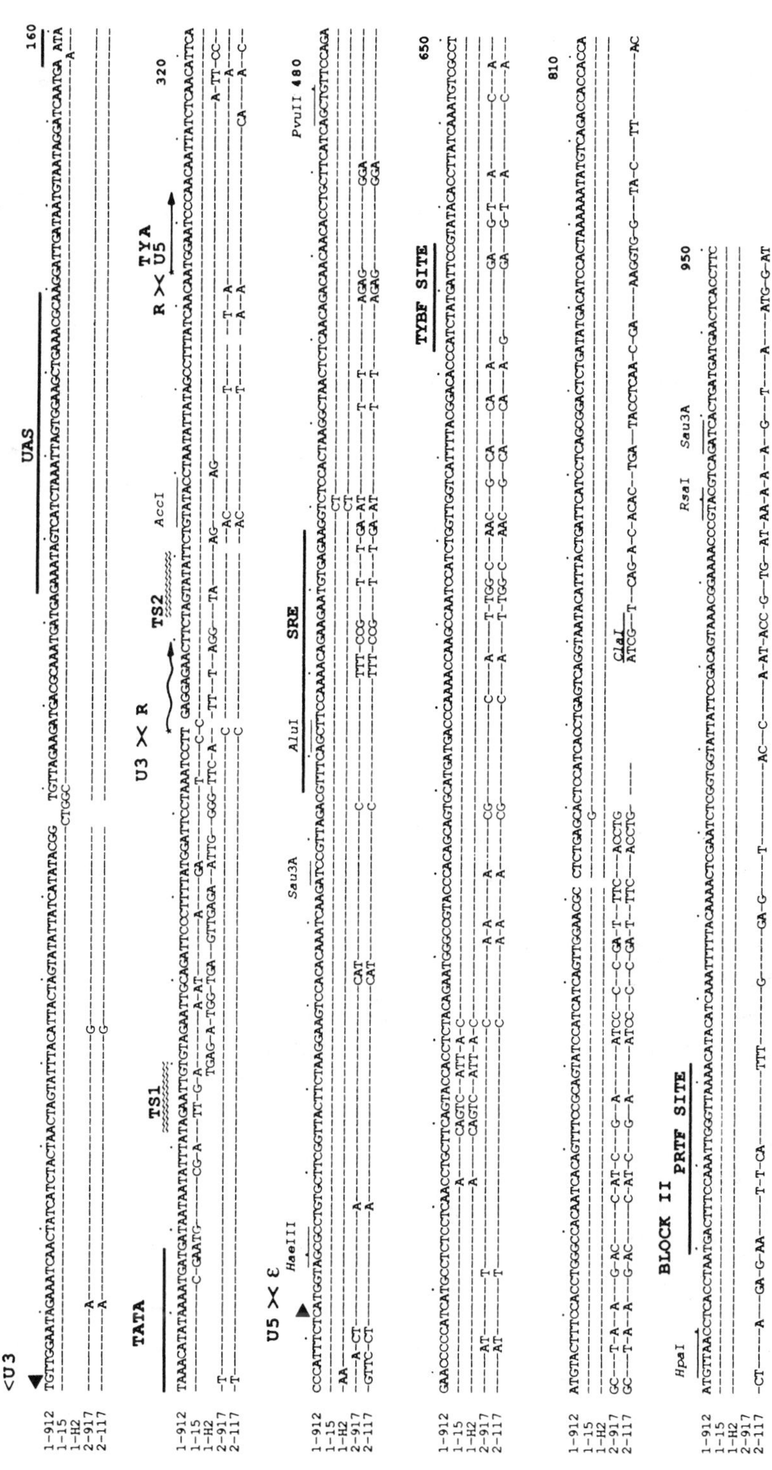

Figure 3 (See facing page for legend.)

retrovirus-like replication. The major 5′ end points map at positions 118 (Ty3 nucleotide 223), 114, and 106 from the right end of the upstream σ element. S1 and reverse primer extension analyses show that some of the 5′-end points are farther upstream, possibly outside of the σ element. The major 3′ ends of the transcripts map to a cluster of four sites, 75–100 bp from the right end of the downstream σ element. Some of the transcripts extend to the end of the downstream σ or beyond, implying that termination is inefficient. These end points partition the inserted Ty3 LTRs into U3-R-U5: nucleotides 1–222, 223–240, and 241–340, respectively. Inspection of the sequences at the U5 and U3 junctions with the internal domain reveals functional similarities with the regions involved in priming the replication reaction. At the upstream end of the internal domain in the RNA, the dinucleotide sequence CC is followed by 8 nucleotides that are complementary to the 3′ end of yeast tRNA^{i-met}. If GU pairing is allowed, this can be extended to 10 nucleotides. Immediately upstream of U3 is a tract of 14 purines, which resembles the polypurine tract that functions as the retrovirus plus-strand primer. Thus, the 5.2-kb Ty3 RNA is functionally analogous to the RNA genome of a retrovirus (Hansen et al. 1988).

C. Protein-coding Regions

Currently, characterized yeast retrotransposons contain two long overlapping open reading frames (ORFs) that correspond to retroviral *gag* and *pol* genes. These genes encode proteins required for retrovirus capsid formation and for replication and integration, respectively. In retroviruses, the *pol*-encoded proteins are synthesized as part of a *gag/pol* polyprotein and include a protease, a reverse transcriptase, an RNase H, and an endonuclease. Comparisons of the ORFs and encoded domains in each element are shown in Figure 2 and Table 1.

1. Ty1 and Ty2

The two long overlapping ORFs in Ty1 and Ty2 have been designated *TYA* and *TYB*. Where clarification is necessary, these will be distinguished further as *TYA1*, *TYA2*, etc. These ORFs are nearly coextensive in a total of three sequenced Ty1 elements (Clare and Farabaugh 1985; Hauber et al. 1985; Boeke et al. 1988c) and two Ty2 elements (Warmington et al. 1985; P.J. Farabaugh, pers. comm.). In Ty1 elements, *TYA* and *TYB* are 440 and 1328 or 1329 codons, respectively. In Ty2 elements, the ORFs are 438 and 1349 codons in length. The initiator codon of the *TYA* genes in Ty1 and Ty2 elements is preceded by an adenine residue at position −3, characteristic of eukaryotic genes (Kozak 1984). Ty1 and Ty2 are unusual among retrotransposons in that translation of the first ORF begins within the δ (LTR) sequence.

2. Ty3

The internal domain of a transpositionally active Ty3 element, Ty3-1, contains two long overlapping ORFs of 307 and 1270 codons, designated *TYA3* and *TYB3*, respectively. The internal domain of Ty3-2, a transpositionally inactive element, contains ORFs of 307, 1221, and 110 codons (Hansen et al. 1988; Hansen and Sandmeyer 1990). The differences between Ty3-1 and Ty3-2 can be accounted for by insertion of 26 codons into the upstream portion of the second ORF of Ty3-2 and deletion of a single nucleotide causing division of *TYB3-2* into two ORFs. *TYA3* contains an ATG in the initiator context starting 76 bp inside the internal domain. The *TYB3* ORF overlaps *TYA3* by 13 codons in the plus-one frame. In Ty3, *TYB3* overlaps the downstream LTR.

III. REGULATION OF TY ELEMENT TRANSCRIPTION AND TRANSPOSITION

A. Transcription Initiation and Termination

In retrotransposons and retroviruses, the upstream LTR acts as a promoter and the downstream LTR acts as a polyadenylation site and/or terminator. The basis of differential activities of the identical LTR sequences is not understood for the yeast elements. A promoter competition mechanism has been suggested for retroviruses, but it is not well defined on the molecular level (Cullen et al. 1984). Preliminary evidence suggests that this model does not apply to Ty1 elements (Yu and Elder 1989; G. Natsoulis and J.D. Boeke, unpubl.).

In yeast, the transcription initiation site is typically specified by the TATA sequence (Struhl 1982; Hahn et al. 1985; Nagawa and Fink 1985), together with one of two conserved initiation (I) sites (PuPuPyPuPu or TCPuA) (Hahn et al. 1985). It is known from extensive mutagenesis of TATA elements that many sequences can perform the function of this element (Chen and Struhl 1988). The sequences required to direct polyadenylation and transcription termination have recently been determined for a Ty1 element (Yu and Elder 1989).

B. Transcriptional Regulation of Ty1 and Ty2

1. Initiation

The signals that specify the position of Ty1 and Ty2 transcription initiation lie within the δ element (see Fig. 3). The major start site of Ty1 transcription was mapped by Elder et al. (1983) to the second pyrimidine (Ty1-912 nucleotide 240) within the sequence TPyGA; this sequence resembles the I site consensus and coincides with an *Xho*I site that is present in many elements. An AT-rich region exists, beginning 83 bp upstream of the I site in Ty1 and Ty2 δ elements (Ty1-912/Ty2-917 nucleotides 157–173, AATAT[A/T]AACATATAAAAT). Analysis of deletions and point mutations suggests that this region contains the TATA element (Roeder et al. 1985a; Liao et al. 1987; Fulton et al. 1988; Hirschman et al. 1988). Deletion of the sequence TATAGAA (nucleotides 190–196) from Ty2-917, a Ty2 element, also decreases transcription, particularly in the absence of the first AT-rich sequence region. The second sequence may therefore function as an auxiliary TATA (Liao et al. 1987). Thus, signals for initiation of Ty1 and Ty2 transcription resemble those for initiation of other yeast genes that are transcribed by polymerase II.

2. 3'-end Formation

Yu and Elder (1989) have analyzed the sequences responsible for formation of the 3' end of Ty1 RNA. Their studies have not distinguished between termination and processing and show that 3'-end formation can occur at at least two sites in the 3' δ; a minor site of 3'-end formation is within the U3 region, giving rise to an RNA nonfunctional for transposition, because it lacks the repeated 45-nucleotide R region found in the major transcript. At least two DNA sequences within the 3' δ are required for 3'-end formation in the major Ty1-D15 transcript. These signals for 3'-end formation, designated TS1 and TS2, are strikingly similar to the Zaret-Sherman (1982, 1984) consensus sequence. We note that the TS1 sequence is not at all well conserved among Ty1 3' LTR sequences, suggesting that 3'-end formation in Ty1 does not require this sequence (Elder et al. 1983). TS2 appears to be well conserved among Ty1 3' LTRs. Both of these sites are also present in the 5' portion of the major transcript, leaving unsolved the enigma of how RNA polymerase ignores the termination signals in the 5' δ but utilizes them efficiently in the 3' δ. No evidence was found of signals required in the U3 region, which is not transcribed in the 5' LTR, nor were short RNA transcripts representing termination within the 5' LTR found in yeast cells.

3. Regulation

a. Carbon Source and Mating Type. Ty transcript levels are regulated in response to signals generated by metabolic state, mating type (Elder et al.

1981), and DNA damage. In log-phase haploid cells growing on a glucose carbon source, Ty RNA constitutes 5–10% of total poly(A) RNA (Elder et al. 1981). Some yeast strains grown on nonfermentable carbon sources, such as glycerol, have lower levels of Ty transcripts. In addition, mating-type control is much more severe in cells that show carbon source regulation of Ty transcription. In one study, Ty transcript levels in *MATa/MATα* cells grown on glycerol were 20 times lower than those observed in haploid or diploid *MATa/MATa* or *MATα/MATα* cells (Taguchi et al. 1984).

b. DNA Damage. Yeast cells treated with various DNA-damaging agents have been shown to have increased levels of Ty transcripts and transposition. UV irradiation results in a tenfold increase in Ty transcript levels (Rolfe et al. 1986; McClanahan and McEntee 1984). Although the sequences required for this effect have not been identified, the hexamer sequence ATGATT, which is repeated upstream of the UV-inducible *CDC9* ligase gene (Barker et al. 1985), is related to sequences found in δ upstream of the TATA element. The mutagen 4-nitroquinoline-1-oxide (4NQO) also increases Ty transcript levels; however, this response is much slower and does not peak until 4–6 hours after treatment (McEntee and Bradshaw 1988). This induction may therefore occur by a mechanism independent from that operating with UV irradiation. 4NQO (McEntee and Bradshaw 1988), as well as γ irradiation (Morawetz 1987), increases frequencies of transposition into *ADH2* and *ADH4*, as monitored by selection of antimycin-A-resistant cells, followed by Southern analysis.

DNA-damaging agents are known to increase transcription from certain promoters and to increase frequencies of transposition in bacteria. In the case of Ty elements, increased transcription of transpositionally competent elements is associated with increased transposition, as would be predicted for a retrotransposon. However, it remains to be determined what the basis of the correlation between damage and transposition is and whether there are direct effects of DNA damage, in addition to changes in transcription.

c. Regulatory Sequences Defined by Activation of Ty1 and Ty2 Promoters. Transcription of most yeast genes is dependent on the presence of one or more upstream activating sequences (UAS), which act in an orientation-independent manner from several hundred base pairs upstream of the I site to increase and/or regulate transcription (for review, see Guarente 1988). For Ty2, there is evidence that sequences upstream of the TATA region contribute to maximal promoter activity. Liao et al. (1987) found that deletion through position 93 or 132 of δ reduced expression of a *his4-917/lacZ* fusion between 60- and 70-fold. Similar analysis of a Ty1 element did not show evidence for a δ UAS (Fulton et al. 1988). However, there are differences between the upstream and downstream sequences of these elements that could account for this discrepancy. Neither the Ty1 nor the Ty2 δ promoter functions in the absence of activation sequences located within the ε domain. By its very nature, a mobile element has rigid constraints on the distance upstream of the I site available for regulatory information. In the case of Ty1 and Ty2, the severity of this constraint is mitigated by overlapping regulatory and coding information within the ε domain.

Regions within ε that are required for activation of the δ promoter have been broadly defined by in-frame fusions of 3′-terminal deletions of Ty2 and Ty1 to reporter genes. Levels of transcripts from Ty2-917 fused to *lacZ* at positions 296, 475, and 559 increased from less than 1% to 33% and 100%, respectively (Liao et al. 1987). Fusions of Ty1-15 to the human α2 interferon gene produced qualitatively similar results (Fulton et al. 1988). Studies on both classes of elements are in agreement that sequences within the first 1 kbp are sufficient to produce high levels of transcripts.

C. Transcriptional Regulation of Ty3

1. Promoter Sequences

The σ element contains sequences corresponding to conserved promoter elements in yeast genes. The sequence GATAA occurs at the major start site of σ transcription (Van Arsdell et al. 1987; Clark et al. 1988). The central T (nucleotide 223) corresponds to the Py start position in the conserved I site context PuPuPyPuPu. Two sequences that have homology with the TATA(A/T)A(A/T) consensus occur in σ upstream of the start site, one at nucleotides 138–144 (TATAAAT) and one much closer to the start site at nucleotides 209–220 (TATAATATATAA).

Regulation of Ty3 transcription differs in several respects from Ty1 and Ty2 regulation. One difference is that transcription from the σ promoter appears to be independent of a requirement for transcription activation by the internal domain. Northern blot analysis of RNA from a strain containing two Ty3 elements shows that the 5.2- and 3.1-kb transcripts hybridize to internal domain-specific probes but that the 6.0-, 1.0-, 0.92-, 0.65-, 0.58-, and 0.40-kb transcripts hybridize only to σ-specific probes (Clark et al. 1988). At least some of these transcripts represent RNAs that initiate in isolated σ elements (Van Arsdell et al. 1987; P. Kinsey and S.B. Sandmeyer, unpubl.).

2. Regulation of the σ Promoter

Transcription from the σ promoter occurs at three different levels: haploid-constitutive, pheromone-induced, and mating-type-repressed. The level of Ty3 transcription in uninduced haploid cells varies considerably among different strains (Van Arsdell et al. 1987; Clark et al. 1988). Transcript levels in uninduced haploid cells are not affected by disruption of the STE2 gene, which encodes the α-factor receptor (V. Bilanchone and S.B. Sandmeyer, in prep.).

Transcription is increased from the σ promoter in MATa cells treated with α-factor (Van Arsdell et al. 1987). Van Arsdell et al. (1987) showed that three σ-hybridizing transcripts (5.2, 0.65, and 0.45 kb) increased in levels after treatment of MATa cells with α-factor. A 50-fold increase was shown in levels of a 0.65-kb transcript, the 5′ end of which mapped within the σ element (Van Arsdell et al. 1987).

Fusions of the σ promoter conferred α-factor inducibility on transcription of the SUC2 gene, showing that this regulation is at the transcriptional level. STE2 and STE5 gene products are required for α-factor induction of the σ promoter (Van Arsdell et al. 1987). Van Arsdell and Thorner (1987) pointed out the occurrence of an 8-nucleotide sequence (consensus ATGAAACA). This sequence, or minor variants, occurs four times in σ (starting at nucleotides 15, 56, 66, and 97). It includes the 7-bp consensus, TGAAACA, that is found in the 5′-flanking DNA of FUS1, CSH1, MFa2, STE6, STE2, BAR1, STE3, and MFa1 (Bulawa et al. 1986; Kronstad et al. 1987; Trueheart et al. 1987), other pheromone-inducible genes.

Levels of Ty3 transcripts are negligible in MATa/MATα strains. In diploid strains in which the MATa1 gene is disrupted, levels of these transcripts are restored to those found in haploid cells. Unlike the case for Ty1 and Ty2, Ty3 mating-type regulation is probably mediated by sequences upstream of the internal domain. Artificial fusions containing only the σ element and about 100 bp of the internal domain are strongly mating-type-regulated (V. Bilanchone and S.B. Sandmeyer, in prep.). There are two sequences within the σ element and one in the upstream portion of the internal domain that bear some resemblance to the a1-α2 protein-binding consensus (PyCPuTGTNN[A/T]NANNTACATCA) (see Miller et al. 1985; Siliciano and Tatchell 1986). Nevertheless, it is not clear whether Ty3 repression is mediated directly by the a1-α2 protein complex or indirectly by virtue of dependence on haploid-specific gene products.

IV. EFFECTS OF YEAST TRANSPOSONS ON EXPRESSION
OF NEIGHBORING GENES

A. Ty1 and Ty2

Ty insertions can affect the expression of neighboring genes in several ways. Activation, mating-type regulation, deregulation, inactivation, and cold sensitivity have been observed separately and in various combinations. The nature of the effect is determined by several variables, including the sequence of the Ty element, the position and orientation of the insertion, and the genotypic background of the strain. Mechanistically, these effects can be explained by replacement of the promoter or regulatory sequences of the disrupted gene with Ty promoter, activator, or mating-type control sequences.

1. Activation, Deregulation, and Mating-type Control

Ty insertion can change the expression of a neighboring gene by substitution of Ty activation sequences for the 5′ regulatory sequences of that gene. Ty insertions within the promoter regions of host genes can result in the loss of normal induction or repression and the gain of mating-type repression. A dramatic example of this effect is transcription of the promoterless *his3Δ4* allele, which in *SPT* strains is activated by Ty1 and Ty2 insertions up to 175 bp upstream (Scherer et al. 1982; Boeke et al. 1985, 1986; Curcio et al. 1989). In these activation events, Ty activation sequences are proximal to *HIS3*, and in most cases, the Ty element is oriented so that its transcription is divergent to that of *HIS3*. Insertions that are divergently oriented are usually regulated by mating type and therefore have been referred to as ROAM mutations for *r*egulated *o*verproducing *a*lleles responding to *m*ating signals (Errede et al. 1980). *CYC7-H2* (Rothstein and Sherman 1980; Errede et al. 1980, 1981), *cargA*+O^h (Dubois et al. 1978, 1982; Errede et al. 1980; Jauniaux et al. 1981, 1982), *cargB*+O^h (Deschamps et al. 1979; Errede et al. 1980; Dubois et al. 1982; Degols et al. 1987), constitutive alleles of *DUR1*, *DUR2* (Lemoine et al. 1978; Errede et al. 1980; Cooper et al. 1984), and *ADH2-1^c–ADH2-9^c* (Ciriacy 1976; Ciracy and Williamson 1981; Williamson et al. 1981, 1983) alleles of the genes encoding iso-2-cytochrome *c*, arginase, ornithine deaminase, urea amidolyase, and the glucose-repressible alcohol dehydrogenase (II) are mutations of this type. In the following discussion, the gene neighboring the Ty insertion will be referred to as the target gene. Nothing is implied about Ty insertion specificity.

ROAM expression correlates with Ty transcription under several different conditions. In ROAM mutations, the Ty is oriented with its promoter and activation sequences proximal to the target gene promoter. ROAM expression levels range from similar to levels produced under inducing or derepressing conditions to 20-fold higher. Expression of the target gene is relatively insensitive to normal regulation. In *MATa/MATα*, but not in *MATa/MATa* or *MATα/MATα* cells, expression of ROAM genes is decreased 4- to 20-fold.

The effects of Ty insertions at *ADH2* and *CYC7* have been studied relatively extensively, and the results shed some light on the basis of Ty-mediated gene activation. The seven constitutive *ADH2* mutants result from insertion of six Ty1 elements and one Ty2 element all in the divergent orientation 125–210 bp upstream of the structural sequence of *ADH2*. The 5′ ends of the *ADH2* transcripts in these mutants map to the wild-type start site (Williamson et al. 1983). *ADH2* transcription is no longer sensitive to catabolite repression by glucose, and, as is the case for Ty1, *ADH2* expression is under mating-type control in cells grown on glycerol (Taguchi et al. 1984).

The ROAM phenotype of the *CYC7-H2* allele is attributable to the presence of a rearranged Ty1 element 107 bp upstream of the position to which the 5′ end of the wild-type *CYC7* transcript maps (184 bp upstream of the translation start site). The putative δ UAS and the TATA sequence(s) are not present, making it unlikely that this Ty is transcriptionally active (Errede et al. 1981, 1986). Cells with the *CYC7-H2* allele overexpress iso-2-cytochrome *c* in haploid, *MATa/MATa*, and *MATα/MATα* cells by 20-fold and repress expression in *MATa/MATα* cells by 10-fold (Rothstein and Sherman 1980; Errede et al. 1980).

Several generalizations about Ty-mediated gene activation can be made from the studies of ROAM mutations at *ADH2* (Williamson et al. 1983; Taguchi et al. 1984) and *CYC7-H2* (Errede et al. 1981, 1986). First, loss of regulation correlates with displacement of regulatory sequences by the 6-kbp length of the Ty. Second, activation occurs for insertions that are oriented so that Ty activation sequences in the ε domain are proximal to the target gene. Third, Ty activation sequences, rather than Ty transcription per se, are likely to be critical for Ty-mediated gene activation. However, the existence of Ty insertions, such as *his4-917*, which are in the ROAM orientation and are transcribed but inactivate the target, shows that activation does not result in every case from simply substituting Ty activation sequences for target gene UAS sequences. At the very least, particular Ty elements vary widely in their effects on target genes. The *cis*- and *trans*-acting mutations described below shed some additional light on interactions of the Ty and the target gene promoter.

2. Internal, cis-Acting Ty Sequences That Affect Transcription of Neighboring Genes

Several studies have exploited the effects of Ty insertions on target genes in order to identify transcriptional activation or regulatory sequences within ε. To facilitate comparison of results with different elements for this discussion, positions within all elements are given relative to the comparable coordinate in the Ty1-912 sequence. An alignment of DNA sequences from three Ty1 elements (Ty1-912 [Clare and Farabaugh 1985]; Ty1-15 [Fulton et al. 1985]; and Ty1-CYC7-H2 [Errede et al. 1985]) and two Ty2 elements (Ty2-917 [Roeder et al. 1985a] and Ty2-117 [Warmington et al. 1985]) is shown in Figure 3. As is apparent in this alignment, sequences of Ty1 and Ty2 elements diverge after about 600 bp. For neighboring gene activation, as in the case of Ty transcription itself, the upstream 2 kb of the Ty is essential (Errede et al. 1984). From studies of Ty1 and Ty2, there is evidence for at least three internal regions that independently bind proteins and affect Ty transcription.

The promoter-proximal region responsible for activation of CYC7-H2 has been localized between the AccI and PvuII sites within ε (Ty1-912 nucleotides 266–477) (Errede et al. 1984, 1985; Company and Errede 1987; Company et al. 1988). This fragment forms several different complexes with yeast cell extracts. One of these complexes forms over a 40-bp region termed the sterile-responsive region (SRE) (Company and Errede 1988). This complex is absent in extracts from ste12 cells and reduced in ste7 cells, possibly explaining the reduced transcription of Ty1 in MATa/MATα cells (Company and Errede 1988). The STE12 gene product can be phosphorylated; the phosphorylated form is active in complex formation. These data suggest that the STE7 and STE11 gene products, which have been shown to be protein kinases, may play a role in activating Ste12p (Teague et al. 1986; B. Errede, pers. comm.).

The analogous Sau3A–PvuII fragment from Ty1-15, when substituted for the PGK UAS in the ROAM orientation, stimulated phosphoglycerate kinase (PGK) transcription in a mating-type-dependent manner. The region defined in these studies was referred to as Ty activator sequence 1 (TAS1) (Rathjen et al. 1987). It is likely that the SRE and TAS1 activities are equivalent.

Selection of His[+] revertants of the his4-917 mutation has identified an A–G transition within ε (Ty1-912 nucleotide 614), which increases transcript levels from Ty2 about twofold and transcript levels from HIS4 about eightfold, as measured with lacZ fusions. When this point mutation is coupled with a mutation in the Ty TATA sequence that virtually eliminates Ty transcription but does not affect HIS4 transcription, there is a 19-fold increase in HIS4 activity (Coney and Roeder; 1988; Goel and

Pearlman 1988). Goel and Pearlman (1988) have demonstrated, by gel retardation and methylation interference assays over two regions of the mutant Ty, that a complex forms in extracts from cells of all three mating types and *spt3* mutants. No complex is formed with fragments from the same region in *his4-917*. The protein in this complex has been designated Ty-binding factor (TyBF) (Goel and Pearlman 1988). The protected sequence is present in at least one other Ty2 element (see Fig. 3). The related sequence in Ty1 does not form a complex with TyBF. This is consistent with independent observations suggesting that Ty1 and Ty2 are not regulated identically.

The region between the *Hpa*I (Ty1-912 nucleotide 815) and *Rsa*I (Ty1-912 nucleotide 920) or *Sau*3A (Ty1-912 nucleotide 925) sites in Ty1 has also been demonstrated to activate transcription of heterologous promoters (Errede et al. 1987; Fulton et al. 1988). Errede et al. (1985) noted a sequence (block II) within this region of the *CYC7-H2* Ty1 insert that is similar to the SV40 enhancer core, a 12/16 match to the palindromic pheromone receptor/gene transcription factor (PRTF) consensus (Bender and Sprague 1987; Company and Errede 1988), and overlaps an imperfect inverted repeat of a sequence with similarity to the a*1*-α*2* consensus (Miller et al. 1985; Siliciano and Tatchell 1986). A 57-bp oligonucleotide (region H, Ty1-912 nucleotide 815–872), containing the regions with PRTF-site and mating-type sequence similarities, stimulated expression in an orientation-independent, mating-type-regulated manner when inserted alone upstream of *CYC7* (Company and Errede 1988). Stimulation mediated by this sequence is independent of *STE12*. Gel-shift analysis has identified two complexes within H. The DNase and methylation interference patterns of the upstream complex map over the region with PRTF site similarity, and this complex has been shown to include PRTF (B. Errede, pers. comm.). Protein complexes with the H fragment could not be competed with SV40 enhancer sequences, but they could be competed with known PRTF and a*1*-α*2* sequences.

The position of the block II activation element coincides with a *Hpa*I–*Rsa*I fragment containing an activation sequence from Ty1-15 (Fulton et al. 1988). This fragment, termed TAS2, was shown to activate *PGK* transcription when substituted for the natural UAS. In this assay, activation was found only in the non-ROAM orientation and was not mating-type-regulated.

3. Inactivation of Target Genes

Ty insertions that inactivate target gene function have been identified at *HIS4*, *LYS2* (Eibel and Philippsen 1984; Simchen et al. 1984; Natsoulis et al. 1989), *CAN1* (Wilke et al. 1989), and *URA3* (Rose and Winston 1984; Natsoulis et al. 1989). Insertions of Ty1 elements in either orientation into the region just upstream of the *CYR1* initiator codon result in decreased expression of adenylate cyclase and confer constitutive stress resistance. This effect is under mating-type control in the case of the divergently transcribed Ty1, which causes a further reduction in adenylate cyclase activity in *MATa/MATα* compared to haploid cells (Iida 1988). Although selections for loss-of-function mutations have recovered Ty insertions in structural sequences, these constitute a minority of the Ty-mediated mutations. Ty insertions that inactivate nearby genes occur in either orientation and occur both upstream and downstream from the target gene TATA sequence.

Two Ty insertion mutations at the *HIS4* locus, *his4-912* and *his4-917*, have been studied extensively (Chaleff and Fink 1980; Roeder and Fink 1980; Roeder et al. 1980). The *his4-912* mutation is a Ty1 insertion at position −97, with respect to the transcription start site (−162, with respect to the initiator ATG). Ty1-912 is in the same orientation as *HIS4*. The *his4-917* mutation is a Ty2 insertion at position −7 with respect to the transcription start site and is transcribed away from *HIS4*. The *HIS4* TATA element at position −60 is thus displaced in the *his4-917* mutation but not in *his4-912*. No *HIS4* transcripts are produced in cells containing the *his4-912* mutation, and barely detectable levels are produced in cells carrying the *his4-917* mutation (Silverman and Fink 1984).

Insertions of Ty1 and Ty2 at *LYS2* and, to a lesser extent, *LYS5* (Chattoo et al. 1979; Garfinkel et al. 1988), are recovered as a small percentage of spontaneous α-aminoadipate-resistant cells (see Eibel and Philippsen 1984; Simchen et al. 1984; Natsoulis et al. 1989). Most of the insertions are just upstream of the *LYS2* structural sequence, but within the transcribed region or close to the 5′ end of the transcript. These insertions are in both orientations, and insertions in the divergent orientation frequently display residual *LYS2* activity.

cis- and *trans*-acting mutations that suppress the effects of Ty insertions at *HIS4* (Chaleff and Fink 1980; Roeder and Fink 1980; Winston et al. 1984b, 1987; Roeder et al. 1985a; Fassler and Winston 1988), *LYS2* (Simchen et al. 1984), and *URA3* (G. Natsoulis and J.D. Boeke, in prep.) have been recovered. Five classes of revertants of the *his4-912* and *his4-917* insertions have been identified, including translocations and inversions, deletions fusing *HIS4* sequences to the Ty promoter, conversion or point mutation within the Ty1-912 or Ty2-917 sequence (Roeder and Fink 1982), deletion of ε mediated by δ-δ recombination (*his4-912δ* and *his4-917δ*) (Chaleff and Fink 1980; Roeder and Fink 1980, 1982; Roeder 1983), and mutations in genes for *trans*-acting factors required for Ty-mediated disruption (*SPT* for *suppressors* of *Ty*; see Table 2) (Winston et al. 1984b, 1987; Fassler and Winston 1988). With the exception of the Ty/*HIS4* fusion, reversion is accompanied by an increase in the levels of the wild-type *HIS4* transcript (Silverman and Fink 1984; Hirschman et al. 1988). Phenotypes of revertants of Ty insertions at *LYS2* have not been studied in as much detail, but reversion seems to occur by qualitatively similar mutations (Simchen et al. 1984).

Solo δ insertions constitute a major class of revertants of Ty insertions at *HIS4* or *LYS2* (*his4-912δ*, *his4-917δ*, or the analogous *lys2-128δ* mutations). The *his4-912δ* mutation confers a cold-sensitive, His⁺ phenotype. The basis of δ inactivation at low temperature is probably quite different from that of Ty inactivation. In addition to being cold sensitive, the *his4-912δ* allele is also regulated by general amino acid biosynthetic control. Selection for revertants of *his4-912δ* cold sensitivity yields point mutations that inactivate the δ TATA sequence and point mutations that may increase the activity of the *HIS4* TATA region (Hirschman et al. 1988). The His⁻ phenotype caused by *his4-912δ* could be explained by "competition" between the upstream δ promoter and the parallel, downstream *HIS4* promoter for the *HIS4* UAS or the polymerase, or both. This model is in agreement with previous findings that adjacent promoters compete for enhancer activity and that UAS-promoter interactions occur across large distances (for review, see Cullen et al. 1984; Emerman and Temin 1986; Ptashne 1986, 1988; Guarente 1988).

Reversion analyses, together with the activation studies described above, suggest that Ty elements that insert in the same orientation as the target gene displace normal regulatory sequences without substituting Ty activation sequences. This could mean either that these activation sequences do not act on foreign promoters in this orientation or that they cannot act from a distance of 5 kbp. Although the behavior of δ mutations can be explained by promoter competition, there is clearly not a universal inverse correlation between transcription of a full Ty insert and target gene transcription. Studies of *trans*-acting suppressors discussed below show that Ty-target interactions involve the products of many genes.

4. trans-*Acting Genes That Affect the Interaction between Ty1 and Ty2 and Nearby Promoters*

Five classes of *trans*-acting genes that affect the expression of genes flanked by Ty1 or Ty2 insertions or the expression of Ty1 or Ty2 directly have been described: (1) *STE* genes, required for expression of mat-

Table 2 Suppression of Ty and δ insertions at *HIS4* and *LYS2*

spt	Suppression — *lys2-128δ*	*his4-912δ* (cs)	*his4-917δ*	*his4-917(Ty2)*	Additional phenotypes	Map position (chromosome)	References
wt	−	−	−	+/−		—	
1	n.a.[a]	+	−	−		—	Winston et al. (1984b)
2 (*SIN1*)	+	+	−	−	largest group of alleles; some dominant; gene encodes 333-amino-acid protein	rt arm, V	Winston et al. (1984b); Simchen et al. (1984); Roeder et al. (1985b); Kruger and Herskowitz (1989)
3	+	+	+	+	weak sterile; deficient sporulation; decreased level of 5.7-kb Ty transcript; gene encodes 337-amino-acid protein; decreased exp of *MFa1*, *MFa2*, *MFα1*	rt arm, IV	Winston et al. (1984a,b); Simchen et al. (1984); Winston and Minehart (1986); Hirschhorn and Winston (1988)
4	+	+	+/−	−	some alleles mms sensitive; synthetic lethal with *spt5*, *spt6*	—	Winston et al. (1984b)
5	+	+	−	−	synthetic lethal with *spt4*, *spt6*	—	Winston et al. (1984b)
6 (*SSN20*)	+	+	−/+	−/+	essential gene; suppresses fermentation defect of *snf2*, *snf4*, and *snf5*	rt arm, VII	Winston et al. (1984b); Clark-Adams and Winston (1987); Neigeborn et al. (1987)
7	+	+	+	+	deficient sporulation; decreased 5.7-kb Ty transcript	rt arm, II	Winston et al. (1984b, 1987)

					Phenotype	Chromosome	Reference
8	+	+	+	+	deficient sporulation; decreased 5.7-kb Ty transcript	rt arm, XII	Winston et al. (1984b, 1987)
9	+	n.a.	n.a.	+	essential gene	—	Fassler and Winston (1988)
10	+	+	−	+	synthetic lethal with spt11, spt12; activates transcription from 3' LTR	left arm, X	Fassler and Winston (1988); F. Winston (pers. comm.); G. Natsoulis and J.D. Boeke (unpubl.)
11 (HTA1)	+	+	−	+	histone gene; synthetic lethal with spt10	rt arm, IV	Fassler and Winston (1988); Clark-Adams et al. (1988)
12 (HTB1)	+	+	−	+	histone gene, synthetic lethal with spt10	rt arm, IV	Fassler and Winston (1988); Clark-Adams et al. (1988)
13 (GAL11)	−	−	−	+	weak sterile; deficient sporulation	—	Fassler and Winston (1988)
14	−	−	−	+	essential gene	—	Fassler and Winston (1988)
15	+	+	+	+	TFIID (TATA-binding factor) gene	—	Winston et al. (1987); Eisenmann et al. (1989)
21	+	+	−	+	activates transcription from 3' LTR; synthetic lethal with spt11, spt12	XIII	G. Natsoulis and J.D. Boeke (unpubl.); C. Dollard and F. Winston (pers. comm.)

[a]Information not available.

ing-type-regulated genes; (2) *SPT* genes, mentioned above; (3) *TYE* genes (for Ty expression), defined by mutations that suppress activating Ty insertion mutations at *ADH2*; (4) *ROC* or *TEC* genes (Dubois et al. 1982; Laloux et al. 1990), defined by mutations that suppress activating Ty insertion effects on *DUR1* and *DUR2*; and (5) *SNF2*, *SNF5*, and *SNF6* genes, defined as mutations that result in the inability to ferment sucrose due to the failure of *SUC2* induction. *snf2*, *snf5*, and *snf6* strains have recently been shown to have reduced levels of Ty transcripts (A. Happel et al., pers. comm.). The phenotypes of mutations in these genes are summarized in Table 3.

Ty1 and Ty2 are transcribed in haploid cells of both mating types and are repressed in diploid *MATa/MATα* cells (Elder et al. 1981). They are therefore classed as haploid-specific genes. Alleles of *STE4*, *STE5*, *STE7*, *STE11*, and *STE12* have been tested for their effects on various ROAM mutations (Errede et al. 1980, 1981; Dubois et al. 1982; Taguchi et al. 1984). Mutations in three of these, *STE7*, *STE11*, and *STE12*, result in significant reductions in Ty-mediated activation (Fields and Herskowitz 1985, 1987; Fields et al. 1988). Genes required for ROAM activation are probably also required for Ty1 expression; however, this has only been shown specifically for *STE7* (Dubois et al. 1982).

Mutations in *SPT* genes have been identified as suppressors of the negative effects of Ty and/or δ insertion mutations at *HIS4* (Winston et al. 1984b), *LYS2* (Eibel and Philippsen 1984; Simchen et al. 1984b), and *URA3* (G. Natsoulis and J.D. Boeke, unpubl.). A total of 16 complementation groups (*SPT1–SPT15, SPT21*) have been defined (Winston et al. 1984b, 1987; Fassler and Winston 1988; G. Natsoulis and J.D. Boeke, in prep.).

The *SPT* genes can be described in groups defined by the pattern of suppression exhibited by alleles of each toward different Ty or δ insertions. Even though some other mutant phenotypes differ within these suppression groups (F. Winston, pers. comm.), the suppression patterns are a convenient way to consider these genes. The first group suppresses *his4-917*, *his4-912δ*, and *his4-917δ*, and *lys2-128δ*—a divergent Ty and δ insertions in both orientations. Mutations in *SPT3*, *SPT7*, and *SPT8* yield suppressors of this class. In cells with mutations in these genes, the level of the 5.7-kb transcript is decreased, and the level of the 5.0-kb transcript initiated in ε is increased or unaffected. It is the only *SPT* class in which the overall pattern of Ty transcripts is affected (Winston et al. 1984b, 1987; Winston and Minehart 1986).

The phenotype of cells with *his4δ* insertions and mutations in *SPT3*, *SPT7*, and *SPT8* is consistent with reversal of promoter competition between δ and *his4* sequences. The same mechanism can also be invoked to explain suppression of *his4-917*. The *his4-917* insertion displaces the *HIS4* TATA element by 6 kbp. Inspection of the sequence upstream of *his4-917*, within the δ of *Ty2-917*, reveals the sequence of TATAAA, 200 bp upstream of the *HIS4* transcription initiation site. In Spt$^+$ cells, *his4-917* transcription is not observed. Mutations in *SPT3*, *SPT7*, or *SPT8*, by abolishing transcription from the normal Ty2 promoter, would reduce competition for transcription of *HIS4* from this fortuitous TATA sequence.

Suppression of δ insertions in the same orientation as *HIS4* can also be understood as release from promoter interference. Inactivation of the δ promoter in *his4-912* results in *cis*-acting reversion of *his4-912*δ (Hirschman et al. 1988). Similarly, mutations in *SPT3*, *SPT7*, and *SPT8* are likely to functionally inactivate the δ promoter because of the absence of required transcription factors.

The largest category of *SPT* genes are suppressors of *his4-917*, and *his4-912*δ or *lys2-128*δ—a divergent Ty and two convergent δ insertions. This class includes alleles of *SPT1*, *SPT2*, *SPT6*, *SPT9*, *SPT10*, *SPT11*, *SPT12*, and *SPT21*. Mutations in these genes do not cause any striking change in the pattern of Ty transcripts. Whether Ty2-917 transcription is affected specifically has not been reported. *SPT2* and *SPT6* alleles constituted the largest complementation groups isolated in the original selection for suppressors of *his4-912*δ (Winston et al. 1984b). Both of these genes have been cloned and sequenced (Roeder et al. 1985b; Clark-Adams and Winston 1987). *SPT2* is the same as *SIN1*, mutations which make *HO* expression independent of the positive regulatory genes *SWI1*, *SWI2*, *SWI3*, or *SWI5* (Sternberg et al. 1987; Kruger and Herskowitz 1989). *SPT6* is the same as *SSN20*, mutations which suppress *snf2*, *snf5*, and *snf6* mutations (Clark-Adams and Winston 1987; Neigeborn et al. 1987). Thus, *SPT2* and *SPT6* gene products could be regulators of several classes of genes.

Perturbation of chromatin structure can also result in suppression of the effects of inactivating Ty insertions. The *SPT11* and *SPT12* genes have been cloned and shown to be identical to *HTA1-HTB1*, one of the pairs of genes for histones *H2A* and *H2B*. The *HTA1-HTB1* gene pair, as well as *HTA2-HTB2*, has been shown to be capable of suppression of δ mutations when overexpressed on high-copy plasmids. High copy number of either of the pairs of genes for histones *H3* and *H4*, *HHT1-HHF1* and *HHT2-HHF2*, causes weak suppression of *his4-912*δ (Clark-Adams et al. 1988). Mutations in two other *SPT* genes, *SPT10* and *SPT21*, are lethal when combined with mutations in *SPT11* or *SPT12*. This suggests that the *SPT10* and *SPT21* gene products might function in histone modification or nucleosome assembly to mitigate the effect of an imbalance in

Table 3 Genes that affect the ability of Ty1 and Ty2 to activate neighboring genes

Mutation	Ty-related phenotype	Additional phenotypes	References
tye1	reduced expression of Ty-activated, *ADH2c* alleles	inefficient mating	Ciriacy and Williamson (1981)
tye2	reduced expression of Ty-activated, *ADH2c* alleles	reduced ADHII and isocitrate lyase; inefficient mating	Ciriacy and Williamson (1981)
tye3 (snf2)	low expression of *ADH2c* alleles caused by Ty insertion; low levels of Ty transcripts	unable to derepress secreted invertase on sucrose or raffinose; unable to grow on galactose or glycerol; unable to derepress acid phosphatase; partially constitutive for *PRB1* expression	Ciriacy and Williamson (1981); M. Ciriacy (pers. comm.); A. Happel et al. (pers. comm.); Neigeborn and Carlson (1984); Abrams et al. (1986); Moehle and Jones (1990)
tye4 (snf5)	low expression of *ADH2c* alleles caused by divergent Ty insertions; low levels of Ty transcripts	reduced ADHII and isocitrate lyase activity; unable to derepress secreted invertase on sucrose or raffinose; unable to derepress acid phosphatase; partially constitutive for *PRB1* expression	Ciriacy and Williamson (1981); M. Ciriacy (pers. comm.); A. Happel et al. (pers. comm.); Neigeborn and Carlson (1984); Abrams et al. (1986); Moehle and Jones (1990)

snf6	low levels of Ty transcripts	unable to derepress secreted invertase on sucrose or raffinose; unable to grow on galactose or glycerol	A. Happel et al. (pers. comm.) Neigeborn and Carlson (1984)
tec1 (*roc1*)	reduced expression of *durO*^{h}, *cargA*$^+O^h$, and *cargB*$^+O^h$ alleles caused by Ty insertion	mating competent	Dubois et al. (1982) Laloux et al. (1990)
tec2 (*roc2*)	reduced expression of *durO*^{h}, *cargA*$^+O^h$, and *cargB*$^+O^h$ alleles caused by Ty insertion	mating competent	Dubois et al. (1982)

histone subunits (Fassler and Winston 1988; C. Dollard et al., unpubl.). Thus, changes in gene products that are unlikely to interact specifically with Ty1 or Ty2 can suppress insertion mutations.

Mutations of genes in the third class, *SPT4* and *SPT5*, suppress *his4-912δ* and *lys2-128δ* – δ insertions in the same orientation as the target gene. The products of these *SPT* genes may interact with the product of the *SPT6* gene because double mutations are inviable.

The fourth class of *spt* mutations causes suppression of divergent Ty insertions at *HIS4* and *LYS2*, such as *his4-917* and *lys2-61*, but members of this class do not cause suppression of δ insertions (Fassler and Winston 1988). Two complementation groups have been identified in this class, *SPT13* and *SPT14*. *spt13* mutations not only fail to suppress δ insertions, but also block the ability of alleles of several other *SPT* genes, including *SPT1*, *SPT3*, *SPT7*, and *SPT8*, to suppress the His$^+$ phenotype of δ insertions. Mutations in *SPT13* and *SPT14* have pleiotropic effects. Their suppression of Ty, but not δ insertions, implies that the *SPT13* and *SPT14* gene products may interact directly with ε sequences. The epistasis of *spt13* alleles, however, suggests a more complex situation.

The *spt* mutations define a most interesting set of gene products. Because of the strategies used in their selection, they are not strictly positive or negative regulators of Ty or target gene function. Global regulators, transcription factors, and chromatin assembly functions are likely to be identified by these mutations. This was shown by the fact that the *SPT15* gene encodes the TATA-binding factor (TFIID) required for RNA polymerase II initiation (Buratowski et al. 1988; Eisenmann et al. 1989). In addition, because mutant phenotypes reflect competition between parallel and divergent promoters, these mutations should provide insights into the process through which chromatin structure is set to favor the activation of certain promoters over others.

Other genes have been identified by suppression of the activation caused by divergent Ty element insertions. The fact that some members of this group encode positive regulators of Ty expression is not surprising. *tye1-4* mutant alleles cause reduced expression of *ADH2^c* mutations caused by divergent Ty insertions. *tye3* and *tye4* have been shown to be the same as *snf2* and *snf5*, respectively (M. Ciriacy, pers. comm.), and these two genes have been shown to be required for Ty expression (A. Happel et al., pers. comm.). *ROC* alleles, now renamed *TEC* for *Ty expression control*, cause reduced expression from constitutive mutations at *DUR1*, *DUR2*, *cargA*, and *cargB* caused by Ty insertions. *TEC1* and *TEC2* are required for normal levels of Ty transcripts.

The activities of several tRNA genes associated with Ty, δ, τ, and σ, have been monitored by injection of DNA into *Xenopus* oocytes. Two tRNA$^{Glu}_3$ alleles, differing by the presence of a τ element almost 200 bp upstream of the tRNA, differed in the level of tRNA produced, with the τ-containing variant having a somewhat higher level of expression. Different tRNA genes associated with full-length Ty elements are quite active. The transciptional activity of a tRNA$^{Lys}_1$ gene and a polymorphic variant with a 5′-flanking Ty1 insertion were compared by injection into frog oocytes. The results of these experiments suggested that a region including the δ and about 140 bp of the internal domain was responsible for sixfold activation of the Ty-containing variant. This activation occurred over distances of greater than 7 kb and, in some cases, was orientation independent (Hauber et al. 1986).

B. Effects of Ty3 Insertions on Flanking Genes

Insertion specificity makes it unlikely that Ty3 insertions will interrupt structural sequences of genes transcribed by polymerase II, and no spontaneous Ty3 insertions have been recovered in selections for Ty1 and Ty2 transposition. Thus, the effects of spontaneous Ty3 insertion may be limited to tRNA genes. The tRNA gene promoter is internal and located downstream from the actual Ty3 insertion site. Nevertheless, there is potential for effects of Ty3 insertion on tRNA gene expression because the 5′-flanking sequence plays a role in the degree to which these genes are expressed (for review, see Geiduschek and Tocchini-Valentini 1988). In addition, there are now several examples of interactions between polymerase II and polymerase III promoters.

The effect of Ty3 and σ insertion on tRNA gene expression has been addressed at the *SUP2* locus, which encodes a tRNA^{Tyr} gene. σ-containing and -lacking alleles of tRNA^{Tyr} *SUP2* have been identified and cloned (Sandmeyer and Olson 1982). The σ element at *SUP2* is in the more common orientation in that the direction of transcription of the σ is divergent to that of the tRNA gene. A Ty3-containing *SUP2* locus was created artificially by insertion of an *Xho*I fragment (containing the internal domain and one σ element) into the *Xho*I site of a single σ element. Expression from *SUP2* was monitored by hybridization to a probe that was specific for a small insertion in the intervening sequence of *SUP2*. The *SUP2* gene is active in the presence of both the σ and the Ty3 insertions. Activation of the σ promoter did not have significant effects on pre-tRNA levels from *SUP2*, even when the *SUP2* gene was immediately downstream from a solo σ element (P. Kinsey and S.B. Sandmeyer, in prep.).

V. TRANSPOSON-ENCODED GENE PRODUCTS AND THEIR EXPRESSION

A. Ty1 and Ty2

1. TYA/gag

The *TYA* ORF is predicted to encode a protein of 49 kD. The product of this gene is the main structural component of the VLPs made by Ty1 elements (Dobson et al. 1984; Mellor et al. 1985b; Adams et al. 1987b; Muller et al. 1987; Youngren et al. 1988) and is sufficient for VLP formation in the absence of the *TYB* protein products (Adams et al. 1987b; Muller et al. 1987; Youngren et al. 1988). The unusually basic composition of these proteins, together with an overrepresentation of proline, appears to cause anomalous varying electrophoretic migration. In addition, there are differences in the molecular-weight estimates of the *TYA* proteins from different laboratories. Two potential sources of differences are a peculiar sensitivity of these proteins to electrophoretic conditions or differential proteolysis. Laboratories agree, however, that overexpression of *TYA*, in the absence of the protease function, results in the appearance of a protein of between 56 and 62 kD (pro-TYA, p1, or p58-TYA) (Adams et al. 1987b; Muller et al. 1987; Boeke and Garfinkel 1988; Youngren et al. 1988). This species is present as a minor form in wild-type mature particles, along with a major species with an apparent molecular mass between 51 and 55 kD (p2 or p54). The p54 protein is probably derived from p58 by endoproteolytic cleavage performed by the Ty1 protease (see below). Although several lower-molecular-weight species are present in some preparations, they are thought to be artifacts of degradation. No low-molecular-weight protein has yet been observed of a size or stoichiometry that would suggest derivation from p58 by the cleavage that generates p54 (S. Youngren and D.J. Garfinkel, pers. comm.).

The structural role played by *TYA* proteins makes them analogous to the retroviral Gag-encoded proteins. Because the Gag proteins have been characterized, this comparison suggests some properties of the *TYA* proteins. The retroviral *gag* gene can encode as many as five separate proteins, which include the matrix protein, the major phosphoprotein of unknown function; a moderately hydrophobic major core protein; and the highly basic nucleic-acid-binding protein (NC). A protease is also encoded in the *gag* gene of avian viruses, whereas it is encoded in the *pol* gene of most retroviruses and in Ty elements (Varmus and Swanstrom 1984).

The *TYA* protein-coding region reveals a carboxy-terminal concentration of basic residues similar to that found in the carboxy-terminal nucleic-acid-binding protein of retroviruses. In addition, both p54 and p58 have been shown to bind DNA nonspecifically, although the significance of this is uncertain (Mellor et al. 1985b). The retroviral nucleic-acid-binding protein is quite small and contains one or two copies of the highly conserved motif, $CX_2CX_4HX_4C$ (Covey 1986), which mediates nucleic acid binding in vitro (J. Berg, pers. comm.). This motif is not present in the predicted *TYA* protein sequence, but some homology within p54 to the helix-turn-helix motif of some DNA-binding proteins has been reported (Clare and Farabaugh 1985). In addition, in Ty1 and Ty2, *TYA* sequences predict proteins containing 9.0% and 7.3% proline, respectively, compared to an average value of 4.4% in a random collection of more than 60 yeast protein sequences. Proline is also overrepresented in some Gag proteins.

As is the case for some Gag proteins, the Ty1 capsid proteins are phosphoproteins (Mellor et al. 1985b; H. Xu and J.D. Boeke, in prep.). A greater degree of *TYA* protein phosphorylation is observed in *MATa* cells that have been treated with α-factor, and transposition frequencies in these cells are also greatly reduced. The kinase responsible for this phosphorylation does not appear to be encoded by *STE7* or *STE11* (H. Xu and J.D. Boeke, unpubl.). Consensus sites for phosphorylation by casein kinase II and protein kinase C are present in the *TYA* translation product.

Overall, the coding capacity of *TYA* is smaller than that of retroviral *gag* genes, so some *gag*-encoded functions are presumably missing or condensed into fewer mature proteins in Ty elements. VLPs do not appear to interact with membranes, so a matrix protein equivalent may be absent. The protease that is encoded by the *gag* gene of some retroviruses is encoded by *TYB* in the Ty elements. It is likely that the functions performed by the *TYA*-encoded proteins correspond most closely to those performed by the core and nucleic-acid-binding proteins of retroviruses.

2. TYB/pol

TYB encodes protein domains with some similarity to characterized retroviral gene products encoded by the *pol* gene. In the *pol* gene, these occur in the order protease, reverse transcriptase, RNase H (part of the reverse transcriptase protein in retroviruses) and integrase. In Ty1 and Ty2 and in the related *Drosophila* element *copia*, the order of the reverse transcriptase and integrase domains are reversed (Clare and Farabaugh 1985; Mount and Rubin 1985). Like the *pol* gene products, *TYB* proteins are expressed as translational readthrough products of the first ORF. The primary translation product, a *TYA*/*TYB* polyprotein, is processed, probably by the Ty1 protease, to give rise to the mature Ty proteins.

3. Translational Frameshifting

Ribosomal frameshifting to generate a *TYA/TYB* fusion protein has been demonstrated for Ty1 and Ty2 (Clare and Farabaugh 1985). The nucleotide sequences of the overlap regions are shown in Figure 4. The first and second ORFs overlap by 38 nucleotides in the Ty1 transcript and by 44 nucleotides in the Ty2 transcript; therefore, in both cases, a plus-one frameshift is required for translational readthrough. The efficiency of this event for Ty1 is about 20% (Clare et al. 1988). Alignment of the overlap regions of Ty1 and Ty2 shows some similarity (Wilson et al. 1986). A 14-nucleotide region of similarity within the overlap mediates frameshifting in a completely heterologous context (the yeast *HIS4* gene; Clare et al. 1988). Recent studies have further delimited the region sufficient for Ty1 and Ty2 frameshifting in a heterologous context to the sequence CTTAGGC (Belcourt and Farabaugh 1990) and have shown that it must be in the proper reading frame to effect frameshifting. The related sequence GTAAGGC is found in the Ty3 overlap region but occurs in a different reading frame. Inspection of this heptanucleotide in the Ty1 and Ty2 frame shows that the rarely used codon AGG occurs within the conserved sequence. The tRNA^{Arg} corresponding to this codon is encoded by a single-copy gene. Overexpression of this tRNA gene results in a decrease in frameshifting efficiency and in transposition frequency (Belcourt and Farabaugh 1990; Xu and Boeke 1990b). Belcourt and Farabaugh (1990) also presented data suggesting that tRNA$^{Leu}_{UAG}$ was required for frameshifting and that it reads the overlapping CUU and UUA leucine codons, thereby effecting the frameshift in the +1 direction. These data suggest that ribosome pausing may be essential for proper frameshifting. This mechanism has been invoked to explain some prokaryotic frameshift events, which are strongly influenced by the availability of tRNAs required to decode the frameshift region (Weiss and Gallant 1983).

Although the human immunodeficiency virus (HIV) *gag-pol* overlap can effect frameshifting in yeast (Kramer et al. 1986; Wilson et al. 1988), there are some apparently fundamental differences between the mechanisms of retroviral and Ty frameshifting. Retroviral genomes cause minus-one frameshifting, whereas all three Ty RNAs cause plus-one (or minus-two) frameshifting. Within the retrovirus frameshift region, two different, but characteristic, consensus sequences, consisting of short homopolymer runs, which do not occur in the yeast overlaps, have been directly implicated in ribosomal slipping. Finally, the frameshifting of at least some retroviral genomes (although probably not HIV [Wilson et al. 1988]) requires a well-defined stem-loop structure just downstream from the frameshift position (Jacks et al. 1988). No such structure could occur

```
Ty1   TGAACAATAAGCACGACCTTCACCTTAGGCCAGGAACTTACTGA
      *  *  *  **  *   ****  ***    **********  ***   *   **
Ty2   TAAGCGATGACAACGAACTTAGTCTTAGGCCAGCAACAGAAAGAATCTAA
                *  *            *  ****  **
Ty3   TGAACGAATGTAGAGCACGTAAGGCGAGTTCTAACCGATCTTGA
```

Figure 4 Frameshift regions in Ty1–Ty3. The nucleotide sequences from the *TYA/TYB* overlaps of the three classes of elements are aligned by similarity. (Line 1) Ty1-912 (nucleotides 1573–1616; Clare and Farabaugh 1985); (line 2) Ty2-117 or Ty2-917 (nucleotides 1559–1608; Warmington et al. 1985; P.J. Farabaugh, pers. comm.); (line 3) Ty3-1 (nucleotides 1245–1288; Hansen et al. 1988). The sequences begin with the stop codon preceding the *TYB* ORF and end with the stop codon terminating the *TYA* ORF. Stars between Ty1 and Ty2 indicate common nucleotides between those two sequences; stars between Ty3 and Ty2 indicate only nucleotides that are common among all three sequences.

within the short region shown to mediate Ty frameshifting or is predicted downstream from this region in the complete Ty RNAs.

4. TYB *Proteins*

In Ty1, the primary translation product of *TYA/TYB* is a 190- to 200-kD protein called p190-TYA/TYB. Two mature *TYB* protein products of approximately 90 kD (p90-TYB) and 65 kD (p65-TYB) have been identified. The exact termini of these proteins have not been determined, but immunomapping experiments show that the p90-TYB protein contains the integrase homology domain (Eichinger and Boeke 1988) and that p65-TYB contains the reverse transcriptase and RNase H domains (S. Youngren and D.J. Garfinkel, pers. comm.). A single protein with reverse transcriptase/RNase H domains is similar to the arrangement found in retroviruses. A separate Ty1 protease protein has recently been identified (S. Youngren and D.J. Garfinkel, pers. comm.).

a. Protease. The Ty1 protease domain contains the sequence DSG found in the active sites of all retrovirus and retrovirus-like element proteases described to date, and in the cellular aspartyl proteases (Toh et al. 1985a; Doolittle et al. 1989; Skalka 1989) (some proteases contain the related DTG sequence). Mutagenesis experiments using Ty1 overexpression systems have shown that the protease region is required for maturation of Ty1 VLPs, cleavage of primary translation products of *TYA* and *TYB* in vivo, and transposition (Adams et al. 1987b; Muller et al. 1987; Youngren et al. 1988).

b. Integrase. The integrase domains of retroviruses and retrotransposons, including Ty elements, contain a well-conserved motif that has features reminiscent of the Zn^{++} finger of some DNA-binding proteins (Johnson et al. 1986). This motif has the consensus sequence $HX_{3-5}HX_{23-33}CX_2C$. However, additional residues occurring between the histidine and cysteine pairs that are conserved among demonstrated DNA-binding proteins containing the Zn^{++}-finger motif are not found, and the order of histidine and cysteine residues is also diffferent (Frankel and Pabo 1988).

Three mutations mapping within the Ty1 integrase domain greatly reduce transposition frequencies. The first, *tyb-2098*, is a point mutation that converts a leucine codon, lying between the two histidine codons in the region encoding the Zn^{++}-finger-like domain, to an isoleucine codon. This point mutation reduces Ty transposition about 60-fold. The exact nature of the transposition defect is unclear. Levels of the p90-TYB protein are very low, and a precursor species seems to be present in its place. Levels of VLPs and reverse transcriptase activity, however, appear normal (Boeke et al. 1988c; J.D. Boeke, unpubl.). Two in-frame linker insertion mutations within the integrase homology region, *tyb-M2600* and *tyb-M2725*, are easier to interpret phenotypically. They have normal amounts of VLPs, reverse transcriptase activity, Ty DNA in the VLPs, and p90-TYB but fail to function in the in vitro transposition system described below.

Although it is clear that mutations in the integrase domain disrupt integration, the reactions that the integrase catalyzes have not been defined. Experiments indicate that the integration reaction of retroviruses requires a small number of base pairs at the 5′ and 3′ ends of the reverse transcript, which are probably a substrate for the integrase (Panganiban and Temin 1983). The same is true of Ty1 (Eichinger and Boeke 1990). Presumably, the integrase is also responsible for the staggered nicks at the integration site (Craigie et al. 1990), although this has not yet been directly shown for the Ty1 integrase. Many of these mechanistic questions can now be addressed by in vitro studies.

c. Reverse Transcriptase. Ty1 and Ty2 elements contain a reverse-transcriptase-like coding sequence within *TYB*. The reverse transcriptase and RNase H are probably part of the same protein molecule, p65-TYB (S. Youngren and D.J. Garfinkel, pers. comm.). In Ty1 and Ty2, the reverse transcriptase/RNase-H-coding region is considerably 3′ of the integrase-coding region. In the predicted protein sequence, these domains are separated by a long stretch of protein sequence that shows no detectable homology with retroviral sequences. The reverse transcriptase and

RNase-H-coding regions of most retroid elements contain a highly conserved sequence, including the YXDD (X = M, V, or L) box characteristic of reverse transcriptase. Ty1 and Ty2 have the related sequence FVDD.

Reverse transcriptase activity has now been demonstrated in cells overproducing Ty1 (Garfinkel et al. 1985; Mellor et al. 1985a) and Ty2 elements (J.D. Boeke, unpubl.). The reverse transcriptase is assayed in intact VLP preparations with added poly(C) and oligo(dG) or with the Ty RNA as an endogenous substrate. It is in the class of magnesium-dependent reverse transcriptases. The Ty1 enzyme activity has a low-temperature optimum in vitro, possibly explaining the temperature sensitivity of the Ty1 transposition process (Paquin and Williamson 1984, 1986; Garfinkel et al. 1985; Boeke et al. 1986). The specific activity of VLP reverse transcriptase is quite low compared to that of a retrovirus preparation. Attempts to solubilize the VLP do not result in increased activity with the exogenous primer template. Ty1 reverse transcripts, even in cells overproducing a transposition-competent element, are estimated to be present at less than one to ten copies per cell (Eichinger and Boeke 1990). The low level of reverse transcriptase activity may reflect some disadvantage to cells having higher levels or it may be that assay conditions are not yet fully optimized.

d. RNase H. The evidence for the involvement of RNase H in Ty element transposition is limited to the sequence homology carboxy-terminal to the reverse-transcriptase-coding domain. A mutation in this domain that affects transposition negatively has been isolated (D.J. Garfinkel, pers. comm.). Evidence for enzymatic activity corresponding to this protein has not been obtained, although a yeast RNase H with associated reverse transcriptase activity has been described and partially characterized (Karwan et al. 1983, 1986). The relationship of this enzyme to Ty elements is uncertain, but it is clearly different from the "cellular" RNase H gene product encoded by the *RNH1* locus, which lacks reverse transcriptase activity.

e. Regions of Unknown Function. A region of about 1.5 kb in the *TYB* genes of Ty1 and Ty2, which lies between the integrase- and reverse-transcriptase-coding regions, lacks coding sequences similar to retroviral proteins. This region is also one of great divergence between Ty1 and Ty2 elements (Kingsman et al. 1981). Its function is unknown, but antipeptide and antifusion protein antibodies directed against this region react with p90-TYB, suggesting that this region encodes domains of the integrase protein (Garfinkel et al. 1985; Youngren et al. 1988).

B. Ty3

1. TYA3/gag

The translated portion of the first ORF, *TYA3*, is predicted to encode a protein of about 34 kD. The products of *TYA3* are the structural proteins of the VLP. They include low amounts of a protein with an apparent M_r of 38 kD and higher amounts of proteins of 26 kD and 9 kD (L.J. Hansen and S.B. Sandmeyer, unpubl.). The predicted *TYA3* protein sequence has several similarities to the sequences of Gag proteins. The amino-terminal 30 amino acids of *TYA* are 20% proline. In addition, the carboxy-terminal region is rich in basic residues and contains one copy of the $CX_2CX_4HX_4C$ motif.

2. TYB3/pol

TYB3 is homologous to the retroviral *pol* gene. The predicted size of the *TYA3/TYB3* fusion polyprotein is 178 kD. Although a protein of this size has not been observed, it is known that translation of *TYB3* is dependent on ribosomal frameshifting (D. Forrest and S.B. Sandmeyer, unpubl.). *TYB3* encodes a multidomain protein that has homology with retroviral protease, reverse transcriptase, RNase H, and integrase, in that order (Hansen et al. 1988). The domains in this protein are in the same order as the homologous domains in proteins encoded by most retroviral *pol* genes.

The first domain in *TYB3* is predicted to encode a protease. The Ty3 sequence aligns slightly better with the cellular aspartyl proteases than with retroviral aspartyl proteases and contains the active site sequence DSG. Cells in which Ty3 is expressed at high levels show elevated reverse transcriptase activity. Alignments of the sequence of the polymerase and RNase H domains of Ty3 reverse transcriptase with retroviral reverse transcriptases show that the polymerase and RNase H domains are closely juxtaposed in Ty3, compared to retroviruses in which these domains are separated by a tether region (Johnson et al. 1986).

The peculiar insertion specificity of Ty3 makes the integrase domain of particular interest. Experimental evidence suggests that the integrase is required for retroviral and Ty1 insertion. Comparison of the Ty3 integrase to integrases from a set of seven other elements and retroviruses revealed that of eight residues that are absolutely conserved among the others, Ty3 contained all but one. This suggests that despite the position specificity of the element, the integrase is likely to function by a mechanism fundamentally similar to that of retroviruses and retrovirus-like elements. The predicted *TYB3* protein sequence extends from the position of alignment with the amino terminus of the Rous sarcoma virus integrase for 450 codons. Antibodies generated against synthetic peptides designed from the *TYB3* protein domain with integrase homology react with proteins of 61 and 58 kD. Interruption of the ORF immediately upstream of the LTR polypurine tract causes severe reductions in transposition efficiency.

C. Relationships between Ty and Other Retroid Elements

The Ty1 and Ty2 elements are closely related to each other, the *Arabidopsis* element Ta-1 (Voytas and Ausubel 1988), the *Nicotiana tabacum* Tnt-1 element (Grandbastien et al. 1989), the *Drosophila copia* element, and, perhaps, to the *Hpa*II repeat element of *Physarum*, but are only distantly related to Ty3. Comparisons of the polymerase domains show that this group is only distantly related to vertebrate retroviruses and many other classes of retrotransposons (Xiong and Eickbush 1988; Doolittle et al. 1989).

Alignment of the polymerase domain of the Ty3 reverse transcriptase (the most conserved domain among retroviruses) shows that Ty3 belongs to the *gypsy*-like group of retroid elements (Toh et al. 1983, 1985b; Saigo et al. 1984; Marlor et al. 1986; Hansen et al. 1988; Doolittle et al. 1989), which includes the *gypsy*-like elements of *Drosophila*; caulimoviruses, a group of plant DNA viruses that replicate via reverse transcription; and *del1*, a highly repeated sequence in lily (Smyth et al. 1989). Within the polymerase domain, Ty3 is about 43% identical with members of the *gypsy* group and *del1*, and 30–33% identical with the plant DNA viruses. This region has only 12% and 11% identity with Ty1 and *copia* sequences, respectively. Comparison of the Ty3 polymerase domain to several animal retrovirus polymerase domains shows that it is most similar to Moloney murine leukemia virus at 26% identity (Hansen et al. 1988). These similarities among retroid elements of plants, insects, and fungi have prompted the suggestion that horizontal transfer may have occurred among progenitors of these organisms (Doolittle et al. 1989). Relationships among these elements and retroviruses plotted with the progressive alignment programs of Feng and Doolittle (1987) are shown in Figure 5 (kindly provided by R.F. Doolittle).

VI. TRANSPOSITION MECHANISM

The mechanism of Ty element transposition, or retrotransposition, has many parallels with the life cycle of retroviruses (Fig. 1). The major difference is that retroviruses have an obligatory extracellular phase.

Retroviruses transpose from one cell to another via an infectious process, whereas Ty elements transpose intracellularly. Efforts to demonstrate infectivity of the Ty1 transposition process directly have been fruitless so far.

The Ty element transposition process (based on what is known about Ty1 transposition) consists of several discrete stages: transcription of the

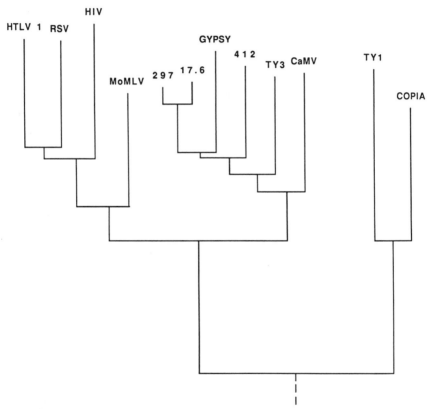

Figure 5 Relatedness of retroviruses and Ty elements. This diagram is based on a tree plot generated by the progressive alignment programs of Feng and Doolittle (1987). Vertical distance is proportional to the percentage of identical amino acid residues within the protein sequence of the conserved portion of the RT domains of the reverse transcriptases of the indicated retroid elements. (HTLV 1) Human T-cell leukemia virus; (RSV) Rous sarcoma virus; (HIV) human immunodeficiency virus; (MoMLV) Moloney murine leukemia virus; (CaMV) cauliflower mosaic virus. HTLV, RSV, HIV, and MoMLV are animal retroviruses; *297, 17.6, gypsy, 412,* and *copia* are *Drosophila* retrotransposons; CaMV is a plant DNA virus.

Ty element, expression of the Ty-encoded proteins, assembly of VLPs, reverse transcription of the genomic RNA to form a full-length Ty DNA, and integration of the DNA copy into a target site on a yeast chromosome or plasmid.

The transposition process has been studied by use of plasmids bearing a 2-micron plasmid origin of replication and a Ty element fused at the Ty transcriptional initiation site to the yeast *GAL1* promoter or *GAL1* UAS. In cells carrying these plasmids under inducing conditions, transposition frequency is increased and amounts of Ty proteins are elevated (Boeke et al. 1985; Garfinkel et al. 1988; Hansen et al. 1988). Various heterologous genes can be inserted into Ty1, Ty2, and Ty3 elements (Boeke et al. 1988b; Garfinkel et al. 1988; Jacobs et al. 1988; Chalker and Sandmeyer 1990), allowing transposition to be followed phenotypically (Boeke et al. 1988b). Much of our knowledge of the mechanism of Ty element transposition is inferred from the behavior of marked Ty1 elements. Unless otherwise specified, the information in this section refers to Ty1 elements. We suspect that in most ways, Ty2 and Ty3 will behave similarly.

A. Ty1

1. Transcription and Gene Expression

Transcription of Ty elements by the host RNA polymerase II (Jerome and Jaehning 1986; Nonet et al. 1987) represents the first step in the retrotransposition process. Ty RNA is required both as a source of structural and catalytic proteins and as the genetic material required for transposition, making transcription a point of potential regulation of transposition. Despite the observation that transposition levels are elevated in cells in which Ty1 or Ty2 elements are overexpressed, there is not a direct correlation between the levels of Ty RNA and the frequency of transposition. In Ty1 overexpression experiments, the levels of Ty RNA are increased about twofold, whereas the frequency of transposition increases by about two orders of magnitude. This discrepancy can be partially explained by the high number of inactive elements in the genome (Boeke et al. 1988a,c; Curcio et al. 1988).

Ty RNA may have restricted translational potential. Although at least some yeast mRNAs are actively translated in wheat germ extracts, Ty RNAs do not seem to be good templates. Secondary structure in the 5' end of the transcript could occlude the AUG, causing Ty RNA to be a poor template. The translational activity of Ty RNA templates could also be affected by TYA structural proteins. These proteins are thought to bind to a packaging signal in the RNA not far from the AUG (during

VLP formation) and could have an autoregulatory role if they compete with the translation machinery for RNA templates.

2. Particle Assembly

The Ty1-VLP has been visualized by electron microscopy in cells that overproduce Ty1 transcripts and proteins. It is a roughly oval, electron-dense structure about 60 nm in diameter. Several events must occur to assemble a complete Ty-VLP. The order in which these events occur is not known.

One of us (Boeke 1989) has suggested a simple model for VLP assembly, in which unprocessed p58-TYA monomers initially associate around Ty RNA at a packaging site. The model proposes that the monomers associate in a way such that the amino termini extend outward and the carboxyl termini compress inward toward the center of the particle. This orientation of the protein would provide an economical mechanism for the incorporation of the *TYB* protein precursor, p190-TYA/TYB, into the VLP: Because p190-TYA/TYB is identical to a *TYA* protein monomer at its amino terminus, the carboxyl terminus of p190, which contains the protease, integrase, reverse transcriptase, and RNase H domains, would be on the inside of the particle, where the RNA is expected to be. This orientation is also consistent with the basicity of the carboxy-terminal portion of the *TYA* protein. This hypothetical precursor assembly, composed of unprocessed Ty proteins and Ty RNA, is referred to as the pre-Ty-VLP.

Some progress has been made in defining regions of Ty sequence required for encapsidation into VLPs. Deletion of the region between nucleotides 478 and 620 from mini-Ty1 plasmids renders them unable to transpose, and the RNA from such mini-Ty elements is packaged at a greatly reduced efficiency. Sequences downstream from nucleotide 620 appear to be dispensable for packaging (Xu and Boeke 1990a).

Proteolytic processing of the *TYA* and *TYA/TYB* precursor proteins into mature species would convert the pre-Ty-VLP into a VLP capable of reverse transcription. In Ty1 protease mutants, particle assembly takes place, but replication, assayed by the "endogenous reaction" (see next section), is blocked. Nevertheless, these protease mutant particles contain active reverse transcriptase as assayed by the addition of an exogenous template/primer complex (Youngren et al. 1988). This implies that processing precedes replication. Reverse transcription of the Ty RNA first into an RNA/DNA hybrid and later into a double-stranded DNA molecule would give rise to a mature DNA-containing Ty-VLP. Testing of this or alternative models for the VLP assembly process will require fractionation of the different forms of Ty-VLP.

VLPs are readily penetrated by outside macromolecules. The ability of exogenous primer template complexes to act as substrates for reverse transcriptase suggests that VLPs may consist of a cage-like structure, rather than an intact shell. Furthermore, Ty nucleic acids are accessible to pancreatic RNase and micrococcal nuclease (Garfinkel et al. 1985; Eichinger and Boeke 1988). The same property is characteristic of Moloney murine leukemia virus integration complexes (Bowerman et al. 1989). The Moloney particles are also permeable to the restriction enzymes *Bam*HI and *Bgl*II (Bowerman et al. 1989).

3. Reverse Transcription

Reverse transcription of Ty RNA has not yet been studied in detail, but the structure of Ty RNA, Ty reverse transcripts, and the presence of initiator methionine tRNA in the VLPs suggest that the mechanism closely resembles the mechanism thought to be used by retroviruses (Gilboa et al. 1979; for more detailed reviews of the model, see Varmus and Swanstrom 1984). tRNA$^{\text{i-met}}$ is thought to prime synthesis from the minus-strand pbs. A minus-strand, strong-stop species of the appropriate length has been identified for Ty1. Plus-strand strong-stop DNA has not been identified for Ty elements but has been isolated from some *Drosophila* retrotransposons (Arkhipova et al. 1986). The primer for these molecules has not yet been identified.

Studies of Ty1 suggest that the process of reverse transcription is associated with high frequencies of recombination. In wild-type cells, the transposed progeny of marked Ty donor elements are heterogeneous in sequence. However, in *spt3* strains, in which transcription of chromosomal Ty elements is essentially abolished, but transcription of the *GAL*/Ty is unaffected, no such sequence heterogeneity is observed. This argues that "recombination" during reverse transcription of different Ty RNA templates is responsible for much of the observed heterogeneity (Boeke et al. 1986). Although direct evidence for a retrovirus-like, dimeric genome has not been obtained for Ty elements, VLPs are likely to contain more than one RNA. Evidence for a recombination-like reaction occurring during reverse transcription has also been provided by studies of Ty elements with tandem copies of a marker segment. A high percentage of transposed progeny elements have undergone precise deletion of one copy of the repeat. Examination of the nucleic acids within the particles reveals the presence of RNA bearing two copies of the marker, but the reverse transcripts contain only a single copy (Xu and Boeke 1987). This association of reverse transcription with recombination-like phenomena has also been observed in a retroviral system (Rhode et al. 1987).

4. Integration of Reverse Transcripts

On the basis of electron microscopy of cells expressing high levels of Ty1 transcripts, VLP assembly probably occurs in the cytoplasm (Garfinkel et al. 1985; Mellor et al. 1985c; Adams et al. 1987b; Muller et al. 1987). An unsolved problem is the means by which the extrachromosomal Ty DNA is delivered to its target in the nucleus. In vitro transposition studies suggest that the active transposition intermediate is too large to fit through a nuclear pore without distortion (Eichinger and Boeke 1988). Because the yeast nucleus does not break down and reassemble during mitosis, nuclear transport of Ty DNA cannot occur by that route. Several possibilities can be imagined: (1) Nuclear DNA is somehow accessible to the VLP within the cytoplasmic space (by looping out through the pores?); (2) Ty VLPs can deform themselves or nuclear pores to gain access to the nucleoplasm; (3) a subassembly of the VLP is actually transported across the nuclear pore; or (4) a VLP-specific transport mechanism exists.

Integration of the Ty element reverse transcript is another step in the life cycle of Ty that closely resembles the retroviral process. Like proviruses and other integrated retrotransposons, genomic copies of Ty DNA are associated with short duplications of the insertion site, which are generated by the transposition process (Farabaugh and Fink 1980; Gafner and Philippsen 1980). The size of these target site duplications is characteristic of the particular virus or retrotransposon, suggesting that an element-specific function is involved.

We have recently developed an in vitro system to study the integration process (Fig. 6). A similar system was independently developed for Moloney murine leukemia virus (Brown et al. 1987, 1989). These systems make use of a powerful prokaryotic selection scheme to detect rare events. The target DNA is derived from bacteriophage λgtWES, which carries amber mutations in three essential genes. This bacteriophage essentially never reverts to ability to grow on a nonsuppressing host strain of *Escherichia coli*. Naked bacteriophage DNA is mixed with extracts (or purified VLPs) prepared from a yeast strain carrying a *GAL1*/Ty1 fusion plasmid marked with the πan7 miniplasmid, which itself carries a copy of the *E. coli supF* gene, a tRNA suppressor gene capable of suppressing the three amber mutations in the phage DNA. The reaction mixture is then deproteinized, and the resulting DNA is concentrated and packaged into bacteriophage λ in vitro. Plaques obtained on nonsup-

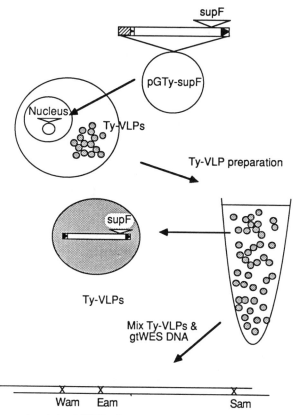

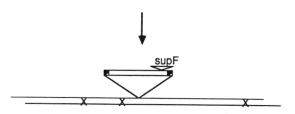

Figure 6 In vitro transposition assay. The basis of an in vitro assay for Ty1 transposition is shown. Cells containing a pGTy1-H3 plasmid marked with a bacterial suppressor tRNA gene (*supF*) are the source of transposition intermediates, which copurify with Ty-VLPs. Only recombinant bacteriophage λ bearing the transposon and its associated *supF* gene are able to form plaques on nonsuppressing *E. coli* host strains. (Reprinted, with permission, from Eichinger and Boeke 1988.)

pressing hosts consist of recombinant λ derivatives carrying πan7(*supF*)-marked Ty elements. Insertion of the Ty elements is associated with a 5-bp duplication of the target λ DNA, showing that the in-vitro-generated events have the same structure as in vivo transposition events and are not the product of illegitimate recombination. The reaction requires only Ty-VLPs, a divalent cation, and target DNA. Polyethylene glycol is stimulatory, but not absolutely required. When Ty-VLPs are made from an integrase mutant construct, no in vitro integration occurs (Eichinger and Boeke 1988). Recently, we have shown that Ty1 VLPs will effect in vitro integration of cloned DNA molecules containing as little as 12 bp of Ty1 sequence at each end. Analysis of such substrates indicates that the $3'OH$ group of Ty1 is required, but the $5'PO_4$ group is not (Eichinger and Boeke 1990). These experiments provide the strongest evidence that the Ty-VLPs are direct intermediates in the transposition process.

B. Ty3

1. Transposition

Ty3 is not highly transcribed without pheromone induction and spontaneous transposition is not observed. Fusion of the *GAL1–GAL10* UAS upstream of the Ty3 TATA element has been used to circumvent this transcriptional limitation on transposition. This system is essentially similar to that described for Ty1 (Boeke et al. 1985). In the original experiments, an unmarked, plasmid-borne inducible Ty3 element was the source of the transposing Ty3 (Hansen et al. 1988). In subsequent experiments, a genetically marked, plasmid-borne inducible Ty3 element was complemented by an unmarked, inducible helper Ty3 element (Chalker and Sandmeyer 1990). In both types of experiments, the cells that had lost the plasmid source of the transposing Ty3 element were selected. Transposed Ty3 elements were detected by colony hybridization in a null Ty3 background or by genetic selection. Under conditions of galactose induction for 10 days, 3–6% of cells sustained transposition events. Because of the dependence of transposition on transcription, it is predicted that pheromone induction would increase the rate of Ty3 transposition.

2. Particles

Extracts of cells containing induced Ty3 elements fractionated on sucrose step gradients yielded Ty3 particles at the 30–70% sucrose interface. These particles are analogous to those described for Ty1 in that they are composed of structural protein reverse transcriptase and integrase in addition to Ty3 5.2-kb RNA and 5.4-kb full-length linear DNA (D. Chalker and S.B. Sandmeyer, unpubl.). Electrophoretic fractionation of concentrated particle preparations, followed by staining with Coomassie blue, reveals proteins of apparent molecular masses of 38, 26, and ~9 kD, which are not found in extracts of null strains. The largest protein is a minor component and may be a precursor species. These proteins are inferred to be the particle structural proteins encoded by *TYA3* (L. Hansen and S.B. Sandmeyer, unpubl.).

VII. DISTRIBUTION OF TRANSPOSONS IN THE GENOME

Insertion site determination is not clearly understood for any retrovirus-like element. Because the position and orientation of retrovirus or retrotransposon insertions determine the extent of disruption sustained by the target genome, this is a problem of considerable interest. Evidence from several sources suggests that retrotransposon and retrovirus insertions are not a random process. Three *gypsy*-like elements, *297*, *17.6*, and *tom* (Ikenaga and Saigo 1982; Inouye et al. 1984; Tanda et al. 1988), insert preferentially into the sequence (T)ATAT. In several cases, this sequence coincides with known TATA elements of other genes. The data concerning retrovirus insertion preferences are incomplete. Within certain loci, insertions are associated with DNase-hypersensitive sites in the chromatin (Vijaya et al. 1986; Rohdewohld et al. 1987). Analysis of randomly chosen, independent Rous sarcoma virus insertion sites (Shih et al. 1988) showed that a small percentage of insertions occur into precisely the same sites, making them highly preferred overall; why these sites are preferred has not been determined. The insertion patterns of the Ty element have been considered by analysis of locations of preexisting genomic elements and by selection for transposition into particular targets.

A. Insertion Specificity

1. Ty1 and Ty2

Ty elements have been mapped to many different chromosomal locations (Klein and Petes 1984). Multiple Ty insertions have been observed at a number of chromosomal loci (Cameron et al. 1979; Liebman et al. 1981; Eigel and Feldmann 1982; Gafner et al. 1983; Genbauffe et al. 1984; Nelbock et al. 1985; Warmington et al. 1986; Stucka et al. 1986, 1987; Van Arsdell et al. 1987; Sandmeyer et al. 1988). A common feature of all of these clusters of Ty or LTR sequences is the presence of at least one tRNA gene within several hundred base pairs. For Ty1, Ty2, and Ty4 and their respective LTRs, the distance of the element from the tRNA gene varies from quite close to several hundred base pairs. The presence of localized concentrations of unselected insertions is consistent with regional hot spots. Unfortunately, the circumstantial nature of these data does not allow one to address the more interesting questions of whether Ty elements accumulate because of AT-rich sequence or particular target features, or for the somewhat less interesting reason that insertions in these particular regions are not disruptive.

The location of Ty1 and Ty2 insertions recovered in selective schemes also suggests that specific positions are strongly preferred over others. Because only insertions within a given locus that result in a particular phenotype can be compared, these data are subject to the caveat that they may not be representative overall. Nevertheless, the results from these studies are quite interesting. As already described, Ty insertions have been recovered at *ADH2*, *CAN1*, *his3Δ4*, *LYS2*, *HIS4*, *URA3*, *CYR1*, and *SUP4* loci. In each case, the data suggest both regions and positions that are preferred for insertion. Selections at *CAN1*, *LYS2*, *URA3*, *CYR1*, and *SUP4* were for loss of function and were therefore less restrictive concerning the types of events recovered. At *LYS2*, *CAN1*, and *URA3*, insertions were concentrated within the 5′ portion of the genes, and multiple insertions occurred at particular positions (Eibel and Philippsen 1984; Simchen et al. 1984; Natsoulis et al. 1989; Wilke et al. 1989). Interestingly, evidence has been presented that the apparent target site preference for 5′ regions (at the *CAN1* locus) is present only in some genetic backgrounds (Wilke et al. 1989). The Ty insertions recovered at *URA3* on the basis of 5-fluoroorotic acid (5-FOA) resistance were almost exclusively within coding sequence, probably because promoter insertions do not survive the selection. The loose consensus (A/C/G) N (A/T) N (T/G/C) was deduced from the *URA3* insertion analysis. Evidence for preferred integration was also obtained for Ty1 insertions into *SUP4*, a tRNA[Tyr] gene (Giroux et al. 1988). In a selection for loss of *SUP4-o* suppression, Ty1 insertions were recovered in 6.1% of mutants. Of the 12 Ty insertions, 10 were at the same position near the middle of the gene (Giroux et al. 1988). No Ty3 insertions were recovered in this selection for *SUP4-o* reversion. These data argue strongly that there are preferred positions of Ty1, and probably Ty2, insertion in the vicinity of genes transcribed by polymerases II and III.

2. Ty3

To date, of more than 30 σ elements and several Ty3 elements for which flanking sequence has been determined, in no case is the insertion separated by more than 19 bp from the 5'end of mature tRNA-coding sequence (summarized in Sandmeyer et al. 1988). These numbers include sequence analysis of 17 clones selected solely on the basis of σ hybridization. In addition, no Ty3 insertion has been recovered in selections that recover spontaneous Ty1 and Ty2 transpositions. These data do not rule out the possibility that Ty3 could insert elsewhere but do argue that insertion is highly specific. Recently, a plasmid target assay has been used to show that 5S and U6 genes, which are also transcribed by polymerase III, can be targets for Ty3 insertion. Other experiments with plasmid targets show that a transcriptionally active tRNA gene is a much more efficient target than a transcriptionally inactive gene (D.L. Chalker and S.B. Sandmeyer, unpubl.).

Several kinds of target features could be proposed to make the Ty3 insertion site distinctive: conserved patterns in the DNA sequence within or upstream (undetected) of the tRNA-coding domain, a protein-induced conformation of the DNA, a protein bound to the tRNA gene, or some combination of these three. Conserved promoter sequences, referred to as A and B boxes, lie within tRNA genes; these boxes could provide a target for insertion upstream. In addition, a comparison of the sequences flanking σ elements showed that in half of the cases, the tRNA gene proximal nick (of the staggered nick) would have occurred after the sequence TCA (Sandmeyer et al. 1988).

It is tempting to speculate as to the nature of protein/DNA interactions that could make the region upstream of the tRNA gene a target for transposition. An obvious question is whether there is a connection between transcription and target formation. Polymerase III transcription complexes are quite stable and, when visualized by electron microscopy, appear to lend a distinct conformation to the DNA. Transcription of tRNAs typically initiates 4–10 bp upstream of the mature coding sequence (for review, see Geiduschek and Tocchini-Valentini 1988). This places initiation quite close to the 5-bp staggered cleavage proposed to accompany Ty3 insertion. In addition, TFIIIB, a polymerase III transcription factor required for tRNA gene expression, has been shown to protect from DNase digestion the tRNA-like strands of *SUP4* and *SUP6*, two other tRNA^{Tyr} genes, from positions −53 to −15 and −56 to −14, respectively (Kassavetis et al. 1989). The position of transcription initiation has been mapped for the *SUP2* alleles containing and lacking Ty3. There is a major start at position −10 and a minor start at position −12. The tRNA gene proximal nick for insertion is inferred to have occurred between positions −11 and −12 on the basis of the position of the 5-bp repeat in the σ-containing *SUP2* allele (P. Kinsey and S.B. Sandmeyer, in prep.). Thus, insertion is physically very close to the transcription initiation site for at least one gene and probably most others.

The mechanism of insertion of some prokaryotic mobile elements is relatively well understood. These systems have demonstrated the role played by DNA supercoiling, host transcription factors, and scaffolding proteins in creating an insertion target for the endonuclease (Berg and Howe 1989). In yeast, the absence of strong consensus target sequences, together with the existence of preferred regions of insertion near promoter elements, suggests that targets could be similarly composed of DNA that is conformationally constrained by transcription factors and chromatin scaffolding.

B. Genomic Rearrangements Involving Ty Elements

Homologous recombination between allelic and nonallelic repeats occurs in yeast (for review, see Petes et al., this volume). Because mitotic and meiotic gene conversion occur at high frequencies and are accompanied by crossing-over about half of the time, the existence of a repeated element in the genome is potentially detrimental. If the 60 or so Ty elements in a typical diploid strain were to recombine during meiosis at the 0.5% observed for some other markers, spore lethality in excess of 30% of tetrads would be predicted. Although deletions, inversions, and translocations involving Ty elements have been documented, high frequencies of lethality are not observed. Several mechanisms have been proposed to account for this discrepancy.

Mitotic recombination occurs much less frequently than meiotic recombination. Nevertheless, mitotic recombination of Ty elements at several loci has been studied. Recombination between Ty2-917 at *HIS4* and Ty2-117 at *LEU2*, two directly repeated Ty2 elements separated by 21 kb on chromosome III, occurs at less than 10^{-7} in haploid cells (Roeder 1983). Recombination between two directly oriented Ty1 elements on chromosome X deletes the intervening 13 kb at a frequency of 10^{-2} to 10^{-6}, depending on the strain. This recombination event explains the phenotype of strains carrying the *DEL1* allele (Liebman et al. 1981; Picologlou et al. 1988). The more similar and the closer together two Ty elements are, the more likely they are to recombine (Roeder et al. 1984). Mitotic recombination between elements on nonhomologous chromosomes is also documented. These events include gene conversion and crossing-over (Chaleff and Fink 1980; Roeder and Fink 1980; Scherer and Davis 1980; Breilmann et al. 1985; Kupiec and Petes 1988). Kupiec and Petes (1988) reported that mitotic removal of a *URA3* insertion in a Ty2 (itself inserted at *LYS2* on chromosome II) must occur at less than 1.25×10^{-7}. At least at this locus, a marked Ty2 does not undergo mitotic ectopic gene conversion (i.e., with a Ty at another genetic locus) at high frequencies, and crossing-over was not observed. Other data, not necessarily in conflict with the foregoing data, suggest that the recombination events between the Ty elements would be even more likely to result in gene conversion than in crossing-over between flanking markers. For very closely spaced *HIS4* alleles on chromosome III, gene conversion was found to be the predominant result of recombination. If similar recombination occurred between closely spaced Ty elements, it would not be disruptive. The data are consistent with the intuitive conclusion that mitotic recombination between Ty elements is not highly detrimental.

Several factors could be invoked to account for the low level of meiotic ectopic Ty recombination. The frequency of meiotic recombination has been shown to be affected both by the chromosomal position of the genes undergoing recombination and by whether the positions are allelic or ectopic. In one study, allelic recombination between repeats of a sequence in different chromosomal positions varied by as much as 38-fold (Lichten et al. 1987). The magnitude of an additional contribution from ectopic events is difficult to estimate in the case of Ty elements. Ectopic recombination frequencies that have been measured for other genes vary from similar to allelic to 17-fold reduced (Jinks-Robertson and Petes 1985, 1986; Lichten et al. 1987). These observations all suggest that tremendous differences might be observed in the frequency with which any pair of nonallelic Ty elements would recombine. Allelic, meiotic recombination between Ty elements has been reported at 0.5–0.7% (Roeder 1983; Kupiec and Petes 1988), which is within the range of meiotic recombination frequencies observed for other genes. Kupiec and Petes (1988) measured the frequency with which 5-FOA-resistant cells arose meiotically in a strain carrying one copy of a *URA3*-marked Ty2 element. Although an identical Ty element was available on another chromosome, the frequency of overall ectopic recombination was about 2.7×10^{-6}. This was much lower than the frequency of recombination (about 10^{-5} of nonallelic *lys2* genes in the isogenic background (Kupiec and Petes 1988). Almost all of the Ty recombination events observed in the study by Kupiec and Petes were gene conversions by closely related Ty2 elements. Surprisingly, δ-δ recombination was not observed.

These data suggest several reasons why the genomic damage sustained from recombination between Ty elements is not as great as that predicted by the thumbnail calculation. First, the results of the Ty recombination examined argue that for each Ty, recombination is only within a pool of highly similar elements. Unfortunately, studies have been conducted mostly with Ty2 elements, which are fewer and more homogeneous in sequence than Ty1 elements. If these results can be extrapolated, however, the variation within the Ty1 class of elements would limit the size of recombinogenic pools to some number below the total number of elements. Second, the most frequent nonallelic Ty recombination events are gene conversions without associated crossing-over. Such events do not yield substantive genome rearrangments, such as inversions, deletions, and reciprocal translocations. For other nonallelic genes that have been studied, reciprocal interchromosomal recombination events are almost equally represented compared to gene conversion events (Jinks-Robertson and Petes 1986; Lichten et al. 1987). Gene conversion has been shown to predominate among intrachromosomal meiotic events, however (Klein and Petes 1981), and much of Ty recombination appears to be with nearby sequences on the same chromosome. Third, even gene conversion events between nonallelic Ty elements do not occur as frequently as would be predicted based on data from non-Ty heteroalleles.

The interesting (and unanswered) questions are, why is gene conversion predominant, and why is the measured frequency even lower (for at least one set of nonallelic Ty elements) than predicted? We would like to speculate that although Ty elements may insert into various chromosomal regions, they are rapidly lost from regions of the genome active in δ-δ recombination and tend to accumulate in nonrecombinogenic regions. Thus, naturally occurring Ty elements would tend to be found in regions that are not recombinogenic. There is accumulating evidence correlating recombinogenicity with transcriptional activity in yeast (Keil and Roeder 1984; Nicolas et al. 1989; Thomas and Rothstein 1989). Although Ty elements are clearly transcriptionally active, in diploid cells where meiotic recombination takes place, Ty transcription is repressed to some extent by mating-type control. In addition, even when transcriptionally active, Ty1 and Ty2 elements have internal activation sequences and may not present a substrate for recombination that is equivalent to other transcriptionally active sequences. Transcriptional repression limits the ability of the silent mating-type loci to participate in gene conversion. A potentially unrelated, but interesting, observation is that the *SIR2* locus, which is required for repression of the silent mating-type genes, is involved in suppression of recombination among rRNA repeats (Gottlieb and Esposito 1989). There is also some evidence consistent with specific regulators of Ty recombination. Rothstein (1984) reported that mutations in the *EDR* gene result in increased δ-δ recombination in the *SUP4* region. One possible explanation of the phenotype is that *EDR* encodes a repressor of δ recombination. However, the recent finding that this gene encodes a topoisomerase-like protein (Wallis et al. 1989) suggests that this is not a simple case of repression by DNA binding. In summary, Ty meiotic recombination does not occur at 30 times the frequency of al-lelic, single-copy recombination, but whether any Ty-specific suppression of recombination accounts for this is not known.

VIII. TY ELEMENTS AS MOLECULAR TOOLS

Ty overexpression systems have direct technical applications in other areas of yeast biology. These applications exploit the ability of Ty elements to mobilize other genes for insertion in many genomic locations and the property of some Ty proteins to direct self-assembly into discrete particles.

A. Transposon Mutagenesis

High-frequency transposition of marked Ty elements makes transposon tagging an attractive strategy. The target gene is first disrupted by Ty insertion and is then cloned from genomic DNA by using selectable markers carried by the element. There are several variations on this application involving both yeast and bacterial markers. The bacterial neo^r gene (from Tn903) confers G418 resistance in yeast and neomycin resistance in bacteria when inserted in a nonessential region of Ty1 (Boeke et al. 1988b). The yeast TRP1 and HIS3 genes, expressed from within a Ty element, functionally complement the corresponding auxotrophic yeast and E. coli mutants. Thus, the disrupted gene can be followed genetically in yeast or bacteria. A Ty2 element containing a cassette consisting of the neo^r gene, together with an E. coli origin of replication, offers the additional advantage of Ty insertions that can be selected in yeast and in E. coli. Thus, one can retrieve genes of interest by simply cleaving and religating total genomic DNA and transforming into bacteria (Garfinkel et al. 1988).

B. Transposons as Retrovirus-like Vectors

Ty elements can be used to construct yeast strains bearing multiple copies of a gene of interest (Boeke et al. 1988b). Although the efficiency of transposition is reduced with insertions larger than 1.5 kbp, foreign insertions in Ty of up to 3.0 kb can be transposed. Mini-Ty elements may allow much larger inserts to be mobilized. Derivatives of GAL1/Ty plasmids containing cassettes composed of a strong yeast promoter linked to a reporter gene, e.g., E. coli galK, have been constructed and transposed into chromosomal targets (Jacobs et al. 1988). A special advantage of Ty1 and Ty2 is that the marker can be inserted downstream from the second ORF without affecting transposition function. The internal promoter in this system produces amounts of galactokinase similar to that produced by plasmids bearing an expression cassette alone. This level of activity would seem to distinguish the Ty cassettes from retroviral vectors in which viral LTR promoters compete with the expression of inserted genes (Emerman and Temin 1986).

C. Ty-VLPs as Epitope Expression Systems

VLPs produced from *TYA* fusions to other protein-coding sequences have been used to produce self-assembling, particulate immunogens. These heterologous VLPs can be readily purified and have been shown to present antigenic determinants very effectively. Hybrid VLPs carrying *TYA* protein fused to domains from the HIV envelope protein (Adams et al. 1987a) and to interferon (Malim et al. 1987) have each been used to generate antibodies. The success of these experiments implies that epitopes which are carboxy-terminal relative to the *TYA* protein are on the outside of the particle. This result is superficially at odds with inferences from Ty particle assembly studies that the *TYB* protein component of the *TYA/TYB* fusion, together with the RNA template, is inside the *TYA* protein coat. These differences may be reconciled by considering the different stoichiometries involved. In the Ty particle, the ratio of *TYA:TYA-TYB* fusion proteins is perhaps 10:1 (Clare et al. 1988), whereas in the antigenic construct, only fusion protein is made. Thus, these immunogenic particles probably have three-dimensional structures that are distinct from the wild-type VLPs.

IX. PREVIEWS OF COMING ATTRACTIONS

The study of transposition in yeast has been facilitated recently with the development of overproducer plasmids, transposon tagging vectors, and in vitro transposition systems. The close connections between the transposition processes of Ty elements and the replication processes of vertebrate retroviruses have also stimulated research in this area.

Substantial effort is currently directed to deciphering biochemically the discrete steps in the retrotransposition process. The surface has barely been scratched in understanding the pathway and mechanism of assembly of VLPs, the packaging signal(s) in Ty RNA, the mechanism of packaging of primer tRNAs, the role of proteolytic processing in assembly, the role of phosphorylation and other postsynthetic modifications (if any), and the roles of the different Gag-like proteins. Much is still unknown about the structure of the VLP itself: Is it a shell, tightly enclosing the nucleic acids and enzymes or a more relaxed, cage-like structure? Does the genomic RNA(s) have a defined structure or location within the VLP or is its structure inside the VLP unordered? Is the genome dimeric, as in retroviruses? Can VLPs be assembled in vitro from defined components? Once a particle is assembled, how is the reverse transcription process activated and how are the "jumps" of strong-stop DNAs effected? What primes plus-strand synthesis in Ty elements? How does reverse transcriptase know at which point to stop reverse transcription to generate the appropriate substrates for integration? Is the reverse-transcript-containing particle transported into the nucleus intact by a special mechanism or does the VLP perch outside the nuclear pore, "spitting" its reverse transcript/integrase complex into the nucleus? Can an integration-proficient (DNA-containing) VLP be assembled from defined components?

The unusual target site specificities of Ty elements will also be investigated in greater detail, particularly with Ty3. Development of an in vitro system to study its transposition will be invaluable for understanding its apparently fastidious requirement for DNA adjacent to tRNA-coding regions as substrate. A greater understanding of the target specificities of Ty elements could also be gained from studying specificity of transposition over the whole genome or, perhaps, single chromosomes.

Our ability to put Ty elements to good use will surely be extended, as Ty elements are engineered for different tasks, as has been done with bacterial transposons. In addition to refinement of Ty tagging vectors and techniques and use of Ty elements as epitope delivery systems, as described above, modified Ty elements designed to generate gene fusions (similar to bacterial Mud-*lac* phages) throughout the yeast genome can be easily imagined. Similarly, libraries of insertions bearing a controllable UAS sequence near one terminus might be a useful tool.

The newly discovered Ty4 and Tf1 elements will extend our knowledge of the properties of yeast retrotransposons much further. Tf1 of *S. pombe* will enable a "compare-and-contrast" approach to be taken; the great phylogenetic distance between the two yeasts should be instructive. Other transposable elements will undoubtedly be isolated from other yeasts (e.g., *Candida*) as their genomes are explored. Perhaps some yeasts will actually bear members of other classes of transposons, such as the P-element type, which apparently does not transpose via RNA, and the processed pseudogene-type (e.g., LINE element), which lacks LTR sequences.

Perhaps the potentially biggest contribution the study of yeast Ty (and Tf) elements will make to obtaining a deeper understanding of retrotransposons and retroviruses rests on our ability to manipulate the host genome readily. To this end, interactions between Ty elements and host genes are being sought by various approaches. Genes that have a negative or positive effect on transposition rates, when mutated, are being sought. Similarly, genes that affect transposition, when carried on a high-copy-number vector, would be of interest. Once such genes are identified, their role(s) in transposition and their other roles can be rapidly ascertained by using the many old and new techniques of genetics and molecular genetics that are available in the yeasts.

ACKNOWLEDGMENTS

We thank F. Winston and B. Errede for helpful discussions and B. Errede, H. Feldmann, R. Pearlman, and F. Winston for communicating results prior to publication. We gratefully acknowledge the support of research in our laboratories by grants from the National Institutes of Health.

REFERENCES

Abrams, E., L. Neigeborn, and M. Carlson. 1986. Molecular analysis of *SNF2* and *SNF5* genes required for expression of glucose-repressible genes in *Saccharomyces cerevisiae*. *Mol. Cell. Biol.* **6**: 3643.

Adams, S.E., K.M. Dawson, K. Gull, S.M. Kingsman, and A.J. Kingsman. 1987a. The expression of hybrid HIV:Ty virus-like particles in yeast. *Nature* **329**: 68.

Adams, S.E., J. Mellor, K. Gull, R.B. Sim, M.F. Tuite, S.M. Kingsman, and A.J. Kingsman. 1987b. The functions and relationships of Ty-VLP proteins in yeast reflect those of mammalian proteins. *Cell* **49**: 111.

Arkhipova, I.R., A.M. Mazo, V.A. Cherkasova, T.V. Gorelova, N.G. Schuppe, and Y.V. Ilyin. 1986. The steps of reverse transcription of *Drosophila* mobile dispersed genetic elements and U3-R-U5 structure of their LTRs. *Cell* **44**: 555.

Baltimore, D. 1985. Retroviruses and retrotransposons: The role of reverse transcription in shaping the eukaryotic genome. *Cell* **40**: 481.

Barker, D.G., J.H.M. White, and L.H. Johnston. 1985. The nucleotide sequence of the DNA ligase gene (*CDC9*) from *Saccharomyces cerevisiae*: A gene which is cell-cycle regulated and induced in response to DNA damage. *Nucleic Acids Res.* **13**: 8323.

Belcourt, M. and P.J. Farabaugh. 1990. Ribosomal frameshifting in the yeast retrotransposon Ty: tRNAs induce slippage on a 7 nucleotide minimal site. *Cell* **62**: 339.

Bender, A. and G.F. Sprague. 1987. MATα1 protein, a yeast transcription activator, binds synergistically with a second protein to a set of cell-type-specific genes. *Cell* **50**: 681.

Berg, D.E. and M.M. Howe. 1989. *Mobile DNA*. American Society for Microbiology, Washington, D.C.

Boeke, J.D. 1988. Retrotransposons. In *RNA Genetics*, vol. II: *Retroviruses, viroids and RNA recombination* (ed. E. Domingo et al.), p. 59. CRC Press, Boca Raton, Florida.

—————. 1989. Transposable elements in *Saccharomyces cerevisiae*. In *Mobile DNA* (ed. D. Berg and M. Howe), p. 335. American Society for Microbiology, Washington, D.C.

Boeke, J.D. and V.G. Corces. 1989. Transcription and reverse transcription in retrotransposons. *Annu. Rev. Microbiol.* **43**: 403.

Boeke, J.D. and D.J. Garfinkel. 1988. Yeast Ty elements as retroviruses. In *Viruses of fungi and lower eukaryotes* (ed. M.J. Leibowitz and Y. Koltin), p. 15. Marcel Dekker, New York.

Boeke, J.D., D. Eichinger, and G.R. Fink. 1988a. Regulation of yeast Ty element transposition. *Banbury Rep.* **30**: 169.

Boeke, J.D., C.A. Styles, and G.R. Fink. 1986. *Saccharomyces cerevisiae SPT3* gene is required for transposition and transpositional recombination of chromosomal Ty elements. *Mol. Cell. Biol.* **6**: 3575.

Boeke, J.D., H. Xu, and G.R. Fink. 1988b. A general method for the chromosomal amplification of genes in yeast. *Science* **239**: 280.

Boeke, J.D., D. Eichinger, D. Castrillon, and G.R. Fink. 1988c. The yeast genome contains functional and nonfunctional copies of transposon Ty1. *Mol. Cell. Biol.* **8**: 1432.

Boeke, J.D., D.J. Garfinkel, C.A. Styles, and G.R. Fink. 1985. Ty elements transpose through an RNA intermediate. *Cell* **40**: 491.

Bowerman, B., P.O. Brown, J.M. Bishop, and H.E. Varmus. 1989. A nucleoprotein complex initiates the integration of retroviral DNA. *Genes Dev.* **3**: 467.

Breilmann, D., J. Gafner, and M. Ciriacy. 1985. Gene conversion and reciprocal exchange in a Ty-mediated translocation in yeast. *Curr. Genet.* **9**: 553.

Brodeur, G.M., S.B. Sandmeyer, and M.V. Olson. 1983. Consistent association between sigma elements and tRNA genes in yeast. *Proc. Natl. Acad. Sci.* **80**: 3292.

Brown, P.O., B. Bowerman, H.E. Varmus, and J.M. Bishop. 1987. Correct integration of retroviral DNA in vitro. *Cell* **49**: 347.

—————. 1989. Retroviral integration: Structure of the initial covalent product and its precursor, and a role for the viral IN protein. *Proc. Natl. Acad. Sci.* **86**: 2525.

Bulawa, C.E., M. Slater, E. Cabib, J. Au-Young, A. Sburlati, L. Adair, and P.W. Robbins. 1986. The *Saccharomyces cerevisiae* structural gene for chitin synthetase (*CHS1*) is not required for chitin synthesis in vivo. *Cell* **46**: 213.

Buratowski, S., S. Hahn, P.A. Sharp, and L. Guarente. 1988. Function of a yeast TATA-binding element in a mammalian transcription system. *Nature* **334**: 37.

Cameron, R.R., E.Y. Loh, and R.W. Davis. 1979. Evidence for transposition of dispersed repetitive DNA families in yeast. *Cell* **16**: 739.

Chaleff, D.T. and G.R. Fink. 1980. Genetic events associated with an insertion mutation in yeast. *Cell* **21**: 227.

Chalker, D. and S.B. Sandmeyer. 1990. Transfer RNA genes are genomic targets for de novo transposition of the yeast retrotransposon Ty3. *Genetics* (in press).

Chattoo, B.B., F. Sherman, D.A. Azubalis, T.A. Fjellstedt, D. Mehvert, and M. Ogur. 1979. Selection of *lys2* mutants of the yeast *Saccharomyces cerevisiae* by the utilization of α-amino-adipate. *Genetics* **93**: 51.

Chen, W. and K. Struhl. 1988. Saturation mutagenesis of a yeast *HIS3* "TATA element": Genetic evidence for a specific TATA-binding protein. *Proc. Natl. Acad. Sci.* **85**: 2691.

Chisholm, G.E., F.S. Genbauffe, and T.G. Cooper. 1984. Tau, a repeated DNA sequence in yeast. *Proc. Natl. Acad. Sci.* **81**: 2965.

Cigan, M. and T.F. Donahue. 1986. The methionine initiator tRNA genes of yeast. *Gene* **41**: 343.

Ciriacy, M. 1976. *Cis*-dominant regulatory mutations affecting the formation of glucose-repressible alcohol dehydrogenase (ADHII) in *Saccharomyces cerevisiae. Mol. Gen. Genet.* **145**: 327.

Ciriacy, M. and V.M. Williamson. 1981. Analysis of mutations affecting Ty-mediated gene expression in *Saccharomyces cerevisiae. Mol. Gen. Genet.* **182**: 159.

Clare, J. and P. Farabaugh. 1985. Nucleotide sequence of a yeast Ty element: Evidence for an unusual mechanism of gene expression. *Proc. Natl. Acad. Sci.* **82**: 2829.

Clare, J.J., M. Belcourt, and P.J. Farabaugh. 1988. Efficient translational frameshifting occurs within a conserved sequence of the overlap between the two genes of a yeast Ty1 transposon. *Proc. Natl. Acad. Sci.* **85**: 6816.

Clark, D.J., V.W. Bilanchone, L.J. Haywood, S.L. Dildine, and S.B. Sandmeyer. 1988. A yeast sigma composite element, Ty3, has properties of a retrotransposon. *J. Biol. Chem.* **263**: 1413.

Clark-Adams, C.D. and F. Winston. 1987. The *SPT6* gene is essential for growth and is required for δ-mediated transcription in *Saccharomyces cerevisiae. Mol. Cell. Biol.* **7**: 679.

Clark-Adams, C.D., D. Norris, M.A. Osley, J.S. Fassler, and F. Winston. 1988. Changes in histone gene dosage alter transcription in yeast. *Genes Dev.* **2**: 150.

Company, M. and B. Errede. 1987. Cell-type-dependent gene activation by yeast transposon Ty1 involves multiple regulatory determinants. *Mol. Cell. Biol.* **7**: 3205.

———. 1988. A Ty1 cell-type-specific regulatory sequence is a recognition element for a constitutive binding factor. *Mol. Cell. Biol.* **8**: 5299.

Company, M., C. Adler, and B. Errede. 1988. Identification of a Ty1 regulatory sequence responsive to *STE7* and *STE12. Mol. Cell. Biol.* **8**: 2545.

Coney, L.R. and G.S. Roeder. 1988. Control of yeast gene expression by transposable elements: Maximum expression requires a functional Ty activator sequence and a defective Ty promoter. *Mol. Cell. Biol.* **8**: 4009.

Cooper, T.G. and G. Chisholm. 1984. Position-dependent, Ty-mediated enhancement of *DUR1,2* gene expression. In *Genome rearrangement* (ed. M. Simon and I. Herskowitz), p. 289. A.R. Liss, New York.

Covey, S.N. 1986. Amino acid sequence homology in *gag* region of reverse transcribing elements and the coat protein of cauliflower mosaic virus. *Nucleic Acids Res.* **14**: 623.

Craigie, R., T. Fujiwara, and F. Bushman. 1990. The IN protein of Moloney murine leukemia virus processes the viral DNA ends and accomplishes their integration in vitro. *Cell* **62**: 829.

Cullen, B.R., P.T. Lomedico, and G. Ju. 1984. Transcriptional interference in avian retroviruses—Implications for the promoter insertion model of leukemogenesis. *Nature* **307**: 241.

Curcio, M.J., N.J. Sanders, and D.J. Garfinkel. 1988. Transpositional competence and transcription of endogenous Ty elements in *Saccharomyces cerevisiae*: Implications for regulation of transposition. *Mol. Cell. Biol.* **8**: 3571.

Curcio, M.J., A.M. Hedge, J.D. Boeke, and D.J. Garfinkel. 1989. Ty RNA levels determine the spectrum of retrotransposition events that activate gene expression in *Saccharomyces cerevisiae. Mol. Gen. Genet.* **220**: 213.

Degols, G., J.-C. Jauniaux, and J.-M. Wiame. 1987. Molecular characterization of transposable-element-associated mutations that lead to constitutive L-ornithine aminotransferase expression in *Saccharomyces cerevisiae. Eur. J. Biochem.* **165**: 289.

del Rey, F.J., T.F. Donahue, and G.R. Fink. 1982. Sigma, a repetitive element found adjacent to tRNA genes of yeast. *Proc. Natl. Acad. Sci.* **79**: 4138.

Deschamps, J., E. Dubois, and J.M. Wiame. 1979. L-ornithine transaminase synthesis in

Saccharomyces cerevisiae: Regulation by inducer exclusion. *Mol. Gen. Genet.* **174**: 225.

Dobson, M.J., J. Mellor, A.M. Fulton, N.A. Roberts, B.A. Bowen, S.M. Kingsman, and A.J. Kingsman. 1984. The identification and high level expression of a protein encoded by the yeast Ty element. *EMBO J.* **3**: 1115.

Doolittle, R.F., D.F. Feng, M.S. Johnson, and M.A. McClure. 1989. Origins and evolutionary relationships of retroviruses. *Q. Rev. Biol.* **64**: 1.

Dubois, E., E. Jacobs, and J.-C. Jauniaux. 1982. Expression of the ROAM mutations in *Saccharomyces cerevisiae*: Involvement of *trans*-acting regulatory elements and relation with the Ty1 transcription. *EMBO J.* **1**: 1133.

Dubois, E., D. Hiernaux, M. Grenson, and J.M. Wiame. 1978. Specific induction of catabolism and its relation to repression of biosynthesis in arginine metabolism of *Saccharomyces cerevisiae*. *J. Mol. Biol.* **122**: 3836.

Eibel, H. and P. Philippsen. 1984. Preferential integration of yeast transposable element Ty1 into a promoter region. *Nature* **307**: 386.

Eibel, H., J. Gafner, A. Stotz, and P. Philippsen. 1981. Characterization of the yeast mobile genetic element Ty1. *Cold Spring Harbor Symp. Quant. Biol.* **45**: 609.

Eichinger, D.J. and J.D. Boeke. 1988. The DNA intermediate in yeast Ty1 element transposition copurifies with virus-like particles: Cell-free Ty1 transposition. *Cell* **54**: 955.

———. 1990. A specific terminal structure is required for Ty1 transposition. *Genes Dev.* **4**: 324.

Eigel, A. and H. Feldmann. 1982. Ty1 and delta elements occur adjacent to several tRNA genes in yeast. *EMBO J.* **1**: 1245.

Eisenmann, D.M., C. Dollard, and F. Winston. 1989. *SPT15*, the gene encoding the yeast TATA binding factor TFIID, is required for normal transcription initiation in vivo. *Cell* **58**: 1183.

Elder, R.T., E.Y. Loh, and R.W. Davis. 1983. RNA from the yeast transposable element Ty1 has both ends in the direct repeats, a structure similiar to retrovirus RNA. *Proc. Natl. Acad. Sci.* **80**: 2432.

Elder, R.T., T.P. St. John, D.T. Stinchcomb, and R.W. Davis. 1981. Studies on the transposable element Ty1 of yeast. I. RNA homologous to Ty1. *Cold Spring Harbor Symp. Quant. Biol.* **45**: 581.

Emerman, M. and H. Temin. 1986. Quantitative analysis of gene expression in integrated retrovirus vectors. *Mol. Cell. Biol.* **6**: 792.

Errede, B., M. Company, and C.A. Hutchison III. 1987. Ty1 sequence with enhancer and mating-type-dependent regulatory activities. *Mol. Cell. Biol.* **7**: 258.

Errede, B., M. Company, and R. Swanstrom. 1986. An anomalous Ty1 structure attributed to an error in reverse transcription. *Mol. Cell. Biol.* **6**: 1334.

Errede, B., T.S. Cardillo, M.A. Teague, and F. Sherman. 1984. Identification of regulatory regions within the Ty1 transposable element that regulate iso-2-cytochrome C production in the *CYC7-H2* yeast mutant. *Mol. Cell. Biol.* **4**: 1393.

Errede, B., T.S. Cardillo, G. Wever, and F. Sherman. 1981. Studies on transposable elements in yeast. I. ROAM mutations causing increased expression of yeast genes: Their activation by signals directed toward conjugation functions and their formation by insertion of Ty1 repetitive elements. *Cold Spring Harbor Symp. Quant. Biol.* **45**: 593.

Errede, B., M. Company, J.D. Ferchak, C.A. Hutchison III, and W.S. Yarnell. 1985. Activation regions in a yeast transposon have homology to mating type control sequences and to mammalian enhancers. *Proc. Natl. Acad. Sci.* **82**: 5423.

Errede, B., T.S. Cardillo, F. Sherman, E. Dubois, J. Deschamps, and J.M. Wiame. 1980.

Mating signals control expression of mutations resulting from insertion of a transposable repetitive element adjacent to diverse yeast genes. *Cell* **22**: 427.

Farabaugh, P.J. and G.R. Fink. 1980. Insertion of the eukaryotic transposable element Ty1 creates a 5 base pair duplication. *Nature* **286**: 352.

Fassler, J.S. and F. Winston. 1988. Isolation and analysis of a novel class of suppressor of Ty insertion mutations in *Saccharomyces cerevisiae*. *Genetics* **118**: 203.

Feng, D.-F. and R.F. Doolittle. 1987. Progressive sequence alignment as a prerequisite to correct phylogenetic trees. *J. Mol. Evol.* **25**: 351.

Fields, S. and I. Herskowitz. 1985. The yeast *STE12* product is required for expression of two sets of cell-type-specific genes. *Cell* **42**: 923.

————. 1987. Regulation by the yeast mating-type locus of *STE12*, a gene required for cell-type-specific expression. *Mol. Cell. Biol.* **7**: 3818.

Fields, S., D.T. Chaleff, and G.F. Sprague. 1988. Yeast *STE7*, *STE11*, and *STE12* genes are required for expression of cell-type-specific genes. *Mol. Cell. Biol.* **8**: 551.

Fink, G.R., J.D. Boeke, and D.J. Garfinkel. 1986. The mechanism and consequences of retrotransposition. *Trends Genet.* **2**: 118.

Frankel, A.D. and C.O. Pabo. 1988. Fingering too many proteins. *Cell* **53**: 675.

Fulton, A.M., P.D. Rathjen, S.M. Kingsman, and A.J. Kingsman. 1988. Upstream and downstream transcriptional control signals in the yeast retrotransposon, Ty. *Nucleic Acids Res.* **16**: 5439.

Fulton, A.M., J. Mellor, M.J. Dobson, J. Chester, J.R. Warmington, K.J. Indge, S.G. Oliver, P. de la Paz, W. Wilson, A.J. Kingsman, and S.M. Kingsman. 1985. Variants within the yeast Ty sequence family encode a class of structurally conserved proteins. *Nucleic Acids Res.* **13**: 4097.

Gafner, J. and P. Philippsen. 1980. The yeast transposon Ty1 generates duplications of target DNA on insertion. *Nature* **286**: 414.

Gafner, J., E.M. DeRobertis, and P. Philippsen. 1983. Delta sequences in the 5' noncoding region of yeast tRNA genes. *EMBO J.* **2**: 583.

Garfinkel, D.J., J.D. Boeke, and G.R. Fink. 1985. Ty element transposition: Reverse transcriptase and virus-like particles. *Cell* **42**: 507.

Garfinkel, D.J., M.F. Mastrangelo, N.J. Sanders, B.K. Shafer, and J.N. Strathern. 1988. Transposon tagging using Ty elements in yeast. *Genetics* **120**: 95.

Geiduschek, E.P. and G.P. Tocchini-Valentini. 1988. Transcription by RNA polymerase III. *Annu. Rev. Biochem.* **57**: 873.

Genbauffe, F.S., G.E. Chisholm, and T.G. Cooper. 1984. Tau, sigma, and delta. *J. Biol. Chem.* **259**: 10518.

Gilboa, E., S.W. Mitra, S. Goff, and D. Baltimore. 1979. A detailed model of reverse transcription and a test of crucial aspects. *Cell* **18**: 93.

Giroux, C.N., J. Mis, M.K. Pierce, S.E. Kohalmi, and B.A. Kunz. 1988. DNA sequence analysis of spontaneous mutations in the *SUP-o* gene of *Saccharomyces cerevisiae*. *Mol. Cell. Biol.* **8**: 978.

Goel, A. and R.E. Pearlman. 1988. Transposable element-mediated enhancement of gene expression in *Saccharomyces cerevisiae* involves sequence-specific binding of a *trans*-acting factor. *Mol. Cell. Biol.* **8**: 2572.

Gopinathan, K.P., L.A. Weymouth, T.A. Kunkel, and LA. Loeb. 1979. Mutagenesis in vitro by DNA polymerase from an RNA tumor virus. *Nature* **289**: 857.

Gottlieb, S. and R.E. Esposito. 1989. A new role for a yeast transcriptional silencer gene, *SIR2*, in regulation of recombination in ribosomal DNA. *Cell* **56**: 771.

Grandbastien, M.A., A. Spielmann, and M. Caboche. 1989. Tnt1, a mobile retroviral-like element of tobacco isolated by plant cell genetics. *Nature* **337**: 676.

Guarente, L. 1988. UASs and enhancers: Common mechanisms of transcriptional activation in yeast and mammals. *Cell* **52**: 303.

Hahn, S., E.T. Hoar, and L. Guarente. 1985. Each of three "TATA elements" specifies a subset of the transcription initiation sites at the CVC-1 promoter of *Saccharomyces cerevisiae. Proc. Natl. Acad. Sci.* **82**: 8562.

Hansen, L.J. and S.B. Sandmeyher. 1990. Characterization of a transpositionally active Ty3 element and identification of the Ty3 integrase protein. *J. Virol.* **64**: 2599.

Hansen, L.J., D.L. Chalker, and S.B. Sandmeyer. 1988. Ty3, a yeast tRNA-gene retrotransposon associated with tRNA genes, has homology to animal retroviruses. *Mol. Cell. Biol.* **8**: 5245.

Hauber, J., P. Nelbock-Hochstetter, and H. Feldmann. 1985. Nucleotide sequence and characteristics of a Ty element from yeast. *Nucleic Acids Res.* **13**: 2745.

Hauber, J., P. Nelbock, U. Pilz, and H. Feldmann. 1986. Enhancer-like stimulation of yeast tRNA gene expression by a defined region of the Ty element micro-injected into *Xenopus* oocytes. *Biol. Chem. Hoppe-Seyler* **367**: 1141.

Hirschhorn, J.N. and F. Winston. 1988. *SPT3* is required for normal levels of a-factor and α-factor expression in *Saccharomyces cerevisiae. Mol. Cell. Biol.* **8**: 822.

Hirschman, J.E., K.J. Durbin, and F. Winston. 1988. Genetic evidence for promoter competition in *Saccharomyces cerevisiae. Mol. Cell. Biol.* **8**: 4608.

Iida, H. 1988. Multistress resistance of *Saccharomyces cerevisiae* is generated by insertion of retrotransposon Ty into the 5 ' coding region of the adenylate cyclase gene. *Mol. Cell. Biol.* **8**: 5555.

Ikenaga, H. and K. Saigo. 1982. Insertion of a movable genetic element, *297*, into the TATA box for the H3 histone in *Drosophila melanogaster. Proc. Natl. Acad. Sci.* **79**: 4143.

Inouye, S., S. Yuki, and K. Saigo. 1984. Sequence-specific insertion of the *Drosophila* element, 17.6. *Nature* **310**: 332.

Jacks, T., H.D. Madhani, F.R. Masiarz, and H.E. Varmus. 1988. Signals for ribosomal frameshifting in the Rous sarcoma virus *gag-pol* region. *Cell* **55**: 447.

Jacobs, E., M. Dewerchin, and J.D. Boeke. 1988. Retrovirus-like vectors for *Saccharomyces cerevisiae*: Integration of foreign genes controlled by efficient promoters into yeast chromosomal DNA. *Gene* **67**: 259.

Jauniaux, J.C., E. Dubois, M. Crabeel, and J.M. Wiame. 1981. DNA and RNA analysis of arginase regulatory mutants in *Saccharomyces cerevisiae. Arch. Int. Physiol. Biochim.* **89**: B111.

Jauniaux, J.C., E. Dubois, S. Vissers, M. Crabeel, and J.M. Wiame. 1982. Molecular cloning, DNA structure, and RNA analysis of the arginase gene in *Saccharomyces cerevisiae*. A study of *cis*-dominant regulatory mutants. *EMBO J.* **1**: 1125.

Jerome, J.F. and J.A. Jaehning. 1986. mRNA transcription in nuclei isolated from *Saccharomyces cerevisiae. Mol. Cell. Biol.* **6**: 1633.

Jinks-Robertson, S. and T.D. Petes. 1985. High-frequency meiotic gene conversion between repeated genes on nonhomologous chromosomes in yeast. *Proc. Natl. Acad. Sci.* **82**: 3350.

———. 1986. Chromosomal translocations generated by high-frequency meiotic recombination between repeated yeast genes. *Genetics* **114**: 731.

Johnson, M.S., M.A. McClure, D.-F. Feng, J. Gray, and R.F. Doolittle. 1986. Computer analysis of retroviral *pol* genes: Assignment of enzymatic functions to specific sequences and homologies with nonviral enzymes. *Proc. Natl. Acad. Sci.* **83**: 7648.

Karwan, R., H. Blutsch, and U. Wintersberger. 1983. Physical association of a DNA polymerase stimulating activity with a ribonuclease H purified from yeast. *Bio-*

chemistry **22**: 5500.

Karwan, R., C. Kuhne, and U. Wintersberger. 1986. Ribonuclease H(70) from *Saccharomyces cerevisiae* possesses cryptic reverse transcriptase activity. *Proc. Natl. Acad. Sci.* **83**: 5919.

Kassavetis, G.A., D.L. Riggs, R. Negri, N.H. Nguyen, and E.P. Geiduschek. 1989. Transcription factor IIIB generates extended DNA interactions in RNA polymerase III transcription complexes on tRNA genes. *Mol. Cell. Biol.* **9**: 2551.

Keil, R.L. and G.S. Roeder. 1984. *Cis*-acting recombination-stimulating activity in a fragment of the ribosomal DNA of *S. cerevisiae*. *Cell* **39**: 377.

Kingsman, A.J., S.M. Kingsman, and K.F. Chater. 1989. *Transposition*. Cambridge University Press, Cambridge, England.

Kingsman, A.J., R.L. Gimlich, L. Clarke, A.C. Chinault, and J. Carbon. 1981. Sequence variation in dispersed repetitive sequences in *Saccharomyces cerevisiae*. *J. Mol. Biol.* **145**: 619.

Klein, H.L. and T.D. Petes. 1981. Intrachromosomal gene conversion in yeast. *Nature* **289**: 144.

———. 1984. Genetic mapping of Ty elements in *Saccharomyces cerevisiae*. *Mol. Cell. Biol.* **4**: 329.

Kozak, M. 1984. Compilation and analysis of sequences upstream from the translation start site in eukaryotic mRNAs. *Nucleic Acids Res.* **12**: 857.

Kramer, R.A., M.D. Schaber, A.M. Skalka, K. Ganguly, F. Wong-Staal, and E.P. Reddy. 1986. HTLV-III *gag* protein is processed in yeast cells by the virus *pol*-protease. *Science* **231**: 1580.

Kronstad, J.A., J.A. Holly, and V.L. MacKay. 1987. A yeast operator overlaps an upstream activation site. *Cell* **50**: 369.

Kruger, W. and I. Herskowitz. 1989. The role of the *SIN1* gene product in regulating expression of *HO* and other genes in *Saccharomyces cerevisiae*. In *Abstracts from Yeast Genetics and Molecular Biology Meeting*, Atlanta, Georgia, p.31. Genetics Society of America, Bethesda, Maryland.

Kupiec, M. and T.D. Petes. 1988. Meiotic recombination between repeated transposable elements in *Saccharomyces cerevisiae*. *Mol. Cell. Biol.* **8**: 2942.

Laloux, I., E. Dubois, M. Dewerchin, and E. Jacobs. 1990. *TEC1*, a gene involved in the activation of Ty1 and Ty1-mediated gene expression in *Saccharomyces cerevisiae*: Cloning and molecular analysis. *Mol. Cell. Biol.* **10**: 3541.

Lemoine, Y., E. Dubois, and J.M. Wiame. 1978. The regulation of urea amidolyase of *Saccharomyces cerevisiae*: Mating type influence on a constitutivity mutation acting in *cis. Mol. Gen. Genet.* **166**: 251.

Levin, H.L., D.C. Weaver, and J.D. Boeke. 1990. Two related families of retrotransposons from fission yeast. *Mol. Cell. Biol.* **10**: (in press).

Liao, X-b., J.J. Clare, and P.J. Farabaugh. 1987. The UAS site of a Ty2 element of yeast is necessary but not sufficient to promote maximal transcription of the element. *Proc. Natl. Acad. Sci.* **84**: 8520.

Lichten, M., R.H. Borts, and J.E. Haber. 1987. Meiotic gene conversion and crossing over between dispersed homologous sequences occurs frequently in *Saccharomyces cerevisiae*. *Genetics* **115**: 233.

Liebman, S., P. Shalit, and S. Picologlou. 1981. Ty elements are involved in the formation of deletions in *DEL1* strains of *Saccharomyces cerevisiae*. *Cell* **26**: 401.

Malim, M.H., S.E. Adams, K. Gull, A.J. Kingsman, and S.M. Kingsman. 1987. The production of hybrid Ty:IFN virus-like particles in yeast. *Nucleic Acids Res.* **15**: 7571.

Marlor, R.L., S.M. Parkhurst, and V.G. Corces. 1986. The *Drosophila melanogaster gypsy* transposable element encodes putative gene products homologous to retroviral proteins. *Mol. Cell. Biol.* **6**: 1129.

McClanahan, T.A. and K. McEntee. 1984. Specific transcripts are elevated in *Saccharomyces cerevisiae* in response to DNA damage. *Mol. Cell. Biol.* **4**: 2356.

McEntee, K. and V. Bradshaw. 1988. Effects of DNA damage on transcription and transposition of Ty retrotransposons in yeast. *Banbury Rep.* **30**: 245.

Mellor, J., A.M. Fulton, M.J. Dobson, W. Wilson, S.M. Kingsman, and A.J. Kingsman. 1985a. A retrovirus-like strategy for the expression of a fusion protein encoded by yeast transposon Ty1. *Nature* **313**: 243.

Mellor, J., A.M. Fulton, M.J. Dobson, N.A. Roberts, W. Wilson, A.J. Kingsman, and S.M. Kingsman. 1985b. The Ty transposon of *Saccharomyces cerevisiae* determines the synthesis of at least three proteins. *Nucleic Acids Res.* **13**: 6249.

Mellor, J., M.H. Malim, K. Gull, M.F. Tuite, S. McCready, T. Dibbayawan, S.M. Kingsman, and A.J. Kingsman. 1985c. Reverse transcriptase activity and Ty RNA are associated with virus-like particles in yeast. *Nature* **318**: 583.

Miller, A.M., V.L. MacKay, and K.A. Nasmyth. 1985. Identification and comparison of two sequence elements that confer cell-type specific transcription in yeast. *Nature* **314**: 598.

Moehle, C.M. and E.W. Jones. 1990. Consequences of growth media, gene copy number and regulatory mutations on the expression of the *PRB1* gene of *Saccharomyces cerevisiae*. *Genetics* **124**: 39.

Morawetz, C. 1987. Effect of irradiation and mutagenic chemicals on the generation of ADH2-constitutive mutants in yeast. Significance for the inducibility of Ty transposition. *Mutat. Res.* **177**: 53.

Mount, S.M. and G.M. Rubin. 1985. Complete nucleotide sequence of the *Drosophila* transposable element *copia*: Homology between *copia* and retroviral proteins. *Mol. Cell. Biol.* **5**: 1630.

Muller, F., K.-H. Bruhl, K. Freidel, K.V. Kowallik, and M. Ciriacy. 1987. Processing of Ty1 proteins and formation of Ty1 virus-like particles in *Saccharomyces cerevisiae*. *Mol. Gen. Genet.* **207**: 421.

Nagawa, F. and G.R. Fink. 1985. The relationship between the "TATA" sequence and transcription initiation sites at the *HIS4* gene of *Saccharomyces cerevisiae*. *Proc. Natl. Acad. Sci.* **82**: 8557.

Natsoulis, G., W. Thomas, M.C. Roghmann, F. Winston, and J.D. Boeke. 1989. Ty1 transposition in *Saccharomyces cerevisiae* is nonrandom. *Genetics* **123**: 269.

Neigeborn, L. and M. Carlson. 1984. Genes affecting the regulation of *SUC2* gene expression by glucose repression in *Saccharomyces cerevisiae*. *Genetics* **108**: 845.

Neigeborn, L., J.L. Celenza, and M. Carlson. 1987. *SSN20* is an essential gene with mutant alleles that suppress defects in *SUC2* transcription in *Saccharomyces cerevisiae*. *Mol. Cell. Biol.* **7**: 672.

Nelbock, P., R. Stucka, and H. Feldmann. 1985. Different patterns of transposable elements in the vicinity of tRNA genes in yeast: A possible clue to transcriptional modulation. *Biol. Chem. Hoppe-Seyler* **366**: 1041.

Nicolas, A., D. Treco, N.P. Schultes, and J.W. Szostak. 1989. An initiation site for meiotic gene conversion in the yeast *Saccharomyces cerevisiae*. *Nature* **338**: 35.

Nonet, M., C. Scafe, J. Sexton, and R. Young. 1987. Eucaryotic RNA polymerase conditional mutant that rapidly ceases mRNA synthesis. *Mol. Cell. Biol.* **7**: 1602.

Panganiban, A.T. and H.M. Temin. 1983. The terminal nucleotides of retrovirus DNA are required for integration but not virus production. *Nature* **306**: 155.

Paquin, C.E. and V.M. Williamson. 1984. Temperature effects on the rate of Ty transposition. *Science* **226:** 53.

———. 1986. Ty insertions at two loci account for most of the spontaneous antimycin A resistance mutations during growth at 15°C of *Saccharomyces cerevisiae* strains lacking *ADH1. Mol. Cell. Biol.* **6:** 70.

Picologlou, S., M. Dicig, P. Kovarik, and S. Liebman. 1988. The same configuration of Ty elements promotes different types and frequencies of rearrangements in different yeast strains. *Mol. Gen. Genet.* **211:** 272.

Ptashne, M. 1986. Gene regulation by proteins acting nearby and at a distance. *Nature* **322:** 697.

———. 1988. How eukaryotic transcriptional activators work. *Nature* **335:** 683.

Rathjen, P.D., A.J. Kingsman, and S.M. Kingsman. 1987. The yeast ROAM mutation—Identification of the sequences mediating host gene activation and cell-type control in the yeast retrotransposon, Ty. *Nucleic Acids Res.* **15:** 7309.

Rhode, B.W., M. Emerman, and H.M. Temin. 1987. Instability of large direct repeats in retrovirus vectors. *J. Virol.* **61:** 925.

Roberts, J.D., K. Bebenek, and T.A. Kunkel. 1988. The accuracy of reverse transcriptase from HIV-1. *Science* **242:** 1171.

Roeder, G.S. 1983. Unequal crossing over between yeast transposable elements. *Mol. Gen. Genet.* **190:** 117.

Roeder, G.S. and G.R. Fink. 1980. DNA rearrangements associated with a transposable element in yeast. *Cell* **2l:** 239.

———. 1982. Movement of yeast transposable elements by gene conversion. *Proc. Natl. Acad. Sci.* **79:** 5621.

Roeder, G.S., A.B. Rose, and R.E. Perlman. 1985a. Transposable element sequences involved in the enhancement of yeast gene expression. *Proc. Natl. Acad. Sci.* **82:** 5428.

Roeder, G.S., M. Smith, and E.J. Lambie. 1984. Intrachromosomal movement of genetically marked *Saccharomyces cerevisiae* transposons by gene conversion. *Mol. Cell. Biol.* **4:** 703.

Roeder, G.S., C. Beard, M. Smith, and S. Keranen. 1985b. Isolation and characterization of the *SPT2* gene, a negative regulator of Ty-controlled yeast gene expression. *Mol. Cell. Biol.* **5:** 1543.

Roeder, G.S., P.J. Farabaugh, D.T. Chaleff, and G.R. Fink. 1980. The origins of gene instability in yeast. *Science* **209:** 1375.

Rohdewohld, H., H. Weiher, W. Reik, R. Jaenisch, and M. Breindl. 1987. Retrovirus integration and chromatin structure: Moloney murine leukemia proviral integration sites map near DNase I hypersensitive sites. *J. Virol.* **61:** 336.

Rolfe, M., A. Spanos, and G. Banks. 1986. Induction of yeast Ty element transcription by ultraviolet light. *Nature* **319:** 339.

Rose, M. and F. Winston. 1984. Identification of a Ty insertion within the coding sequence of the *S. cerevisiae URA3* gene. *Mol. Gen. Genet.* **193:** 557.

Rothstein, R.J. 1984. Double-strand break repair, gene conversion, and postdivision segregation. *Cold Spring Harbor Symp. Quant. Biol.* **49:** 629.

Rothstein, R.J. and F. Sherman. 1980. Dependence on mating type for the overproduction of iso-cytochrome C in the yeast mutant *CYC7-H2. Genetics* **94:** 891.

Rothstein, R., C. Helms, and N. Rosenberg. 1987. Concerted deletions and inversions are caused by mitotic recombination between delta sequences in *Saccharomyces cerevisiae. Mol. Cell. Biol.* **7:** 1198.

Saigo, K., W. Kugimiya, Y. Matsuo, S. Inouye, K. Yoshioka, and S. Yuki. 1984. Identification of the coding sequence for a reverse transcriptase-like enzyme in a transposable genetic element in *Drosophila melanogaster. Nature* **312:** 659.

Sandmeyer, S.B. and M.V. Olson. 1982. Insertion of a repetitive element at the same position in the 5'-flanking regions of two dissimilar yeast tRNA genes. *Proc. Natl. Acad. Sci.* **79:** 7674.

Sandmeyer, S.B., V.W. Bilanchone, D.J. Clark, P. Morcos, G.F. Carle, and G.M. Brodeur. 1988. Sigma elements are position-specific for many different yeast tRNA genes. *Nucleic Acids Res.* **16:** 1499.

Scherer, S. and R.W. Davis. 1980. Recombination of dispersed repeated DNA sequences in yeast. *Science* **209:** 1380.

Scherer, S., C. Mann, and R.W. Davis. 1982. Reversion of a promoter deletion in yeast. *Nature* **298:** 815.

Shapiro, J.A. 1983. *Mobile genetic elements.* Academic Press, New York.

Shih, C.C., J.P. Stoye, and J.M. Coffin. 1988. Highly preferred targets for retrovirus integration. *Cell* **53:** 531.

Siliciano, P.G. and K. Tatchell. 1986. Identification of the DNA sequences controlling the expression of the *MATα* locus of yeast. *Proc. Natl. Acad. Sci.* **83:** 2320.

Silverman, S.J. and G.R. Fink. 1984. Effects of Ty insertions on HIS4 transcription in *Saccharomyces cerevisiae. Mol. Cell. Biol.* **4:** 1246.

Simchen, G., F. Winston, C.A. Styles, and G.R. Fink. 1984. Ty-mediated gene expression of the *LYS2* and *HIS4* genes of *Saccharomyces cerevisiae* is controlled by the same *SPT* genes. *Proc. Natl. Acad. Sci.* **81:** 2431.

Simsek, M. and U.L. RajBhandary. 1972. The primary structure of yeast initiator methionine tRNA. *Biochem. Biophys. Res. Commun.* **49:** 508.

Skalka, A.M. 1989. Retroviral proteases: First glimpses at the anatomy of a processing machine. *Cell* **56:** 911.

Smyth, D.R., P. Kalitsis, J.L. Joseph, and J.W. Sentry. 1989. Plant retrotransposon from *Lilium henryi* is related to Ty3 of yeast and the *gypsy* group of *Drosophila. Proc. Natl. Acad. Sci.* **86:** 5015.

Sternberg, P.W., M.J. Stern, I. Clark, and I. Herskowitz. 1987. Activation of the yeast *HO* gene by release from multiple regulators. *Cell* **48:** 567.

Struhl, K. 1982. Regulatory sites for *his3* gene expression in yeast. *Nature* **300:** 284.

Stucka, R., J. Hauber, and H. Feldmann. 1986. Conserved and non-conserved features among the yeast Ty elements. *Curr. Genet.* **11:.** 193.

————. 1987. One member of the tRNA (Glu) gene family in yeast codes for a minor GAGtRNA(Glu) species and is associated with several short transposable elements. *Curr. Genet.* **12:** 323.

Stucka, R., H. Lochmuller, and H. Feldmann. 1989. Ty4, a novel low copy number element in *Saccharomyces cerevisiae:* One copy is located in a cluster of Ty element and tRNA genes. *Nucleic Acids Res.* **17:** 4993.

Taguchi, A.K.W., M. Ciriacy, and E.T. Young. 1984. Carbon source dependence of transposable element-associated gene activation in *Saccharomyces cerevisiae. Mol. Cell. Biol.* **4:** 61.

Tanda, S., A.E. Shrimpton, C. Ling-Ling, H. Itayama, H. Matsubayashi, K. Saigo, Y.N. Tobari, and C.H. Langley. 1988. Retrovirus-like features and site specific insertions of a transposable element, *tom*, in *Drosophila ananassae. Mol. Gen. Genet.* **214:** 405.

Teague, M.A., D.T. Chaleff, and B. Errede. 1986. Nucleotide sequence of the yeast regulatory gene *STE7* predicts a protein homologous to protein kinases. *Proc. Natl. Acad. Sci.* **83:** 7371.

Thomas, B.J. and R. Rothstein. 1989. Elevated recombination rates in transcriptionally active DNA. *Cell* **56:** 619.

Toh, H., H. Hayashida, and T. Miyata. 1983. Sequence homology between retroviral reverse transcriptase and putative polymerases of hepatitis B virus and cauliflower mosaic virus. *Nature* **305:** 827.

Toh, H., M. Ono, K. Saigo, and T. Miyata. 1985a. Retroviral protease-like sequence in the yeast transposon Ty1. *Nature* **315:** 691.

Toh, H., R. Kikuno, H. Hayashida, T. Miyata, W. Kugimiya, S. Inouye, S. Yuki, and K. Saigo. 1985b. Close structural resemblance between putative polymerase of a *Drosophila* transposable genetic element 17.6 and pol gene product of Moloney murine leukaemia virus. *EMBO J.* **4:** 1267.

Trueheart, J., J.D. Boeke, and G.R. Fink. 1987. Two genes required for cell fusion during yeast conjugation: Evidence for a pheromone-induced surface protein. *Mol. Cell. Biol.* **7:** 2316.

Van Arsdell, S.W. and J. Thorner. 1987. Hormonal regulation of gene expression in yeast. In *Transcriptional control mechanisms* (ed. D.K. Granner et al.), p. 325. A.R. Liss, New York.

Van Arsdell, S.W., G.L. Stetler, and J. Thorner. 1987. The yeast repeated element sigma contains a hormone-inducible promoter. *Mol. Cell. Biol.* **7:** 749.

Varmus, H.E. 1985. Reverse transcriptase rides again. *Nature* **314:** 583.

Varmus, H.E. and R. Swanstrom. 1984. Replication of retroviruses. In *RNA tumor viruses 1/Text* (ed. R. Weiss et al.), p. 369. Cold Spring Harbor Laboratory, Cold Spring Harbor, New York.

Vijaya, S., D.L. Steffen, and H.L. Robinson. 1986. Acceptor sites for retroviral integration map near DNAse hypersensitive sites in chromatin. *J. Virol.* **60:** 683.

Voytas, D.F. and F.M. Ausubel. 1988. A *copia*-like transposable element family in *Arabidopsis thaliana*. *Nature* **336:** 242.

Wallis, J.W., G. Chrebet, G. Brodsky, M. Rolfe, and R. Rothstein. 1989. A hyper-recombination mutation in *S. cerevisiae* identifies a novel eukaryotic topoisomerase. *Cell* **58:** 409.

Warmington, J.R., R.B. Waring, C.S. Newlon, K.J. Indge, and S.G. Oliver. 1985. Nucleotide sequence characterization of Ty 1-17, a class II transposon from yeast. *Nucleic Acids Res.* **13:** 6679.

Warmington, J.R., R. Anwar, C.S. Newlon, R.B. Waring, R.W. Davies, K.J. Indge, and S.G. Oliver. 1986. A "hot spot" for Ty transposition on the left arm of yeast chromosome III. *Nucleic Acids Res.* **14:** 3475.

Weiner, A.M., P.L. Deininger, and A. Efstratiadis. 1986. Nonviral retroposons: Genes, pseudogenes, and transposable elements generated by the reverse flow of genetic information. *Annu. Rev. Biochem.* **55:** 631.

Weiss, R. and J. Gallant. 1983. Mechanism of ribosome frameshifting during translation of the genetic code. *Nature* **302:** 389.

Wilke, C.M., S.H. Heidler, N. Brown, and S. Liebman. 1989. Analysis of yeast transposon Ty1 insertions at the CAN1 locus. *Genetics* **123:** 655.

Williamson, V.M., E.T. Young, and M. Ciriacy. 1981. Transposable elements associated with constitutive expression of yeast alcohol dehydrogenase II. *Cell* **23:** 605.

Williamson, V.M., D. Cox, E.T. Young, D.W. Russell, and M. Smith. 1983. Characterization of transposable element-associated mutations that alter yeast alcohol dehydrogenase II expression. *Mol. Cell. Biol.* **3:** 20.

Wilson, W., M.H. Malim, J. Mellor, A.J. Kingsman, and S.M. Kingsman. 1986. Expression strategies of the yeast retrotransposon Ty: A short sequence directs ribosomal frameshifting. *Nucleic Acids Res.* **14:** 7001.

Wilson, W., M. Braddock, S.E. Adams, P.D. Rathjen, S.M. Kingsman, and A.J. Kingsman. 1988. HIV expression strategies: Ribosomal frameshifting is directed by a short sequence in both mammalian and yeast systems. *Cell* **55:** 1159.

Winston, F. and P.L. Minehart. 1986. Analysis of the yeast *SPT3* gene and identification of its product, a positive regulator of Ty transcription. *Nucleic Acids Res.* **14:** 6885.

Winston, F., K.J. Durbin, and G.R. Fink. 1984a. The *SPT3* gene is required for normal transciption of Ty elements in *S. cerevisiae. Cell* **39:** 675.

Winston, F., D.T. Chaleff, B. Valent, and G.R. Fink. 1984b. Mutations affecting Ty-mediated expression of the *HIS4* gene of *Saccharomyces cerevisiae. Genetics* **107:** 179.

Winston, F., C. Dollard, E.A. Malone, J. Clare, J.G. Kapakos, P. Farabaugh, and P.L. Mineheart. 1987. Three genes required for *trans*-activation of Ty element transcription in yeast. *Genetics* **115:** 649.

Xiong, Y. and T.H. Eickbush. 1988. Similarity of reverse transcriptase-like sequences of viruses, transposable elements, and mitochondrial introns. *Mol. Biol. Evol.* **5:** 675.

Xu, H. and J.D. Boeke. 1987. High frequency deletion between homologous sequences during retrotransposition of Ty elements in *Saccharomyces cerevisiae. Proc. Natl. Acad. Sci.* **84:** 8553.

————. 1990a. Localization of sequences required in *cis* for yeast Ty1 element transposition near the long terminal repeats: Analysis of mini-Ty1 elements. *Mol. Cell. Biol.* **10:** 2695.

————. 1990b. Host genes that influence transposition in yeast: The abundance of a rare tRNA regulates Ty1 transposition frequency. *Proc. Natl. Acad. Sci.* **87:** (in press).

Youngren, S.D., J.D. Boeke, N.J. Sanders, and D.J. Garfinkel. 1988. Functional organization of the retrotransposon Ty from *Saccharomyces cerevisiae*: The Ty protease is required for transposition. *Mol. Cell. Biol.* **8:** 1421.

Yu, K. and R.T. Elder. 1989. Some of the signals for 3′ end formation in transcription of the *Saccharomyces cerevisiae* Ty-D15 element are immediately downstream of the initiation site. *Mol. Cell. Biol.* **9:** 2431.

Zaret, K.S. and F. Sherman. 1982. DNA sequence required for efficient transcription in yeast. *Cell* **28:** 563.

————. 1984. Mutationally altered 3′ ends of yeast *CYC1* mRNA affect transcript stability and translational efficiency. *J. Mol. Biol.* **176:** 107.

5

Yeast RNA Virology:
The Killer Systems

Reed B. Wickner

Section on Genetics of Simple Eukaryotes
Laboratory of Biochemical Pharmacology, National Institute
of Diabetes, Digestive and Kidney Diseases,
National Institutes of Health, Bethesda, Maryland 20892

I. INTRODUCTION

Most strains of *Saccharomyces cerevisiae* harbor one or more double-stranded RNA (dsRNA) viruses. These viral species were first discover-

Volume I. *The Molecular and Cellular Biology of the Yeast* Saccharomyces: *Genome Dynamics, Protein Synthesis, and Energetics.* Copyright 1991 Cold Spring Harbor Laboratory Press 0-87969-355-X/91 $3 + 00   **263**

ed as the agents responsible for the killer phenotype of yeast. Killer strains of yeast are those that secrete a protein toxin that is lethal to most nonkiller *Saccharomyces* strains. The viral particles are infectious only by cell-to-cell fusion, a route less commonly used by animal viruses. In other respects, these yeast viruses are quite similar to dsRNA viruses found in larger eukaryotic organisms. As a result of this similarity, the viruses of *Saccharomyces* provide a useful model for the central mechanisms of virus replication, such as RNA-dependent transcription, replication, packaging, ribosomal frameshifting, and the interrelationship of host and viral functions. The killer toxin, encoded by a satellite virus, has also been useful for the study of protein processing and secretion and toxin action and receptors.

Yeast RNA viruses include three families with dsRNA genomes (L-A, L-BC, and M), two other families of dsRNA replicating in yeast that may yet prove to be viral (T and W), and four families of retroviruses (Ty1–Ty4) (Table 1). This paper focuses on the dsRNA viruses and the single-stranded circular RNA replicon, 20S RNA; retroviruses are reviewed by Boeke and Sandmeyer (this volume), although several striking parallels between these systems will be mentioned here as well. This review also includes several speculative models designed to relate the molecular, biochemical, and genetic information on dsRNA viruses to their biological context. Detailed descriptions of the basic genetics, physical organization, and physiology of the yeast dsRNA viruses and their associated killer phenotypes are available in the earlier edition of this book (Wickner 1981). The emphasis in this review will be on developments since that time. Other recent reviews of this system include those by Tipper and Bostian (1984), Wickner (1986, 1988), and Bussey (1988).

II. YEAST dsRNA VIRUSES ARE INFECTIOUS ONLY BY CELL-TO-CELL FUSION

Most viruses of animals, plants, and bacteria spread horizontally. They multiply in one cell, leave that cell, and enter another cell, where they resume multiplication. Vertical transmission of viruses is also encountered—either as part of the host genome (λ and integrated retroviruses) or by stable replication separate from, but coordinated with, the host genome (phage P1). The σ virus of *Drosophila* is a rhabdovirus that is *only* transmitted vertically (Brun and Plus 1980). Some animal viruses rely occasionally on direct cell fusion for transmission. This is true of some herpesviruses (Hooks et al. 1976), as well as human immunodeficiency virus type 1 (HIV-1; Gupta et al. 1989). In general, however, an animal or bacterial virus that never left its original host cell (or

Table 1 Families of yeast viruses

Name	Size (kb)	Encoded proteins and genetic functions
		dsRNA Replicons
L-A	4.6	80-kD major coat protein
		180-kD minor ssRNA-binding protein, RNA polymerase; [HOK], [NEX], [EXL], [B][a]
M_1, M_2, M_{28}	1.8	32-kD killer toxin/immunity precursor protein
L-BC[b]	4.6	77-kD, 73-kD major coat proteins
T	2.7	?
W	2.25	?
		Single-stranded Circular RNA Replicon
20S RNA	~3.0	probable RNA polymerase
		Retroviruses (Retrotransposons)
Ty1, Ty2	5.9	*TYA* = precursor of major coat proteins
		TYA/TYB (*gag/pol*) fusion protein = precursor of protease, integrase, reverse transcriptase, and RNase H
Ty3	5.4	*TYA3* = *gag* precursor
		TYB3 = precursor of protease, reverse transcriptase, RNase H, and integrase
Ty4	6.3	

[a][HOK], [NEX], [EXL], and [B] are genetic activities found on various natural variants of L-A. They are defined in terms of their interactions with M_1 and M_2 dsRNAs and the chromosomal *MAK* and *MKT* genes (see text).

[b]L-BC is the same size as L-A but is unrelated in sequence (Sommer and Wickner 1982). It is found in intracellular viral particles with a coat distinct from that of L-A (Sommer and Wickner 1982; El-Sherbeini et al. 1984; Thiele et al. 1984a).

the progeny of that cell) would not have spread far and might never have had the privilege of being studied by molecular biologists.

In contrast, fungal viruses apparently spread exclusively by cell-to-cell fusion. No natural extracellular infectious cycle of a fungal virus has been demonstrated despite numerous attempts (for review, see Buck 1988). In *S. cerevisiae*, a majority of strains carry both L-A and L-BC dsRNA viruses, although these viruses spread only by cytoplasmic mixing. This probably reflects the fact that fungi mate or fuse hyphae very frequently in nature, allowing passage between strains of such viruses.

The yeast retroviruses (called Ty1, Ty2, Ty3, and Ty4) also spread only by an intracellular route, called retrotransposition (Boeke and Sandmeyer, this volume). They activate production of viral particles (an inter-mediate in the transposition process) in response to haploidy or mating pheromones. The adaptive value of this regulation may be to prepare the virus to hop into the genome of a potentially unoccupied host (the genome of the cell with which its host is about to mate).

Recently, the introduction of viral particles along with plasmid DNA in transformation of *S. cerevisiae* has been reported (El-Sherbeini and Bostian 1987). This novel method has already been used to show that one genetic element, [B], affecting M_1 dsRNA replication, is located on L-A dsRNA (Uemura and Wickner 1988) and that another, [D], is not (Esteban and Wickner 1987). Further application of this technology will undoubtedly prove quite valuable in dissecting various aspects of the viral life cycle and expression.

III. dsRNA VIRUS REPLICATION IN VIVO, IN VIRO, AND IN VITRO

A. L-A Replication Cycle

In vivo studies suggested that the overall process of L-A replication is conservative; i.e., like transcription of DNA, but unlike DNA replication, the parental strands of L-A remain together during replication (Newman et al. 1981; Sclafani and Fangman 1984). In addition, in pulse-label ex-periments, the (+) strands of L-A dsRNA become labeled before the (−) strands (Newman and McLaughlin 1986). These results suggest that (+) strands are made from dsRNA genomes and (−) strands are made on a (+) strand template to form L-A dsRNA.

Isolated viral particles have been shown to carry out both (+) and (−) strand synthesis with endogenous templates. We refer to such experi-ments as in viro (in the virus) to distinguish them from experiments in which the template and enzymes are added separately (in vitro). In viro experiments were used to establish the viral particle replication cycles for L-A and M dsRNAs (see Fig. 1) (Esteban and Wickner 1986; Fujimura and Wickner 1986). Viral particles that make (+) single-stranded RNA (ssRNA) on a dsRNA template have been separated from those that make (−) strands on a (+) strand template. The fact that the former do not label dsRNA at any time in the reaction demonstrates that replication is conservative (Fujimura and Wickner 1986). This is also supported by the fact that label in the L-A dsRNA template did not significantly enter the (+) strand product (Williams and Leibowitz 1987). This mode of replica-tion is similar to that used by reovirus (Joklik 1983), whereas *Penicillium stoloniferum* virus (Buck 1978), *Aspergillus foetidus* virus S (Ratti and Buck 1979), and *Ustilago maydis* virus segment M_2 (Yie et al. 1989) all replicate by a semiconservative mechanism.

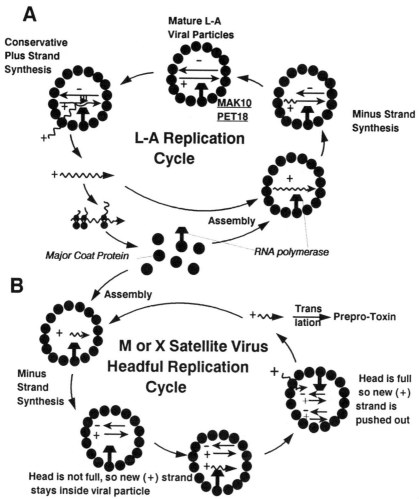

Figure 1   Replication cycles of L-A and M or X dsRNAs. In both cycles, replication is conservative (parental strands stay together), intraviral (both [+] and [−] strands are synthesized within the viral particles), and sequential ([+] strands are made at the transcription step and [−] strands are made at the replication step). Headful replication means that replication within the head continues until the head is full. New (+) strands are then all extruded to be translated and to be encapsidated to form new viral particles.

The finding that particles making L-A (+) ssRNA on a dsRNA template were distinct and separable from those making L-A (−) strands on a (+) ssRNA template (to yield dsRNA) confirmed the in vivo evidence that L-A replication is asynchronous; i.e., (+) strands and (−) strands are not made at the same time. This observation also showed that the replication is intraviral (Fujimura et al. 1986; Fujimura and Wickner 1987). Finally, the existence of viral particles containing only a (+) single strand showed that (+) ssRNA, rather than dsRNA, is the form recognized by the packaging machinery for encapsidation into particles.

B. M_1 and X dsRNA Replication Cycles: Headful Replication

Each L-A particle contains either one L-A dsRNA or one L-A (+) ssRNA molecule. In contrast, M_1 particles can contain either one M_1 dsRNA molecule or two M_1 dsRNA molecules (Esteban and Wickner 1986). Both types of M_1 particles have the same diameter and the same protein composition as each other and as L-A dsRNA-containing particles. However, although L-A dsRNA particles and those with two M_1 dsRNA molecules per particle extrude all of their (+) ssRNA transcripts from the particle, particles with only one M_1 dsRNA molecule per particle retain 30–60% of their transcripts within the particle. This retained (+) ssRNA is subsequently converted to a second M_1 dsRNA molecule in the same particle.

To explain these results, Esteban and Wickner (1986) proposed that the structure of the L-A-encoded capsid is designed to hold one L-A dsRNA molecule. When M_1 (+) strands are encapsidated in this coat and converted to dsRNA, the heads are not full, because M_1 (1.8 kb) is less than half the size of L-A (4.6 kb). Thus, new M_1 (+) strand transcripts often remain inside the viral particles, where they are converted to a second M_1 dsRNA molecule. This "headful replication model" (Esteban and Wickner 1986) is distinct from the "headful packaging" model for phage T4 and other phage, in which replication occurs outside the particle and packaging is limited by the head size. In the dsRNA system, packaging of (+) ssRNA is followed by replication until the head is full. Only then are new (+) strands extruded to be translated and/or packaged to make new viral particles.

The headful packaging model predicted that a deletion mutant of L-A less than half the size of L-A would also show multiple viral particle species. Such mutant genomes should be present in particles at a number up to the reciprocal of its fractional length of the L-A genome; i.e., a mutant viral genome that is one-quarter the length of the L-A genome should yield particles containing up to four copies of the mutant genome. This prediction has been fulfilled. A mutant was isolated and called X dsRNA (Esteban and Wickner 1988; Esteban et al. 1988). X dsRNA is 530 bp in length and is replicated and transcribed. The fact that particles having one to eight X dsRNA molecules per particle were observed confirmed the headful replication model.

The headful replication model predicts that only one viral (+) single strand is packaged per particle. This has been directly verified in vivo using a DNA vector expressing transcripts containing only the viral packaging signal (Fujimura et al. 1990). Just one transcript was packaged per particle.

The headful replication model also predicted that although L-A dsRNA synthesized by isolated particles is only labeled in the (–) strands (Fujimura et al. 1986), M_1 dsRNA or its deletion derivatives (called S dsRNAs) should be labeled in both strands. This was shown to be the case by Williams and Leibowitz (1987). This fact has made it difficult to prove that in viro transcription of M or its derivatives is conservative. Nevertheless, Williams and Leibowitz (1987) have obtained evidence that this reaction is conservative, and recent data using the in vitro transcription system (see below) confirm this conclusion.

In the *U. maydis* killer systems, a similar pattern of one large dsRNA molecule per particle or one to three smaller molecules per particle is seen (Bozarth et al. 1981). These results may also be explained by the headful replication model, but the data do not yet rule out other possibilities.

Headful replication implies that the L-A capsid size and structure are determined by the capsid protein structure, not by the size of the genome encapsidated. Indeed, particles containing one L-A dsRNA or one M_1 dsRNA molecule per particle were apparently the same size (Esteban and Wickner 1986). One might expect that intraviral replication in particles of fixed dimensions would automatically result in headful replication unless there were a specific mechanism to force out new (+) strands.

C. In Vitro Site-specific Binding of Viral (+) Strand RNA to Viral Particles

Detailed analysis of the dsRNA replication cycle has been made possible by the development of techniques to isolate opened empty virions that are capable of specifically binding (+) ssRNA (Esteban et al. 1988; Fujimura and Wickner 1988a,b), converting it to the dsRNA form (Fujimura and Wickner 1988a), and transcribing dsRNA to yield new (+) strands (Fujimura and Wickner 1989). Our current appreciation of each of these steps is described below.

Binding of RNA to opened empty capsids is specific for viral (+) strands. L-A, M_1, and X (+) strands are all capable of binding (Esteban et al. 1988; Fujimura and Wickner 1988b). Detailed studies of X (+) strand binding has identified the recognition site as a 24-base sequence with a stem-loop structure that has an A residue protruding on the 5 ′ side of the stem (see Fig. 2) (Esteban et al. 1988, 1989). The structure, but not the

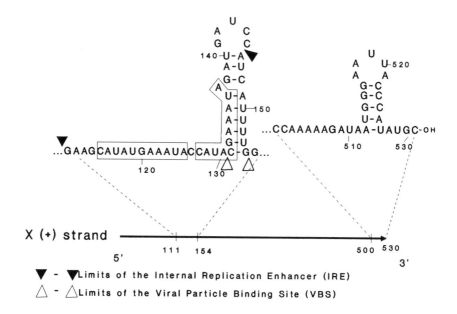

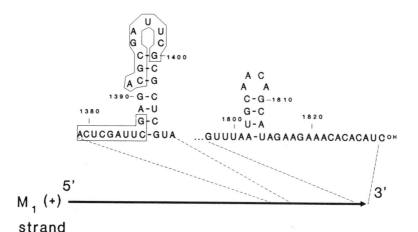

Figure 2 Sites on X (L-A) and M_1 (+) strands for in vitro replication and encapsidation. The 3'-end 4 bases and the adjacent stem-loop structure are required for the replication of X (+) strands, and, by inference, L-A (+) strands. The IRE increases the efficiency of replication five- to tenfold. The VBS is necessary for binding of viral (+) strands to open particles in vitro and is the encapsidation signal. The corresponding sites on M_1 are shown and have been demonstrated by synthesis and insertion into vectors or substitution in the X sequence.

sequence, of the stem is essential. The presence of a protruding A residue and the sequence of the 5 bases in the loop are also essential for RNA binding (Fujimura et al. 1990). This structure has been designated the virus-binding site (VBS) and is located about 400 bases from the 3'end of X (+) ssRNA (or L-A). A similar structure can be found at approximately the same distance from the 3'end of M_1 (+) strands (Fig. 2). This region is retained in all M_1 deletion mutants whose structures have been determined (Thiele et al. 1984b; Lee et al. 1986). These structures in M and L-A are sufficient for the binding reaction (Fujimura et al. 1990).

Insertion of either of these viral particle binding sites into a yeast DNA expression vector resulted in the packaging of the transcripts in L-A particles in vivo. Thus, these are the encapsidation signals for L-A and M_1 (Fujimura et al. 1990).

Western blots of proteins of purified L-A virions showed that a 180-kD viral protein bound ssRNA (Fujimura and Wickner 1988b). The protein bound dsRNA very poorly, and ssRNA binding was not competed by tRNA. These results argue that this 180-kD protein is responsible for the viral particle binding of (+) ssRNA. This 180-kD protein is encoded by L-A itself, and its novel structure forms the basis for a model of RNA packaging discussed below (Fujimura and Wickner 1988b).

D. In Vitro Replication: (–) Strand Synthesis on a (+) Strand Template

Empty L-A viral particles can replicate added L-A, M_1, or X (+) ssRNA, converting it to dsRNA by the synthesis of (–) strands. (–) Strand synthesis occurs completely de novo, and the replication machinery is absolutely specific for viral (+) strands as template. In vitro replication requires added host factor(s), in the form of a crude fraction of an extract from virus-free cells, as well as a low concentration of polyethylene glycol (Fujimura and Wickner 1988a).

Detailed studies of the template requirements of X (+) ssRNA have shown that two regions of the X (+) strands are important (Esteban et al. 1989). The first region encompasses the 3'-terminal 3 bases and a stem-loop structure immediately adjacent to the 3'end. This stem-loop structure is actually present in L-A (+) strands (Thiele et al. 1984a). The structure, but not the sequence, of the stem is critical, as is the sequence of the loop. The second important region overlaps the VBS that is the packaging signal (Fig. 2). This 35-base region is called the internal replication enhancer (IRE) and contains a 10 of 11 base, direct repeat sequence, which may be related to its activity. Deletions within this region lower template activity five- to tenfold.

Although the IRE and the viral packaging signal are distinct, their proximity may not be coincidental. M_1 (+) strands, like those of L-A, have a direct repeat (9 of 10 bases) overlapping the 5' side of its packaging signal. This could serve as an IRE. Perhaps the particles first bind (+) strands by the packaging site, then change their grip on the (+) strands (using the IRE) so that they can recognize the 3' end and begin (–) strand synthesis.

E. In Vitro Transcription: (+) Strand Synthesis on a dsRNA Template

The same opened empty particles that can bind and replicate (+) ssRNA to make dsRNA can also use viral dsRNA to make (+) strands in a conservative reaction (Fujimura and Wickner 1989). The reaction is template-specific, but the precise signals for synthesis have not yet been determined. Transcription requires 20% polyethylene glycol, presumably to increase the effective concentration of the dsRNA template. The opened empty particles indeed have a very low affinity for dsRNA templates. In vivo, the dsRNA is present inside the particles where it was synthesized. Its effective concentration to the transcriptase in the particles is thus very high, so the transcriptase need not have a high affinity for dsRNA (Fujimura and Wickner 1989).

F. L-A *gag/pol* Fusion Protein: Parallels with Retroviruses and (+) Strand Viruses

L-A dsRNA is a 4579-bp molecule with two long open reading frames (ORFs), both in the (+) strand (Fig. 3) (Icho and Wickner 1989). ORF1 (680 amino acids) lies at the 5' end of the genome and encodes the 76-kD L-A viral particle major coat protein. This was shown by comparing the ORF1 sequence with that of several cyanogen bromide peptides of the major coat protein determined by Dihanich et al. (1989). ORF2 (868 amino acids) occupies the 3' part of the L-A (+) strand and overlaps ORF1 in the –1 frame by 130 bases.

1. ORF2 Apparently Encodes an RNA Polymerase Similar to That of (+) Strand RNA Viruses

The amino acid sequence of ORF2 (Icho and Wickner 1989) includes a consensus sequence for RNA-dependent RNA polymerases found in more than 20 (+) strand RNA viruses of animals and plants (Kamer and Argos 1984). This consensus sequence is largely shared by all dsRNA viruses whose RNA polymerase-encoding segment has been sequenced (Fig. 4) (Icho and Wickner 1989; Wickner 1989). In addition to the per-

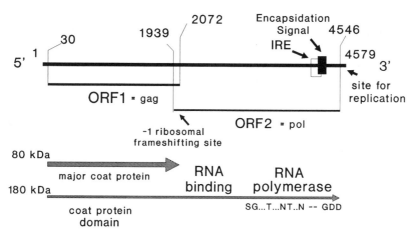

Figure 3 Structure and expression of L-A dsRNA. L-A has two ORFs, both on the (+) strand. ORF1 encodes the major coat protein (Icho and Wickner 1989) and a fusion of ORF1 and ORF2, formed by ribosomal frameshifting, encodes the 180-kD minor coat protein (Fujimura and Wickner 1988b; Icho and Wickner 1989; Dinman et al. 1991) that has ssRNA-binding activity and is therefore thought to be involved in packaging (see Fig. 6) (Fujimura and Wickner 1988b). The sequence of ORF2 includes a pattern diagnostic of viral RNA-dependent RNA polymerases (see Fig. 4) (Icho and Wickner 1989).

```
(+) Strand RNA Viruses:
Carnat    524  GCRMSGDMNTALGNCLLACLITKH.............LMKIRSRLINNGDDCVLI
TobEch   2580  KGNNSGQPSTVVDNTLMVIIAMLY........TCEKCGINKEEIVYYVNGDDLLIA

Mdlbrg    804  AMMKSGMFLTLFVNTMLNMTIAS.........RVLEERLTNSKCAAFIGDDNIVH
WNileF   3120  DQRGSGQVVTYALNTFTNLAVQKV.25 aa.RTWLFENGEERLSRMAVSGDDCVVK
YellowF  3100  DQRGSGQVVTYALNTITNLKVQLI.25 aa.EAWLTEHGCDRLKRMAVSGDDCVVR

Rhino14  2006  GGMPSGCSGTSIFNSMINNIIIRT.....LILDAYKGIDLD.KLKILAYGDDLIVS
Hepat A  1191  GSMPSGSPCTALLNSIINNVNLYY...VFSKIFGKSPVFFCQALKILCYGDDVLIF
Polio    2025  GGMPSGCSGTSIFNSMINNLIIRT.....LLLKTYKGIDLD.HLKMIAYGDDVIAS
Coxsac B  300  GGMPSGCSGTSIFNSMINNIIIRT.....LMLKVYKGIDLD.QFRMIAYGDDVIAS
                   **   *   **  *                            ***
dsRNA Viruses:
PHI6      390  VGLSSGQGATDLMgTLLmSITYLVMQLD.24 A.A.s.QGHEEIRQISKsDDAILG
IBDV      476  YGQGSGNAATFINNhLLsTLVLDQWNL...20 A.A.s....NFKIERSiDDIRGK
L-A       544  GTLLSGWRLTTFMNTVLNWAYMKLAGV.........FDLDDVQDSVHNGDDVMIS
BTV       715  DTHLSGENSTLIANSMHNMAIGTLIQRA....VGREQPGILTFLSEQYVGDDTLFY
REOVIRUS  676  TTFPSGSTATSTEhTANNSTMMETFLTV...20 A.A.s...QRNYVCQGDDGLMI
```

Figure 4 dsRNA viruses share the sequence pattern characteristic of (+) ssRNA viral RNA-dependent RNA polymerases. The RNA-dependent RNA polymerases of more than 20 (+) ssRNA viruses of animals and plants share the consensus pattern, SG...T...N[ST]..N—20–50 amino acids—GGD (Kamer and Argos 1984). L-A, Bluetongue virus, reovirus, and, to a lesser extent, bacteriophage φ6 and infectious bursal disease virus (all dsRNA viruses), share the same pattern.

fect matches found in L-A and the Bluetongue virus L1 segment (Fig. 4), a near-perfect match is present in the reovirus L1 segment, and less complete matches are found in bacteriophage φ6 and infectious bursal disease virus of chickens. The evidence that this sequence is part of the RNA-dependent RNA polymerases of the (+) strand RNA viruses is quite convincing, although the precise role of these residues in the enzyme reaction has not been defined for either (+) strand or dsRNA viruses. The fact that these two classes of RNA viruses, but not (–) strand RNA viruses, share this pattern suggests that the (+) and dsRNA groups are more closely related than may have been supposed.

2. Ribosomal Frameshifting Fuses ORF1 and ORF2

ORF2 is expressed as a fusion protein with ORF1 to yield the 180-kD minor viral coat protein that has ssRNA-binding activity (Fujimura and Wickner 1988b; see above). ORF1 and ORF2 overlap by 130 bp, with ORF2 in the –1 frame relative to ORF1 (Icho and Wickner 1989). Jacks et al. (1988) have shown that –1 ribosomal frameshifting in Rous sarcoma virus (RSV) and other retroviruses occurs by a simultaneous slippage of peptidyl and aminoacyl tRNAs on the mRNA, as diagramed in Figure 5. This model suggests that when ribosomes encounter an mRNA sequence ...*NNX XXY YYN NNN*... (where *N* represents any base, *X*s are all the same base, and *Y*s are all the same base), then the aminoacyl tRNA bound to *YYN* can slip back 1 base to *YYY* and still have its non-wobble bases properly paired. Likewise, the peptidyl tRNA bound to *XXY* can frameshift by –1 and still have its nonwobble bases correctly paired to *XXX* on the mRNA. This simultaneous –1 frameshift of both tRNAs on the ribosome is favored by a strong stem-loop or stem-loop-pseudoknot structure just after the site of the shift, perhaps due to pausing of the ribosomes at this point (Jacks et al. 1988; Brierley et al. 1989). As shown in Figure 5, L-A has such a structure within the 130-bp overlap of ORF1 and ORF2 (Icho and Wickner 1989).

The region necessary and sufficient for –1 frameshifting by L-A is just the slippery site and stem-loop-pseudoknot identified by inspection of the ORF1-ORF2 overlap region (Dinman et al. 1991). A detailed analysis of the slippery site showed that its requirements are as predicted by the simultaneous slippage model but suggest that a low probability of unpairing of the aminoacyl-tRNA at the ribosomal A site reduces the efficiency of frameshifting (Dinman et al. 1991). The efficiency of the L-A frameshift site is 1.8%, similar to the observed molar ratio in viral particles of the 180-kD fusion protein to the major coat protein.

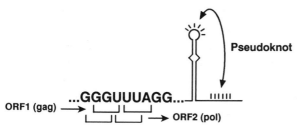

Figure 5 Simultaneous shift model of ribosomal frameshifting (Jacks et al. 1988) applied to L-A (Icho and Wickner 1989; Dinman et al. 1991). The occurrence of a sequence of the type ...*X XXY YYZ*....stem-loop-pseudoknot in the ORF1 frame allows the peptidyl tRNA and the aminoacyl tRNA bound to the ribosome to simultaneously slip back by 1 base on the mRNA and still have their nonwobble bases properly paired. The presence of a stem-loop (Jacks et al. 1988) or a pseudoknot (Brierley et al. 1989) probably slows the movement of the ribosome and thereby favors the shift. L-A has a stem-loop-pseudoknot structure (Icho and Wickner 1989; Dinman et al. 1991) immediately following the "shifty site" shown here. The sequence requirements for L-A's frameshifting are those predicted by the simultaneous slippage model (Dinman et al. 1991).

One reason (Icho and Wickner 1989) for using ribosomal frameshifting is the disadvantage of using splicing or some other modification of the mRNA to create the fused polypeptide. As in retroviruses and (+) strand RNA viruses, the (+) strands of dsRNA viruses serve not only as mRNAs, but also as the template for replication and packaging. As a result, any modification or editing of (+) strands in their capacity as mRNA must remove packaging or replication signals to prevent formation of mutant genomes. Thus, splicing has not been encountered among dsRNA viruses or (+) strand RNA viruses. This is true even though a number of (–) strand RNA viruses use splicing and mRNA editing, and splicing is ubiquitous among DNA viruses. Splicing in retroviruses always removes the packaging signal and thus prevents the propagation of this spliced message. All of the packaging, transcription, and replication signals for L-A are within 25 bp of the 5′ end of the (+) strand and within 500 bp of the 3′ end (Esteban et al. 1988, 1989; Fujimura et al. 1990). These signals could not be spliced out easily with part of the overlap region and still leave a message large enough to encode a 180-kD protein.

G. A Model for Packaging

Retroviruses synthesize their reverse transcriptase as a *gag/pol* fusion protein whose amino-terminal domain is a major coat protein monomer and whose carboxy-terminal domain includes reverse transcriptase, RNase H, protease, and integrase domains. Likewise, the amino-terminal domain of the 180-kD protein is a major coat protein monomer, whereas its carboxy-terminal domain is apparently the RNA polymerase. The packaging model depicted in Figure 6 suggests that the carboxy-terminal ssRNA-binding domain of the 180-kD protein specifically binds (+) ssRNA, whereas the amino-terminal major coat protein domain primes polymerization of the coat protein by interacting with major coat protein monomers (Fujimura and Wickner 1988b). When the polymerization of coat proteins is complete, the structure of the 180-kD protein has ensured packaging of both a viral (+) strand and the polymerase.

Because the structure of the 180-kD protein is analogous to that of the retroviral *gag/pol* fusion proteins, this packaging model could also apply to retroviruses. The retrovirus *gag/pol* fusion protein is cleaved by a virus-encoded protease but not until after the capsid has been formed and the viral RNA encapsidated. Thus, if the *pol* domain has specific affinity for the viral RNA, the *gag/pol* fusion protein could have the same role in retroviral assembly, as has been hypothesized for the 180-kD protein in L-A viral assembly.

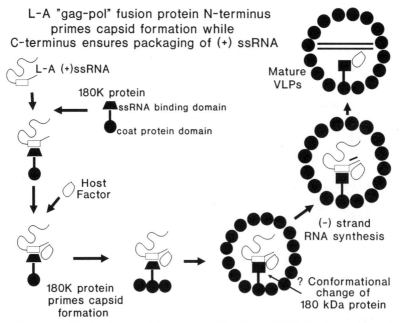

Figure 6 Packaging model suggested by the ssRNA-binding and major coat protein domain structure of the 180-kD fusion protein of L-A viral particles (Fujimura and Wickner 1988b).

IV. Control of dsRNA Replication in Yeast

The copy numbers of L-A, L-BC, and M dsRNAs, and even their ability to replicate at all, are affected by both chromosomal genes and non-chromosomal factors. The chromosomal genes include those necessary for dsRNA replication (*MAK* genes, *PET18*, and *CLO*) and those that repress dsRNA replication (*SKI* genes). Recently, the products of some of the *MAK* and *SKI* genes have been identified (Table 2). Most *MAK* genes for which a determination has been made are essential for host viability. This prevents the host from eliminating dsRNAs by simply deleting a gene apparent for propagation of the virus but nonessential to the host. The sole essential function of both of the two *SKI* genes tested is the repression of virus replication. This suggests that the *SKI* genes comprise a system dedicated to suppression of virus replication, rather than genes whose function incidentally affects dsRNA copy number.

A. *MAK* Genes and dsRNA Replication and Maintenance

The *MAK* gene products include a ribosomal protein (*MAK8* = *TCM1*; Wickner et al. 1982), DNA topoisomerase I (*MAK1* = *TOP1*; Thrash et al. 1984), a membrane-associated protein that has homology with *CDC4* and β-transducin (*MAK11*; Icho and Wickner 1988; M. Geobl, pers.

Table 2 Chromosomal genes essential for dsRNA replication

	Host function	Viral function
MAK8	ribosomal protein L3	?
SPE1, SPE2	polyamine biosynthesis	?
MAK1 = *TOP1*	DNA topoisomerase I	?
MAK11	essential membrane-associated protein	?
GCD1, GCD10, GCD11, GCD13	translation factors	?
MAK16	G_1 to S transition nuclear protein	?
MAK10	?	viral particle structural stability
PET18A	?	viral particle structural stability
Host factor (gene unknown)	?	replication reaction (in vitro [−] strand synthesis)

comm.), a nuclear protein required to transit G_1 (*MAK16*; Wickner 1988), enzymes in polyamine biosynthesis (*SPE2, SPE1*; Tyagi et al. 1984), *GCD* proteins that may be translation factors (Harashima and Hinnebusch 1986; see Hinnebusch and Liebman, this volume), and a product needed by L-A and M only at temperatures above 30°C (*PET18*; Toh-e and Sahashi 1985). In addition, mutations in *MAK10* or *PET18* causing temperature dependence for maintenance of L-A and M yield viral particles (produced at the permissive temperature) that are structurally unstable (Fujimura and Wickner 1986, 1987). This suggests that these gene products are components of the viral particles or are involved in the packaging process.

Although all *MAK* gene products are needed for M_1 dsRNA replication or maintenance, only *PET18* (>30°C), *MAK10*, and *MAK3* are required for maintenance of L-A dsRNA. Unexpectedly, X dsRNA, which is derived directly from L-A, requires all of the *MAK* genes. What do X and M_1 have in common that distinguishes them from L-A? Both X and M_1 replicate by the headful mechanism. However, this feature is unlikely to be the relevant, distinguishing characteristic, because headful packaging is simply a mechanical consequence of the rigid structure of the L-A-encoded head. The other common feature of M_1 and X is that both must appropriate coat proteins from L-A to replicate. This suggests that many *MAK* gene products could be involved in (1) the appropriation process, (2) protecting M_1 or X (e.g., from the *SKI* products) until they are packaged (Icho and Wickner 1988), or (3) determining the rate of production of coat proteins.

Several facts bear on this question. A natural variant of L-A, called L-A-B (B for bypass) suppresses the usual need of M_1 for a group of *MAK* genes (Toh-e and Wickner 1980; Uemura and Wickner 1988). Furthermore, the expression of the L-A ORF1 from a cDNA clone suppresses the *mak11-1*, *mak18-1*, and *mak27-1* mutations (Wickner et al. 1990). Clearly, further work will be needed to determine the role of *MAK* genes in viral maintenance.

B. *SKI* Genes: An Intracellular Antiviral System

Mutations in *SKI2, SKI3, SKI4, SKI6, SKI7*, or *SKI8* result in derepression of L-A, L-BC, and M dsRNA replication (Toh-e et al. 1978; Ball et al. 1984; Ridley et al. 1984). In strains with M_1 dsRNA, this produces a superkiller phenotype, as well as cold sensitivity for growth (Ridley et al. 1984). Cold sensitivity of *ski* strains requires the presence of an M replicon. However, S dsRNA deletion mutants, which lack nearly the entire coding region of M_1, produce the same growth defect in *ski* strains as does M_1 itself. This shows that overproduction of the killer toxin or immunity protein is not the cause of cytopathology (Ridley et al. 1984). In addition, the growth defect cannot be ascribed simply to an excess load of dsRNA in the cell. M dsRNA represses the copy number of L-A dsRNA, and this effect is particularly marked in *ski* strains (Ball et al. 1984). Accordingly, elimination of M from a *ski* killer results in a net increase in total cellular dsRNA (on either a mass or molar basis), because derepression of L-A more than makes up for the loss of M. Nevertheless, the *ski* M-o strains with higher total dsRNA are healthy, whereas the *ski* M_1 strains are sick. Thus, cold sensitivity of *ski* strains results specifically from the presence of elevated levels of M replicons but in some capacity other than its coding potential (Ridley et al. 1984).

In confirmation of this view, the level of secreted toxin from a cDNA-based plasmid is not increased in a *ski3* strain, showing that the toxin overproduction phenotype is dependent on an increased level of M_1 dsRNA (Hougan et al. 1989).

The cytopathology of *ski* M strains is enhanced by a new non-Mendelian genetic element called [D] (for disease). [D] is not located on M or L-A but seems to depend on L-A for its maintenance or replication (Esteban and Wickner 1987). *ski* M [D] strains not only are inviable at 8–12ºC, but also grow poorly at 30ºC and fail to grow at 37ºC. Cytoplasmic elements that interfere with [D] and chromosomal genes that are necessary for the maintainance of [D] (*MAD* genes) have also been described. The molecular identity of [D] is not yet known.

Using null mutations of *SKI3* or *SKI8* genes, it has been shown that the only essential function of either gene is prevention of the viral cytopathology discussed above (Sommer and Wickner 1987; Rhee et al. 1989). The *SKI3* product is a 165-kD protein located in the nucleus (Rhee et al. 1989), but the mechanism by which the *SKI* genes function remains completely unknown. Evidence also exists for an anti-retroviral system in yeast cells (Curcio et al. 1988), possibly involving the *RAD6*-encoded ubiquitin-conjugating enzyme (Picologlou et al. 1990).

C. Mitochondrial Functions and L-A Regulation

Several recent observations illuminate a connection between mitochondrial function and L-A maintenance, although the precise nature of this connection is unclear. None of the dsRNAs are necessary for mitochondrial function in wild-type strains, nor is the mitochondrial genome needed for dsRNA replication. However, growth on glycerol of cells carrying L-A results in a threefold higher L-A copy number than does growth on glucose. In addition, elimination of the mitochondrial genome prevents the loss of L-A and M dsRNAs from *mak10* mutants. Growth on glycerol of strains lacking the mitochondrial outer-membrane general diffusion pore protein (porin) results in massive overproduction of L-A viral particles (Dihanich et al. 1987, 1989). Furthermore, certain *mak10* and *mak3* strains grow poorly on glycerol, suggesting that these genes are involved in mitochondrial function (Dihanich et al. 1989).

Similar overproduction of L-A viral particles accompanies mutation of the *NUC1* gene, which encodes the major mitochondrial nonspecific nuclease (Liu and Dieckmann 1989). *ski* mutations also result in elevation of L-A (and L-BC) levels (Ball et al. 1984); but, unlike *ski* mutations, *nuc1* increases L-A levels without affecting M_1 levels (Liu and Dieckmann 1989). *nuc1* mutants are not superkillers nor are they cold-sensitive (R.B. Wickner, unpubl.). As in the case of the *ski* mutants (Ball et al. 1984), L-A overproduction in *nuc1* strains is eliminated by introduction of M dsRNA (Liu and Deickmann 1989). Reduction of L-A copy number by M_1 is epistatic to the elevation of L-A copy number by the *ski* or *nuc1* mutation.

D. L-A Genetics

A number of natural variants of L-A have been recognized on the basis of their interactions with each other, with M_1 and M_2, and with various chromosomal genes (Wickner 1980, 1983, 1987; Sommer and Wickner 1982; Wickner and Toh-e 1982; Ridley et al. 1984; Hannig et al. 1985; Uemura and Wickner 1988). M_1 and M_2 are distinct M dsRNA genes responsible for K_1 and K_2 killer phenotypes, respectively. They encode distinct killer toxins and toxin resistances, which have nonoverlapping specificities. L-A-E is unable to support replication of M_1 or M_2 in a wild-type strain but can do so in a *ski* host. The function missing in L-A-E is called [HOK], or H, for *h*elper *o*f *k*iller. L-A-H carries [HOK] and, accordingly, can support M_1 or M_2 in a wild-type host, although L-A-H was originally present in a particular K_2 strain. The fact that *ski* mutations make the [HOK] function dispensable for M replication suggests that the role of [HOK] is to antagonize the antiviral *SKI* products. Expression of L-A's ORF1 alone from a cDNA clone is sufficient to provide [HOK] (Wickner et al. 1990).

When L-A-E and L-A-H coexist in the same strain, L-A-E dramatically lowers the copy number of L-A-H, without actually excluding it. If the strain also carries either M_1 or M_2, the M genome is lost because of this reduction of available [HOK]. This property of L-A-E is called *exclusion* ([EXL], or E).

Some L-A-H variants are resistant to copy number depression by L-A-E. Accordingly, these resistant variants render M_1 or M_2 in the strain *nonex*cludable ([NEX], or N) by L-A-E. Such L-A variants are said to carry [NEX]. A majority of wild-type killer strains carry L-A-HN, an L-A with both helper and nonexcludable properties.

The properties of L-A discussed to this point do not distinguish M_1 and M_2; however, in certain genetic backgrounds, M_1 and M_2 respond differently to L-A. Specifically, L-A-HN cannot maintain M_2 at temperatures above 30°C in a host defective in *mkt1* or *mkt2* (*m*aintenance of *K two*). M_1, on the other hand, is stable in such hosts at temperatures up to 37°C. Because the same phenotype is obtained in strains containing null alleles of *mkt1*, the *MKT1* product apparently renders some temperature-dependent aspect of M_2 replication or maintenance temperature-independent.

Certain L-A variants make M_1 replication independent of a subset of *MAK* genes (M_2 has not been tested for this effect). These L-A variants are said to possess a bypass property (B). Thus, a *mak11-1* L-A-HN strain initially harboring M_1 loses M_1 rapidly, whereas a *mak11-1* L-A-HNB strain carrying M_1 has as strong a killer phenotype and as high an M_1 copy number as does a *MAK11* L-A-HNB strain. The L-A-HNB studied also maintains M_1 at a higher copy number than does the L-A-HN studied.

E. L-BC, T, and W dsRNAs

Elimination of L-A dsRNA by heat curing or by mutation of *mak3*, *mak10*, or *pet18* revealed other dsRNA species of the same size as L-A. T1 RNase fingerprints of these non-L-A dsRNAs showed that they comprise two families of dsRNA, and they were designated L-B and L-C. These two species were completely unrelated in sequence to L-A (Sommer and Wickner 1982) but were related to each other by hybridization. Since these species in most strains have not been completely characterized as L-B or L-C, they are currently referred to as the L-BC family. L-BC dsRNA is encapsidated in viral particles with a coat protein distinct from that of L-A (Sommer and Wickner 1982), and these particles have transcriptase and replicase activity (Sommer and Wickner 1982; Fujimura et al. 1986). L-BC has also been shown to encode its own major coat protein (El-Sherbeini et al. 1984).

Elimination of L-BC, by mutation of the *clo* gene, revealed several minor species of dsRNA, the two most abundant of which have been partially characterized (Wesolowski and Wickner 1984). T dsRNA (2.7 kb) and W dsRNA (2.25 kb) are not homologous to L-A, L-B, L-C, or each other, are inherited in a non-Mendelian fashion, and are increased in copy number in cells grown at 37°C.

V. STRUCTURE, PROCESSING, SECRETION, AND ACTION OF KILLER TOXINS

A. Structure of the K_1 Preprotoxin Gene and the K_1 Toxin

Killer toxin and immunity to killer toxin are encoded by M dsRNA. M_1 dsRNA possesses a single ORF that extends from an AUG codon at base 14 from the 5′ end of the M_1 (+) strand to a UAG codon at base 963. The 3′ region of the M_1 genome lacks coding information but encompasses the sites for packaging and replication enhancement discussed above. Lying between the packaging/replication region and the coding region is a stretch of poly(A)/poly(U) of about 200 bp, whose length varies from generation to generation. Comparison of the protein sequence of purified toxin with the nucleotide sequence of the M_1-coding region revealed that killer toxin is composed of two subunits, α (9.5 kD) and β (9.0 kD), which originate from the amino-terminal and carboxy-terminal domains of the preprotoxin, respectively (Fig. 7) (Thiele and Leibowitz 1982; Bostian et al. 1984; Skipper et al. 1984; Zhu et al. 1987). The α and β domains flank a segment called γ, which is not part of the mature toxin. Protoxin is glycosylated at several sites within the γ segment (Bostian et al. 1984), but glycosylation appears to be dispensable for both toxin secretion and immunity (Lolle et al. 1984).

Both toxin production and immunity can be expressed from cDNA clones of the preprotoxin gene, and site-specific mutagenesis of these clones has localized the toxin and immunity domains within the preprotoxin gene. Toxin binds to cells, recognizing a (1-6)-β-D-glucan linkage in cell walls, and causes cell death by creating cation-permeable pores in the membranes of sensitive cells (de la Pena et al. 1980, 1981; Martinak et al. 1990). Mutants in either α or β lose toxin activity against whole cells. Several β mutants remain able to kill spheroplasts, although they cannot kill whole cells (Lolle et al. 1984; Boone et al. 1986; Sturley et al. 1986). This is explained by assuming that the β-subunit is involved in binding to the cell-wall receptor and the α toxin functions as the ionophore.

Strains carrying M_1 dsRNA or a preprotoxin gene cDNA clone are immune to the toxin that they secrete (Lolle et al. 1984). Mutations in the α and γ regions from residues 85 to 177 show loss of immunity. Deletions of all of β and the carboxy-terminal half of γ leave immunity largely unaffected (Boone et al. 1986; Sturley et al. 1986). The immunity domain defined by these mutations is shown in Figure 7. As discussed below, the products of the *KEX* (*k*iller *ex*pression) genes are responsible for proteolytic processing of preprotoxin, except for signal peptide cleavage. In *kex2* mutants, no processing of the protoxin is observed, yet these strains are fully immune to the toxin. This observation, in conjunction with the extent of the immunity domain determined by localized

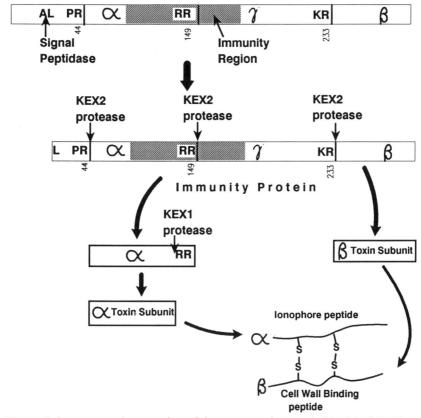

Figure 7 Structure and processing of the preprotoxin encoded by M$_1$ dsRNA.

mutagenesis, suggests that the protoxin itself is responsible for immunity (Boone et al. 1986). One model suggests that the protoxin confers immunity by binding to the membrane receptor for the toxin in such a way that active toxin, itself, cannot bind.

A chromosomal gene, *REX1*, is necessary for the immunity of M$_1$-containing cells (Wickner 1974). Accordingly, *rex1* mutants have a K$^+$R$^-$ suicide phenotype. Douglas et al. (1988) have detected a 22-kD peptide derived from the protoxin that reacts with antitoxin and with anti-γ antibodies. This product is not seen in a *rex1* mutant but reappears in Rex$^+$ revertants. This suggests that the 22-kD peptide may be responsible for immunity. Further studies by these workers showed that *pep6* (*vpl3*) and *pep12* (*vpl6*), defective in vacuolar protein localization, as well as *pep5* (*end1*), defective in vacuolar biogenesis, fail to express full immunity.

B. Toxin Processing: *KEX1* and *KEX2* Proteases

The *kex* mutants were first isolated on the basis of their inability to secrete the K_1 killer toxin. *kex2* mutants were also found to be deficient in mating and sporulation (Leibowitz and Wickner 1976). The basis of these defects became apparent when *KEX1* and *KEX2* were cloned and shown to encode proteases necessary for processing both killer toxin and α-factor precursor proteins (Julius et al. 1984; Dmochowska et al. 1987).

The Kex2 protein (Kex2p) is an endopeptidase that cleaves to the immediate carboxyl side of pairs of basic residues (Julius et al. 1984; for review, see Fuller et al. 1988). The kex1 protein (Kex1p) (Dmochowska et al. 1987; Wagner and Wolf 1987; Cooper and Bussey 1989) is a carboxypeptidase that removes pairs of basic residues. As shown in Figure 7, the combined action of these two proteases yields mature toxin from protoxin. Similarly, mature α-pheromone is derived from a precursor protein consisting of tandem repeats of the mature pheromone sequence, separated by spacer peptides each initiated by a pair of basic amino acids (see Sprague and Thorner, these volumes). Accordingly, Kex2p is absolutely required for maturation of α-factor. However, the last of the α-factor repeat units lies at the absolute carboxyl ends of the prohormone and therefore is not dependent on Kex1p for its maturation.

KEX2 shows clear homology with the subtilisin family (Mizuno et al. 1988), as does a similar gene from *Kluyveromyces lactis* that is necessary for processing of the toxin encoded by the pGKl1 linear DNA plasmid of that organism (Wesolowski-Louvel et al. 1988). Although the protease activity is Ca^{++}-dependent, no Ca^{++}-binding domain is evident in the sequences of the protein. The enzyme is membrane-associated and the sequence shows an amino-terminal signal peptide-like region and a carboxy-terminal hydrophobic domain that might serve to anchor the protein to the membrane.

Kex1p is an 82-kD membrane-associated polypeptide to which N-linked carbohydrate chains are attached. Its sequence resembles yeast carboxypeptidase Y and possesses a typical serine protease active site. The serine in this sequence is essential for the *KEX1* complementing activity (Dmochowska et al. 1987; Cooper and Bussey 1989). Kex1p has been detected in membrane fractions (Achstetter and Wolf 1985; Wagner and Wolf 1987; Cooper and Bussey 1989) and exhibits the expected specificity for the removal of basic residues from the carboxyl terminus of synthetic substrates.

Proteolysis is widely used in the endocrine and nervous systems to release active hormones or neuropeptides from larger precursor polypeptides. The signals for such cleavages are most often pairs of basic amino acid residues, with the cleavage made on the immediate carboxy-terminal side. This is followed by removal of the two basic residues. One example of this is the processing of prepro-opiomelanocortin to produce ACTH, β-lipotropin, γ-lipotropin, β-endorphin, α-MSH, and γ-MSH. Despite extensive efforts, the enzymes responsible for this type of processing have not yet been positively identified in any mammalian system (for review, see Bussey 1988; Fuller et al. 1988).

The processing of the K_1 killer toxin and α-pheromone requires the same cleavages as those necessary for mammalian prohormone processing. Perhaps the most dramatic demonstration of the parallel between the yeast *KEX* system and mammalian prohormone processing was the demonstration by G. Thomas et al. (1988) and L. Thomas et al. (1990) that expressing *KEX1* and *KEX2* from plasmids in a mammalian cell line allowed correct processing of prepro-opiomelanocortin and incorporation of the products into secretory granules. Thus, Kex1p and Kex2p are not only functionally homologous to their mammalian counterparts, they are actually interchangeable. A possible human homolog of *KEX2*, "furin," has recently been identified by sequence similarity (Fuller et al. 1989; Brake et al. 1990; Bresnahan et al. 1990).

kex2 mutations also unexpectedly suppress temperature-sensitive mutations in RNA polymerase II and also allow cells to grow at a higher than normal growth temperature (Martin and Young 1989).

Processing of the killer toxin also requires the products of the *SEC* genes, responsible for general *sec*retion (Bussey et al. 1983). Mutants in *sec18* or *sec7* accumulate the 43-kD glycosylated protoxin in the endoplasmic reticulum, whereas *sec1* mutants accumulate it in secretory vesicles.

C. *KRE* Genes and the Mechanism of Action of Killer Toxins

The K_1 toxin binds specifically to (1-6)-β-D-glucan and can be purified rapidly by using (1-6)-β-D-glucan columns (Hutchins and Bussey 1983). Mutants in the *KRE1* gene (*k*iller *re*sistant), selected from a sensitive strain by their ability to resist the K_1 toxin, are defective in their ability to synthesize the (1-6)-β-D-glucan linkage. Thus, the first K_1 toxin receptor is in the cell wall. However, spheroplasts of *kre1* strains are killed by the K_1 toxin. In addition, spheroplasts of *Candida*, *Kluyveromyces*, and *Schwannomyces* strains can also be killed by *Saccharomyces* killer toxins, even though intact cells of these strains are resistant to killing (Zhu and Bussey 1989). These results suggest the existence of a second membrane-associated receptor, a notion also supported by the fact that spheroplasts of strains carrying M_1 dsRNA are resistant to the effects of the K_1 toxin.

The study of genes responsible for glucan synthesis has been approached by isolation of killer-toxin-resistant mutants. In addition to *KRE1*, mutants in *KRE5* and *KRE6* are also defective in (1-6)-β-D-glucan biosynthesis (Boone et al. 1990). The *KRE5* gene encodes a probable endoplasmic reticulum protein that is required for normal cell growth (Meaden et al. 1990). The *KRE1* gene has been isolated and shown to encode a nonessential, highly *O*-glycosylated, threonine/serine-rich protein, apparently localized on the cell surface (Boone et al. 1990).

Although the actual cell membrane receptor for the K_1 toxin has not yet been identified, studies by de la Pena et al. (1980, 1981) have shown that the killer toxin has almost immediate effects on proton transport accompanying either K^+ or amino acid uptake. Membrane patches from toxin-treated spheroplasts and liposomes into which the K_1 toxin was incorporated have a cation channel that is almost always open, independent of the transmembrane voltage (Martinac et al. 1990). These results indicate that the toxin acts by forming a pore in the membrane that is cation permeable.

D. New Killer Yeasts

Recent descriptions of killer yeasts include strains of *Hansenula* and *Kluyveromyces*, as well as other *Saccharomyces* strains. A *Hansenula mrakii* killer secretes an 8.9-kD protein toxin that is extremely stable to heat or pH and has a broad spectrum of activity. An 8.5-kD toxin called HYI produced by *Hansenula saturnus* has been purified to homogeneity and shown to have a very broad spectrum of activity against yeasts (Ashida et al. 1983; Nomoto et al. 1984; Ohta et al. 1984).

Hara et al. (1983) have described several new types of killer strains of *Saccharomyces* isolated from wine yeasts. One of these strains, called Y-1, has L and M dsRNAs, another, Y-9, has only L dsRNA, and a third, Y-6, has no dsRNA. Genetic analysis of Y-9 showed that its killer trait is controlled by two chromosomal genes: *KHS* on chromosome V distal to *rad3* and controlling secretion of a heat-sensitive toxin, and *KHR* on chromosome IX very close to *cdc29* and controlling secretion of a heat-resistant toxin. The absense of an M dsRNA in Y-9 and the finding of two toxins with different properties controlled by two different chromosomal genes suggest that these genes may encode the toxins (Kitano et al. 1986).

Naumov (1985) has described a cytoplasmically inherited killer trait in *Saccharomyces paradoxus* strain CBS5829 that is eliminated by ethidium treatment and by an *ade* mutation. The strain appears to lack dsRNA, and Naumov suggests that the killer trait is encoded by the mitochondrial genome. Its toxin has the remarkable property of killing strains with the K_1 or K_2 killer trait but not strains sensitive to these other killers. Thus, this new killer might be called a "killer killer."

The KT28 toxin was produced by a single strain of *S. cerevisiae* and has been purified to homogeneity (Pfeiffer and Radler 1982). This toxin is also determined by an M dsRNA (Schmitt and Tipper 1990) and is a glycoprotein. Unlike the K_1 toxin that binds to cell-wall (1-6)-β-D-glucan, the KT28 toxin binds first to cell-wall mannan (Schmitt and Radler 1988). Thus, the toxin can be purified on mannan/Sepharose affinity columns (Schmitt and Radler 1989), and among KT28 toxin-resistant mutants are those with an altered mannan biosynthesis (Schmitt and Radler 1988).

Gunge discovered a killer system in *K. lactis* determined by two linear cytoplasmic DNA plasmids, pGKl1 and pGKl2. The extensive studies of this system are beyond the scope of this review, but they are reviewed by Gunge (1986) and by Stark et al. (1990).

VI. APPLICATIONS

A. Brewing Strains

Introduction of the killer trait into fermentation strains can be used to prevent contaminating killer (or nonkiller) strains from spoiling the brew. The killer virus can be introduced by cytoduction without changing either the nuclear genotype or the mitochondrial genotype of the brewing strain (Ouchi et al. 1979). This method is most efficient if the brewing strain is capable of mating; most brewing strains are sterile polyploids. Rare matings can be used, however, to accomplish the same purpose (Hammond and Eckersley 1984), and the resulting strains are indistinguishable as brewing strains for beer (Hammond and Eckersley 1984) or for saki (Ouchi et al. 1979). The "transfection" of isolated viral particles accompanying DNA transformation of the brewing strain is one way to avoid the mating requirement (El-Sherbeini and Bostian 1987). Yet another approach uses a cDNA clone of M dsRNA to express toxin production (Bussey and Meaden 1985; Boone et al. 1989). This last approach has the unique advantage of allowing construction of a strain stably expressing both K_1 and K_2 killer toxins, avoiding the usual exclusion of M_2 dsRNA by M_1 dsRNA. Either the self-selection of K^+ cells by killing of K^- segregants or the integration of the M cDNA sequences ensures stability.

B. Secretion Vectors

The toxin cDNA clone has also been used to construct a secretion vector with the signal region of the toxin. This was applied to construct a yeast strain secreting the carboxymethylcellulase from *Cellomonas fimi* (Skipper et al. 1985).

C. Production of Foreign Proteins

The extremely high copy number of L-A dsRNA, particularly in a *ski* strain lacking M dsRNA, has been crucial in developing the biochemistry of replication of the L-A system. Because such strains (R. Esteban et al., unpubl.) and the porin-defective mutants (Dihanich et al. 1989) may have 10–20% of their total protein as major coat protein, it seems likely that the cDNA clones of L-A (Icho and Wickner 1989) or those of its deletion mutant, X (Esteban et al. 1988), can be used to produce large amounts of a desired protein in yeast.

D. Biotyping of Fungi and Other Microorganisms

Polonelli, Morace, and their co-workers have developed methods of classifying and conducting epidemiologic studies on pathogenic fungi, particularly *Candida* isolates, by using their sensitivity or resistance to a wide range of yeast killer toxins, especially those from *Saccharomyces*, *Pichia*, *Kluyveromyces*, and *Candida* (Polonelli et al. 1983, 1987; Polonelli and Morace 1988). The same methods have also proved applicable to classifying actinomycetes (Morace et al. 1988).

VI. 20S RNA IS A CIRCULAR ssRNA REPLICON

20S RNA of *S. cerevisiae* was first described as a species whose synthesis was induced by shifting cells to potassium acetate medium, the same medium used to induce meiosis and sporulation (Kadowaki and Halvorson 1971a,b). That the induction of 20S RNA synthesis was unrelated to these processes was shown by the ability of many haploid strains to induce 20S synthesis, although they cannot undergo meiosis or sporulation, and by the inability of some sporulation-competent strains to induce 20S RNA (Garvik and Haber 1978). Mating strains that could induce 20S RNA synthesis with those that could not produced meiotic progeny, all of whom could induce 20S RNA synthesis (Garvik and Haber 1978). This non-Mendelian segregation pattern suggested that 20S RNA was an independent replicon, but whether the 20S RNA was inherited in a non-Mendelian fashion, or only its inducibility, was not clear. 20S RNA was found to be present in 32S intracellular particles associated with 18–20 copies of a 23-kD protein (Wejksnora and Haber 1978). This protein partially protects the RNA from nuclease attack.

Recently, clones of 20S RNA have been isolated and used to show that 20S RNA is an autonomous replicon that replicates by an RNA-RNA pathway. It is amplified 10,000-fold on shifting to potassium acetate medium. Its replication is also repressed three- to fivefold by the *SKI* antiviral system (Matsumoto et al. 1990).

Electron microscopic examination of purified 20S RNA showed that most of the molecules were circular. Both 3′ and 5′ ends were inaccessible to labeling, and the molecule showed the anomalous migration on two-dimensional gel electrophoresis typical of circular RNA molecules (Matsumoto et al. 1990). Although 20S RNA is clearly circular in some sense, the possibility of an unusual linkage of 3′ and 5′ ends of a linear RNA molecule was not completely ruled out.

Almost all of the 20S RNA genome has now been cloned and sequenced (Y. Matsumoto and R.B. Wickner, unpubl.). Unlike all previously described circular RNA replicons, it appears to encode its own RNA polymerase. The presence of two typical cAMP-dependent phosphorylation sites in the putative RNA polymerase suggests that amplification of the 20S RNA genome might be regulated by cAMP, a signal known to transmit the same nutritional status information to the sporulation-control system.

ACKNOWLEDGMENTS

The author thanks his co-workers Jon Dinman, Rosa Esteban, Tsutomu Fujimura, Tateo Icho, Yang Ja Lee, Yutaka Matsumoto, Juan Carlos Ribas, Juan Carlos Tercero, Rosaura Valle, and Bill Widner.

REFERENCES

Achstetter, T. and D.H. Wolf. 1985. Hormone processing and membrane-bound proteases in yeast. *EMBO J.* **4:** 173.

Ashida, S., T. Shimazaki, K. Kitano, and S. Hara. 1983. New killer toxin of *Hansenula mrakii. Agric. Biol. Chem.* **47:** 2953.

Ball, S.G., C. Tirtiaus, and R.B. Wickner. 1984. Genetic control of L-A and L-(BC) dsRNA copy number in killer systems of *Saccharomyces cerevisiae. Genetics* **107:** 199.

Boone, C., S.S. Sommer, and H. Bussey. 1990. Yeast *KRE* genes provide evidence for a pathway of cell wall β-glucan assembly. *J. Cell Biol.* **110:** 1833.

Boone, C., H. Bussey, D. Greene, D.Y. Thomas, and T. Vernet. 1986. Yeast killer toxin: Site-directed mutations implicate the precursor protein as the immunity component. *Cell* **46**: 105.

Boone, C., A.-M. Sdicu, J. Wagner, R. Degre, C. Sanchez, and H. Bussey. 1989. Integration of the yeast K_1 killer toxin gene into the genome of marked wine yeasts, and its effect on vinification. *Am. J. Enol. Vitic.* **41**: 37.

Bostian, K.A., Q. Elliot, H. Bussey, V. Burn, A. Smith, and D.J. Tipper. 1984. Sequence of the preprotoxin dsRNA gene of type I killer yeast: Multiple processing events produce a two-component toxin. *Cell* **36**: 741.

Bozarth, R.F., Y. Koltin, M.B. Weissman, R.L. Parker, R.E. Dalton, and R. Steinlauf. 1981. The molecular weight and packaging of dsRNAs in the mycovirus from *Ustilago maydis* killer strains. *Virology* **113**: 492.

Brake, A.J., R.S. Fuller, and J. Thorner. 1990. Identification of human prohormone-processing endoproteases by homology to the *KEX2* gene product. *Proc. Alfred Benzon Symp.* **29**: 197.

Bresnahan, P.A., R. Ledric, L. Thomas, J. Thorner, H.L. Gibson, A.J. Brake, P.J. Barr, and G. Thomas. 1990. Human *fur* gene encodes a yeast *KEX2*-like endoprotease that cleaves pro-β-NGF in vivo. *J. Cell Biol.* (in press).

Brierley, I., P. Digard, and S.C. Inglis. 1989. Characterization of an efficient coronavirus ribosomal frameshifting signal: Requirement for an RNA pseudoknot. *Cell* **57**: 537.

Brun, G. and N. Plus. 1980. The viruses of *Drosophila*. In *The genetics and biology of Drosophila* (ed. M. Ashburner and T.F. Wright), vol. 2, p. 625. Academic Press, New York.

Buck, K.W. 1978. Semi-conservative replication of double-stranded RNA by a virion-associated RNA polymerase. *Biochem. Biophys. Res. Commun.* **84**: 639.

———. 1988. From interferon induction to fungal viruses. *Eur. J. Epidemiol.* **4**: 395.

Bussey, H. 1988. Proteases and the processing of precursors to secreted proteins in yeast. *Yeast* **4**: 17.

Bussey, H. and P. Meaden. 1985. Selection and stability of yeast transformants expressing cDNA of an M_1 killer toxin-immunity gene. *Curr. Genet.* **9**: 285.

Bussey, H., D. Saville, D. Greene, D.J. Tipper, and K.A. Bostian. 1983. Secretion of *Saccharomyces cerevisiae* killer toxin: Processing of the glycosylated precursor. *Mol. Cell. Biol.* **3**: 1362.

Cooper, A. and H. Bussey. 1989. Characterization of the yeast *KEX1* gene product: A carboxypeptidase involved in processing secreted precursor proteins. *Mol. Cell. Biol.* **9**: 2706.

Curcio, M.J., N.J. Sanders, and D.J. Garfinkel. 1988. Transposition competence and transcription of endogenous Ty elements in *Saccharomyces cerevisiae*: Implications for regulation of transposition. *Mol. Cell. Biol.* **8**: 3571.

de la Pena, P., F. Barros, S. Gascon, P.S. Lazo, and S. Ramos. 1981. Effect of yeast killer toxin on sensitive cells of *Saccharomyces cerevisiae*. *J. Biol. Chem.* **256**: 10420.

de la Pena, P., F. Barros, S. Gascon, S. Ramos, and P. Lazo. 1980. Primary effects of yeast killer toxin. *Biochem. Biophys. Res. Commun.* **96**: 544.

Dihanich, M., K. Suda, and G. Schatz. 1987. A yeast mutant lacking mitochondrial porin is respiratory deficient, but can recover respiration with simultaneous accumulation of an 86 kDa extramitochondrial protein. *EMBO J.* **6**: 723.

Dihanich, M., E. Van Tuinen, J.D. Lambris, and B. Marshallsay. 1989. Accumulation of viruslike particles in a yeast mutant lacking a mitochondrial pore protein. *Mol. Cell. Biol.* **9**: 1100.

Dinman, J.D., T. Icho, and R.B. Wickner. 1991. A −1 Ribosomal frameshift in a double-stranded RNA virus of yeast forms a gag-pol fusion protein *Proc. Natl. Acad. Sci.* **88:** 174.

Dmochowska, A., D. Dignard, D. Henning, D.Y. Thomas, and H. Bussey. 1987. Yeast *KEX1* gene encodes a putative protease with a carboxypeptidase B-like function involved in killer toxin and alpha factor precursor processing. *Cell* **50:** 573.

Douglas, C.M., S.L. Sturley, and K.A. Bostian. 1988. Role of protein processing, intracellular trafficking and endocytosis in production of and immunity to yeast killer toxin. *Eur. J. Epidemiol.* **4:** 400.

El-Sherbeini, M. and K.A. Bostian. 1987. Viruses in fungi: Infection of yeast with the K_1 and K_2 killer virus. *Proc. Natl. Acad. Sci.* **84:** 4293.

El-Sherbeini, M., D.J. Tipper, D.J. Mitchell, and K.A. Bostian. 1984. Virus-like particle capsid proteins encoded by different L double-stranded RNAs of *Saccharomyces cerevisiae*: Their roles in maintenance of M double-stranded killer plasmids. *Mol. Cell. Biol.* **4:** 2818.

Esteban, R. and R.B. Wickner. 1986. Three different M_1 RNA-containing viruslike particle types in *Saccharomyces cerevisiae*: *In vitro* M_1 double-stranded RNA synthesis. *Mol. Cell. Biol.* **6:** 1552.

––––––. 1987. A new non-Mendelian genetic element of yeast that increases cytopathology produced by M_1 double-stranded RNA in *ski* strains. *Genetics* **117:** 399.

––––––. 1988. A deletion mutant of L-A double-stranded RNA replicates like M_1 dsRNA. *J. Virol.* **62:** 1278.

Esteban, R., T. Fujimura, and R.B. Wickner. 1988. Site-specific binding of viral (+) single-stranded RNA to replicase-containing open virus-like particles of yeast. *Proc. Natl. Acad. Sci.* **85:** 4411.

––––––. 1989. Internal and terminal cis-acting sites are necessary for in vitro replication of the L-A double-stranded RNA virus of yeast. *EMBO J.* **8:** 947.

Fujimura, T. and R.B. Wickner. 1986. Thermolabile L-A virus-like particles from *pet18* mutants of *Saccharomyces cerevisiae*. *Mol. Cell. Biol.* **6:** 404.

––––––. 1987. L-A double-stranded RNA virus-like particle replication cycle in *Saccharomyces cerevisiae*: Particle maturation *in vitro* and effects of *mak10* and *pet18* mutations. *Mol. Cell. Biol.* **7:** 420.

––––––. 1988a. Replicase of L-A virus-like particles of *Saccharomyces cerevisiae*: *In vitro* conversion of exogenous L-A and M_1 single-stranded RNAs to double-stranded form. *J. Biol. Chem.* **263:** 454.

––––––. 1988b. Gene overlap results in a viral protein having an RNA binding domain and a major coat protein domain. *Cell* **55:** 663.

––––––. 1989. Reconstitution of template-dependent in vitro transcriptase activity of a yeast double-stranded RNA virus. *J. Biol. Chem.* **264:** 10872.

Fujimura, T., R. Esteban, and R.B. Wickner. 1986. In vitro L-A double-stranded RNA synthesis in virus-like particles from *Saccharomyces cerevisiae*. *Proc. Natl. Acad. Sci.* **83:** 4433.

Fujimura, T., R. Esteban, L.M. Esteban, and R.B. Wickner. 1990. Portable encapsidation signal of the L-A double-stranded RNA virus of *S. cerevisiae*. *Cell* **62:** 819.

Fuller, R.S., A.J. Brake, and J. Thorner. 1989. Intracellular targeting and structural conversation of a prohormone-processing endopeptidase. *Science* **246:** 482.

Fuller, R.S., R.E. Stearne, and J. Thorner. 1988. Enzymes required for yeast prohormone processing. *Annu. Rev. Physiol.* **50:** 345.

Garvik, B. and J.E. Haber. 1978. New cytoplasmic genetic element that control 20S RNA synthesis during sporulation in yeast. *J. Bacteriol.* **134:** 261.

Gunge, N. 1986. Linear DNA killer plasmids from the yeast *Kluyveromyces*. *Yeast* **2:**

153.

Gupta, P., B. Balachandran, M. Ho, A. Enrico, and C. Rinaldo. 1989. Cell-to-cell transmission of human immunodeficiency virus type 1 in the presence of azidothymidine and neutralizing antibody. *J. Virol.* **63**: 2361.

Hammond, J.R.M. and K.W. Eckersley. 1984. Fermentation properties of brewing yeast with killer character. *J. Inst. Brew.* **90**: 167.

Hannig, E.M., M.J. Leibowitz, and R.B. Wickner. 1985. On the mechanism of exclusion of M_2 double-stranded RNA by L-A-E double-stranded RNA in *Saccharomyces cerevisiae. Yeast* **1**: 57.

Hara, S., Y. Iimura, and K. Otsuka. 1983. Development of a new pure culture winemaking method using killer wine yeasts and prevention of film-forming spoilage of wines. *Nippon Nogeikagaku Kaishi* **57**: 897.

Harashima, S. and A.G. Hinnebusch. 1986. Multiple *GCD* genes required for repression of *GCN4*, a transcriptional activator of amino acid biosynthetic genes in *Saccharomyces cerevisiae. Mol. Cell. Biol.* **6**: 3990.

Hooks, J.J., W. Burns, K. Hayashi, S. Geis, and A.L. Notkins. 1976. Viral spread in the presence of neutralizing antibody: Mechanisms of persistence in foamy virus infection. *Infect. Immun.* **14**: 1172.

Hougan, L., D.Y. Thomas, and M. Whiteway. 1989. Cloning and characterization of the *SKI3* gene of *Saccharomyces cerevisiae* demonstrated allelism to *SKI5. Curr. Genet.* **16**: 139.

Hutchins, K. and H. Bussey. 1983. Cell wall receptor for yeast killer toxin: Involvement of $(1\rightarrow6)$-β-D-glucan. *J. Bacteriol.* **154**: 161.

Icho, T. and R.B. Wickner. 1988. The *MAK11* gene is essential for cell growth and replication of M double-stranded RNA and is apparently a membrane-associated protein. *J. Biol. Chem.* **263**: 1467.

―――. 1989. The double-stranded RNA genome of yeast virus L-A encodes its own putative RNA polymerase by fusing two open reading frames. *J. Biol. Chem.* **264**: 6716.

Jacks, T., H.D. Madhani, F.R. Masiarz, and H.E. Varmus. 1988. Signals for ribosomal frameshifting in the Rous sarcoma virus *gag-pol* region. *Cell* **55**: 447.

Joklik, W. 1983. *The reoviridae.* Plenum Press, New York.

Julius, D., A. Brake, L. Blair, R. Kunisawa, and J. Thorner. 1984. Isolation of the putative structural gene for the lysine-arginine-cleaving endopeptidase required for the processing of yeast prepro-alpha-factor. *Cell* **36**: 309.

Kadowski, K. and H.O. Halvorson. 1971a. Appearance of a new species of ribonucleic acid during sporulation in *Saccharomyces cerevisiae. J. Bacteriol.* **105**: 826.

―――. 1971b. Isolation and properties of a new species of ribonucleic acid synthesized in sporulating cells of *Saccharomyces cerevisiae. J. Bacteriol.* **105**: 831.

Kamer, G. and P. Argos. 1984. Primary structural comparison of RNA-dependent polymerases from plant, animal and bacterial viruses. *Nucleic Acids Res.* **12**: 7269.

Kitano, I., A. Totsuka, and S. Hara. 1986. Yeast killer phenomena that do not depend on an RNA plasmid. In *Yeast biotechnology: Basic and applied* (ed. T. Hirano), p. 1. Gakkai Shuppan Center, Tokyo.

Lee, M., D.F. Pietras, M.E. Nemeroff, B.J. Corstanje, L.J. Field, and J.A. Bruenn. 1986. Conserved regions in defective interfering viral double-stranded RNAs from a yeast virus. *J. Virol.* **58**: 402.

Leibowitz, M.J. and R.B. Wickner. 1976. A chromosomal gene required for killer plasmid expression, mating, and spore maturation in *Saccharomyces cerevisiae. Proc. Natl. Acad. Sci.* **73**: 2061.

Liu, Y. and C.L. Dieckmann. 1989. Overproduction of yeast virus-like particle coat protein genome in strains deficient in a mitochondrial nuclease. *Mol. Cell. Biol.* **9**: 3323.

Lolle, S., N. Skipper, H. Bussey, and D.Y. Thomas. 1984. The expression of cDNA clones of yeast M_1 double-stranded RNA in yeast confers both killer and immunity phenotypes. *EMBO J.* **3**: 1383.

Martin, C. and R.A. Young. 1989. *KEX2* mutations supress RNA polymerase II mutants and alter the temperature range of yeast cell growth. *Mol. Cell. Biol.* **9**: 2341.

Martinac, B., H. Zhu, A. Kubalski, X. Zhou, M. Culbertson, H. Bussey, and C. Kung. 1990. Yeast K1 killer toxin forms ion channels in sensitive yeast spheroplasts and in artificial lipsomes. *Proc. Natl. Acad. Sci.* **87**: 6228.

Matsumoto, Y., R. Fishel, and R.B. Wicker. 1990. Circular single-stranded RNA replicon in *Saccharomyces cerevisiae. Proc. Natl. Acad. Sci.* **87**: (in press).

Meaden, P., K. Hill, J. Wagner, D. Slipetz, S.S. Sommers, and H. Bussey. 1990. The yeast *KRE5* gene encodes a probable endoplasmic reticulum protein required for $(1\rightarrow6)$-β-D-glucan synthesis and normal cell growth. *Mol. Cell. Biol.* **10**: 3013.

Mizuno, K., T. Nakamura, T. Ohshima, S. Tanaka, and H. Matsuo. 1988. Yeast *KEX2* gene encodes an endopeptidase homologous to subtilisin-like serine proteases. *Biochem. Biophys. Res. Commun.* **156**: 246.

Morace, G., G. Dettori, M. Sanguinetti, S. Manzara, and L. Polonelli. 1988. Biotyping of aerobic actinomycetes by modified killer system. *Eur. J. Epidemiol.* **4**: 99.

Naumov, G.I. 1985. Comparative genetics of yeast. XXIII. Unusual inheritance of toxin formation in *Saccharomyces paradoxus* Batschinskaia. *Genetika* **21**: 1794.

Newman, A.M. and C.S. McLaughlin. 1986. The replication of double-stranded RNA. In *Extrachromosomal elements in lower eukaryotes* (ed. R.B. Wickner et al.), p. 173. Plenum Press, New York.

Newman, A.M., S.G. Elliott, C.S. McLaughlin, P.A. Sutherland, and R.C. Warner. 1981. Replication of dsRNA of the virus-like particles in *Saccharomyces cerevisiae. J. Virol.* **38**: 263.

Nomoto, H., K. Kitano, T. Shimazaki, K. Kodama, and S. Hara. 1984. Distribution of killer yeasts in the genus *Hansenula. Agric. Biol. Chem.* **48**: 807.

Ohta, Y., Y. Tsukada, and T. Sugimori. 1984. Production, purification and characterization of HYI, an anti-yeast substance, produced by *Hansenula saturnus. Agric. Biol. Chem.* **48**: 903.

Ouchi, K., R.B. Wickner, A. Toh-e, and H. Akiyama. 1979. Breeding of killer yeasts for sake brewing by cytoduction. *J. Ferment. Technol.* **57**: 483.

Pfeiffer, P. and F. Radler. 1982. Purification and characterization of extracellular and intracellular killer toxin of *Saccharomyces cerevisiae* strain 28. *J. Gen. Microbiol.* **128**: 2699.

Picologlou, S., N. Brown, and S.W. Liebman. 1990. Mutations in *RAD6*, a yeast gene encoding a ubiquitin-conjugating enzyme, stimulate retrotransposition. *Mol. Cell. Biol.* **10**: 1017.

Polonelli, L. and G. Morace. 1988. Killer systems and pathogenic fungi. *Eur. J. Epidemiol.* **4**: 415.

Polonelli, L., C. Archibusacci, M. Sestito, and G. Morace. 1983. Killer system: a simple method for differentiating *Candida albicans* strains. *J. Clin. Microbiol.* **17**: 774.

Polonelli, L., G. Dettori, C. Cattel, and G. Morace. 1987. Biotyping of mycelial fungus cultures by the killer system. *Eur. J. Epidemiol.* **3**: 237.

Ratti, G. and K.W. Buck. 1979. Transcription of double-stranded RNA in virions of *Aspergillus foetidus* virus S. *J. Gen. Virol.* **42**: 59.

Rhee, S.K., T. Icho, and R.B. Wickner. 1989. Structure and nuclear localization signal of the SKI3 antiviral protein of *Saccharomyces cerevisiae*. *Yeast* 5: 149.

Ridley, S.P., S.S. Sommer, and R.B. Wickner. 1984. Superkiller mutations in *Saccharomyces cerevisiae* supress exclusion of M_2 double-stranded RNA by L-A-HN and confer cold-sensitivity in the presence of M and L-A-HN. *Mol. Cell. Biol.* 4: 761.

Schmitt, M. and F. Radler. 1988. Molecular structure of the cell wall receptor for killer toxin K28 in *Saccharomyces cerevisiae*. *J. Bacteriol.* 170: 2192.

———. 1989. Purification of yeast killer toxin KT28 by receptor-mediated affinity chromatography. *J. Chromatogr.* (in press).

Schmitt, M. and D.J. Tipper. 1990. K_{28}, a unique double-stranded RNA killer virus of *Saccharomyces cerevisiae*. *Mol. Cell. Biol.* 10: 4807.

Sclafani, R.A. and W.L. Fangman. 1984. Conservative replication of double-stranded RNA in *Saccharomyces cerevisiae* by displacement of progeny strands. *Mol. Cell. Biol.* 4: 1618.

Skipper, N., D.Y. Thomas, and P.C.K. Lau. 1984. Cloning and sequencing of the preprotoxin-coding region of the yeast M_1 double-stranded RNA. *EMBO J.* 3: 107.

Skipper, N., M. Sutherland, R.W. Davies, D. Kilburn, R.C. Miller, A. Warren, and R. Wong. 1985. Secretion of a bacterial cellulase by yeast. *Science* 230: 958.

Sommer, S.S. and R.B. Wickner. 1982. Yeast L dsRNA consists of at least three distinct RNAs; evidence that the non-Mendelian genes [HOK], [NEX] and [EXL] are on one of these dsRNAs. *Cell* 31: 429.

———. 1987. Gene disruption indicates that the only essential function of the SKI8 chromosomal gene is to protect *Saccharomyces cerevisiae* from viral cytopathology. *Virology* 157: 252.

Stark, M.J.R., A. Boyd, A.J. Mileham, and M.A. Romanas. 1990. The plasmid-encoded killer system of *Kluyveromyces lactis:* A review. *Yeast* 6: 1.

Sturley, S.L., Q. Elliot, J. LeVitre, D.J. Tipper, and K.A. Bostian. 1986. Mapping of functional domains within the *Saccharomyces cerevisiae* type 1 killer preprotoxin. *EMBO J.* 5: 3381.

Thiele, D.J. and M.J. Leibowitz. 1982. Structural and functional analysis of separated strands of killer double-stranded RNA of yeast. *Nucleic Acids Res.* 10: 6903.

Thiele, D.J., E.M. Hannig, and M.J. Leibowitz. 1984a. Multiple L double-stranded RNA species of *Saccharomyces cerevisiae*: Evidence for separate encapsidation. *Mol. Cell. Biol.* 4: 92.

———. 1984b. Genome structure and expression of a defective interfering mutant of the killer virus of yeast. *Virology* 137: 20.

Thomas, G., B.A. Thorne, L. Thomas, R.G. Allen, D.E. Hruby, R. Fuller, and J. Thorner. 1988. Yeast KEX2 endopeptidase correctly cleaves a neuroendocrine prohormone in mammalian cells. *Science* 241: 226.

Thomas, L., A. Cooper, H. Bussey, and G. Thomas. 1990. Yeast *KEX1* protease cleaves a prohormone processing intermediate in mammalian cells. *J. Biol. Chem.* 265: 10821.

Thrash, C., K. Voelkel, S. DiNardo, and R. Sternglanz. 1984. Identification of *Saccharomyces cerevisiae* mutants deficient in DNA topoisomerase I activity. *J. Biol. Chem.* 259: 1375.

Tipper, D.J. and K.A. Bostian. 1984. Double-stranded ribonucleic acid killer systems in yeasts. *Microbiol. Rev.* 48: 125.

Toh-e, A. and Y. Sahashi. 1985. The *PET18* locus of *Saccharomyces cerevisiae*: A complex locus containing multiple genes. *Yeast* 1: 159.

Toh-e, A. and R.B. Wickner. 1980. "Superkiller" mutations suppress chromosomal mutations affecting double-stranded RNA killer plasmid replication in *Saccharomyces cerevisiae. Proc. Natl. Acad. Sci.* **77**: 527.

Toh-e, A., P. Guerry, and R.B. Wickner. 1978. Chromosomal superkiller mutants of *Saccharomyces cerevisiae. J. Bacteriol.* **136**: 1002.

Tyagi, A., R.B. Wickner, C.W. Tabor, and H. Tabor. 1984. Specificity of polyamine requirements for the replication and maintenance of different double-stranded RNA plasmids in *Saccharomyces cerevisiae. Proc. Natl. Acad. Sci.* **81**: 1149.

Uemura, H. and R.B. Wickner. 1988. Suppression of chromosomal mutations affecting M_1 virus replication in *Saccharomyces cerevisiae* by a variant of a viral RNA segment (L-A) that encodes coat protein. *Mol. Cell. Biol.* **8**: 938.

Wagner, J.-C. and D.H. Wolf. 1987. Hormone (pheromone) processing enzymes in yeast. The carboxy-terminal processing enzyme of the mating pheromone α-factor, carboxypeptidase ysc-α is absent in α-factor maturation-defective *kex1* mutant cells. *FEBS Lett.* **221**: 423.

Wejksnora, P.J. and J.E. Haber. 1978. Ribonucleoprotein particle appearing during sporulation in yeast. *J. Bacteriol.* **134**: 246.

Wesolowski, M. and R.B. Wickner. 1984. Two new double-stranded RNA molecules showing non-Mendelian inheritance and heat-inducibility in *Saccharomyces cerevisiae. Mol. Cell. Biol.* **4**: 181.

Wesolowski-Louvel, M., C. Tanguy-Rougeau, and H. Fukuhara. 1988. A nuclear gene required for the expression of the linear DNA-associated killer system in the yeast *Kluyveromyces lactis. Yeast* **4**: 71.

Wickner, R.B. 1974. Chromosomal and nonchromosomal mutations affecting the "killer character" of *Saccharomyces cerevisiae. Genetics* **76**: 423.

———. 1980. Plasmids controlling exclusion of the K_2 killer double-stranded RNA plasmid of yeast. *Cell* **21**: 217.

———. 1981. Killer systems in *Saccharomyces cerevisiae*. In *The molecular biology of the yeast* Saccharomyces: *Life cycle and inheritance* (ed. J.N. Strathern et al.), p. 415. Cold Spring Harbor Laboratory, Cold Spring Harbor, New York.

———. 1983. Killer systems in *Saccharomyces cerevisiae*: Three distinct modes of exclusion of M_2 double-stranded RNA by three species of double-stranded RNA, M1, L-A-E, and L-A-HN. *Mol. Cell. Biol.* **3**: 654.

———. 1986. Double-stranded RNA replication in yeast: The killer system. *Annu. Rev. Biochem.* **55**: 373.

———. 1987. *MKT1*, a nonessential *Saccharomyces cerevisiae* gene with a temperature-dependent effect on replication of M_2 double-stranded RNA. *J. Bacteriol.* **169**: 4941.

———. 1988. Host function of *MAK16*: G1 arrest by a *mak16* mutant of *Saccharomyces cerevisiae. Proc. Natl. Acad. Sci.* **85**: 6007.

———. 1989. Yeast virology. *FASEB J.* **3**: 2257.

Wickner, R.B. and A. Toh-e. 1982. [HOK], a new yeast non-Mendelian trait, enables a replication-defective killer plasmid to be maintained. *Genetics* **100**: 159.

Wickner, R.B., T. Icho, T. Fujimura, and W.R. Widner. 1990. Expression of yeast L-A dsRNA virus proteins produces derepressed replication — A ski ⁻ phenocopy. *J. Virol.* (in press).

Wickner, R.B., S.P. Ridley, H.M. Fried, and S.G. Ball. 1982. Ribosomal protein L3 is involved in replication or maintenance of the killer double-stranded RNA genome of *Saccharomyces cerevisiae. Proc. Natl. Acad. Sci.* **79**: 4706.

Williams, T.L. and M.J. Leibowitz. 1987. Conservative mechanism of the in vitro transcription of killer virus of yeast. *Virology* **158**: 231.

Yie, S.W., G.K. Podila, and R.F. Bozarth. 1989. Semiconservative strand-displacement transcription of the M2 dsRNA segment of *Ustilago maydis* virus. *Virus Res.* **12**: 221.

Zhu, H. and H. Bussey. 1989. The K1 killer toxin of *Saccharomyces cerevisiae* kills spheroplasts of many yeast species. *Appl. Environ. Microbiol.* **55:** 2105.

Zhu, H., H. Bussey, D.Y. Thomas, J. Gagnon, and A.W. Bell. 1987. Determination of the carboxyl termini of the α and β subunits of yeast K_1 killer toxin. Requirement of a carboxypeptidase B-like activity for maturation. *J. Biol. Chem.* **262:** 10728.

6

Circular DNA Plasmids of Yeasts

James R. Broach
Department of Biology
Princeton University
Princeton, New Jersey 08544

Fredric C. Volkert
Department of Microbiology and Immunology
SUNY Health Science Center
Brooklyn, New York 11203

  I. **Introduction**
 II. **Structure of 2-micron Circle**
      A. Genome Organization
      B. Structural Variants
III. **Biology of 2-micron Circle**
      A. Localization and Replication of Plasmid
      B. Phenotypes Associated with 2-micron Circle
 IV. ***FLP1*-mediated Site-specific Recombination**
      A. The *FLP1* Recombination Target (*FRT*) Site
      B. *FLP1* Recombinase
      C. Mechanism of Recombination
  V. **Mechanism of Plasmid Persistence**
      A. Plasmid Stability
      B. Plasmid Partitioning
         1. Components
         2. Mechanism of Plasmid Partitioning
      C. Plasmid Amplification
      D. Control of Plasmid Gene Expression
      E. A Model for Plasmid Copy-number Control
 VI. **Other Circular Double-stranded Plasmids in Yeasts**
      A. Functional Analyses of 2-micron-like Plasmids
         1. *ARS* Elements
         2. Site-specific Recombination
         3. Stability Functions
      B. Origins of Yeast Plasmids
VII. **Conclusions and Perspectives**
      A. Use of 2-micron Circle in Genetic Analysis and Strain Manipulations
      B. Double Rolling Circle as a Mechanism of Chromosomal Amplification
      C. Future Directions

I. INTRODUCTION

In addition to chromosomes and mitochondrial DNA, the genomes of most yeasts encompass a variety of stable extrachromosomal elements.

Volume I. *The Molecular and Cellular Biology of the Yeast* Saccharomyces: *Genome Dynamics, Protein Synthesis, and Energetics.* Copyright 1991 Cold Spring Harbor Laboratory Press 0-87969-355-X/91 $3 + 00 **297**

These include double-stranded DNA plasmids, such as 2-micron circle and 2-micron-circle-like plasmids described in this chapter; double-stranded RNA viruses, described by Wickner (this volume); and a variety of retrotransposons, reviewed by Boeke and Sandmeyer (this volume). Because these elements appear neither to confer any selective advantages to the cells in which they are resident nor to inflict any particular harm, they most likely represent various forms of benign molecular parasitism. As such, each type of extrachromosomal element has evolved various strategies to subvert aspects of the host's reproductive machinery to its own ends without placing undue stress on the host itself.

Recent attempts to appreciate the molecular basis for the delicate tightrope act underlying the successful persistence of 2-micron circle have revealed novel and elegant processes of genome replication and segregation. Probing these processes has provided new and valuable perspectives on the analysis of DNA metabolism, cellular and nuclear architecture, and segregation of nuclear components during division. Not only have these studies uncovered unexpected mechanisms for selective amplification of specific DNA sequences and control of DNA replication, but they also have afforded the development of specialized vectors that have facilitated the use of yeast in the scientific and commercial application of genetic engineering.

In this paper, we summarize our current understanding of the molecular biology of 2-micron circle, with special emphasis on those processes responsible for its successful persistence. Other general reviews of 2-micron circle have appeared recently (Volkert and Broach 1987; Armstrong et al. 1988; Futcher 1988; Volkert et al. 1989). In addition, reviews that emphasize the use of 2-micron circle as a model system for site-specific recombination (Sadowski 1986; Cox 1989) or as a substrate for vector design (Armstrong et al. 1989; Rose and Broach 1990) have also appeared. These reviews serve as valuable adjuncts to and elaborations of the overview provided in this paper.

II. STRUCTURE OF 2-MICRON CIRCLE

A. Genome Organization

The canonical 2-micron plasmid found in most strains of *Saccharomyces cerevisiae* and designated Scp1 is a 6318-bp double-stranded circular DNA species. The genome is organized as two unique segments, 2774 and 2346 bp in length, separated by two regions of 599 bp each, which are precise inverted repeats of each other (Fig. 1) (Hartley and Donelson 1980). In yeast, recombination occurs readily between the two repeated regions to yield inversion of the two unique segments relative to one an-

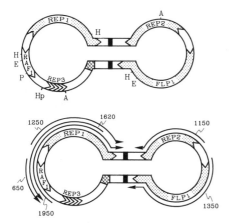

Figure 1 Structural organization of 2-micron circle. (*Top*) Genome organization. A diagram of the genomic organization of the yeast plasmid is shown, drawn to distinguish the inverted repeats (horizontal lines) from unique sequences (circular regions). On the diagram are indicated the locations of the open reading frames (*dotted regions*, arrows signifying the 5′ to 3′ orientation), the origin of replication (*crosshatched region*), the *FRT* site (solid region), the *cis*-acting stability locus (*chevrons*), and various restriction sites. The D-coding region is sometimes referred to as *RAF1*. (*Bottom*) Transcription map. The map positions of the major 2-micron circle transcripts are indicated on a diagram of the B form of the 2-micron circle genome. Each transcript is designated by its length in bases, as determined by previous Northern analysis of 2-micron circle transcription (Broach et al. 1979). The precise locations of the 3′ ends of all the transcripts shown (indicated by the arrowheads) and the 5′ ends of all but the 1950- and 1620-base transcripts are known (Sutton and Broach 1985). The locations of the four open reading frames are indicated by the heavy lines on the diagram of the genome, with tapers lying at the 3′ ends of the genes.

other. As a consequence, 2-micron circle isolated from yeast consists of an equal mixture of two molecular species, which differ solely in the orientation of one unique region with respect to the other (see Broach 1981).

The plasmid contains four genes, whose positions and orientations are indicated in Figure 1 (top) and Table 1 (see p. 319). Each unique region encompasses a pair of divergently transcribed genes. As discussed in subsequent sections, the largest open reading frame, *FLP1*, encodes a protein that catalyzes site-specific recombination between the plasmid inverted repeats. The intermediate-size genes, *REP1* and *REP2*, encode proteins required for plasmid segregation and regulation of plasmid gene expression. The smallest coding region, *RAF1*, encodes a protein also involved in regulation of gene expression. Each of these genes is represented in vivo by a discrete mRNA species that encompasses the coding region. Several additional 2-micron-specific RNA species of unknown function are also found in plasmid-bearing cells (Fig. 1, bottom). These include a 1950-bp transcript whose 5'end lies 3'of the *RAF1*-coding region and whose 3'end is coterminal with that of *REP1* mRNA, a 1620-bp transcript whose 3'end is coterminal with that of *RAF1* mRNA, and a 600-bp transcript corresponding to the noncoding strand of *RAF1* (Broach et al. 1979; Sutton and Broach 1985; Murray et al. 1987).

Several other structural features of the plasmid are worth noting. The plasmid contains a single autonomous replication sequence (*ARS*), lying at the junction between the larger unique region and one of the inverted repeats (Broach et al. 1983b). This site serves as the sole in vivo origin of replication within the plasmid (Brewer and Fangman 1987; Huberman et al. 1987). Located in the larger unique region, approximately 500 bp from the origin of replication, are a series of inexact tandem repeats of a 62-bp sequence. Five of these repeats, lying almost precisely between the *Hpa*I and *Ava*I sites in the large unique region, are highly homologous; they are bracketed by two additional repeats exhibiting somewhat reduced homology. These repeats correspond to the region required in *cis* for plasmid segregation, which has been designated variously as *REP3* or *STB* (Jayaram et al. 1983, 1985; Kikuchi 1983).

B. Structural Variants

Several structural variants of 2-micron circle have been observed in various *S. cerevisiae* strains. Although none of these variants have been sequenced, restriction analysis suggests that they are closely related to Scp1, differing only by the number of repeated elements within the *REP3* locus and by several scattered restriction site polymorphisms (Cameron et al. 1977; Livingston 1977; Kikuchi 1983). In all cases, the variant plasmid constitutes the sole plasmid species in a strain, indicating that each variant is fully competent in all essential plasmid functions.

III. BIOLOGY OF 2-MICRON CIRCLE

A. Localization and Replication of Plasmid

To a first approximation, the properties of 2-micron circle plasmid are indistinguishable from those of a chromosomal replicon. Plasmid molecules reside in the nucleus and are packaged into chromatin with the normal complement and phasing of nucleosomes (Livingston 1977; Livingston and Hahne 1979; Nelson and Fangman 1979; Taketo et al. 1980). During normal mitotic growth, plasmid molecules are replicated once and only once during each S phase of the cell cycle (Zakian et al. 1979). Replication initiates at a single site within the molecule, precisely within the region identified previously as the sole *ARS* within the plasmid genome, and is dependent on the same set of gene products as required for replication of chromosomal DNA (Broach et al. 1979, 1983b; Williamson 1985; Brewer and Fangman 1987; Huberman et al. 1987).

Two facets distinguish 2-micron circle from other chromosomal replicons, and these features relate to its mode of persistence. First, the plasmid is present in most cells at high copy. Haploid cells contain approximately 60–100 copies of 2-micron circle, and copy number increases fairly linearly with increasing ploidy (Clark-Walker and Miklos 1974; Futcher and Cox 1983; Mead et al. 1986). Second, the plasmid is not physically attached to a centromere and thus does not directly engage the mitotic apparatus during nuclear division. Rather, the plasmid encodes a partitioning system that promotes random dispersal of the multiple copies of plasmid molecules to mother and daughter cells during mitosis or to the four spores during meiosis. The components and mechanism of this dispersal system are discussed below.

B. Phenotypes Associated with 2-micron Circle

Examinations of isogenic strains that either possess a full complement of 2-micron circle (designated [cir $^+$]) or completely lack 2-micron circle (designated [cir ^{0}]) have documented that the plasmid contributes little to the phenotype of the cell in which it is resident. Such isogenic pairs of strains have been produced either by curing [cir $^+$] strains of endogenous 2-micron circle (Dobson et al. 1980) or by introducing 2-micron circle into a formerly plasmid-free cell (Volkert and Broach 1986; Armstrong et al. 1989). The fact that [cir ^{0}] strains grow normally under standard culture conditions certifies that the plasmid encodes no functions essential to the growth of the yeast cell (Dobson et al. 1980; Mead et al. 1986). In addition, previous attribution of oligomycin resistance (Guerineau et al. 1974) or other drug resistance phenotypes to 2-micron circle has been discounted using isogenic strains. Finally, the fact that all four identifiable plasmid genes are involved in plasmid maintenance makes it unlikely that the plasmid will be found to contribute directly to a phenotype conferring some selective advantage to the plasmid-bearing cell.

The presence of 2-micron circle appears to inflict a slight growth inhibition to cells in which it is resident. By careful comparison of the relative growth rates of isogenic [cir $^+$] and [cir ^{0}] strains, Mead et al. (1986) concluded that 2-micron-circle-bearing strains have a 1.5–3% longer generation time than the corresponding plasmid-free strains. These results confirmed similar results obtained previously with less isogenic pairs of strains (Futcher and Cox 1983).

A curious phenotype associated with 2-micron circle, perhaps related to the slight inhibitory effect of the plasmid on growth of wild-type yeasts, is a nibbled colony morphology. *Saccharomyces carlsbergensis* strains carrying the recessive *nib1* allele exhibit clonal lethality in the presence, but not the absence, of endogenous 2-micron circle (Holm 1982; see also Broach 1981). This clonal lethality is readily evident as a nibbled morphology of colonies on agar. Cultures of *nib1* [cir +] strains contain large and small cells. Large cells appear to carry 2-micron circle at an unusually high copy number and generally fail to form colonies. Small cells yield nibbled colonies that contain a mixture of small and large cells. Thus, high copy levels of 2-micron circle are associated with inviability of *nib1* cells, and the characteristic colony morphology probably reflects a steady-state production of dead cells. However, whether the loss of *NIB1* function induces higher than normal levels of 2-micron circle or whether its loss simply renders the cell more sensitive to high copy levels has not been determined. In addition, the *nib1* phenotype is not strictly dependent on 2-micron circle: Aspects of the nibbled phenotype can be produced in the presence of any high copy plasmid, whether or not it contains 2-micron circle sequences.

An extreme version of the nibbled morphology phenotype can be induced in wild-type *S. cerevisiae* strains by massive overproduction of 2-micron circle. As described below, plasmid copy number can be driven to exceptionally high copy levels by constitutive expression of the plasmid amplification system (Reynolds et al. 1987; Som et al. 1988). Strains induced for plasmid amplification yield very small, ragged colonies and contain a large percentage of cells in the medial nuclear division stage of the cell cycle (Reynolds et al. 1987; Rose and Broach 1990; J.R. Broach, unpubl.). Whether such cells are terminally or only transiently arrested has not been clarified. A reasonable explanation for the observed phenotype is that the extremely high plasmid copy levels subvert much of the chromosomal replication machinery, resulting in significant under-replication of chromosomal DNA and arrest at mitosis. Less likely, but not excluded, is the possibility that high-level expression of a plasmid-encoded protein is toxic to the cell. In either event, the slight mitotic growth disadvantage of [cir +] strains, the nibbled morphology of *nib1* strains, and the severe growth defects of strains with exceptionally high copy levels may simply reflect differing degrees of the same 2-micron-circle-induced load on mitotic growth of the cell.

The results presented above raise an obvious conundrum: How can an extrachromosomal element persist in the population if those cells that carry it are at a selective disadvantage compared to those cells that lack it? A solution to this paradox, proposed by Futcher et al. (1988), emerges by considering 2-micron circle as a sexually transmitted infection; i.e., the measured stability of 2-micron circle and the ability of the plasmid to amplify copy number following matings between a [cir $^+$] strain and a [cir ^{0}] strain (see below) can compensate fully for the slightly deleterious effect on mitotic growth, as long as one assumes a reasonable frequency of outbreeding. Thus, the known behavior of 2-micron circle can completely account for its indefinite persistence in a yeast population, even in the absence of any selectable phenotype.

IV. *FLP1*-MEDIATED SITE-SPECIFIC RECOMBINATION

Interconversion of the two inversion isomers of 2-micron circle is catalyzed by the product of the *FLP1* gene acting at specific sites, designated *FRT*, which lie near the center of the plasmid inverted repeats (Broach et al. 1982). This site-specific recombination system has been subject to extensive examination, both as a model system for specialized recombination and as a means of inducing specific rearrangements in eukaryotic cells.

The *FLP1/FRT* system represents a completely unadorned version of DNA recombination and, as such, presents distinct advantages as a model system. The only protein required for the reaction is Flp1p itself: The *FLP1*-mediated reaction proceeds efficiently in *Escherichia coli* (Cox 1983; Vetter et al. 1983), in *Drosophila* (Golic and Lindquist 1989), or in a completely purified in vitro system (Vetter et al. 1983; Meyer-Leon et al. 1984; Prasad et al. 1986). Recombination in vitro requires only Flp1p, a substrate, a buffer, and appropriate concentrations of salt. No divalent cations or high-energy cofactors are needed. The reaction is completely nonspecific with regard to the topology of the substrate. Supercoiled circles, relaxed circles, or linear molecules are all substrates, although supercoiling confers a measurable rate enhancement in in vitro reactions. Finally, the nature of the reaction is completely general. Depending only on the relative orientation and location of the interacting *FRT* sites, Flp1p will catalyze intermolecular recombination or intramolecular inversion or deletion (Broach and Hicks 1980; McLeod et al. 1984; Meyer-Leon et al. 1987; Gates and Cox 1988). These features have substantially facilitated the biochemical dissection of *FLP1*-mediated recombination, as summarized in the following sections.

A. The *FLP1* Recombination Target (*FRT*) Site

The *FRT* site consists of an 8-bp core element bracketed by two 13-bp repeated elements in inverted orientation. To one side, the 13-bp element

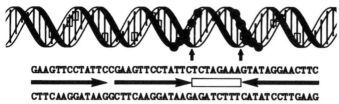

GAAGTTCCTATTCCGAAGTTCCTATTCTCTAGAAAGTATAGGAACTTC

CTTCAAGGATAAGGCTTCAAGGATAAGAGATCTTTCATATCCTTGAAG

Figure 2 FRT site. The sequence of the *FRT* site is shown under a scale drawing of a B-form DNA helix on which are indicated the phosphate residues (*solid circles*) and guanosine residues (*open boxes*) protected by Flp1p and the sites of Flp1p-induced strand cleavage (arrows). The core region (*open rectangle*) and symmetry elements (heavy arrows) are designated within the sequence.

is repeated in a direct orientation after one intervening base pair (see Fig. 2) (Broach et al. 1982). Although Flp1p binds to each of the repeated elements, only the two immediately adjacent to the core are required for recombination (Gronostajski and Sadowski 1985b; Senecoff et al. 1985). However, the third element may contribute to the efficiency of the reaction in yeast (Jayaram 1985) and to the particular order of cleavage and exchange events during recombination (Beatty and Sadowski 1988). Flp1p makes staggered cleavages within the *FRT* site at the boundaries of the spacer and the adjacent repeats to yield an 8-base 5′overhang (Andrews et al. 1985; Senecoff et al. 1985). The protein binds covalently through a tyrosyl linkage to the exposed 3′phosphate at the cleavage site (Gronostajski and Sadowski 1985a).

The core region of *FRT* is not involved in Flp1p binding but establishes the specificity and directionality of the site. To a first approximation, the specific sequence of the core region is unimportant. The only requirement for reactivity of the site is that the core regions of the two interacting sites be homologous (McLeod et al. 1986; Senecoff and Cox 1986; Andrews et al. 1987). Thus, single-base-pair changes within the core abolish recombination with a wild-type site, but recombination between two identical mutant sites occurs readily. This requirement for homology imposes an orientation on the *FRT* site and thereby determines the directionality of the reaction, i.e., whether intramolecular recombination between two sites will yield inversion or deletion of the intervening sequences. Mutant core regions with symmetrical sequences are completely functional but do not exhibit directionality (Senecoff and Cox 1986).

Although single-base-pair changes within the core do not affect the recombination proficiency of *FRT*, more substantial alteration of the core region can diminish its ability to undergo recombination. In particular, increasing or decreasing the size of the core severely compromises recombination competence (Senecoff et al. 1985). In addition, substantially increasing the G:C content or disrupting the purine-rich tract extending from the middle of the core through the first 9 bases of the adjacent repeats has a deleterious effect on recombination (Umlauf and Cox 1988).

The nature of the interaction of Flp1p with the repeated elements within *FRT* has been subjected to extensive biochemical and mutational analyses. A single Flp1p protomer binds to each of the repeat elements, making contact with several purine residues within the repeat and several phosphates at the core/repeat junction (see Fig. 2) (Andrews et al. 1985, 1987; Bruckner and Cox 1986; Prasad et al. 1986). Most point mutations within the binding site have little effect on Flp1p binding and subsequent recombination (Andrews et al. 1986; McLeod et al. 1986; Prasad et al. 1986; Senecoff et al. 1988). However, any alteration of the G residue located 7 bp from the junction or a transversion of any of the 3 bases immediately adjacent to the core causes a substantial diminution of recombinase activity, presumably reflecting a reduction in Flp1p binding (Fig. 2).

B. *FLP1* Recombinase

The *FLP1* product, a 423-amino-acid polypeptide with a predicted molecular mass of 48,622, has been purified to near homogeneity by affinity chromatography (Meyer-Leon et al. 1987; Gates and Cox 1988). The purified protein has been shown to function as an enzyme that catalyzes recombination between two *FRT* sites (Gates and Cox 1988). In the absence of any additional proteins or high-energy cofactor, Flp1p will promote intramolecular recombination between two *FRT* sites to yield inversion or deletion of the intervening sequences, depending on the orientation of the two sites. In this highly purified system, intramolecular recombination predominates over intermolecular recombination. However, addition of histones, *E. coli* histone-like protein, or reagents that concentrate the substrate, such as polyethylene glycol, enhances the relative rate of intermolecular recombination (Bruckner and Cox 1986; Meyer-Leon et al. 1987).

In its ability to catalyze site-specific recombination, Flp1p resembles a number of recombinases from various prokaryotic systems (Argos et al. 1986). These include integrase protein (*int*) encoded by bacteriophage λ or the related phages 80, P22, 186, and P4; resolvases from the Tn*3* class of transposable elements; and the *cre* protein of phage P1. Whereas these enzymes exhibit little sequence homology, all members share three conserved amino acids near their carboxyl end, to which correspond His-305, Arg-308, and Tyr-343 of Flp1p. Mutational analysis of two of these three sites (His-305 and Tyr-343) suggests that each plays an essential role in Flp1p function (Prasad et al. 1987; Parsons et al. 1988). In particular, mutations that change Tyr-343 to serine or phenylalanine inactivate the ability of Flp1p to promote recombination but do not decrease its binding to the *FRT* site, an observation consistent with the possibility that Tyr-343 is the residue that becomes covalently linked to the target site DNA during recombination (Parsons et al. 1988). Additional mutational analysis of Flp1p suggests that the DNA-binding domain of the protein is not restricted to a discrete region of the protein but may be composed of several disparate portions of the protein spatially juxtaposed in a specific tertiary structure (Amin and Sadowski 1989). The facility with which specific *FLP1* mutants can be isolated (Govind and Jayaram 1987) and analyzed (Amin and Sadowski 1989) suggests that mutational analysis of this system will continue to prove informative.

C. Mechanism of Recombination

The mechanism of *FLP1*-mediated recombination has been thoroughly dissected, and a synopsis of this process as we currently appreciate it is provided in the following. The reaction is initiated by cooperative binding of Flp1p protomers to the *FRT* site (Prasad et al. 1986; Andrews et al. 1987; Beatty and Sadowski 1988; Senecoff et al. 1988). Binding of Flp1p to the *FRT* site induces substantial bending of the DNA, which is critical for recombination to proceed. Bending is apparently associated with formation of a complex in which DNA of the recombination site is wrapped around a protein core of Flp1p protomers (Schwartz and Sadowski 1989). Once this complex is formed, recombination can proceed through several rounds without intervening dissociation of the complex.

Subsequent to complex formation, two Flp1p/*FRT* complexes align in close juxtaposition in a process of synapsis (Amin et al. 1990). Since the organization of the complex is symmetrical even though the underlying *FRT* sequence is not, synapsis can occur either with the two *FRT* core sequences oriented in the same direction (a productive orientation) or with the two *FRT* core sequences oriented in opposite directions (a nonproductive orientation). Once a synapse is formed, actual recombination is initiated by Flp1p-mediated cleavage and ligation at one end of the core region to generate a cross-strand Holliday intermediate (Meyer-Leon et al. 1988, 1990; Parsons et al. 1988). If interacting chromatids are in a nonproductive orientation, then branch migration of the Holliday intermediate through the core region is impeded by lack of homology (Andrews et al. 1986; Senecoff and Cox 1986). In this case, the reverse reaction restores the initial configuration of the strands, the chromatids separate, and the process begins again. If interacting chromatids are in a productive orientation, branch migration of the Holliday structure proceeds through the 8-bp core to the opposite side. During this process, Flp1p protomers in the complex undergo an as yet undefined isomerization (Meyer-Leon et al. 1990). The reaction is then completed by a second set of cleavage and ligation reactions to resolve the Holliday structure either into recombined products or back into the initial chromatid configuration.

Although currently available information has allowed development of this reasonably detailed view of the *FLP1* recombination process, a number of questions remain unanswered. The three-dimensional structure of the complex and the dynamics of this structure during the recombination process are completely unexplored. In addition, the appearance of unusual products arising from *FLP1*-mediated recombination in vivo suggests that other undefined elements can influence the course of the reaction (Jayaram 1986). Finally, the degree to which site availability and partner selection can be modulated in vivo as a means of regulating *FLP1*-mediated amplification of copy number (see below) has yet to be addressed. These issues will clearly be directions for further exploration.

V. MECHANISM OF PLASMID PERSISTENCE

A. Plasmid Stability

During normal mitotic growth, the stability of 2-micron circle approaches that of a chromosome. In haploid [cir $^+$] strains, plasmid-free cells arise at a rate of approximately 10^{-4} to 10^{-5} per cell per generation (Futcher and Cox 1983; Mead et al. 1986). In diploid cells, the loss rate is even less, most likely reflecting the higher plasmid copy number in diploid versus haploid cells. This rate of loss stands in marked contrast to that of artificial autonomously replicating plasmids, composed of bacterial plasmid sequences and a yeast chromosomal origin of replication (YRp plasmids). Such plasmids are lost at a rate of 10^{-1} to 10^{-2} per cell per generation, even though the copy level of these plasmids in the subset of plasmid-bearing cells in a culture can be as high as that of 2-micron circle plasmids (Murray and Szostak 1983; Williamson 1985).

Two factors contribute to the high level of 2-micron circle stability. The primary component of the stability system of the plasmid is its ability to disperse itself fairly uniformly between mother and daughter cells during mitosis and among all four spore cells during meiosis

(Kikuchi 1983; Murray and Szostak 1983; Jayaram et al. 1985). A second factor contributing to the stability of the plasmid is its ability to assess its own copy number in the cell and induce plasmid copy number amplification in those cells that contain a sub-optimum level of plasmid (Futcher 1986; Volkert and Broach 1986; Murray et al. 1987; Reynolds et al. 1987; Som et al. 1988). Although the contribution of this latter process to plasmid stability is not documented as readily as that of plasmid partitioning, the fact that this amplification system is a component of all circular double-stranded DNA plasmids identified to date in yeasts (see below) argues forcefully that it is critical to the successful persistence of the plasmids.

B. Plasmid Partitioning

1. Components

Mutational analysis and subcloning experiments have defined three elements that together constitute the 2-micron circle partitioning system. These include the proteins encoded by the plasmid genes *REP1* and *REP2* and a *cis*-acting site designated *REP3* or *STB* (Broach and Hicks 1980; Jayaram et al. 1983, 1985; Kikuchi 1983). Most of the genetic analyses of plasmid stability have been conducted with hybrid 2-micron circle plasmids, which are inherently less stable than 2-micron circle itself. Nevertheless, loss of any of these components reduces plasmid stability 10- to 100-fold and renders such plasmids as unstable as YRp plasmids. Conversely, addition of the *REP3* locus to a YRp plasmid increases its stability 10- to 100-fold when propagated in a [cir $^+$] strain, i.e., in the presence of Rep1p and Rep2p.

The *REP3* locus corresponds to the series of direct tandem repeats located in the *Hpa*I-*Ava*I segment of the large unique region. A thorough functional dissection of this region has not been performed to date. However, analyses of both hybrid plasmids and natural variants document that the presence of all of the repeats is not required for efficient partitioning activity (Cameron et al. 1977; Kikuchi 1983; Jayaram et al. 1985). However, whether the repeat units are essentially interchangeable or represent functionally distinct entities has not been determined. In addition, whether a single repeat is sufficient for activity or whether increasing numbers of repeats enhance partitioning activity has not been addressed.

REP3 activity is affected by its context. In particular, unrestricted transcription through the locus appears to diminish function, a property shared with other *cis*-acting elements, such as centromeres and origins of replication (Hill and Bloom 1987; Snyder et al. 1988). The 250-bp

*Pst*I–*Hpa*I region of the 2-micron circle genome immediately adjacent to the repeated elements of *REP3* encompasses the transcriptional terminator for the *RAF1*-coding region, as well as the promoter and regulatory sequences for the 1950-base transcript (Sutton and Broach 1985). Deletion of these sequences impairs *REP3*-promoted stability, presumably as a consequence of read-through transcription from *RAF1* (Murray and Cesarini 1986). The potential for transcriptional inactivation of *REP3* may explain the variable results obtained from subcloning experiments used in defining the functional domains of the locus.

The products of both the *REP1* and *REP2* genes are required in *trans* for plasmid stability. Rep1p is a 43-kD protein that appears to comprise an as yet ill-defined intranuclear structure. In cell fractionation studies, Rep1p copurifies with an insoluble subnuclear fraction, which has been previously designated as the nuclear matrix/lamina/pore component (Wu et al. 1987). Immunoelectron microscopy indicates that the protein is dispersed throughout the nucleus (J.R. Broach, unpubl.). Since it is not restricted to the periphery of the nucleus, Rep1p cannot be a component of a nuclear lamina (see Fig. 3). Immunofluorescence analysis reveals punctate staining of Rep1p (Fig. 3), which suggests that Rep1p is not a freely diffusible molecule but is restricted to dispersed aggregates or a fixed spatial architecture. The sequence of Rep1p fortifies the suggestion that it participates in some aggregated structure. Secondary structure analysis predicts a bimorphic arrangement for a Rep1p protomer, with the amino-terminal half of the protein assuming a globular form and the carboxy-terminal half assuming an extended α-helix. In addition, the α-helical portion of the molecule exhibits the heptad repeat pattern characteristic of proteins that form extended coiled-coil structures, such as vimentin, myosin heavy chain, and nuclear lamins A and C. Finally, some evidence has been obtained to suggest that Rep1p possesses DNA-binding activity with some specificity for 2-micron circle sequences (L.C.C. Wu and J.R. Broach, unpubl.).

Rep2p is also a nuclear protein. However, immunofluorescence staining of Rep2p reveals a reasonably uniformly dispersed pattern throughout the nucleus, clearly distinct from the staining of Rep1p (Fig. 3). Although Rep2p is thus unlikely to be a component of the same intranuclear structure in which Rep1p participates, the two proteins appear to interact in the cell. In [cir ^{0}] strains in which either or both of the Rep proteins are expressed from an inducible promoter, Rep1p is found in the nucleus, whether or not Rep2p is coexpressed. However, Rep2p is found in the nucleus only when coexpressed with Rep1p (K.A. Armstrong and J.R. Broach, unpubl.). Thus, Rep2p nuclear transport or stability appears to be dependent on Rep1p.

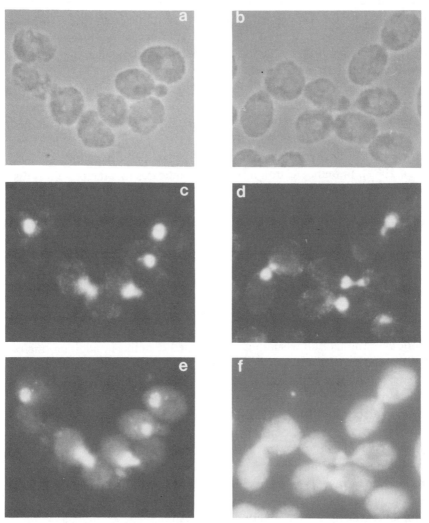

Figure 3 Cellular localization of Rep1p and Rep2p by immunofluorescence microscopy. Strain GF2 (*leu2::LEU2-GAL10-FLP1* [cir ⁺]) was grown overnight in galactose-containing medium and prepared for immunofluorescence by standard procedures. Cells were incubated with DAPI and either anti-Rep1p (*a, c, e*) or anti-Rep2p (*b, d, f*) rabbit serum, followed by fluorescein-conjugated goat anti-rabbit IgG. Cells were then visualized by phase-contrast (*a, b*) or by fluorescence microscopy under conditions specific for DAPI (*c, d*) or fluorescein (*e, f*). *a, c,* and *e* are different images of the same field, as are *b, d,* and *f*.

2. Mechanism of Plasmid Partitioning

How does the *REP* system function to promote plasmid stability? This question is difficult to answer since we do not yet understand why plasmids without a centromere or partitioning system are so unstable. The context of the problem is fairly clear: Plasmids lacking a partitioning or segregation system are transmitted inefficiently to daughter cells prior to mitotic cell division and are rarely incorporated into spore cells at meiosis (see Zakian and Scott 1982; Murray and Szostak 1983). Because of this asymmetric segregation, plasmid-free cells arise frequently and plasmids accumulate to very high copy number in the remaining plasmid-bearing cells (mother cell lineage) (see Futcher and Cox 1984). However, even at high copy number, such plasmids are transmitted inefficiently.

We can account for the asymmetric segregation of these unstable plasmids if we make two assumptions. First, plasmid molecules must be restricted in their migration through the nucleus. Second, newly synthesized nuclear material (e.g., envelope and nuclear matrix) must be preferentially transmitted to the daughter cells at mitosis and to the spore cells at meiosis. These assumptions merit consideration not only because they provide a molecular basis for the behavior of artificial plasmids in yeast, but also because we can marshal some experimental support for them. First, very small (2 kb) circular plasmids and small- to medium-sized linear plasmids exhibit reasonably stable propagation. This is consistent with the notion that stability is inversely proportional to the cross-sectional area presented by a plasmid and, accordingly, that larger circular plasmids confront some restriction to their diffusion. Second, bacterial plasmids with lower copy number than those obtained by yeast YRp plasmids rely successfully on random diffusion as the sole means of partitioning. Thus, if plasmids were unrestricted in their diffusion, then their copy would be sufficient to ensure their efficient transmission. Finally, although the distribution of bulk nuclear material at mitosis has not been examined, at least one nuclear protein (the *KAR1* product) shows segregation of newly synthesized material to the daughter cell (M. Rose, pers. comm.). This at least demonstrates that segregation of some nuclear material can be asymmetrical.

Why might plasmids in yeast be restricted in their diffusion? One possibility is that they are attached to some fixed nuclear structure. Consistent with this hypothesis is the observation that a region near the origin of 2-micron circle is preferentially associated with the insoluble "nuclear matrix" fraction of yeast, raising the possibility that plasmid molecules might remain attached to the matrix at their sites of synthesis (Amati and Gasser 1988). A second possibility is that the internal architecture of the nucleus is reasonably rigid and presents a barrier to the free migration of plasmid molecules. With regard to plasmid segregation, we might envision the internal milieu of the nucleus as a gel matrix, not dissimilar from agarose or polyacrylamide. The relative transmissibility of small versus large plasmids and linear versus circular plasmids during growth mirrors their relative migration in such gel systems.

Flowing from this consideration of the possible barriers to plasmid segregation by diffusion, we can suggest several models to explain how the *REP* system promotes plasmid stability. One possibility is that the Rep proteins form an extended, intranuclear structure that provides a dispersed set of attachment sites for plasmid molecules. The *REP3* locus would be the site by which the plasmid bound to the structure. The assembly of this structure would serve as a driving force to distribute newly synthesized plasmid molecules throughout the growing nucleus and thereby ensure their presence in the daughter nucleus. Consistent with this model is the fact that Rep1p appears to form a reasonably distinct intranuclear structure. In addition, an insoluble fraction of the yeast nucleus specifically binds a restriction fragment spanning the *REP3* locus and origin of replication (Amati and Gasser 1988). Whether this binding is actually relevant to partitioning awaits further experimentation.

A second class of model would postulate that the *REP* system simply promotes plasmid diffusion. For instance, the Rep proteins could initiate plasmid condensation from the *REP3* locus to yield a compact structure that could migrate more readily through the nucleus. Alternatively, these proteins could induce local depolymerization of the yeast nuclear matrix, perhaps triggered by binding to *REP3*. Distinguishing experimentally among these various models of *REP*-mediated partitioning will undoubtedly be afforded high priority in the continuing analysis of 2-micron circle.

C. Plasmid Amplification

The second component of the stability system of 2-micron circle is amplification of plasmid copy number. This property initially seemed paradoxical. On the one hand, plasmid replication is stringently regulated. During steady-state growth, greater than 95% of all plasmid molecules replicate once and only once during the cell cycle (Zakian et al. 1979). Thus, the plasmid is normally precluded from more than one initiation event per cell cycle. On the other hand, under certain circumstances, the average plasmid copy number can increase significantly during the growth of a clone of cells. One situation in which this is evident are crosses of a [cir ^{0}] and a [cir $^+$] strain. The resulting diploids rapidly attain the same relative plasmid copy number as that of the haploid [cir $^+$]

parent, an outcome that requires the complement of plasmids in the zygote to double relative to the genomic DNA content during the first several generations of outgrowth (Mead et al. 1986; Futcher et al. 1988). How the plasmid can increase its copy number without abrogating cell-cycle control of replication initially presented an intriguing dilemma.

An ingenious solution to this paradox was suggested by Futcher (1986) and is diagramed in Figure 4. Under normal conditions, plasmid replication forks proceed in a bidirectional manner away from the single origin of replication to converge at the opposite end of the plasmid to yield two copies of the parent plasmid. Futcher noted that if an *FLP1*-mediated recombination between an unreplicated repeat and one of the replicated repeats occurred during the process of plasmid replication, the relative orientation of the two forks would be inverted. Rather than proceeding in opposite directions around the circle, they would now chase each other around an inverted form of the parent plasmid, spinning off an ever-increasing concatemer of plasmid multimers. This process would continue until another specific recombination event reinstated the bidirectional orientation of the forks. This amplification event can be viewed as a transient shift in the mode of plasmid replication from the normal θ process to rolling circle, generating extra copies of the plasmid genome from a single initiation event. Two aspects of this model rendered it immediately compelling: (1) Amplification is achieved without abrogation of the stringent control of replication initiation and (2) the mode of amplification accounts for the unusual structural organization of the plasmid and the otherwise apparently pointless inversion process.

The Futcher model predicts the existence of unusual DNA structures during the process of amplification. These include the double rolling circle intermediate shown in Figure 4, as well as more convoluted structures. Attempts to visualize such structures by electron microscopy have been unavailing, most likely due to the large size and low abundance of the predicted molecules. However, indirect evidence for the existence of such intermediates has been reported (Futcher 1988).

A more productive approach to testing the Futcher model for plasmid amplification has been to examine the structural requirements for amplification. This has been achieved experimentally by examining the fate in vivo of a single plasmid molecule following its synchronous excision from the chromosome in an appropriately tailored yeast strain (Volkert and Broach 1986; Reynolds et al. 1987). These studies have shown that amplification requires the plasmid to be an extrachromosomal circle with a pair of intact *FRT* sites in inverted orientation. In addition, amplification requires the continuous presence of Flp1p and ongoing

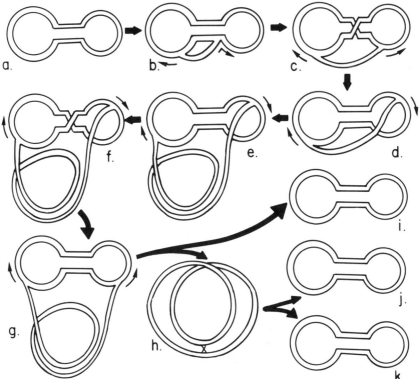

Figure 4 Futcher model for plasmid amplification. (*a, b*) Semiconservative DNA replication proceeds bidirectionally from the plasmid origin. Arrows indicate replication fork movement. (*c, d*) An Flp1p-mediated recombination reaction reorients the forks so that they no longer converge. (*e*) Continuing replication in this mode yields a multimeric replication intermediate. (*f, g*) Another Flp1p-mediated recombination event restores the converging orientation of the replication forks. Completion of replication yields a 2-micron circle monomer (*i*) and a multimer (*h*). Further Flp1p-mediated or general recombination resolves the multimer to monomers (*j, k*).

cellular growth. Finally, plasmid genes other than *FLP1* are neither necessary (Reynolds et al. 1987) nor sufficient (Volkert and Broach 1986) for copy number amplification. These observations are entirely consistent with the Futcher model and provide compelling confirmation of its essential features.

D. Control of Plasmid Gene Expression

As noted in the preceding section, 2-micron circle can increase its copy number by a novel recombination-mediated amplification pathway, which is activated under conditions of reduced plasmid copy number. Because plasmid amplification cannot be detected during normal mitotic growth, 2-micron circle must be able to assess its copy level and activate or inhibit amplification in response to that assessment. Control of plasmid gene expression most likely embodies the means by which this regulation of amplification is accomplished.

A summary of the transcriptional regulatory circuitry of 2-micron circle is presented in Figure 5. These interactions have been deduced from studies in which 2-micron circle gene products were produced in vivo from high-level inducible promoters, and their effects on plasmid transcription were assessed both by using *lacZ* fusions to various plasmid-coding regions and by directly measuring transcript levels from various 2-micron circle genes (Murray et al. 1987; Reynolds et al. 1987; Som et al. 1988). These studies yielded the following conclusions. Rep1p and Rep2p coordinately repress transcription from most of the 2-micron circle genes studied, including *FLP1*, *REP1*, and *RAF1*, but not *REP2*. In addition, the Rep1p/Rep2p complex represses production of the 1950-base transcript. Repression of the *FLP1* and 1950-base transcripts can be as much as 100-fold. Finally, hyperexpression of Raf1p antagonizes the

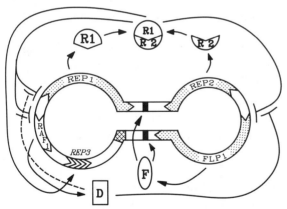

Figure 5 Regulatory circuitry underlying plasmid copy-number control. Interactions of 2-micron circle plasmid products with the plasmid genome are indicated. Flp1p (F) catalyzes recombination between specific sites (*solid regions*) within the inverted repeats, a process required for plasmid copy number amplification. The products of the *REP1* and *REP2* genes (R1 and R2) act in concert, perhaps as a heterologous dimeric complex, via the *REP3* locus to promote equipartitioning at cell division. In addition, these products repress transcription of *FLP1*, *REP1*, and *RAF1*. Raf1p (D) antagonizes *REP1/REP2*-mediated repression of *FLP1* gene expression, and perhaps of *REP1* and *RAF1* gene expression as well.

repression effect of Rep1p/Rep2p on the *FLP1* promoter (Murray et al. 1987). This last conclusion confirmed indications of a role for Raf1p in control of *FLP1* expression (Cashmore et al. 1988).

E. A Model for Plasmid Copy-number Control

The genetic analysis of plasmid gene function and the pattern of transcriptional regulation of plasmid gene expression can be synthesized into a coherent model for stable maintenance of 2-micron circle copy levels. This model has been described in detail previously (Armstrong et al. 1988) and is only briefly outlined here. The primary stabilizing factor of the plasmid is equal mitotic partitioning, mediated by Rep1p and Rep2p acting on *REP3*. Rep1p and Rep2p also repress transcription of *REP1*, *FLP1*, *RAF1*, and the 1950-base transcript. When plasmid copy number falls—due to unequal partitioning or errors in plasmid replication, or as a result of a [cir $^+$] strain mating with a [cir ^{0}] strain—concentration of the Rep proteins also would fall, relieving repression of the regulated promoters. *FLP1* expression would increase and the copy number deficit would be corrected by the amplification scheme described above. As the copy number rises, so would Rep protein concentration, resulting in reestablishment of repression and an end to amplification. As discussed previously (see Armstrong et al. 1988), autogenous regulation of *REP1* expression by the Rep1p/Rep2p complex and antagonism of *FLP1* and *REP1* repression by Raf1p serve to sharpen the responsiveness of plasmid gene expression to changes in repressor concentration. In theory, this enhances the sensitivity of the plasmid to fluctuations in plasmid copy levels.

Several observations are consistent with this model of plasmid copy-number control. First, by using an in vivo recombination assay, more Flp1p activity could be detected in cells with a single copy of the gene, chromosomally inserted in a [cir ^{0}] strain, than in [cir $^+$] cells containing some 60 copies of this gene (Reynolds et al. 1987; Som et al. 1988). This strongly suggests that *FLP1* expression is substantially repressed by plasmid-encoded components. Second, high-level *FLP1* expression raises the copy number of endogenous 2-micron circle by severalfold. This indicates that the availability of Flp1p limits copy number under ordinary conditions (Murray et al. 1987; Reynolds et al. 1987; Som et al. 1988). Third, mutation of *REP1* or *REP2* increases the copy level of 2-micron-circle-based plasmids with an intact *FLP1/FRT* recombination system, compared to plasmids with wild-type *REP1* and *REP2* or to *rep1$^-$/rep2$^-$* plasmids in which the *FLP1/FRT* system has been inactivated (Volkert and Broach 1986; Veit and Fangman 1988). In addition, Veit and Fangman (1988) observed shifts in the nuclease protection patterns in nucleosome-assembled plasmids upon mutational inactivation of *REP1* or *REP2*. These mutations changed the nucleosome phasing patterns around *REP3* and in the intergenic region immediately to the 5′ side of *FLP1*, indicating that 2-micron circle stabilization and expression regulation might share a common mechanism. Finally, the *REP3* locus and the regions immediately 5′ to the genes regulated by Rep1p/Rep2p all encompass one or more copies of the specific nononucleotide sequence 5′-TGCATTTTT-3′ (Armstrong et al. 1988). This conserved element could represent the site through which these proteins effect partitioning and repression.

Despite the fact that 2-micron circle can regulate its copy number and induce copy number amplification, direct demonstration of the adaptive significance of these processes has been difficult to obtain. One possible role for *FLP1*-mediated amplification may be to prevent dilution of the plasmid in matings between [cir $^+$] and [cir ^{0}] cells. Certainly, plasmid amplification is induced following such matings. In addition, Futcher et al. (1988) have demonstrated that 2-micron circle can in fact spread in mixed populations of [cir $^+$] and [cir ^{0}] cells via repeated rounds of induced sporulation and mating. A second function of amplification could be to readjust copy levels during mitotic growth to compensate for inexact plasmid segregation or occasional plasmid replication failures. If partitioning involves random dispersal of plasmids to daughter cells, the process is likely to be imprecise. In this case, daughter cells with below average plasmid copy levels would arise occasionally during normal mitotic growth. Additionally, because the majority of plasmid copies are monomers and thus rely on a single origin to initiate replication, a finite rate of replication failure may occur. This would necessitate copy number correction as well. A significant effort will likely be expended in testing genetically the contribution of amplification and copy number control to plasmid survival.

VI. OTHER CIRCULAR DOUBLE-STRANDED PLASMIDS IN YEASTS

An often informative approach to determining the functional significance of particular features of a biological system is to examine their conservation among phylogenetically distinct examples of that system. The identification of circular, double-stranded DNA plasmids other than 2-micron circle in various yeasts has afforded an opportunity to perform such cross-species comparisons. Accordingly, in this section, we present a summary of our current knowledge of all the known 2-micron-circle-like plasmids in yeasts.

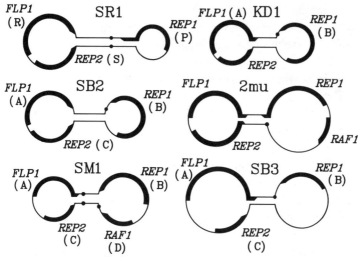

Figure 6   Structural organization of 2-micron-circle-like plasmids. A diagram of the genomic organizations of the known double-stranded DNA circular yeast plasmids, drawn approximately to scale, indicates the relative positions of the inverted repeats (horizontal lines) and unique sequences (circular regions). On the diagram are indicated the locations of the open reading frames (*solid regions*, tapers signifying the 5′ to 3′ orientation) and the origins of replication (*solid circles*). The open reading frames are labeled using 2-micron circle gene nomenclature, identifying the open reading frames in each plasmid homologous to Flp1p and Rep1p. The third largest open reading frame in each plasmid has been designated *REP2*, and the fourth largest, if present, has been designated *D* (*RAF1*). Because the sequence of plasmid pSB1 has not been obtained, a diagram of this plasmid is not included.

To date, only six distinct, circular, double-stranded DNA plasmids have been isolated from yeasts. The genome organizations of these plasmids are diagramed in Figure 6, and several features of these plasmids are listed in Table 1. Five of these plasmids were isolated in two separate screens, one a random survey of various yeast species and one focused specifically on osmophilic yeasts (Toh-e et al. 1982, 1984). The sixth plasmid was adventitiously identified in studies of *Kluyveromyces drosophilarum* (Chen et al. 1986). Five of these plasmids have been sequenced and all have been genetically characterized to some extent (Araki et al. 1985; Toh-e and Utatsu 1985; Chen et al. 1986; Utatsu et al. 1987).

The most striking feature of these plasmids is the similarity of their genome organization to that of 2-micron circle. All of the plasmids are

Table 1 Structural features of 2-micron-circle-like plasmids from yeasts

Plasmid	Source[a]	Size[b]				Open reading frames[c]				References[h]
		U1	U2	IR	T	FLP1[d]	REP1[e]	REP2[f]	RAF1[g]	
pSR1[i]	Z. rouxii	2654	1679	959	6251	490	410	233		1,2
pSB1	Z. bailii	2300	2900	675	6550					3
pSB2	Z. bailii	2457	2004	477	5415	474	408[j]	158		3,4
pSB3	Z. rouxii[k]	3168	2665	391	6615	568	322	178		3,5
pSM1	Z. fermentati	2552	2160	352	5416	372	394[j]	260		4
pKD1	K. drosophilarum	2137	1928	346	4757	447	415	212	200	6
2-micron	S. cerevisiae	2774	2346	599	6318	387	373	295	180	7

[a] (S) Saccharomyces; (Z) Zygosaccharomyces; (K) Kluyveromyces. Prior to 1985, Zygosaccharomyces rouxii and Zygosaccharomyces bailii were classified in Saccharomyces rouxii and Saccharomyces bailii (see Murray et al. 1988).

[b] Lengths in base pairs of the larger unique region (U1), smaller unique region (U2), inverted repeat (IR), and the entire plasmid (T).

[c] Sizes in amino acid codons of the open reading frames identified by sequence analysis of the plasmids; plasmid pSB1 has not been sequenced.

[d] The gene encoding the plasmid site-specific recombinase, as judged either from mutational analysis (for plasmids pSR1 and pSB3) or from sequence homology with 2-micron circle Flp1p and the site-specific recombinases of pSR1 and pSB3.

[e] The gene encoding a protein whose predicted amino acid sequence is homologous to Rep1p of 2-micron circle.

[f] Designated by analogy to 2-micron circle gene organization (no homology is evident among the predicted protein sequences). For plasmids pSR1 and PSB3, this gene is required for stable plasmid propagation (Jearnpipatkul et al. 1987a,b).

[g] Designated by analogy to 2-micron circle gene organization (no homology is evident between the predicted protein sequences).

[h] (1) Araki et al. 1985; (2) Toh-e et al. 1982; (3) Toh-e et al. 1984; (4) Utatsu et al. 1987; (5) Toh-e and Utatsu 1985; (6) Chen et al. 1986; (7) Hartley and Donelson 1980.

[i] Plasmids pSR2 (Toh-e et al. 1982) and pSB4 (Toh-e et al. 1984) have identical restriction maps to plasmid pSR1.

[j] Published sequences have been modified by single-base-pair addition within the predicted open reading frame, inserted to optimize predicted amino acid homology with Rep1p of 2-micron circle.

[k] Formerly classified as Zygosaccharomyces bisporus and Saccharomyces bisporus (see footnote a above).

319

small (4757–6615 bp) and contain two regions that are precise inverted repeats of each other and that divide the genome into approximately equal halves. All of the plasmids encompass either three or four extended open reading frames, one of which encodes a protein that has extensive sequence homology with Flp1p and likely promotes recombination between the repeats (Toh-e et al. 1984; Utatsu et al. 1986, 1987). All of the plasmids carry one or more *ARS* elements, either within the inverted repeat or just adjacent to it (Toh-e et al. 1984; Toh-e and Utatsu 1985; Chen et al. 1986). These similarities exist even though the sequences of these seven plasmids are generally quite dissimilar.

All of the features shared by yeast circular plasmids are precisely those that are required for amplification via the recombination mechanism described above for 2-micron circle. This provides strong circumstantial evidence that all of these plasmids exploit recombination-promoted plasmid amplification as an integral component of their life cycle. In addition, the fact that these are the only circular DNA plasmids identified in yeast to date argues that site-specific recombination is essential to the survival of a circular plasmid in yeast, most likely as a means of amplification.

A. Functional Analyses of 2-micron-like Plasmids

1. ARS *Elements*

One or more *ARS* elements functional in *S. cerevisiae* have been identified and mapped in every plasmid examined to date (Toh-e et al. 1984; Toh-e and Utatsu 1985; Chen et al. 1986). The approximate locations of these elements are shown in Figure 6. *ARS* elements have also been mapped for plasmids pKD1 and pSB3, using strains of the same genus as those from which the plasmid was originally isolated (*Kluyveromyces lactis* for pKD1 and *Zygosaccharomyces rouxii* for pSB3). In both cases, the site of the *ARS*, as judged from propagation in *S. cerevisiae*, is exactly the same as that determined by propagation in the normal host species. This suggests some degree of conservation in the replication apparatuses among these yeast.

Surprisingly, the *ARS* elements from pKD1 and pSB3 do not encompass the consensus sequence characteristic of most *ARS* elements isolated from *S. cerevisiae* (Williamson 1985). In addition, although all six of these plasmids will replicate in *S. cerevisiae*, 2-micron circle does not replicate in *Z. rouxii* (Araki et al. 1985) or *K. lactis* (Das and Hollenberg 1982).

2. Site-specific Recombination

All of the plasmids identified to date undergo high-frequency intramolecular recombination between inverted repeat domains (Toh-e et al. 1984; Chen et al. 1986; Utatsu et al. 1987). This recombination can occur both in the normal cellular environment of the plasmids and in *S. cerevisiae*. Each of the plasmids encompasses a gene whose predicted protein product exhibits striking homology over its entire length with the Flp1p of 2-micron circle. For pSR1 and pSB3, this product has been shown to be required for plasmid recombination. In addition, recombination within pSR1 has been shown to occur at a specific site within the inverted repeat (Utatsu et al. 1986; Matsuzaki et al. 1988). This site exhibits a sequence organization similar to that of the *FRT* site in 2-micron circle (Matsuzaki et al. 1988). An analogous sequence can also be seen in the inverted repeats of all the plasmids sequenced to date, suggesting that initiation of recombination is restricted to a specific site in each of these plasmids as well (Utatsu et al. 1987). Although the sequences of the recombinases encoded by 2-micron circle and the five other sequenced plasmids are highly homologous, plasmid-encoded recombination systems are generally not cross-functional. The 2-micron circle *FLP1* product will not catalyze recombination between the inverted repeats of any of these plasmids and vice versa (Toh-e et al. 1984; Utatsu et al. 1986). Similarly, the pSB3 and pSR1 Flp proteins will not act on the heterologous recombination sites (Utatsu et al. 1986). The cross-functionality of the two most closely related recombinases, encoded by pSR1 and pSB2, has not been reported to date.

3. Stability Functions

All of the plasmids sequenced to date encompass a gene whose predicted product has extensive homology with the 2-micron circle Rep1p. This suggests that for all of these plasmids, either partitioning or copy-number control, or both, is effected by a mechanism not dissimilar to that used by 2-micron circle. Confirmation of the functional significance of this sequence homology has been derived by mutational analysis of stability functions of plasmids pSR1 and pSB3 in *Z. rouxii* (Toh-e and Utatsu 1985; Jearnpipatkul et al. 1987a,b). Insertional inactivation of either *REP1* homologous gene (gene *P* for pSR1; gene *B* for pSB3) yields diminished stability of the hybrid plasmid in a [cir ^{0}] *Z. rouxii* strain, compared to that of the same plasmid with all reading frames intact. Stability is restored to these mutant plasmids by propagation in a *Z. rouxii* strain harboring the cognate wild-type plasmid. This indicates that the destabilizing mutations are in *trans*-acting functions.

The stability of pSR1 and pSB3 is also dependent on the integrity of the third largest coding region in each plasmid. Mutation of this gene (genes *S* and *C* for pSR1 and pSB3, respectively) diminishes stability in *Z. rouxii*, a phenotype that can be reversed by *trans*-complementation with a wild-type copy of the respective gene. This functional role of the *C* (or *S*) gene, plus its genomic location in all the plasmids examined to date, suggests that these genes are functional analogs to the 2-micron circle *REP2* gene. However, unlike the *REP1* homologous genes present in all of the circular double-stranded DNA plasmids, the *REP2*-equivalent genes and their products display no extensive sequence homology.

Jearnpipatkul et al. (1987b) have identified a site within pSR1 that is required in *cis* for stable propagation. This site encompasses a series of repeated elements in direct and inverted orientation with low (65–80%) cross homology. Whether these repeated elements are responsible for the stability-enhancing function of the site has not been determined.

Even less is known at present about the mechanism of plasmid-promoted stability enhancement for pSR1 and pSB3 than is known for the 2-micron circle system. For instance, whether mutant plasmids with reduced stability exhibit mother cell bias in segregation, as is the case for 2-micron-circle-based plasmids, has not been determined. To a first approximation, plasmid stability systems are host-cell-specific; i.e., pSR1 and pSB3 propagate stably in *Z. rouxii* but exhibit relatively rapid loss when propagated in *S. cerevisiae*. This would suggest that at least one component of the stability machinery interacts with some component of the host cell. To date, neither the plasmid-encoded products nor the host-encoded products that participate in this host-plasmid interaction have been identified.

B. Origins of Yeast Plasmids

The origin and evolution of the yeast 2-micron-circle-like plasmids are quite puzzling. Several considerations support the assumption that these plasmids have all derived from a common ancestral species. First, the genomic architectures of all the plasmids exhibit extensive correspondence. Not only are those features of the genome organization required for *FLP1*-mediated amplification retained, but all of the plasmids display the same relative locations and orientations of the three major coding regions (cf. Fig. 6). In addition, the *FLP1*-coding regions and the *REP1*-coding regions from the six plasmids exhibit extensive sequence similarities (Table 2) (Utatsu et al. 1987; Murray et al. 1988). These correlations provide compelling evidence for a common origin for these plasmids.

Despite the likelihood that all of the 2-micron-circle-like plasmids evolved from a common ancestral plasmid, several aspects of the relationships among these plasmids are not easily explained by standard models of molecular evolution. First, different regions of the plasmids appear to have evolved at significantly different rates; i.e., the *FLP1*-coding regions and portions of the *REP1*-coding regions are well conserved, whereas the remainder of the plasmid genomes have diverged so extensively as to render any relatedness unrecognizable. Second, the phylogenetic relationships among the plasmids as determined by sequence similarity do not correlate with the phylogenetic relationships of the hosts in which these plasmids are resident. Plasmids pSB3 and pSR1 are the most distantly related of the six plasmids, as judged from *REP1* homologies, and yet they both were isolated from and propagate equally well in *Z. rouxii* strains. Conversely, the most closely related plasmid species, pSR1 and pSB2, were obtained from different species, albeit of the same genus.

These conundrums regarding the molecular evolution of yeast plasmids have been highlighted and explored by Murray et al. (1988), who suggest that several factors may contribute to the anomalous phylogenetic relationships among these plasmids. One factor that may account for the enhanced rate of evolution of the plasmids may be the particular mode of persistence used by these plasmids; i.e., the ability of individual plasmid molecules to expand rapidly by *FLP1*-mediated amplification, the tendency for plasmids to partition nonrandomly (a process enhanced in partitioning-deficient mutants), and the high rates of intra- and intermolecular recombination and gene conversion could accelerate the fixation of base-pair changes and other mutations in the population. A competing influence on the rate of evolution of these plasmids may be constraints imposed by coevolution of DNA target sites and the proteins that interact with them; i.e., both the *FLP1* product and the *FRT*

Table 2 Structural similarities of the *FLP1* and *REP1* genes of 2-micron-circle-like plasmids

	pSR1	pSB2	pSM1	pSB3	2-micron	pKD1
pSR1	—	1415/1030	735/365	663/143	614/239	592/271
pSB2	—	—	757/316	698/137	565/240	565/224
pSM1	—	—	—	750/134	594/381	614/245
pSB3	—	—	—	—	584/143	550/112
2-micron	—	—	—	—	—	550/208

Shown are the optimized similarity scores from the FASTA program of Pearson and Lipman (1988) for the predicted amino acid sequences of Flp1p (first score) and Rep1p (second score) obtained for each pairwise combination of the six sequenced plasmids.

site on which it acts have to evolve in concert to maintain the functional integrity of the site-specific recombination system. Similarly, the plasmid *REP3* locus and whichever of the Rep proteins binds to it must also maintain mutual recognition. This constraint may account for the conservation in sequence of *FLP1* and of portions of *REP1*. The facts that the DNA-binding domain of Flp1p appears to be dispersed throughout the protein (Amin and Sadowski 1989) and that Rep1p, rather than Rep2p, likely binds to *REP3* are consistent with this hypothesis. Further exploration of the connection between sequence conservation and functional domains of plasmid products may prove informative.

VII. CONCLUSIONS AND PERSPECTIVES

A. Use of 2-micron Circle in Genetic Analysis and Strain Manipulations

The yeast plasmid has been exploited extensively and effectively as the basis of vectors for propagation and expression of cloned genes in yeast (Broach 1983; Broach et al. 1983a; A.B. Rose and Broach 1990; M.D. Rose and Broach 1991). The ability of the plasmid to promote both stable and high-copy-level propagation in yeast has established it as the vector of choice in harnessing yeast to the production of biologically and commercially relevant proteins. In addition, these properties have also fostered development of a novel and effective tool that has enhanced the armamentarium available to the yeast geneticist: high-copy suppression. This technique has been productively applied to identify novel genes involved in such diverse processes as gene regulation (Hinnebusch and Fink 1983), cell-cycle control (Toda et al. 1987; Bender and Pringle 1989), and chromosome mechanics (Kuo and Campbell 1983).

The *FLP1/FRT* system has been used successfully as a means of inducing specific genome rearrangements in yeast and other eukaryotes. Given the absence of any topological constraints on *FLP1*-mediated recombination, judicious placement of pairs of *FRT* sites can be used to promote specific, *FLP1*-dependent intrachromosomal deletions or inversions or to direct integration of a circular molecule into a specific chromosomal site. Because *FLP1* can be placed under stringent but inducible expression systems, these rearrangements can be induced in a cell culture or organism at a particular time and in a synchronous fashion. Finally, because the only components needed for recombination are the target sites and Flp1p, these rearrangements can potentially be performed in any eukaryotic or prokaryotic cell. In addition, the ability to alter the specificity of the target site without losing *FLP1* responsiveness means that multiple independent recombination events can be promoted simultaneously in the same cell. To date, this system has been used to launch circular plasmids from chromosomal sites in *S. cerevisiae* (Volkert and Broach 1986; Reynolds et al. 1987), to induce specific deletions of the *white* locus of *Drosophila* (Golic and Lindquist 1989), and to construct conditional, high-copy vectors (F. Volkert, unpubl.). This is a tool that is still relatively underutilized but whose possible applications are only restricted by imagination and implementation. We anticipate that the successful application of the *FLP1/FRT* system to *Drosophila* genetics will undoubtedly spawn interest in the use of this system to promote specific rearrangements in other large eukaryotes.

B. Double Rolling Circle as a Mechanism of Chromosomal Amplification

The amplification mechanism used by 2-micron circle and its relatives is an elegant solution to the problem of obtaining multiple copies of a defined segment of DNA under conditions in which replication of the sequence is restricted to a single initiation event per cell cycle. We have noted elsewhere that this process may be operative in amplification of other specific DNA species, such as chloroplast or other organelle genomes. In addition, amplification of some chromosomal sequences could proceed by a variation of this mechanism.

A number of loci in mammalian cells can, under certain circumstances, exhibit local amplification. One mechanism often invoked to account for this process involves activation of a single replication origin multiple times within a single cell cycle (Schimke 1984). Although this may account for some programmed amplification events, such as chorion gene amplification (Kafatos et al. 1985; Kalfayan et al. 1985), it may not be universally applicable. Cox (1989) has suggested that reiterated tandem repeats could arise by formation of a rolling circle intermediate by recombination between repeated sequences lying in direct orientation on either side of a chromosomal replication fork. Amplified loci in higher eukaryotes have been shown frequently to feature DNA segments reiterated in inverted orientation (see Passananti et al. 1987). These molecules could result from double rolling circle replication of DNA that had been excised from a chromosome as a circle with an inverted repeat. In both these cases, it is not generally assumed that the double rolling circle is formed by site-specific recombination; general homologous recombination would have the same effect, albeit at a lower efficiency. Although the double rolling circle mechanism does not account for all examples of specific gene amplification, it warrants inclusion in the panoply of processes by which selective expansion of chromosomal domains occurs.

C. Future Directions

The 2-micron circle has presented yeast molecular biologists with a set of intriguing problems, exploration of which not only has revealed fascinating and often novel biological mechanisms, but also has provided valuable entrées into fundamental issues of molecular and cellular biology. As noted in previous sections, investigation of the plasmid-encoded, site-specific recombinases continues to refine our appreciation of the molecular details of recombination, at a resolution that is rapidly approaching the atomic level. The mechanism of plasmid partitioning still remains essentially unsolved. Further progress in this area will likely move hand in hand with accumulation of an increasingly sophisticated knowledge of the structure and dynamics of nuclear architecture and nuclear biogenesis. Finally, analysis of other 2-micron-circle-like plasmids has opened the door to a productive interplay between the investigation of phylogenetic relationships among plasmid species and molecular dissections of the basic mechanisms of plasmid persistence. In this light, we can anticipate that the fruits emerging from the continued investigation of this benign parasite will continue to enrich our knowledge of the biology of the host in which it resides.

REFERENCES

Amati, B.B. and S.M. Gasser. 1988. Chromosomal *ARS* and *CEN* elements bind specifically to the yeast nuclear scaffold. *Cell* **54:** 967.

Amin, A.A. and P.D. Sadowski. 1989. Synthesis of an enzymatically active FLP recombinase in vitro: Search for a DNA-binding domain. *Mol. Cell. Biol.* **9:** 1987.

Amin, A.A., L.G. Beatty, and P.D. Sadowski. 1990. Synaptic intermediates promoted by the FLP recombinase. *J. Mol. Biol.* **214:** 55.

Andrews, B.J., L.G. Beatty, and P.D. Sadowski. 1987. Isolation of intermediates in the binding of the FLP recombinase of the yeast plasmid 2-micron circle to its target sequence. *J. Mol. Biol.* **193:** 345.

Andrews, B.J., M. McLeod, J. Broach, and P.D. Sadowski. 1986. Interaction of the FLP recombinase of the *Saccharomyces cerevisiae* 2μm plasmid with mutated target sequences. *Mol. Cell. Biol.* **6:** 2482.

Andrews, B.J., G.A. Proteau, L.G. Beatty, and P.D. Sadowski. 1985. The FLP recombinase of the 2μ circle DNA of yeast: Interaction with its target sequences. *Cell* **40:** 795.

Araki, H., A. Jearnpipatkul, H. Tatsumi, T. Sakurai, K. Ushio, T. Muta, and Y. Oshima. 1985. Molecular and functional organization of yeast plasmid pSR1. *J. Mol. Biol.* **182:** 191.

Argos, P., A. Landy, K. Abremski, J.B. Egan, E. Haggard-Ljungquist, R.H. Hoess, M.L. Kahn, B. Kalionis, S.V.L. Narayana, L.S. Pierson, N. Sternberg, and J.M. Leong. 1986. The integrase family of site-specific recombinases: Regional similarities and global diversity. *EMBO J.* **5:** 433.

Armstrong, K.A., T. Som, F.C. Volkert, and J.R. Broach. 1988. Regulation of yeast plasmid amplification. *Cancer Cells* **6**: 213.

Armstrong, K.A., T. Som, F.C. Volkert, A. Rose, and J.R. Broach. 1989. Propagation and expression of genes in yeast using 2-μm circle vectors. In *Yeast genetic engineering* (ed. P.J. Barr et al.), p.165. Butterworth, Stoneham, Massachusetts.

Bender, A. and J.R. Pringle. 1989. Multicopy suppression of the *cdc24* budding defect in yeast by *CDC42* and three newly identified genes including the *ras*-related gene *RSR1*. *Proc. Natl. Acad. Sci.* **86**: 9976.

Beatty, L.G., and P.D. Sadowski. 1988. The mechanism of loading of the FLP recombinase onto its DNA target sequence. *J. Mol. Biol.* **204**: 283.

Brewer, B.J. and W.L. Fangman. 1987. The localization of replication origins on *ARS* plasmids in *S. cerevisiae*. *Cell* **51**: 463.

Broach, J.R. 1981. The yeast plasmid 2μ circle. In *The molecular biology of the yeast Saccharomyces: Life cycle and inheritance* (ed. J.N. Strathern et al.), p. 445. Cold Spring Harbor Laboratory, Cold Spring Harbor, New York.

———. 1983. Construction of high copy yeast vectors using 2-μm circle sequences. *Methods Enzymol.* **101**: 307.

Broach, J.R. and J.B. Hicks. 1980. Replication and recombination functions associated with the yeast plasmid, 2μ circle. *Cell* **21**: 501.

Broach, J.R., V.R. Guarascio, and M. Jayaram. 1982. Recombination within the yeast plasmid 2μ circle is site-specific. *Cell* **29**: 227.

Broach, J.R., J.F. Atkins, C. McGill, and L. Chow. 1979. Identification and mapping of the transcriptional and translational products of the yeast plasmid, 2μ circle. *Cell* **16**: 827.

Broach, J.R., Y.-Y. Li, L.C.-C. Wu, and M. Jayaram. 1983a. Vectors for high-level inducible expression of cloned genes in yeast. In *Experimental manipulation of gene expression* (ed. M. Inouye), p. 83. Academic Press, New York.

Broach, J.R., Y.-Y. Li, J. Feldman, M. Jayaram, J. Abraham, K.A. Nasmyth, and J.B. Hicks. 1983b. Localization and sequence analysis of yeast origins of DNA replication. *Cold Spring Harbor Symp. Quant. Biol.* **47**: 1165.

Bruckner, R.C., and M.M. Cox. 1986. Specific contacts between the FLP protein of the yeast 2-micron plasmid and its recombination site. *J. Biol. Chem.* **261**: 11798.

Cameron, J.R., P. Philippsen, and R.W. Davis. 1977. Analysis of chromosomal integration and deletions of yeast plasmids. *Nucleic Acids Res.* **4**: 1429.

Cashmore, A.M., M.S. Albury, C. Hadfield, and P.A. Meacock. 1988. The 2 μm *D* region plays a role in yeast plasmid maintenance. *Mol. Gen. Genet.* **212**: 426.

Chen, X.J., M. Saliola, C. Falcone, M.M. Bianchi, and H. Fukuhara. 1986. Sequence organization of the circular plasmid pKD1 from the yeast *Kluyveromyces drosophilarum*. *Nucleic Acids Res.* **14**: 4471.

Clark-Walker, G.D. and G.L.C. Miklos. 1974. Localization and quantitation of circular DNA in yeast. *Eur. J. Biochem.* **41**: 359.

Cox, M.M 1983. The FLP protein of the yeast 2-μm plasmid: Expression of a eukaryotic genetic recombination system in *Escherichia coli. Proc. Natl. Acad. Sci.* **80**: 4223.

———. 1989. DNA inversion in the 2 μm plasmid of *Saccharomyces cerevisiae*. In *Mobile DNA* (ed. D.E. Berg and M.M. Howe), p. 661. American Society for Microbiology, Washington, D.C.

Das, S. and C.P. Hollenberg. 1982. A high-frequency transformation system for the yeast *Kluyveromyces lactis. Curr. Genet.* **6**: 123.

Dobson, M.J., A.B. Futcher, and B.S. Cox. 1980. Loss of 2 μm DNA from *Saccharomyces cerevisiae* transformed with the chimaeric plasmid pJDB219. *Curr. Genet.* **2**: 201.

Futcher, A.B. 1986. Copy number amplification of the 2 micron circle plasmid of *Saccharomyces cerevisiae. J. Theoret. Biol.* **119**: 197.

———. 1988. The 2 μm circle plasmid of *Saccharomyces cerevisiae. Yeast* **4**: 27.

Futcher, A.B. and B.S. Cox. 1983. Maintenance of the 2 μm circle plasmid in populations of *Saccharomyces cerevisiae. J. Bacteriol.* **154**: 612.

———. 1984. Copy number and the stability of 2-μm circle-based artificial plasmids of *Saccharomyces cerevisiae. J. Bacteriol.* **157**: 283.

Futcher, B., E. Reid, and D.A. Hickey. 1988. Maintenance of the 2 μm circle plasmid of *Saccharomyces cerevisiae* by sexual transmission: An example of a selfish DNA. *Genetics* **118**: 411.

Gates, C.A. and M.M. Cox. 1988. FLP recombinase is an enzyme. *Proc. Natl. Acad. Sci.* **85**: 4628.

Golic, K.G. and S. Lindquist. 1989. The FLP recombinase of yeast catalyzes site-specific recombination in the *Drosophila* genome. *Cell* **59**: 499.

Govind, N.S. and M. Jayaram. 1987. Rapid localization and characterization of random mutations within the 2 μ circle site-specific recombinase: A general strategy for analysis of protein function. *Gene* **51**: 31.

Gronostajski, R.M. and P.D. Sadowski. 1985a. The FLP recombinase of the *Saccharomyces cerevisiae* 2 μm plasmid attaches covalently to DNA via a phosphotyrosyl linkage. *Mol. Cell. Biol.* **5**: 3274.

———. 1985b. The FLP protein of the 2-micron plasmid of yeast. Inter- and intramolecular reactions. *J. Biol. Chem.* **260**: 12328.

Guerineau, M., P.P. Slonimski, and P.R. Avner. 1974. Yeast episome: Oligomycin resistance associated with a small covalently closed non-mitochondrial circular DNA. *Biochem. Biophys. Res. Commun.* **61**: 462.

Hartley, J.L. and J.E. Donelson. 1980. Nucleotide sequence of the yeast plasmid. *Nature* **286**: 860.

Hill, A. and K. Bloom. 1987. Genetic manipulations of centromere function. *Mol. Cell. Biol.* **7**: 2397.

Hinnebusch, A.G. and G.R. Fink. 1983. Positive regulation in the general amino acid control of *Saccharomyces cerevisiae. Proc. Natl. Acad. Sci.* **80**: 5374.

Holm, C. 1982. Clonal lethality caused by the yeast plasmid 2μ DNA. *Cell* **29**: 585.

Huberman, J.A., L.D. Spotila, K.A. Nawotka, S.M. El-Assouli, and L.R. Davis. 1987. The in vivo replication origin of the yeast 2 μm plasmid. *Cell* **51**: 473.

Jayaram, M. 1985. Two-micrometer circle site-specific recombination: The minimal substrate and the possible role of flanking sequences. *Proc. Natl. Acad. Sci.* **82**: 5875.

———. 1986. Mating type-like conversion promoted by the 2 μm circle site-specific recombinase: Implications for the double-strand-gap repair model. *Mol. Cell. Biol.* **6**: 3831.

Jayaram, M., Y.-Y. Li, and J.R. Broach. 1983. The yeast plasmid 2 micron circle encodes components required for its high copy propagation. *Cell* **34**: 95.

Jayaram, M., A. Sutton, and J.R. Broach. 1985. Properties of *REP3*: A *cis* acting locus required for stable propagation of the *Saccharomyces cerevisiae* plasmid 2 μm circle. *Mol. Cell. Biol.* **5**: 2466.

Jearnpipatkul, A., H. Araki, and Y. Oshima. 1987a. Factors encoded by and affecting the holding stability of yeast plasmid pSR1. *Mol. Gen. Genet.* **206**: 88.

Jearnpipatkul, A., R. Hutacharoen, H. Araki, and Y. Oshima. 1987b. A *cis*-acting locus for the stable propagation of yeast plasmid pSR1. *Mol. Gen. Genet.* **207**: 355.

Kafatos, F., S.A. Mitsialis, N. Spoerel, B. Mariani, J.R. Lingappa, and C. Delidakis. 1985. Studies on the developmentally regulated expression and amplification of insect chorion genes. *Cold Spring Harbor Symp. Quant. Biol.* **50:** 537.

Kalfayan, L., J. Levine, T. Orr-Weaver, S. Parks, B. Wakimoto, D. deCicco, and A. Spradling. 1985. Localization of sequences regulating *Drosophila* chorion gene amplification and expression. *Cold Spring Harbor Symp. Quant. Biol.* **50:** 527.

Kikuchi, Y. 1983. Yeast plasmid requires a *cis*-acting locus and two plasmid proteins for its stable maintenance. *Cell* **35:** 487.

Kuo, C.-L. and J.L. Campbell. 1983. Cloning of *Saccharomyces cerevisiae* DNA replication genes: Isolation of the *CDC8* gene and two genes that compensate for the *cdc8-1* mutation. *Mol. Cell. Biol.* **3:** 1730.

Livingston, D.M. 1977. Inheritance of the 2 μm DNA plasmid from *Saccharomyces*. *Genetics* **86:** 73.

Livingston, D.M. and S. Hahne. 1979. Isolation of a condensed, intracellular form of the 2-μm DNA plasmid of *Saccharomyces cerevisiae*. *Proc. Natl. Acad. Sci.* **76:** 3727.

Matsuzaki, H., H. Araki, and Y. Oshima. 1988. Gene conversion associated with site-specific recombination in yeast plasmid pSR1. *Mol. Cell. Biol.* **8:** 955.

McLeod, M., S. Craft, and J.R. Broach. 1986. Identification of the crossover site during *FLP*-mediated recombination in the yeast plasmid 2 μm circle. *Mol. Cell. Biol.* **6:** 3357.

McLeod, M., F.C. Volkert, and J.R. Broach. 1984. Components of the site-specific recombination system encoded by the yeast plasmid 2-micron circle. *Cold Spring Harbor Symp. Quant. Biol.* **49:** 779.

Mead, D.J., D.C.J. Gardner, and S.G. Oliver. 1986. The yeast 2 μ plasmid: Strategies for the survival of a selfish DNA. *Mol. Gen. Genet.* **205:** 417.

Meyer-Leon, L., R.B. Inman, and M.M. Cox. 1990. Characterization of Holliday structures in FLP protein-promoted site-specific recombination. *Mol. Cell. Biol.* **10:** 235.

Meyer-Leon, L., J.F. Senecoff, R.C. Bruckner, and M.M. Cox. 1984. Site-specific genetic recombination promoted by the FLP protein of the yeast 2-micron plasmid in vitro. *Cold Spring Harbor Symp. Quant. Biol.* **49:** 797.

Meyer-Leon, L., C.A. Gates, J.M. Attwood, E.A. Wood, and M.M. Cox. 1987. Purification of the FLP site-specific recombinase by affinity chromatography and re-examination of basic properties of the system. *Nucleic Acids Res.* **15:** 6469.

Meyer-Leon, L., L.C. Huang, S.W. Umlauf, M.M. Cox, and R.B. Inman. 1988. Holliday intermediates and reaction by-products in FLP protein-promoted site-specific recombination. *Mol. Cell. Biol.* **8:** 3784.

Murray, A.W. and J.W. Szostak. 1983. Pedigree analysis of plasmid segregation in yeast. *Cell* **34:** 961.

Murray, J.A.H. and G. Cesarini. 1986. Functional analysis of the yeast plasmid partition locus *STB. EMBO J.* **5:** 3391.

Murray, J.A.H., G. Cesareni, and P. Argos. 1988. Unexpected divergence and molecular coevolution in yeast plasmids. *J. Mol. Biol.* **200:** 601.

Murray, J.A.H., M. Scarpa, N. Rossi, and G. Cesareni. 1987. Antagonistic controls regulate copy number of the yeast 2μ plasmid. *EMBO J.* **6:** 4205.

Nelson, R.G., and W.L. Fangman. 1979. Nucleosome organization of the yeast 2-μm DNA plasmid: A eukaryotic minichromosome. *Proc. Natl. Acad. Sci.* **76:** 6515.

Parsons, R.L., P.V. Prasad, R.M. Harshey, and M. Jayaram. 1988. Step-arrest mutants of FLP recombinase: Implications for the catalytic mechanism of DNA recombination. *Mol. Cell. Biol.* **8:** 3303.

Passananti, C., B. Davies, M. Fork, and M. Fried. 1987. Structure of an inverted duplication formed as a first step in a gene amplification event: Implications for a model of gene amplification. *EMBO J.* **6:** 1697.

Pearson, W.R. and D.J. Lipman. 1988. Improved tools for biological sequence comparison. *Proc. Natl. Acad. Sci.* **85:** 2444.

Prasad, P.V., L.J. Young, and M. Jayaram. 1987. Mutations in the 2-μm circle site-specific recombinase that abolish recombination without affecting substrate recognition. *Proc. Natl. Acad. Sci.* **84:** 2189.

Prasad, P.V., D. Horensky, L.J. Young, and M. Jayaram. 1986. Substrate recognition by the 2 μm circle site-specific recombinase: Effect of mutations within the symmetry elements of the minimal substrate. *Mol. Cell. Biol.* **6:** 4329.

Reynolds, A., A. Murray, and J. Szostak. 1987. Function of the yeast plasmid gene products. *Mol. Cell. Biol.* **7:** 3566.

Rose, A.B. and J.R. Broach. 1990. Propagation and expression of cloned genes in yeast 2-μm circle-based vectors. *Methods Enzymol.* **185:** 234.

Rose, M.D. and J.R. Broach. 1991. Cloning genes by complementation in yeast. *Methods Enzymol.* **194:** 195.

Sadowski, P. 1986. Site-specific recombinases: Changing partners and doing the twist. *J. Bacteriol.* **165:** 341.

Schimke, R.T. 1984. Gene amplification in cultured animal cells. *Cell* **37:** 705.

Schwartz, C.J. and P.D. Sadowski. 1989. FLP recombinase of the 2 μm circle plasmid of *Saccharomyces cerevisiae* bends its DNA target. Isolation of FLP mutants defective in DNA bending. *J. Mol. Biol.* **205:** 647.

Senecoff, J.F. and M.M. Cox. 1986. Directionality in FLP protein-promoted site-specific recombination is mediated by DNA-DNA pairing. *J. Biol. Chem.* **261:** 7380.

Senecoff, J.F., R.C. Bruckner, and M.M. Cox. 1985. The FLP recombinase of the yeast 2-μm plasmid: Characterization of its recombination site. *Proc. Natl. Acad. Sci.* **82:** 7270.

Senecoff, J.F., P.J. Rossmeissl, and M.M. Cox. 1988. DNA recognition by the FLP recombinase of the yeast 2 μ plasmid. A mutational analysis of the FLP binding site. *J. Mol. Biol.* **201:** 405.

Snyder, M., R.J. Sapolsky, and R. Davis. 1988. Transcription interferes with elements important for chromosome maintenance in *Saccharomyces cerevisiae. Mol. Cell. Biol.* **8:** 2184.

Som, T., K.A. Armstrong, F.C. Volkert, and J.R. Broach. 1988. Autoregulation of 2-μm circle gene expression provides a model for maintenance of stable plasmid copy levels. *Cell* **52:** 27.

Sutton, A. and J.R. Broach. 1985. Signals for transcription initiation and termination in the *Saccharomyces cerevisiae* plasmid 2 μm circle. *Mol. Cell. Biol.* **5:** 2770.

Taketo, M., S.M. Jazwinski, and G.M. Edelman. 1980. Association of the 2-μm DNA plasmid with yeast folded chromosomes. *Proc. Natl. Acad. Sci.* **77:** 3144.

Toda, T., S. Cameron, P. Sass, M. Zoller, and M. Wigler. 1987. Three different genes in *S. cerevisiae* encode the catalytic subunits of the cAMP-dependent protein kinase. *Cell* **50:** 277.

Toh-e, A. and I. Utatsu. 1985. Physical and functional structure of a yeast plasmid, pSB3, isolated from *Zygosaccharomyces bisporus. Nucleic Acids Res.* **13:** 4267.

Toh-e, A., S. Tada, and Y. Oshima. 1982. 2-μm DNA-like plasmids in the osmophilic haploid yeast *Saccharomyces rouxii. J. Bacteriol.* **151:** 1380.

Toh-e, A., H. Araki, I. Utatsu, and Y. Oshima. 1984. Plasmids resembling 2-μm DNA in the osmotolerant yeasts *Saccharomyces bailii* and *Saccharomyces bisporus. J. Gen. Microbiol.* **130:** 2527.

Umlauf, S.W. and M.M. Cox. 1988. The functional significance of DNA sequence structure in a site-specific genetic recombination reaction. *EMBO J.* **7:** 1845.

Utatsu, I., A. Utsunomiya, and A. Toh-e. 1986. Functions encoded by the yeast plasmid pSB3 isolated from *Zygosaccharomyces rouxii* IFO 1730 (formerly *Saccharomyces bisporus* var. mellis). *J. Gen. Microbiol.* **132:** 1359.

Utatsu, I., S. Sakamoto, T. Imura, and A. Toh-e. 1987. Yeast plasmids resembling 2 μm DNA: Regional similarities and diversities at the molecular level. *J. Bacteriol.* **169:** 5537.

Veit, B.E. and W.L. Fangman. 1988. Copy number and partition of the *Saccharomyces cerevisiae* 2 μm plasmid controlled by transcription regulators. *Mol. Cell. Biol.* **8:** 4949.

Vetter, D., B.J. Andrews, L. Roberts-Beatty, and P.D. Sadowski. 1983. Site-specific recombination of yeast 2-μm DNA in vitro. *Proc. Natl. Acad. Sci.* **80:** 7284.

Volkert, F.C. and J.R. Broach. 1986. Site-specific recombination promotes plasmid amplification in yeast. *Cell* **46:** 541.

————. 1987. The mechanism of propagation of the yeast 2-micron circle plasmid. In *The biochemistry and molecular biology of industrial yeasts* (ed. G.G. Stewart et al.), p. 145. CRC Press, Boca Raton, Florida.

Volkert, F.C., D.W. Wilson, and J.R. Broach. 1989. Deoxyribonucleic acid plasmids in yeasts. *Microbiol. Rev.* **53:** 299.

Williamson, D.H. 1985. The yeast ARS element, six years on: A progress report. *Yeast* **1:** 1.

Wu, L.-C.C., P.A. Fisher, and J.R. Broach. 1987. A yeast plasmid partitioning protein is a karyoskeletal component. *J. Biol. Chem.* **262:** 883.

Zakian, V.A. and J.F. Scott. 1982. Construction, replication and chromatin structure of TRP1-RI circle, a multiple-copy synthetic plasmid derived from *Saccharomyces cerevisiae* chromosomal DNA. *Mol. Cell. Biol.* **2:** 221.

Zakian, V.A., B.J. Brewer, and W.L. Fangman. 1979. Replication of each copy of the yeast 2 micron DNA plasmid occurs during the S phase. *Cell* **17:** 923.

7

Biogenesis of Yeast Mitochondria

Liza Pon[1] and Gottfried Schatz

Biocenter, University of Basel,
CH-4056 Basel, Switzerland

I. INTRODUCTION

The question of how mitochondria are formed has fascinated biologists ever since Altmann (1890) first described these organelles 100 years ago. Initially, mitochondria were regarded as intracellular parasites. In the 1950s, when the resolving power of the electron microscope revealed the

---

[1]Present address: Department of Anatomy and Cell Biology, Columbia University College of Physicians & Surgeons, New York, New York 10032.

multitude of intracellular membranes in eukaryotes, mitochondria were frequently assumed to represent just another intracellular membrane system composed of unit membranes. In the early 1960s, the discovery of mitochondrial DNA (mtDNA) led some investigators to propose that mitochondria were self-replicating units. Today, we view mitochondria as bona fide organelles that are very much controlled by the nucleus yet have retained telltale earmarks of their endosymbiotic past.

The first meeting devoted specifically to mitochondrial biogenesis was held in 1967 (Slater et al. 1968). Since then, the field has grown explosively, aided to a large measure by the unique experimental possibilities offered by the mitochondrial system of *Saccharomyces cerevisiae*. It is now clear that the processes by which mitochondria are formed vary little between different eukaryotic cells; most of the differences found so far are not fundamental but variations on a common theme. Although in this article we emphasize mitochondrial biogenesis in *S. cerevisiae*, we consider results obtained in other systems as well. Historical details and additional information can be found in the various reviews cited throughout this chapter, as well as in several books (Slonimski 1953; Lehninger 1964; Roodyn and Wilkie 1968; Keilin 1970; Sager 1972; Gillham 1978; Ernster and Schatz 1981; Tzagoloff 1982). The lipid composition of yeast mitochondria and mitochondrial lipid metabolism are covered elsewhere in these volumes (Paltauf and Henry).

II. MORPHOLOGY

To understand how yeast mitochondria are made, one must know what these organelles look like. The morphology of *S. cerevisiae* mitochondria has been studied extensively, but the information derived from these studies is not widely known. This may stem from the fact that, until recently, yeast mitochondria were mainly the playground of molecular geneticists, who often considered the mitochondrial membranes to be bothersome wrappings around the mtDNA, to be dissolved as quickly as possible by detergents or chloroform/phenol. As yeast is becoming a favorite pet of cell biologists, this attitude is changing rapidly. We are beginning to appreciate that mitochondria of yeast, like those of other cells, are highly dynamic structures whose morphological changes may affect the behavior of the mitochondrial genome to mutations and during heteromitochondrial crosses.

The previous edition of this monograph featured an excellent chapter on the morphology of yeast mitochondria (Stevens 1981). Additional useful information can be found in several reviews and books (Matile et al. 1969; Lloyd 1974; Munn 1974; Gillham 1978; Tzagoloff 1982). This section summarizes the main points of these reviews, draws attention to some earlier studies that deserve further emphasis, and discusses some recent work that has not yet been reviewed.

A. Methods

The morphology of yeast mitochondria has been studied by three main methods. The oldest of these is vital staining of mitochondria (e.g., with Janus Green) and examination by light microscopy. Since the late 1950s, the favorite method has been electron microscopic examination of thin-sectioned cells after fixation with glutaraldehyde and permanganate and embedding in organic resins (Fig. 1) (Agar and Douglas 1957; Hagedorn 1957; Hashimoto et al. 1959; Hirano and Lindegren 1961; Vitols et al. 1961; Yotsuyanagi 1962a,b). Such studies revealed that mitochondria of yeast, like those of mammals, possess two distinct membranes, the outer and inner membranes, which define and enclose two soluble compartments, the matrix and the intermembrane space. The inner membrane is highly invaginated, forming pleats referred to as cristae, and is tightly associated with regions of outer membrane at membrane-contact sites (Hackenbrock 1968; Kellems et al. 1975; Pon et al. 1989).

More recently, light microscopy has made an impressive comeback with the introduction of fluorescent dyes that specifically stain the mitochondria of living cells (Bereiter-Hahn 1976; Johnson et al. 1980; Pringle et al. 1989). These dyes are lipophilic cations; because the membrane potential of respiring mitochondria is positive toward the cytosol (in contrast to that of other eukaryotic membranes), these dyes are specifically accumulated in the mitochondrial matrix. They may therefore be less effective for visualizing mutant or incompletely developed yeast mitochondria, which may maintain only a weak transmembrane potential.

B. Effects of Growth Phase

By examining cells collected at various points during their growth on glucose, Yotsuyanagi (1962a) showed that the changes in cellular respiration (see Fig. 2A) were paralleled by distinctive changes in mitochondrial morphology (Fig. 2B,C). During mid- and mid-to-late logarithmic growth, cells contained only few mitochondria with poorly developed cristae (points A–I); during the late logarithmic phase, the exhaustion of glucose induced nearly full mitochondrial development: Mitochondrial mass per cell increased, the mitochondria acquired typical cristae, and, as shown by light microscopy, appeared to fuse into a mitochondrial reticulum composed of one big and several smaller tubular organelles (point J). As the cells switched from glucose fermentation to ethanol oxidation during early stationary phase (Fig. 2A, point K), the

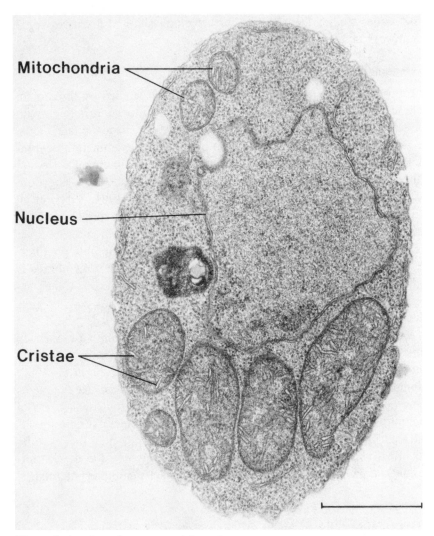

Figure 1 A spheroplast prepared from *S. cerevisiae* strain S41 grown under conditions favoring the development of respiring mitochondria. Spheroplasts were fixed with glutaraldehyde and OsO_4. Note the clearly visible nucleus and the striking arrangement of cristae-rich mitochondria. Bar, 1 μm. (Courtesy of B. Stevens.)

mitochondrial reticulum broke up into many small, regular mitochondria that were preferentially arranged at the periphery of the cell. During extended starvation in the late stationary phase (point L and later), mitochondria changed slowly into degenerative forms composed of parallel or concentric lamellae (Fig. 2C).

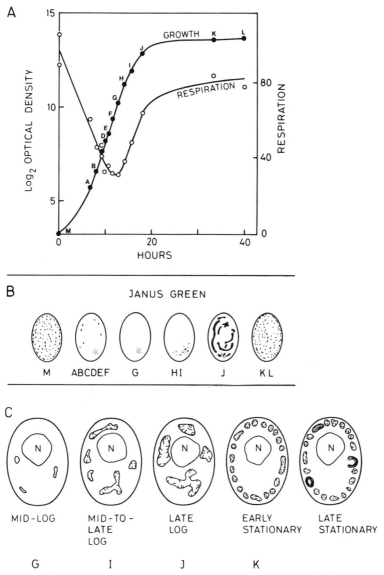

Figure 2 Morphology of yeast mitochondria as a function of the growth phase in a rich medium containing 5.4% glucose. Uppercase letters identify the growth stages (see *A*) at which cells were analyzed. (*A*) Cell number (measured by optical density) and respiratory activity (expressed as Q_{o2}) as a function of time. (*B*) Staining of mitochondria with Janus Green. Drawings are based on appearance of cells in the light microscope. (*C*) Examination of thin sections of permanganate-fixed, embedded cells in the electron microscope. Drawings are based on electron micrographs. The growth stage corresponding to late stationary phase is beyond the time range of the experiment depicted in *A*. (N) Nucleus. (Adapted from Yotsuyanagi 1962a.)

Later studies (Hoffmann and Avers 1973; Stevens 1977, 1981; Damsky et al. 1969; Damsky 1976; Stevens and White 1979) confirmed and extended these findings significantly. Stevens showed that the number of individual mitochondria per yeast cell could vary from 1–2 (in early logarithmic phase on glucose) to 44 (in stationary phase during glucose limitation) and that the percentage of cell volume contributed by mitochondria varied accordingly from 3.2% to as much as 12.6% (Stevens 1981). The variability of the yeast mitochondrial system in response to growth conditions probably accounts for differing earlier reports on the number of mitochondria per cell.

During meiosis, the morphology of yeast mitochondria changes even more dramatically. With the fluorescent dyes DAPI and dimethylaminostyrylmethylpyridinium iodine (DASPMI), Miyakawa et al. (1984) visualized mitochondria of diploid spheroplasts in early stages of meiosis as many discrete, small bodies (Fig. 3A). Four hours later (approximately during early prophase), these mitochondria had fused into a highly convoluted mitochondrial reticulum (Fig. 3B); upon the second nuclear division, this reticulum yielded four mitochondrial tubules, each of them associated with one of the four nuclei in a strikingly oriented pattern (Fig. 3C). Fusion and fragmentation of mitochondria were also observed during cell division and germ tube formation of the yeast *Candida albicans* (Tanaka et al. 1985).

C. Effects of Mutations

The morphology of yeast mitochondria can also be altered by mutations. As already noted by Yotsuyanagi (1962b), yeast cells lacking functional mtDNA (ρ^- mutants) have aberrant mitochondrial profiles characterized by irregular shapes, few (if any) cristae, and frequent multilamellar configurations (see Figs. 4 and 5). Because ρ^- mutants lack mitochondrial protein synthesis (Kûzela and Grecná 1969; Schatz and Saltzgaber 1969), it is not surprising that similar morphological changes are induced by growing ρ^+ cells in the presence of chloramphenicol, an inhibitor of mitochondrial protein synthesis (Clark-Walker and Linnane 1967). In contrast, most nuclear (*pet*) mutations causing respiratory deficiency (see Section IV) have little effect on mitochondrial morphology as long as these mutations do not also cause the loss of the mitochondrial genome (Fig. 5).

Figure 3 Changes in mitochondrial morphology during meiosis. *S. cerevisiae* spheroplasts were inoculated into sporulation medium and stained with DAPI immediately after (*A*), 4 hr after (*B*), and 12 hr after (*C*) inoculation. (Photographs kindly supplied by I. Miyakawa, N. Sando and T. Kuroiwa.)

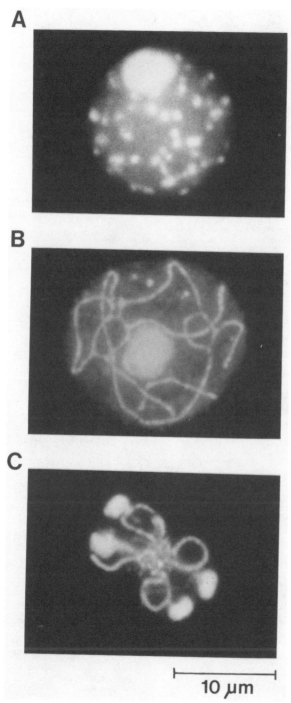

Figure 3 *(See facing page for legend.)*

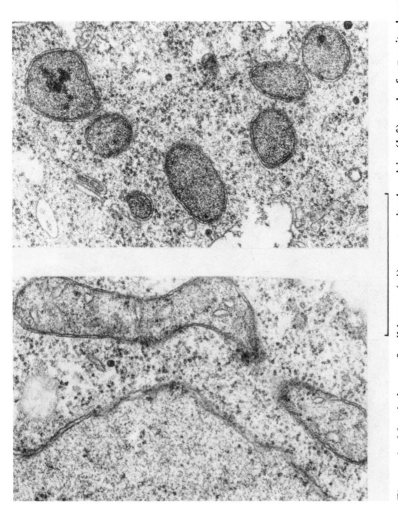

Figure 4 Morphology of wild-type (ρ^+) yeast mitochondria (*left*) and of ρ^- mitochondria lacking functional mtDNA (*right*). Spheroplasts from early stationary-phase cells were fixed with glutaraldehyde and OsO_4. Bar, 1 μm. (Courtesy of B. Stevens.)

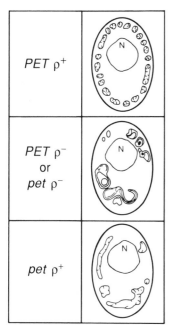

PET ρ^+	
PET ρ^- or *pet* ρ^-	
pet ρ^+	

Figure 5   Mitochondrial morphology of different respiratory-deficient yeast mutants in early stationary phase of growth. (*Top*) *PET* ρ^+ (wild type); (middle) *PET* ρ^-, *pet7* ρ^-, or *pet5* ρ^-; (*bottom*) *pet5* ρ^+ or *pet7* ρ^+. (N) Nucleus. Note that mitochondria of double mutants (e.g., *pet7* ρ^-) have essentially the same morphology as those of a *PET* ρ^- strain. (Adapted from Yotsuyanagi 1962b.)

D. Promitochondria of Anaerobically Grown Yeast

In the 1960s, the question of whether anaerobically grown yeast cells contain mitochondria generated considerable discussion. By that time, it was well established that aerobically grown yeast contains mitochondria (see above) and a typical respiratory chain, including cytochromes b, c_1, c, and aa_3, and that respiratory activity and cytochrome content decrease drastically upon anaerobic growth (Slonimski 1953). In 1964, examination of anaerobically grown *S. cerevisiae* by conventional electron microscopy suggested that the cells lacked mitochondria; however, the cells appeared to regain mitochondria upon aeration (Polakis et al. 1964; Wallace and Linnane 1964). This led to the suggestion that during respiratory adaptation, mitochondria are generated de novo, rather than from preexisting mitochondria. The issue became more complex when Morpurgo et al. (1964) detected mitochondrial profiles in anaerobically grown cells, but only if the growth medium had been supplemented with ergosterol and oleic acid. These essential membrane lipids require oxygen for their synthesis and must therefore be added to the medium for long-term anaerobic growth of yeast (see Andreasen and Stier 1953).

The supposed absence of mitochondria from anaerobically grown cells became questionable when subfractionation of these cells identified subcellular particles that were respiration-deficient, yet contained several biochemical markers typical of mitochondria (Schatz 1965). These promitochondria were suggested to differentiate into typical mitochondria upon respiratory adaptation of the cells. In subsequent studies, promitochondria were shown to contain mtDNA, a typical mitochondrial protein-synthesizing system, and even part of the oxidative phosphorylation machinery (Schatz et al. 1968; Criddle and Schatz 1969; Damsky et al. 1969; Rabinowitz and Swift 1970; Groot et al. 1972). The main difference between these organelles and typical yeast mitochondria was the absence of a respiratory chain. Examination of the anaerobically grown cells invariably revealed mitochondrial structures (Damsky et al. 1969; Plattner and Schatz 1969); however, if anaerobic growth had occurred in the absence of added oleic acid and ergosterol, the promitochondria could not be visualized reliably by conventional fixation methods, probably because the low content of unsaturated lipids in mitochondrial membranes renders these methods ineffective. These promitochondria required freeze-etching for proper visualization in yeast cells. Promitochondria may contain typical cristae (Fig. 6), even though these are less frequent and less well developed than in mitochondria from cells grown aerobically on a nonfermentable carbon source.

Promitochondria are the structural precursors to respiring yeast mitochondria: When their membrane proteins were specifically labeled with ^{3}H in vivo, these labeled proteins were recovered with respiring mitochondria after respiratory adaptation. Quantitative autoradiography and cytochemical staining for cytochrome oxidase, both at the electron microscope level, confirmed that membrane proteins formerly present in promitochondria were present in mitochondrial particles exhibiting cytochrome oxidase activity (Plattner et al. 1970). This brief survey may serve as a reminder that the morphological changes of mitochondria are one of the least understood aspects of these organelles. What are the molecular mechanisms that decide whether mitochondria grow, divide, or fuse with each other? Does the cytoskeleton play a role in these changes or in the distribution of mitochondria to the bud in mitosis? All of these questions are still unanswered.

III. MITOCHONDRIAL GENETIC SYSTEM

The 27 years since the discovery of mtDNA in yeast (Schatz et al. 1964) have witnessed an almost explosive growth of research on this topic. The

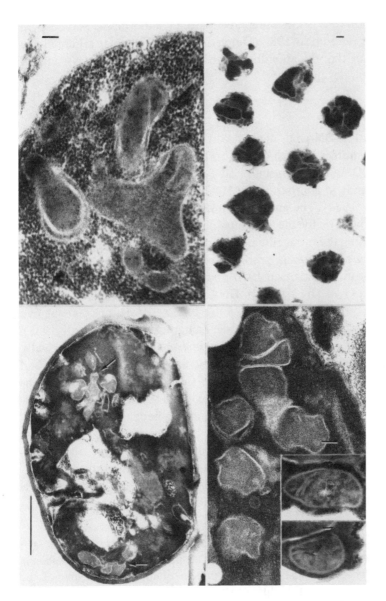

Figure 6 Promitochondria of *S. cerevisiae* (strain D273-10B) grown anaerobically under different conditions. (*Left*) Spheroplasts; (*right*) isolated promitochondria; (*top*) growth in presence of oleic acid and ergosterol; (*bottom*) growth in absence of added lipids. (*Bottom left*) Arrows identify promitochondrial profiles; some promitochondria (e.g., those shown at *bottom right*) exhibit cristae. Bars: (*top* and *bottom right*) 0.1 μm; (*bottom left*) 1 μm. Fixation was performed by freeze substitution with glutaraldehyde, followed by post-fixation staining with uranyl acetate and lead citrate. (Reprinted, with permission, from Plattner et al. 1970.)

previous edition of this monograph contains an authoritative and detailed review of the field up to 1981 (Dujon 1981). This chapter reiterates the key features of the mitochondrial genetic system and presents some of the more recent developments.

A. Mitochondrial DNA

The amount of mtDNA per haploid cell appears to vary among different *S. cerevisiae* strains; in most strains, mtDNA accounts for 5–15% of the total cellular DNA (Williamson et al. 1971; Nagley and Linnane 1972). This corresponds to approximately 10–30 molecules of mtDNA per cell (Grimes et al. 1974). The amount of mtDNA per cell is little affected by anaerobic growth or by the ρ^- mutation (Dujon 1981), but the mechanisms controlling the intracellular level of mtDNA are unknown.

S. cerevisiae mtDNA is probably circular (Hollenberg et al. 1970; Christiansen and Christiansen 1976), but it is not certain whether this is always true in vivo. The size of mtDNA varies among strains; it is usually about 75 kb (corresponding to 25 µm or 5×10^7 daltons), about five times larger than human mtDNA (de Zamaroczy and Bernardi 1985). One of the most striking features of *S. cerevisiae* mtDNA is its unusually low G + C content of 18% and its extreme subdivision into GC-rich and AT-rich regions (Grossman et al. 1971). Because of this extreme compositional heterogeneity, yeast mtDNA exhibits several abnormal physical properties, such as a highly atypical melting behavior (Bernardi et al. 1970).

Irrespective of its size, *S. cerevisiae* mtDNA encodes seven proteins of the oxidative phosphorylation machinery, one protein of the mitochondrial ribosomal small subunit, two rRNAs, 24–25 tRNAs, and one 9S RNA participating in the maturation of tRNAs (Figs. 7 and 8). In addi-

Figure 7 Protein-coding genes in mtDNAs from different organisms. (*b*) Apocytochrome *b*; (I–III) subunits I–III of cytochrome oxidase; (6,8,9) subunits 6, 8, and 9 of ATP synthase; (Var1p) protein of the small subunit of the mitochondrial ribosome; (bI2–bI4 and aI1–aI6) introns 2–4 in the gene for apocytochrome *b* (*COB*) and introns 1–6 in the gene for cytochrome oxidase subunit I (*COX I*), respectively. These introns contain ORFs that partly encode processing enzymes (maturases) acting upon mitochondrial introns (see Section III. C). (r1) ORF within the intron of the gene for the large rRNA; the ORF encodes a DNA endonuclease. (This intron is only present in some [ω^+] strains of *S. cerevisiae*.) (ORFs 1–5) ORFs encoding proteins that may interact with RNA; (9) an apparently nonfunctional gene for ATP synthase subunit IX; (S5 and S13) proteins 5 and 13, respectively, of the small subunit of the mitochondrial ribosome; (α) ATP synthase α-subunit. (Adapted from Attardi and Schatz 1988.)

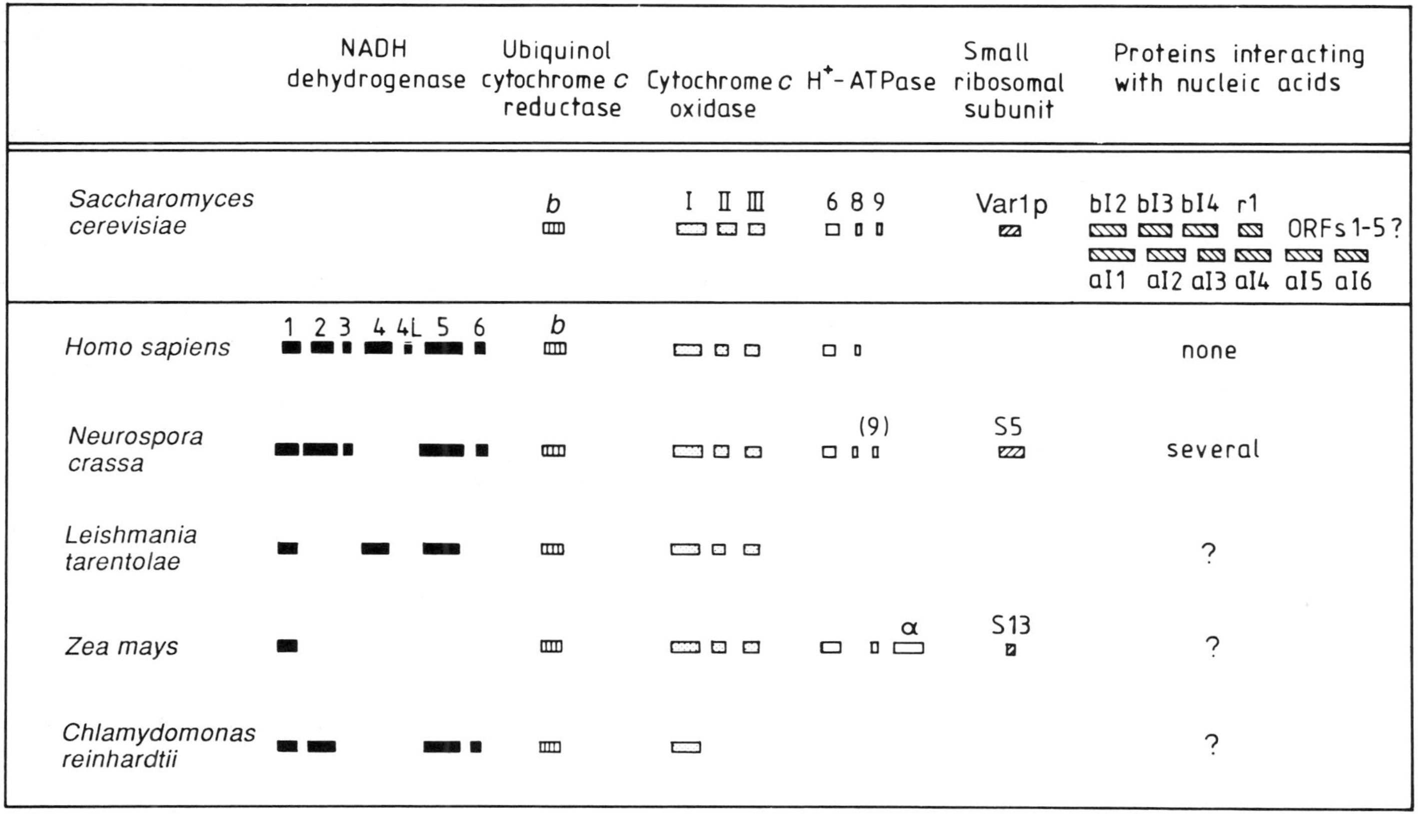

NADH dehydrogenase | Ubiquinol cytochrome c reductase | Cytochrome c oxidase | H⁺-ATPase | Small ribosomal subunit | Proteins interacting with nucleic acids

Column sub-labels: 1 2 3 4 4L 5 6 — b — I II III — 6 8 9 — Var1p — bI2 bI3 bI4 r1 ORFs 1-5? bI1 aI2 aI3 aI4 aI5 aI6 aI1

Saccharomyces cerevisiae

Homo sapiens — none

Neurospora crassa — (9) — S5 — several

Leishmania tarentolae — ?

Zea mays — α — S13 — ?

Chlamydomonas reinhardtii — ?

Figure 7 (See facing page for legend.)

345

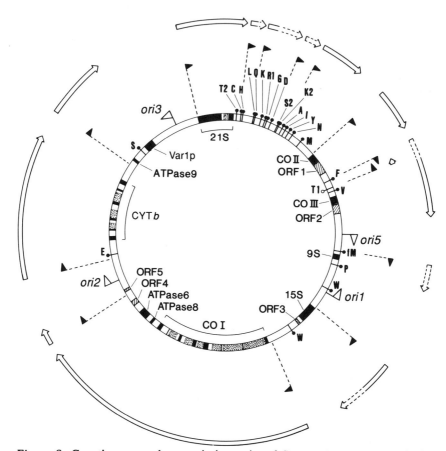

Figure 8 Genetic map and transcription units of *S. cerevisiae* mtDNA. (*Solid bars*) Genes or their exons; (*hatched bars*) ORFs of unknown function (ORF 1–ORF 5); (*dotted bars*) ORFs within introns. (*Solid circles*) tRNA genes transcribed from the main coding strand, identified by the one-letter codes of the cognate amino acids; (*open circle*) the single tRNA gene (tRNA Thr1) transcribed from the opposite strand. (*Solid flags*) The positions and orientations of transcription initiation sites; (*open flags*) functional origins of replication (*ori1*, *ori2*, *ori3*, and *ori5*) and the 5′ to 3′ direction of the corresponding RNA primers (de Zamaroczy et al. 1984). In the outer circle, arrows identify primary transcripts (in the 5′ to 3′ direction), and dashed lines indicate probable extensions of some of these transcripts. (21S and 15S) Large and small rRNAs; (9S) RNA component of mitochondrial RNase; (Var1p) protein of small ribosomal subunit; (CO) cytochrome oxidase; (CYT *b*) apocytochrome *b*. (Reprinted, with permission, from Attardi and Schatz 1988.)

tion, it may contain up to 20 additional open reading frames (ORFs), most of which are located within introns of the genes encoding the large rRNA, cytochrome oxidase subunit I (*COX I*), and cytochrome *b* (*COB*).

Genes are separated by long AT-rich intergenic regions (sometimes termed spacers; Fonty et al. 1978), whose function is not clear. The size differences between mtDNAs of different *S. cerevisiae* strains result from the variable presence of some introns and size variations of the intergenic regions.

The gene content of yeast mtDNA differs in two important respects from that of most other mtDNAs. First, it lacks genes for subunits of NADH dehydrogenase; this is paralleled by the fact that electron flow through the NADH dehydrogenase of *S. cerevisiae* is not coupled to ATP synthesis (Ohnishi et al. 1966; Schatz and Racker 1966). Second, many of the ORFs of yeast mtDNA are absent in human mtDNA and may or may not be present in mtDNAs from other species (Fig. 7). These ORFs encode proteins performing one or more of the following functions: (1) excision, at the level of mRNA, of the intron encoding them (maturase function; see below); (2) duplicative transposition, at the DNA level, of the corresponding intron (Lazowska et al. 1980; Jacquier and Dujon 1985; Colleaux et al. 1986; Wenzlau et al. 1989); (3) deletion, at the DNA level, of introns (reverse transcriptase function) (Michel and Lang 1985); and (4) induction of genetic recombination between homologous exons (Kotylak et al. 1985).

B. Transcription and Replication of mtDNA

Transcription of mitochondrial genes appears to be catalyzed by a nucleus-encoded RNA polymerase composed of at least two nonidentical subunits (Winkley et al. 1985; Kelly and Lehman 1986; Schinkel et al. 1987, 1988; Ticho and Getz 1988): A 145-kD subunit functions as a core enzyme, and a specificity factor allows the core subunit to recognize one of the approximately 20 specific mitochondrial transcription initiation signals (5'-TATAAGTA-3') on mtDNA (Baldacci and Bernardi 1982; Osinga et al. 1982; Christianson and Rabinowitz 1983). The core subunit exhibits little sequence similarity to the corresponding *Escherichia coli* protein but is similar to the RNA polymerases of bacteriophages T3 and T7 (Masters et al. 1987). The size of the specificity factor has been given as 43 kD (Schinkel et al. 1987) or 70 kD (Ticho and Getz 1988); it is not clear whether this reflects the existence of two different proteins or whether the 43-kD polypeptide is a functionally active proteolytic degradation product of the 70-kD protein. The synthesis of both subunits is repressed by glucose, and tentative evidence suggests that the core subunit is repressed more strongly than the specificity-conferring subunit (Wilcoxen et al. 1988).

Disruption of the nuclear *RPO41* gene encoding the core subunit causes loss of mtDNA, suggesting that the enzyme also functions as a primase in the replication of mtDNA (Greenleaf et al. 1986; Kelly et al. 1986; Schinkel and Tabak 1989). The mtDNA polymerase, itself, has only been characterized incompletely. The presence of such an enzyme was already demonstrated more than 20 years ago (Wintersberger 1966; Wintersberger and Wintersberger 1970a,b), and this DNA polymerase activity was characterized further in later studies (Wintersberger and Blutsch 1976). However, the enzyme has not yet been purified. Genga et al. (1986) have identified a nuclear gene (*MIP1*) whose mutation can cause loss of mtDNA without apparent inhibition of nuclear DNA synthesis. However, it is not clear whether this gene encodes a subunit of the mtDNA polymerase. It is also not clear why inhibition of mitochondrial protein synthesis either by prolonged growth of cells in chloramphenicol (Weislogel and Butow 1970; Williamson et al. 1971) or by mutations (Myers et al. 1985) causes instability or loss of functional mtDNA. Because ρ^- cells (which lack mitochondrial protein synthesis) replicate their defective mtDNA, either these defective molecules replicate by a different mechanism than ρ^+ genomes or the absence of mitochondrial translation induces deletions in mtDNA by indirect means.

C. Maturation of Mitochondrial mRNAs

In *S. cerevisiae*, the mitochondrial gene for the large rRNA, *COB*, and *COX I* may contain one, up to five, and up to ten introns, respectively. It appears that introns are not inherently essential for the formation of respiring mitochondria, because *S. cerevisiae* strains have been constructed that lack all introns in their mtDNA yet have an apparently normal oxidative phosphorylation system (Séraphin et al. 1987). However, deletion or mutational alteration of *individual* introns may completely block the formation of respiring mitochondria (see below). This apparent paradox is explained by the fact that many of the introns in the *COB* and *COX I* genes encode proteins that mediate excision of intron-encoded sequences from mitochondrial mRNAs. This maturase function of intron-encoded proteins is one of the most bizarre, but also one of the most fascinating, aspects of the yeast mitochondrial genetic system. The literature on this subject has become immense. We shall not attempt to cover it in detail but only outline the essential features.

The *COB* gene of most laboratory strains exists in one of two forms (Fig. 9). The long form contains six exons (B1–B6) that are separated by five introns (bI1–bI5); the short form lacks the first three introns. Introns bI2, bI3, and bI4 contain long ORFs in-frame with the preceding exons.

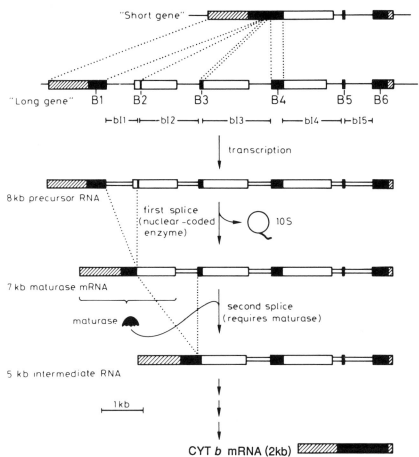

Figure 9   Initial steps in the maturation of mRNA for *S. cerevisiae* apo-cytochrome *b*. (B1–B6, *solid bars*) Exons; (bI1–bI5) introns. (*Large open bars*) Intronic ORFs in-frame with the preceding exons. (*Cross-hatched bars*) Regions corresponding to the 5'- and 3'-untranslated regions of the mRNA. The diagram does not show the endonucleolytic removal of glutamyl tRNA from the 5' end of the pre-mRNA (see text). (Adapted from Borst and Grivell 1981.)

Transcription of the long gene initially yields an 8-kb mRNA precursor. The region corresponding to the first intron is then excised as a 10S lariat molecule by one or more proteins imported from the cytoplasm; this creates an in-frame fusion between the mRNA sequences corresponding to the first two exons B1 and B2 and the ORF of intron bI2. Translation of this incompletely processed mRNA yields a 42-kD fusion protein whose amino-terminal part is identical to that of apocytochrome *b* and whose carboxy-terminal part is the product of the bI2 ORF. This chimeric protein now acts upon its own mRNA by excising the region corresponding to intron bI2, creating a 5-kb mRNA intermediate. This second splice now connects the ORF of bI3 in-frame with the fused exons B1–B3, resulting in the transient synthesis of a 50-kD chimeric maturase. Once again, this maturase acts upon its own mRNA by excising the region corresponding to intron bI3, creating an in-frame fusion between the fused exons B1–B4 and the ORF of intron bI4; translation of this mRNA intermediate yields another chimeric maturase that splices out the mRNA region corresponding to intron bI4. The final maturation steps of *COB* mRNA have not yet been worked out in detail; presumably, they are mediated by nucleus-encoded proteins because the last intron of the *COB* gene appears to lack an extended ORF.

This model has several key features. First, each intron-encoded maturase regulates its own synthesis by destroying its own mRNA. Second, the action of each maturase permits synthesis of the maturase for the next downstream intron. Third, inactivation of a maturase (e.g., by nonsense mutation in one of the upstream exon-derived sequences or by mutations in the intronic ORF) causes accumulation of incompletely processed *COB* mRNA and of unusual mitochondrially synthesized proteins antigenically related to apocytochrome *b*.

A similar splicing cascade appears to govern the maturation of *COX I* mRNA; most of the *COX I* introns have long ORFs in-frame with the preceding exons, and at least one of these intronic ORFs (that of aI4) has sequence homology with bI4 of the *COB* gene. However, the 56-kD protein encoded by *COX I* exons A1–A4 and the ORF of aI4 cannot function as a maturase for either *COX I* pre-mRNA or *COB* pre-mRNA; instead, the aI4-encoded regions are excised from *COX I* pre-mRNA by the chimeric maturase encoded by exons B1–B4 and the ORF of bI4. Thus, mutational inactivation of this maturase blocks not only synthesis of apocytochrome *b* (as discussed above), but also synthesis of cytochrome oxidase subunit I. This *trans*-effect of mutations in the *COB* gene on expression of the *COX I* gene (Cobon et al. 1976; Kotylak and Slonimki 1976) has been termed the box phenotype. A *trans*-acting box mutation in bI4 that produces a truncated maturase can be suppressed in at least three different ways (Fig. 10): (1) by a glutamic acid to lysine point mutation (termed *mim2-1*) in aI4 (Dujardin et al. 1982); (2) by mutation or overexpression of the nuclear gene (*NAM2*) coding for a leucyl tRNA synthetase (Labouesse at al. 1987; Benne 1988); and (3) by alteration of a piece of mtDNA encoding an active, carboxy-terminal 27-kD fragment of the bI4 maturase to allow expression in the cytosol and import of the corresponding maturase protein into mitochondria (Banroques et al. 1986, 1987).

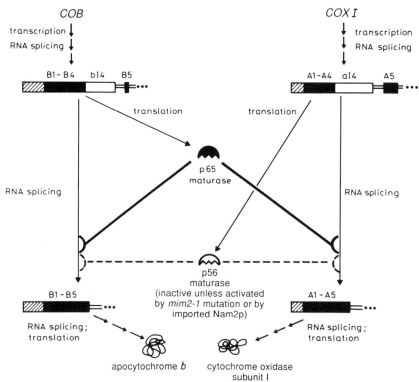

Figure 10 Interactions among *COB* (encoding apocytochrome *b*), *COX I* (encoding cytochrome oxidase subunit I), and *NAM2* (encoding a leucyl tRNA synthetase). (*Solid bars*) The first five exons of *COB* (B1–B5) and *COX I* (A1–A5), respectively; (*large open bars*) ORFs in the fourth intron of *COB* (bI4) and *COX I* (aI4), respectively; (*small open bars*) intron regions not coding for protein; (*crosshatched bars*) 5'-untranslated leader regions of *COB* and *COX I* pre-mRNAs; (p65 maturase) 65-kD protein encoded by partially processed *COB* pre-mRNA in which B1–B4 and the ORF of bI4 have been fused in-frame; the RNA maturase activity of this protein appears to reside entirely in its intron-encoded 27-kD carboxy-terminal domain, which can sometimes be detected as a distinct proteolytic fragment. (p56 maturase) 56-kD protein (silent maturase) encoded by partially processed *COX I* pre-mRNA in which A1–A4 and the ORF of aI4 have been fused in-frame; (Nam2p) product of the nuclear *NAM2* gene. (For further details, see text; adapted from a drawing kindly supplied by Dr. C. Jacq.)

These intriguing experiments suggest that the normally inactive maturase encoded by aI4 can be activated by the *mim2-1* mutation so that it can now excise not only aI4-encoded sequences from its own mRNA, but also bI4-encoded sequences from *COB* pre-mRNA. The experiments also show that a nucleus-encoded tRNA synthetase participates in mito-chondrial mRNA splicing, a fact also established for *Neurospora crassa* (Akins and Lambowitz 1987). Because mitochondrially encoded RNA maturases may act on more than one pre-mRNA or may be active only under special conditions, the total number of these maturases in yeast is still unknown.

Although correct splicing in vivo invariably appears to require proteins encoded by nuclear and/or mtDNA, at least some of the mitochondrial mRNAs discussed above can self-splice in vitro in the ab-sence of proteins (Tabak and Grivell 1986; van der Veen et al. 1986). It is not known how maturases perform their function. They may serve as scaffolds that facilitate correct folding of the precursor mRNAs and thereby suppress side reactions and/or allow the autocatalytic splice reac-tion to occur under physiological conditions (Grivell and Schweyen 1989). Such a role has been suggested for the protein encoded by the nuclear *CBP2* gene, which mediates the in vivo splicing of the "self-splicing" intron bI5 (Hill et al. 1985). Some of these proteins may also act as bona fide catalysts. Still others may act via indirect mechanisms that may prove difficult to sort out. For example, some nuclear mutations blocking splicing of mitochondrial pre-mRNAs also impair the formation of respiring mitochondria in *S. cerevisiae* strains devoid of mitochondrial introns (Séraphin et al. 1988, 1989).

In addition to splicing, maturation of mitochondrial RNAs (mtRNAs) also involves endonucleolytic processing of polycistronic transcripts. In yeast, at least three nucleus-encoded proteins mediate cleavage of mitochondrial transcripts to smaller fragments. The best-characterized example is once again the *COB* pre-mRNA. The short form of the *COB* gene (and perhaps also the long one) is cotranscribed with the 5′-located gene for glutamyl tRNA (see Fig. 8). Removal of the 5′-terminal region from this polycistronic transcript is mediated by the protein encoded by the nuclear gene *CBP1* (Dieckmann et al. 1982, 1984a,b). At least two other proteins participate in the 5′ and 3′ maturation of pre-tRNAs. 5′ maturation depends on an RNase-P-like activity that involves a mitochondrially encoded 9S RNA and a nucleus-encoded protein, or set of proteins (Martin et al. 1985; Miller et al. 1985; Hollingsworth and Martin 1986); 3′ maturation requires only nucleus-encoded proteins (Chen and Martin 1988; Martin et al. 1989). However, none of these tRNA-processing proteins have been purified, and none of their genes have been identified. It is surprising how little information exists on the proteins acting upon mitochondrial polycistronic transcripts even though these proteins are probably central for regulating the expression of mitochondrial genes in yeast.

D. Mutations of mtDNA

An extensive literature exists on the various mutations of yeast mtDNA. Because this topic has been treated comprehensively in the previous edition of this monograph (Dujon 1981), we summarize only the most important facts.

Mutations affecting only a single mitochondrial gene (termed mit^- mutations) generally produce the expected phenotype, particularly if the gene lacks introns: The corresponding protein is altered or no longer accumulated. This has served as the basis for the demonstration that a genetically defined locus on mtDNA (named *oxi1* by Slonimski and Tzagoloff [1976], but now better termed *COX II*) encodes subunit II of cytochrome oxidase (Cabral et al. 1978). Mutations altering an intron-encoded maturase (see above) may cause more complex effects. As some maturases act on more than one mRNA, mutations may alter more than one protein.

In syn^- mutants, the mutation inactivates an essential component of the mitochondrial protein synthesis machinery (such as a tRNA or the product of the *var1* gene, Var1p). The mutation is thus pleiotropic: The mutant lacks all proteins encoded by mtDNA. As the Var1p of the small ribosomal subunit is made on mitochondrial ribosomes (Douglas and Butow 1976; Groot et al. 1979; Terpstra and Butow 1979; Maheshwari and Marzuki 1985; Hibbs et al. 1987), all syn^- mutants lack functional mitochondrial ribosomes. As loss of mitochondrial protein synthesis results in loss of intact mtDNA (see above), the true phenotype of syn^- mutants can only be studied if they are leaky or conditional. Otherwise, most of them would be scored as ρ^- mutants.

In ρ^- mutants, mtDNA has extensive deletions, and the remaining sequences are amplified so that the normal size and amount of mtDNA are retained. Because ρ^- mutants cannot respire, they form smaller colonies than wild-type cells on glucose-poor media and are referred to as cytoplasmic petite mutants. As the genes that are essential for mitochondrial protein synthesis are spread all over the mtDNA (see Fig. 8), any sizable deletion inactivates mitochondrial protein synthesis. Although inactive ribosomal subunits may be retained if the genes for the corresponding rRNAs are retained (Maheshwari and Marzuki 1984), functional mitochondrial ribosomes are always absent from ρ^- mutants because Var1p is missing. Some tRNA genes may be retained; however, as the gene for the 9S RNA of mitochondrial RNase P is often lost, ρ^- mutants frequently accumulate unprocessed tRNA precursors (Martin and Underbrink-Lyon 1981; Martin et al. 1985). Moreover, as ρ^- mutants cannot make maturases encoded by mtDNA, they accumulate unprocessed transcripts of those mosaic genes that are still present (see above). Finally, because ρ⁻ mutants have lost extensive portions or all of their mtDNA, they cannot revert to ρ⁺. The properties of mitochondria isolated from ρ⁺ cells grown under different conditions, from ρ⁰ mutants (which have no mtDNA), and from ρ⁻ mutants are summarized in Table 1.

In most ρ⁻ mtDNAs, the retained sequences represent less than one third of wild-type mtDNA; they are generally amplified head to tail, although more complex rearrangements are observed as well (Morimoto et al. 1975; Lazowska and Slonimski 1976; Lewin et al. 1978; Heyting et al. 1979). One of the most extreme documented cases is a ρ⁻ mtDNA consisting of tandem repeats of a 66-bp sequence (Mol et al. 1974; Bos et al. 1978). If the sequences amplified in ρ⁻ mtDNA include one of the major origins of replication (*ori* sequences), the ρ⁻ mtDNA might be expected to replicate faster than ρ⁺ mtDNA. When such a ρ⁻ mutant is crossed to a ρ⁺ partner, the diploid progeny will be partly, largely, or completely ρ⁻: The corresponding ρ⁻ strain is suppressive. If the ρ⁻ mtDNA is not enriched in *ori* sequences or even deficient in them, it does not outgrow ρ⁺ mtDNA and all diploid progeny will be ρ⁺: The ρ⁻ mutant is neutral. Although this model is still tentative, it explains why ρ⁰ mutants are invariably neutral and offers a relatively straightforward means of rationalizing the large amount of data on these bizarre mutants.

E. Transformation of Mitochondria

Until recently, mitochondrial genes could be isolated, amplified, and modified, but they could not be put back into mitochondria. Despite intensive efforts (T. Fox, pers. comm.) and a reported success (Atchison et al. 1980), reproducible methods for transforming mitochondria were not available; this has now changed (for review, see Howe 1988). Mitochondrial genes can be stably reintroduced into mitochondria of living yeast cells by attaching them to metal microprojectiles that are shot into yeast cells by a particle gun. In the approach chosen by Johnston et al. (1988), a recipient *mit⁻* mutant carrying a deletion in the *COX I* gene (as well as a mutation in the nuclear *URA3* gene) was bombarded with pieces of DNA containing functional *COX I* and *URA3* genes. *URA3* transformants, which represented the subset of cells that remained viable subsequent to introduction of the foreign DNA, were screened for respiratory competence (Fig. 11). Fox and his colleagues (1988) introduced a functional *COX II* gene into a recipient ρ⁰ strain and screened transformants for their ability to yield respiring diploids upon mating to a *mit⁻* mutant carrying a deletion in the *COX II* gene. This second ap-

Table 1 Properties of mitochondria from ρ^0 and ρ^- mutants and from ρ^+ *S. cerevisiae* cells grown under different conditions

Type of mitochondria	Cristae	mtDNA	Mitochondrial ribosomes	Mitochondrial protein synthesis	Respiration	Cytochrome				Energy coupling system
						aa_3	c	c_1	b	
Wild-type, aerobic growth, glucose derepressed	+++	+++	+++	+++	+++	+++	+++	+++	+++	++
Wild-type, aerobic growth, glucose repressed	+	++	+	+	+	+	+	+	+	+
Wild-type, anaerobic growth in presence of unsaturated lipids	+	++	+	+	−	−	−	−	−	+
Wild-type, anaerobic growth in absence of unsaturated lipids	+/−	+[a]	+[a]	+[a]	−	−	−	−	−	−
ρ^- mutant, aerobic growth, glucose derepressed	+/−	++	+/−[b]	−	−	−	+++	+++	−	−
ρ^0 mutant, aerobic growth, glucose derepressed	+/−	−	−	−	−	−	+++	+++	−	−
ρ^- mutant, anaerobic growth in presence of unsaturated lipids	+/−	++	+/−[b]	−	−	−	−	−	−	−
Wild-type, aerobic growth in presence of chloramphenicol	+/−	+[a]	++[a]	−	−	−	+++	+++	−	−

For further details and references, see text.

[a] Unstable; lost upon prolonged growth.

[b] Absent or present, depending on whether the regions encoding the large and small mitochondrial rRNAs are retained in the ρ^- strain. Even if both ribosomal subunits are present, they are inactive in protein synthesis because one protein of the small subunit (Var1p) is encoded by mtDNA and made inside the mitochondria.

355

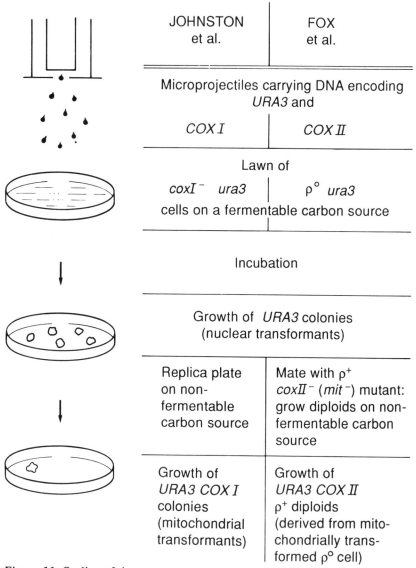

JOHNSTON et al.	FOX et al.
Microprojectiles carrying DNA encoding *URA3* and	
COX I	*COX II*
Lawn of	
coxI⁻ ura3	*ρ° ura3*
cells on a fermentable carbon source	

Incubation

Growth of *URA3* colonies
(nuclear transformants)

Replica plate on non-fermentable carbon source	Mate with ρ⁺ *coxII⁻* (*mit⁻*) mutant: grow diploids on non-fermentable carbon source
Growth of *URA3 COX I* colonies (mitochondrial transformants)	Growth of *URA3 COX II* ρ⁺ diploids (derived from mito-chondrially trans-formed ρ° cell)

Figure 11 Outline of the two approaches used for mitochondrial transformation of *S. cerevisiae* (for details, see text). (Adapted from Fox et al. [1988] and Johnston et al. [1988].)

proach is particularly versatile, as the synthetic ρ⁻ mutant created by the bombardment appeared to replicate even nonmitochondrial plasmid sequences inside its mitochondria. This should make it possible to introduce foreign DNA sequences into yeast mitochondria and check expression of these sequences upon mating the ρ^- cells to ρ^+ partners.

The mechanism of this transformation remains to be clarified; it seems to involve multiple crossovers between homologous regions and/or gene conversion, because Johnston et al. (1988) did not observe integration of nonmitochondrial vector sequences into their recipient mtDNA. Mitochondrial transformation has ushered in a new exciting phase in research on yeast mtDNA.

IV. NUCLEAR CONTROL OF MITOCHONDRIAL BIOGENESIS

Mitochondria of *S. cerevisiae*, like those of other organisms, are built up from hundreds of different macromolecules, only a small fraction of which are made inside the mitochondria; as discussed in Section II, yeast mtDNA contains, at most, 50 genes (~27 genes for RNAs and 8–23 genes for proteins). All of the other mitochondrial components, including the small molecules, are direct or indirect products of nuclear genes. As most of the mitochondrially encoded polypeptides are not complete enzymes, but only function in association with nucleus-encoded proteins, nuclear gene products participate in almost every intramitochondrial reaction (Grivell 1989). (An exception to this may be the transposase encoded by the intron of the large mitochondrial rRNA; Jacquier and Dujon 1985.) All of the nucleus-encoded proteins and many of the small molecules are made in the cytosol and subsequently imported into the mitochondria. Import is generally mediated by specific transport systems, most of which are still poorly characterized. The best-characterized systems are those for importing adenine nucleotides, P_i, and dicarboxylic acids (for review, see Aquila et al. 1987). Although there is no solid information to indicate that mitochondria export macromolecules, it is now clear that they import proteins, lipids, and possibly even several RNAs.

The nucleus-encoded proteins imported by mitochondria are not only building blocks and housekeeping enzymes, but also signals by which the nucleus controls mitochondrial formation (Fig. 12). The 10 years since the first edition of this monograph have witnessed impressive advances in deciphering some of these protein signals.

Many of the imported proteins were first identified through *S. cerevisiae* mutants carrying specific lesions in one of the corresponding nuclear genes. These mutants are usually defective in oxidative phosphorylation and can grow only by fermentation. They resemble cytoplasmic petite mutants in that they form only small colonies on glucose-poor media but differ in that they can usually revert to respiring cells and in that their phenotypes are inherited according to normal Men-

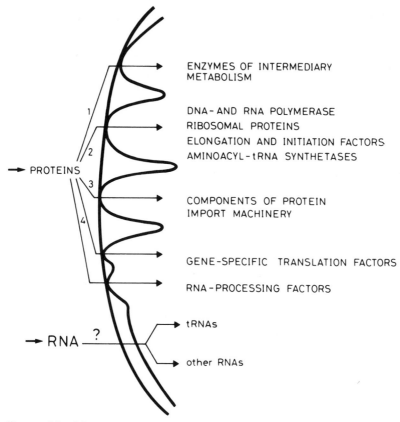

ENZYMES OF INTERMEDIARY
METABOLISM

DNA- AND RNA POLYMERASE
RIBOSOMAL PROTEINS
ELONGATION AND INITIATION FACTORS
AMINOACYL-tRNA SYNTHETASES

COMPONENTS OF PROTEIN
IMPORT MACHINERY

GENE-SPECIFIC TRANSLATION FACTORS

RNA-PROCESSING FACTORS

tRNAs

other RNAs

PROTEINS

RNA ?

Figure 12 Macromolecules imported into *S. cerevisiae* mitochondria (for details, see text).

delian laws. Initially, they were termed segregational petite mutants or chromosomal petite mutants (Chen et al. 1950; Ephrussi 1953); they are now usually referred to as nuclear petite or simply *pet* mutants. Thousands of them have been isolated, and approximately 300 distinct complementation groups have been identified (Michaelis et al. 1982). Molecular and biochemical analyses of these mutants have not only told us much about how mitochondria are made, but also revealed some unexpected features of mitochondrial metabolism. Each of the major classes of imported proteins is discussed in detail below.

A. Proteins of Mitochondrial Intermediary Metabolism

A large fraction of the imported mitochondrial proteins mediate respiration, the Krebs cycle, the biosynthesis of heme and amino acids, respiration-driven ATP synthesis, or transport of metabolites. Indeed, the *S. cerevisiae CYC1* gene, encoding iso-1-cytochrome *c*, was the first nuclear gene that was rigorously proved to encode a mitochondrial protein (Sherman et al. 1966). Inactivation of such nuclear genes usually causes the specific lesions expected from the known function of the affected protein. Thus, mutants lacking a functional gene for fumarase have cytochromes yet fail to grow on a nonfermentable carbon source (Wu and Tzagoloff 1987). Similarly, mutants lacking the nucleus-encoded subunit IV of cytochrome oxidase will lack only cytochrome oxidase activity and retain the other cytochromes as well as an energy-coupling system (Dowhan et al. 1985).

In some cases, however, mutations in the nuclear gene for a mitochondrial protein may fail to cause an obvious phenotype, or cause multiple phenotypic changes. The former case often reflects the presence of two or more functionally interchangeable isoforms of the affected mitochondrial protein. For example, yeast synthesizes an isoform of cytochrome *c*, termed iso-2-cytochrome *c*, encoded by the nuclear *CYP3* gene (later renamed *CYC7*; Verdiere and Petrochilo 1975; Downie et al. 1977; Petrochilo and Verdiere 1977; Sherman et al. 1978). Similar work has identified three nuclear genes encoding closely related adenine nucleotide translocators (Lawson and Douglas 1988; Kolarov et al. 1990), an extramitochondrial citrate synthase (Kim et al. 1986; Rosenkrantz et al. 1986), and a second pore-forming protein of the mitochondrial outer membrane (Dihanich et al. 1987, 1989). On the other hand, complex phenotypes resulting from a single nuclear mutation may reflect an interruption of regulatory circuits. For example, mutational inactivation of δ-aminolevulinic acid synthase (which catalyzes the first committed step of heme synthesis) causes not only the expected absence of heme, but also lowered transcription of other nuclear genes (Guarente and Mason 1983), defective assembly of cytochrome oxidase subunits (Saltzgaber-Muller and Schatz 1978), and incomplete proteolytic maturation of apocytochrome c_1 (Ohashi et al 1982). Although such complex phenotypes may initially baffle the investigator, they ultimately offer unique opportunities for insights into mitochondrial metabolism or the assembly of oligomeric mitochondrial enzymes.

B. Proteins Mediating Global Expression of mtDNA

Proteins of this group include mtDNA and mtRNA polymerases, all but one of the proteins of mitochondrial ribosomes, and the aminoacyl tRNA synthetases (for review, see Tzagoloff and Myers 1986). If the corresponding *pet* mutations are conditional or slightly leaky, strains with these mutations exhibit the expected phenotype, i.e., they have lowered levels of all mitochondrially synthesized proteins (Michaelis et al. 1982; Nobrega and Nobrega 1986; Hibbs et al. 1987). However, as discussed in Section III.D, strains bearing tight mutations of this class are normally difficult to analyze for at least three reasons: (1) Complete inhibition of mitochondrial protein synthesis induces loss of intact mtDNA; (2) inhibition of mitochondrial protein synthesis prevents synthesis of mitochondrially made mRNA maturases and thus indirectly interferes with mRNA splicing; and (3) some aminoacyl tRNA synthetases may participate directly in the splicing of mitochondrial mRNA.

C. Proteins Mediating Import of Precursor Proteins into Mitochondria

The known proteins of this class will be discussed in Section V. Some of them are essential for cell viability even in the absence of respiration; thus, mutations in the corresponding genes have not been isolated as *pet* mutations, but as temperature-sensitive lethal mutations or by reverse genetics using diploid cells. This underscores the fact that mitochondrial function is not restricted to oxidative phosphorylation but is essential for the life of yeast even under fermentative conditions.

D. Nucleus-encoded Proteins Controlling Expression of Individual Mitochondrial Genes

The expression of many mitochondrial genes is regulated by nucleus-encoded proteins that mediate processing of polycistronic mRNA, excision of intron sequences from pre-mRNA, or translation of individual mRNAs inside the mitochondria (Table 2). The only known nuclear gene controlling processing of polycistronic mitochondrial transcripts is *CBP1* (see Section III.C). Much more is known about nucleus-encoded proteins participating in the splicing of mitochondrial pre-mRNAs. Mutant genes for these proteins have been identified mainly by either of two approaches. First, strains with nuclear *pet* mutations causing defects in *COB* and/or *COX I* expression were checked for aberrant processing of *COB* or *COX I* mRNAs. Second, nuclear mutations were identified that suppressed a splicing defect of *mit⁻* mutants carrying restricted mutations in or near mitochondrial intron sequences.

To estimate the number of nuclear genes controlling splicing of mitochondrial pre-mRNAs, Séraphin et al. (1987) screened a collection of 300 nuclear *pet* mutations (comprising ~180 different complementation groups) for their effects in yeast cells having either intron-containing or intron-free mtDNA. This approach suggested that mitochondrial mRNA

Table 2 Nuclear genes controlling expression of specific mitochondrial genes in *S. cerevisiae*

Nuclear gene	Target (m)RNA[a]	Putative action	Reference[b]
CBP1	tRNA Glu-*COB*	5´-processing or mRNA stability	1, 2, 13
MSS51	*COX I* and *COB*	splicing	3
MSS116	*COX I* and *COB*	splicing	4
MSS8	*COX I*	splicing or stability	5
MSS18	*COX I* (aI5β)	splicing	5, 6
NAM2	*COX I* and *COB*	splicing	7, 8
CBP2	*COB*	splicing (aI5)	9, 10
MRS1	*COB*	splicing (aI3)	11
MRS3	*COB*	splicing? (aI1)	12
PET494	*COX III*	translation	14–18
PET122	*COX III*	translation	19–22
PET54 (wild-type allele of *ts2341*)	*COX III*	translation	23, 24, 20
	COX I	splicing	35
PET111	*COX II*	translation	14, 15, 25, 26
PET112	*COX II*	translation	27
SCO1	*COX I* and *COX II*	translation or assembly	28, 29
CBS1 (MK2)	*COB*	translation [c]	30–32
CBS2	*COB*	translation [c]	31
CBP3	*COB*	translation?	33
CBP6	*COB*	translation [d]	34

[a] (*COB*) Mitochondrial gene for apocytochrome *b*; (*COX I, COX II, COX III*) mitochondrial genes for cytochrome oxidase subunits I, II, and III, respectively; these genes were formerly termed OXI III, OXI I, and OXI II, respectively.

[b] (1) Dieckmann et al. 1982; (2) Dieckmann et al. 1984a,b; (3) Faye and Simon 1983; (4) Séraphin et al. 1989; (5) Faye and Simon 1983; (6) Séraphin et al. 1988; (7) Labouesse et al. 1987; (8) Herbert et al. 1988a,b; (9) McGraw and Tzagoloff 1983; (10) Hill et al. 1985; (11) Kreike et al. 1986; (12) Schmidt et al. 1987; (13) Dieckmann and Mittelmeier 1987; (14) Ebner et al. 1973a,b; (15) Cabral and Schatz 1978; (16) Müller et al. 1984; (17) Costanzo and Fox 1986; (18) Costanzo et al. 1986b; (19) McEwen et al. 1986; (20) Kloeckener-Gruissem et al. 1987; (21) Costanzo and Fox 1988; (22) Ohmen et al. 1988; (23) Costanzo et al. 1986a; (24) Costanzo et al. 1989; (25) Poutre and Fox 1987; (26) Strick and Fox 1987; (27) Seaver and Fox, cited in Costanzo et al. 1986a; (28) Schulze and Roedel 1988; (29) Schulze and Roedel 1989; (30) Roedel et al. 1985; (31) Roedel et al. 1986; (32) Roedel 1986; (33) Crivellone et al., cited in Tzagoloff and Myers 1986; (34) Dieckmann and Tzagoloff 1985; (35) Valencik et al. 1989.

[c] Also necessary for processing of *COB* and *COX I* pre-mRNAs.

[d] With the mutant allele characterized so far, translation of *COB* mRNA is probably only partly inhibited, as processing of *COB* pre-mRNA is not impaired.

splicing may require the action of approximately 18 nuclear genes, a number not too different from that listed in Table 2. However, this screen would have missed *pet* alleles that also cause loss of functional mtDNA. Furthermore, several nuclear mutations affecting splicing also impair respiration-driven growth of yeast cells lacking mitochondrial introns (Séraphin et al. 1989), suggesting that some of these 18 nuclear genes affect mitochondrial mRNA splicing indirectly. Nevertheless, the study by Séraphin et al. (1987) raises the possibility that a major fraction of the nuclear genes affecting mitochondrial splicing has already been found.

Nucleus-encoded proteins can also mediate the translation of specific mitochondrial mRNAs (for review, see Fox 1986; and references in Table 2). Mutations in the nuclear genes encoding these proteins generate *pet* mutants lacking specific proteins encoded by mtDNA. This was first demonstrated by Ebner et al. (1973a,b), who isolated *pet* mutants specifically lacking either subunit III (mutant *pet494*) or subunit II (mutant *petE11*, later renamed *pet111*) of cytochrome oxidase. The *pet494* mutation could be suppressed by a nucleus-encoded suppressor tRNA known to act on cytosolic protein synthesis (Ono et al. 1975); this strongly suggested that the wild-type *PET494* gene encodes a protein that is made in the cytosol, imported into mitochondria, and promotes expression of the mitochondrial *COX III* gene. Fox and his group have provided a detailed characterization of this control circuit (Fig. 13). They showed that *PET494* encodes a 56-kD protein with a typical amino-terminal matrix-targeting sequence (Costanzo et al. 1986b), that this protein is imported into mitochondria, and that it is required for translation of the *COX III* mRNA (Costanzo and Fox 1986). In the absence of functional Pet494p, *COX III* mRNA is made and is quite stable, but it is not translated. Translation can be restored by an mtDNA rearrangement in which the untranslated 5′region of the *COX III* gene is replaced by that of another mitochondrial gene, for example, *COB* or that encoding subunit IX of ATPase (Müller et al. 1984).

Pet494p thus must normally promote translation by interacting with the untranslated 5′-leader region of *COX III* mRNA. Translation of *COX III* mRNA requires not only Pet494p, but also proteins encoded by the nuclear *PET54* and *PET122* genes (Costanzo et al. 1986b; McEwen et al. 1986; Kloeckener-Gruissem et al. 1987, 1988; Costanzo and Fox 1988). All of these proteins appear to be imported into mitochondria in vivo. A closely analogous interaction was discovered for the untranslated 5′-leader region of *COB* mRNA and several nucleus-encoded proteins (Roedel et al. 1985, 1986; Roedel 1986; Roedel and Fox 1987; Michaelis et al. 1988). It is tempting to speculate that the different activator proteins function as an oligomeric complex, perhaps as part of the mito-

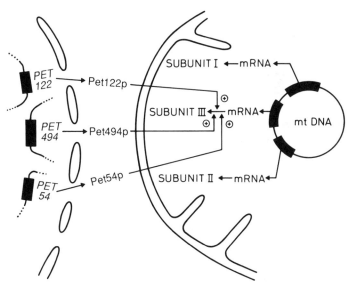

Figure 13   The translation of mitochondrial *COX III* mRNA in *S. cerevisiae* depends on proteins that are encoded by the nuclear genes *PET54*, *PET122*, and *PET494*, translated in the cytosol, and imported into mitochondria. The *PET54* product may, in addition, participate in the splicing of *COX I* pre-mRNA.

chondrial translation machinery. At least some of these proteins may also participate in the splicing of mitochondrial mRNAs: Valencik et al. (1989) discovered that Pet54p promotes not only translation of *COX III* mRNA, but also excision of one of the introns of *COX I* pre-mRNA. This raises the possibility that mitochondrial pre-mRNAs are spliced while bound to a component of the translation machinery, such as the ribosome.

As with mitochondrial splicing, many effects of nuclear genes on mitochondrial translation are difficult to rationalize at present. For example, overproduction of the wild-type Tif1p (which is encoded in the nucleus) can suppress missense mutations in the mitochondrial *COX III* gene (Linder and Slonimski 1989), even though Tif1p itself is apparently not imported into mitochondria. Tif1p appears to be eIF-4A, one of the initiation factors for protein synthesis in the yeast cytosol. As it suppresses mitochondrial missense mutations weakly, and only if over-expressed, it may act by some indirect mechanism. Nevertheless, this interaction may serve as a reminder that present knowledge on nuclear/mitochondrial interactions may deal only with the "tip of the iceberg."

E. Control of Nuclear Genes for Mitochondrial Proteins

Many, and probably most, nuclear genes controlling mitochondrial formation in yeast are themselves controlled by other nuclear genes. As this topic is covered elsewhere in these volumes, the present discussion only touches on a few central points.

Many nuclear genes encoding mitochondrial proteins have one or more upstream activator sites (UASs). These UASs lie 5′ to the TATA boxes of the genes and bind *trans*-acting proteins encoded by a variety of regulatory genes (Guarente 1984, 1987; Pinkham and Guarente 1985; Pinkham et al. 1987). UASs are functionally similar to enhancer sequences found in genes of higher eukaryotes in that their exact positions relative to the TATA boxes, and sometimes even their orientations, are not crucial (Guarente and Hoar 1984). The first detailed analysis of a UAS was published for *CYC1*, the nuclear gene encoding yeast iso-1-cytochrome *c* (Guarente and Mason 1983; Guarente et al. 1984; Lalonde et al. 1986). Two distinct UASs were identified that bind distinct, but overlapping, sets of regulatory proteins. The proteins binding to UAS1 control the induction of *CYC1* by oxygen and heme; they include RC2 and the proteins encoded by the *HAP1* and *RAF1* genes. The proteins binding to UAS2 control repression of the *CYC1* gene by glucose; they include proteins encoded by the *HAP2* and *HAP3* genes. As synthesis of heme in yeast requires molecular oxygen, induction by oxygen probably occurs via formation of heme. Hap1p (Guarente et al. 1984; Verdiere et al. 1986; earlier termed the CYP1 protein by Clavilier et al. 1969) appears to be necessary also for the heme-dependent transcriptional activation of many other genes encoding mitochondrial proteins, including iso-2-cytochrome *c*, Mn^{++} superoxide dismutase, cytochrome b_2, and cytochrome oxidase subunit Va (for review, see Verdier et al. 1988). It also regulates transcription of genes for certain nonmitochondrial proteins such as catalase (Spevak et al. 1986). Some genes, such as *ANB1* and the gene encoding cytochrome oxidase subunit Vb, contain UASs that cause heme-mediated *repression* of the downstream transcription units. These UASs interact with the *ROX1* gene product, whose expression is itself stimulated by heme (Zitomer et al. 1987; Cerdan and Zitomer 1988; Lowry and Zitomer 1988; Trueblood and Poyton 1988; Trueblood et al. 1988). Furthermore, some genes (such as *HEM1*, encoding the first enzyme of heme biosynthesis) contain at least two UASs, one of which mediates repression, and the other induction, by heme (Keng and Guarente 1987). The expression of such a gene may therefore appear to be unaffected by intracellular heme levels. As heme also controls import (Nicholson et al. 1987; Nargang et al. 1988), processing (Ohashi et al. 1982), and assembly (Saltzgaber-Muller and Schatz 1978) of some mitochondrial proteins, it occupies a central role in coordinating mitochondrial biogenesis. This role is complex and will occupy investigators for many years to come.

F. Import of RNAs into Yeast Mitochondria?

A substantial body of evidence suggests that mitochondria from mammals and some unicellular organisms import RNAs (for review, see Attardi and Schatz 1988). As yet, only a single observation indicates that this may also be true for yeast mitochondria. In their extensive studies on tRNAs associated with yeast mitochondria, Martin et al. (1977, 1979) detected a lysine tRNA (anticodon CUU) that was not encoded by mtDNA yet was inaccessible to externally added RNase in isolated intact mitochondria. This tRNA probably does not function in mitochondrial protein synthesis, because it is not charged by intramitochondrial aminoacyl tRNA synthetases. Conceivably, this tRNA could be involved in controlling intramitochondrial mRNA splicing reactions or other regulatory functions (see above). To identify its role (if any), it will probably be necessary to disrupt its nuclear gene and check for the effect of this disruption on mitochondrial functions.

How RNAs can move across mitochondrial membranes is unknown. Suyama (1986) has suggested that tRNAs may be imported in association with their cognate aminoacyl tRNA synthetases. Recent experiments have shown that nucleic acids can be imported into isolated yeast mitochondria if covalently attached to the carboxyl terminus of an imported precursor protein (Vestweber and Schatz 1989). These model studies show that, in principle, nucleic acids can be transported via the mitochondrial protein import machinery.

V. IMPORT OF PROTEINS INTO MITOCHONDRIA

Mitochondria are built up from two separate membranes that enclose two distinct spaces. Thus, a protein destined for import into mitochondria must carry not only a signal for routing it to mitochondria, but also one or more signals for sorting it to its correct intramitochondrial location. The mechanisms by which this is achieved are not yet fully understood, but many important features are now known. Although the present discussion will focus mainly on protein import into yeast mitochondria, most, and perhaps all, mitochondria appear to import proteins by the same basic mechanisms. However, because of their many experimental advantages, yeast mitochondria have been a favorite experimental system for studying mitochondrial protein import. A more extensive discussion of results obtained with other mitochondrial systems can be found in several recent reviews (Hay et al. 1984; Douglas et al. 1986; Attardi and Schatz 1988; Verner and Schatz 1988; Hartl et al. 1989).

A. General Features of Mitochondrial Protein Import

Every protein transported into mitochondria must be sorted into one of the four major intramitochondrial compartments: matrix, inner membrane, intermembrane space, and outer membrane. In addition, proteins transported into one of the two mitochondrial membranes must assume a specific orientation within that membrane. Of these sorting processes, import into the matrix space is characterized best. Most proteins imported into the matrix are initially synthesized as larger precursors carrying amino-terminal extensions (referred to as signal sequences, targeting sequences, or presequences). These extensions are usually 15–35 residues long but may be considerably longer. The precursor proteins bind to one or more receptor proteins on the mitochondrial surface and insert their presequences across both membranes, probably at sites where the two membranes are in close contact. This process requires an electric potential ($\Delta\psi$) across the inner membrane. It is followed by translocation of the mature portion of the precursor into the matrix space. Translocation usually requires cleavage of ATP and a loose conformation of the precursor protein. Finally, the presequence is removed by a soluble metalloendoprotease in the matrix.

This process has been studied mainly by two approaches. In one approach, yeast cells are transformed with a plasmid-borne gene for an imported mitochondrial protein, and the intracellular fate of this protein is then followed by pulse labeling or immune blotting coupled with subcellular fractionation (Hase et al. 1984; Hurt et al. 1985a,b; Keng et al. 1986). In the second approach, a radioactively labeled precursor protein (either synthesized in a cell-free system or purified from *E. coli* cells transformed with the corresponding gene) is presented to isolated mitochondria, and its import is assessed by monitoring the generation of mature protein inaccessible to externally added protease (Maccecchini et al. 1979; Eilers and Schatz 1986).

Both of these approaches have shown that many precursor proteins can be imported into mitochondria posttranslationally. This does not exclude the possibility that some proteins are imported cotranslationally in vivo or that a given precursor may be imported either co- or posttranslationally, depending on the physiological conditions. The possibility that import may occur cotranslationally was first raised by Butow and his colleagues, who found that isolated yeast mitochondria contain bound cytosolic ribosomes that appear to be anchored to the mitochondrial surface by nascent chains of proteins destined to be imported into mitochondria (Kellems and Butow 1972, 1974; Kellems et al. 1974, 1975; Ades and Butow 1980a,b). More recently, Pon et al. (1989) subfractionated yeast mitochondria containing bound cytoplasmic ribosomes and found cofractionation of these ribosomes with a submitochondrial membrane fraction containing the protein import sites. However, electron micrographs of yeast cells fixed during active growth have thus far failed to detect mitochondrion-associated cytosolic ribosomes (Stevens 1981). Most likely, the relative contribution of post- and cotranslational mechanisms to overall import is governed by the relative rates of protein synthesis and import: If protein synthesis is rapid and import is slow (as, e.g., in glucose-repressed cells), import would be predominantly posttranslational. Conversely, if protein synthesis is slow and import is rapid (as in derepressed cells in early stationary phase), import would be largely cotranslational (Reid and Schatz 1982a; Schatz and Butow 1983).

B. Matrix-targeting Signals

In all cases that have been studied so far, the signal for targeting a protein into the mitochondrial matrix is a contiguous, short stretch of amino acids at or near the amino terminus of the precursor. If this stretch of amino acids is fused to a passenger protein that normally resides in the cytosol (such as mouse dihydrofolate reductase [DHFR]), the resulting fusion protein is imported into the mitochondrial matrix (Hurt et al. 1984, 1985b). If the imported protein is synthesized as a larger precursor, the matrix-targeting signal is generally located within the cleavable presequence (Hurt et al. 1985b; Emr et al. 1986; Keng et al. 1986). In most cases, it is not known how much of a cleavable presequence is occupied by the actual targeting signal. In the few cases where this has been studied, the targeting sequence was shorter than the presequence and located at or near the amino terminus of the presequence. The targeting signal need not be proteolytically removed to act. For example, some precursors are only cleaved *after* they have been completely translocated (Reid et al. 1982b). Also, if cleavage is prevented by mutating the precursor near its normal cleavage site or by inactivating the matrix-localized processing protease in isolated mitochondria, translocation into the matrix is not prevented (Zwizinski and Neupert 1983; Hurt et al. 1985b; Vassarotti et al. 1987). The minimal length of a matrix-targeting signal is not known but appears to be inversely related to the tightness of folding of the attached passenger protein. If progressively shorter amino-terminal stretches of a matrix-targeting sequence are fused to normal mouse DHFR, the shortest sequence that still directs transport of the fusion protein into mitochondria contains between 10 and 12 residues (Hurt et al. 1985b). However, if the DHFR domain is labilized by point mutations, only seven residues of the same presequence are sufficient (Verner and Lemire 1989).

Figure 14 shows the matrix-targeting sequence of cytochrome oxidase subunit IV, a protein that is imported into the matrix and subsequently assembled within the mitochondrial inner membrane. Like most matrix-targeting sequences, this sequence is rich in positively charged, hydrophobic, and hydroxyl amino acids; contains no negatively charged amino acids; and lacks extended hydrophobic stretches. Matrix-targeting sequences are thus quite different from the hydrophobic presequences of proteins transported across the endoplasmic reticulum or the bacterial plasma membrane (for review, see Roise and Schatz 1988; Gierasch 1989). However, they are similar to the sequences that target proteins from the cytoplasm into the stroma of chloroplasts. For example, the amino-terminal two thirds of a stroma-targeting sequence can direct an attached protein into yeast mitochondria in vivo, although at low efficiency (Hurt et al. 1986). In a living plant cell, the two types of sequences are apparently sufficiently different to discriminate effectively between mitochondria and chloroplasts (Boutry et al. 1987).

Apart from the general features mentioned above, the matrix-targeting sequences of different precursors lack significant sequence similarities. However, they share the potential to form an amphiphilic structure (generally an α-helix). Thus, the chemically synthesized prepeptides of cytochrome oxidase subunit IV (Roise et al. 1986) and mammalian ornithine *trans*-carbamylase (Epand et al. 1986) are water soluble, yet interact strongly with phospholipid mono- and bilayers. In a membrane-like environment, these peptides fold into amphiphilic α-helices with

$$
\overset{\oplus}{\text{Met}}-\text{Leu}-\text{Ser}-\text{Leu}-\overset{\oplus}{\text{Arg}}-\text{Gln}-\text{Ser}-\text{Ile}-\overset{\oplus}{\text{Arg}}-\text{Phe}-\text{Phe}-\overset{\oplus}{\text{Lys}}-\text{Pro}-
$$

$$
-\text{Ala}-\overset{\oplus}{\text{Thr}}-\overset{\oplus}{\text{Arg}}-\text{Thr}-\text{Leu}-\text{Cys}-\text{Ser}-\text{Ser}-\overset{\oplus}{\text{Arg}}-\text{Tyr}-\text{Leu}-\text{Leu}-
$$

Figure 14 Cleavable presequence of subunit IV of yeast cytochrome oxidase (Maarse et al. 1984). It contains two cleavage sites for mitochondrial proteases: one after residue 17 and the other after residue 25 (arrows). The precursor protein is cleaved at the second site during import into mitochondria; purified matrix protease cleaves only at the first site (Hurt et al. 1985b). The significance of the two cleavages is not known. The matrix-targeting signal is located within the first 12 residues.

polar, positively charged, and apolar surfaces (Fig. 15). The membrane-perturbing properties of matrix-targeting sequences might provide a rationale for the fact that these sequences are usually proteolytically removed (and presumably degraded) upon import, even though this cleavage is not essential for translocation itself. They could also explain why yeast mutants temperature-sensitive for presequence cleavage in the matrix die at the nonpermissive temperature (Yaffe and Schatz 1984; Yaffe et al. 1985).

The hypothesis that matrix-targeting sequences are positively charged amphiphilic domains is further supported by the following observations. First, a computer-aided analysis of all known matrix-targeting sequences predicted that most of them can form amphiphilic helices (von Heijne 1986). Second, completely synthetic sequences composed of only arginine, leucine, glutamine, and serine can function as matrix-targeting sequences; their targeting efficiency is roughly proportional to their experimentally determined amphiphilicity (Allison and Schatz 1986; Roise et al. 1988). Third, an amphiphilic, positively charged helix that is normally buried inside a cytosolic protein functions as a matrix-targeting sequence if placed in front of a passenger protein (Hurt and Schatz 1987). Fourth, a large percentage (>10%) of random fragments derived from *E. coli* DNA encode functional matrix-targeting sequences (Baker and Schatz 1987). These sequences resemble authentic presequences in

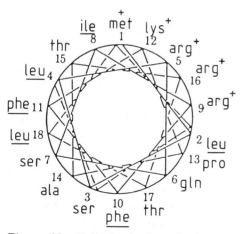

Figure 15 Helical wheel projection of the first 18 residues of the yeast cytochrome oxidase subunit IV precursor. Each residue is displaced from the next by 100°, giving 3.6 residues for each complete turn of the helix. Strongly hydrophobic residues are underlined. Note the clustering of positive charges in the upper right quadrant. (Reprinted, with permission, from Roise et al. 1986.)

overall amino acid composition and degree of hydrophobicity. Fifth, about 25% of the sequences in a large pool of random 27-bp oligodeoxyribonucleotides encode functional matrix-targeting sequences if placed between an ATG start codon and the gene for a passenger protein (Lemire et al. 1989). The in vivo targeting efficiencies of these sequences correlate well with predicted α-helical amphiphilicity. A similar result is obtained if an authentic matrix-targeting sequence is subjected to saturation mutagenesis and the resulting mutant sequences are analyzed (Bedwell et al. 1989). These nonbiased tests involving analysis of many separate peptide sequences provide compelling support for the view that the signal targeting a protein into the mitochondrial matrix is a positively charged amphiphilic domain (usually an α-helix) of degenerate sequence that is exposed on the surface of the precursor.

C. The Importance of Protein Conformation

Proteins appear to be unable to pass through biological membranes if tightly folded. This was first suggested by several observations indicating that bacterial or mitochondrial precursor proteins are folded more loosely than the corresponding mature proteins and that mitochondrial precursors trapped during import have an extended conformation at their amino termini (for review, see Zimmermann and Meyer 1986; Eilers and Schatz 1988). Proof that a tightly folded native conformation is translocation-incompetent came from the demonstration that import of precursors into isolated mitochondria is specifically blocked by ligands that stabilize the tight native conformation of the mature domain of a precursor (Eilers and Schatz 1986; Chen and Douglas 1987). Thus, import of a purified artificial precursor protein (containing a yeast matrix-targeting sequence fused to mouse DHFR) is completely blocked by the folate analog methotrexate. Conversely, import of the purified artificial precursor is accelerated by almost two orders of magnitude if the protein is presented to mitochondria in a denatured state (Eilers et al. 1988). Import of this precursor is also enhanced if its DHFR moiety is labilized by point mutations (Vestweber and Schatz 1988a).

Although it is clear that the translocation-competent conformation of a precursor is different from the tightly folded native conformation of the corresponding mature protein, the translocation-competent conformation is unknown. Two observations suggest that a protein need not be completely unfolded to enter mitochondria. First, mitochondria can import even artificially created, branched polypeptide chains (Vestweber and Schatz 1988b). Second, they can import double-stranded DNA if this DNA is covalently attached to the carboxyl terminus of a precursor protein (Vestweber and Schatz 1989). Because the diameter of double-stranded DNA is approximately 20 Å, the mitochondrial protein import channel can apparently accommodate structures more bulky than a fully extended polypeptide chain. This channel, however, fails to accommodate bovine pancreatic trypsin inhibitor, a tightly folded 6-kD protein with three internal disulfide bonds (Vestweber and Schatz 1988c); if this protein is attached to the carboxyl terminus of a mitochondrial precursor protein, the resulting chimeric protein starts to enter mitochondria but becomes stuck in the import machinery (see below).

No definitive information exists on how the translocation-competent structures of precursors are generated in vivo; tentative evidence suggests that several mechanisms may be involved. First, import may occur cotranslationally so that the growing polypeptide chain can only fold on the *trans*-side of the membrane barrier. Second, the nascent polypeptide may bind to a cytosolic antifolding component that prevents tight folding of the precursor; interaction of this complex with the mitochondrial surface would deliver the precursor to the import machinery in a loose conformation. Third, completely folded precursors may be unfolded in the cytosol or on the mitochondrial surface (Rothman and Kornberg 1986). Each of these possibilities is supported by some experimental evidence.

A cotranslational import mode is suggested by the observations that cytosolic ribosomes bound to the surface of isolated yeast mitochondria are enriched in mRNAs for imported mitochondrial proteins (Ades and Butow 1980a; Suissa and Schatz 1982) and that steady-state cytosolic pools of several mitochondrial precursors are low or undetectable, at least under certain conditions (Ades and Butow 1980b; Reid and Schatz 1982c).

The antifolding model is supported by the observation that mitochondrial precursors made in a cell-free protein-synthesizing system become slightly more protease-resistant upon depletion of ATP, suggesting that ATP prevents tight folding (Pfanner et al. 1987). Furthermore, protein translocation into mitochondria or the endoplasmic reticulum in yeast is stimulated by 70-kD heat-shock proteins (termed Hsp70) in the cytosol. The *SSA* subgroup of Hsp70 genes in *S. cerevisiae* contains four homologous members that perform overlapping functions (Werner-Washburne et al. 1987; Craig, these volumes). A triple mutant carrying disruptions in the genes *SSA1*, *SSA2*, and *SSA4* is inviable, but can be kept alive by a plasmid-borne intact *SSA1* gene that has been placed under the control of a *GAL1* UAS element. When expression of *SSA1* is turned off by growth on glucose, death of the cells is preceded by the accumulation of unprocessed precursors to proteins normally transported across the endoplasmic reticulum or the mitochondrial membranes. In these experiments, lack of processing probably reflects lack of translocation across the appropriate target membrane. No such accumulation is seen if death of the cells is caused by intracellular depletion of calmodulin (Deshaies et al. 1988). Stimulation by Hsp70 was also demonstrated for import of in-vitro-synthesized precursors into isolated yeast mitochondria (Murakami et al. 1988). However, neither of these studies shows that protein import into mitochondria is absolutely dependent on Hsp70; it is not even firmly established that mitochondrial precursors bind to these proteins or, if they do, whether this binding requires additional components. Nevertheless, these results are compatible with the suggestion that Hsp70 binds (and thereby rescues) partly unfolded proteins and releases them again in the presence of ATP (Pelham 1986). At least part of the ATP requirement for protein import into mitochondria (see Section V.D) might thus reflect ATP-dependent release of incompletely folded precursors from cytosolic Hsp70.

Although there is no evidence to suggest that Hsp70 can unfold proteins, there is nevertheless some support for the unfolding model. Thus, some precursor proteins can be unfolded by acidic phospholipids on the mitochondrial surface (Eilers et al. 1989; Endo et al. 1989). This was shown with a fusion protein consisting of a matrix-targeting sequence fused to mouse DHFR. Physicochemical and enzymatic measurements showed that the DHFR moiety of this artificial precursor protein was folded nearly as tightly as the authentic, presequence-free mouse enzyme. However, interaction of the fusion protein (but not of the presequence-free enzyme) with ATP-depleted mitochondria, with isolated mitochondrial outer membranes, or even with liposomes rich in acidic phospholipids induced a slow and temperature-dependent partial unfolding of the DHFR moiety. This partial unfolding was inhibited either by methotrexate, which stabilizes DHFR (see above), or by adriamycin, which binds to acidic phospholipids. Because both drugs also block import of the fusion protein into isolated mitochondria, this import appears to require a lipid-mediated conformational change. A possible involvement of acidic phospholipids in mitochondrial protein import was also suggested by Ou et al. (1988), who found that binding of in-vitro-synthesized mitochondrial precursor proteins to liposomes was greatly stimulated by the presence of cardiolipin in the liposomes. However, because the lipid-mediated conformational change observed by Endo et al. (1989) was ATP-independent, it does not help to explain why protein import into mitochondria requires ATP (see Section V.D).

In summary, we cannot yet define the translocation-competent structure of a protein or how this structure is established. However, progress in this area is now quite rapid; at least some of the catalysts involved in these conformational changes should soon be identified.

D. Energy Requirements

Transport of proteins into the mitochondrial matrix requires two forms of energy, ATP and an electrical potential across the mitochondrial inner membrane (for review, see Eilers and Schatz 1988). The electrical potential (positive toward the cytosol) is required for insertion of the positively charged presequence of the precursor across the outer and inner membranes (Schleyer and Neupert 1985). This insertion appears to occur at sites of close contact between the two membranes (Schwaiger et al. 1987; Pon et al. 1989), suggesting that the membrane junctions allow the precursor access to the energized inner membrane. The role of ATP is less well defined: ATP appears to be needed outside, as well as inside, the mitochondria. For example, the ATP-dependent insertion of porin into the outer membrane (Pfanner et al. 1988) requires ATP outside the mitochondria (Hwang and Schatz 1989). Moreover, low levels of external ATP may promote binding of some precursors to the mitochondrial surface (Pfanner et al. 1987), perhaps by inducing or maintaining a loose conformation of these precursors (see above). However, transport of proteins into the matrix space depends predominantly on ATP (or some other nucleoside triphosphate) inside the inner membrane (Hwang and Schatz 1989). This internal nucleoside triphosphate has at least two distinct effects: It stimulates translocation of precursors across the inner membrane and it induces the release of precursors from "chaperone" proteins, such as Hsp60, that promote refolding of the newly imported precursor in the matrix (see below). However, it is not clear why transport of incompletely folded precursor proteins into the matrix requires less, or even no, nucleoside triphosphate (Verner and Schatz 1987; Chen and Douglas 1988).

Chaperone proteins, whose role in organelle assembly was first recognized by Ellis (1987), bind proteins that are newly translocated into, or synthesized within, chloroplasts and mitochondria; they appear to prevent premature aggregation or misfolding of these proteins and to promote their proper assembly into oligomeric complexes. The first organellar "chaperonin" to be discovered was a chloroplast stroma protein that binds the chloroplast-encoded large subunit of ribulose-1,5-bisphosphate carboxylase. The sequence of this chaperonin was significantly similar to that of groEL, an abundant protein of the E. coli cytoplasm that mediates phage assembly and proper folding of endogenous E. coli proteins (Hemmingsen et al. 1988). A distinctive property of E. coli groEL is its assembly into a 20S particle composed of 14 identical subunits (Hendrix 1979; Hohn et al. 1979). A closely similar heat-inducible protein (mitochondrial Hsp60) was identified by McMullin and Hallberg (1987, 1988) in mitochondria from diverse organisms, including yeast. The yeast gene encoding this protein proved to be the same as a gene found to be necessary for growth and for the assembly of mitochondrial ornithine transcarbamylase subunits in a temperature-sensitive yeast mutant (Cheng et al. 1989). The matrix-located, groEL-like Hsp60 binds proteins imported into the matrix and releases them in a controlled manner, thereby ensuring their correct folding. This release requires at least one additional protein as well as cleavage of ATP (Ostermann et al. 1989; Hartl and Neupert 1990). In the *mif4* (*hsp60*) mutant isolated by Cheng et al. (1989), translocation of precursor proteins still operates at the non-permissive temperature, at least in short-term pulse-labeling experiments.

Hsp60 may not be the only mitochondrial heat-shock protein with a chaperone function. Recent experiments have shown that yeast mitochondria also contain at least one Hsp70 that is essential for growth. The protein is encoded by the nuclear *SSC1* gene (Craig et al. 1989); it is located in the matrix, binds specifically to ATP/Sepharose, and is similar in sequence to that of the dnaK protein, another heat-shock protein of *E. coli* (Bardwell and Craig 1984). As this mitochondrial Hsp70 (mHsp70) associates tightly with precursors as these enter the matrix, it appears to catalyze proper refolding of these precursors (Scherer et al. 1990).

The discovery of chaperone proteins has led to the startling realization that protein folding in organelle biogenesis is enzyme-catalyzed.

Figure 16 Key steps in the import of a protein into the internal mitochondrial compartments (for a detailed description, see text). The cleavable matrix-targeting signal is depicted as a helical domain with three positive charges. (Hsp70) Cytosolic 70-kD heat shock protein; (*dotted arrow*) some precursor proteins may not bind to cytosolic Hsp70 for their import; (Mas70p) one of several receptor proteins on the mitochondrial surface; (X) one or more additional receptor proteins (such as MOM19 of *N. crassa*); the association between receptor proteins shown here is hypothetical; (Y) hypothetical protein participating with ISP42 in creating a transport channel across the outer membrane; (Z and Z') hypothetical proteins creating a channel across the inner membrane; ($\Delta\psi$) electrochemical potential across the inner membrane (positive outside); (Hsp60) groEL-like 60-kD heat-shock protein in the matrix; upon release from Hsp60, the cleaved precursor is usually cleaved by the protease encoded by the nuclear *MAS1* and *MAS2* genes; it may then remain in the matrix (*a*), assemble with partner proteins (*shaded* and *crosshatched*) in the inner membrane (*b*), or perhaps move across the inner membrane into the intermembrane space (*c*). (P) Membrane-associated protease ("inner membrane protease I") removing sorting signals from some proteins imported to the intermembrane space. Note that the two possible cleavage steps (one in the matrix and one presumably on the outer face of the inner membrane) generate different amino termini (N' and N''). This figure does not include the recently discovered role of mHsp70 in the import process (see text) (Scherer et al. 1990).

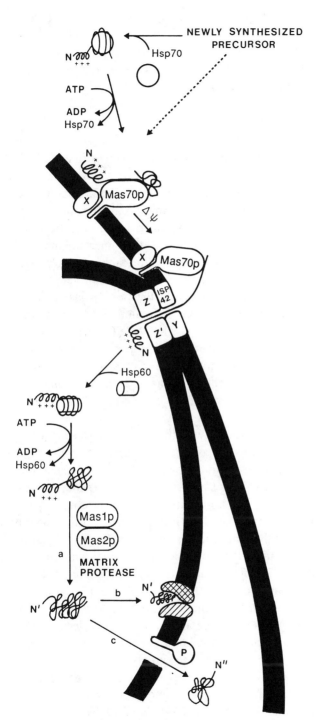

Figure 16 (See facing page for legend.)

This suggests, *a fortiori*, that higher-order interactions are also involved in more complex events, such as the translocation of a polypeptide chain across membranes. Figure 16 summarizes our current understanding of the distinct steps in the import of a precursor protein into yeast mitochondria. Some of the components shown in Figure 16 will be discussed more fully below.

E. Intramitochondrial Sorting of Proteins

Sorting of proteins to the mitochondrial outer membrane was first studied with the 29-kD major pore-forming protein (porin) of *Neurospora* and yeast mitochondria (Freitag et al. 1982; Mihara et al 1982; Gasser and Schatz 1983). In contrast to import of proteins into the matrix, import of porin into the outer membrane is not accompanied by proteolytic processing and does not require a potential across the inner membrane. However, it does require ATP (Kleene et al. 1987; Pfanner et al. 1987; Hwang and Schatz 1989). A targeting sequence directing a protein to the outer membrane was first identified in studying the in vivo insertion of Mas70p, a major 70-kD protein of the yeast mitochondrial outer membrane (Hase et al. 1983, 1984; Riezman et al. 1983a,b). This protein, which functions as a receptor for mitochondrial protein import (see below), is anchored to the outer membrane through its amino-terminal region, which contains a long stretch of uncharged and hydrophobic residues (Fig. 17). The remaining 60 kD of the protein protrudes into the cytosol.

Deletion and gene fusion studies showed that the amino-terminal sequence also targets the protein to the outer membrane in vivo. Some deletions within the hydrophobic region shown in Figure 17 cause the protein to enter the matrix space. This suggests that the outer-membrane-targeting signal consists of an amino-terminal matrix-targeting signal (which targets the proteins to mitochondria) followed by a sorting signal (which retains the protein in the outer membrane). If the sorting signal is inactivated, the matrix-targeting signal transports the protein completely into the matrix. This view is supported by the demonstration that the amino-terminal 12–21 residues of the 70-kD protein target an attached passenger protein across the inner membrane (Hurt et al. 1985a; Hase et al. 1986; Nakai et al. 1989). Although the amino-terminal dodecapeptide is not predicted to form an amphiphilic α-helix, its amino acid composition resembles that of typical matrix-targeting signals.

A distinct sorting signal downstream from an amino-terminal matrix-targeting signal also effects the sorting of some proteins to the intermembrane space. This was first indicated by the findings that yeast cytochrome c_1 (a protein anchored to the inner membrane, but exposed

Imported protein	Imported into	Functional domains in amino-terminal (pre) sequences
subunit IV of cytochrome *c* oxidase	matrix / inner membrane	matrix-targeting 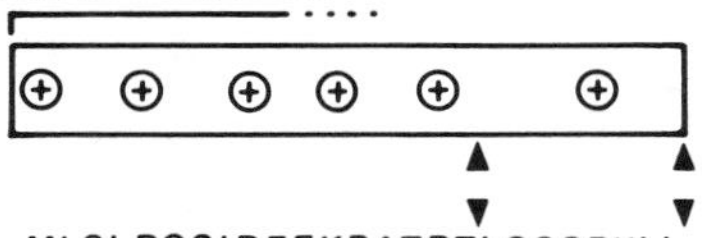 MLSLRQSIRFFKPATRTLCSSRYLL (1, 10, 20)
cytochrome c_1	inter-membrane space	matrix-targeting; sorting (inner membrane) 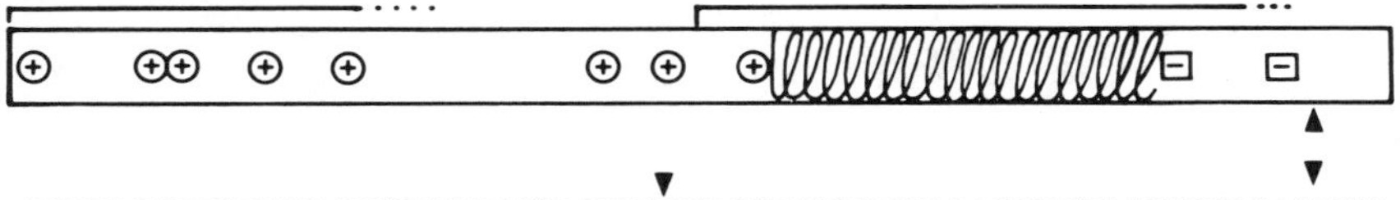 MFSNLSKRWAQRTLSKSFYSTATGAASKSGKLTQKLVTAGVAAAGITASTLLYADSLTAEMTA (1, 10, 20, 30, 40, 50, 60)
70 kDa outer membrane protein (Mas70p)	outer membrane	matrix-targeting; sorting (outer membrane) 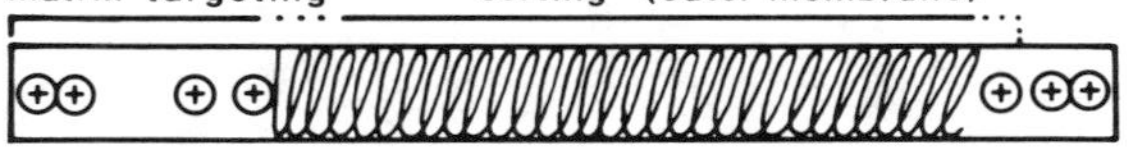 MKSFITRNKTAILATVAATGTAIGAYYYYNQLQQQQRGKK (1, 10, 20, 30, 40)

Figure 17 Sequences targeting proteins to distinct intramitochondrial locations (for details, see text). Amino acids are given in the single-letter code and are numbered from the amino terminus of each precursor. (+ and –) Positively and negatively charged amino acids, respectively. (*Arrowheads*) Cleavage sites; (*helical lines*) uncharged stretches of amino acids. (*Dotted lines*) Exact borders of the respective signal are not known. (Adapted from Hurt and van Loon [1986] and Schatz [1987].)

to the intermembrane space) as well as cytochrome c peroxidase and cytochrome b_2 (soluble proteins of the intermembrane space) carry unusually long presequences (Kaput et al. 1982; Sadler et al. 1984; Guiard 1985) that are removed by two successive cuts (Daum et al. 1982; Gasser et al. 1982; Ohashi et al. 1982; Reid et al. 1982). The first cut is catalyzed by the *MAS*-encoded protease in the matrix (see below), and the second one is catalyzed by other proteases. The second cleavage of the cytochrome b_2 precursor is mediated by "inner membrane protease I," which contains a 21.4-kD subunit with sequence similarity to *E. coli* leader peptidase (see Pratje and Guiard 1986; M. Behrens et al.; A. Schneider et al.; both in prep.). With cytochrome c_1, the second cut requires the presence of heme, whose covalent addition to the intermediate form of the apocytochrome may induce the conformation necessary for the second cleavage.

The presequence of cytochrome c_1 is composed of three distinct domains: a typical amino-terminal matrix-targeting signal comprising up to 32 residues, a stretch of 19 uncharged residues, and a brief domain containing 2 acidic residues (Fig. 17). This presequence contains all the necessary information for intracellular sorting because it directs attached foreign "passenger proteins" to the intermembrane space (Hurt and van Loon 1986; van Loon et al. 1986, 1987). Mutations within the two carboxy-terminal domains convert this presequence into a matrix-targeting sequence. van Loon and Schatz (1987) initially proposed that this sorting sequence functioned as a stop-transfer sequence similar to that of the outer membrane protein Mas70p, except that it was specific for the inner membrane. However, Hartl et al. (1986, 1987) suggested that the routing may be more complex. They proposed that precursors to cytochrome b_2 and cytochrome c_1 are initially transported completely into the matrix, where an amino-terminal part of the presequence is removed. The resulting intermediate forms are then re-exported across the inner membrane to the intermembrane space. According to this model, the sorting sequences of cytochrome c_1 and cytochrome b_2 are re-export sequences that are structurally and functionally similar to bacterial signal sequences. This controversy deserves further study (Kaput et al. 1989).

Apocytochrome c follows a completely different pathway to the intermembrane space (Figs. 18 and 19). Extensive studies by Neupert et al. in *Neurospora* have established that the apocytochrome c (which is amphiphilic; Rietveld and de Kruijff 1984; Rietveld et al. 1986) directly inserts partway across the lipid bilayer of the outer membrane in a process that does not require a potential across the inner membrane. The partly translocated protein is then covalently attached to heme by heme

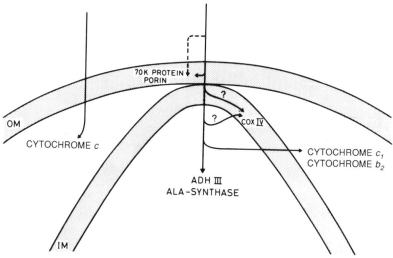

Figure 18 Intramitochondrial sorting of proteins. (ADH III) Mitochondrial isozyme of yeast alcohol dehydrogenase; (ALA) δ-aminolevulinic acid; (COX IV) subunit IV of cytochrome oxidase; (70K protein) major 70-kD protein (Mas70p) of the yeast mitochondrial outer membrane. (OM and IM) Outer and inner membrane, respectively. It is not yet clear whether outer-membrane proteins insert exclusively at membrane contact sites (*solid arrow*) or also into outer membrane areas that are not in contact with the inner membrane (*dashed arrow*). (Adapted from Hartl et al. 1989.)

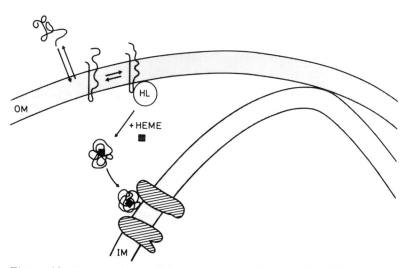

Figure 19 Import of apocytochrome *c* into mitochondria. (HL) Heme lyase; (OM and IM) outer and inner membranes, respectively. The holocytochrome *c* generated in the intermembrane space (for details, see text) binds electrostatically to cytochrome oxidase and the cytochrome bc_1 complex (*hatched areas*) in the inner membrane.

lyase in the intermembrane space. This probably causes the protein to move completely across the outer membrane (Nicholson et al. 1987, 1988; Korb and Neupert 1978). As expected from this model, yeast or *Neurospora* mutants lacking functional heme lyase fail to accumulate apocytochrome *c* in the mitochondrial intermembrane space (Dumont et al. 1988, 1990; Nargang et al. 1988).

It is still not clear whether proteins are targeted into the inner membrane by specific inner-membrane-targeting signals. Present evidence suggests that they are first targeted to either the matrix or the intermembrane space by one of the signals discussed above and then inserted into the inner membrane. Insertion may be mediated by as yet unknown insertion signals or by assembly into oligomeric complexes with partner proteins in the inner membrane.

In summary, intramitochondrial sorting of proteins represents a fascinating field in which almost every major question is still unanswered. What distinguishes sorting sequences for outer membrane and intermembrane space proteins? How do proteins lacking typical targeting sequences, such as heme lyase (Dumont et al. 1987) or the 17-kD subunit of the cytochrome bc_1 complex (van Loon et al. 1984), reach the intermembrane space? Contrary to some optimistic predictions (Kolata 1985), this topic may still vex investigators for many years to come.

F. Protein Import Machinery

In addition to the cytosolic heat-shock proteins described above, eight mitochondrial polypeptides have been identified as components of the yeast mitochondrial protein import system: four matrix proteins (the two subunits of processing protease, Hsp60 and mHsp70) and four proteins of the mitochondrial outer membrane (ISP42, Mas70p, a 32-kD protein, and Mps1p) (Table 3). The protease that removes the matrix-targeting sequences from precursors imported into the matrix space was initially identified in yeast by Böhni et al. (1980) and was later partially purified in two laboratories (McAda and Douglas 1982; Böhni et al. 1983; Cerletti et al. 1983). Similar enzymes were identified in mitochondria from other species (for review, see Pfanner and Neupert 1989) and purified from *N. crassa* (Hawlitschek et al. 1988) and rat liver (Ou et al. 1989). The yeast enzyme is soluble in the matrix, inhibited by metal chelators such as EDTA or 1,10-phenanthroline, and only acts upon mitochondrial precursor proteins. It seems to recognize some conformational feature of its substrates, because cleavage is not sequence-specific and is blocked by mutations of the precursors at regions distant from the actual cleavage site (Hurt et al. 1987). The pure enzymes from *N. crassa* (Hawlitschek et

Table 3 Proteins mediating mitochondrial protein import in yeast

Protein (Gene)	Location[a]	Function	Essentiality	Reference[b]
Hsp 70 (*SSA1, 2,* and *4*)	cytosol	maintenance of import-competent protein conformation	yes[c]	1
ISP42 (*ISP42*)	OM	component of import machinery in the OM	yes	2, 3
Mas70p (*MAS70*)	OM	receptor	no	4, 5
32-kD protein	OM	interacts with targeting signals; probable receptor	no	6
Cytochrome *c*-heme lyase (*CYC3*)	IMS	mediates import of apocytochrome *c* by attaching heme	no	7, 8
Mas1p (*MAS1, MIF1*)	matrix	subunit of matrix protease; removes matrix-targeting signals	yes	9, 10, 11
Mas2p (*MAS2, MIF2*)	matrix	subunit of matrix protease; removes matrix-targeting signals	yes	9, 10, 11
Hsp60 (*MIF4, HSP60*)	matrix	controls refolding of imported precursor proteins	yes	12, 13, 14
mHsp70 (*SSC1*)	matrix	controls refolding of imported precursor proteins?	yes	15, 16
Inner membrane protease I	IMS/IM	removes sorting sequences from cytochrome b_2 and cytochrome c_1	no	17, 18, 19
Mps1p (*MPS1*)	OM	mediates protein sorting between OM and IM	no	20

[a] (OM) Outer membrane; (IMS) intermembrane space; (IM) inner membrane.

[b] (1) Deshaies et al. 1988; (2) Vestweber et al. 1989; (3) Baker et al. 1990; (4) Hase et al. 1983; (5) Hines et al. 1990; (6) Pain et al. 1990; (7) Dumont et al. 1988; (8) Nicholson et al. 1988; (9) Böhni et al. 1983; (10) Witte et al. 1988; (11) Pollock et al. 1988; (12) McMullin and Hallberg 1987; (13) Cheng et al. 1989; (14) Reading et al. 1989; (15) Scherer et al. 1990; (16) Craig et al. 1989; (17) Pratje and Guiard 1986; (18) M. Behrens et al., in prep.; (19) A. Schneider et al., in prep.; (20) M. Nakai et al., in prep.

[c] Lethality is only observed if three cytosolic members of the Hsp70 family (those encoded by *SSA1, SSA2,* and *SSA4*) are deleted.

al. 1988) and yeast (Yang et al. 1988) consist of two nonidentical subunits. Interestingly, the two subunits exhibit significant sequence similarity over their entire lengths. Both subunits (particularly the smaller
one, encoded in yeast by *MAS1*, see below) are also similar but not
identical in sequence to the largest subunit of the cytochrome bc_1 complex (encoded in yeast by *COR1*). In *Neurospora*, the smaller protease
subunit and the subunit of the cytochrome bc_1 complex are actually
identical (Schulte et al. 1989), raising the question of how this polypeptide is correctly apportioned between these two oligomeric complexes.

   The nuclear genes (*MAS1* and *MAS2*) for the yeast protease subunits
were initially identified and cloned with the aid of temperature-sensitive
mutants that accumulate uncleaved mitochondrial precursors at the nonpermissive temperature (Yaffe and Schatz 1984; Yaffe et al. 1985;
Jensen and Yaffe 1988; Witte et al. 1988). These two genes were also
identified independently by Pollock et al. (1988) on the basis of a different screen. Overproduction of either subunit alone does not increase
the processing activity of isolated matrix fractions; however, overproduction of both subunits does (Geli et al. 1990; Yang et al. 1991). Heat inactivation of either the *MAS1* or *MAS2* product in intact cells bearing a
temperature-sensitive *mas* mutation blocks processing of all mitochondrial precursors tested, indicating that both subunits are essential for
activity and that this protease cleaves most, and perhaps all, cleavable
precursors imported into the matrix. Unexpectedly, inactivation of the
protease also blocks translocation of precursors across the inner membrane in vivo (Yaffe and Schatz 1984), suggesting some coupling between processing and import under these conditions. No such coupling is
seen with isolated mitochondria (see Section V.B). Both subunits are
made as larger precursors, whose processing is blocked by the *mas1*
mutation (Witte et al. 1988); the protease thus catalyzes its own processing.

   The role of mitochondrial surface proteins in the import of proteins
into the organelle was first demonstrated by the findings that treatment of
intact mitochondria with proteases (Riezman et al. 1983a; Zwizinski et
al. 1983, 1984; Ohba and Schatz 1987a) or antibodies against mitochondrial outer membrane proteins (Ohba and Schatz 1987b) inhibits
protein import. Recently, four of these proteins have been identified.

   A 42-kD outer membrane protein (termed *i*mport *s*ite *p*rotein 42, or
ISP42) was characterized as a component of the import machinery by
specifically tagging mitochondrial protein import sites with the chimeric
precursor protein shown in Figure 20. When this protein construct is
added to energized mitochondria, it becomes stuck across the two mitochondrial membranes with the DHFR moiety in the matrix, the trypsin

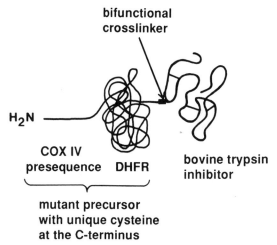

Figure 20 Chimeric precursor protein that "jams" the protein import sites of isolated yeast mitochondria (Vestweber and Schatz 1988c).

inhibitor moiety still outside the mitochondria, and the matrix-targeting signal cleaved off by the matrix protease (Vestweber and Schatz 1988c).

The protein appears to be stuck specifically in the import sites, as it completely blocks the import of authentic precursors. Quantitation of this inhibition indicates that each mitochondrial particle contains between 100 and 1000 import sites (Vestweber and Schatz 1988c). When this experiment was repeated with a modified version of the chimeric precursor that allows photo-cross-linking to adjacent mitochondrial proteins, the stuck protein was specifically cross-linked to ISP42 (Vestweber et al. 1989). As an antiserum monospecific for this protein blocks import of some precursors into isolated mitochondria, and as depletion of ISP42 from cells causes accumulation of uncleaved precursor proteins outside the mitochondria, ISP42 appears to be a component of the mitochondrial import machinery. ISP42 is an integral membrane protein even though the nucleotide sequence of its gene predicts a 41.9-kD hydrophilic, serine-rich protein (Baker et al. 1990). The protein is essential for cell viability, but its exact role remains to be clarified.

Pain et al. (1990) have recently identified three yeast mitochondrial proteins that react with an anti-idiotypic antibody against a chemically synthesized matrix-targeting peptide. One of these proteins, a 32-kD hydrophobic outer membrane protein, was enriched at sites where the two membranes are closely apposed (see below). An antibody specific for this protein inhibited import of some, but not all, precursors into isolated mitochondria, suggesting that this protein mediates import of only a subclass of precursors. Indeed, disruption of its nuclear gene was not lethal but rendered the cells respiration-deficient. The exact role of this protein is not known, but it may function as a receptor for some matrix-targeting signals.

Mas70p, a major 70-kD outer membrane protein, is yet another component of the yeast mitochondrial protein import machinery. Hase et al. (1983) and Riezman et al. (1983b) characterized this protein and its gene (see Section V.E). Mitochondria from a *mas70* null mutant import the precursors to the F_1-ATPase β-subunit or the adenine nucleotide translocator much more slowly than do wild-type mitochondria both in vitro and in vivo. Moreover, import of these precursors (but not that of several others) into wild-type mitochondria is strongly inhibited by anti-Mas70p serum (Hines et al. 1990). As Mas70p is linked to the outer membrane only by a short amino-terminal domain, it is most likely a surface receptor for a subclass of imported precursor proteins. Interestingly, Mas70p belongs to a protein family exhibiting several 34-residue repeats (Sikorski et al. 1990); the members of this family participate in mitosis, neurogenesis, and transcription control in yeast, fungi, and *Drosophila*.

Cross-linking studies with heterobifunctional membrane-permeant cleavable cross-linkers have shown that an intramitochondrial 70-kD heat-shock protein binds to a precursor protein as soon as the precursor's amino terminus penetrates into the matrix space (Scherer et al. 1990). This mHsp70 is similar in sequence to the *E. coli* dnaK protein and the cytosolic members of the yeast 70-kD heat-shock proteins. It is encoded by the essential nuclear gene *SSC1* (Craig et al. 1989). mHsp70 is thus the fifth component of the protein import machinery that is essential for viability of the yeast cell (see Table 3). At present, it is not clear whether mHsp70 functions before, separately, or in parallel with Hsp60. Most likely, both proteins mediate the refolding of newly imported precursors in the matrix space.

Figure 21 Contacts between the inner and outer membranes of *S. cerevisiae* mitochondria. The matrix space of isolated mitochondria was shrunk by incubation under hyperosmotic conditions; the mitochondria were then fixed with glutaraldehyde and paraformaldehyde, dehydrated, and embedded in Lowicryl. Thin sections were decorated with antibody against outer-membrane porin (*top*) or the inner-membrane marker holocytochrome oxidase (*bottom*), and the primary antibodies were then visualized with gold-labeled second antibody. (O) Outer membrane; (I) inner membrane; (CS) contact site. Bar, 0.1 μm. (Reprinted, with permission, from Pon et al. 1989.)

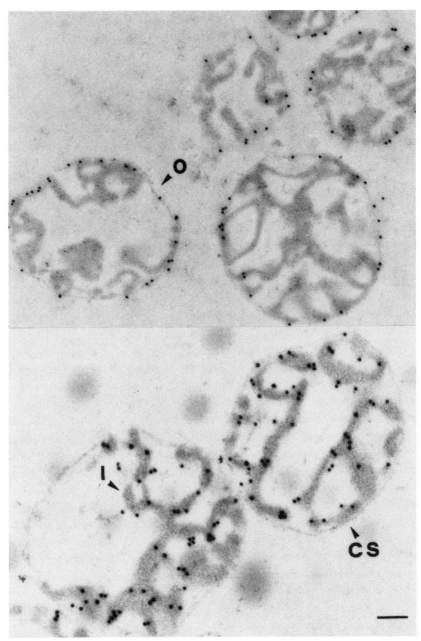

Figure 21 (See facing page for legend.)

M. Nakai et al. (in prep.) have identified a 41-kD outer membrane protein (termed Mps1p, for *m*itochondrial *p*rotein *s*orting) that ensures correct sorting of proteins among the two mitochondrial membranes. A mutation in the *MPS1* gene causes certain fusion proteins destined for the outer membrane to be misrouted to the inner membrane. The predicted sequence of the Mps1p reveals an ATP-binding motif and an amino-terminal domain resembling that of Mas70p. The role of Mps1p in mitochondrial protein import is unknown.

Proteins seem to enter mitochondria at sites of close contact between the two membranes (Kellems et al. 1975; Schleyer and Neupert 1985; Schwaiger et al. 1987; Vestweber and Schatz 1988c). The apparent attachment sites are revealed especially clearly when the inner membrane-matrix compartment has been shrunk by hyperosmotic conditions (Fig. 21).

To isolate these membrane contact sites, Pon et al. (1989) specifically labeled them in isolated yeast mitochondria with the chimeric precursor protein shown in Figure 20, which becomes irreversibly stuck as a transmembrane intermediate in the import sites. When the tagged mitochondria were subfractionated by ultrasonic disruption and density gradient centrifugation, the stuck precursor was specifically recovered in a vesicle population whose buoyant density was intermediate between those of outer and inner membranes. This fraction consisted largely of closed inner membrane vesicles attached to leaky outer membrane vesicles and was enriched in several mitochondrial proteins. Surprisingly, ISP42 was not enriched in these membrane vesicles. This raises the possibility that precursors initially interact with ISP42 or other outer membrane components (such as acidic phospholipids; see Section V.C) all over the mitochondrial surface and are only subsequently recruited by the import machinery located at the membrane contact sites.

It is not known whether mitochondrial membrane contact sites are fixed structures or form only transiently (e.g., when a mitochondrial precursor protein binds to an outer membrane receptor). Some support for the second possibility is provided by the finding that isolated inner membrane vesicles essentially free of outer membrane components can still import precursor proteins into their lumens (Hwang et al. 1989). This import activity cannot be explained by residual contact sites in the vesicle preparations, because it is also found if the vesicles are prepared from mitochondria whose import sites were inactivated by antibodies, or treatment with protease, or with the partly translocatable chimeric precursor protein (see above). Import directly across the inner membrane resembles that into intact mitochondria in requiring a functional mitochondrial targeting signal on the precursor protein, an energized inner membrane, and ATP. However, it differs in being insensitive to an antibody against ISP42. Most of the import sites of the inner membrane are probably cryptic in vivo, perhaps being recruited into functional contact sites only when these sites are formed.

VI. OUTLOOK

In the 100 years since Altman first described mitochondria, our understanding of the formation and functions of this organelle has grown enormously. We now have detailed information on the genes present in mtDNA, are beginning to understand how expression of these genes is controlled by the nucleus, and have made progress in describing the pathways by which nucleus-coded proteins enter the mitochondria. However, there are still many lacunae in our knowledge. We do not know how lipids, RNA, or charged coenzymes are imported into mitochondria, nor can we explain how mitochondria transmit signals to the nuclear genome. Insertion of mitochondrially synthesized proteins into the inner membrane is still a black box to us. We are ignorant about the mechanisms that cause mitochondria to divide or fuse with each other. There is also no direct information on how mitochondria move within the cell and how this movement is controlled.

Some advances, however, may be just around the corner. The recent successes in introducing DNA into mitochondria and in reintroducing isolated mitochondria into cells (King and Attardi 1988; Sulo et al. 1989) offer the hope that the power of recombinant DNA methods can now be exploited to study the expression of mitochondrial genes and the mechanisms governing mitochondrial mutations. The cloning from yeast of the first gene encoding a phospholipid exchange protein, the finding that this protein is essential for life (Aitken et al. 1990), and the surprising discovery (Bankaitis et al. 1990) that it is identical to the *SEC14* (Schekman and Novick 1982) gene product may help to answer the longstanding question of how lipids are inserted into the mitochondrial membranes. As some components of the mitochondrial protein import machinery are now known, identification of additional components should soon follow. In all of these developments, the yeast *S. cerevisiae* will undoubtedly continue to feature prominently as a powerful experimental system.

REFERENCES

Ades, I.Z. and R.A. Butow. 1980a. The transport of proteins into yeast mitochondria. Kinetics and pools. *J. Biol. Chem.* **255:** 9925.

————. 1980b. The products of mitochondria-bound cytoplasmic polysomes in yeast. *J. Biol. Chem.* **255**: 9918.

Agar, H.D. and H.C. Douglas. 1957. Studies on the cytological structure of yeast: Electron microscopy of thin sections. *J. Bacteriol.* **73**: 365.

Aitken, J.F., G.P.H. van Heusden, M. Temkin, and W. Dowhan. 1990. The gene encoding the phosphatidylinositol transfer protein is essential for cell growth. *J. Biol. Chem.* **265**: 4711.

Akins, R.A. and A.M. Lambowitz. 1987. A protein required for splicing group I introns in *Neurospora* mitochondria is mitochondrial tyrosyl-tRNA synthetase or a derivative thereof. *Cell* **50**: 331.

Allison, D.S. and G. Schatz. 1986. Artificial mitochondrial presequences. *Proc. Natl. Acad. Sci.* **83**: 9011.

Altmann, R. 1890. *Die Elementarorganismen und ihre Beziehungen zu den Zellen.* Veit, Leipzig.

Andreasen, A.A. and T.J.B. Stier. 1953. Anaerobic nutrition of *Saccharomyces cerevisiae. J. Cell. Comp. Physiol.* **41**: 23.

Aquila, H., T.A. Link, and M. Klingenberg. 1987. Solute carriers involved in energy transfer of mitochondria form a homologous protein family. *FEBS Lett.* **212**: 1.

Atchison, B.A., R.J. Devenish, A.W. Linnane, and P. Nagley. 1980. Transformation of *Saccharomyces cerevisiae* with yeast mitochondrial DNA linked to two micron circular yeast plasmid. *Biochem. Biophys. Res. Commun.* **96**: 580.

Attardi, G. and G. Schatz. 1988. The biogenesis of mitochondria. *Annu. Rev. Cell Biol.* **4**: 289.

Baker, A. and G. Schatz. 1987. Sequences from a prokaryotic genome or the mouse dihydrofolate reductase gene can restore the import of a truncated precursor protein into yeast mitochondria. *Proc. Natl. Acad. Sci.* **84**: 3117.

Baker, K.P., A. Schaniel, D. Vestweber, and G. Schatz. 1990. ISP42 a protein of the yeast mitochondrial outer membrane is essential for protein import and cell viability. *Nature* (in press).

Baldacci, G. and G. Bernardi. 1982. Replication origins are associated with transcription initiation sequences in the mitochondrial genome of yeast. *EMBO J.* **1**: 987.

Bankaitis, V.A., J.R. Aitken, A.E. Cleves, and W. Dowhan. 1990. An essential role for a phospholipid transfer protein in yeast Golgi function. *Nature* **347**: 561.

Banroques, J., A. Delahodde, and C. Jacq. 1986. A mitochondrial RNA maturase gene transferred to the yeast nucleus can control mitochondrial mRNA splicing. *Cell* **46**: 837.

Banroques, J., J. Perea, and C. Jacq. 1987. Efficient splicing of two yeast mitochondrial introns controlled by a nuclear-encoded maturase. *EMBO J.* **6**: 1085.

Bardwell, J.C.A. and E.A. Craig. 1984. Major heat-shock gene of *Drosophila* and the *Escherichia coli* heat-inducible *dnaK* gene are homologous. *Proc. Natl. Acad. Sci.* **81**: 848.

Bedwell, D.M., S.A. Strobel, K. Yun, G.D. Jongeward, and S.D. Emr. 1989. Sequence and structural requirements of a mitochondrial protein import signal defined by saturation cassette mutagenesis. *Mol. Cell. Biol.* **9**: 1014.

Benne, R. 1988. Aminoacyl-tRNA synthetases are involved in RNA splicing in fungal mitochondria. *Trends Genet.* **4**: 181.

Bereiter-Hahn, J. 1976. Dimethylaminostyrylmethylpyridium iodine (DASPMI) as a fluorescent probe for mitochondria in situ. *Biochim. Biophys. Acta* **423**: 1.

Bernardi, G., M. Faures, G. Piperno, and P.P. Slonimski. 1970. Mitochondrial DNA's from respiratory-sufficient and cytoplasmic respiratory-deficient mutant yeast. *J. Mol.*

Biol. **48:** 23.

Böhni, P.C., G. Daum, and G. Schatz. 1983. Import of proteins into mitochondria: Partial purification of a matrix-located protease involved in cleavage of mitochondrial precursor polypeptides. *J. Biol. Chem.* **258:** 4937.

Böhni, P.C., S. Gasser, C. Leaver, and G. Schatz. 1980. A matrix-localized mitochondrial protease processing cytoplasmically-made precursors to mitochondrial proteins. In *The organization and expression of the mitochondrial genome* (ed. A.M. Kroon and C. Saccone), p. 423. Elsevier/North Holland, Amsterdam.

Borst, P. and L.A. Grivell. 1981. One gene's intron is another gene's exon. *Nature* **289:** 439.

Bos, J.L., C.F. van Kreijl, F.H. Ploegaert, J.N. Mol, and P. Borst. 1978. A conserved and unique (AT)-rich segment in yeast mitochondrial DNA. *Nucleic Acids Res.* **5:** 4563.

Boutry, M., F. Hagy, C. Poulsen, K. Aoyagi, and N.-H. Chua. 1987. Targeting of bacterial chloramphenicol acetyltransferase to mitochondria in transgenic plants. *Nature* **328:** 340.

Cabral, F. and G. Schatz. 1978. Identification of cytochrome *c* oxidase subunits in nuclear yeast mutants lacking the functional enzyme. *J. Biol. Chem.* **253:** 4396.

Cabral, F., M. Solioz, Y. Rudin, and G. Schatz. 1978. Identification of the structural gene for yeast cytochrome *c* oxidase subunit II on mitochondrial DNA. *J. Biol. Chem.* **253:** 297.

Cerdan, M.E. and R.S. Zitomer. 1988. Oxygen-dependent up-stream activation sites of *Saccharomyces cerevisiae* cytochrome *c* genes are related forms of the same sequence. *Mol. Cell. Biol.* **8:** 2275.

Cerletti, N., P.C. Böhni, and K. Suda. 1983. Import of proteins into mitochondria: Isolated yeast mitochondria and a solubilized matrix protease correctly process cytochrome *c* oxidase subunit V precursor at the NH_2 terminus. *J. Biol. Chem.* **258:** 4944.

Chen, J.Y. and N.C. Martin. 1988. Biosynthesis of tRNA in yeast mitochondria. An endonuclease is responsible for the 3′-processing of tRNA precursors. *J. Biol. Chem.* **263:** 13677.

Chen, S.-Y., B. Ephrussi, and H. Hottinguer. 1950. Nature génétique des mutants à déficience respiratoire de la souche B-II de la levure de boulangerie. *Heredity* **4:** 337.

Chen, W.-J. and M.G. Douglas. 1987. The role of protein structure in the mitochondrial import pathway. Unfolding of mitochondrially bound precursors is required for membrane translocation. *J. Biol. Chem.* **262:** 15605.

―――. 1988. An F_1-ATPase beta-subunit precursor lacking an internal tetramer-forming domain is imported into mitochondria in the absence of ATP. *J. Biol. Chem.* **263:** 4979.

Cheng, M.Y., F.U. Hartl, J. Martin, R.A. Pollock, F. Kalousek, W. Neupert, E.M. Hallberg, R.L. Hallberg, and A.L. Horwich. 1989. Mitochondrial heat-shock protein hsp60 is essential for assembly of proteins into yeast mitochondria. *Nature* **337:** 620.

Christiansen, G. and C. Christiansen. 1976. Comparison of the fine structure of mitochondrial DNA from *Saccharomyces cerevisiae* and S. *carlsbergensis*: Electron microscopy of partially denatured molecules. *Nucleic Acids Res.* **3:** 465.

Christianson, T. and M. Rabinowitz. 1983. Identification of multiple transcriptional initiation sites on the yeast mitochondrial genome by *in vitro* capping with guanyltransferase. *J. Biol. Chem.* **258:** 14025.

Clark-Walker, G.D. and A.W. Linnane. 1967. The biogenesis of mitochondria in *Saccharomyces cerevisiae*. A comparison between cytoplasmic respiratory-deficient mutant yeast and chloramphenicol-inhibited wild type cells. *J. Cell Biol.* **34:** 1.

Clavilier, L., G. Pere, and P.P. Slonimski. 1969. Demonstration of several independent loci involved in the synthesis of iso-2-cytochrome *c* in yeast. *Mol. Gen. Genet.* **104:** 195.

Cobon, G.S., D.J. Groot-Obink, R.M. Hall, R. Maxwell, J. Murphy, J. Pytka, and A.W. Linnane. 1976. Mitochondrial genes determining cytochrome *b* (complex III) and cytochrome oxidase function. In *Genetics and biogenesis of chloroplasts and mitochondria* (ed. T. Bücher et al.), p. 453. Elsevier/North-Holland, Amsterdam.

Colleaux, L., L. d'Auriol, M. Betermier, G. Cottarel, A. Jaquier, F. Galibert, and B. Dujon. 1986. Universal code equivalent of a yeast mitochondrial intron reading frame is expressed in *E. coli* as a specific double strand endonuclease. *Cell* **44:** 521.

Costanzo, M.C. and T.D. Fox. 1986. Product of *S. cerevisiae* nuclear gene *PET494* activates translation of a specific mitochondrial mRNA. *Mol. Cell. Biol.* **6:** 3694.

————. 1988. Specific translational activation by nuclear gene products occurs in the 5'-untranslated leader of a yeast mitochondrial mRNA. *Proc. Natl. Acad. Sci.* **85:** 2677.

Costanzo, M.C., E.C. Seaver, and T.D. Fox. 1986a. At least two nuclear gene products are specifically required for translation of a single mitochondrial mRNA. *EMBO J.* **5:** 3637.

————. 1989. The *PET54* gene of *Saccharomyces cerevisiae*: Characterization of a nuclear gene encoding a mitochondrial translational activator and subcellular localization of its product. *Genetics* **122:** 297.

Costanzo, M.C., P.P. Mueller, C.A. Strick, and T.D. Fox. 1986b. Primary structure of wild-type and mutant alleles of the *PET494* gene of *Saccharomyces cerevisiae*. *Mol. Gen. Genet.* **202:** 294.

Craig, E.A., J. Kramer, J. Shilling, M. Werner-Washburne, S. Holmes, J. Kosic-Smithers, and C.M. Nicolet. 1989. *SSC1*, an essential member of the yeast HSP70 multigene family, encodes a mitochondrial protein. *Mol. Cell. Biol.* **9:** 3000.

Criddle, R.S. and G. Schatz. 1969. Promitochondria of anaerobically grown yeast. I. Isolation and biochemical properties. *Biochemistry* **8:** 322.

Damsky, C.H. 1976. Environmentally induced changes in mitochondria and endoplasmic reticulum of *Saccharomyces carlsbergensis* yeast. *J. Cell Biol.* **71:** 123.

Damsky, C.H., W.M. Nelson, and A. Claude. 1969. Mitochondria in anaerobically grown, lipid-limited brewer's yeast. *J. Cell Biol.* **43:** 174.

Daum, G., S. M. Gasser, and G. Schatz. 1982. Import of proteins into mitochondria. Energy-dependent, two-step processing of the intermembrane space enzyme cytochrome b_2 by isolated yeast mitochondria. *J. Biol. Chem.* **257:** 13075.

Deshaies, R.J., B.D. Koch, M. Werner-Washburne, E.A. Craig, and R. Schekman. 1988. A subfamily of stress proteins facilitates translocation of secretory and mitochondrial precursor polypeptides. *Nature* **332:** 800.

de Zamaroczy, M. and G. Bernardi. 1985. Sequence organization of the mitochondrial genome of yeast—A review. *Gene* **37:** 1.

de Zamaroczy, M., G. Faugeron Fonty, G. Baldacci, R. Goursot, and G. Bernardi. 1984. The ori sequences of the mitochondrial genome of a wild-type yeast strain: Number, location, orientation and structure. *Gene* **32:** 439.

Dieckmann, C.L. and T.M. Mittelmeier. 1987. Nuclearly-encoded *CBP1* interacts with the 5' end of mitochondrial cytochrome *b* pre-mRNA. *Curr. Genet.* **12:** 391.

Dieckmann, C. L. and A. Tzagoloff. 1985. Assembly of the mitochondrial membrane system. *CBP6*, a yeast nuclear gene necessary for synthesis of cytochrome *b*. *J. Biol. Chem.* **260:** 1513.

Dieckmann, C.L., G. Homison, and A. Tzagoloff. 1984a. Assembly of the mitochondrial membrane system. Nucleotide sequence of a yeast nuclear gene (*CBP1*) involved in

5′ end processing of cytochrome *b* pre-mRNA. *J. Biol. Chem.* **259:** 4732.

Dieckmann, C.L., T.J. Koerner, and A. Tzagoloff. 1984b. Assembly of the mitochondrial membrane system. *CBP1*, a yeast nuclear gene involved in 5′ end processing of cytochrome *b* pre-mRNA. *J. Biol. Chem.* **259:** 4722.

Dieckmann, C.L., L.K. Pape, and A. Tzagoloff. 1982. Identification and cloning of a yeast nuclear gene (*CBP1*) involved in expression of mitochondrial cytochrome *b*. *Proc. Natl. Acad. Sci.* **79:** 1805.

Dihanich, M., K. Suda, and G. Schatz. 1987. A yeast mutant lacking mitochondrial porin is respiratory-deficient, but can recover respiration with simultaneous accumulation of an 86 kD extramitochondrial protein. *EMBO J.* **6:** 723.

Dihanich, M., A. Schmid, W. Oppliger, and R. Benz. 1989. Identification of a new pore in the mitochondrial outer membrane of a porin-deficient yeast mutant. *Eur. J. Biochem.* **181:** 703.

Douglas, M.G. and R.A. Butow. 1976. Variant forms of mitochondrial translation products in yeast: Evidence for location of determinants on mitochondrial DNA. *Proc. Natl. Acad. Sci.* **73:** 1083.

Douglas, M.G., T.M. McCammon, and A. Vassarotti. 1986. Targeting proteins into mitochondria. *Microbiol. Rev.* **50:** 166.

Dowhan, W., C.R. Bibus, and G. Schatz. 1985. The cytoplasmically-made subunit IV is necessary for assembly of cytochrome *c* oxidase in yeast. *EMBO J.* **4:** 179.

Downie, J.A., J.W. Stewart, N. Brockman, A. Schweingruber, and F. Sherman. 1977. Structural gene for iso-2-cytochrome *c*. *J. Mol. Biol.* **113:** 369.

Dujardin, G., C. Jacq, and P.P. Slonimski. 1982. Single base substitution in an intron of oxidase gene compensates splicing defects of the cytochrome *b* gene. *Nature* **298:** 628.

Dujon, B. 1981. Mitochondrial genetics and functions. In *Molecular biology of the yeast Saccharomyces: Life cycle and inheritance* (ed J.N. Strathern et al.), p. 505. Cold Spring Harbor Laboratory, Cold Spring Harbor, New York.

Dumont, M.E., J.F. Ernst, and F. Sherman. 1988. Coupling of heme attachment to import of cytochrome *c* into yeast mitochondria. Studies with heme lyase-deficient mitochondria and altered apocytochromes *c*. *J. Biol. Chem.* **263:** 15928.

Dumont, M.E., J.F. Ernst, D.M. Hampsey, and F. Sherman. 1987. Identification and sequence of the gene encoding cytochrome *c* heme lyase in the yeast *Saccharomyces cerevisiae*. *EMBO J.* **6:** 235.

Dumont, M.D., A.J. Mathews, B.T. Nall, S.B. Baim, D.C. Eustice, and F. Sherman. 1990. Differential stability of two apo-isocytochromes *c* in the yeast *Saccharomyces cerevisiae*. *J. Biol. Chem.* **265:** 2733.

Ebner, E., T.L. Mason, and G. Schatz. 1973a. Mitochondrial assembly in respiration-deficient mutants of *Saccharomyces cerevisiae*. II. Effect of nuclear and extrachromosomal mutations on the formation of cytochrome *c* oxidase. *J. Biol. Chem.* **248:** 5369.

Ebner, E., L. Mennucci, and G. Schatz. 1973b. Mitochondrial assembly in respiration-deficient mutants of *Saccharomyces cerevisiae*. I. Effect of nuclear mutations on mitochondrial protein synthesis. *J. Biol. Chem.* **248:** 5360.

Eilers, M. and G. Schatz. 1986. Binding of a specific ligand inhibits import of a purified precursor protein into mitochondria. *Nature* **322:** 228.

————. 1988. Protein unfolding and the energetics of protein translocation across biological membranes. *Cell* **52:** 481.

Eilers, M., T. Endo, and G. Schatz. 1989. Adriamycin, a drug interacting with acidic phospholipids, blocks import of precursor proteins by isolated yeast mitochondria. *J. Biol. Chem.* **264:** 2945.

Eilers, M., S. Hwang, and G. Schatz. 1988. Unfolding and refolding of a purified precursor protein during import into isolated mitochondria. *EMBO J.* **7:** 1139.

Ellis, J. 1987. Proteins as molecular chaperones. *Nature* **328:** 378.

Emr, S.D., A. Vassarotti, J. Garrett, B.L. Geller, M. Takeda, and M.G. Douglas. 1986. The amino terminus of the yeast F_1-ATPase beta-subunit precursor functions as a mitochondrial import signal. *J. Cell Biol.* **102:** 523.

Endo, T., M. Eilers, and G. Schatz. 1989. Binding of a tightly folded artificial mitochondrial precursor protein to the mitochondrial outer membrane involves a lipid-mediated conformational change. *J. Biol. Chem.* **264:** 2951.

Epand, R.M., S.W. Hui, C. Argan, L.L. Gillespie, and G.C. Shore. 1986. Structural analysis and amphiphilic properties of a chemically synthesized mitochondrial signal peptide. *J. Biol. Chem.* **261:** 10017.

Ephrussi, B. 1953. *Nucleo-cytoplasmic relationships in microorganisms*. Clarendon Press, Oxford.

Ernster, L. and G. Schatz. 1981. Mitochondria: A historical review. *J. Cell Biol.* **91:** 227s.

Faye, G. and M. Simon. 1983. Analysis of a yeast nuclear gene involved in the maturation of mitochondrial pre-messenger RNA of the cytochrome oxidase subunit I. *Cell* **32:** 77.

Fonty, G., R. Goursot, D. Wilkie, and G. Bernardi. 1978. The mitochondrial genome of wild-type yeast cells. VII. Recombination in crosses. *J. Mol. Biol.* **119:** 213.

Fox, T.D. 1986. Nuclear gene products required for translation of specific mitochondrially coded mRNAs in yeast. *Trends Genet.* **2:** 97.

Fox, T.D., J.C. Sanford, and T.W. McMullin. 1988. Plasmids can stably transform yeast mitochondria lacking endogenous mtDNA. *Proc. Natl. Acad. Sci.* **85:** 7288.

Freitag, H., M. Janes, and W. Neupert. 1982. Biosynthesis of mitochondrial porin and insertion into the outer mitochondrial membrane of *Neurospora crassa*. *Eur. J. Biochem.* **126:** 197.

Gasser, S.M. and G. Schatz. 1983. Import of proteins into mitochondria. In vitro studies on the biogenesis of the outer membrane. *J. Biol. Chem.* **258:** 3427.

Gasser, S.M., A. Ohashi, G. Daum, P.C. Böhni, J. Gibson, G.A. Reid, T. Yonetani, and G. Schatz. 1982. Imported mitochondrial proteins cytochrome b_2 and cytochrome c_1 are processed in two steps. *Proc. Natl. Acad. Sci.* **79:** 267.

Geli, V., M. Yang, K. Suda, A. Lustig, and G. Schatz. 1990. The *MAS*-encoded processing protease of yeast mitochondria: Overproduction and characterization of the two non-identical subunits. *J. Biol. Chem.* **265:** 19216.

Genga, A., L. Bianchi, and F. Foury. 1986. A nuclear mutant of *Saccharomyces cerevisiae* deficient in mitochondrial DNA replication and polymerase activity. *J. Biol. Chem.* **261:** 9328.

Gierasch, L.M. 1989. Signal sequences. *Biochemistry* **28:** 923.

Gillham, N.W. 1978. *Organelle heredity*. Raven Press, New York.

Greenleaf, A.L., J.L. Kelly, and I.R. Lehman. 1986. Yeast *RPO41* gene product is required for transcription and maintenance of the mitochondrial genome. *Proc. Natl. Acad. Sci.* **83:** 3391.

Grimes, G.W., H.R. Mahler, and P.S. Perlman. 1974. Mitochondrial morphology. *Science* **185:** 630.

Grivell, L.A. 1989. Nucleo-mitochondrial interactions in yeast mitochondrial biogenesis. *Eur. J. Biochem.* **182:** 477.

Grivell, L.A. and R.J. Schweyen. 1989. RNA splicing in yeast mitochondria: Taking out the twists. *Trends Genet.* **5:** 39.

Groot, G.S., T.L. Mason, and N. van Harten-Loosbroek. 1979. Var1 is associated with the small ribosomal subunit of mitochondrial ribosomes in yeast. *Mol. Gen. Genet.* **174:** 339.

Groot, G.S., W. Rouslin, and G. Schatz. 1972. Promitochondria of anaerobically grown yeast. VI. Effect of oxygen on pro-mitochondrial protein synthesis. *J. Biol. Chem.* **247:** 1735.

Grossman, L.I., D.R. Cryer, E.S. Goldring, and J. Marmur. 1971. The petite mutation in yeast. 3. Nearest-neighbor analysis of mitochondrial DNA from normal and mutant cells. *J. Mol. Biol.* **62:** 565.

Guarente L. 1984. Yeast promoters: Positive and negative elements. *Cell* **36:** 799.

——. 1987. Regulatory proteins in yeast. *Annu. Rev. Genet.* **21:** 425.

Guarente, L. and E. Hoar. 1984. Upstream activation sites of the *CYC1* gene of *Saccharomyces cerevisiae* are active when inverted but not when placed downstream of the "TATA box". *Proc. Natl. Acad. Sci.* **81:** 7860.

Guarente, L. and T. Mason. 1983. Heme regulates transcription of the *CYC1* gene of *S. cerevisiae* via an upstream activation site. *Cell* **32:** 1279.

Guarente, L., B. Lalonde, P. Gifford, and E. Alani. 1984. Distinctly regulated tandem upstream activation sites mediate catabolite repression of the *CYC1* gene of *Saccharomyces cerevisiae*. *Cell* **36:** 503.

Guiard, B. 1985. Structure, expression and regulation of a nuclear gene encoding a mitochondrial protein: The yeast L(+)-lactate cytochrome *c* oxidoreductase (cytochrome b_2). *EMBO J.* **4:** 3265.

Hackenbrock, C.R. 1968. Chemical and physical fixation of isolated mitochondria in low-energy and high-energy states. *Proc. Natl. Acad. Sci.* **61:** 598.

Hagedorn, H. 1957. Der elektronenmikroskopische Nachweis der Mitochondrien in *Saccharomyces cerevisiae*. *Naturwissenschaften* **24:** 641.

Hartl, F.-U. and W. Neupert. 1990. Protein sorting to mitochondria: Evolutionary conservations of folding and assembly. *Science* **247:** 930.

Hartl, F.-U., J. Ostermann, B. Guiard, and W. Neupert. 1987. Successive translocation into and out of the mitochondrial matrix: Targeting of proteins to the intermembrane space by a bipartite signal peptide. *Cell* **51:** 1027.

Hartl, F.-U., N. Pfanner, D.W. Nicholson, and W. Neupert. 1989. Mitochondrial protein import. *Biochim. Biophys. Acta* **988:** 1.

Hartl, F.-U., B. Schmidt, E. Wachter, H. Weiss, and W. Neupert. 1986. Transport into mitochondria and intramitochondrial sorting of the Fe/S protein of ubiquinol-cytochrome *c* reductase. *Cell* **47:** 939.

Hase, T., M. Nakai, and H. Matsubara. 1986. The *N*-terminal 21 amino acids of a 70 kDa protein of the yeast mitochondrial outer membrane direct *E. coli* β-galactosidase into the mitochondrial matrix space in yeast cells. *FEBS Lett.* **197:** 199.

Hase, T., U. Mueller, H. Riezman, and G. Schatz. 1984. A 70 kD protein of the yeast mitochondrial outer membrane is targeted and anchored via its extreme amino terminus. *EMBO J.* **3:** 3157.

Hase, T., H. Riezman, K. Suda, and G. Schatz. 1983. Import of proteins into mitochondria: Nucleotide sequence of the gene for a 70 kD protein of the yeast mitochondrial outer membrane. *EMBO J.* **2:** 2169.

Hashimoto, T., S.F. Conti, and H.B. Naylor. 1959. Studies on the fine structure of microorganisms. IV. Observations on budding *Saccharomyces cerevisiae* by light and electron microscopy. *J. Bacteriol.* **77:** 344-354.

Hawlitschek, G., H. Schneider, B. Schmidt, M. Tropschug, F.-U. Hartl, and W. Neupert. 1988. Mitochondrial protein import: Identification of processing peptidase and of PEP, a processing enhancing protein. *Cell* **53:** 795.

Hay, R., P.C. Böhni, and S. Gasser. 1984. How mitochondria import proteins. *Biochim. Biophys. Acta* **779:** 65.

Hemmingsen, S.M., C. Woolford, S.M. van der Vies, K. Tilly, D.T. Dennis, C.P. Georgopoulos, R.W. Hendrix, and R.J. Ellis. 1988. Homologous plant and bacterial proteins chaperone oligomeric protein assembly. *Nature* **333:** 330.

Hendrix, R.W. 1979. Purification and properties of groE, a host protein involved in bacteriophage assembly. *J. Mol. Biol.* **129:** 375.

Herbert, C.J., B. Dujardin, M. Labouesse, and P.P. Slonimski. 1988a. Divergence of the mitochondrial leucyl tRNA synthetase genes in two closely related yeasts *Saccharomyces cerevisiae* and *Saccharomyces douglasii*: A paradigm of incipient evolution. *Mol. Gen. Genet.* **213:** 297.

Herbert, C.J., M. Labouesse, G. Dujardin, and P.P. Slonimski. 1988b. The *NAM2* proteins from *S. cerevisiae* and *S. douglasii* are mitochondrial leucyl-tRNA synthetases, and are involved in mRNA splicing. *EMBO J.* **7:** 473.

Heyting, C., J.L. Talen, P.J. Weijers, and P. Borst. 1979. Fine structure of the 21S ribosomal RNA region on yeast mitochondrial DNA. II. The organization of sequences in petite mitochondrial DNAs carrying genetic markers from the 21S region. *Mol. Gen. Genet.* **168:** 251.

Hibbs, A.R., K.K. Maheshwari, and S. Marzuki. 1987. Assembly of the mitochondrial ribosomes in a temperature-conditional mutant of *Saccharomyces cerevisiae* defective in the synthesis of the var1 protein. *Biochim. Biophys. Acta.* **908:** 179.

Hill, J., P. McGraw, and A. Tzagoloff. 1985. A mutation in yeast mitochondrial DNA results in a precise excision of the terminal intron of the cytochrome *b* gene. *J. Biol. Chem.* **260:** 3235.

Hines, V., A. Brandt, G. Griffiths, H. Horstmann, H. Brütsch, and G. Schatz. 1990. Protein import into yeast mitochondria is accelerated by the outer membrane protein MAS70. *EMBO J.* **9:** 3191.

Hirano, T. and C.C. Lindegren. 1961. Electron microscopy of mitochondria in *Saccharomyces. J. Ultrastruct. Res.* **5:** 321.

Hoffmann, J.P. and C.J. Avers. 1973. Mitochondrion of yeast: Ultrastructural evidence for one giant, branched organelle per cell. *Science* **181:** 749.

Hohn, T., B. Hohn, A. Engel, and M. Wurtz. 1979. Isolation and characterization of the host protein groE involved in bacteriophage lambda assembly. *J. Mol. Biol.* **129:** 375.

Hollenberg, C.P., P. Borst, and E.F. van Bruggen. 1970. Mitochondrial DNA. V. A 25 micron closed circular duplex DNA molecule in wild-type yeast mitochondria. Structure and genetic complexity. *Biochim. Biophys. Acta* **209:** 1.

Hollingsworth, M.J. and N.C. Martin. 1986. RNase P activity in the mitochondria of *Saccharomyces cerevisiae* depends on both mitochondrion and nucleus-encoded components. *Mol. Cell. Biol.* **6:** 1058.

Howe, C.J. 1988. Organelle transformation. *Trends Genet.* **4:** 150.

Hurt, E.C. and G. Schatz. 1987. A cytosolic protein contains a cryptic mitochondrial targeting signal. *Nature* **325:** 499.

Hurt, E.C. and A.P.G.M. van Loon. 1986. How proteins find mitochondria and intramitochondrial compartments. *Trends Biochem. Sci.* **11:** 204.

Hurt, E.C., U. Mueller, and G. Schatz. 1985a. The first twelve amino acids of a yeast mitochondrial outer membrane protein can direct a nuclear-coded cytochrome oxidase subunit to the mitochondrial inner membrane. *EMBO J.* **4:** 3509.

Hurt, E.C., B. Pesold-Hurt, and G. Schatz. 1984. The cleavable prepiece of an imported mitochondrial protein is sufficient to direct cytosolic dihydrofolate reductase into the mitochondrial matrix. *FEBS Lett.* **178:** 306.

Hurt, E.C., D.S. Allison, U. Mueller, and G. Schatz. 1987. Amino-terminal deletions in the presequence of an imported mitochondrial protein block the targeting function and proteolytic cleavage of the presequence at the carboxy terminus. *J. Biol. Chem.* **262:** 1420.

Hurt, E.C., B. Pesold-Hurt, K. Suda, W. Oppliger, and G. Schatz. 1985b. The first twelve amino acids (less than half of the pre-sequence) of an imported mitochondrial protein can direct mouse cytosolic dihydrofolate reductase into the yeast mitochondrial matrix. *EMBO J.* **4:** 2061.

Hurt, E.C., N. Soltanifar, M. Goldschmidt-Clermont, J.-D. Rochaix, and G. Schatz. 1986. The cleavable presequence of an imported chloroplast protein directs attached polypeptides into yeast mitochondria. *EMBO J.* **5:** 1343.

Hwang, S.T. and G. Schatz. 1989. Translocation of proteins across the mitochondrial inner membrane, but not into the outer membrane, requires nucleoside triphosphates in the matrix. *Proc. Natl. Acad. Sci.* **86:** 8432.

Hwang, S., T. Jascur, D. Vestweber, L. Pon, and G. Schatz. 1989. Disrupted yeast mitochondria can import precursor proteins directly through their inner membrane. *J. Cell Biol.* **109:** 487.

Jacquier, A. and B. Dujon. 1985. An intron-encoded protein is active in a gene conversion process that spreads an intron into a mitochondrial gene. *Cell* **41:** 383.

Jensen, R.E. and M.P. Yaffe. 1988. Import of proteins into yeast mitochondria: The nuclear *MAS2* gene encodes a component of the processing protease that is homologous to the *MAS1*-encoded subunit. *EMBO J.* **7:** 3863.

Johnson, L.V., M.L. Walsh, and L.B. Chen. 1980. Localization of mitochondria in living cells with rhodamine 123. *Proc. Natl. Acad. Sci.* **77:** 990.

Johnston, S.A., P.Q. Anziano, K. Shark, J.C. Sanford, and R.A. Butow. 1988. Mitochondrial transformation in yeast by bombardment with microprojectiles. *Science* **240:** 1538.

Kaput, J., M.C. Brandriss, and T. Prussak Wieckowska. 1989. *In vitro* import of cytochrome *c* peroxidase into the intermembrane space: Release of the processed form by intact mitochondria. *J. Cell. Biol.* **109:** 101.

Kaput, J., S. Goltz, and G. Blobel. 1982. Nucleotide sequence of the yeast nuclear gene for cytochrome *c* peroxidase precursor. Functional implications of the pre-sequence for protein transport into mitochondria. *J. Biol. Chem.* **257:** 15054.

Keilin, D. 1970. *The history of cell respiration and cytochrome.* Cambridge University Press, Cambridge, England.

Kellems, R.E. and R.A. Butow. 1972. Cytoplasmic 80S-type ribosomes associated with yeast mitochondria. I. Evidence for ribosome binding sites on yeast mitochondria. *J. Biol. Chem.* **247:** 8043.

———. 1974. Cytoplasmic type 80S ribosomes associated with yeast mitochondria. III. Changes in the amount of bound ribosomes in response to changes in metabolic state. *J. Biol. Chem.* **249:** 3304.

Kellems, R.E., V.F. Allison, and R.A. Butow. 1974. Cytoplasmic type 80S ribosomes associated with yeast mitochondria. II. Evidence for the association of cytoplasmic ribosomes with the outer mitochondrial membrane in situ. *J. Biol. Chem.* **249:** 3297.

———. 1975. Cytoplasmic type 80S ribosomes associated with yeast mitochondria. IV. Attachment of ribosomes to the outer membrane of isolated mitochondria. *J. Cell Biol.* **65:** 1.

Kelly, J.L. and I.R. Lehman. 1986. Yeast mitochondrial RNA polymerase. Purification and properties of the catalytic subunit. *J. Biol. Chem.* **261:** 10340.

Kelly, J.L., A.L. Greenleaf, and I.R. Lehman. 1986. Isolation of the nuclear gene encoding a subunit of the yeast mitochondrial RNA polymerase. *J. Biol. Chem.* **261:** 10348.

Keng, T. and L. Guarente. 1987. Constitutive expression of the yeast *HEM1* gene is actually a composite of activation and repression. *Proc. Natl. Acad. Sci.* **84:** 9113.

Keng, T., E. Alani, and L. Guarente. 1986. The nine amino-terminal residues of delta-aminolevulinate synthase direct beta-galactosidase into the mitochondrial matrix. *Mol. Cell. Biol.* **6:** 355.

Kim, K.S., M.S. Rosenkrantz, and L. Guarente. 1986. *Saccharomyces cerevisiae* contains two functional citrate synthase genes. *Mol. Cell. Biol.* **6:** 1936.

King, M.P. and G. Attardi 1988. Injection of mitochondria into human cells leads to a rapid replacement of the endogenous mitochondrial DNA. *Cell* **52:** 811.

Kleene, R., N. Pfanner, R. Pfaller, T.A. Link, W. Sebald, W. Neupert, and M. Tropschug. 1987. Mitochondrial porin of *Neurospora crassa*: cDNA cloning, in vitro expression and import into mitochondria. *EMBO J.* **6:** 2627.

Kloeckener-Gruissem, B., J.E. McEwen, and R.O. Poyton. 1987. Nuclear functions required for cytochrome *c* oxidase biogenesis in *S. cerevisiae*: Multiple trans-acting nuclear genes exert specific effects on expression of each of the cytochrome *c* oxidase subunits encoded on mitochondrial DNA. *Curr. Genet.* **12:** 311.

————. 1988. Identification of a third nuclear protein-coding gene required specifically for posttranscriptional expression of the mitochondrial *COX3* gene in *Saccharomyces cerevisiae. J. Bacteriol.* **170:** 1399.

Kolarov, J., N. Kolarova, and N. Nelson. 1990. A third ADP/ATP translocation gene in yeast. *J. Biol. Chem.* **265:** 12711.

Kolata G. 1985. How do proteins find mitochondria? *Science* **228:** 1517.

Korb, H. and W. Neupert. 1978. Biogenesis of cytochrome *c* in *Neurospora crassa*. Synthesis of apocytochrome *c*, transfer to mitochondria and conversion to holocytochrome *c. Eur. J. Biochem.* **91:** 609.

Kotylak, Z. and P.P. Slonimski. 1976. Joint control of cytochrome *a* and *b* by a unique mitochondrial DNA region comprising four genetic loci. In *The genetic function of mitochondrial DNA* (ed. C. Saccone and A.M. Kroon), p. 143. Elsevier, Amsterdam.

Kotylak, Z., J. Lazowska, D. Hawthorne, and P. Slonimski. 1985. Intron encoded proteins of mitochondria: Key elements of gene expression and genomic evolution. In *Achievements and perspectives of mitochondrial research* (ed. E. Quagliariello et al.), vol. 2, p. 1. Elsevier, Amsterdam.

Kreike, J., M. Schulze, T. Pillar, A. Korte, and G. Roedel. 1986. Cloning of a nuclear gene *MRS1* involved in the excision of a single group I intron (bI3) from the mitochondrial *COB* transcript in *S. cerevisiae. Curr. Genet.* **11:** 185.

Kûzela, S. and E. Grecná. 1969. Lack of amino acid incorporation by isolated mitochondria from respiratory-deficient cytoplasmic yeast mutants. *Experientia* **25:** 776.

Labouesse, M., C.J. Herbert, G. Dujardin, and P.P. Slonimski. 1987. Three suppressor mutations which cure a mitochondrial RNA maturase deficiency occur at the same codon in the open reading frame of the nuclear *NAM2* gene. *EMBO J.* **6:** 713.

Lalonde, B., B. Arcangioli, and L. Guarente. 1986. A single *Saccharomyces cerevisiae* upstream activation site (UAS1) has two distinct regions essential for its activity. *Mol. Cell. Biol.* **6:** 4690.

Lawson, J.E. and M.G. Douglas. 1988. Separate genes encode functionally equivalent ADP/ATP carrier proteins in *Saccharomyces cerevisiae*. Isolation and analysis of *AAC2. J. Biol. Chem.* **263:** 14812.

Lazowska, J. and P.P. Slonimski. 1976. Electron microscopy analysis of circular repetitive mitochondrial DNA molecules from genetically characterized *rho⁻*-mutants of *Sac-*

charomyces cerevisiae. Mol. Gen. Genet. **146**: 61.

Lazowska, J.C. Jacq, and P.P. Slonimski. 1980. Sequence of introns and flanking exons in wild-type and *box3* mutants of cytochrome *b* reveals an interlaced splicing protein coded by an intron. *Cell* **22**: 333.

Lehninger, A.L. 1964. *The mitochondrion.* Benjamin, New York.

Lemire, B.D., C. Fankhauser, A. Baker, and G. Schatz. 1989. The mitochondrial targeting function of randomly generated sequences correlates with predicted helical amphiphilicity. *J. Biol. Chem.* **264**: 20206.

Lewin, A., R. Morimoto, and M. Rabinowitz. 1978. Restriction enzyme analysis of mitochondrial DNAs of petite mutants of yeast: Classification of petites, and deletion mapping of mitochondrial genes. *Mol. Gen. Genet.* **163**: 257.

Linder, P. and P.P. Slonimski. 1989. An essential yeast protein, encoded by duplicated genes *TIF1* and *TIF2* and homologous to the mammalian translation initiation factor eIF-4A, can suppress a mitochondrial missense mutation. *Proc. Natl. Acad. Sci.* **86**: 2286.

Lloyd, D. 1974. *The mitochondria of microorganisms.* Academic Press, New York.

Lowry, C.V. and R. Zitomer. 1988. *ROX1* encodes a heme-induced repression factor regulating *ANB1* and *CYC7* of *Saccharomyces cerevisiae. Mol. Cell. Biol.* **8**: 4651.

Maarse, A.C., A.P.G.M. van Loon, H. Riezman, I. Gregor, G. Schatz, and L.A. Grivell. 1984. Subunit IV of yeast cytochrome *c* oxidase: Cloning and nucleotide sequencing of the gene and partial amino acid sequencing of the mature protein. *EMBO J.* **3**: 2831.

Maccecchini, M.L., Y. Rudin, G. Blobel, and G. Schatz. 1979. Import of proteins into mitochondria: Precursor forms of the extramitochondrially made F_1-ATPase subunits in yeast. *Proc. Natl. Acad. Sci.* **76**: 343.

Maheshwari, K.K. and S. Marzuki. 1984. The formation of a defective small subunit of the mitochondrial ribosomes in petite mutants of *Saccharomyces cerevisiae. Biochim. Biophys. Acta* **781**: 153.

―――. 1985. Defective assembly of the mitochondrial ribosomes in yeast cells grown in the presence of mitochondrial protein synthesis inhibitors. *Biochim. Biophys. Acta* **824**: 273.

Martin, N.C. and K. Underbrink-Lyon. 1981. A mitochondrial locus is necessary for the synthesis of mitochondrial tRNA in the yeast *Saccharomyces cerevisiae. Proc. Natl. Acad. Sci.* **78**: 4743.

Martin, N.C., D.L. Miller, K. Underbrink, and X. Ming. 1985. Structure of a precursor to the yeast mitochondrial tRNAMetf. Implications for the function of the tRNA synthesis locus. *J. Biol. Chem.* **260**: 1479.

Martin, R.P., J.M. Schneller, A.J. Stahl, and G. Dirheimer. 1977. Study of yeast mitochondrial tRNAs by two-dimensional polyacrylamide gel electrophoresis: Characterization of isoaccepting species and search for imported cytoplasmic tRNAs. *Nucleic Acids Res.* **4**: 3497.

―――. 1979. Import of nuclear deoxyribonucleic acid coded lysine-accepting transfer ribonucleic acid (anticodon C-U-U) into yeast mitochondria. *Biochemistry* **18**: 4600.

Martin, R.P., J.C. Amé, R. Bordonné, C. Desliens, M. Alt, E. Lemerle, and G. Dirheimer. 1989. In *Proceedings of the 13th International tRNA Workshop,* Vancouver, Canada.

Masters, B.S., L.L. Stohl, and D.A. Clayton. 1987. Yeast mitochondrial RNA polymerase is homologous to those encoded by bacteriophages T3 and T7. *Cell* **51**: 89.

Matile, P., H. Moor, and C.F. Robinow. 1969. Yeast cytology. In *The yeasts: Biology of yeasts* (ed. A.H. Rose and J.S. Harrison), p. 219. Academic Press, London.

McAda, P. and M.G. Douglas. 1982. A neutral metallo-endoprotease involved in the processing of an F_1-ATPase subunit precursor in mitochondria. *J. Biol. Chem.* **257**: 3177.

McEwen, J.E., C. Ko, B. Kloeckener Gruissem, and R.O. Poyton. 1986. Nuclear functions required for cytochrome *c* oxidase biogenesis in *Saccharomyces cerevisiae*. Characterization of mutants in 34 complementation groups. *J. Biol. Chem.* **261:** 11872.

McGraw, P. and A. Tzagoloff. 1983. Assembly of the mitochondrial membrane system. Characterization of a yeast nuclear gene involved in the processing of the cytochrome *b* pre-mRNA. *J. Biol. Chem.* **258:** 9459.

McMullin, T.W. and R.L. Hallberg. 1987. A normal mitochondrial protein is selectively synthesized and accumulated during heat shock in *Tetrahymena thermophila*. *Mol. Cell. Biol.* **7:** 4414.

————. 1988. A highly evolutionarily conserved mitochondrial protein is structurally related to the protein encoded by the *E. coli groEL* gene. *Mol. Cell. Biol.* **8:** 371.

Michaelis, G., G. Mannhaupt, E. Pratje, E. Fischer, J. Naggert, and E. Schweizer 1982. Mitochondrial translation products in nuclear respiration deficient *pet* mutants of *S. cerevisiae*. In *Mitochondrial genes* (ed. G. Attardi et al.), p. 311. Cold Spring Harbor Laboratory, Cold Spring Harbor, New York.

Michaelis, U., T. Schlapp, and G. Roedel. 1988. Yeast nuclear gene *CBS2*, required for translational activation of cytochrome *b*, encodes a basic protein of 45 kDa. *Mol. Gen. Genet.* **214:** 263.

Michel, F. and B.F. Lang. 1985. Mitochondrial class II introns encode proteins related to the reverse transcriptases of retroviruses. *Nature* **316:** 641.

Mihara, K., G. Blobel, and R. Sato. 1982. *In vitro* synthesis and integration into mitochondria of porin, a major protein of the outer mitochondrial membrane of *Saccharomyces cerevisiae*. *Proc. Natl. Acad. Sci.* **79:** 7102.

Miller, D.L., J.L. Krupp, H.H. Shu, and N.C. Martin. 1985. Polymorphism in the structure of the yeast mitochondrial tRNA synthesis locus. *Nucleic Acids Res.* **13:** 859.

Miyakawa, I., H. Aoi, N. Sando, and T. Kuroiwa. 1984. Fluorescence microscopic studies of mitochondrial nucleoids during meiosis and sporulation in the yeast, *Saccharomyces cerevisiae*. *J. Cell Sci.* **66:** 21.

Mol, J.N., P. Borst, F.G. Grosveld, and J.H. Spencer. 1974. The size of the repeating unit of the repetitive mitochondrial DNA from a "low-density" petite mutant of yeast. *Biochim. Biophys. Acta* **374:** 115.

Morimoto, R., A. Lewin, H.J. Hsu, M. Rabinowitz, and H. Fukuhara. 1975. Restriction endonuclease analysis of mitochondrial DNA from grande and genetically characterized cytoplasmic petite clones of *Saccharomyces cerevisiae*. *Proc. Natl. Acad. Sci.* **72:** 3868.

Morpurgo, G., G. Serlupi-Crescenzi, G. Tecce, F. Valente, and D. Venettacci. 1964. Influence of ergosterol on the physiology and the ultrastructure of *Saccharomyces cerevisiae*. *Nature* **201:** 897.

Müller, P., M.K. Reif, S. Zonghou, C. Sengstag, T.L. Mason, and T.D. Fox. 1984. A nuclear mutation that post-transcriptionally blocks accumulation of a yeast mitochondrial gene product can be suppressed by a mitochondrial gene rearrangement. *J. Mol. Biol.* **175:** 431.

Munn, E.A. 1974. *The structure of mitochondria*. Academic Press, London.

Murakami, H., D. Pain, and G. Blobel. 1988. 70-kD heat shock-related protein is one of at least two cytosolic factors stimulating protein import into mitochondria. *J. Cell Biol.* **107:** 2051.

Myers, A.M., L.K. Pape, and A. Tzagoloff. 1985. Mitochondrial protein synthesis is required for maintenance of intact mitochondrial genomes in *Saccharomyces cerevisiae*. *EMBO J.* **4:** 2087.

Nagley, P. and A.W. Linnane. 1972. Biogenesis of mitochondria. XXI. Studies on the nature of the mitochondrial genome in yeast: The degenerative effects of ethidium bromide on mitochondrial genetic information in a respiratory competent strain. *J. Mol. Biol.* **66**: 181.

Nakai, M., T. Hase, and H. Matsubara. 1989. Precise determination of the mitochondrial import signal contained in a 70kDa protein of yeast mitochondrial outer membrane. *J. Biochem.* **105**: 513.

Nargang, F. E., M. E. Drygas, P. L. Kong, D. W. Nicholson, and W. Neupert. 1988. A mutant of *Neurospora crassa* deficient in cytochrome *c* heme lyase activity cannot import cytochrome *c* into mitochondria. *J. Biol. Chem.* **263**: 9388.

Nicholson, D.W., C. Hergersberg, and W. Neupert. 1988. Role of cytochrome *c* heme lyase in the import of cytochrome *c* into mitochondria. *J. Biol. Chem.* **263**: 19034.

Nicholson, D.W., H. Kohler, and W. Neupert. 1987. Import of cytochrome *c* into mitochondria. Cytochrome *c* heme lyase. *Eur. J. Biochem.* **164**: 147.

Nobrega, M.P. and F.G. Nobrega. 1986. Mapping and sequencing of the wild-type and mutant (G 116-40) alleles of the tyrosyl-tRNA mitochondrial gene in *Saccharomyces cerevisiae. J. Biol. Chem.* **261**: 3054.

Ohashi, A., J. Gibson, I. Gregor, and G. Schatz. 1982. Import of proteins into mitochondria. The precursor of cytochrome c_1 is processed in two steps, one of them heme-dependent. *J. Biol. Chem.* **257**: 13042.

Ohba, M. and G. Schatz. 1987a. Disruption of the outer membrane restores protein import to trypsin-treated yeast mitochondria. *EMBO J.* **6**: 2117.

———. 1987b. Protein import into yeast mitochondria is inhibited by antibodies raised against 45 kD proteins of the outer membrane. *EMBO J.* **6**: 2109.

Ohmen, J.D., B. Kloeckener Gruissem, and J.E. McEwen. 1988. Molecular cloning and nucleotide sequence of the nuclear *PET122* gene required for expression of the mitochondrial *COX3* gene in S. *cerevisiae. Nucleic Acids Res.* **16**: 10783.

Ohnishi, T., K. Kawaguchi and B. Hagihara. 1966. Preparation and some properties of yeast mitochondria. *J. Biol. Chem.* **241**: 1797.

Ono, B.I., G. Fink, and G. Schatz. 1975. Mitochondrial assembly in respiration-deficient mutants of *Saccharomyces cerevisiae.* IV. Effects of nuclear amber suppressors on the accumulation of a mitochondrially made subunit of cytochrome *c* oxidase. *J. Biol. Chem.* **250**: 775.

Osinga, K. A., M. de Haan, T. Christianson, and H.F. Tabak. 1982. A nonanucleotide sequence involved in promotion of ribosomal RNA synthesis and RNA priming of DNA replication in yeast mitochondria. *Nucleic Acids Res.* **10**: 7993.

Ostermann, J., A.L. Horwich, W. Neupert, and F.-U. Hartl. 1989. Protein folding in mitochondria requires complex formation with hsp60 and ATP hydrolysis. *Nature* **341**: 125.

Ou, W.-J., A. Ito, H. Okazaki, and T. Omura. 1989. Purification and characterization of a processing protease from rat liver mitochondria. *EMBO J.* **8**: 2605.

Ou, W.-J., A. Ito, M. Umeda, K. Inoue, and T. Omura. 1988. Specific binding of mitochondrial protein precursors to liposomes containing cardiolipin. *J. Biochem.* **103**: 589.

Pain, D., H. Murakami, and G. Blobel. 1990. Identification of a receptor for protein import into mitochondria. *Nature* **347**: 444.

Pelham, H.R.B. 1986. Speculations on the functions of the major heat shock and glucose-regulated proteins. *Cell* **46**: 959.

Petrochilo, E. and J. Verdiere. 1977. A glycine to serine substitution identifies the *CYP3* locus as the structural gene of iso-2-cytochrome *c* in *Saccharomyces cerevisiae. Biochem. Biophys. Res. Commun.* **79**: 364.

Pfanner, N. and W. Neupert. 1989. Transport of proteins into mitochondria. *Curr. Opinion Cell Biol.* **1:** 624.

Pfanner, N., M. Tropschug, and W. Neupert. 1987. Mitochondrial protein import: Nucleoside triphosphates are involved in conferring import-competence to precursors. *Cell* **49:** 815.

Pfanner, N., R. Pfaller, R. Kleene, M. Ito, M. Tropschug, and W. Neupert. 1988. Role of ATP in mitochondrial protein import. Conformational alteration of a precursor protein can substitute for ATP requirement. *J. Biol. Chem.* **263:** 4049.

Pinkham, J.L. and L. Guarente. 1985. Cloning and molecular analysis of the *HAP2* locus: A global regulator of respiratory genes in *Saccharomyces cerevisiae. Mol. Cell. Biol.* **5:** 3410.

Pinkham, J.L., J.T. Olesen, and L.P. Guarente. 1987. Sequence and nuclear localization of the *Saccharomyces cerevisiae HAP2* protein, a transcriptional activator. *Mol. Cell. Biol.* **7:** 578.

Plattner, H. and G. Schatz. 1969. Promitochondria of anaerobically grown yeast. III. Morphology. *Biochemistry* **8:** 339.

Plattner, H., M.M. Salpeter, J. Saltzgaber, and G. Schatz. 1970. Promitochondria of anaerobically grown yeast. IV. Conversion into respiring mitochondria. *Proc. Natl. Acad. Sci.* **66:** 1252.

Polakis, E.S., W. Bartley, and G.A. Meek. 1964. Changes in the structure and enzyme activity of *Saccharomyces cerevisiae* in response to changes in the environment. *Biochem. J.* **90:** 369.

Pollock, R.A., F.-U. Hartl, M.Y. Cheng, J. Ostermann, A. Horwich, and W. Neupert. 1988. The processing peptidase of yeast mitochondria: The two cooperating components MPP and PEP are structurally related. *EMBO J.* **7:** 3493.

Pon, L., T. Moll, D. Vestweber, B. Marshallsay, and G. Schatz. 1989. Protein import into mitochondria: ATP-dependent protein translocation activity in a submitochondrial fraction enriched in membrane contact sites and specific proteins. *J. Cell Biol.* **109:** 2603.

Poutre, C.G. and T.D. Fox. 1987. *PET111* a *Saccharomyces cerevisiae* nuclear gene required for translation of the mitochondrial mRNA encoding cytochrome *c* oxidase subunit II. *Genetics* **115:** 637.

Pratje, E. and B. Guiard. 1986. One nuclear gene controls the removal of transient presequences from two yeast proteins: One encoded by the nuclear the other by the mitochondrial genome. *EMBO J.* **5:** 1313.

Pringle, J.R., R.A. Preston, A.E.M. Adams, T. Stearns, D.G. Drubin, B.K. Haarer, and E.W. Jones. 1989. Fluorescence microscopy methods for yeast. *Methods Cell Biol.* **31:** 357.

Rabinowitz, M. and H. Swift. 1970. Mitochondrial nucleic acids and their relation to the biogenesis of mitochondria. *Physiol. Rev.* **50:** 376.

Reading, D.S., R.L. Hallberg, and A.M. Myers. 1989. Characterization of the yeast *HSP60* gene coding for a mitochondrial assembly factor. *Nature* **337:** 665.

Reid, G.A. and G. Schatz. 1982a. Import of proteins into mitochondria. Yeast cells grown in the presence of carbonyl cyanide m-chlorophenylhydrazone accumulate massive amounts of some mitochondrial precursor polypeptides *J. Biol. Chem.* **257:** 13056.

———. 1982b. Import of proteins into mitochondria. Extramitochondrial pools and post-translational import of mitochondrial protein precursors in vivo. *J. Biol. Chem.* **257:** 13062.

Reid, G.A., T. Yonetani, and G. Schatz. 1982. Import of proteins into mitochondria. Import and maturation of the mitochondrial intermembrane space enzymes cytochrome b_2 and cytochrome *c* peroxidase in intact yeast cells. *J. Biol. Chem.* **257:** 13068.

Rietveld, A. and B. de Kruijff. 1984. Is the mitochondrial precursor protein apocytochrome c able to pass a lipid barrier? *J. Biol. Chem.* **259:** 6704.

Rietveld, A., T.A. Berkhout, A. Roenhorst, D. Marsch, and B. de Kruijff. 1986. Preferential association of apocytochrome c with negatively charged phospholipids in mixed model membranes. *Biochim. Biophys. Acta* **858:** 38.

Riezman, H., R. Hay, C. Witte, N. Nelson, and G. Schatz. 1983a. Yeast mitochondrial outer membrane specifically binds cytoplasmically-synthesized precursors of mitochondrial proteins. *EMBO J.* **2:** 1113.

Riezman, H., T. Hase, A.P.G.M. van Loon, L.A. Grivell, K. Suda, and G. Schatz. 1983b. Import of proteins into mitochondria: A 70 kilodalton outer membrane protein with a large carboxy-terminal deletion is still transported to the outer membrane. *EMBO J.* **2:** 2161.

Roedel, G. 1986. Two yeast nuclear genes, *CBS1* and *CBS2*, are required for translation of mitochondrial transcripts bearing the 5′-untranslated *COB* leader. *Curr. Genet.* **11:** 41.

Roedel, G. and T.D. Fox. 1987. The yeast nuclear gene *CBS1* is required for translation of mitochondrial mRNAs bearing the *COB* 5′-untranslated leader. *Mol. Gen. Genet.* **206:** 45.

Roedel, G., A. Koerte, and F. Kaudewitz. 1985. Mitochondrial suppression of a yeast nuclear mutation which affects the translation of the mitochondrial apocytochrome b transcript. *Curr. Genet.* **9:** 641.

Roedel, G., U. Michaelis, V. Forsbach, J. Kreike, and F. Kaudewitz. 1986. Molecular cloning of the yeast nuclear genes *CBS1* and *CBS2*. *Curr. Genet.* **11:** 47.

Roise, D. and G. Schatz. 1988. Mitochondrial presequences. *J. Biol. Chem.* **263:** 4509.

Roise, D., S.J. Horvath, J.M. Tomich, J.H. Richards, and G. Schatz. 1986. A chemically synthesized pre-sequence of an imported mitochondrial protein can form an amphiphilic helix and perturb natural and artificial phospholipid bilayers. *EMBO J.* **5:** 1327.

Roise, D., F. Theiler, S.J. Horvath, J.M. Tomich, J.H. Richards, D.S. Allison, and G. Schatz. 1988. Amphiphilicity is essential for mitochondrial presequence function. *EMBO J.* **7:** 649.

Roodyn, D.B. and D. Wilkie. 1968. *The biogenesis of mitochondria.* Methuen, London.

Rosenkrantz, M., T. Alam, K.S. Kim, B.J. Clark, P.A. Srere, and L. Guarente. 1986. Mitochondrial and nonmitochondrial citrate synthases in *Saccharomyces cerevisiae* are encoded by distinct homologous genes. *Mol. Cell. Biol.* **6:** 4509.

Rothman, J.E. and R.D. Kornberg. 1986. An unfolding story of protein translocation. *Nature* **322:** 209.

Sadler, I., K. Suda, G. Schatz, F. Kaudewitz, and A. Haid. 1984. Sequencing of the nuclear gene for the yeast cytochrome c_1 precursor reveals an unusually complex amino-terminal presequence. *EMBO J.* **3:** 2137.

Sager, R. 1972. *Cytoplasmic genes and organelles.* Academic Press, New York.

Saltzgaber-Muller, J. and G. Schatz. 1978. Heme is necessary for the accumulation and assembly of cytochrome c oxidase subunits in *Saccharomyces cerevisiae*. *J. Biol. Chem.* **253:** 305.

Schatz, G. 1965. Subcellular particles carrying mitochondrial enzymes in anaerobically grown cells of *Saccharomyces cerevisiae*. *Biochim. Biophys. Acta* **96:** 342.

———. 1987. Signals guiding proteins to their correct locations in mitochondria. *Eur. J. Biochem.* **165:** 1.

Schatz, G. and R.A. Butow. 1983. How are proteins imported into mitochondria? *Cell* **32:** 316.

Schatz, G. and E. Racker. 1966. Stable phosphorylating submitochondrial particles from baker's yeast. *Biochem. Biophys. Res. Commun.* **22:** 579.

Schatz, G. and J. Saltzgaber. 1969. Protein synthesis by yeast promitochondria in vivo. *Biochem. Biophys. Res. Commun.* **37:** 996.

Schatz, G., R.S. Criddle, and F. Paltauf. 1968. The oxygen-induced synthesis of respiratory enzymes. In *Biochemie des Sauerstoffs* (ed. B. Hess and H. Staudinger), p. 318. Springer Verlag, Berlin.

Schatz, G., E. Haslbrunner, and H. Tuppy. 1964. Deoxyribonucleic acid associated with yeast mitochondria. *Biochem. Biophys. Res. Commun.* **15:** 127.

Schekman, R. and R. Novick. 1982. The secretory process and yeast cell-surface assembly. In *The molecular biology of the yeast* Saccharomyces: *Metabolism and gene expression* (ed. J.N. Strathern et al.), p. 361. Cold Spring Harbor Laboratory, Cold Spring Harbor, New York.

Scherer, P.E., U.C. Krieg, S.T. Hwang, D. Vestweber, and G. Schatz. 1990. A precursor protein partly translocated into yeast mitochondria is bound to a 70 kd mitochondrial stress protein. *EMBO J.* **9:** 4315.

Schinkel, A.H. and H.F. Tabak. 1989 Mitochondrial RNA polymerase—Dual role in transcription and replication. *Trends Genet.* **5:** 149.

Schinkel, A.H., M.J. Koerkamp, and H.F. Tabak. 1988. Mitochondrial RNA polymerase of *Saccharomyces cerevisiae*: Composition and mechanism of promoter recognition. *EMBO J.* **7:** 3255.

Schinkel, A.H., M.J. Koerkamp, E.P. Touw, and H.F. Tabak. 1987. Specificity factor of yeast mitochondrial RNA polymerase. Purification and interaction with core RNA polymerase. *J. Biol. Chem.* **262:** 12785.

Schleyer, M. and W. Neupert. 1985. Transport of proteins into mitochondria: Translocational intermediates spanning contact sites between outer and inner membranes. *Cell* **43:** 339.

Schmidt, C., T. Söllner, and R.J. Schweyen. 1987. Nuclear suppression of a mitochondrial RNA splice defect: Nucleotide sequence and disruption of the *MRS3* gene. *Mol. Gen Genet.* **210:** 145.

Schulte, U., M. Arretz, H. Schneider, M. Tropschug, E. Wachter, W. Neupert, and H. Weiss. 1989. A family of mitochondrial proteins involved in bioenergetics and biogenesis. *Nature* **339:** 147.

Schulze, M. and G. Roedel. 1988. *SCO1*, a yeast nuclear gene essential for accumulation of mitochondiral cytochrome *c* oxidase subunit II. *Mol. Gen. Genet.* **211:** 492.

———. 1989. Accumulation of the cytochrome *c* oxidase subunits I and II in yeast requires a mitochondrial membrane-associated protein, encoded by the nuclear *SCO1* gene. *Mol. Gen. Genet.* **216:** 37.

Schwaiger, M., V. Herzog, and W. Neupert. 1987. Characterization of translocation contact sites involved in the import of mitochondrial proteins. *J. Cell Biol.* **105:** 235.

Séraphin, B., M. Simon, and G. Faye. 1988. *MSS18*, a yeast nuclear gene involved in the splicing of intron aI5 beta of the mitochondrial *cox1* transcript. *EMBO J.* **7:** 1455.

Séraphin, B., A. Boulet, M. Simon, and G. Faye. 1987. Construction of a yeast strain devoid of mitochondrial introns and its use to screen nuclear genes involved in mitochondrial splicing. *Proc. Natl. Acad. Sci.* **84:** 6810.

Séraphin, B., M. Simon, A. Boulet, and G. Faye. 1989. Mitochondrial splicing requires a protein from a novel helicase family. *Nature* **337:** 84.

Sherman, F., J.W. Stewart, C. Helms, and J.A. Downie. 1978. Chromosome mapping of the *CYC7* gene determining yeast iso2-cytochrome *c*: Structural and regulatory regions. *Proc. Natl. Acad. Sci.* **75:** 1437.

Sherman, F., J.W. Stewart, E. Margoliash, J. Parker, and W. Campbell. 1966. The structural gene for yeast cytochrome *c*. *Proc. Natl. Acad. Sci.* **55**: 1498.

Sikorski, R. S., M. S. Buguski, M. Goeble, and P. Hieter. 1990. A repeating amino acid motif in *CDC23* defines a family of proteins and a new relationship among genes required for mitosis and RNA synthesis. *Cell* **60**: 307.

Slater, E.C., J.M. Tager, S. Papa, and E. Quagliariello, eds. 1968. *Biochemical aspects of the biogenesis of mitochondria*. Adriatica Editrice, Bari.

Slonimski, P.P. 1953. *La formation des enzymes respiratoires chez la levure*. Masson, Paris.

Slonimski, P.P. and A. Tzagoloff. 1976. Localization in yeast mitochondrial DNA of mutations expressed in a deficiency of cytochrome oxidase and/or coenzyme QH_2-cytochrome *c* reductase. *Eur. J. Biochem.* **61**: 27.

Spevak, W., A. Harting, P. Meindl, and H. Ruis. 1986. Heme control region of the catalase T gene of the yeast *Saccharomyces cerevisiae*. *Mol. Gen. Genet.* **203**: 73.

Stevens, B.J. 1977. Variation in number and volume of the mitochondria in yeast according to growth conditions. A study based on serial sectioning and computer graphics reconstruction. *Biol. Cell.* **28**: 37.

————. 1981. Mitochondrial structure. In *The molecular biology of the yeast* Saccharomyces: *Life cycle and inheritance* (ed. J.N. Strathern et al.), p. 471. Cold Spring Harbor Laboratory, Cold Spring Harbor, New York.

Stevens, B.J. and J.G. White. 1979. Computer reconstruction of mitochondria from yeast. *Methods Enzymol.* **56**: 718.

Strick, C.A. and T.D. Fox. 1987. *Saccharomyces cerevisiae* positive regulatory gene *PET111* encodes a mitochondrial protein that is translated from an mRNA with a long 5′ leader. *Mol. Cell. Biol.* **7**: 2728.

Suissa, M. and G. Schatz. 1982. Import of proteins into mitochondria. Translatable mRNAs for imported mitochondrial proteins are present in free as well as mitochondria-bound cytoplasmic polysomes. *J. Biol. Chem.* **257**: 13048.

Sulo, P., P. Griac, V. Klubocnikova,, and L. Kovác. 1989. A method for the efficient transfer of isolated mitochondria into yeast protoplasts. *Curr. Genet.* **15**: 1.

Suyama, Y. 1986. Two-dimensional polyacrylamide gel electrophoresis analysis of *Tetrahymena* mitochondrial tRNA. *Curr. Genet.* **10**: 411.

Tabak, H.F. and L.A. Grivell. 1986. RNA catalysis in the excision of yeast mitochondrial introns. *Trends Genet.* **2**: 51.

Tanaka, K., T. Kanbe, and T. Kuroiwa. 1985. Three-dimensional behaviour of mitochondria during cell division and germ tube formation in the dimorphic yeast *Candida albicans. J. Cell Sci.* **73**: 207.

Terpstra, P. and R.A. Butow. 1979. The role of var1 in the assembly of yeast mitochondrial ribosomes. *J. Biol. Chem.* **254**: 12662.

Ticho, B.S. and G.S. Getz. 1988. The characterization of yeast mitochondrial RNA polymerase. A monomer of 150,000 daltons with a transcription factor of 70,000 daltons. *J. Biol. Chem.* **263**: 10096.

Trueblood, C.E. and R.O. Poyton. 1988. Identification of *REO1*, a gene involved in negative regulation of *COX5b* and *ANB1* in aerobically grown *Saccharomyces cerevisiae*. *Genetics* **120**: 671.

Trueblood, C.E., R.M. Wright, and R.O. Poyton. 1988. Differential regulation of the two genes encoding *Saccharomyces cerevisiae* cytochrome *c* oxidase subunit V by heme and the *HAP2* and *REO1* genes. *Mol. Cell. Biol.* **8**: 4537.

Tzagoloff, A. 1982. *Mitochondria*. Plenum Press, New York.

Tzagoloff, A. and A.M. Myers. 1986. Genetics of mitochondrial biogenesis. *Annu. Rev.*

Biochem. **55:** 249.

Valencik, M. L., B. Kloeckener-Gruissem, R.O. Poyton, and J.E. McEwen. 1989. Disruption of the yeast nuclear *PET54* gene blocks excision of mitochondrial intron aI5 beta from pre-mRNA for cytochrome *c* oxidase subunit I. *EMBO J.* **8:** 3899.

van der Veen, R., A.C. Arnberg, G. van der Horst, L. Bonen, H.F. Tabak, and L.A. Grivell. 1986. Excised group II introns in yeast mitochondria are lariats and can be formed by self-splicing in vitro. *Cell* **44:** 225.

van Loon, A.P.G.M. and G. Schatz. 1987. Transport of proteins to the mitochondrial intermembrane space: The "sorting" domain of the cytochrome c_1 presequence is a stop-transfer sequence specific for the mitochondrial inner membrane. *EMBO J.* **6:** 2441.

van Loon, A.P., A.W. Brändli, and G. Schatz. 1986. The presequences of two imported mitochondrial proteins contain information for intracellular and intramitochondrial sorting. *Cell* **44:** 801.

van Loon, A.P., A.W. Brändli, B. Pesold-Hurt, D. Blank, and G. Schatz. 1987. Transport of proteins to the mitochondrial intermembrane space: The "matrix-targeting" and the "sorting" domains in the cytochrome c_1 presequence. *EMBO J.* **6:** 2433.

van Loon, A.P., R.J. de Groot, M. de Haan, A. Dekker, and L.A. Grivell. 1984. The DNA sequence of the nuclear gene coding for the 17 kD subunit VI of the yeast ubiquinol-cytochrome *c* reductase: A protein with an extremely high content of acidic amino acids. *EMBO J.* **3:** 1039.

Vassarotti, A., W.J. Chen, C. Smagula, and M.G. Douglas. 1987. Sequences distal to the mitochondrial targeting sequences are necessary for the maturation of the F_1-ATPase β-subunit precursor in mitochondria. *J. Biol. Chem.* **262:** 411.

Verdiere, J. and E. Petrochilo. 1975. *CYP3* a likely candidate for the structural gene of iso-2-cytochrome *c* in *Saccharomyces cerevisiae. Biochem. Biophys. Res. Commun.* **67:** 1451.

Verdiere, J., F. Creusot, L. Guarente, and P.P. Slonimski. 1986. The overproducing *CYP1* and the underproducing *HAP1* mutations are alleles of the same gene which regulates in trans the expression of the structural genes encoding isocytochromes *c. Curr. Genet.* **10:** 339.

Verdiere, J., M. Gaisne, B. Guiard, N. Defranoux, and P.P. Slonimski. 1988. *CYP1* (*HAP1*) regulator of oxygen-dependent gene expression in yeast. II. Missense mutations suggest alternative Zn fingers as discriminating agents of gene control. *J. Mol. Biol.* **204:** 277.

Verner, K. and B.D. Lemire. 1989. Tight folding of a passenger protein can interfere with the targeting function of a mitochondrial presequence. *EMBO J.* **88:** 1491.

Verner, K. and G. Schatz. 1987. Import of an incompletely folded precursor protein into isolated mitochondria requires an energized inner membrane, but no added ATP. *EMBO J.* **6:** 2449.

———. 1988. Protein translocation across membranes. *Science* **241:** 1307.

Vestweber, D. and G. Schatz. 1988a. Point mutations destabilizing a precursor protein enhance its post-translational import into mitochondria. *EMBO J.* **7:** 1147.

———. 1988b. Mitochondria can import artificial precursor proteins containing a branched polypeptide chain or a carboxy-terminal stilbene disulfonate. *J. Cell Biol.* **107:** 2045.

———. 1988c. A chimeric mitochondrial precursor protein with internal disulfide bridges blocks import of authentic precursors into mitochondria and allows quantitation of import sites. *J. Cell Biol.* **107:** 2037.

———. 1989. DNA-protein conjugates can enter mitochondria via the protein import pathway. *Nature* **338:** 170.

Vestweber, D., J. Brunner, A. Baker, and G. Schatz. 1989. A 42K outer membrane protein is a component of the yeast mitochondrial import site. *Nature* **341**: 205.

Vitols, E., R.J. North, and A.W. Linnane. 1961. Studies on the oxidative metabolism of *Saccharomyces cerevisiae*. I. Observations on the fine structure of the yeast cell. *J. Biochem. Biophys. Cytol.* **9**: 689.

von Heijne, G. 1986. Mitochondrial targeting sequences may form amphiphilic helices. *EMBO J.* **5**: 1335.

Wallace, P.G. and A.W. Linnane. 1964. Oxygen-induced synthesis of yeast mitochondria. *Nature* **201**: 1191.

Weislogel, P.O. and R.A. Butow. 1970. Low temperature and chloramphenicol induction of respiratory deficiency in a cold-sensitive mutant of *Saccharomyces cerevisiae*. *Proc. Natl. Acad. Sci.* **67**: 52.

Wenzlau, J.M., R.J. Saldanha, R.A. Butow, and P.S. Perlman. 1989. A latent intron-encoded maturase is also an endonuclease needed for intron mobility. *Cell* **56**: 421.

Werner-Washburne, M., D.E. Stone, and E.A. Craig. 1987. Complex interactions among members of an essential subfamily of hsp70 genes in *Saccharomyces cerevisiae*. *Mol. Cell. Biol.* **7**: 2568.

Wilcoxen, S.E., C.R. Peterson, C.S. Winkley, M.J. Keller, and J.A. Jaehning. 1988. Two forms of *RPO41*-dependent RNA polymerase. Regulation of the RNA polymerase by glucose repression may control yeast mitochondrial gene expression. *J. Biol. Chem.* **263**: 12346.

Williamson, D.H., N.G. Maroudas, and D. Wilkie. 1971. Induction of the cytoplasmic petite mutation in *Saccharomyces cerevisiae* by the antibacterial antibiotics erythromycin and chloramphenicol. *Mol. Gen. Genet.* **111**: 209.

Winkley, C.S., M.J. Keller, and J.A. Jaehning. 1985. A multicomponent mitochondrial RNA polymerase from *Saccharomyces cerevisiae*. *J. Biol. Chem.* **260**: 14214.

Wintersberger, E. 1966. Occurrence of a DNA polymerase in isolated yeast mitochondria. *Biochem. Biophys. Res. Commun.* **25**: 1.

Wintersberger, U. and H. Blutsch. 1976. DNA-dependent DNA polymerase from yeast mitochondria. Dependence of enzyme activity on conditions of cell growth, and properties of the highly purified polymerase. *Eur. J. Biochem.* **68**: 199.

Wintersberger, U. and E. Wintersberger. 1970a. Studies on deoxyribonucleic acid polymerases from yeast. I. Partial purification and properties of two DNA polymerases from mitochondria-free cell extracts. *Eur. J. Biochem.* **13**: 11.

―――. 1970b. Studies on deoxyribonucleic acid polymerases from yeast. II. Partial purification and characterization of mitochondrial DNA polymerase from wild type and respiration-deficient yeast cells. *Eur. J. Biochem.* **13**: 20.

Witte, C., R.E. Jensen, M.P. Yaffe, and G. Schatz. 1988. *MAS1*, a gene essential for yeast mitochondrial assembly, encodes a subunit of the mitochondrial processing protease. *EMBO J.* **7**: 1439.

Wu, M. and A. Tzagoloff. 1987. Mitochondrial and cytoplasmic fumarases in *Saccharomyces cerevisiae* are encoded by a single nuclear gene *FUM1*. *J. Biol. Chem.* **262**: 12275.

Yaffe, M.P. and G. Schatz. 1984. Two nuclear mutations that block mitochondrial protein import in yeast. *Proc. Natl. Acad. Sci.* **81**: 4819.

Yaffe, M.P., S. Ohta, and G. Schatz. 1985. A yeast mutant temperature-sensitive for mitochondrial assembly is deficient in a mitochondrial protease activity that cleaves imported precursor polypeptides. *EMBO J.* **4**: 2069.

Yang, M., R.E. Jensen, M.P. Yaffe, and G. Schatz. 1988. Import of proteins into yeast mitochondria: The purified matrix processing protease contains two subunits which are encoded by the nuclear *MAS1* and *MAS2* genes. *EMBO J.* **7**: 3857.

Yang, M., V. Geli, W. Oppliger, K. Suda, P. James, and G. Schatz. 1991. The *MAS*-encoded processing protease of yeast mitochondria: Interaction of the purified enzyme with signal peptides and a purified precursor protein. *J. Biol. Chem.* (in press).

Yotsuyanagi, Y. 1962a. Etudes sur le chondriome de la levure. I. Variation de l'ultrastructure du chondriome au cours du cycle de la croissance aérobic. *J. Ultrastruct. Res.* **7**: 121.

―――. 1962b. Etudes sur le chondriome de la levure. II. Chondriomes des mutants á deficience respiratoire. *J. Ultrastruct. Res.* **7**: 141.

Zimmermann, R. and D.I. Meyer. 1986. A year of new insights into how proteins cross membranes. *Trends Biochem. Sci.* **11**: 512.

Zitomer, R.S., J.W. Sellers, D.W. McCarter, G.A. Hastings, P. Wick, and C.V. Lowry. 1987. Elements involved in oxygen regulation of the *Saccharomyces cerevisiae CYC7* gene. *Mol. Cell. Biol.* **7**: 2212.

Zwizinski, C. and W. Neupert. 1983. Precursor proteins are transported into mitochondria in the absence of proteolytic cleavage of the additional sequences. *J. Biol. Chem.* **258**: 13340.

Zwizinski, C., M. Schleyer, and W. Neupert. 1983. Transfer of proteins into mitochondria. Precursor to the ADP/ATP carrier binds to receptor sites on isolated mitochondria. *J. Biol. Chem.* **258**: 4071.

―――. 1984. Proteinaceous receptors for the import of mitochondrial precursor proteins. *J. Biol. Chem.* **259**: 7850.

8
Recombination in Yeast

Thomas D. Petes
Department of Biology
University of North Carolina
Chapel Hill, North Carolina 27599-3280

Robert E. Malone
Department of Biology
University of Iowa
Iowa City, Iowa 52242

Lorraine S. Symington
Institute of Cancer Research
Columbia University College of Physicians & Surgeons
New York, New York 10032

Volume I. *The Molecular and Cellular Biology of the Yeast* Saccharomyces: *Genome Dynamics, Protein Synthesis, and Energetics.* Copyright 1991 Cold Spring Harbor Laboratory Press 0-87969-355-X/91 $3 + 00   **407**

I. INTRODUCTION

The exchange of information between DNA molecules is a process found in every organism that has been investigated in detail. Because recombination in eukaryotes occurs most frequently during meiosis after chromosome replication, much of our understanding of eukaryotic recombination is based on organisms (such as *Saccharomyces cerevisiae*) in which all products of an individual meiosis can be analyzed. In this chapter, we summarize studies of homologous recombination (meiotic and mitotic) in the yeast *S. cerevisiae*. We discuss less comprehensively recombination in the fission yeast *Schizosaccharomyces pombe*, emphasizing areas of research representing either particularly important contributions (e.g., the analysis of hot spots for meiotic recombination) or interesting contrasts with *S. cerevisiae.*

The two goals of this chapter are (1) to introduce genetic studies of recombination and issues of interest in this area to yeast workers outside the immediate field and (2) to discuss recent recombination studies in the context of earlier work summarized in previous reviews (Esposito and Wagstaff 1981; Fogel et al. 1981; Orr-Weaver and Szostak 1985; Resnick 1987; Hastings 1988; Roeder and Stewart 1988; Thaler and Stahl 1988). This review is limited to general homologous recombination and does not concern exchanges catalyzed by specialized enzyme systems.

As with other areas of yeast research, recombinant DNA procedures have had a major impact on studies of recombination (Haber et al. 1988; Petes and Hill 1988; Petes et al. 1989). Before we discuss meiotic and mitotic recombination in detail, we outline some of the general applications of this methodology to the analysis of recombination.

II. ANALYSIS OF RECOMBINATION USING RECOMBINANT DNA PROCEDURES

Classical studies of recombination in yeast usually involved analysis of exchange in diploid strains heterozygous for one or more single-copy genes. In general, the exact nature of the mutations and their locations within the genes were unknown. Recombinant DNA techniques, coupled to the yeast transformation procedure, allow the construction of yeast strains with precisely defined mutational changes at defined sites within the genome. As discussed below, these methods not only improve the resolution of earlier studies, but also allow the investigation of a new repertoire of recombinational interactions.

The construction of strains with mutations at defined sites in the genome is relatively easy because in *S. cerevisiae*, transforming DNA usually integrates at sites of extended sequence homology (Hinnen et al. 1978). (However, integration events involving very limited homology [4 bp] have recently been detected in *S. cerevisiae* [R. Schiestl and T. Petes, unpubl.], and nonhomologous integration events have been observed in *S. pombe* [Wright et al. 1986].) A single crossover integrating a circular plasmid results in a duplication of DNA sequences in which the repeats are separated by vector sequences (Fig. 1a). In a second type of manipulation (transplacement), the sequences in the transforming DNA replace those on the chromosome by protocols involving either one (Rothstein 1983) (Fig. 1b) or two steps (Scherer and Davis 1979). Because cotransformation in yeast occurs at a relatively high frequency (Hicks et al. 1979), it is sometimes possible to screen for transplacement events in strains that have been transformed at an unlinked locus.

a. Integration of plasmid creating intrachromosomal repeat.

b. Insertion of gene on non-homologous chromosome by one-step transplacement

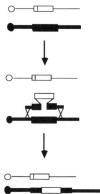

Figure 1 Construction by transformation of yeast strains with duplicated yeast genes. (*a*) Integrative transformation resulting in an intrachromosomal repeat. (*Thick lines*) Bacterial plasmid DNA; (*thin lines*) yeast DNA sequences; (*hatched rectangles*) a selectable yeast gene; (*open rectangles*) a second yeast gene; (*short vertical line*) a modification of the second gene. (*b*) One-step transplacement resulting in duplicated genes on nonhomologous chromosomes. Sequences of the two chromosomes are indicated by lines of different widths. In this version of the procedure, a wild-type gene (*open rectangle*) is inserted into a gene derived from a different chromosome (*closed rectangle*). The resulting DNA fragment is used for transformation.

These methods have been useful in recombination studies in a variety of different ways: (1) constructing defined mutational changes at defined positions, (2) physically monitoring recombination, (3) constructing isogenic strains to examine the effects of various mutations on recombination, (4) examining the properties of recombination between repeated genes, and (5) constructing strains that are useful for the isolation of mutations affecting recombination. Each of these uses is discussed briefly below.

To study recombination genetically, one needs strains containing heterozygous markers. Until recently, most such markers were obtained by in vivo mutagenesis. With the exception of mutations in the gene encoding iso-1-cytochrome *c*, analyzed in the laboratory of Fred Sherman (for review, see Moore et al. 1988), these mutations were not well defined. Recombinant DNA procedures allow the quick generation of defined mutations of cloned DNA sequences. For example, one can cut a plasmid with an enzyme that leaves protruding 5′ ends, "fill in" the recessed 3′ ends using DNA polymerase, and ligate the blunt ends. The resulting plasmid cannot be cleaved with the original restriction enzyme at the same site and has a small duplication (usually 4 bp). The plasmid containing the altered sequence can then be transplaced into the genome of a diploid cell such that the alteration is present on only one of the two homologous chromosomes (Borts and Haber 1987; Symington and Petes 1988a,b). The meiotic segregation of the markers into spores can then be monitored by Southern analysis. The advantages of this type of marker over conventional mutations are (1) the marker can be placed precisely on a physical map; (2) markers can be densely spaced along the chromosome; (3) the nature of the change (a small insertion) can be defined; (4) the mutations do not readily revert; and (5) changes can be made in regions of the genome outside of expressed genes. Other defined mutational alterations of cloned sequences (large insertions, single-base-pair changes, deletions) are also useful in recombination studies. Because most "natural" mutations represent base substitutions, it is possible that some "artificial" mutations may have marker effects on recombination that are different from those caused by single-base-pair changes.

Strains with heterozygous restriction sites can also be used to monitor meiotic recombination physically (Borts et al. 1984, 1986). Recombination between two heterozygous sites results in the appearance of a restriction fragment that differs in size from either parental fragment. The timing of the appearance of the recombinant fragment during meiosis can be assayed by Southern analysis. This method is also useful in examining recombination in strains that do not produce viable spores.

One technical difficulty in studying *trans*-acting mutations that affect recombination has been the problem of constructing strains that are isogenic except for the single mutation whose effect is being tested. Both differences in genetic background in different strains and extraneous mutations resulting from the mutagenesis contribute to this problem; differences in genetic backgrounds are much more pronounced for *S. cerevisiae* strains than for *S. pombe* strains, which were derived from a single homothallic isolate. If the gene encoding the wild-type allele of the mutated gene has been cloned, it is easy to construct strains that differ only at the locus being tested. The relevant phenotype can then be assayed in two strains: the original mutant strain and a strain in which the mutant gene has been replaced (by two-step transplacement) with the wild-type allele. Alternatively, the wild-type gene can be disrupted by the one-step transplacement procedure.

Another important use of recombinant DNA technology has been the construction of yeast strains for the study of recombination between repeated genes (for review, see Petes and Hill 1988). With the exception of an analysis of recombination between dispersed tRNA genes in *S. pombe* (Munz and Leupold 1981), such recombination events were not analyzed prior to the development of recombinant DNA procedures. One obstacle for this type of study by classical techniques is that genetic analysis requires heterozygous markers, and quite often, different members of a gene family are very similar or identical in DNA sequence. Even when there were differences (such as restriction site polymorphisms) between members of a gene family, selecting or screening for recombination events was difficult.

Recombinant DNA procedures have allowed the construction of repeated gene families (usually two-member families) in which each repeat has a different, easily scorable phenotype. As diagrammed in Figure 1a, one method of constructing strains with a tandem duplication of yeast sequences (a two-member family) is by transforming cells with a recombinant plasmid. By using methods described below, recombination between the duplicated sequences in such strains is easily detected. Recombinant plasmids containing two noncontiguous yeast genes can be used to construct strains that have duplicated genes located far apart on one chromosome or located on nonhomologous chromosomes (Fig. 1b) (Scherer and Davis 1980). Such plasmids can also be used to "tag" a naturally occurring family of repeats with a selectable yeast gene. This approach was used to study recombination within the rRNA gene cluster (Petes 1980; Szostak and Wu 1980).

Finally, haploid strains containing duplicated genes are useful in the isolation of mutations affecting recombination. Most classical studies of recombination have involved recombination in diploid strains. Because most mutations are recessive, it is difficult to screen directly for mutations affecting recombination. Consequently, most existing recombination mutants were first detected by a different phenotype (e.g., failure to repair radiation- or chemical-induced damage) that can be observed in a haploid. Some mutations affecting recombination (e.g., *pms1* [Williamson et al. 1985] and *rec46* [Esposito et al. 1986]) were isolated in haploid strains that were disomic for a chromosome containing heteroalleles. An alternative approach is to use haploid strains that contain a duplication of genes with different mutant alleles. By using such strains, mutations affecting mitotic or meiotic recombination have been identified.

In summary, recombinant DNA procedures have expanded and re-

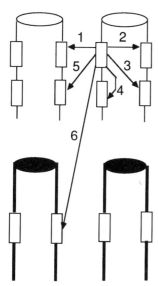

Figure 2   Different types of recombination (from Petes and Hill 1988). Chromosomes are depicted after DNA synthesis. (*Open rectangles*) Homologous DNA sequences. Arrows indicate the possible recombinational interactions (see text). Nonhomologous chromosomes are shown by lines of different widths.

fined the methods available for the study of recombination. Below, we describe our current understanding of recombination in yeast based on both classical and new methods of genetic analysis.

III. MEIOTIC RECOMBINATION IN YEAST

The general principles of genetic analysis in yeast and in other eukaryotes are very similar. Most of the differences in technology reflect two features of the yeast life cycle: the stable haploid cycle (in *S. cerevisiae)* and the recovery of all four meiotic products in a single ascus.

Before discussing meiotic exchange, we need to define a number of different types of recombination (Fig. 2). If meiotic recombination occurs after DNA synthesis and involves only genes represented in one copy per haploid genome, recombination can occur either between genes located on nonsister chromatids of homologous chromosomes (pathway 1) or between genes located on sister chromatids (pathway 2). Almost all early studies involved the first type of exchange, which we term classical recombination. The second type is called sister-strand recombination. If a gene is duplicated on one pair of homologs, the recombinational interactions illustrated by pathways 3, 4, and 5 could occur, called unequal sister strand, intrachromatid, and quasiclassical recombination, respectively. Pathway 6 illustrates recombination between repeated genes on nonhomologous chromosomes, which we term heterochromosomal recombination. Pathways 3–6 are collectively described as ectopic recombination, and recombination events involving either sister strands or intrachromatid interactions (pathways 2–4) are termed intrachromosomal recombination. If the type of recombination is not specified in the terms described above, it can be assumed to be classical recombination.

A. Meiotic Crossing-over

Most studies of meiotic recombination (prior to 1980) were done in diploid strains that were heterozygous for one or more single-copy genes. The heterozygous markers usually represented either one wild-type gene and one mutant gene or two different mutant genes. In a typical experiment, a diploid was constructed by crossing one haploid strain with the linked wild-type alleles A, B, and C to a haploid strain containing mutant alleles a, b, and c. When this diploid was sporulated and tetrads were dissected, in most tetrads, two of the spores contained the wild-type allele and two contained the mutant allele for each gene pair. In considering the segregation of two different genetic markers (e.g., A and B) that show 2:2 segregation, three different patterns are possible: parental ditype ($2AB$:$2ab$), nonparental ditype ($2Ab$:$2aB$), and tetratype ($1AB$:$1Ab$:$1aB$:$1ab$). The genetic definition of linkage in yeast is that the frequency of parental ditype tetrads significantly exceeds the frequency of nonparental ditype tetrads. As shown in Figure 3, a single crossover between linked genes A and B yields an ascus that is tetratype; crossing-over is also termed reciprocal recombination, because the same tetrad that has the recombinant spore Ab also has the reciprocal recombinant product aB. A four-strand double crossover would yield a nonparental ditype tetrad.

Genetic distances between pairs of linked markers are usually established by tetrad analysis and conversion of tetrad data to map lengths using standard equations (Perkins 1949; Mortimer and Schild 1981). To make map distances derived from tetrad data equivalent to those in organisms in which only individual products of meiosis can be analyzed, distance is calculated on a per chromosome basis, rather than a per tetrad basis; thus, 1 centiMorgan (cM) distance between two markers indicates that about 2% of the tetrads (or 1% of the chromosomes) have had a crossover in that interval. The total length of the yeast genetic map is

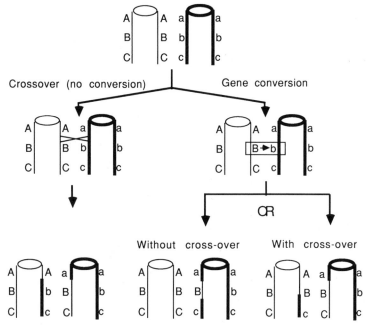

Figure 3 Meiotic crossing-over and gene conversion. Meiotic recombination is shown between replicated nonsister homologous chromatids. The left side of the figure shows a crossover that is not associated with a detectable gene conversion event. The right side of the figure shows a conversion event at the *B* locus. Slightly fewer than half of such conversion events are associated with reciprocal exchange of flanking markers.

about 4500 cM, and there are about 0.37 cM/kb (estimated by correlating genetic and physical distances on chromosome III; Strathern et al. 1979). One can calculate from this map length that there are about 100 cross-overs per yeast cell per meiosis. This is a very high rate of recombination per unit of physical length relative to those observed in most eukaryotes. For example, the genetic map lengths for humans and yeast are very similar despite a 200-fold difference in genome size.

1. Distribution of Crossover Events

Until physical distances between genes were known, it was difficult to determine whether such distances were proportional to genetic map lengths. It was clear, however, that the genetic maps established by meiotic recombination experiments were different from those established by mitotic experiments both in *S. cerevisiae* (Malone et al. 1980) and in *S. pombe* (Minet et al. 1980). An analysis of these differences is likely to yield useful information about the mechanics of the recombination process. It is now clear that chromosomal regions exist with more (hot spots) or fewer (cold spots) exchanges than an average genomic interval. For example, Larkin and Woolford (1983) found that the 21-kb interval between *MAT* and *CRY1* had a genetic distance of only 2.2 cM, making this region three- to fourfold colder than average. Two other regions, *CDC7-CEN4* (Meddle et al. 1984) and *TRP1-CEN4* (Stinchcomb et al. 1982), were reported to be about fivefold colder than average. An extreme example of a cold region is the rRNA gene cluster (150 tandemly arranged 9-kb repeats located on chromosome XII), which is about 100-fold colder for classical recombination than an average interval of this size (Petes and Botstein 1977); however, unequal sister-strand recombination events between these clusters occur frequently (Petes 1980). Conversely, the interval between *CDC24* and *PYK1* on chromosome I is fivefold hotter (Coleman et al. 1986) and the *LEU2-HIS4* interval on chromosome III is about threefold hotter (Newlon et al. 1986) than the average. Kaback et al. (1989) have argued that chromosome I (the smallest yeast chromosome) has more crossing-over than expected for its physical length (see also below). In *S. pombe*, the region between the *mat2* and *mat3* loci has at least tenfold fewer meiotic recombination events than expected from its physical length (Egel 1984).

The correlation between genetic and physical maps of yeast chromosomes has also been examined at higher resolution (Symington and Petes 1988a). In this study, diploid strains were constructed that contained ten heterozygous markers located between *LEU2* and the centromere of chromosome III (a distance of 6 cM and 22 kb). Tetrads with a crossover between *LEU2* and *CEN3* were detected using a centromere-linked marker on another chromosome. The genetic map distance between *LEU2* and *CEN3* was the same in the diploid containing the heterozygous markers as in an isogenic strain lacking the markers. In crossover tetrads, Southern analysis was done on DNA isolated from spore cultures. Of the ten genetic intervals examined, there was one interval with significantly more (twofold) and one interval with significantly fewer (fivefold) crossovers than expected. The 5-kb interval that was closest to the centromere had approximately the expected number of crossovers, indicating that the centromere did not strongly suppress meiotic crossing-over.

Lambie and Roeder (1986) investigated the effect of the centromere on meiotic crossing-over by constructing strains in which *CEN3* was relocated on chromosome III. They found that crossing-over was increased in the interval from which they removed the centromere and decreased (three- to fivefold) in the interval into which they inserted the centromere. Because Symington and Petes (1988a) found an average frequency of crossing-over near the centromere, it is possible that the centromere slightly suppresses meiotic recombination of DNA sequences that would have a higher than average frequency of exchange in the absence of the centromere. Alternatively, the rate of crossing-over may be artifactually increased or decreased by the chromosomal alterations used in the two studies. It is clear that in some other organisms (e.g., *Drosophila*), meiotic crossing-over is strongly (200-fold) repressed by the centromere (Dobzhansky 1931). In *S. pombe,* in which centromeric DNA is larger and more complex than in *S. cerevisiae,* meiotic recombination is suppressed in a region of 60 kb that includes the centromere (Nakaseko et al. 1986).

In addition to the study of natural genomic sequences, crossing-over within DNA sequences introduced into the yeast genome by transformation has also been examined. The *URA3* gene (normally located on chromosome V), when inserted into chromosome III, was a hot spot for crossing-over in some locations (Borts et al. 1984) but not in others (Borts and Haber 1987). This type of context effect suggests that formation of the hot spot requires two sites (or sequences), one provided by the *URA3* gene and one provided by certain chromosomal regions. A mini-Tn3 element (Seifert et al. 1986) containing the *amp*[r] gene of pBR322, a *loxP* site and a *URA3* gene, when integrated in chromosome III as a homozygous insertion, strongly (13-fold) stimulated meiotic crossing-over at three different sites examined (Stapleton and Petes 1991). Insertions of poly(G,T)/poly(C,A) sequences into the genome appeared to stimulate both simple crossovers and complex patterns of recombination (Treco and Arnheim 1986).

Another interesting meiotic crossover hot spot results from the M26 mutation in the *ade6* gene of *S. pombe* (Gutz 1971). This mutation is a G to T change within the coding region; this change stimulates locally both meiotic gene conversion and crossing-over (Gutz 1971; Ponticelli et al. 1988; Schuchert and Kohli 1988; Szankasi et al. 1988). The effect of the mutation is context-dependent; when a 3-kb fragment containing the mutant sequences was placed in a different chromosomal location (within *ura4*), no stimulation of meiotic recombination was observed (A. Ponticelli and G. Smith, pers. comm.). This result demonstrates that one or more nucleotides more than 1 kb from M26 are required for hot spot activity.

It is unclear by what mechanism hot spots or cold spots exert their effects. Perhaps the most obvious possibility is that the enzymes that catalyze homologous recombination have DNA sequence preferences, reflecting a preference for a specific site (analogous to a restriction site), structure (e.g., Z-DNA), or base composition. Alternatively, because the DNA is complexed with proteins in the chromosome, hot spots and cold spots might reflect the accessibility of the DNA to the enzymes that catalyze recombination. This accessibility could be affected by transcription, specific DNA-binding proteins, local regions of DNA winding and unwinding, and other factors. A third possibility is that these regions reflect a more global aspect of chromosome structure. In one model of chromosome packing, the chromosome is arranged in a series of radial loops, the base of each loop being connected to a proteinaceous scaffold (for review, see Gasser and Laemmli 1987). The ability of homologous sequences to pair and recombine could be affected by the position of the sequences within the loops. Alternatively, the rate of recombination may be dependent on the interactions of two sites, one site acting at a distance to promote recombination at a second site. Although such possibilities may seem needlessly baroque, two-site interactions are one way of explaining context-specific effects on recombination. In addition, such interactions have been demonstrated clearly in *Neurospora* (Angel et al. 1970) and in prokaryotes (for review, see Smith 1987). The various possible mechanisms should eventually be distinguished by sequence and deletion analyses of various hot spots together with experiments in which hot spots are moved from one chromosomal location to another.

One complication in analyzing the distribution of crossover events is the evidence that these events may sometimes be affected by the heterozygous markers that are used to detect them. Borts and Haber (1987) examined meiotic recombination in a diploid that was homozygous for an insertion of plasmid (pBR322 and *URA3*) sequences located between duplicated *MAT* loci (duplicated *MAT*a loci on one chromosome and duplicated *MAT*α loci on the other). Using multiple heterozygous markers, Borts and Haber did a detailed analysis of crossing-over in these insertions. Simple (allelic) crossovers in this interval produced tetrads that had pairs of nonmating spores. However, the frequency of such crossovers was a function of the number of heterozygous markers used. If nine heterozygous markers were used (in a 9-kb region), 11% of the unselected tetrads had simple crossovers. When these same markers were made homozygous, the frequency of simple crossovers doubled. These authors also found a number of other classes of recombinant tetrads, most of which involved either ectopic crossovers of the flanking *MAT* loci, double recombination events, or classic conversion of the *MAT* loci. The frequencies of these other classes were higher in the strains with multiple heterozygous markers than in those with few or no heterozygous markers.

On the basis of these results, Borts and Haber (1987) suggested that heteroduplexes formed between DNA sequences containing multiple mismatches are repaired by a process that generates additional recombination events (Hastings 1984). If the rate of initiation of recombination is not affected by the heterozygous markers, the proposed mechanism would lead to a loss of simple crossovers and an increase in other classes of recombinants. An alternative interpretation of their data is that classical recombination events in the plasmid sequences compete with other types of recombination (classical and ectopic recombination involving the *MAT* loci). Thus, if reduction in the level of homology between plasmid sequences (as a result of introducing heterozygous mutations) reduces the level of recombination between these sequences, recombination events involving the flanking *MAT* loci might be enhanced.

It is not yet clear why heterozygous markers affected recombination in the experiments of Borts and Haber (1987) but had no obvious effect in the experiments of Symington and Petes (1988a). Borts and Haber (1989) subsequently showed that only certain combinations and spacings of markers affected recombination; such specific arrangements of markers may not have been present in the other experiment. Alternatively, because Borts and Haber (1987) used constructions in which a duplication flanked the region in which they were monitoring recombination, this duplication may have allowed the redirection of recombination events into ectopic pathways of recombination unavailable in a strain containing no duplications.

An issue related to the distribution of crossovers is the total number of crossovers per chromosome. In one direct measurement of crossovers on chromosome III, a strain was constructed that was heterozygous for seven markers distributed from one end to the other (Haber et al. 1984). The average number of crossovers detected in this chromosome per meiosis was 2.8, and in only 2 of the 62 tetrads examined were no crossovers detected. In a similar study of chromosome I (the shortest yeast chromosome), an average of about three crossovers per meiosis was detected, and the frequency of noncrossover chromosomes was less than 1% (Kaback et al. 1989). Because meiotic crossing-over in most organisms appears to be required for proper chromosome disjunction at meiosis I (for review, see Hawley 1988), it is reassuring that the frequency of tetrads with no crossovers for chromosomes I and III is low. In both studies, the proportion of chromosomes with no crossovers was less than that expected assuming a Poisson distribution for the number of exchanges per chromosome. Because crossovers appear to interfere with each other (as described below), this skewed distribution is not unexpected.

S. cerevisiae may have a "backup" system to allow a degree of proper chromosome disjunction even in the absence of crossing-over; a similar system (distributive pairing) was described previously for *Drosophila* (for review, see Hawley 1988). Centromere-containing plasmids that had not apparently recombined and/or shared no significant sequence homology usually segregated away from each other in meiosis (Dawson et al. 1986; Mann and Davis 1986; Kaback 1989). In addition, although no recombination was detected between two copies of chromosome V that had diverged in sequence (by about 15%), the level of normal disjunction was 96% (Nilsson-Tillgren et al. 1986).

2. Interference

Since Muller's (1916) studies of *Drosophila*, it has been clear that the occurrence of a crossover can influence the probability of a second crossover nearby (chiasma interference). A number of methods exist for estimating interference (Snow 1979), using data derived from two-point crosses or crosses involving multiple linked markers. Whatever method is used, if the observed multiple crossovers are fewer than expected, based on the frequency of single crossovers, there is positive interference; if there are more than expected, there is negative interference. Although positive chiasma interference is very common, it is not universal (for review, see Whitehouse 1982); for example, neither *Aspergillus nidulans* (Strickland 1958) nor *S. pombe* (Snow 1979) displays chiasma interference. Even in organisms that have interference, chiasma interference is usually not observed for adjacent intervals if one of the intervals includes the centromere (Whitehouse 1982).

In the fairly limited number of studies of chiasma interference in *S. cerevisiae,* positive interference has usually been observed for adjacent intervals that do not span the centromere (Hawthorne and Mortimer 1960; Fogel and Hurst 1967; Mortimer and Fogel 1974; Snow 1979). For crosses in which one of the intervals includes the centromere, no interference has been observed in some experiments (Hawthorne and Mortimer 1960), whereas in most other crosses, positive interference has been seen (for a summary, see Snow 1979). There are two somewhat unexpected features of these interference studies. First, there is no clear correlation in *S. cerevisiae* between the degree of positive interference and the genetic length of the interval (Snow 1979). Second, some intervals appear to have very little positive interference; many of these exceptional intervals are located on chromosome VII (Snow 1979). No systematic study of the correlation between the physical length of the intervals and the degree of interference has been undertaken. In the analysis of crossing-over between *LEU2* and *CEN3* described above (Symington and

Petes 1988a), there appeared to be negative interference. This interval is considerably smaller genetically (6 cM) than most of the intervals considered by Snow.

There is no convincing molecular model for chiasma interference in any organism. The following are among the models that have been suggested: (1) Recombination at one site on the chromosome results in local depletion of factors necessary to catalyze a second event nearby (Holliday 1977; Hastings 1987b); (2) binding of one recombination nodule (a spherical particle observed by electron microscopy that appears to be associated with recombining chromosomes; Carpenter 1975) covers a sufficient amount of the chromosome to prevent a second nodule from occupying the nearby chromosome; and (3) the topological movements necessary to generate a crossover prevent an adjacent second crossover. In considering the last point, it should be mentioned that in yeast, as in other fungi, gene conversion events not associated with crossovers do not interfere with crossovers in nearby intervals (Fogel and Hurst 1967). More extensive high-resolution studies of the distribution of crossovers along the chromosome would be useful in developing better-defined models for interference.

A second potential type of interference is chromatid interference, a specific inhibition of two-strand double crossovers relative to three- and four-strand double crossovers (for review, see Whitehouse 1982). If there is no chromatid interference and sister-strand crossing-over is infrequent relative to nonsister exchanges (see below), the expected ratio of 2:3:4-strand double crossovers is 1:2:1. This approximate ratio has been observed in those studies involving many tetrads (Hawthorne and Mortimer 1960; Fogel and Hurst 1967). Thus, in *S. cerevisiae,* as in most other organisms (Whitehouse 1982), there appears to be no chromatid interference.

3. Sister-strand Crossing-over

Because sister strands (chromatids originally connected to the same centromere) contain identical DNA sequences, recombination between them is not examined easily by conventional techniques. Three different approaches have been used to examine this type of recombination. In one study (Fig. 4), unequal sister-strand recombination within the rRNA gene cluster was detected by observing the segregation of an inserted *LEU2* gene (Petes 1980). At least 15% of unselected tetrads had an unequal sister-strand exchange. Another estimate of sister-strand recombination was also based on studies of unequal crossing-over between duplicated genes constructed by integrating a plasmid (see Fig. 1a). In a study in

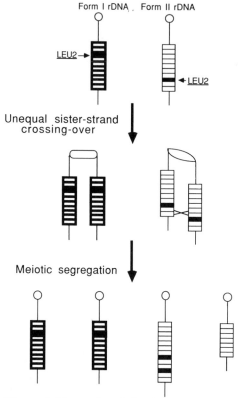

Figure 4 Unequal meiotic sister-strand crossing-over. A haploid strain that had form I rRNA genes (seven *Eco*RI sites per repeat) and a *LEU2* gene integrated in the rRNA gene cluster was mated to a haploid that had form II rRNA genes (six *Eco*RI sites per repeat) that also had an integrated *LEU2* gene. The resulting diploid was sporulated and tetrads were dissected; 15% of the tetrads contained a Leu⁻ spore. In such tetrads, the Leu⁻ spore had the same type of rRNA genes as the Leu⁺ spore that contained a duplication of the *LEU2* sequences.

which both homologs contained a 19-kb duplication, Jackson and Fink (1985) found unequal sister-strand crossovers in about 1.5% of un- selected tetrads.

Meiotic sister-strand exchanges were also observed by using a diploid containing one linear and one circular derivative of chromosome III (Haber et al. 1984). A single or odd number of sister-strand crossovers between circular sister chromatids would result in a dicentric double cir- cle, which would presumably not be found in viable spores. From the fre- quency of tetrads with two viable spores containing linear chromosomes III, Haber et al. (1984) concluded that in at least 12.5% of the tetrads, there was a sister-strand exchange between the circular sister strands. Because there were about 2.8 crossovers in chromosome III involving non-sister strands, sister-strand crossovers were about 20-fold less common than classical exchanges. This conclusion is generalizable only if the rate of sister-strand crossovers is the same for circular and linear chromosomes.

Game et al. (1989) have used diploid strains heterozygous for circular derivatives of chromosome III for the physical detection of sister-strand exchanges. Intact chromosomal DNA molecules were gently isolated from meiotic cells of such strains and examined by using pulsed-field gel electrophoresis. Both crossovers between a circular and a linear chromosome III (classical crossovers) and crossovers between two circular chromosomes (sister-strand exchanges) were detected. The level of sister-strand recombination was about one per chromosome, a level only threefold less than that of recombination between nonsisters.

In summary, meiotic sister-strand crossing-over occurs at a substantial frequency, although for most chromosomal regions, classical exchange appears to predominate. One exception to this generalization is crossing-over in the rRNA gene cluster, where sister-strand exchange is more frequent than nonsister-strand recombination. The coupled suppression of nonsister-strand exchange and enhancement of sister-strand recombination in the rRNA gene cluster suggests that these two processes may be in competition, a possibility supported by the observation that sister-strand exchange appears to be elevated in $spo13$ haploid yeast cells that undergo meiosis (Wagstaff et al. 1985).

It is unclear how the recombination system distinguishes between sister and nonsister strands. Some mutations affecting recombination influence these two types of recombination differently (see Section III.H). It is possible that the synaptonemal complex (discussed in Section III.G) has some role in establishing this distinction. Because no synaptonemal complex is seen in the nucleolus during meiosis (Byers and Goetsch 1975; Dresser and Giroux 1988), the complex may be involved in inhibiting sister or promoting classical recombination. Alternatively, it is possible that the ratio of sister to classical exchange is affected by the organization (single-copy DNA sequences vs. tandemly arranged repeats), the sequence, or the level of transcription of the interacting chromosomal DNA sequences.

4. Timing of Crossing-over

The time of crossing-over during meiosis in yeast was first investigated by experiments in which vegetative diploid cells were shifted into sporu-

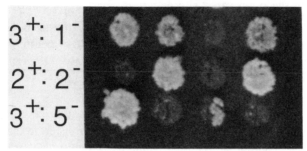

$3^+ : 1^-$

$2^+ : 2^-$

$3^+ : 5^-$

Figure 5 Spore colonies depicting normal Mendelian (2:2), gene conversion (3:1), and postmeiotic segregation (3:5) patterns (photograph provided by D. Nag, University of North Carolina). A diploid strain that was heterozygous for a mutation at the *HIS4* locus was sporulated and dissected onto plates containing a rich nutrient medium. After 3 days of growth, the resulting colonies were replica plated onto plates containing medium that lacked histidine. After 1 day of incubation, the replica-plated colonies were photographed. Each row of colonies is derived from a different tetrad.

lation medium, and then transferred back to media allowing vegetative growth and selection of recombinant products (for references, see Esposito and Klapholz 1981). Such experiments demonstrated that cells become committed to meiotic levels of recombination (both gene conversion and crossing-over) before becoming committed to haploidization. Commitment to recombination occurred at about the same time as commitment to meiotic DNA synthesis. This type of experiment does not directly measure the time at which crossovers occur.

With the development of methods to monitor the kinetics of crossing-over by Southern analysis (Borts et al. 1984; Haber et al. 1988; Cao et al. 1990), it was shown that crossing-over occurs after meiotic DNA synthesis. It is not yet clear whether different regions of the genome cross over at different times, as suggested by earlier commitment studies (Esposito and Klapholz 1981).

B. Meiotic Gene Conversion

For all heterozygous genetic loci that have been examined in yeast, occasional (1–20% of unselected tetrads) departures from 2:2 segregation have been observed (Fig. 5). For example, Figure 3 shows a tetrad in which the *B* gene is segregating 3*B*:1*b*. Because it appears that one allele has been converted to the other allele in such tetrads, these events are called gene conversions. If the *A* and the *B* genes are examined together in the convertant tetrad, it is noted that one spore contains the recombinant product *aB*, but the other three spores contain parental combinations of markers. For this reason, gene conversion events are also called nonreciprocal recombination events. Much of our knowledge of meiotic gene conversion in *S. cerevisiae* is the result of research done by S. Fogel, R.K. Mortimer, and collaborators (for review, see Fogel et al. 1981).

1. General Features of Gene Conversion

As described above, gene conversion represents the nonreciprocal transfer of information between two homologous DNA sequences, the replacement of two DNA strands from one chromosome with two DNA strands derived from the homologous chromosome. Gene conversion is observed for every locus and every allele that has been examined in detail at frequencies that vary from about 0.5% (for *trp1*) to 30% (for *his4*) (Fogel et al. 1981; Nag et al. 1989). Conversion events are observed for every type of allele that has been examined, including point mutations, small (1–4 bp) insertions and deletions, and large insertions and deletions (Fink and Styles 1974; Fogel et al. 1981; McKnight et al. 1981; Pukkila et al. 1986; Borts and Haber 1987; Symington and Petes 1988a). In general, the frequency of gene conversion does not appear to be related specifically to the type of mutational change.

Although strains containing one wild-type and one mutant allele are often used to study gene conversion, these events can also be detected in strains with two different mutant alleles. In such intragenic recombination experiments, recombinant spores (either a wild-type spore or a spore with both mutations) usually (about 90% of the time) reflect gene conversion events rather than crossovers (Fogel and Hurst 1967).

Meiotic gene conversion is a faithful process. In convertant tetrads that contain an extra mutant spore (1*A*:3*a*), all three mutant spores contain the same DNA sequence. One strong test of the fidelity of conversion was the analysis of conversion events involving an *ochre* mutation (Fogel and Mortimer 1970). These workers showed that in convertant tetrads containing three mutant spores, all mutants were suppressible by *ochre* suppressors. In addition, in studies involving heterozygous restriction sites, DNA sequences are transferred faithfully by gene conversion (Borts and Haber 1987; Judd and Petes 1988; Symington and Petes 1988a,b).

For every heterozygous marker that has been examined extensively in *S. cerevisiae*, both 3*A*:1*a* and 1*A*:3*a* conversion events have been detected. Alleles that have the same frequency of these two types of conversions show parity (Fogel et al. 1981). Although significant departures from parity have been detected for certain alleles in yeast, the difference between the two classes of tetrads is generally less than a factor of two. Insertions of 4 bp show approximate parity of conversion (Borts and Haber 1987; Symington and Petes 1988a), although large (>100 bp) deletions frequently show small departures from parity, having an excess of tetrads in which the deletion is lost by conversion (Fogel et al. 1981; McKnight et al. 1981; Pukkila et al. 1986). In two of three heterozygous insertions of Ty elements (5.7 kb), approximate parity of conversion was detected (Vincent and Petes 1989).

The amount of DNA transferred during a single gene conversion event varies between less than several hundred base pairs to greater than 12 kb. From studies involving closely linked sites within the *ARG4* gene, Fogel and Mortimer (1969) concluded that several hundred base pairs were transferred in a typical conversion event. In contrast, DiCaprio and Hastings (1976) observed coconversion events involving genes 10 cM apart at a frequency of about 0.5% of unselected tetrads, suggesting that some conversion tracts were very long. In studies involving strains with heterozygous restriction sites, the average minimum conversion tract lengths (calculated by averaging the distances between the sites known to have been coconverted) were 0.4 kb (Borts and Haber 1987), 1.6 kb (Symington and Petes 1988a), 0.9 kb (Judd and Petes 1988), and 1.4 kb (Borts and Haber 1989). The average tract lengths (calculated by averaging the minimum and maximum possible values for each tract) were 1.5 kb (Borts and Haber 1987), 3.7 kb (Symington and Petes 1988a), and 2.3 kb (Judd and Petes 1988). There are three conclusions from these studies. First, meiotic gene conversion tracts often exceed several hundred base pairs in length; about 20% of the tracts in one study were in excess of 5 kb (Symington and Petes 1988a). Second, the sizes of the tracts vary according either to the strain or to the position in the genome. Third, when flanking markers showed gene conversion, an intervening marker almost always converted in the same direction. Therefore, information is usually transferred continuously from one chromosome to the other.

2. Preferred Sites for the Initiation of Gene Conversion

In fungi, the frequency with which a mutation converts is often related to its position within the gene: Markers located near one end of the gene convert more frequently than those located at the other end (for review, see Whitehouse 1982). This phenomenon is called polarity. We refer to the end of the gene that has a high rate of conversion as the high-conversion end. It is not clear whether most yeast genes show polarity, because a demonstration of polarity involves extensive tetrad analysis with many well-mapped mutant alleles. To our knowledge, there is no example of a yeast gene without a polarity gradient, and there are four yeast genes that do have polarity gradients: *HIS1* (Fogel and Hurst 1967), *ARG4* (Fogel et al. 1981), *HIS4* (M. White et al., unpubl.), and *HIS2* (R. Malone, unpubl.). The high-conversion end of the gene is the 5′ end for *ARG4* and *HIS4*, but the 3′ end for *HIS2*; the transcriptional orientation of the *HIS1* gene has not yet been established. The polarity gradient at *ARG4* ranges from 8.6% at the 5′ end of the gene to 1.4% at the 3′ end (Fogel et al. 1981). Because a similar polarity gradient at *ARG4* was found using mutations generated in vitro by filling in restriction sites (Nicolas et al. 1989), polarity is not specific to the type of allele used to assay conversion.

Polarity gradients could reflect (1) an initiation site for conversion that is located in the high-conversion end of the gene, and from which conversion tracts are propagated in a distance-dependent manner (for review, see Whitehouse 1982; Hastings 1988) or (2) a site located at the low-conversion end of the gene that represses gene conversion in a distance-dependent manner. If the first possibility is correct, it should be possible to delete DNA sequences at the high-conversion end of the gene and lower the rate of conversion within the gene. This approach was used by Nicolas et al. (1989) to analyze gene conversion at *ARG4*. These workers found that several overlapping deletions that removed the *ARG4* promoter lowered the rate of conversion of alleles near the 5′ end of the gene by two- to ninefold. When these deletions (abbreviated Δ) were heterozygous, there was a disparity in conversion, with more tetrads having 1 wild type:3Δ spores than 3 wild type:1Δ spore. This disparity is the opposite of that observed for deletions that do not contain an initiation site (see Section III.B.1). These results suggest that the chromosome containing the initiation site tends to receive rather than donate information during the conversion event.

The data described above suggested the possibility that the site of initiation of gene conversion may be the promoter and that gene conversion is therefore related to transcription (Nicolas et al. 1989). An alternative interpretation (Nicolas et al. 1989) is that the initiation site is close to but separable from the promoter. Consistent with the latter interpretation are the observations that neither a promoter deletion (reducing gene expression 15-fold) nor a deletion of a *trans*-acting positive activator (reducing gene expression 4-fold) strongly affected gene conversion at the *HIS4* locus (M. White et al., unpubl.). The conclusion that there is a single initiation site for gene conversion at *ARG4* is also complicated by the findings that one of the deletions that removed the promoter did not reduce the level of conversion (Nicolas et al. 1989) and that one deletion that removed only DNA sequences 3′ to the gene reduced the level of conversion (N. Schultes and J. Szostak, pers. comm.). It is not clear whether these results indicate a requirement for two sites for the initiation of conversion or complicating interactions between flanking DNA sequences. For example, the 3′ deletion may move a DNA sequence that regionally represses recombination closer to *ARG4*.

Similar complexities were also observed in an analysis of a hot spot for gene conversion at the *HIS2* locus (R. Malone et al., unpubl.). Both insertion and deletion studies indicated that DNA sequences to the 3′ side of the gene were required for gene conversion; but when the 3′ region of *HIS2* was moved near *LYS2*, no stimulation of *LYS2* conversion was observed. A 3-kb DNA fragment containing *HIS2* with flanking 3′ and 5′ sequences stimulated gene conversion at *LYS2* about sevenfold. These results support the conclusion that the initiation site is bipartite and/or only functions in certain contexts. Interestingly, the mini-Tn3 element that dramatically stimulated meiotic crossing-over in a different chromosomal region in a different strain (discussed in Section III.A.1) reduced gene conversion at *HIS2* when inserted near the 3′ end of the gene.

The *ade6-M26* mutation of *S. pombe,* which has already been discussed as a hot spot for crossing-over, is also a conversion hot spot (Gutz 1971). The rate of conversion of *ade6* was about tenfold the rates for other closely linked *ade6* mutations. Conversion events involving the heterozygous *M26* allele showed disparity, 3 *ade6+*:1 *ade6-M26* tetrads exceeding 1 *ade6+*:3 *ade6-M26* tetrads by about 12-fold. These results suggest that the *M26* mutation creates an initiation site for conversion and that the chromosome containing the initiation site tends to receive rather than donate information during conversion (Gutz 1971). The stimulation of conversion by *M26* does not require heterozygosity (Goldman and Smallets 1979; Ponticelli et al. 1988). Because the *M26* mutation is a nonsense mutation within the coding sequence of *ade6* (Gutz 1971; Ponticelli et al. 1988; Szankasi et al. 1988), the stimulation of conversion is presumably not connected with transcription of *ade6*. Any interpretation of the effects of *M26* is complicated by the observation that the effect of the mutation is dependent on the chromosomal context (A. Ponticelli and G. Smith, pers. comm.).

Nonrandomness in the frequency of gene conversion was also observed in the 22-kb interval between *LEU2* and *CEN3* (Symington and Petes 1988a). In this study, in 82 tetrads with a crossover in the interval, one heterozygous marker (located ~7 kb from the centromere) converted in 27 tetrads, whereas a marker located in the middle of the interval converted only four times. In general, the regions of high conversion bordered the regions of high crossing-over. Large (38-fold) differences in the rate of gene conversion were also observed between *leu2* heteroalleles inserted at different positions in the genome (Lichten et al. 1987).

In considering the effects of chromosomal context on gene conversion, one interesting question is whether the centromere affects the rate of these events. The DNA sequences that define the *S. cerevisiae* centromere are about 100 bp in length (for review, see Fitzgerald-Hayes 1987; Newlon 1988), and mutations within the centromere have been constructed (Gaudet and Fitzgerald-Hayes 1987). Some of these mutant centromeres contain a changed restriction site but are wild type for centromere function. A diploid strain was constructed that was heterozygous for a selectable gene (*URA3*) inserted about 1 kb from *CEN3*, for a restriction site change within the centromeric sequences, and for a restriction site change on the opposite side of the centromere from *URA3* (Symington and Petes 1988b). Approximately 2% of unselected tetrads had conversion events involving *URA3*, half of which were also convertant for the altered site within the centromere; moreover, in about half of the tetrads in which the centromere mutation was converted, the mutation on the opposite side of the centromere was coconverted. Thus, the yeast centromere does not act as a barrier to the extension of a conversion tract, and meiotic gene conversion occurs within the centromere at a rate (at least 1% of unselected tetrads) similar to that observed for other genomic regions.

The effect of the centromere on gene conversion was also examined by Lambie and Roeder (1988). These workers examined heteroallelic gene conversion at the *HIS4* locus in a strain with *CEN3* at its normal position (~45 kb from *HIS4*) and in a strain in which *CEN3* was relocated to within 500 bp of *HIS4*. They found that gene conversion was about fourfold less frequent in the strain containing *CEN3* near *HIS4*. The results from these two studies can be reconciled if it is assumed that a modest negative effect of the centromere on gene conversion is balanced by hot spots for conversion near *CEN3*; interestingly, Lambie and Roeder (1988) found that nonfunctional centromeres appeared to stimulate gene conversion. Alternatively, the results of the two studies may have been influenced by technical differences (measuring gene conversion rates by tetrad analysis [Symington and Petes 1988b] vs. random spores [Lambie and Roeder 1988]) or details of the constructions of the experimental strains. In particular, it is possible that altering the chromosome by insertions or deletions of any DNA sequence affects the normal pattern of recombination.

In summary, the conclusions that can be drawn about initiation sites for meiotic gene conversion are currently somewhat limited. Polarity studies (for review, see Whitehouse 1982) indicate that most putative natural initiation sites are likely to be between genes rather than within genes, although analysis of the *M26* mutation suggests that an initiation site can also be within a gene. In those crosses in which putative initiation sites are heterozygous, the chromosome containing the site is more often the recipient than the donor of information. Mechanisms proposed to account for conversion initiation sites are similar to those proposed for crossover hot spots in Section III.A.1. In our view, it is unlikely that a single factor, such as transcription, will be responsible for controlling the initiation of gene conversion. Further fine-structure deletion analysis of several different conversion initiation sites will be required to resolve these issues.

3. Association between Gene Conversion and Crossing-over

Gene conversion and crossing-over are associated events in yeast (Fogel and Hurst 1967) and other fungi (for review, see Whitehouse 1982). This association has been demonstrated in two ways. First, in tetrads in which a heterozygous marker has converted, there is often an associated crossover of flanking markers. The degree of association appears to vary somewhat for different genetic intervals. In one compilation (Fogel et al. 1981), the frequency of conversions associated with an exchange varied from 0.18 to 0.66, with an average association of about 0.35.

In one study, Fogel and Hurst (1967) examined conversion-associated exchange at the *HIS1* locus by constructing a diploid that was heteroallelic at *HIS1* and heterozygous for two flanking genes. Tetrads with one His$^+$ spore were analyzed. Most of these His$^+$ spores were the result of gene conversion, and about 39% had a conversion-associated exchange. In 75% of the tetrads that had a conversion event involving only the centromere-proximal marker and that had an associated crossover, the crossover occurred between the centromere-proximal flanking marker and the centromere-distal *his1* mutation. In the remaining 25%, the crossover occurred between the centromere-distal *his1* mutation and the centromere-distal flanking marker. The implications of this result are discussed further elsewhere in this review. It should be pointed out that crossovers can be mapped only if the flanking markers segregate 2:2.

Just as conversion events are associated with crossovers, crossovers (selected in a region with a high density of heterozygous markers) are often associated with gene conversion of one or more adjacent markers (Borts and Haber 1987; Symington and Petes 1988a). Because the number of conversion events detected is a function of the density of markers, the observed associations (57–59%) are minimal estimates. Among tetrads without a crossover between *LEU2* and *CEN3*, only 16% of the tetrads had a detectable conversion event (Symington and Petes 1988a). Although most of the conversion events were adjacent to the crossover in these studies, the conversion tract was separated from the position of the crossover by a marker segregating 2:2 in about 10% of the tetrads.

The simplest interpretation of the association between gene conversion and crossing-over is that these events are mechanistically connected, and such a connection is explicit in most models of recombination (Holliday 1964; Meselson and Radding 1975; Szostak et al. 1983). In these models, gene conversion events reflect an intermediate in the formation of crossovers. The alternative possibility—that conversion and crossing-over represent mechanistically distinct events that require a common condition (e.g., "accessible" DNA)—has not been ruled out (for review, see Fink and Petes 1984; Hastings 1987b; Ray et al. 1988). The experimental data that are relevant to this heterodox alternative come from several sources. First, conversion events in different regions of the genome show a variable association with crossing-over. Second, in experiments involving recombination between repeated genes within a chromosome, gene conversion is only weakly (~10%) associated with crossing-over (Klein and Petes 1981; Klar and Strathern 1984; Klein 1984; Jackson and Fink 1985; Ray et al. 1988). Third, as described above, the associated crossover is not necessarily adjacent to the conversion tract (Fogel and Hurst 1967; Borts and Haber 1987; Symington and Petes 1988a). Finally, *trans*-acting yeast mutations (discussed below) have been isolated that differentially affect conversion and crossing-over. In summary, gene conversion events may represent an intermediate in the formation of a crossover whose resolution is under sophisticated regulation. Alternatively, these events may be mechanistically distinct processes that tend to occur in the same region of the chromosome, perhaps reflecting a preference for a similar substrate.

C. Postmeiotic Segregation

In some fungi, the ascus contains eight rather than four spores. In such fungi, following the two meiotic divisions, there is a mitotic division of each meiotic product. Usually, the spores derived from a single meiotic product have the same genotype. Thus, in strains heterozygous for gene *A*, normal Mendelian segregation is 4*A*:4*a*, and gene conversion events are either 6*A*:2*a* or 2*A*:6*a*. At a frequency that is somewhat organism- and allele-dependent, segregation ratios of 5*A*:3*a*, 3*a*:5*A*, or other patterns are also observed (Kitani et al. 1961). In organisms with ordered asci (the order of spores in the ascus reflecting the segregation of the meiotic products), it is clear that such patterns are the result of segregation of the A and a alleles into two spores derived from the same meiotic product, a postmeiotic segregation (PMS). Because the two alleles segregate from a single haploid meiotic product, the simplest interpretation of a PMS event is that it reflects replication of a chromosome that contains an unrepaired mismatch in a heteroduplex that has one DNA strand of A and one DNA strand of a information.

To detect PMS events in yeast (which has only four spores per ascus), Esposito (1971) dissected tetrads directly on petri dishes containing a rich nutrient medium (Fogel et al. 1979). The colonies derived from the individual spores were replica plated to omission media (media lacking one amino acid or base), and PMS events for heterozygous auxotrophic markers were signaled by sectored colonies (one half of the colony growing and one half not growing as shown in Fig. 5). For describing PMS events in yeast, the nomenclature of eight-spored fungi is used. Thus, if the segregation pattern for A and a is 2A spore colonies:1a spore colony:1 sectored A/a spore colony, the segregation pattern is 5A:3a. The segregation pattern 1A:1a:2 sectored A/a spore colonies is called an aberrant 4:4.

1. General Features of PMS in Yeast

For most alleles, PMS events are considerably less frequent than gene conversion events (Fogel et al. 1981). The fraction of aberrant (non-2:2) segregations that represents PMS events varies from 0% to 54% for various mutant alleles; the median frequency of PMS as a fraction of the total aberrant segregations is 3.5%. Because the average frequency of aberrant segregation is about 4.5%, PMS events represent only 0.15% of unselected tetrads for an average allele. The low frequency of PMS in yeast indicates either that heteroduplex formation between two chromosomes occurs infrequently or that most mismatches existing in heteroduplexes are efficiently repaired. As some genes have both high and low PMS alleles (alleles with high or low frequencies of PMS relative to other types of aberrant segregation) at similar map positions, the latter interpretation appears to be correct for at least some loci. Because PMS events in yeast are rare for most alleles, it is not clear whether most alleles show equal numbers of 5:3 and 3:5 tetrads. The *arg4-16* allele, which shows frequent PMS, deviates strongly from parity—95 5:3 tetrads and 289 3:5 tetrads (Fogel et al. 1981).

Although for some mutant alleles in yeast, the 5:3 and 3:5 classes of tetrads are observed at substantial frequencies, the aberrant 4:4 class is almost always extremely rare (Fogel et al. 1981). The frequency of aberrant 4:4 tetrads is not greater than that expected as a result of two independent 5:3 and 3:5 events. As discussed later, this observation indicates that heteroduplex formation in yeast usually involves an asymmetric transfer of a single DNA strand from one chromosome to another, rather than a symmetric exchange of single strands between two chromosomes.

PMS events, like conversion events, represent faithful transfers of information from one chromosome to another (Hurst et al. 1972). In studies of crosses with multiple closely linked heterozygous markers (DiCaprio and Hastings 1976; Fogel et al. 1981), PMS events including more than one marker were detected. From such studies, it appears that the amount of DNA transferred in a single PMS event is similar to that observed for gene conversion.

2. High and Low PMS Alleles

Although the frequency of non-2:2 segregation is often related to the position of the mutant allele in the gene, the fraction of such segregations that are PMS does not have any obvious relationship to location. For example, the *arg4-16* mutation is only 200 bp from the *arg4-17* mutation (White et al. 1985), yet about half of the non-2:2 segregations involving *arg4-16* are PMS compared to only 5% for *arg4-17*. In the following discussion, all estimates of the PMS frequency are given as a proportion of total non-2:2 segregations.

Sequence analysis of a number of *arg4* mutations indicated that the specific base-pair mismatches were likely to be important in determining which point mutations would have high levels of PMS (White et al. 1985). The sequence changes in *arg4-16*, *arg4-17*, and an amber derivative of *arg4-17* (*arg4-17*[*]) were G to C, A to T, and A to G, respectively. Heteroduplexes formed between a wild-type and mutant DNA strand would be expected to result in two different types of mismatches, depending on whether the heteroduplex is formed between a mutant non-transcribed and wild-type transcribed strand or vice versa. Thus, in diploids with one mutant and one wild-type gene, the *arg4-16*, *arg4-17*, and *arg4-17*[*] mutations would be expected to result in mismatches of G-G and C-C, A-A and T-T, and C-A and G-T, respectively. The *arg4-16* allele had a PMS frequency about sixfold higher than the other two alleles, suggesting that C-C and/or G-G mismatches were inefficiently repaired in meiosis. Other mutations creating C-C and G-G mismatches at the *ARG4* (Lichten et al. 1990) and *HIS4* loci (Detloff et al. 1991) have also been shown to result in high PMS frequencies. At the *HIS4* locus, heterozygous mutations representing all possible mismatches were examined, and only the C-C, G-G mismatch pair exhibited high PMS frequencies.

Several experiments indicate that the C-C, rather than the G-G, mismatch is inefficiently repaired. First, when heteroduplexes with specific mismatches were constructed in vitro and transformed into mitotically dividing cells, Bishop et al. (1989) and B. Kramer et al. (1989) found that the C-C mismatch was poorly repaired relative to the G-G mismatch. In a second approach, Lichten et al. (1990) physically examined spore DNA derived from a diploid that was heterozygous for an *arg4* mutation that was expected to generate C-C and G-G mismatches. Using a denaturing gel technique, these workers found that unrepaired C-C mismatches were more common than unrepaired G-G mismatches. Using a different type of analysis, Detloff et al. (1991) also found that C-C mismatches were repaired more poorly than G-G mismatches at the *HIS4* locus.

Interestingly, similar specificities in mismatch repair are also found in other systems. In *S. pombe,* Hofer et al. (1979) and Thuriaux et al. (1980) showed that G-C to C-G transversion mutations in two different serine tRNA genes showed unusually high PMS frequency. In *Streptococcus pneumoniae,* transformation specificities (which reflect mismatch repair) indicate that C-C and/or G-G mismatches are repaired inefficiently (Lacks et al. 1982), and, in *Escherichia coli,* C-C mismatches are inefficiently repaired by the methyl-directed mismatch repair system (Kramer et al. 1984). One complication to the generalization that C-C mismatches are inefficiently repaired is the observation that in a cell-free mismatch repair system derived from mitotic cells (Muster-Nassal and Kolodner 1986), C-C mismatches were repaired slightly more efficiently than the G-G mismatches. Whether this result is an artifact of the in vitro system or reflects other complicating factors (such as strain differences in repair specificity or an effect of the context of the mismatch on repair) is not yet clear.

One indication that the rules governing PMS in yeast are not likely to be simple is the results obtained with insertion and deletion alleles. Early experiments indicated that large (at least 100 bp) insertions or deletions (Fogel 1981), 4-bp insertions (Borts and Haber 1987; Symington and Petes 1988a), and 1-bp insertions (White et al. 1985) showed no PMS, suggesting that the corresponding mismatches were efficiently repaired. Similarly, in studies done by transforming mitotic yeast cells with heteroduplexes formed in vitro, Bishop (1989) found that mismatches involving a 1-bp insertion were repaired efficiently. Because these insertion and deletion mismatches were repaired efficiently, it was surprising that *ade8-18*, which has a 54% PMS frequency (Esposito 1971; Fogel et al. 1981), is a 38-bp deletion (White et al. 1985). From the analysis of the PMS behavior of a number of other small deletions, White et al. (1988)

suggested that the junction formed as a result of the deletion could create a protein-binding site and binding of a protein at this site could prevent mismatch repair. High PMS alleles involving small insertions have also been observed. For example, a 26-bp insertion (containing the *lexA* operator site of *E. coli*) in the *HIS4* gene of yeast has a very high (65%) frequency of PMS (Nag et al. 1989); this result is observed in the absence of the LexA protein. Because the *lexA* operator is a perfect palindrome (Brent and Ptashne 1984), in a heteroduplex formed between a wild-type *HIS4* gene and the mutant gene, one would expect the operator sequence to show intrastrand pairing, forming a hairpin. The high level of PMS indicates that this structure is poorly recognized by the yeast repair system. This effect is not specific to the *lexA* operator, as another palindrome with an unrelated sequence had similar properties (Nag et al. 1989). The DNA sequences that were deleted in *ade8-18* were not palindromic, suggesting that the inefficient repair of the *ade8-18* deletion is not directly related to the inefficient repair of palindromic insertions.

If high PMS alleles reflect inefficient repair of mismatches (and not a stimulation of heteroduplex formation), it is likely that low PMS alleles in the same region of the gene reflect efficient repair of mismatches existing within the heteroduplex; the repair could result in a gene conversion event (Holliday 1964). This conclusion is further supported by studies showing that alleles with low PMS generally reduce the level of PMS of closely linked high PMS alleles. For example, in one cross, the *arg4-16* allele, when heterozygous by itself, had a PMS frequency of 48%; if both the *arg4-16* and a closely linked low PMS *arg4* allele were heterozygous, the frequency of PMS at *arg4-16* was reduced to 28% and the frequency of conversion of *arg4-16* was elevated proportionally (Fogel et al. 1979). Similar results were also obtained at the *HIS1* locus (Hastings 1984) and at several loci in other fungi (Hastings 1987a, 1988). These results suggest that inefficiently repaired mismatches are recognized poorly by the mismatch repair system, rather than being intrinsically difficult to repair.

3. Association between PMS and Crossing-over

In crosses that are heterozygous for a high PMS allele and flanking markers, the association between PMS and crossing-over can be examined. Both in yeast (Fogel et al. 1979) and in other fungi (Whitehouse 1982; Rossignol et al. 1988), a significant association is found. For example, about half of tetrads displaying PMS for *arg4-16* also had an associated crossover. As discussed below, this observation has been interpreted as indicating that a region of heteroduplex (reflected by the PMS event) is an intermediate in the formation of a crossover (Holliday 1964; Meselson and Radding 1975). Alternatively, as discussed above with respect to gene conversion, PMS and crossovers may require a common substrate (e.g., chromatin in an "open" configuration) but follow mechanistically different pathways.

D. DNA Intermediates in Recombination

To look at intermediates in recombination, investigators have pursued three different approaches: examination of the meiotic DNA for single-strand nicks or double-strand breaks, analysis of meiotic recombination after induction of double-strand breaks, and electron microscopic examination of meiotic DNA. Using sucrose gradients, Resnick et al. (1984) failed to find either nicks or breaks in meiotic DNA isolated from wild-type cells. The investigators estimated that a steady-state level of greater than ten breaks per genome could have been detected. Indeed, as described in Section III.H, meiosis-specific single-strand nicks were observed in strains with *rad52* or *rad57* mutations. Using the OFAGE procedure on DNA isolated from meiotic cells, Game et al. (1989) found a low level (less than one per chromosome at a given time) of double-strand breaks on chromosome III.

Meiotic double-strand breaks have been detected at specific sites in two studies. Sun et al. (1989) found double-strand breaks at the 5' end of *ARG4* in both *ARG4*-containing plasmids and chromosomal DNA molecules. The broken DNA molecules had S1-sensitive tails, indicating terminal single-stranded regions of several hundred nucleotides. The position of the break at *ARG4* was close to the site required for high rates of gene conversion at this locus (Nicolas et al. 1989), and a deletion that reduced the level of gene conversion at *ARG4* eliminated formation of the break. The interpretation of these results is complicated by the observation that a deletion at the 3' end of *ARG4* that reduces gene conversion has little effect on the formation of double-strand breaks in *ARG4*-containing plasmids (J. Szostak, pers. comm.); this deletion has not yet been tested for its effects on the double-strand break in chromosomal DNA. Cao et al. (1990) also found evidence for a meiosis-specific discontinuity in the chromosome. The endpoint was mapped in *LEU2* sequences that had been introduced near *HIS4* on chromosome III, as well as in the normal *LEU2* locus in the same strain. The breaks occurred after meiotic DNA replication but preceded formation of the synaptonemal complex and crossover products (measured using heterozygous restriction sites) in the same chromosomal region.

There are at least three possible interpretations of these experiments. First, the breaks may represent the initiation of meiotic recombination events, as predicted by the double-strand-break repair model of recombination (Szostak et al. 1983), although it is unclear why such breaks were not also detected in wild-type strains by Resnick et al. (1984). Second, the breaks may reflect artifactual processing of some other type of DNA replication or recombination intermediate (e.g., nuclease degradation of a gapped molecule into a broken molecule). Third, these structures may represent a recombination intermediate that is different from a simple chromosomal break (Cao et al. 1990).

An alternative method of defining recombination intermediates is to induce a break in the chromosome during meiosis and to determine whether the break is recombinogenic. Kolodkin et al. (1986) constructed a strain in which a specific double-strand break was induced in meiosis at the *MAT*a locus by using a galactose-inducible HO endonuclease. They found that the break stimulated crossing-over of markers flanking the *MAT* locus but that conversion events at *MAT* were usually 4:0 instead of 3:1. This result suggests that meiotic recombination can be induced by double-strand breaks but that the effects of these HO-induced breaks are not the same as those observed for normal meiotic recombination. In contrast to these results, Klar and Miglio (1986) found that double-strand breaks at the *mat1* locus of *S. pombe* stimulated both crossing-over and the expected type of meiotic conversion (3:1). One interpretation of these results is that double-strand breaks may be the normal initiator of recombination in *S. pombe* but not in *S. cerevisiae*. Alternatively, double-strand breaks may also be the normal initiator of recombination in *S. cerevisiae*, but the timing, number, or structure of the HO-induced breaks in the study of Kolodkin et al. (1986) is different from that of the breaks induced normally in meiosis. For example, the observed 4:0 segregations may reflect breaking of the chromosome before DNA replication or breaks in both sister chromatids after DNA replication.

Bell and Byers (1979) used electron microscopy to detect meiosis-specific DNA structures that may represent recombination intermediates resulting from crossing-over between the inverted repeats of the 2-micron plasmid. These workers examined 2-micron plasmids that migrated anomalously in an electric field and observed DNA molecules in which the inverted repeats were joined at a single point. It is not completely clear that these structures reflect the action of enzymes catalyzing general homologous recombination, because yeast has a site-specific enzyme (Flp1p) that catalyzes recombination between the inverted repeats of the plasmid (Broach et al. 1982; Broach and Volkert, this volume). In subsequent studies using a similar approach, Bell and Byers (1983) identified meiosis-specific X-shaped intermediates derived from yeast chromosomal DNA.

E. Models of Recombination

In most models of recombination, the three processes of crossing-over, gene conversion, and PMS are considered to be intimately related mechanistically. We restrict detailed discussion to three particularly influential models proposed by Holliday (1964), Meselson and Radding (1975), and Szostak et al. (1983).

1. General Features of the Holliday and Meselson-Radding Models

In the Holliday model (Fig. 6), recombination is initiated by single-strand nicking of homologous chromosomes at the same position. Strand exchange then occurs symmetrically, leading to formation of the Holliday junction. If these symmetric heteroduplexes include a heterozygous marker, there will be mismatches on both chromatids. Repair of both mismatches in the same direction (both to mutant or both to wild type) would result in gene conversion, whereas failure to repair both mismatches would result in an aberrant 4:4 segregation. Repair of only one mismatch would result in a 5:3 or 3:5 segregation. To allow segregation of the connected chromatids, the Holliday junction is cleaved. As shown in Figure 6, cleavage in one plane leads to crossing-over of flanking markers, whereas cleavage in the other plane fails to generate a crossover. In this model, as both gene conversion and PMS reflect formation of a heteroduplex and a heteroduplex is an obligatory intermediate in the formation of a crossover, these events are associated.

Although the Holliday model was consistent with most of the available data at the time it was conceived, several experiments done subsequently in yeast were difficult to explain. For example, Fogel et al. (1981) found that the *ade8-18* allele showed a high frequency of 5:3 and 3:5 segregation but a very low frequency of aberrant 4:4 segregation (about that expected for two independent events, each involving a single heteroduplex). This result suggests either that symmetric heteroduplexes are rare in yeast or that repair of mismatches in symmetric heteroduplexes is constrained in peculiar ways. In addition, repair of both mismatches within a symmetric heteroduplex would sometimes lead to tetrads that appeared to have had a two-strand double crossover. For example, in Figure 6, if the top mismatch is repaired to *b* and the bottom mismatch is repaired to *B*, the resulting tetrad (assuming no associated crossover) would have spores with the genotypes *ABC*, *AbC*, *aBc*, and

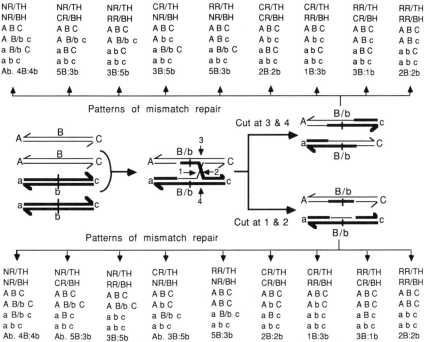

Figure 6 Patterns of meiotic segregation expected as the result of symmetric heteroduplex formation (Holliday 1964). A symmetric heteroduplex is formed at the *B* locus. The mutant base in the *b* allele is indicated by the short vertical line. Resolution of the structure by cleavage at positions 1 and 2 results in tetrads with no associated crossover (*bottom*), whereas cleavage at positions 3 and 4 results in tetrads with an associated crossover (*top*). Nine different possible repair patterns exist for each pathway. (NR) No repair; (CR) conversional repair: repair of mismatch in the direction of the invading strand; (RR) restorational repair: repair of mismatch in the direction of the invaded strand; (TH) top heteroduplex; (BH) bottom heteroduplex; (Ab.) aberrant.

abc. Two-strand, three-strand, and four-strand double crossovers are expected in a ratio of 1:2:1 from "true" crossover events. No excesses of two-strand double crossovers altering this ratio have been detected in yeast for a number of different chromosomal intervals (Fogel et al. 1981). Another class of tetrad expected for symmetric heteroduplexes is obtained if the top mismatch in Figure 6 is not repaired and the bottom mismatch is repaired to *B*. If there is not an associated crossover, the spores would show an "aberrant 5:3" segregation pattern: *ABC, A(B/b)C, aBc, abc.* This class of tetrad is almost never found in yeast (DiCaprio and Hastings 1976). However, in some other fungi, aberrant 4:4 and aberrant 5:3 segregation patterns are much more common, consistent

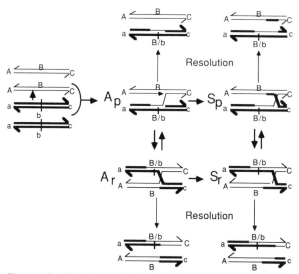

Figure 7 Meselson-Radding model of recombination. Recombination is in-itiated by nicking one of the chromosomes. The nicked strand invades the homologous chromatid, displacing the resident strand, which is degraded. Repair synthesis fills in the gap on the donor chromatid. The resulting intermediate (A_p) has an asymmetric heteroduplex with the flanking markers in the parental con-figuration. A_p can isomerize into a structure with the flanking markers in the recombinant configuration (A_r) or branch migration can occur, resulting in a structure with symmetric (as well as asymmetric) heteroduplex and markers still in the parental configuration (S_p). Either branch migration of A_r or isomerization of S_p can yield the S_r structure. Resolution of the intermediates occurs by cleavage of the "crossing" strand(s), followed by ligation of the free ends.

with the patterns expected for the Holliday model (for review, see Whitehouse 1982).

To explain the presence of 5:3 and 3:5 patterns and the near absence of aberrant 4:4 and 5:3 patterns, Meselson and Radding (1975) suggested that heteroduplex formation in yeast was often asymmetric (Fig. 7). In their model, heteroduplex formation is initiated by a single-strand nick on one chromosome, followed by displacement of the nicked strand, which invades the neighboring chromatid. This results in an asymmetric heteroduplex with flanking markers in the parental configuration (A_p). Rearrangement (isomerization) of this structure and/or associated branch migration of the region of heteroduplex can generate structures with asymmetric heteroduplexes and flanking markers in the recombinant configuration (A_r) or various structures with symmetric heteroduplexes (S_r and S_p). These structures are resolved by cutting the crossing DNA

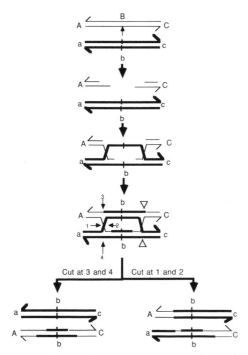

Figure 8 Double-strand-break repair model of recombination (after Fig. 10 in Szostak et al. 1983). Recombination is initiated by a double-strand break of the DNA that is expanded into a gap. The resulting ends invade the neighboring chromatid and the gap is repaired. One of the junctions of the double Holliday structure is resolved by cutting the outer strands (*open up and down arrowheads*), whereas the other junction is cut either at positions 3 and 4 (no crossover results) or at positions 1 and 2 (a crossover results).

strand(s). Meselson and Radding (1975) suggested that in organisms like *S. cerevisiae*, recombination is restricted to the asymmetric modes (A_p and A_r), whereas in organisms with more abundant aberrant 4:4 segregations, all four modes might occur.

2. General Features of the Double-strand-break Repair Model of Recombination

Szostak et al. (1983) proposed three variant models, one of which is shown in Figure 8. In this model, the initiating event for recombination is a double-strand break that is extended into a gap with single-strand ends on the same strand. The free single-strand ends invade the neighboring chromatid, and the gap is repaired by DNA synthesis using the invaded DNA as a template. If a heterozygous marker is in the region covered by the gap, it will undergo gene conversion without involvement in a heteroduplex. PMS events will occur if a heterozygous marker is included in the region of heteroduplex near the borders of the gap. Because heteroduplex DNA occurs on only one chromatid, aberrant 4:4 tetrads would not be expected. Repair of a mismatch in the heteroduplex region is a second way of generating a gene conversion event. Following gap repair, the resulting double Holliday structure is cleaved to yield chromosomes in which the gap-repaired site is flanked by markers in the parental (if both junctions are cleaved in the same sense) or recombinant configurations (if the two junctions are cleaved in different senses).

3. Experimental Evidence Relevant to the Models

Some of the distinguishing features of the Meselson-Radding and the double-strand-break repair models include (1) the mode of initiating the recombination event, (2) the position of the crossover associated with PMS or conversion, and (3) the nature of the repair process that results in gene conversion. Data relevant to each of these issues are described below. Other reviews treating these issues in detail include those of Orr-Weaver and Szostak (1985), Hastings (1987a,b, 1988), Rossignol et al. (1988), and Thaler and Stahl (1988).

a. Initiation of Recombination. One of the most intuitively disagreeable features of the double-strand-break repair model is the disruption in the continuity of the chromosome required to initiate recombination. Because there are about 100 crossover events and 200 conversion events per meiosis in yeast and a single unrepaired double-strand break would be lethal, this mode of initiating recombination seems to be intrinsically risky. Kolodkin et al. (1986) found that induction of a single double-strand break in meiosis had no obvious effect on spore viability, indicating that the meiotic repair process is reasonably efficient. However, a small effect of the double-strand break on spore viability would not have been detected in these studies. It would be useful to induce multiple breaks in the genome during meiosis by similar techniques and to monitor spore viability.

As described in Section III.D, two groups have obtained evidence for meiosis-specific double-strand breaks in the DNA extracted from wild-type strains, and the introduction of double-strand breaks stimulated conversion and crossing-over, although (at least in *S. cerevisiae*) they were abnormal. No evidence for specific single-strand nicks in wild-type strains has been reported, although some *rad* mutants did accumulate nicks (Resnick et al. 1984) (also see Section III.H).

As described in Section III.B, analysis of several putative gene-conversion initiation sites indicates that the chromosome containing the site acts as the recipient for information. This result is explained most easily in the context of the double-strand-break repair model (Szostak et al. 1983). By this model, the initiation site is a DNA sequence that is susceptible to being cut; once the chromosome containing the site is broken, the break is expanded into a gap that is repaired by using the chromosome that does not have the site as a donor. These observations are explained less easily by the Meselson-Radding model, because if the initiation site represented a region of DNA that was easily nicked, the chromosome containing the site would be expected to be the donor rather than the recipient of information. One modification of the Meselson-Radding model (Radding 1982) is that the chromosome containing the site is nicked and the nick is expanded into a single-strand gap. The chromosome containing this gap then acts as a recipient for a single strand transferred from the other chromatid (Fig. 9). A related possibility is that initiation sites represent regions of DNA that are susceptible (although not necessarily nicked or gapped) to invasion by single strands of DNA from the homologous chromatid.

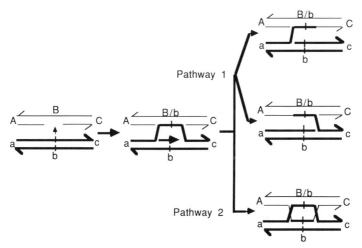

Figure 9 Radding's (1982) model of recombination. Recombination is initiated by a nick that is extended into a single-strand gap. The gap is then invaded by a strand derived from the homologous chromatid. The resulting D loop can be processed by either of two pathways. In pathway 1, one of the crossing strands is cleaved to form an intermediate similar to the A_p structure of Fig. 7. This intermediate can isomerize and be resolved to produce crossovers. Alternatively, a double Holliday structure may be formed (pathway 2); this structure can be processed as described in Fig. 8.

b. Position of the Crossover Associated with PMS or Conversion. For both the Meselson-Radding and double-strand-break repair models, polarity of conversion or PMS reflects the occurrence of an initiation event at one site, followed by the propagation of a conversion tract, with decreasing probability, away from this site. In the Meselson-Radding model, the propagated event is the extent of asymmetric or symmetric heteroduplex; in the double-strand-break repair model, the propagated event is the extent of the double-strand gap or of the flanking single-strand regions. In the original formulation of the Meselson-Radding model, the crossovers associated with PMS or conversion events are expected to be located on the low-conversion side of the tract (as determined by the polarity gradient) because the strand connecting the two chromosomes is located at this position. In the double-strand-break repair model or the modified version of the Meselson-Radding model (Radding 1982), the crossovers can occur either on the low-conversion side or on the high-conversion side of the conversion tract because it is flanked by two Holliday junctions.

If a diploid strain is heterozygous for three linked markers (*A*, *B*, and *C*) and a gene conversion event occurs at *B* (for this example, resulting in a 3*B*:1*b* segregation), it is not possible to determine on which side of the conversion tract the associated crossover occurred. Associated crossovers between *A* and *B*, as well as those between *B* and *C*, would result in tetrads with the segregation patterns *ABC*, *ABc*, *aBC*, and *abc*. Crossovers can only be mapped between markers when neither marker exhibits gene conversion. If the *B* gene exhibits PMS, however, it is possible to map the crossover. Thus, a crossover between *A* and *B*, associated with a 5*B*:3*b* segregation, would result in meiotic products with the segregation pattern *ABC*, *A(B/b)c*, *aBC*, *abc*; an associated crossover between *B* and *C* would yield patterns *ABC*, *ABc*, *a(B/b)C*, and *abc*. Using this experimental rationale, Fogel et al. (1979) found that 20 tetrads had a crossover on the low-conversion side of *arg4-16*, and 44 tetrads had a crossover on the high-conversion side. These results are explained more easily by the double-strand-break repair model than by the Meselson-Radding model, although Radding's (1982) modification of the original model allows crossing-over on either the low-conversion or high-conversion side of the conversion tract (Fig. 9).

c. Nature of the Repair Process that Results in Gene Conversion. For most alleles in yeast, gene conversion is much more common than PMS. In the Meselson-Radding model, the explanation of this result is that repair of mismatches in yeast is usually very efficient. In the double-strand-break repair model, the explanation is that most conversion events result from double-strand gap repair rather than from mismatch repair of heteroduplexes; in the original model, the gap was large relative to the size of the single-strands.

Several lines of evidence support the notion that conversion is often the result of heteroduplex repair rather than gap repair. In particular, as discussed in Sections III.C and III.H, (1) high PMS alleles tend to be ones that would yield poorly repaired mismatches; (2) the *pms1* mutation increases PMS and decreases gene conversion of many (although not all) alleles; and (3) low PMS alleles tend to decrease the frequency of PMS for high PMS alleles located nearby. Another observation that is more easily explained by the Meselson-Radding model is the virtual absence of aberrant 5:3 and 3:5 segregations (Section III.E.1). Such segregations would not be expected to occur in the asymmetric phase of the Meselson-Radding model. In contrast, by the double-strand-break repair model, these segregations would be expected to occur frequently unless the double Holliday junctions are constrained to be cut in specific ways (Szostak et al. 1983; Thaler and Stahl 1988). For example, if the double Holliday junctions (see Fig. 8) were both cut in the horizontal plane and there was an unrepaired mismatch at the *B* locus within one of the heteroduplex regions, an aberrant 3:5 tetrad (*ABC*, *AbC*, *a(B/b)c*, and *abc*) would be obtained.

There are also, however, two findings that are difficult to explain by heteroduplex repair of conversion tracts. First, gene conversion of large (>1 kb) deletions and insertions occurs readily in yeast and often with approximate parity (Fogel et al. 1981; Szostak et al. 1983). Because in vitro formation of heteroduplex DNA catalyzed by the RecA protein is blocked by a heterology of about 800 bp (for review, see Radding 1982), one might expect that the conversion rate of such mutations would be less than that for point mutations. Second, gene conversion events are usually continuous (Fogel et al. 1981). Even in conversion tracts extending several kilobases in which multiple markers are heterozygous, markers are almost invariably converted in the same direction and it is unusual to observe a marker segregating 2:2 with flanking markers segregating 3:1 or 1:3 (Borts and Haber 1987; Symington and Petes 1988a). If conversion is due to repair of mismatches in a heteroduplex, these mismatches must be repaired coordinately, either because there is a single excision tract or because one of the strands is "tagged" for repair.

A controversial issue concerning mismatch repair is whether conversion and restoration occur to equal extents. If a DNA strand containing one allele invades a chromosome containing another, repair of the mismatch in the heteroduplex in the direction of the invading allele results in a conversion event; repair in the direction of the invaded allele is called a restoration event. Hastings and Savage (Savage and Hastings 1981; Hastings and Savage 1983) examined gene conversion and associated crossovers at the *HIS1* locus and argued that conversion is more common than restoration. To reach this conclusion, these authors assumed that all associated crossovers occurred toward the low-conversion side of the conversion tract. Although they presented suggestive evidence in support of this assumption (Savage and Hastings 1981), more direct evidence indicates that the crossovers can be on either the low-conversion side or the high-conversion side of the heteroduplex region (as discussed above). Thus, the relative frequency of conversion and restoration in yeast is not known. If further research shows that no restoration events occur, this would support the double-strand-break repair model, as there is no restoration possible with gap repair.

4. Alternative Models and Future Prospects

As described above, neither the Meselson-Radding model nor the double-strand-break repair model, as originally formulated, easily explains all of the available data on yeast meiotic recombination. The most prominent difficulties with the Meselson-Radding model concern the positions of the crossovers, the continuity of the conversion tract, and the observation that the chromosome containing the initiation site for conversion acts as a recipient. The major difficulty with the original version of the double-strand-break repair model is the increasing body of evidence that a substantial fraction of gene-conversion events are the result of heteroduplex repair.

Hastings (1984) proposed a model that had some features of both of the earlier models. In the Hastings model, heteroduplex formation occurs as proposed by Meselson and Radding. Although some of the resulting mismatches are then repaired by excision and replacement of a single strand, others trigger a double-strand break. The broken chromosome is repaired by using the homologous chromosome as a template. The experimental results predicted by this model depend on the efficiency of mismatch repair and whether mismatches are more frequently repaired following excision of a single strand or a double strand. In this model, the continuity of the conversion tracts and the position of the associated crossover relative to the conversion tract are explained in the same way as for the double-strand-break repair model. As for the Meselson-Radding model, the Hastings model allows the specific mismatch to control the relative amounts of PMS and gene conversion and predicts that the chromosome with the initiation site will be a donor in the conversion event. Another clear prediction is that recombination will be qualitatively altered by increasing the number of heterozygous markers. In two studies involving multiple heterozygous markers, such an alteration was detected in one analysis (Borts and Haber 1987) but not in the other (Symington and Petes 1988a).

In our view, the two models of recombination that appear to have the fewest conflicts with the available yeast data are other variations on the Meselson-Radding and double-strand-break repair models. One is Radding's (1982) model in which the initiating event is a nick that is expanded into a gap, followed by formation of an asymmetric heteroduplex with the gapped molecule as the recipient (Fig. 9). Although the steps following this initiation event were not detailed by Radding, one possibility is that a double Holliday junction is formed (Fig. 9, pathway 2). This structure could then be resolved by cleaving each junction as described for the double-strand-break repair model. Alternatively, formation of the asymmetric heteroduplex could be followed by cleavage of one of the two crossing strands (Fig. 9, pathway 1); the remaining structure would then be processed as described in the original Meselson-Radding model. Radding's model allows the crossover to occur on either side of the conversion tract and accounts for the observation that the chromosome on which the conversion event initiates is the recipient. A variant double-strand-break repair model that also accounts for most of the observations is that the initiation event is a double-strand break, which is then followed by extensive degradation of single strands, allowing extensive heteroduplex formation (Thaler and Stahl 1988; Sun et al. 1989). This model, in which most conversion is the result of mismatch repair in the heteroduplex, is quite similar to that proposed by Resnick (1976), as noted by Thaler and Stahl (1988).

These two models are clearly closely related. The major difference is that the Radding model has the appealing feature that the continuity of the chromosomal DNA is not disrupted during recombination. Whether this feature of the model is as appealing to the yeast cell as it is to us will require considerably more research.

Further progress in defining the mechanisms of meiotic recombination in yeast is likely to require progress in two areas: the physical characterization of recombination intermediates and the analysis of additional mutations affecting meiotic recombination. For the physical analysis, a description of the DNA lesion (nick or break) is needed, coupled with evidence that this lesion is correlated with crossing-over or gene conversion in its timing, its response to mutations affecting recombination, and its response to deletions and point mutations that alter the rate of recombination. The analysis of additional mutations affecting meiotic recombination should reveal, among other things, whether there are multiple pathways of recombination in yeast, as there are in *E. coli* (for review, see Smith 1987). It is possible that the primary difficulty in designing a single model of recombination that is consistent with all of the available data is that recombination in yeast involves multiple pathways utilizing different mechanisms.

F. Ectopic Recombination

Because *S. cerevisiae* and *S. pombe* contain repeated genes, these organisms have regions of DNA sequence homology that do not occupy equivalent positions on homologous chromosomes. In this section, we describe briefly the different types of ectopic recombination involving such genes (see Fig. 2). More detailed reviews of such events are given by Klein (1988a) and Petes and Hill (1988).

As discussed in Section II, ectopic recombination studies often utilize yeast strains in which selectable single-copy genes have been duplicated by using recombinant DNA and transformation techniques (see Fig. 1a). If the two copies of the selectable gene are different (either one mutant and one wild type or two different mutants), meiotic gene conversion between the two genes can be monitored. In such studies (Klein and Petes 1981; Klein 1984; Jackson and Fink 1985), it was shown that intrachromosomal gene conversion events occurred at a frequency (several percent of unselected tetrads) comparable to "normal" gene conversion events. Thus, pairing between homologous chromosomes at equivalent positions is not necessary for high levels of meiotic conversion. Interestingly, fewer than 10% of the intrachromosomal conversion events were associated with crossing-over. This association is significantly lower than that observed for most allelic recombination. A similar lack of associated crossovers was observed when intrachromosomal gene conversion was induced by a specific double-strand break (Ray et al. 1988). These results suggest that the resolution of the conversion event as a crossover is more difficult (perhaps for topological reasons) in intrachromosomal interactions, that crossing-over requires more extended homology than conversion, or that conversion and crossing-over are not mechanistically associated.

For intrachromosomal conversion events that are not associated with a crossover, it is not possible to determine whether the conversion involved an intrachromatid or unequal sister-strand interaction. If the conversion event is associated with a crossover, a distinction between the two types of interactions can often be made. Intrachromatid crossing-over between repeats leads to either loss of the duplication (if the repeats are in direct orientation) or inversion of the DNA sequences located between the repeats (if the repeats are in inverted orientation). Unequal sister-strand crossing-over between repeats leads to a loss of the repeat from one sister strand and a triplication of the repeat in the other sister. Both intrachromatid (Willis and Klein 1987) and unequal sister-strand crossovers (Jackson and Fink 1985; Wagstaff et al. 1985; Maloney and Fogel 1987) (see Section III.A.3) occur at reasonably high rates. A high rate of unequal sister-strand exchange was observed even when the repeats were separated by 100 kb (Borts et al. 1984; Lichten et al. 1987). In addition, frequent unequal sister-strand crossovers have been detected between different members of natural gene families, including the Ty repeats (Roeder 1983) and the rRNA genes (Petes 1980; Section III.A.3).

Quasiclassical (see Fig. 2) recombination events also occur at high rates in meiosis. For example, a 20-kb duplication of the region of the chromosome containing the *HIS4* gene had quasiclassical crossovers in about 17% of unselected tetrads (Jackson and Fink 1985); in the same strain, unequal sister-strand exchanges were about tenfold less frequent. In similar experiments, Maloney and Fogel (1987) found a fivefold difference in the same direction.

High rates of meiotic recombination are detected even when the duplicated segments are located on nonhomologous chromosomes (heterochromosomal recombination). In one such study, meiotic gene conversion in unselected tetrads was detected between the copy of *HIS3* at its normal position (on chromosome XV) and a 1.7-kb duplication of this gene (located on chromosome IX) at a frequency of 0.5% (Jinks-Robertson and Petes 1985). The rate of gene conversion between the same genes at allelic positions was remarkably similar (1.5%). In another study (Jinks-Robertson and Petes 1986), the heterochromosomal frequency of recombination between duplicated *ura3* genes was 17-fold less than the classical (allelic) level. Lichten et al. (1987) examined recombination between duplicated *leu2* genes, one located at *URA3* on chromosome V and one at its normal position on chromosome III. They found that the rate of recombination was about tenfold less than the classical rate for the same sequences located on chromosome III but about twofold more than the classical rate for the same sequences located at *URA3*. Thus, although it cannot be stated that the rates of heterochromosomal and classical recombination are the same (because the chromosome context influences the rate of recombination), at some loci the rates are roughly comparable. Unlike intrachromosomal conversion events, heterochromosomal conversion is frequently (43% and 22%, in two studies) associated with crossing-over (Jinks-Robertson and Petes 1986; Lichten et al. 1987). The products of such crossovers were reciprocally translocated chromosomes. Thus, high levels of meiotic gene conversion and crossing-over do not require extensive regions of sequence homology or end-to-end chromosome synapsis. It follows that the relative lack of crossovers associated with intrachromosomal gene conversion is unlikely to represent simply an effect of limited homology.

A number of interesting questions are raised by the observed high levels of heterochromosomal recombination. First, what is the role, if any, of end-to-end chromosome synapsis, and hence the synaptonemal complex, in meiotic recombination? This issue is discussed in the following section. Second, are special pairing sites essential for high levels of meiotic exchange? The finding that three different repeated genes all had high levels of heterochromosomal exchange suggests that such sites do not exist, that they allow pairing of nonhomologous and homologous chromosomes, or that they have only a modest (tenfold) stimulatory effect on recombination. Third, because recombination is frequent, even for a single pair of repeated yeast genes, and because there are a number of repeated gene families (such as the Ty family) with many dispersed members, why does yeast have a stable karyotype? Although some degree of polymorphism in the size of chromosomes in different laboratory strains has been reported (Carle and Olson 1985) and one laboratory strain had a naturally occurring chromosome I–III translocation (Mikus and Petes 1982; M. Rose and D. Botstein, pers. comm.), no unselected alterations in the karyotypes of individual strains have been reported. One possibility is that the frequency of meiotic recombination is not proportional to the number of copies of repeated genes (Kupiec and Petes 1988b; Louis and Haber 1990); therefore, karyotypic changes may not be as frequent as expected from an extrapolation of the recombination frequency of a single pair of repeats. In addition, it is possible that meiotic recombination is suppressed between naturally occurring repeats. Kupiec and Petes (1988a,b) found that meiotic gene conversion between Ty elements on nonhomologous chromosomes was suppressed, although recombination between allelic Ty elements occurred at approximately normal rates as did intrachromosomal recombination (Roeder 1983). In addition, very few of the heterochromosomal conversion events involving Ty elements were associated with crossing-over (Kupiec and Petes (1988a,b). In S. pombe, heterochromosomal gene conversion between the dispersed serine tRNA genes occurred at the low frequency of 10^{-6} per meiosis (Munz and Leupold 1981; Heyer et al. 1986), and none of the 554 meiotic conversion events detected were associated with a crossover (Szankasi et al. 1986). Thus, although heterochromosomal conversion events have been detected between natural repeats in yeast, the rate is lower and the association with crossing-over is less than would have been predicted from studies of artificial duplications. Because the tRNA genes are small (<200 bp), the lack of associated crossovers may reflect a requirement for more extensive sequence homology, at least in *S. pombe*. However, as the Ty repeats are 5.7 kb, the lack of associated crossovers in *S. cerevisiae* cannot be explained in this way.

G. Relationship between Meiotic Recombination and Chromosome Pairing

In this section, we discuss the relationship between recombination (a process usually assayed genetically) and cytological events (assayed microscopically) involving the meiotic chromosomes. In yeast, as in most eukaryotes, homologous chromosomes become aligned side by side in meiosis to form a bivalent (for review, see Giroux 1988). In *S. cerevisiae*, several changes in the chromosomes can be detected after meiotic DNA synthesis but prior to chromosome synapsis (Moens and Rapport 1971a,b; Zickler and Olson 1975; Dresser and Giroux 1988). Early in meiosis (at leptotene), linear filaments are first observed; these elements are believed to represent the formation of axial elements (an intermediate in synaptonemal complex assembly) on replicated but unpaired homologous chromosomes. In zygotene, pairing of axial elements and coincident formation of a characteristic tripartite structure between the homologous chromosomes (the synaptonemal complex) are detected. The pairing is initially discontinuous and begins either at the ends or in the middle of the homologous chromosomes. Pachytene is signaled by the formation of a synaptonemal complex extending end to end between the paired chromosomes.

The *S. cerevisiae* synaptonemal complex (see Moens and Rapport 1971a,b; Byers and Goetsch 1975; Dresser and Giroux 1988) appears to be similar to those observed in other eukaryotes (Fig. 10); it consists of a medial element flanked by two lateral elements, which are themselves surrounded by diffusely staining chromatin. No synaptonemal complexes have been detected in *S. pombe* with procedures that allow detection in *S. cerevisiae* (Olson et al. 1978). Sixteen synaptonemal complexes are observed in *S. cerevisiae* (Byers and Goetsch 1975; Dresser and Giroux 1988), correlating with the number of chromosomes detected genetically. One of the largest complexes (presumably that associated with chromosome XII) is bisected by a prominent nucleolus; no complex is evident within the nucleolus (Byers and Goetsch 1975; Dresser and Giroux 1988). Although earlier studies of the complex involved serial-section electron microscopy on meiotic cells, methods have been developed recently that allow visualization of surface-spread preparations of the complex (Dresser and Giroux 1988).

In association with the central region of the synaptonemal complex of

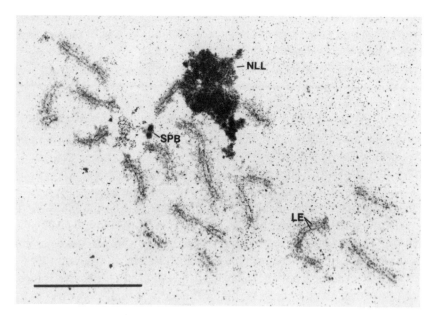

Figure 10 Electron micrograph of surface-spread *S. cerevisiae* synaptonemal complexes (provided by C. Giroux, Wayne State University). Spheroplasts from pachytene yeast cells were surface-spread, fixed, and stained with silver nitrate. The lateral elements (LE) of each of the 16 complexes are evident. The spindle pole body (SPB) and nucleolus (NLL) are indicated. Bar, 3 μm. (Reprinted, by copyright permission, from Dresser and Giroux 1988 [Rockefeller University Press].)

Drosophila, Carpenter (1975) observed densely staining bodies by electron microscopy. Because the number and distribution of these bodies correlate with the number and location of crossovers and respond to events that perturb the number of crossovers (Carpenter 1979), they are called recombination nodules. Byers and Goetsch (1975) observed similar nodules associated with the yeast synaptonemal complex. Studies in other organisms (for review, see Carpenter 1987; Giroux 1988) indicate that there may be two types of recombination nodules, one appearing in zygotene (proposed to be associated with gene conversion but unassociated with crossing-over) and one appearing in pachytene (proposed to be associated with crossing-over). To date, only one type of recombination nodule has been described in *S. cerevisiae.*

   It is obviously important to determine the role, if any, of the synaptonemal complex in meiotic recombination in yeast. The possible relationships are as follows: (1) The synaptonemal complex is required for high levels of recombination in meioisis; (2) meiotic recombination is required to form a synaptonemal complex; (3) the two processes are independent of each other; and (4) the two processes are interdependent. The most widely accepted of these possibilities is the first, based on observations in other organisms correlating the presence of the complex with recombinational proficiency (for review, see Von Wettstein et al. 1984).

In *S. cerevisiae*, a number of observations have been interpreted as indicating the importance of the synaptonemal complex in effecting meiotic recombination. First, all four mutations that eliminate the complex also reduce the levels of classical meiotic recombination (discussed in Section III.H). Because complex formation and recombination may be parts of a single developmental pathway, this result does not prove the dependence of recombination on complex formation. Second, yeast cells transiently blocked at pachytene (the time of the maximum extent of complex formation) by incubating them for various time intervals at elevated temperatures have abnormally high levels of gene conversion, PMS, and crossing-over (Davidow and Byers 1984). In addition, Bell and Byers (1983) observed an accumulation of X-shaped DNA molecules (possibly representing recombination intermediates) during pachytene.

On the other hand, several experiments can be interpreted as indicating that a high level of meiotic recombination does not require formation of the synaptonemal complex. First, heterochromosomal meiotic recombination occurs at rates similar to those observed for classical recombination, as discussed in the preceding section. An alternative interpretation of these results is that complex formation occurs efficiently between repeated sequences on nonhomologous chromosomes. Although we do not favor this view, Carpenter (1987) has pointed out that temporary synaptonemal complex formation between nonhomologous chromosomes is fairly common in other organisms. Second, as discussed in Section III.H, for some of the mutations that eliminate the complex, classical meiotic recombination is only reduced about tenfold and intrachromosomal recombination is not affected. Strains with one such mutation (*red1*) have wild-type levels of classical recombination at the *TRP1* locus, although three other loci have reduced levels of exchange (Rockmill and Roeder 1990).

The results described above are susceptible to a number of alternative interpretations. For example, the synaptonemal complex could be required for all forms of meiotic recombination, but the complexes in some mutant strains, as well as those formed between repeated genes, could be short and/or labile and thus difficult to detect microscopically. Alternatively, the complex could be required for some types of recombination, but not others. More specifically, the complex could be dispensable for intrachromosomal exchange but have an important role in most interchromosomal (involving either homologs or nonhomologs) recombination events. A third (and our favored) view is that the synaptonemal complexes stimulate recombination (perhaps in a sequence-dependent manner) but are not required for high levels of meiotic exchange. A final possibility is that complex formation and meiotic recombination are not intimately related. The relationship between the synaptonemal complex and recombination should eventually be clarified by the isolation and disruption of genes encoding proteins specific to the complex. Even a relatively small enhancing effect of the synaptonemal complex on recombination may be important, as a reduction in recombination would presumably lead to an increase in chromosomal nondisjunction. If the synaptonemal complex proves to be dispensable for meiotic recombination in yeast, then the ubiquity of this complex in eukaryotes would strongly suggest that it has some other important function in meiosis (perhaps related to a role in chromosome segregation during the reductional division) distinct from an effect on recombination.

Another related issue concerns the mechanisms that allow homologous chromosomes to find each other in the cell. Although many possible mechanisms have been proposed (for discussion, see Giroux 1988), the only mechanism yet known to allow specific point-for-point pairing between two DNA duplexes is heteroduplex formation. Powers and Smithies (1986) and Carpenter (1987) have suggested that ectopic gene conversion may reflect the search for extensive homology necessary to allow stable pairing and synaptonemal complex formation between homologous chromosomes. By this interpretation, interactions involving short regions of heteroduplex are unstable, allowing gene conversion events but not crossovers. If the interaction involves longer regions of homology, as in the interactions between homologous chromosomes, then other nearby regions will also form heteroduplexes, stabilizing the pairing and allowing crossovers to occur. It should be noted that heterochromosomal conversion events between natural repeats such as Ty are very rarely associated with crossing-over (Kupiec and Petes 1988a), consistent with the proposed mechanism. The high frequency of ectopic conversion events between artificial repeats in yeast is also consistent with this model, but the frequent association of these events with crossing-over is not.

H. Mutations Affecting Meiotic Recombination

One effective method of analyzing a complex multistep pathway like meiotic recombination is to isolate and characterize mutations affecting various steps of the pathway. In this section, we discuss the identification, classification, and phenotypic analysis of *S. cerevisiae* and *S. pombe* mutations affecting meiotic exchange. A list of mutations that affect meiotic and mitotic recombination in *S. cerevisiae* is given in Table 1.

1. Methods of Identifying Mutations

There are two barriers to the isolation of Rec⁻ meiotic mutants. First, such mutants produce aneuploid, inviable spores, as expected if meiotic recombination is required for normal segregation at meiosis I (for review, see Esposito and Klapholtz 1981). This problem is more acute for *S. cerevisiae,* which has 16 chromosomes, than for *S. pombe,* which has only 3. Second, as only diploid cells normally undergo meiosis, it is difficult to recover recessive mutations affecting this process. Both direct and indirect methods can be used to circumvent these problems.

The first Rec⁻ meiotic mutants were isolated using a protocol that exposes cells to a medium that induces meiosis (sporulation medium) and then returns them to a medium allowing vegetative growth. This protocol (the return to growth or RTG protocol) exploits the observation that cells can be induced to undergo meiotic levels of recombination before becoming committed to the reductional division (Sherman and Roman 1963; Esposito and Esposito 1974). Roth and Fogel (1971) mutagenized a haploid strain that was disomic for chromosome III and looked for mutants that had lower levels of recombination after induction of meiotic recombination by using the RTG protocol. The disomy for chromosome III allowed both expression of **a** and α information (necessary for entry into meiosis) and the use of *leu2* heteroalleles to monitor meiotic recombination. As most wild-type recombinants generated from heteroalleles represent gene conversion events (Section III.B), the mutants obtained were called *con* mutants. The effects of these mutations on crossing-over were not examined. As expected if these mutants were generally Rec⁻, diploid strains homozygous for the *con* mutations produced inviable spores. Most of these mutants have been lost (R. Roth, pers. comm.). The RTG approach was also used to identify mutations that resulted in increased, rather than decreased, rates of prototroph formation (Fogel et al. 1981; Williamson et al. 1985). The mutations represented three complementation groups (designated *PMS1–PMS3*) and were shown to increase both the frequency of mutation in mitotic cells and the frequency of PMS in meiotic cells, as described further in Section III.H.3.

Another direct method of mutant isolation is based on the use of the *spo13-1* mutation, characterized by Klapholz and Esposito (1980a,b). Although other meiotic processes occur normally in *spo13* strains, the re-

Table 1 Mutant phenotypes of genes affecting recombination in *S. cerevisiae*

Gene	spontaneous mitotic recomb.[a]	repair of induced DNA damage	meiotic recomb.[a]	interaction with *spo13*[b]	presence of synaptonemal complex
		Mutant phenotypes			

Gene	spontaneous mitotic recomb.[a]	repair of induced DNA damage	meiotic recomb.[a]	interaction with *spo13*[b]	presence of synaptonemal complex
(a) Genes that affect meiotic recombination and have additional mutant mitotic phenotypes[c]					
CDC5	up	wt	up		yes
CDC40		wt	down		
MMS9	up	ms	down?		
MMS13	up	ms	down?		
MMS21	up	ms	down?		
PMS1	up	wt	PMS up		
PMS2	up	wt	PMS up		
PMS3	up	wt	PMS up		
RAD6[d]	up	us	down	complex[e]	yes
RAD50	up	xs	down[f]	class 1	no
RAD51	down	xs	down	class 2	
RAD52	down[g]	xs	down	class 2	yes
RAD57	down	xs	down	class 2	
REC3	down	wt	down?		
SIR2	up[h]		up[h]		
(b) Genes affecting meiotic but not mitotic recombination[i]					
HOP1	wt		down[j]	class 1	no
MER1	wt		down	class 1	no
MEI4	wt		down	class 1	
REC102	wt		down	class 1	
REC104	wt	wt	down	class 1	
REC107	wt		down	class 1	
REC113	wt		down	class 1	
REC114	wt		down	class 1	
RED1	wt		down	class 1	no
SPO11	wt		down	class 1	no ?
(c) Genes affecting mitotic but not meiotic recombination[k]					
CDC6	up		wt		
CDC8[l,m]	up				
CDC9[n]	up	ms			
CDC13[l]	up				
CDC14[l]	up				
HPR1[o]	up[p]	wt	wt		
HPR2	up	wt	wt		
HPR3/CDC17[q]	up	us	wt		
HPR4	up	wt	wt		
HPR5/RADH[r]	up	us	wt		
HPR6/CDC2[s]	up	us	wt		
HPR7	up	wt	wt		
RAD1	up[t]	us	wt		
RAD3[u]	up	us	wt		
RAD4	up	us	wt		
RAD18	up[v]	us	wt		
RAD54	down	xs	wt		
REC1	down	wt			
REC754[w]	down	wt			
TOP1[x]	up[y]		wt		
TOP2[z]	up[y]		wt		
TOP3[aa]	up[bb]				

This table lists only those genes discussed in the text. No entry in the column indicates that the phenotype is unknown. Increases, decreases, and no significant change from wild-type levels of recombination (or repair) are indicated by up, down, and wt, respectively. Some of the mutations affect only gene conversion or crossing-over. Relative degrees of change are not shown. xs, us, and ms indicate X-ray (or γ-ray), ultraviolet light, or methylmethanesulfonate (MMS) sensitivity, respectively. Only the major or the originally diagnosed repair defect is shown; many mutations affecting DNA repair confer sensitivity to several types of damage. (?) Evidence is not conclusive or conflicting data exist. References for the data presented in this table are given in the text.

[a] Entries refer to classical recombination (events involving allelic loci on homologous chromosomes) unless otherwise noted.

[b] As noted in the text, mutations that reduce meiotic recombination reduce spore viability. If the spore viability is rescued by the *spo13* mutation (i.e., strains with the double mutation have viable spores), the mutation is considered class 1. Mutations with a defect not rescued by *spo13* are considered class 2. Class-2 mutations do not lead to inviable spores if both a class-1 and a *spo13* mutation are present.

[c] Mutations in genes in this category affect meiotic recombination, as well as some mitotic functions (cell growth, DNA repair, etc.).

[d] *RAD6* encodes a ubiquitin-conjugating enzyme.

[e] Although some mutant alleles of *RAD6* prevent an induction of recombination in RTG experiments, these alleles are not rescued by *spo13*, or double mutations of class 1 and *spo13*.

[f] *RAD50* function is required for classical recombination but not intrachromosomal recombination among the rRNA genes.

[g] The *rad52* mutation affects classical recombination but not unequal sister-strand mitotic crossing-over in the rRNA gene cluster.

[h] The *sir2* mutation results in elevated recombination within the rRNA gene cluster but does not affect intrachromosomal recombination between other types of repeats.

[i] Mutations in genes in this category do not appear to affect any mitotic functions, although not all the mutations have been tested for all mitotic phenotypes. Members of this group were isolated as mutations that affected sporulation or meiotic recombination.

[j] Intrachromosomal recombination is not affected by *hop1*.

[k] Mutations in the genes of this group do not appear to affect meiotic recombination, although every member of this group has not been tested completely.

[l] Mutations in these genes confer a defect in sporulation, but it is not known whether there is a defect in meiotic exchange.

[m] *CDC8* encodes thymidylate kinase.

[n] *CDC9* encodes DNA ligase.

[o] *HPR1* is homologous to *TOP1* and may encode a type I topoisomerase.

[p] Mutations at this locus stimulate intrachromosomal but not classical recombination.

[q] *HPR3 (CDC17)* encodes DNA polymerase α.

[r] The *HPR5 (RADH)* product has homology with several bacterial DNA helicases.

[s] *HPR6 (CDC2)* encodes DNA polymerase δ.

[t] Mutations of *RAD1* result in a hyper-Rec phenotype for classical recombination but have lower levels of intrachromosomal exchange.

[u] *RAD3* encodes a DNA helicase.

[v] Intrachromosomal recombination is only marginally affected.

[w] Most of the *rec* mutations isolated by Esposito et al. (1984) are not listed, because they have not been classified into complementation groups. The *rec754* mutation is listed because it is one of the few mutations that leads to a defect in mitotic recombination that is not defective in DNA repair. It is not known whether this mutation also affects meiotic recombination.

[x] *TOP1* encodes a type I topoisomerase.

[y] The effects of the *top1* and *top2* mutations on mitotic recombination appear to be specific for the rRNA genes.

[z] *TOP2* encodes a type II topoisomerase.

[aa] *TOP3* appears to encode a type I topoisomerase.

[bb] The effect of the *top3* mutation on recombination may be limited to intrachromosomal interactions.

ductional division is largely bypassed. Because this bypass prevents the nondisjunction usually associated with a lack of recombination, many Rec⁻ mutants produce viable spores in a *spo13* genetic background (Malone and Esposito 1981; Klapholz et al. 1985). Thus, the effects of putative *rec* mutations on recombination can be tested directly. Moreover, haploid *spo13* strains that express both a and α information can undergo meiosis and produce viable spores (Wagstaff et al. 1982; Malone 1983), allowing the isolation of recessive mutations affecting meiotic recombination. In one such study, Hollingsworth and Byers (1989) constructed a *spo13* haploid strain that was disomic for chromosome III. One copy of this chromosome had an 11-kb duplication flanking an insertion of the *URA3* and *CYH2* genes. An intrachromosomal recombination event involving the duplication would lead to loss of the *CYH2* gene, exposure of the *cyh2* allele on chromosome VII, and thus a cycloheximide-resistant phenotype. These workers looked for meiotic mutants that displayed an increased frequency of Cyhʳ cells. The *hop1-1* mutation, discussed below, was isolated in this way. A *spo13* haploid strain that was disomic for copies of chromosome III containing a number of heteroalleles was used to identify the *mei4* mutation, which results in decreased meiotic gene conversion (Menees and Roeder 1989).

Another method for identifying mutations that affect meiotic recombination has been developed recently (R. Malone, unpubl.). As discussed in Section III.H.2, the *rad52* mutation appears to affect a late stage of meiotic exchange, and *spo13 rad52* haploid strains produce inviable spores. The viability of spores in this genetic background can be rescued by additional mutations that affect the early steps of recombination. Thus far, 208 such mutations have been isolated. In addition to new mutant alleles of *MEI4*, *MER1*, and *SPO11*, at least five new *REC* genes (*REC102*, *REC104*, *REC107*, *REC113*, and *REC114*) have been identified.

A number of less direct methods of obtaining *rec* mutations have also been used. As meiotic exchange appears to be required for normal spore viability, one such method is to analyze mutant strains that do not produce viable spores (*spo* mutants). To recover recessive mutations, Esposito and Esposito (1969) mutagenized the spores of a homothallic strain, allowed the spores to diploidize, and then screened for mutants that failed to sporulate. This approach resulted in the isolation of *spo11*, a *rec* mutation that is discussed in detail below. Similar screens were used to obtain the mutations *red1* (Rockmill and Roeder 1988) and *mer1* (Engebrecht and Roeder 1989).

Another useful nonspecific approach to the identification of meiotic mutants is to screen radiation-sensitive (*rad*) strains. The rationale is that repair of mutagen-induced damage may require some of the same gene products that are necessary for recombination. The many (>100) different *rad* mutations define at least 46 genes, which can be classified into three epistasis groups (each named after a prominent member of the group): *RAD3*, *RAD6*, and *RAD52* (for review, see Game 1983; Friedberg 1987; Friedberg et al., this volume). This classification is based on the radiation sensitivity of double *rad* mutants; if the double mutant strain is no more radiation sensitive than the more sensitive single mutant, the mutations are presumed to affect different steps in the same repair pathway and are thus classified in the same epistasis group. Mutations in the *RAD52* group are deficient in the ability to repair several types of DNA damage, and several of the genes in this group are also required for mitotic recombination (discussed in Section IV.F). In addition, *rad50-1*, *rad52-1*, and *rad57-1* were found to be Rec⁻ in RTG experiments (Game et al. 1980; Prakash et al. 1980); no effect on meiotic recombination was observed for *rad54* (Game 1983 and pers. comm.). It is less clear whether genes in the other epistasis groups are involved in meiotic recombination. Although *rad6-1* reduced the level of meiotic recombination (Game et al. 1980), other mutant alleles of this gene (Montelone 1981a) and other members (*rad5*, *rad9*, and *rad18*) of the *RAD6* epistasis group (Dowling et al. 1985) had no effect on meiotic recombination. Various members (*rad1–rad4*, *rad10*, *rad14*, and *rad16*) of the *RAD3* epistasis group also had no effect on meiotic recombination (Dowling et al. 1985).

The *cdc5* and *cdc40* mutations, originally identified on the basis of their effects on the cell cycle, may also affect meiotic recombination. Although premeiotic DNA synthesis occurred in *cdc40* strains at restrictive temperatures, no induction of meiotic recombination was detected in RTG experiments (Kassir and Simchen 1978). *cdc5* cells that were held at the temperature-sensitive step displayed elevated levels of gene conversion (Simchen et al. 1981); this may be related to the elevated levels of conversion and PMS found when certain wild-type strains were blocked in pachytene by incubation at elevated temperatures during meiosis (Davidow and Byers 1984; see Section II.G).

2. Classification and Phenotypes of rec Mutations

Strains have been classified as Rec⁻ mutants if they exhibit a reduced level of recombination either in an RTG experiment or in a *spo13* genetic background; the latter criterion is a more specific test. One convenient method of classifying Rec⁻ mutants is by measuring their effects on spore viability in a *spo13* genetic background. In this background, good spore viability is displayed by strains with *rad50* (Malone and Esposito

1981), *spo11* (Klapholz et al. 1985), *hop1* (Hollingsworth and Byers 1989), *red1* (Rockmill and Roeder 1988), *mer1* (Engebrecht and Roeder 1989), or *mei4* (Menees and Roeder 1989) mutations. The newly isolated mutations *rec102*, *rec104*, *rec107*, *rec113*, and *rec114* also share this property (R. Malone, unpubl.). In contrast, poor spore viability is displayed by strains with *rad6* (Malone and Esposito 1981; Malone 1983), *rad52* (Malone and Esposito 1981), *rad57* (J. Nitiss and M. Resnick; J. Game; both pers. comm.), or *rad51* (J. Nitiss and M. Resnick; J. Game; both pers. comm.) mutations. One possible interpretation of the two classes of *rec* mutations is that the mutants that yield viable spores in the presence of *spo13* (class-1 mutants) may be blocked at an early step in the recombination process (e.g., in chromosome pairing or the initiation of exchange), whereas the other mutants (class-2 mutants) may be blocked at later stages of recombination (e.g., the resolution of recombination intermediates). Mutants affected in the later steps would not be expected to yield viable spores even in the absence of a reductional division, because unresolved Holliday structures would hold homologous chromosomes together, preventing a proper equational division. Class-1 mutations are epistatic to class-2 mutations in all combinations tested so far.

a. Phenotypes of Class-1 Mutants. Class-1 mutations vary in the degrees to which they affect intrachromosomal and interhomolog recombination. Mutations in *SPO11*, *RAD50*, *MEI4*, and the five new genes affecting recombination described above (R. Malone, pers. comm.) reduced classical meiotic gene conversion and crossing-over to the mitotic background level (Malone and Esposito 1981; Klapholz et al. 1985; Menees and Roeder 1989), whereas mutations in *RED1*, *HOP1*, and *MER1* reduced this type of recombination about tenfold (Engebrecht and Roeder 1989; Hollingsworth and Byers 1989; Rockmill and Roeder 1990). One interpretation of the failure of null mutations in the *RED1*, *HOP1*, and *MER1* genes to eliminate meiotic recombination is that there are multiple pathways of meiotic recombination. This interpretation is supported by the finding that *mer1* and *hop1* null mutations have synergistic effects on classical gene conversion, as though they affect different pathways.

In addition, analysis of the Rec⁻ mutants indicates that classical and intrachromosomal recombination may have somewhat different pathways. For example, the *HOP1* gene product is necessary for wild-type rates of classical crossing-over but does not affect intrachromosomal crossing-over (Hollingsworth and Byers 1989). These workers suggested that Hop1p is involved specifically in pairing and recombination of homologous chromosomes. In contrast, mutations in *MER1* reduced the rates of both intrachromosomal and classical recombination (Engebrecht and Roeder 1989). Although the *spo11* mutation results in a reduced level of unequal sister-strand recombination in the rRNA gene cluster (Wagstaff et al. 1985), the *rad50* mutation does not affect this type of recombination (Gottlieb et al. 1989). The *rad50* mutation, however, does reduce intrachromosomal recombination at other loci (Gottlieb et al. 1989). The different effects of *rad50* on intrachromosomal recombination involving two different types of sequences suggest that it is dangerous to generalize conclusions about the effects of *rec* mutations on different types of recombination without examining several loci.

The synaptonemal complex appears to be absent in several of the class-1 mutants, including *rad50* (Farnet et al. 1988; B. Byers, pers. comm.), *hop1* (Hollingsworth and Byers 1989), *mer1* (J. Engebrecht and S. Roeder, pers. comm.), and *red1* (Rockmill and Roeder 1990). The situation with *spo11* is less clear. Although Klapholz et al. (1985) reported that a strain containing the *spo11-1* mutation (which reduced recombination between homologous chromosomes about 100-fold) had synaptonemal complexes, Giroux et al. (1989) found that strains containing a deletion of the *SPO11* gene had no mature synaptonemal complexes. Thus, either the *spo11-1* mutation is leaky for complex formation (but not recombination) or the complex made in *spo11* mutant strains is unstable under some conditions of sample preparation. Although any of the class-1 genes whose mutants lack synaptonemal complex might encode components of the complex, there is as yet no compelling evidence that they do. There is an indication that antibodies directed against Hop1p react with meiotic bivalents (Hollingsworth et al. 1990).

Most of the class-1 genes have been cloned, and *SPO11* (Atcheson 1987), *RAD50* (Kupiec and Simchen 1984; Alani et al. 1989), *RED1*, and *MER1* (S. Roeder, pers. comm.) have been sequenced. These sequences have not suggested functions for the gene products, although the predicted sequence of the Rad50p contains two regions of heptad repeats that may indicate the presence of α-helical coils (Alani et al. 1989). *SPO11* (Atcheson et al. 1987), *RED1* (Thompson and Roeder 1989), and *MER1* (S. Roeder, pers. comm.) are all induced in meiosis.

b. Phenotypes of Class-2 Mutants.   As suggested above, mutations in class-2 genes may inactivate intermediate- or late-acting functions in the process of meiotic exchange. If class-1 and class-2 genes encode functions that act in a linear dependent pathway, then a *spo13* strain that contains both an early (class 1) and late (class 2) mutation should produce viable spores. This result has been obtained with most combinations that have been tested, including *rad50-1* and *rad52-1* (Malone 1983), *rad50* and *rad51*, and *rad50* and *rad57* (J. Nitiss and M. Resnick; J. Game; both pers. comm.), *spo11* and *rad52* (S. Klapholz and R. Esposito, pers. comm.), and *mer1* and *rad52* (Engebrecht and Roeder 1989). In contrast, strains with the genotype *rad50 rad6 spo13* produced inviable spores (Malone 1983). However, the recent finding that *RAD6* encodes a ubiquitin-conjugating enzyme (Jentsch et al. 1987) indicates that the effect of this mutation on meiotic recombination is likely to be indirect.

The physical state of the DNA in meiosis in *rad52* and *rad57* strains has been examined in several studies. Although wild-type and *rad50* strains failed to accumulate detectable single- or double-strand breaks (consistent with the proposed early role of *RAD50*), *rad52* strains accumulated single-strand nicks but not double-strand breaks (Resnick et al. 1984). One interpretation of this result is that single-strand nicks represent the normal recombination intermediate, but this intermediate is processed too quickly to be detected in wild-type cells. The level of meiotic recombination in *rad52* strains was estimated to be about 5% that of wild type (Resnick et al. 1986). Because a single double-strand break is lethal in haploid mitotic *rad52* strains (Resnick and Martin 1976), it seems unlikely that the meiotic recombination observed in this experiment was initiated by double-strand breaks. J. Nitiss and M. Resnick (pers. comm.) recently observed double-strand breaks, as well as single-strand nicks, in meiotic DNA of *rad57* strains. Because the double-strand breaks appeared later than the single-strand nicks, single-strand nicks appear more likely to represent the lesion that initiates meiotic recombination.

Hogset and Oyen (1984) also examined meiotic DNA from *rad52* strains and observed a limited number of both single-strand and double-strand breaks. The rRNA gene cluster had fewer nicks and breaks than other regions of the genome. Although Hogset and Oyen pointed out that the lack of lesions in the rRNA genes correlates with the lack of classical recombination in this region, it is not clear that such a correlation is expected because of the high rate of unequal sister-strand meiotic recombination within the rRNA gene cluster (Petes 1980).

Meiotic DNA of various class-1 and class-2 mutants was also examined by Borts et al. (1986), who looked for novel restriction fragments formed by classical recombination between sequences marked by heterozygous restriction sites on homologous chromosomes. Although the appearance of the fragments diagnostic of recombination was blocked by the *rad50* and *spo11* mutations, as expected, substantial amounts of the recombinant band were observed in strains with the *rad52* (33% of wild type), *rad57* (nearly wild-type levels), and *rad6*

(65% of wild type) mutations. The connection between these results and the Rec⁻ phenotypes of these strains as determined in other experiments is not yet clear.

The *RAD51*, *RAD52*, *RAD55*, and *RAD57* genes have been cloned (Calderon et al. 1983; Schild et al. 1983b). Strains containing disruptions of these genes were viable (Schild et al. 1983a; Kupiec 1986), although strains with disruptions of *RAD55* or *RAD57* were cold-sensitive in their ability to repair radiation damage in mitosis (Lovett and Mortimer 1987). Although the sequence of the predicted *RAD52* gene product did not suggest a possible function (Adzuma et al. 1984), Chow and Resnick (1988) found that the level of an endonuclease/exonuclease that was induced in meiotic cells appeared to be under the control of *RAD52*.

3. Mutations Affecting Mismatch Repair

As described in Section III.H.1, in screens for mutants that had an increased frequency of meiotic recombination between heteroalleles, mutations (*pms1* through *pms3*) were identified that affect mismatch repair (Fogel et al. 1981; Williamson et al. 1985). Strains carrying the *pms1* mutation had increased levels of PMS and decreased levels of gene conversion for 8 of the 13 single sites tested. One simple explanation of these findings is that both gene conversion and PMS signal the presence of a heteroduplex, with conversion representing a repaired mismatch and PMS representing an unrepaired mismatch within the heteroduplex (Fogel et al. 1981). The *PMS1* function would then be required for efficient repair. However, at several sites, *pms1* had no obvious effect. For *ade8-18*, a mutant allele with high levels of PMS in wild-type strains, the *pms1* mutation did not significantly increase the level of PMS. Thus, the *PMS1* gene product appears to function with different efficiencies, depending on the type of mismatch. Interestingly, the *pms* mutations had relatively little effect on the viability of spores or the frequency of meiotic crossing-over, suggesting that mismatch repair is not essential for the completion of crossing-over.

Strains containing the *pms1* mutation also show mitotic effects, including a 50-fold increased mutation rate at the *CAN1* locus and a 5-fold increased rate of mitotic recombination at the *HIS4* locus. The *pms1* mutation also reduced the efficiency of mitotic mismatch repair as shown by studies involving heteroduplexes transformed into vegetative yeast cells (Bishop et al. 1987, 1989; B. Kramer et al. 1989). In this system, repair was about 50% more efficient in *PMS1* than in *pms1* strains.

Significant sequence homology has been detected between *PMS1* and the *mutL* gene of *E. coli* (W. Kramer et al. 1989). As *mutL* is required for mismatch repair in *E. coli* (for review, see Modrich 1987), this result further implicates the *PMS1* gene product in either the detection or removal of mismatches. As *pms1*-deletion strains still display gene conversion as well as PMS events (B. Kramer et al. 1989), some meiotic conversion events may be the result of mechanisms other than mismatch repair or there may be more than one repair system.

4. Rec⁻ Mutants of S. pombe

A number of mutations affecting meiotic recombination have been identified in *S. pombe*. Thuriaux (1985) isolated three mutants (*rec2*, *rec3*, and *rec5*) that were affected in ectopic recombination between dispersed tRNA genes but not in interhomolog recombination. Grossenbacher-Grunder and Thuriaux (1981) found that none of 18 different mutations affecting radiation repair (*rad* mutations) reduced the level of meiotic recombination. In addition, only one (*swi5*) of eight mutations that affected mating-type switching reduced (about eightfold) the frequency of meiotic intragenic recombination (Schmidt et al. 1987). Using a screen based on meiotic intragenic recombination between a plasmid and a chromosome, Ponticelli and Smith (1989) isolated mutations in six complementation groups that reduced meiotic recombination. Some of the mutations (*rec7* and *rec10*) reduced intragenic recombination (presumably, primarily gene conversion events) more severely than crossing-over. In addition, several of the mutations differentially reduced intragenic recombination at different loci.

5. Conclusions from the Analysis of Meiotic Mutants

Although no specific function has yet been assigned to any of the genes that are mutable to yield a Rec⁻ phenotype, the analysis of the mutant phenotypes has allowed a number of important conclusions. First, different types of meiotic recombination appear to require different gene products. For example, the *HOP1* gene product is required for classical recombination but not intrachromosomal recombination (Hollingsworth and Byers 1989). Second, even null mutations in some genes (such as *RED1*) do not completely eliminate meiotic recombination. This finding either indicates multiple pathways of recombination or suggests a partial requirement (perhaps because of redundancy of function) for a particular gene product. Third, as meiotic recombination occurs (although at reduced levels) in some strains that apparently lack the synaptonemal complex, it may be that not all recombination is complex-dependent. However, this conclusion is not rigorous because it is difficult to rule out a small amount of complex formation in the mutant strains. Finally, in *spo13* strains, in which the reductional division of meiosis is skipped, meiotic recombination is not required for spore viability (Malone and Esposito 1981).

Several types of experiments are likely to help in the mutational analysis of recombination. First, the collection of Rec⁻ mutants is still very small and should be expanded. This would allow extension of the kinds of correlational and epistasis studies discussed above and potentially allow the identification of genes whose sequences provide immediate clues to their biochemical function. Second, isolation of cold-sensitive and heat-sensitive *rec* mutations would allow the ordering of gene functions by the method of Jarvik and Botstein (1973). Third, isolation of the proteins encoded by the genes described above will help to define their functions, for example, by characterizing their DNA-binding activities and intracellular locations. Finally, it should be useful to examine the effects of various *rec* mutations on the in vitro recombination activity (Symington et al. 1983) of cell-free extracts prepared from meiotic cells.

I. Meiotic Recombination in Other Fungi

In this review, we have restricted detailed discussion of meiotic recombination to results obtained with *S. cerevisiae* and, to a lesser extent, with *S. pombe*. This decision was not entirely the result of species chauvinism. It is possible, and perhaps even likely, that mechanisms of recombination in different organisms are fundamentally different. Results obtained by Rossignol and co-workers with one of the most intensively studied fungi, *Ascobolus immersus*, can be interpreted as supporting this view (for review, see Rossignol et al. 1988; Nicolas and Rossignol 1989). In *Ascobolus*, putative 1-bp insertions or deletions often showed large deviations from parity in gene conversion (Leblon 1972a,b); similar deviations were not observed for the 1-bp *his4-519* insertion (Detloff et al. 1991) or for 4-bp insertions at many loci (Borts and Haber 1987; Symington and Petes 1988a) in *S. cerevisiae*. The meiotic products in *Ascobolus* strains heterozygous for the putative 1-bp insertion mutations had a great (20-fold) excess of $2^+:6^-$ segregation over $6^+:2^-$. In strains heterozygous for putative 1-bp deletions, the reciprocal result was observed, an excess of $6^+:2^-$ octads. Thus, for small insertions in *Ascobolus*, conversion favored the longer strand. This result can be explained in the context of heteroduplex models by preferential excision of the shorter strand. A second (and possibly related) difference in recombination between yeast and *Ascobolus* is that in *Ascobolus*, for many point mutations, PMS events represent a substantial fraction of non-2:2 segregations (Rossignol et al.

1988). As the DNA sequences of these mutations are not yet known, it is not clear whether this difference reflects differences in mutagenesis or in mismatch repair of heteroduplexes between the two organisms.

An additional difference is that in *Ascobolus,* unlike yeast, meiotic products indicative of symmetric heteroduplexes (e.g., aberrant 4:4 segregations) were frequently observed. In the *b2* gene of *Ascobolus,* meiotic products suggestive of asymmetric heteroduplex formation were high at the end of the gene with the higher level of non-2:2 segregations and declined toward the other end; the frequency of meiotic products characteristic of symmetric heteroduplex had the opposite polarity (Paquette and Rossignol 1978). This result was used in the development of the Meselson-Radding model, in which asymmetric heteroduplexes initiated at one end of the gene undergo branch migration to form symmetric heteroduplexes. Although such a transition has not yet been observed in *S. cerevisiae,* no high PMS alleles (which are needed to diagnose aberrant 4:4 tetrads) have yet been identified at the low conversion end of a gene.

It is not yet clear whether the differences described above between yeast recombination and *Ascobolus* recombination reflect the operation of fundamentally different mechanisms or adjustments of various parameters of the same basic mechanism. It should be noted that many features of recombination are conserved between the two systems, including the association between conversion and crossing-over and the polarity of conversion. Conversion and PMS have also been described in the fungi *Podospora anserina, Neurospora crassa, Ustilago maydis, Sordaria brevicollis,* and *Sordaria fimicola,* and conversion events have been seen in *Bombaria lunata* and *Aspergillus nidulans* (for review, see Whitehouse 1982). Although most of these systems are not sufficiently well characterized to determine whether they resemble yeast or *Ascobolus,* marker effects on the parity of conversion, similar to those observed with *Ascobolus,* were also observed for *S. brevicollis,* but not for *Neurospora* or *Aspergillus.* In addition, aberrant 4:4 segregation events have been detected in *S. fimicola* (Kitani et al. 1961). It is possible that most fungal genetic systems will be classifiable as either *S. cerevisiae*-like or *Ascobolus*-like.

IV. MITOTIC RECOMBINATION IN YEAST

Spontaneous mitotic recombination between homologous chromosomes occurs at a rate about three to four orders of magnitude lower than the meiotic rate (for review, see Esposito and Wagstaff 1981). Other features of mitotic recombination that appear to be different from meiotic exchange include the following: (1) Some mitotic recombination appears to occur prior to DNA replication; (2) mitotic gene conversion does not show polarity; (3) mitotic gene conversion frequently appears to involve symmetric heteroduplexes; and (4) mitotic cells do not appear to have synaptonemal complexes (Olson and Zimmermann 1978).

In this section, we focus on spontaneous mitotic recombination between allelic sequences on homologous chromosomes, although other topics are also discussed briefly. Previous relevant reviews are those of Esposito and Wagstaff (1981), Orr-Weaver and Szostak (1985), and Roeder and Stewart (1988).

A. Classical Methods of Studying Mitotic Recombination

Because the rates of spontaneous mitotic recombination are low, agents that stimulate recombination (e.g., X-rays and UV light) are often used. One type of study involved constructing a diploid strain heterozygous for multiple linked recessive auxotrophic markers (Nakai and Mortimer 1969). After X-irradiation, cells were plated on a rich nutrient medium and allowed to form colonies. These colonies were then replica-plated onto various omission media, and the patterns of sectoring were scored. Certain markers frequently cosectored. The most common patterns of sectoring were consistent with the expectation that X-rays induced mitotic crossing-over between duplicated chromatids (the four-strand stage). A crossover between duplicated chromatids, as diagramed in Figure 11a, could produce two types of cells: one homozygous for the mutant alleles of genes located centromere-distal to the exchange and one homozygous for the wild-type alleles for the same genes. If these two cells form a single colony on rich growth medium, one would expect that the h lf of the colony in which the mutant alleles are homozygous would be unable to grow after replica-plating to appropriate omission plates, whereas the other half of the colony would be able to grow on these plates. If tetrads are dissected from cells derived from this latter half of the sector, one would expect $4^+:0^-$ segregation of the centromere-distal genes.

Although many of the colonies analyzed in such studies had the sectoring pattern predicted from crossing-over, about 20% of the colonies had a different pattern. In these colonies, sectoring was observed for one of the centromere-proximal markers without sectoring of markers located farther from the centromere (Fig. 11b). In addition, the wild-type sectors of such colonies were not homozygous for the wild-type alleles. In the example given in Figure 11b, there is a net loss of C information and a

a. Mitotic crossing-over

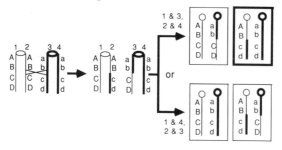

b. Mitotic gene conversion without associated cross-over

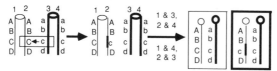

Figure 11 Intergenic mitotic crossing-over and gene conversion. Mitotic recombination is shown between replicated nonsister chromatids. Cells in which the mutant alleles are homozygous are outlined by thick lines. (*a*) Mitotic crossing-over without a detectable gene conversion event. Only one of the two segregation patterns (1 & 3, 2 & 4) during mitosis following the crossover results in the heterozygous markers centromere-distal to the site of recombination (*C* and *D*) becoming homozygous. The other segregation pattern (1 & 4, 2 & 3) leaves all markers still heterozygous but with a new linkage relationship (reverse linkage) between the wild-type and mutant alleles of *A* and *B* versus *C* and *D*. (*b*) Mitotic gene conversion without an associated crossover. As a result of this event, the *c* mutant allele becomes homozygous in one cell and remains heterozygous in its sister cell. In addition, the distal gene *D* remains heterozygous in both sister cells.

net gain of *c* information. Thus, these colonies appear to arise by mitotic gene conversion.

Spontaneous mitotic recombination events are more difficult to analyze than induced mitotic events for several reasons. First, spontaneous events are usually too infrequent to be analyzed by nonselective techniques. Because selective techniques (described below) usually do not allow the recovery of all of the expected products of recombination, there is little direct evidence that spontaneous crossover and conversion events really occur as diagramed in Figures 11 and 12. Second, because spontaneous events can occur at any time during growth of the culture, the frequency of cells containing recombinant products in independent cul-

a.   Intragenic mitotic crossing-over

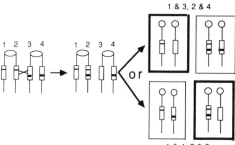

b.   Intragenic mitotic conversion

Distal allele conversion

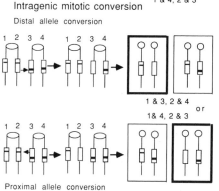

Proximal allele conversion

Figure 12 Intragenic mitotic crossing-over and gene conversion. (*Rectangles*) A single gene; (*thin line in the rectangle*) proximal mutant substitution; (*thick line*) distal mutant substitution. The segregant with a wild-type allele is outlined by thick lines. (*a*) Intragenic crossing-over. If a crossover (unassociated with gene conversion) occurs between the mutant substitutions, about one half of the segregants (cosegregation of chromatids 2 & 3) that contain the wild-type gene will also contain the double mutant allele. Other segregants with one wild-type gene will contain one chromosome with the centromere-proximal mutation. (*b*) Intragenic gene conversion. A conversion event involving the centromere-distal mutation will result in a cell that has the allele containing the centromere-proximal mutation in addition to the wild-type allele. A conversion of the centromere-proximal mutation will result in a cell that has the allele containing the centromere-distal mutation in addition to the wild-type allele.

tures can be very different. Thus, to measure a rate (events per mitotic division), one needs to measure the frequency of recombinants in a number of independent cultures and calculate a rate using the methods developed by Luria and Delbrück (1943) and Lea and Coulson (1949).

One procedure for selecting spontaneous mitotic gene conversion events was developed by Roman (1957). He constructed a diploid that was heterozygous for two noncomplementing mutant alleles of the *ADE5* gene (Fig. 12). He found that Ade⁺ derivatives of this strain arose at a rate much higher than the frequencies of reversion of the individual alleles and hypothesized that these Ade⁺ cells were the result of mitotic recombination. If the *ADE5* genes in the Ade⁺ cells were the result of mitotic crossing-over between the mutant sites and mitotic segregation of the recombined chromosomes occurred randomly, one would expect that about half of the Ade⁺ cells would contain one wild-type gene and one gene with the double mutant allele (Fig. 12a). Roman observed cells of this type very rarely; most Ade⁺ cells had one gene with a single mutation in addition to the wild-type gene (Fig. 12b). Roman therefore suggested that most mitotic recombination events between heteroalleles were gene conversion events.

Spontaneous crossovers are usually detected using strains that are heterozygous for a mutation producing a recessive drug-resistance phenotype (e.g., resistance to the arginine analog canavanine [*can1*ʳ]). About half of the crossover events that occur between this gene and the centromere should result in a cell that is homozygous for the *can1*ʳ allele (represented by the *c* allele in Fig. 11a) and therefore able to grow in medium containing canavanine. Because gene conversion (Fig. 11b) or chromosome loss could also result in a canavanine-resistant phenotype, it is important that the strain have other heterozygous markers located centromere-distal to the *CAN1* locus on the same chromosome, as well as having markers on the other chromosome arm. Chromosome loss would result in all markers on the chromosome becoming hemizygous simultaneously. Gene conversion at the *CAN1* locus would result in the *can1*ʳ allele becoming homozygous but centromere-distal loci remaining heterozygous. Crossing-over between the *CAN1* locus and the centromere is signaled by cells that are homozygous for all loci centromere-distal to *CAN1* but are still heterozygous for loci on the opposite chromosome arm.

B. General Features of Mitotic Crossing-over and Gene Conversion

As discussed in Section III.B, meiotic gene conversion represents the faithful transfer of information (from <100 bp to >10 kb) from one chromosome to the other. Conversion events transfer either mutant or wild-type information, and for most alleles, the two types of information are transferred equally frequently (parity). The frequency of meiotic conversion often declines from one end of the gene to the other (polarity), and meiotic conversion events are correlated with crossing-over of flanking markers. Many of these features have not been demonstrated for mitotic recombination. For example, because most mitotic conversion experiments are done with heteroalleles, the rate of transfer of wild-type information from one gene to the other can be estimated, but the transfer of mutant information between alleles cannot be easily measured; thus, whether mitotic gene conversion displays parity has not yet been established.

1. Lack of Mitotic Conversion Polarity

The available evidence indicates that the rates of spontaneous mitotic conversion are similar at different positions within a gene, in contrast to the situation for meiotic conversion (Kakar 1963; Hurst and Fogel 1964; Golin and Esposito 1984). These studies were done with heteroalleles (Fig. 12b); prototrophic strains with the genotypes expected for conversion of the centromere-proximal alleles appeared at the same rates as did those with the genotypes expected for conversion of the centromere-distal alleles. Conversion events induced by UV, X-rays, or chemical mutagens also showed no polarity (for review, see Esposito and Wagstaff 1981). The lack of polarity in mitotic conversion indicates either that mitotic conversion events initiate randomly (perhaps triggered by randomly located lesions) or that conversion events initiate at specific sites but extend over much longer distances than do meiotic conversion events. Either of these hypotheses is also consistent with the observation that allelic mitotic conversion between the same leu2 heteroalleles at five different loci occurred at about the same rate (Lichten and Haber 1989). The studies cited have dealt primarily with heteroalleles differing by point mutations. However, when one heteroallele contains a point mutation and the other contains a deletion (Chernoff et al. 1984) or insertion (Vincent and Petes 1989), the point mutation is (usually) converted to wild type more easily than is the deletion or insertion, regardless of the positions of these mutations in the gene.

2. Mitotic Conversion Tract Length and Conversion Fidelity

Several lines of evidence indicate that mitotic conversion tracts might be, on average, considerably longer than meiotic tracts. Using strains that were heterozygous for two pairs of heteroallelic markers (trp5 and leu1) 18 cM (and 36 kb) apart, Golin and Esposito (1984) found that Trp+ Leu+ colonies were obtained about 1000-fold more frequently than expected if the mitotic conversion of these two loci were independent; the frequency of the double convertants was only about 50-fold lower than the frequency of single convertants. These results indicated either that mitotic conversion occasionally involves very long conversion tracts or that multiple concerted conversion events sometimes occur along one pair of chromosomes. Subsequent studies in which plasmid sequences were inserted between *LEU1* and *TRP5* to expand the distance between these genes and to provide an intervening genetic marker have been interpreted as indicating that both of these mechanisms may operate (Golin et al. 1986; Golin and Falco 1988; Golin and Tampe 1988). Golin and Falco (1988) also found that conversion of *LEU1* was seldom (7%) associated with conversion of a marker about 9 kb away, indicating that most mitotic recombination conversion tracts are not very long. In addition to concerted mitotic recombination for markers on a single pair of homologous chromosomes, several groups have found that a subpopulation of mitotic cells have high levels of mitotic exchange on different chromosomes (Fogel and Hurst 1963; Minet et al. 1980; Montelone et al. 1981c; Golin and Tampe 1988).

Judd and Petes (1988) compared the lengths of meiotic and mitotic conversion tracts at the *URA3* locus. The diploid strain used was heterozygous for a mutation in *URA3* as well as for a number of restriction site changes within 5 kb of *URA3*. Ura⁻ derivatives of the strain were selected using 5-fluoro-orotate (Boeke et al. 1984) and screened for those derivatives that were still heterozygous for a marker centromere-distal to *URA3*. In 19 of 51 mitotic gene convertants, all markers (spanning a length of 4.7 kb) became homozygous. No examples of conversion tracts in which a nonconverted site was flanked by two converted sites were observed. The distributions of tract lengths for mitotic conversion events induced by UV and methylmethanesulfonate were similar to those observed for spontaneous conversion. If the length of conversion tracts has a normal distribution and there are no defined initiation and termination sites for mitotic conversion, the results obtained in this study indicate that most mitotic conversion tracts at the *URA3* locus are between 4 and 10 kb, compared to 200 bp to 3 kb for meiotic conversion tracts at the same locus in the same strain. These data do not exclude the possibility of rare tracts that are longer, such as those hypothesized by Golin and co-workers (Golin et al. 1986; Golin and Falco 1988; Golin and Tampe 1988).

The lengths of mitotic conversion tracts have also been measured by using repeated genes located either on plasmids (Ahn and Livingston 1986) or on chromosomes (Willis and Klein 1987; Aguilera and Klein 1989). In these studies, the average tract length was less than 500 bp. It is not clear whether the difference between these lengths and those measured by Judd and Petes (1988) reflects locus-specific differences in conversion or differences between intrachromosomal (or intraplasmid) and interchromosomal conversion. In addition, the rate of mitotic recombination between 2-kb repeats within a plasmid was remarkably high, about 2×10^{-3} (Ahn and Livingston 1986), raising the question of whether recombination events on a plasmid may be different from those involving the chromosomes.

Two experiments indicate that mitotic gene conversions represent the faithful transfer of information between homologous DNA sequences. First, Zimmermann (1968) showed that the enzyme produced from wild-type recombinants at the *ilv1* locus had activity that was indistinguishable from that of the true wild-type enzyme. Second, Judd and Petes (1988) found that mitotic conversion events faithfully transferred restriction sites between homologous chromosomes.

3. Association between Gene Conversion and Crossing-over

In strains that are heteroallelic for a centromere-proximal marker and heterozygous for a centromere-distal marker on the same arm, the association between mitotic gene conversion and mitotic crossing-over can be measured. In such studies (for a summary, see Esposito and Wagstaff 1981), between 10% and 55% of the conversion events are associated with crossing-over. Although a similar range is also observed in meiotic recombination, it should be pointed out that only half of the associated mitotic crossovers would be detectable (see Fig. 11) if segregation of chromatids is random (an untested hypothesis). Mitotic crossovers between repeated genes in plasmids (Ahn and Livingston 1986) or on chromosomes (Aguilera and Klein 1989) were preferentially associated with long conversion tracts. In heteroallelic crosses in which one allele was a point mutation and the second was a deletion or insertion, conversion events involving loss of the insertion or deletion were more frequently associated with crossing-over than were conversion events involving the point mutation (Chernoff et al. 1984; Vincent and Petes 1989).

In mitosis, as in meiosis, gene conversion events between repeated genes located on a single chromosome (intrachromosomal gene conversion) are usually (90% of the time) not associated with crossing-over (Jackson and Fink 1981; Klar and Strathern 1984; Ray et al. 1988). Those crossovers that do occur, however, are often associated with conversion (Willis and Klein 1987). Two explanations for these data are possible. First, mitotic gene conversion and crossover events may often be correlated, but conversion events do not represent an intermediate in the production of a crossover. This interpretation would be consistent with the evidence of Roman and Fabre (1983) that conversions and associated crossovers sometimes occur at different times in the cell cycle (see Section IV.B.5). Alternatively, conversion events may represent intermediates in the production of crossovers, but it may be difficult (perhaps for topological reasons) to resolve this intermediate as an intrachromosomal crossover, as also proposed for meiotic intrachromosomal events.

4. Symmetric and Asymmetric Transfer of Information during Mitotic Gene Conversion

As described in Sections III.B and III.C, meiotic gene conversion and PMS in *S. cerevisiae* generally represent the nonsymmetric transfer of information from one chromosome to the other. In the context of the Meselson-Radding/Radding models, this result is interpreted as reflecting repair of mismatches in asymmetric rather than symmetric heteroduplexes (see Figs. 7 and 9). In a heteroallelic diploid (1+ representing one mutant allele and +2 representing the other), mitotic convertants of the types ++/+2 and 1+/++ can also be readily explained as representing mismatch repair of asymmetric heteroduplexes. Although these classes represent about 90% of the convertants, two other classes, ++/++ and ++/12, are also observed (for a summary, see Esposito and Wagstaff 1981). Although the ++/12 class can be explained as a reciprocal mitotic crossover, the ++/++ class, which is found at about the same frequency, cannot. To explain this class as the result of mismatch repair in heteroduplexes, one must postulate that it reflects repair either of symmetric heteroduplexes formed between duplicated chromatids in G_2 or of symmetric heteroduplexes formed between unduplicated chromosomes in G_1 (Esposito 1978). As discussed in the following section, Esposito (1978) also found other evidence for the formation of symmetric heteroduplexes in G_1.

5. Timing of Mitotic Recombination Events

The original interpretation of the results of mitotic crossover experiments was that crossing-over occurs after chromosome replication (in G_2), because crossing-over in G_1 would not be expected to result in heterozygous markers becoming homozygous. In *S. cerevisiae,* several lines of evidence now suggest that mitotic recombination events can occur in both G_1 and G_2. In a series of studies by M. Esposito and co-workers (Esposito 1978; Golin and Esposito 1981, 1984), strains were used that were heteroallelic at the *TRP5* locus and several other markers on chromosome arm VIIL, heterozygous for the centromere-distal *ADE5* gene on the same arm, and homozygous for an *ade2* mutation (chromosome XV). Cells of this genotype, when plated on medium lacking tryptophan, generated red, white, and red/white-sectored Trp+ colonies. Diploids with the genotype *ade2/ade2 ADE5/ADE5* or *ade2/ade2 ADE5/ade5* yield red colonies, whereas diploids with the genotype *ade2/ade2 ade5/ade5* yield white colonies (Roman 1957). The Trp⁺ red/white-sectored colonies were interpreted to represent mitotic gene conversion events at the *TRP5* locus (to generate the wild-type *TRP5* gene) associated with mitotic crossing-over (to generate the red/white sectors). It is important to note that to detect such a colony, both the Ade5⁺ and Ade5⁻ products of crossing-over must have at least one wild-type *TRP5* gene. Thus, neither crossovers without an associated conversion at the *TRP5* locus nor crossovers with a conversion event that generates a single wild-type *TRP5* allele would generate such colonies.

The production of such colonies can be explained as the result of G_2 recombination as shown in Figure 13a. However, by this mechanism, one would expect to find cells with the genotype ++/1+ in one sector and cells with the genotype +2/++ in the other sector. When both sectors were examined by tetrad dissection, Esposito and co-workers (Esposito 1978; Golin and Esposito 1981, 1984) found that this class was in the minority (<5% of the colonies analyzed). The most common classes were those in which both sectors had the same genotype at *TRP5*, either both ++/1+ or both +2/++. Although this configuration cannot be explained by repair of mismatches within a single G_2 heteroduplex, it is readily explained by repair of mismatches within a G_1 heteroduplex. The generation of the +2/++ class is shown in Figure 13b; note that in this model, the resolution of the G_1 recombination event as a mitotic crossover results from replication of the Holliday junction, rather than from strand cleavage (as in the Meselson-Radding [and other] models). Although many of the classes observed in these experiments could be explained by mismatch repair in asymmetric G_1 heteroduplexes (Fig. 13b), some classes appeared to result from mismatch repair in symmetric G_1 heteroduplexes; e.g., in 2 of 70 colonies, both sectors had the ++/++ genotype. It should also be noted that the classes of events that are explicable by repair of mismatches in symmetric G_1 heteroduplexes are difficult to explain if mitotic gene conversion is the result of gap repair of a double-strand break. In particular, repair of gapped molecules in either G_1 or G_2 should not result in sectors with the ++/++ genotype.

The conclusion that spontaneous mitotic recombination events occur in G_1 depends on the assumption that the classes of events analyzed by Esposito do not reflect two or more G_2 recombination events (Whitehouse 1982). Because the rate of formation of sectored Trp⁺ colonies was only tenfold less than the rate of formation of unsectored Trp⁺ red (*ADE5/ade5* or *ADE5/ADE5*) colonies and about the same as the rate of formation of unsectored Trp⁺ white (*ade5/ade5*) colonies, Esposito argued that the sectored colonies were not generated by inde-

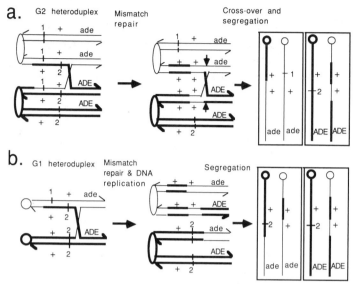

Figure 13 Patterns of heteroduplex formation and repair during mitotic recombination necessary to generate colonies that are sectored for a centromere-distal marker (*ADE*) and have a wild-type allele for a centromere-proximal marker in both sectors (after Esposito 1978). The centromere-proximal marker is heteroallelic before the recombination event. The mutant substitutions are indicated by short vertical lines. (*a*) Mitotic recombination in G_2. A symmetric heteroduplex covering the heteroalleles is formed and repaired to wild type on both chromatids. If this stucture is cleaved as indicated by the arrows, there will be an associated crossover. Appropriate segregation will then yield the daughter cells shown. (*b*) Mitotic recombination in G_1. An asymmetric heteroduplex covering the heteroalleles is formed and repaired to wild type. This structure is resolved by DNA replication. Appropriate segregation will then yield the daughter cells shown.

pendent double events. However, this argument is not unimpeachable, as concerted multiple mitotic conversion events appear to occur on chromosome VII (Golin and Falco 1988; Golin and Tampe 1988).

X-ray- and UV-induced mitotic recombination can apparently also occur in G_1 (Wildenberg 1970; Fabre 1978, 1981). Fabre constructed strains that were heteroallelic for temperature-sensitive mutations at the *CDC4* or the *CDC7* locus; at the restrictive temperature, these cells were arrested in G_1. The formation of *ts*+ cells was inducible by X-rays and UV treatment during the cell-cycle arrest, suggesting that at least the initiation of conversion could be induced in G_1. When a similar experiment was done with *cdc16* cells, which arrested in G_2, very little recombination was induced (Fabre 1981). Because X-ray-induced DNA damage is readily repaired (presumably by recombination) in G_2 (Brunborg et al. 1980), Fabre suggested that induced mitotic recombination events could occur in both G_1 and G_2, but that those occurring in G_2 mainly involved sister strands and therefore did not result in detectable gene conversion. If this interpretation is correct, induced intrachromosomal recombination events would be expected to occur in both G_1 and G_2, a prediction that has not yet been tested.

By analyzing sectored colonies as in the experiments of Esposito (1978), Roman and Fabre (1983) found that some of the G_1 conversion events were associated with crossover events involving at least three homologous chromatids. Such patterns are not simply explained by the model of mitotic recombination shown in Figure 13b and were interpreted as indicating that mitotic gene conversion and crossing-over were temporally (and might be mechanistically) unassociated. Unequal sister-strand recombination events involving repeated genes also clearly represent G_2 mitotic exchanges (see below).

In summary, the available experimental evidence indicates that in mitosis, unlike meiosis, recombination can occur either before or after (and perhaps during) DNA synthesis. The relative amounts of gene conversion and crossing-over that occur in G_1 and G_2 have not been established unambiguously.

C. Chromosomal Distribution of Spontaneous Mitotic Recombination Events

Malone et al. (1980) mapped mitotic crossovers between the *CYH2* locus and the centromere of chromosome VII. Relative to the meiotic map, there was more recombination in the intervals near the centromere. It is not yet clear whether this difference reflects an enhancement of mitotic recombination near the centromere or a possible suppression of meiotic recombination in that region (see Section III.A.1). Neitz and Carbon (1987) found that a 1.5-kb DNA fragment, 500 bp from *CEN14*, enhanced the rate of integration of transforming plasmids. If such integration is a good model for chromosome-chromosome recombination, this result may indicate that a hot spot exists for mitotic exchange near *CEN14*. By constructing strains that were heterozygous for restriction-site markers within and on both sides of the centromere of chromosome III, Liebman et al. (1988) showed that mitotic gene conversion events could include and pass through the centromeric sequences. However, the rate of conversion events involving the centromere was not measured in these studies. Before any generalizations about mitotic (or meiotic) recombination near centromeres can be made, the centromeric regions of additional chromosomes will have to be analyzed.

Several studies have indicated that mitotic recombination may be stimulated by transcription. Keil and Roeder (1984) identified a sequence that increased the rate of mitotic recombination using a plasmid that contained duplicated heteroalleles of *HIS4*. Random *Bgl*II fragments were cloned into this plasmid, and the resulting pool of plasmids was transformed into yeast. Of about 1000 fragments examined, only one (*HOT1*) stimulated (sevenfold) the production of His+ recombinants; this fragment contained the promoter of the 35S rRNA precursor. The effect was orientation-dependent and mitosis-specific. When the rRNA promoter sequences were integrated into the genome, they also stimulated classical gene conversion and intrachromosomal gene conversion and crossing-over (Voelkel-Meiman et al. 1987). In one study (Keil and Roeder 1984), stimulation of classical recombination was observed only if *HOT1* was homozygous, whereas in another similar study (Voelkel-Meiman and Roeder 1990), *HOT1* stimulated recombination even when heterozygous; the reason for this difference is not clear at present. By inserting a transcription terminator, Voelkel-Meiman et al. (1987) showed that the stimulation was the result of RNA polymerase I transcription through the affected region. Subcloning and mutagenesis studies (Stewart and Roeder 1989) indicated that *HOT1* consisted of the promoter for polymerase I as well as an associated enhancer. Voelkel-Meiman et al. suggested that one possible explanation for the effect of *HOT1* is that transcription might lead to unwinding of the DNA, thus exposing single strands to enzymes that catalyze recombination. Because the chromosome containing *HOT1* preferentially receives information during conversion events in heterozygous strains, Voelkel-Meiman and Roeder (1990) suggested that *HOT1*-induced conversion may be initiated by a double-strand break in DNA sequences adjacent to *HOT1* (see Section III.E.3). Alternatively, single-strand gaps in the *HOT1*-containing chromosome would also produce the same effect on the direction of gene conversion (Radding 1982).

Although the *HOT1* sequences can clearly stimulate recombination of nearby non-rRNA gene sequences, it is not clear that the rRNA gene cluster itself has a higher than normal level of mitotic recombination between homologous chromosomes. By measuring the rate of mitotic crossing-over between the centromere and *URA4* (a gene centromere-distal to the rRNA gene cluster) and comparing this rate with those for other genetic intervals, Smolik-Utlaut (1982) estimated that the cluster had approximately the same level of mitotic recombination as did other chromosomal sequences. This observation suggests that the primary effect of *HOT1* may be on sister-strand recombination, which is known to occur at a high mitotic rate (Szostak and Wu 1980). Other possibly relevant data have come from studies of the effects of mutations in the *SIR2* gene (required for repression of transcription of silent *MAT* loci) and *TOP1* or *TOP2* topoisomerase genes on recombination involving chromosomal rRNA gene clusters. Strains with *sir2* mutations have elevated (about tenfold) levels of mitotic and meiotic intrachromosomal recombination in the rRNA genes (Gottlieb and Esposito 1989). Strains with *top1* or *top2* mutations have elevated levels of mitotic recombination, both unequal sister-strand crossovers and crossovers between homologs (Christman et al. 1988). It is not yet clear whether these results reflect effects of *SIR2*, *TOP1*, and *TOP2* on transcription of the rRNA genes.

Thomas and Rothstein (1989) examined the effect of transcription on mitotic recombination between nontandem duplications of *GAL10* heteroalleles. An increased rate of transcription of the plasmid sequences situated between the two *GAL10* genes was correlated with a 15-fold increase in the rate of loss of the duplication (possibly by intrachromosomal crossing-over) without an increase in the rate of conversion of the *GAL10* sequences.

D. Ectopic Recombination

Like meiotic recombination, mitotic recombination is not limited to allelic sequences on homologous chromosomes. Ectopic exchanges (both crossovers and conversion events) occur between repeats within the same chromosome and between repeats located on nonhomologous chromosomes (for detailed review, see Klein 1988a; Petes and Hill 1988). Most of the relevant studies were done using duplications made by transformation. Although these experiments are not discussed in detail here, some important conclusions are mentioned briefly below.

First, the rate of intrachromosomal recombination is often higher (10–100-fold) than the rate of recombination between sequences on homologous chromosomes (Jackson and Fink 1981; Keil and Roeder 1984; Lichten and Haber 1989). However, the rate of intrachromosomal exchange is reduced as the distance between the repeats is increased (Roeder et al. 1984; Lichten and Haber 1989). One plausible interpretation of these results is that the rate of mitotic recombination may be strongly influenced by the probability of two homologous DNA sequences finding each other. Consistent with this interpretation, unequal sister-strand recombination in the rRNA gene cluster occurs at a high rate ($\sim 10^{-2}$ per cell division; Szostak and Wu 1980). Second, for tandemly repeated genes, both intrachromosomal crossovers and unequal sister-strand crossovers occur (see, e.g., Jackson and Fink 1981; Klein, 1988b), although the former events may be rare relative both to the latter events and to unequal sister-strand conversion events (Schiestl et al. 1988). Third, mitotic recombination does not require extensive sequence homology. Intrachromosomal recombination occurs with only about 330 bp of homology (for review, see Liebman and Picologlu 1988), and heterochromosomal recombination occurs between 1.2-kb repeats (Mikus and Petes 1982); 26 bp of homology is enough for a detectable level of conversion between repeats on a plasmid (Ahn and Livingston 1986), and 4–20 bp is sufficient for transforming DNA to recombine with the chromosome (Moerschell et al. 1988; R. Schiestl and T. Petes, unpubl.). Fourth, although ectopic and classical recombination require some of the same gene products, some mutations affect these two processes differentially. For example, although *rad52* mutations affect both conversion and crossing-over between homologous chromosomes (Malone and Esposito 1980; Malone et al. 1988), some types of intrachromosomal crossing-over (e.g., unequal sister-strand exchange in the rRNA genes) do not require *RAD52* function (Prakash and Taillon-Miller 1981; Zamb and Petes 1981). In addition, intrachromosomal gene conversion events appear to be less frequently associated with crossing-over than are classical mitotic gene conversion events (for review, see Fink and Petes 1984); in one recent study, however, about 15% of the intrachromosomal conversion events were associated with crossing-over (Lichten and Haber 1989). Finally, mitotic recombination between naturally occurring repeats is frequent enough that ectopic recombination is likely to be an important cause of chromosomal alterations (deletions, duplications, translocations, etc.). For example, deletions of the *CYC1* locus as a result of recombination between flanking Ty elements occur at a rate of about 10^{-6} per cell generation (Liebman et al. 1981), and recombination between telomeric repeats occurs at a similar rate (Louis and Haber 1990).

E. Induction of Mitotic Recombination

Two lines of evidence indicate a relationship between the repair of DNA damage and mitotic recombination. First, mitotic recombination can be induced by agents that damage DNA, such as X-rays, UV light, and a variety of chemical mutagens (for review, see Esposito and Wagstaff 1981; Haynes and Kunz 1981). Second, as described in Section IV.F, mutations that affect repair often also affect mitotic recombination. In general, both gene conversion and crossing-over are induced by mutagens, although quantitative differences in the rates of induction of these two types of recombination have been observed. For example, at doses that yield the same rate of mitotic crossovers, the rate of induced heteroallelic gene conversion is tenfold higher for UV than for ethylmethanesulfonate (Davies et al. 1975).

It is likely that more than one factor is responsible for the induction of mitotic recombination by DNA-damaging agents. For at least some types of damage, the lesion itself is likely to be responsible for the stimulation. For example, it is known that X-rays induce double-strand breaks (Ho 1975; Resnick and Martin 1976) and that double-strand breaks induced at the *MAT* locus (Strathern et al. 1982) and at other loci (Nickoloff et al. 1986) are recombinogenic. In addition, strains containing *rad52* mutations do not repair double-strand breaks (Ho 1975; Resnick and Martin 1976) and are recombination-defective (Malone and Esposito 1980; Malone et al. 1988). The lesions caused by other recombination-inducing agents (e.g., UV irradiation and a mutational deficiency in DNA ligase) are likely to represent (at least initially) either nicks or single-strand gaps in the DNA, rather than double-strand breaks (for review, see Haynes and Kunz 1981). In addition, some nonmutagenic carcinogens induce intrachromosomal mitotic recombination in yeast (Schiestl 1989). For all agents, the mechanism by which the lesion stimulates recombination is not clear.

In addition to the direct recombinogenic effects of lesions, at least some agents (e.g., UV and X-rays) induce *trans*-acting factors that stimulate mitotic recombination (Fabre and Roman 1977). This induction was detected by mating an irradiated haploid that contained a double mutation in the *ADE6* gene to a diploid (homozygous at the *MAT* locus) that was heteroallelic for the two individual *ade6* mutations. A substantial (tenfold) stimulation of the formation of Ade+ cells was observed, resulting from gene conversion between the unirradiated chromosomes. The mechanism of this stimulation has not yet been established.

Both intragenic and intergenic maps show different patterns for induced and spontaneous mitotic recombination (for review, see Esposito and Wagstaff 1981), although in one study (Judd and Petes 1988), the lengths of gene conversion tracts for spontaneous and UV- or methylmethanesulfonate (MMS)-induced events were similar. It seems clear that conclusions about induced conversion and crossing-over cannot be automatically extended to spontaneous recombination events and vice versa; the mechanisms involved in these two types of recombination may be quite different.

F. Mutations Affecting Mitotic Recombination

Because mitotic recombination is not essential for viability, it is potentially easier to isolate mitotic Rec⁻ mutants than meiotic Rec⁻ mutants. However, a hunt for such mutants faces two difficulties: (1) Mitotic recombination is usually studied in diploids (in which the search for recessive mutations would be difficult) and (2) the rate of spontaneous classical recombination events is low, even in wild-type strains (typically 10^{-6} to 10^{-8} events per locus per mitotic division). Two related approaches for obtaining recessive mutations affecting mitotic exchange are to use strains that are disomic for one chromosome containing heteroalleles and monosomic for all other chromosomes and to use haploid strains containing gene duplications marked by heteroalleles on one chromosome. To circumvent the difficulty of identifying mutations inactivating a process that is already very infrequent (spontaneous mitotic exchange), some researchers have screened for mutations affecting induced mitotic recombination or for mutants that are hyper-Rec. As the repair of DNA damage utilizes some of the same gene products as does recombination, another successful method of obtaining Rec⁻ mutants has been to screen mutants that have increased sensitivity to DNA-damaging agents for defects in recombination. Below, we describe the Rec⁻ and hyper-Rec mutants that have been obtained with these methods. Information about these mutations is summarized in Table 1.

1. Rec⁻ Mutants

As discussed in Section III.H, many mutants defective in the repair of damage induced by UV or X-ray irradiation have been isolated, and most have been classified into the three epistasis groups defined by *RAD3*, *RAD52*, and *RAD6* (for review, see Haynes and Kunz 1981; Game 1983; Friedberg et al., this volume). In the *RAD52* group, most of the mutants have reduced levels of induced mitotic exchange, perhaps related to their defects in the repair of double-strand breaks (Haynes and Kunz 1981; M. Resnick, pers. comm.). In addition, *rad52* (discussed in detail below), *rad51*, and *rad54* mutations appear to reduce spontaneous mitotic exchange; in contrast, the effects of *rad53*, *rad55*, *rad56*, and *rad57* mutations on spontaneous mitotic recombination are less clear (Saeki et al. 1980; J. Game, pers. comm.), and *RAD50* is not required for spontaneous mitotic recombination (Malone and Esposito 1981; Malone 1983; Malone et al. 1985, 1990).

Most *rad52* mutants have reduced levels of spontaneous mitotic gene conversion and crossing-over for at least some intervals. The reductions in the rates of conversion were dramatically different for different loci, varying from 3- to 300-fold (Malone and Esposito 1980; Hoekstra et al. 1986; Malone et al. 1988); these differences were observed even with *rad52* alleles (gene disruptions) that presumably had no *RAD52* activity. Interestingly, the residual levels of conversion in *rad52* strains showed little locus-to-locus or allele-to-allele (Hoekstra et al. 1986) variation, suggesting that there is a constant level of *RAD52*-independent background recombination. The crossover rates in *rad52* strains were reduced by a factor of 10 for two different intervals but not detectably reduced for a third interval (Malone and Esposito 1981).

The *RAD52* product appears to be required for intrachromosomal conversion events (Jackson and Fink 1981) but not for unequal sister-strand crossovers in the rRNA gene cluster (Prakash and Taillon-Miller 1981; Zamb and Petes 1981). In one study (Jackson and Fink 1981), the *RAD52* product did not appear to be required for generation of a triplication from strains containing a duplication (by unequal crossing-over or gene conversion), whereas in two similar studies (Fasullo and Davis 1987; Klein 1988b), triplications were less frequent (4–20-fold) in *rad52* strains. The role of the *RAD52* product in the integration of circular plasmids following transformation is also not clear. Orr-Weaver et al. (1981) found no effect of a *rad52* mutation on the integration of a plasmid containing rRNA gene sequences, whereas Malone et al. (1988) observed a 50-fold decrease in integration of a plasmid containing only single-copy yeast sequences. Although it is possible that this difference reflects the *RAD52* independence of recombination in the rRNA genes (see above), this explanation has not yet been tested. The variability in the effects of *rad52* mutations on different types of recombination indicates that the *RAD52* product is not essential for mitotic recombination in yeast in the same sense that the *recA* gene product is required for all general homologous recombination in *E. coli*.

The association between mitotic conversion and crossing-over appears to be different in *RAD52* and *rad52* strains. Haber and Hearn (1985) found that about 20% of conversion events at the *HIS4* locus were associated with crossovers (half resulting in a heterozygous centromere-distal marker becoming homozygous and half resulting in reverse linkage of the centromere-distal marker) in wild-type strains. In a *rad52* strain, conversion was reduced by a factor of 50, and about one third of the conversion events were associated with chromosome loss. In those convertants retaining two copies of chromosome III, about 80% had an associated crossover. Haber and Hearn (1985) suggested that mitotic conversion events occurring in the absence of *RAD52* are strongly associated with nonreciprocal crossovers in which one of the recombining chromatids is lost. Thus, random mitotic segregation of the chromosomes following recombination would lead to either a monosomic strain or a strain in which the distal markers were homozygous. As *rad52* strains do not repair double-strand breaks, Haber and Hearn suggested that the initiating event might be formation of an asymmetric heteroduplex. These results, as well as others discussed above, suggest that recombination events in *rad52* strains are qualitatively different from those in *RAD52* strains, possibly because different pathways are used.

One gene that may be involved in a *RAD52*-independent pathway for certain types of mitotic recombination is *RAD1*, a member of the *RAD3* epistasis group. Most of the genes (including *RAD1*) in this group are involved in the error-free excision of pyrimidine dimers (for review, see Haynes and Kunz 1981; Friedberg et al., this volume). Klein (1988b) and Schiestl and Prakash (1988) found that *rad1* strains had slightly reduced levels of certain classes of ectopic recombination. In addition, the rate of intrachromosomal recombination in *rad1 rad52* double mutants was depressed synergistically, indicating that these two genes may act in different recombination pathways. However, these conclusions are complicated by several observations on classical recombination. First, *rad1* strains have slightly elevated levels both of crossing-over and of conversion for some genetic loci (Schiestl and Prakash 1988; Montelone et al. 1988). Second, *rad1 rad52* strains have *rad52* levels of conversion at several loci and wild-type levels of crossing-over in the single genetic interval monitored (Montelone et al. 1988). These data indicate either that the *RAD1* gene product plays different roles in ectopic and classical recombination or that the effects of this gene product on recombination are markedly allele- or gene-specific.

Mutants with reduced levels of mitotic recombination have also been isolated by more direct techniques than the analysis of repair-deficient strains. For example, Rodarte-Ramon and Mortimer (1972) screened for Rec⁻ mutants by using a strain that was disomic for chromosome VIII and contained *arg4* heteroalleles. Mutants deficient in the production of prototrophs after UV irradiation were isolated and proved to define four genes. The *rec2* mutation was later found to be allelic with *rad52*. The *rec4* mutation, originally classified as a mutation affecting recombination only at the *arg4* locus, represents a mutation in the arginine biosynthetic pathway and is not actually a Rec⁻ mutant (C. Bruschi, pers. comm.). *rec3* strains fail to produce viable spores, implying that the *REC3* product is likely to be required for both meiotic and mitotic exchange; the *REC1* gene appears to be specific for mitosis. Similarly, using a strain disomic for chromosome VII containing multiple heterozygous and heteroallelic markers, Esposito et al. (1984) isolated several Rec⁻ mutants (as well as hyper-Rec mutants; see below). One of these mutants (*rec754*) was defective for both conversion and crossing-over, seven mutants (*rec396, rec539, rec413, rec336, rec467, rec141,* and *rec100*) were defective for conversion but not for crossing-over, and two (*rec139* and *rec201*) were defective only for crossing-over. These phenotypes offer further support for the possible mechanistic separation of gene conversion and crossing-over. All mutants, except *rec754, rec336, rec141,* and *rec201*, were sensitive to either UV or X-rays. As discussed in Section V, cell extracts prepared from several of these mutants are partially deficient in the ability to catalyze recombination in vitro (Esposito et al. 1984; Symington et al. 1984). Complementation tests between these mutants and the *rad* mutants have not yet been reported.

2. Hyper-Rec Mutants

Mutants with elevated levels of mitotic recombination have been identified indirectly by screening strains with defects in DNA replication or repair. In addition, a number of direct screens for such mutants have been done. The repair-defective strains have been a particularly fruitful source of hyper-Rec mutants, which have been found in all three of the epistasis groups.

In the *RAD3* epistasis group, *rad1, rad3,* and *rad4* mutations have hyper-Rec effects for at least some chromosomal intervals. Snow (1968) did not observe such effects of *rad1* and *rad4* at one diagnostic locus. However, in studies of heteroallelic recombination at four other loci and of crossing-over in two intervals, Montelone et al. (1988) found that *rad1* and *rad4* strains had normal levels of gene conversion but three- to sevenfold elevated levels of crossing-over. These results, like some others described in Sections IV.B.3 and IV.F.1, suggest that gene conversion and crossing over may be formally separable processes.

RAD3 encodes a DNA helicase/ATPase (Sung et al. 1987a,b) and, unlike other members of its epistasis group, is essential for viability (Higgins et al. 1983; Naumovski and Friedberg 1983). Several alleles of *RAD3* that were isolated as UV-sensitive mutations had relatively little effect on mitotic exchange. Snow (1968) reported that *rad3-1* elevated recombination about threefold, and Kern and Zimmermann (1978) reported that the *rad3-2* strains were slightly hyper-Rec. Montelone et al. (1988) subsequently observed no hyper-Rec phenotype for *rad3-2* at four other loci and no effect on crossing-over in two intervals. Mutant alleles of *RAD3* have also been isolated as mutations that increased the frequency of mutation (*rem1-1* = *rad3-101*) (Golin and Esposito 1977) or the level of spontaneous mitotic recombination (*rem1-2* = *rad3-102*) (Malone and Hoekstra 1984). Both of these alleles were semidominant and (when homozygous) resulted in an average 15-fold elevation in gene conversion and crossing-over in every interval examined. The interactions of *rad3-101* and *rad3-102* with other mutations affecting repair and/or recombination have also been examined (Malone and Hoekstra 1984; Montelone et al. 1988). Double-mutant strains with either of these alleles and the *rad52* or *rad50* mutations were inviable, suggesting that the hyper-Rec phenotype of these alleles might be the result of double-strand breaks in the DNA, a suggestion supported by the observation that chromosomal DNA isolated from *rad3-102* strains appeared slightly smaller than that isolated from wild-type strains (Montelone et al. 1988). Because triple-mutant strains with the genotype *rad1 rad3-102 rad52* or *rad4 rad3-102 rad52* were viable, Montelone et al. suggested that other gene products involved in DNA excision repair were necessary to process the primary lesions in *rad3-102* strains into double-strand breaks. If double-strand breaks in *rad3-102* strains are necessary for the hyper-Rec phenotype and mutations in *rad1* (or *rad4*) prevent the processing of DNA damage in *rad3-102* strains into double-strand breaks, strains with the *rad1 rad3-102* genotype should not be hyper-Rec. Montelone et al. found that gene conversion was reduced to wild-type levels in this genetic background, but crossing-over was still elevated. One interpretation of this result is that gene conversion induced by *rad3-102* involves repair of double-strand breaks, but crossovers induced by *rad3-102* do not.

In the *RAD6* epistasis group, two different alleles of *RAD6* itself have modest hyper-Rec effects on both conversion and crossing-over (Kern and Zimmermann 1978; Montelone et al. 1981a); one of these alleles (*rad6-1*) was defective for meiotic recombination, whereas the other (*rad6-2*) was not. Since *RAD6* encodes a ubiquitin-conjugating enzyme and acts on histones (Jentsch et al. 1987), it is possible that its effect is produced by a change in chromatin structure, rather than by a DNA lesion. Mutations in *RAD18* are also hyper-Rec for both classical conversion and crossing-over, although intrachromosomal recombination associated with heterothallic mating-type switching is not affected (Boram and Roman 1976; Schiestl et al. 1990).

In the *RAD52* epistasis group, only *rad50* mutations have been found to confer a hyper-Rec phenotype. Four different mutant alleles (including one null allele) stimulate spontaneous mitotic recombination (both crossing-over and gene conversion) between 5-fold and 20-fold (Malone 1983; Malone et al. 1985, 1990). It is not clear why mutations in a protein that is required for meiotic recombination and repair of X-ray-induced damage of DNA also elevate spontaneous mitotic exchange. One possibility is that Rad50p may be multifunctional, an interpretation supported by the observation that mutations near the amino terminus of the gene eliminate meiotic recombination without affecting DNA repair or the rate of cell growth (Alani et al. 1990). The effects of these mutations on mitotic recombination have not yet been examined.

Mutations that affect the ability to repair MMS-induced damage (Prakash and Prakash 1977) also often confer hyper-Rec phenotypes. Montelone et al. (1981b) showed that mutations in *MMS9*, *MMS13*, and *MMS21* caused elevated levels of gene conversion (25-fold) and crossing-over (10–150-fold). Strains with *mms9* or *mms21* mutations were sensitive to γ-rays (indicating that these mutations might belong to the *RAD52* group), and mutations in all three complementation groups inhibited sporulation (indicating that the gene products may also be involved in meiotic recombination).

Many cell-cycle (*cdc*) mutants, particularly those affected in DNA metabolism, also have hyper-Rec phenotypes. For example, mutations in *CDC9*, encoding DNA ligase, are hyper-Rec for both conversion and crossing-over (Fabre and Roman 1979; Game et al. 1979; Montelone et al. 1981c). Hartwell and Smith (1985) showed that mutations in *CDC2* (a δ-type DNA polymerase; see Section V), *CDC6*, *CDC8* (thymidylate kinase; Sclafani and Fangman 1984), *CDC13*, *CDC17* (DNA polymerase α; Carson 1987), *CDC5*, and *CDC14* all resulted in hyper-Rec phenotypes. In all cases, the enhanced recombination was dependent on *RAD52*. Except for *CDC13, CDC5*, and *CDC14*, temperature-sensitive alleles of all of these genes result in arrest during the S period or have known effects on DNA replication. Hartwell and Smith (1985) suggested that recombination was stimulated by nicks or gaps in the DNA of the mutant strains. Mutations in other cell-cycle genes had little effect on mitotic exchange.

In addition, some mutations originally isolated as affecting meiotic recombination also have hyper-Rec mitotic phenotypes. For example, *pms1* mutants (Williamson et al. 1985; see Section III.H) had a fivefold elevated level of heteroallelic mitotic gene conversion. Since *pms1* strains have a reduced capacity to repair mismatches in heteroduplex DNA, this result suggests that such mismatches formed spontaneously (perhaps as a result of mistakes made during DNA replication) may be an important recombinogenic lesion. A similar suggestion was also made by Montelone et al. (1988) to explain the recombinogenic effects of *rad3-102* (see above).

Hyper-Rec mutants have also been identified by more direct methods in several studies. Maloney and Fogel (1980) used disomic strains (similar to those described by Rodarte-Ramon and Mortimer [1972]) to identify semidominant hyper-*rec* mutations affecting mitotic conversion. Most of these *MIC* mutations stimulated conversion at multiple loci and were also deficient in DNA repair. In addition, several of the *rec* mutants obtained by Esposito et al. (1984, 1986), using methods described above, were hyper-Rec. The mutations *rec276*, *rec952*, and *rec395* resulted in increased crossing-over without increased conversion, whereas the mutations *rec193*, *rec409*, *rec46*, and *rec199* resulted in increased levels of both conversion and crossing-over. The relationships among the hyper-Rec *rec*, *MIC*, and *rad* mutations have mostly not been determined.

In another approach, Aguilera and Klein (1988) used haploid strains containing duplicated heteroallelic genes to look for mutations that elevated the level of intrachromosomal mitotic gene conversion. Mutations in eight different genes (*HPR1–HPR8*) were identified. Mutations in *HPR3*, *HPR5*, or *HPR6* also resulted in increased sensitivity to UV and/or MMS, and all of the mutants required *RAD52* to exhibit the hyper-Rec effect. The *hpr* mutations affected intrachromosomal recombination in different ways: *hpr1* and *hpr6* enhanced "pop-outs" (loss of the duplication as the result of crossing-over or conversion); *hpr4*, *hpr5*, and *hpr8* enhanced gene conversion; and *hpr2*, *hpr3*, and *hpr7* stimulated both conversion and crossing-over. Although the *hpr1* mutation specifically affected intrachromosomal recombination, the mutations *hpr3–hpr5* and *hpr7* were also hyper-Rec for both conversion and crossing-over between homologous chromosomes. *HPR3* proved to be *CDC17* and *HPR6* to be *CDC2*, both of which encode DNA polymerases (see Section V). In addition, DNA sequence analysis of the cloned genes indicates that *HPR1* shares sequence homology with *TOP1*, the yeast gene encoding topoisomerase I (Aguilera and Klein 1990), and *HPR5* is allelic with *SRS2/RADH* (Lawrence and Christiansen 1979; H. Klein, pers. comm.); the *SRS2/RADH* product shares sequence homology with several bacterial DNA helicases (Aboussekhra et al. 1989).

It should be noted that some gene products identified using hyper-Rec mutants may not be directly involved in mitotic recombination, since altering nucleotide pools in a variety of ways also results in a hyper-Rec phenotype (Haynes and Kunz 1986; McIntosh et al. 1986). For example, Kunz et al. (1980) showed that arresting *cdc21* mutants (defective in thymidylate synthetase) at the restrictive temperature stimulated mitotic recombination. From studies of the effects of DNA synthesis inhibitors (such as cytosine arabinoside monophosphate) on mitotic recombination, McIntosh et al. (1986) suggested that any inhibition of DNA replication results in higher levels of mitotic exchange.

3. Conclusions from the Analysis of Mutants

A number of generalizations can be based on the phenotypes of the mutants described above. First, as expected, many (although not all) of the individual mutations that affect meiotic recombination and DNA repair also affect mitotic recombination, indicating that these processes require many common functions. Second, most of the *rad* mutants that are Rec⁻ are defective in double-strand break repair, indicating that this type of repair shares common functions with spontaneous mitotic recombination. Third, several of the *rad* mutants that are hyper-Rec affect error-free excision repair of UV-induced DNA damage, suggesting that the absence of these *RAD* gene products may result in lesions in the DNA that stimulate mitotic exchange. Fourth, the normal association between mitotic gene conversion and crossing-over is altered in some mutant strains, suggesting that gene conversion might not be a precursor in the formation of mitotic crossovers or that the resolution of conversion intermediates as crossovers might be controlled by a number of gene products. Fifth, no single *rec* mutation completely eliminates all types of mitotic exchange, indicating that there are likely to be multiple pathways of mitotic recombination. Sixth, the quantitative effects of particular *rec* mutations on different genetic loci and on different alleles within one locus are different. It is not clear whether this result reflects sequence preferences for the enzymes that catalyze recombination or some other factor, such as the distance between the mutant sites in heteroalleles. Finally, although intrachromosomal recombination events require some of the same gene products as does classical recombination, there are also gene functions that are unique to one pathway or the other.

G. Mechanisms of Mitotic Exchange

Since one can usually analyze only one of the two daughter cells that contain the products of spontaneous mitotic recombination, inferences about the mechanism of mitotic exchange are usually less direct than for meiotic recombination. In particular, the detailed molecular models of recombination are based on meiotic studies almost exclusively. We therefore discuss the mechanism of spontaneous mitotic recombination in the context of the Meselson-Radding and double-strand-break repair models of meiotic recombination.

1. Initiating Event for Spontaneous Mitotic Recombination

It is not known whether spontaneous mitotic recombination is initiated by a double-strand break or by a single-strand nick or gap. One observation consistent with the first possibility is that double-strand breaks induced at random sites by X-rays stimulate mitotic recombination (see above). If recombination is prevented by mutation or by the use of homologous chromosomes with diverged DNA sequences (Resnick et al. 1989), the unrepaired breaks lead to lethality. In addition, site-specific double-strand breaks induced by cleavage with the HO enzyme are recombinogenic at the *MAT* locus (Kostriken 1983) and at other loci containing the appropriate cut site (Nickoloff et al. 1986; Ray et al. 1988; Rudin and Haber 1988). In one study, a double-strand break induced ectopic recombination of a repeated gene located about 10 kb from the break (Ray et al. 1989).

A related argument is that the introduction of a double-strand break into a transforming circular plasmid (in a region that has homology with the chromosome) greatly increases the frequency of plasmid/chromosome recombination (Hicks et al. 1979; Orr-Weaver et al. 1981). When a double-strand gap is made in the plasmid sequences, this gap is filled in using the chromosomal sequences as a template (Orr-Weaver et al. 1981). Thus, the gap-repair process is a gene-conversion-like event in which there is loss of one sequence and duplication of another. Orr-Weaver and Szostak (1983) showed that gap repair was also associated with crossing-over. These observations were used in the formulation of the double-strand-break repair model of recombination (Szostak et al. 1983). However, it is not clear whether the integration of a plasmid into a chromosome following transformation occurs by the same mechanism as does spontaneous recombination between chromosomal sequences. In addition, it is possible that the introduction of transforming DNA sequences may induce novel mitotic recombination pathways.

Although it is clear that double-strand breaks can stimulate mitotic recombination, it has not yet been shown that these lesions are responsible for initiating *spontaneous* mitotic exchange. Support for this possibility comes from the observation that several mutant strains (*rad51*, *rad52*, and *rad54*) defective in their ability to repair double-strand breaks have reduced levels of spontaneous mitotic exchange. However, this argument is weakened by the observation that mutations in *RAD50*, another member of the *RAD52* group that appears defective in the ability to repair double-strand breaks (M. Resnick, pers. comm.), result in a hyper-Rec phenotype. Moreover, the observations that substantial levels of crossovers are observed in *rad52* strains and that some types of intrachromosomal recombination occur at wild-type levels in *rad52* strains suggest that not *all* spontaneous recombination is initiated by double-strand breaks.

The evidence that nicks or single-strand gaps in the DNA are responsible for initiating spontaneous mitotic recombination is also inconclusive. In support of this hypothesis, recombination is stimulated by various treatments (such as UV irradiation) that result in nicked or gapped DNA. Moreover, the observation that many mutations that block cells in S phase are hyper-*rec* (Hartwell and Smith 1985) suggests that the lesions found in replicating DNA (nicks and gaps) are recombinogenic. In particular, the hyper-Rec phenotype of *cdc9* mutants (defective in DNA ligase) is most simply explained by this interpretation. Finally, recent studies (Strathern et al. 1991) have indicated that expression of a site-specific DNA-nicking enzyme (gene 2 protein of phage f1) in yeast can induce mitotic gene conversion near the site recognized by the enzyme. It is not yet clear, however, that the nick expected at this site is not processed into a double-strand break.

One obvious difficulty in detecting initiating lesions is that the low frequency of spontaneous mitotic recombination events at any given locus makes it difficult to obtain a physical description of the lesion. Further progress in identifying the initiating lesions for spontaneous mitotic recombination is likely to require more sensitive physical methods of detection, the analysis of very strong mitotic recombination hot spots, or the improvement of the in vitro recombination system.

2. Mechanisms of Gene Conversion

In the Meselson-Radding model (Meselson and Radding 1975), most conversion is the result of mismatch repair in heteroduplexes, whereas in the double-strand-break repair model (Szostak et al. 1983), most conversion is the result of gap repair. There are a variety of different patterns of mismatch repair in an asymmetric heteroduplex that can produce a wild-type gene in a heteroallelic strain (Fig. 14). As the probability of a conversion event resulting in a wild-type gene is likely to represent a complex function of the lengths of the heteroduplex tracts, the relative rates of DNA repair and replication, and the directions of repair of individual mismatches, hyper-*rec* or *rec* mutations could affect a number of steps other than the frequency of initiating events. For example, a mutant that has longer heteroduplex tracts than the wild-type strain could have less conversion to prototrophy if two mismatches in a heteroduplex were always corrected as a single-excision-repair tract. In contrast, there are fewer different ways of performing gene conversion involving heteroalleles by gap repair. The wild-type gene is constructed by the repair of a DNA molecule that contains a gap that has removed its mutant substitution; to obtain a wild-type allele, the gap cannot include the site that is mutant in the other allele.

Most of the observations on mitotic gene conversion do not bear directly on the question of which of these two types of mechanisms is actually used. The continuity of mitotic conversion tracts involving multi-

a.

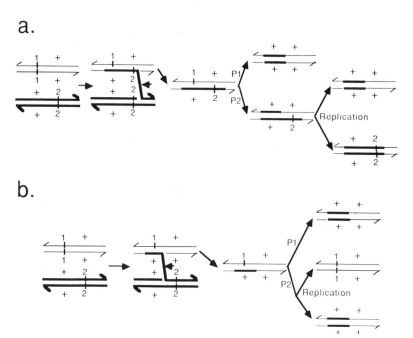

b.

Figure 14 Pathways for generating a wild-type allele by asymmetric heteroduplex formation between heteroalleles in mitosis. The mutant substitutions are
indicated by short vertical lines associated with the DNA strands. (*a*) Repair of a
heteroduplex covering both mutant sites. Following formation of the asymmetric
heteroduplex, there are two ways of generating a wild-type allele: (Pathway 1,
P1) Repair of both mismatches to wild-type information; (Pathway 2, P2) repair
of one mismatch to wild-type information (repair of the [1/+] mismatch is
shown), followed by DNA replication (in the same cell cycle [for a G_1
heteroduplex] or in the next cell cycle [for a G_2 heteroduplex]). (*b*) Processing
of a heteroduplex covering only one mutant site. As in Fig. 14a, a wild-type
gene can be generated either by repair of the mismatch (P1) or by DNA replication (P2).

ple markers (Judd and Petes 1988) is most easily explained by gap repair,
as one might expect repair of a heteroduplex involving multiple mismatches to be "patchy." In addition, it is clear that yeast cells are capable
of repairing double-strand breaks (Strathern et al. 1982) and of performing gap repair (Orr-Weaver et al. 1981). However, certain mitotic recombination events (those that Esposito and co-workers [Esposito 1978;
Golin and Esposito 1981, 1984] suggested were the result of mismatch
repair in symmetric G_1 heteroduplexes) cannot be explained as the result
of a single gap-repair event (see Section IV.B.5). In addition, Rothstein
(1984) showed that when yeast cells (genotypically *ura3 sup4+ cdc8*)

were transformed with fragments of DNA containing multiple markers (*URA3 SUP4-o CDC8*), Ura⁺ colonies were often sectored for one or both of the other two genes. The simplest explanation of this result is that heteroduplexes were formed between the chromosome and the transforming plasmid DNA. In a similar experiment, Ronne and Rothstein (1988) showed that mitotic recombination between direct repeats was associated with heteroduplex formation. Moreover, Singh et al. (1982) and Simon and Moore (1987) found that single-stranded DNA molecules transformed yeast more efficiently than double-stranded molecules. One interpretation of this result is that a region of single-stranded DNA is an intermediate in mitotic exchange. Finally, it is clear from both in vivo (Bishop et al. 1989; B. Kramer et al. 1989) and in vitro (Muster-Nassal and Kolodner 1986) studies that vegetative yeast cells have an effective mismatch repair system.

In summary, mitotic yeast cells appear to have systems that would allow gene conversion either by mismatch repair in heteroduplexes or by gap repair. Consequently, the most important factor determining which pathway is followed is likely to be the lesion that initiates spontaneous recombination; as discussed above, the nature of these lesions is not yet known. It is certainly possible that several types of lesions can initiate spontaneous mitotic recombination.

3. Low Frequency of Mitotic Recombination

Several factors might contribute to the low frequency of spontaneous mitotic recombination relative to meiotic exchange: (1) Most of the enzymes required to catalyze recombination may be quantitatively or qualitatively deficient in vegetative cells; (2) pairing of homologous chromosomes (or homologous sequences), perhaps mediated by the synaptonemal complex (a meiosis-specific structure), may be necessary for high levels of exchange; or (3) although most of the enzymatic machinery necessary to catalyze high levels of recombination may be present in mitotic cells, the lesions (single-strand nicks or double-strand breaks) responsible for the initiation of recombination may be rarer in them than in meiotic cells. Several observations are relevant to the evaluation of these factors. First, efficient mitotic recombination is observed when a double-strand break is introduced either into transforming DNA or into the chromosome. As the level of integrative transformation observed with linear DNA is similar to that observed with autonomously replicating plasmids (Orr-Weaver et al. 1981), the efficiency of the recombination event may be close to one. This conclusion is complicated by the possibility that transformation might induce repair/recombination functions. Second, for many strains, the efficiency of in vitro recombination catalyzed by extracts of mitotic and meiotic cells is not significantly different (Symington et al. 1984). Third, very high levels of mitotic recombination induced by X-ray treatment are not associated with formation of the synaptonemal complex (Olson and Zimmermann 1978). These observations indicate that the major factor limiting the rate of mitotic recombination may be the absence in mitotic cells of the enzymes responsible for making the recombinogenic lesions. If this suggestion is correct, it may be possible to induce near meiotic levels of recombination in mitotic cells by overproduction of a small number of meiotic enzymes. However, the high rate of mitotic intrachromosomal exchange between closely linked repeats, relative to the rate of interchromosomal recombination, suggests that the lack of pairing between homologous chromosomes (although not necessarily the lack of synaptonemal complexes) may also be a significant factor. It is also possible that formation of the initiating lesion and pairing of homologous sequences are related processes.

V. ENZYMOLOGY OF HOMOLOGOUS RECOMBINATION

A complete understanding of homologous recombination will require a complete description of all of the proteins involved. The approaches that are currently being pursued toward this goal include (1) the development and characterization of in vitro recombination systems; (2) the isolation of proteins (strand transferases, single-strand-binding proteins, etc.) that are likely to have a role in recombination; and (3) the isolation and characterization of the products of known recombination genes. Each of these approaches is discussed below.

A. In Vitro Recombination Systems in Yeast

In vitro systems that catalyze recombination have been developed from mitotic (Symington et al. 1983) and meiotic (Symington et al. 1984; Hotta et al. 1985) yeast cells. Extracts were prepared from cells that were converted to spheroplasts and gently lysed. In one study (Symington et al. 1983), recombination was detected between a plasmid containing an XhoI-linker insertion near one end of the gene encoding tetracycline resistance (Tc^r) and a plasmid containing a similar insertion at the other end of the Tc^r gene; both plasmids contained Amp^r genes. After the plasmids were incubated in the extract, the efficiency of recombination was assayed by transforming the DNA into a Tc^s Amp^s recA (recombination-defective) strain of E. coli and determining the ratio of Tc^r transformants (formed by recombination between the Tc genes carried on the two plasmids) to total Amp^r transformants (representing the total number of transforming molecules in the extract). The frequency of Tc^r transformants between two circular plasmids increased about 300-fold during a 2-hour incubation in a mitotic extract. The introduction of a double-strand break into the Tc^r gene of one of the plasmids resulted in a two- to threefold stimulation of the reaction. Plasmid DNAs were isolated from many of the Tc^r transformants and analyzed by gel electrophoresis. In experiments involving two circular plasmids, the most common product (69%) was a monomeric plasmid containing one wild-type Tc^r gene; these molecules presumably resulted from gene conversions between the two mutant Tc^r genes that were not associated with crossing-over. The other products detected were dimeric plasmids with one wild-type Tc^r gene plus a mutant gene with both linker mutations (24% of transformants) or one wild-type Tc^r gene and one mutant Tc^r gene with a single mutation (7% of transformants). The first dimeric class can be explained as resulting from simple crossovers between the mutant Tc^r genes on two plasmids, and the second class may have resulted from conversion events between Tc^r genes on two plasmids that were associated with crossing-over.

Because this assay for recombination involved transformation into *E. coli*, it was possible that some portion of the recombination reaction occurred in vivo after transformation into *E. coli* (despite the *recA* mutation), rather than in vitro. To analyze directly the recombinant products formed in vitro, Symington et al. (1983, 1984, 1985) examined the substrate and product DNA molecules by gel electrophoresis following the in vitro reaction. Because the plasmid substrates contained *Xho*I-linker mutations in the Tc^r gene, the *Xho*I-resistant plasmids detected in these experiments represented formation of wild-type Tc^r genes. In addition, a fraction of the product DNA molecules migrated more slowly than the substrates. When this fraction was examined by electron microscopy, figure-8 molecules (apparently representing the stable association of two circular plasmids) were observed (Fig. 15a). Digestion of these molecules with enzymes that cut once within each plasmid produced Chi forms, confirming the covalent attachment between the two substrate molecules (Fig. 15b).

The abilities of extracts derived from strains containing mutations affecting mitotic recombination were also examined (Symington et al. 1983, 1984). Extracts prepared from *rad52* strains had approximately 20% of wild-type activity (measured by the transformation assay) in a reaction between a linear and a circular substrate; no reduction in activity was seen in the reaction between two circular substrates. Of the other

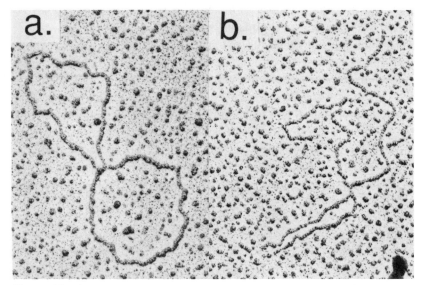

Figure 15 Electron micrographs of recombining plasmids. Two plasmids were incubated with a yeast cell-free extract; recombining DNA was purified by gel electrophoresis, and one of the bands of DNA was eluted from the gel and examined by electron microscopy. (*a*) Recombining circular DNA molecules; (*b*) recombining DNA molecules after treatment with *Eco*RI, which cleaves each circular plasmid once. (Reprinted, with permission, from Symington et al. 1984.)

mutants examined, about one-third affected the activity of the in vitro system and the effects were in the expected directions (hyper-Rec mutants had higher levels and Rec⁻ mutants had lower levels of in vitro recombination). As none of these effects were very large (all were less than fivefold), this system has not been used for purifying the products of these genes. A limitation of this assay for examining mutant strains is that *recA E. coli* strains can resolve recombination intermediates (West et al. 1983); thus, mutations affecting this process might not have an effect on the transformation efficiency.

In vitro recombination has also been examined in meiotic extracts (Symington et al. 1984; Hotta et al. 1985). One problem for such studies is that mitotic extracts from strain SK1 (which has a synchronous and rapid meiosis) have tenfold less recombination activity than extracts from other strains tested. Thus, although a tenfold meiotic induction of in vitro recombination activity was observed in SK1 extracts, this level was not significantly higher than the activity of mitotic extracts in other strains. Hotta et al. (1985) found a 70-fold increase in in vitro recombination activity during meiosis in a strain that had very little mitotic activity. Because a similar increase in activity was also observed in **a/a** and α/α strains, it is not clear that the induced activity observed in vitro is related to the high level of induction of meiotic recombination observed in vivo.

An in vitro assay for repair of DNA mismatches has also been developed (Muster-Nassal and Kolodner 1986). Heteroduplexes constructed in vitro with either single-base-pair mismatches or small (4 and 7 bp) insertions were incubated in mitotic extracts. The insertions and A-C and G-T mismatches were corrected efficiently, whereas six other single-base-pair mismatches were repaired inefficiently. The amount of repair synthesis associated with mismatch correction was less than 20 bp around the mismatch.

B. Isolation of Proteins Likely to Have a Role in Recombination

Our current view of recombination predicts that a number of enzymatic activities will be involved in the recombination process, including endonucleases and exonucleases, strand-transfer proteins, DNA polymerases, helicases, and DNA ligase. Proteins in several of these classes have been purified from *E. coli* and the knowledge of their substrate specificities has provided a useful tool for isolating similar activities from yeast. Although all of these types of enzymes have also been found in yeast, their roles in recombination are not yet established.

1. Endonucleases

This group of enzymes can be subdivided into nonspecific, site-specific, and structure-specific (Holliday junction) endonucleases. The only nonspecific endonuclease that has been well characterized in yeast also has exonuclease activity and is described in the following section. The best-characterized site-specific endonucleases are the products of the *HO* gene (involved in the gene conversion event that leads to mating-type switching; Kostriken et al. 1983) and the *FLP* gene of the 2-micron plasmid (involved in site-specific recombination between the inverted repeats of the plasmid) as reviewed in detail elsewhere (McLeod et al. 1984; Sadowski et al. 1984; Broach and Volkert, this volume). In addition, two other site-specific endonucleases that have no known function have been described. Both Endo.*Sce*I (Watabe et al. 1983) and Endo.*Sce*II (Kostriken et al. 1983) have single recognition sites in pBR322; *Sce*I also has cleavage sites in ϕX174 and in fd RF DNA. Unlike cleavage sites for the *FLP* and *HO* endonucleases and most prokaryotic site-specific endonucleases, five different cleavage sites of the *Sce*I enzyme display no obvious consensus sequence, indicating that this enzyme may actually be a structure-specific endonuclease, rather than a site-specific endonuclease (Watabe et al. 1983).

The endonucleases that are clearly structure-specific are those responsible for recognizing and cleaving Holliday junctions. The configuration of DNA strands in plasmids with extruded cruciforms is similar to the Holliday junction (Fig. 16), except that two arms of the cruciform have homologous sequences (homologous arms), while the other two arms have different DNA sequences (heterologous arms). Enzymes have been isolated from *E. coli* (such as the bacteriophage T4 endonuclease VII [Mizuuchi et al. 1982]) that cut both cruciform plasmids and Holliday junctions. In addition, three groups have isolated enzymes from mitotic yeast cells that cleave cruciform substrates. These enzymes cleave the homologous arms of cruciforms at positions that are within 10 bp of the base of the cruciform and that are symmetrically opposed across the junction, as expected for Holliday junction endonucleases. The enzyme characterized by Symington and Kolodner (1985) also cleaved figure-8 molecules, an intermediate in recombination, to generate the expected monomeric and dimeric circular molecules. Although this protein did not cleave a Holliday junction generated in vitro from oligonucleotides (possibly because of the small size of the arms [<15 bp], the lack of sequence symmetry, or the sequence at the branch point), the oligonucleotide substrate was a competitive inhibitor of cleavage of the cruciform plasmid (Evans and Kolodner 1987). Holliday junctions with longer arms (230–1000 bp) were substrates for this enzyme, although the rates of cleavage, as well as the precise sites that were cut, varied depending on the DNA sequences at the junction (Evans and Kolodner 1988); this result may be relevant to the in vivo observations indicating locus-to-locus variation in the association between gene conversion and crossing-over (as described in Section III.B.3). The enzyme had no activity on three-armed Y-branched substrates.

A protein with the ability to cleave extruded cruciforms has also been isolated by West and co-workers (West and Korner 1985; West et al. 1987). This enzyme cleaved the cruciforms symmetrically in the homologous arms and had little activity in the heterologous arms (Parsons and West 1988). In Holliday junctions in which all four arms were different, cleavages were asymmetric and their positions depended on the DNA sequences of the arms (Parsons and West 1988; Parsons et al. 1989). It is not yet clear whether this enzyme and that isolated by Kolodner and co-workers (Symington and Kolodner 1985; Evans and Kolodner 1987, 1988) are different.

A junction-cleaving activity (called Endo X3) that appears to be dif-

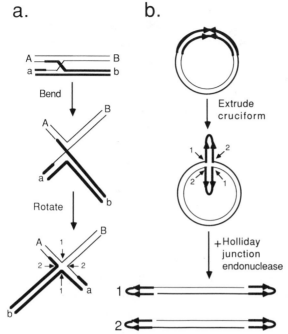

Figure 16 Cleavage of Holliday junctions by Holliday junction endonucleases. (*a*) Expected cleavage patterns for a "natural" Holliday junction connecting homologous chromosomes. The three depictions of the Holliday intermediate are equivalent isomers (Meselson and Radding 1975). Cleavages at the base of the junction at positions 1 would lead to chromosomes with flanking markers in the recombinant configurations, whereas cleavages at positions 2 would yield chromosomes with markers in the parental configurations. (*b*) Assay for cruciform cleavage by Holliday junction endonucleases. In supercoiled plasmids with inverted repeats, the repeat can be induced to extrude a cruciform. This structure is geometrically similar to the "natural" Holliday junction. If the plasmid with the extruded cruciform is incubated with Holliday junction endonuclease, the molecule is cleaved at the base of the cruciform in orientation 1 or 2. The approximate position of cleavage can be mapped by treating the cleaved molecules with a restriction enzyme that cuts at a known position and analyzing the resulting products by gel electrophoresis. DNA sequence analysis of the resulting fragments (see Section V.B.1) indicates that the cleavage of junctions occurs symmetrically in the homologous arms of the cruciform near the junction.

ferent from either of the enzymes described above has also been purified recently (Jensch et al. 1989). This enzyme had cleavage patterns on junction substrates that were very similar to those observed with T4 endonuclease VII (Mizuuchi et al. 1982). Like endonuclease VII, but unlike the enzyme isolated by Kolodner and co-workers (Symington and Kolodner 1985; Evans and Kolodner 1987, 1988), this enzyme cleaves Y-branched DNA as well as a number of other branched structures. This lack of specificity for Holliday junctions may indicate that this enzyme has some general role in DNA metabolism.

As none of the genes encoding these enzymes have been isolated, their role (if any) in recombination has not yet been established. Recently, using cell-free extracts of a heavily mutagenized strain, S. Kleff and R. Sternglanz (pers. comm.) identified a mutant strain that is defective in the ability to cleave DNA containing a cruciform. Surprisingly, this mutation had no detectable effects on either mitotic or meiotic recombination.

2. Exonucleases

Chow and Resnick (1987) purified an endonuclease/exonuclease from yeast on the basis of antigenic cross-reactivity with a *Neurospora crassa* nuclease. This enzyme was present in vegetative cells in log phase and was induced severalfold during meiosis. The nuclease has endonucleolytic activity on linear single-stranded DNA and circular double-stranded DNA and exonucleolytic activity on linear double-stranded molecules. Although the activity of the nuclease appears to be under the control of the *RAD52* gene, the molecular weight of the enzyme is different from that of Rad52p (Chow and Resnick 1988). Several other exonucleases active on single-stranded DNA have also been purified. Two such enzymes have 3' to 5' activity, were associated with DNA polymerase, and had proofreading functions (Chang 1977; Wintersberger 1978; Villadsen et al. 1982; Bauer et al. 1988). The other enzymes have 5' to 3' activity and no known function (Villadsen et al. 1982; Bauer et al. 1988; Burgers et al. 1988). Of the latter class, exonuclease IV also has activity on double-stranded DNA and RNA (Bauer et al. 1988). As with the Holliday junction endonucleases, the possible roles of these exonucleases in recombination have not yet been established.

3. Strand Exchange and Single-strand-binding Proteins

The genetic demonstration of postmeiotic segregation in yeast strongly indicates heteroduplex formation between homologous chromosomes, which is presumably catalyzed by strand-exchange proteins. Two such proteins, Sep1p (Kolodner et al. 1987) and Stpαp (Sugino et al. 1988), have been isolated from yeast using one of the assays employed for characterizing the RecA protein of *E. coli* and Rec1 protein of *Ustilago maydis*. The reaction that is monitored (by gel electrophoresis) is the

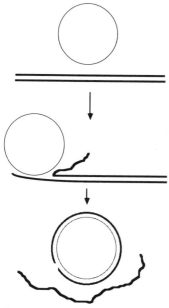

Figure 17 Strand-exchange assay. Addition of a strand-exchange protein stimulates the formation of joint molecules from homologous substrates. A strand is displaced from the linear duplex, coupled with pairing of the single-strand circle to the complementary linear single strand. This reaction is readily monitored by agarose gel electrophoresis.

transfer of a single strand from a linear duplex to a homologous circular single strand (Fig. 17).

Sep1p was isolated from vegetative yeast cells. It catalyzed strand transfer in the same direction as RecA (3′ to 5′) and did not require ATP for activity. The transfer required stoichiometric amounts of protein and produced heteroduplexes several thousand base pairs in length. Like the RecA and Rec1 proteins, Sep1p also catalyzed the renaturation of complementary single-stranded DNA (Heyer et al. 1988). When used in limiting amounts, the protein required a second yeast protein to attain maximal-strand exchange activity. The stimulatory protein was a single-stranded DNA-binding protein (SSB). The yeast SSB substituted for *E. coli* SSB in RecA-catalyzed reactions but the reverse substitution did not work (Heyer and Kolodner 1989). Both Sep1p and the stimulatory SSB have been purified to homogeneity; they have molecular masses of 132 kD and 34 kD, respectively.

As Stpαp has a molecular mass of 38 kD (Sugino et al. 1988), it appears to be different from Sep1p. The specific activity of Stpαp was tenfold higher during meiosis than in vegetative cells and the protein was purified from meiotic cells. Like Sep1p, Stpαp had no nucleotide cofactor requirement, and its activity was greatly stimulated by a 20-kD SSB. The gene encoding Stpαp (*STP1*) was cloned and sequenced (A. Clark et al., pers. comm.). Because Stpαp was expressed to similar extents in mitosis and meiosis, the meiosis-specific activity of the protein may represent a posttranslational modification of the gene product. Strains containing a disruption of *STP1* were viable and had sporulated normally. It is not yet clear whether Sep1p and Stpαp affect recombination in vivo.

4. Replication Proteins

The models of recombination described above require DNA repair synthesis to complete the recombination process. Such synthesis presumably involves one or more DNA polymerases, together with one or more helicases and topoisomerases. Although several different DNA polymerases have been isolated from yeast (for review, see Campbell and Newlon, this volume), it is not clear which is the primary repair polymerase. Certain mutant alleles of *CDC17* (the structural gene for DNA polymerase I [α]; Carson 1987) and *CDC2* (the structural gene for DNA polymerase III [δ]; Sitney et al. 1989; A. Boulet and G. Faye, pers. comm.) have mitotic hyper-Rec phenotypes (Aguilera and Klein 1988). One possible explanation for this result is that the mutant enzymes are less processive than wild type, resulting in recombinogenic gaps. A similar explanation may apply to the increase in mitotic recombination observed in strains containing certain *cdc9* mutations (Game et al. 1979), as *CDC9* encodes DNA ligase (Johnston and Nasmyth 1978).

Two DNA helicases have been purified from yeast. One is the product of the *RAD3* gene, which is required for UV excision repair (Sung et al. 1987a,b). As discussed in Section IV.F.2, *rad3* mutations often result in a hyper-Rec phenotype, possibly because the lack of the *RAD3*-encoded helicase prevents efficient repair of recombinogenic lesions. The second helicase was identified as a DNA-dependent ATPase that stimulates the activity of DNA polymerase I (Sugino et al. 1986). The gene encoding this protein is not known.

Both *S. cerevisiae* and *S. pombe* contain both type I and type II topoisomerases, and mutants deficient in each type of enzyme have been analyzed (for review, see Yanagida and Wang 1987). Type I eukaryotic enzymes relax both positively and negatively supercoiled DNAs by transient breakage of one strand at a time in the absence of a high-energy cofactor, whereas type II enzymes relax both positively and negatively supercoiled DNAs by transient double-strand breaks in the presence of ATP (for review, see Wang 1985). *S. cerevisiae* strains containing mutations in either the *TOP1* (encoding a type I topoisomerase) or *TOP2* (encoding a type II topoisomerase) gene have more than a 50-fold elevated level of mitotic recombination in the rRNA genes (Christman et al. 1988); this effect was not observed for several other types of repeated genes. Similarly, Kim and Wang (1989) found that certain strains with the *top1 top2* genotype had a high level of extrachromosomal rRNA genes, presumably resulting from intrachromatid recombination within the rRNA gene tandem array. Yeast strains with mutations in the *EDR1* gene also have an enhanced level of mitotic recombination between repeated sequences (Rothstein 1984). Wallis et al. (1989) found that the *EDR1* product has significant sequence homology with the active site of the *topA* gene (encoding a type I topoisomerase) of *E. coli* and that expression of *topA* in yeast can suppress the slow growth of *edr1* mutants, indicating that the gene (now called *TOP3*) probably encodes a third topoisomerase. In addition, Aguilera and Klein (1990) found that the *HPR1* gene shares sequence homology with *TOP1*, indicating the possibility of a third type I topoisomerase in yeast; mutations in *HPR1* lead to high levels of intrachromosomal exchange. These observations suggest that increased levels of supercoiling may result directly or indirectly (e.g., by affecting transcription) in increased levels of mitotic recombination in at least some regions of the genome.

C. Isolation and Characterization of the Products of Known Recombination Genes

Analysis of the products of genes known to affect recombination has thus far been restricted mostly to sequence analysis. Although this approach identified interesting homologies for *TOP3*, *HPR1* (see above), and *PMS1* (see Section III.H.3), no other particularly significant homologies that are likely to be relevant to the functions of the proteins have been revealed. Neither *RAD52* (Adzuma et al. 1984) nor *SPO11* (Atcheson et al. 1987) is homologous to any known gene. However, Chen and Bernstein (1988) recently found that expression of *RAD52* in *E. coli* complemented (for growth) mutations in genes 46 and 47 of bacteriophage T4. As these phage genes encode an exonuclease, the *RAD52* product may have a similar function. Since complementation experiments between heterologous elements could be misleading, these data are suggestive, rather than conclusive. The predicted sequence of the *RAD50* product had two regions of heptad repeats characteristic of proteins that form α-helical coiled coils (Alani et al. 1989). It is likely that further progress in assigning functions to gene products affecting recombination will require the isolation of the proteins (from yeast or *E. coli*), followed by various tests of function (DNA-binding properties, DNA-nicking activities, etc.).

A related approach to the identification of yeast recombination proteins is to search for yeast proteins that cross-react with antibodies derived against known recombination proteins from other organisms. For example, two groups used anti-RecA sera to identify a yeast gene encoding a related protein (Elledge and Davis 1987; Hurd et al. 1987). However, the gene isolated by this approach (*RNR2*) was the small subunit of ribonucleotide reductase, rather than a strand transferase.

D. Conclusions

In summary, a variety of approaches have been taken to identify yeast recombination proteins. However, with a few exceptions, as yet, these approaches have identified proteins with a well-defined function but poorly defined effects on recombination or proteins that have no well-defined function but clearly affect recombination. As cloned genes should soon be available for the proteins described in Section V.B, their roles in vivo should soon be clarified. In addition, further development of the in vitro recombination system may eventually allow the characterization of recombination proteins that complement the recombination deficiencies of mutant cell-free extracts.

VI. FINAL WORDS OF INSPIRATION

This is not the end. It is not even the beginning of the end. But it is, perhaps, the end of the beginning. Winston Churchill, 1942.

Give us the tools, and we will finish the job. Winston Churchill, 1941.

The simplest interpretation of the accumulated data concerning yeast recombination is that there are several systems that are capable of catalyzing genetic exchange. These systems show some degree of sequence specificity and appear to be sensitive to alterations in DNA topology. Although a complete understanding of recombination will not come quickly, we now have the appropriate genetic and physical tools to finish the job.

ACKNOWLEDGMENTS

We thank our colleagues for providing unpublished information and Drs. S. Fogel, J. Golin, J. Haber, N. Kleckner, H. Klein, D. Maloney, B.

Montelone, T. Nagylaki, P. Pukkila, G. Smith, and F. Stahl and our co-workers (particularly P. Detloff) for comments on various sections of the review or for valuable discussions. We also thank C. Giroux for providing the photograph for Figure 10 and J.R. Pringle and G.S. Roeder for their exhaustive (and exhausting) comments on the completed manuscript; their efforts greatly improved the final product. Research in our laboratories was supported by the National Institutes of Health.

REFERENCES

Aboussekhra, A., R. Chanet, Z. Zgaga, C. Cassier-Chauvat, M. Heude, and F. Fabre. 1989. *RADH*, a gene of *Saccharomyces cerevisiae* encoding a putative DNA helicase involved in DNA repair. Characteristics of *radH* mutants and sequence of the gene. *Nucleic Acids Res.* **17**: 7211.

Adzuma, K., T. Ogawa, and H. Ogawa. 1984. Primary structure of the *RAD52* gene in *Saccharomyces cerevisiae. Mol. Cell. Biol.* **4**: 2735.

Aguilera, A. and H.L. Klein. 1988. Genetic control of intrachromosomal recombination in *Saccharomyces cerevisiae.* I. Isolation and genetic characterization of hyper-recombination mutants. *Genetics* **119**: 779.

———. 1989. Yeast intrachromosomal recombination: Long gene conversion tracts are preferentially associated with reciprocal exchange and require the *RAD1* and *RAD3* gene products. *Genetics* **123**: 683.

———. 1990. *HPR1*, a novel yeast gene that prevents intrachromosomal excision recombination, shows carboxy-terminal homology to the *Saccharomyces cerevisiae TOP1* gene. *Mol. Cell. Biol.* **10**: 1439.

Ahn, B.-Y. and D.M. Livingston. 1986. Mitotic gene conversion lengths, coconversion patterns, and the incidence of reciprocal recombination in a *Saccharomyces cerevisiae* plasmid system. *Mol. Cell. Biol.* **6**: 3685.

Alani, E., R. Padmore, and N. Kleckner. 1990. Analysis of wild-type and *rad50* mutants of yeast suggests an intimate relationship between meiotic chromosome synapsis and recombination. *Cell* **61**: 419.

Alani, E., S. Subbiah, and N. Kleckner. 1989. The yeast *RAD50* gene encodes a predicted 153-kD protein containing a purine nucleotide-binding domain and two large heptad-repeat regions. *Genetics* **122**: 47.

Angel, T., B. Austin, and D.G. Catcheside. 1970. Regulation of recombination at the *his3* locus in *Neurospora crassa. Aust. J. Biol. Sci.* **23**: 1229.

Atcheson, C.L., B. DiDomenico, S. Frackman, R.E. Esposito, and R.T. Elder. 1987. Isolation, DNA sequence and regulation of a meiosis specific eucaryotic recombination gene. *Proc. Natl. Acad. Sci.* **84**: 8035.

Bauer, G.A., H.M. Heller, and P.M.J. Burgers. 1988. DNA polymerase III from *Saccharomyces cerevisiae. J. Biol. Chem.* **264**: 917.

Bell, L.R. and B. Byers. 1979. Occurrence of crossed strand-exchange forms in yeast DNA during meiosis. *Proc. Natl. Acad. Sci.* **76**: 3445.

———. 1983. Homologous association of chromosomal DNA during yeast meiosis. *Cold Spring Harbor Symp. Quant. Biol.* **47**: 829.

Bishop, D.K., J. Anderson, and R.D. Kolodner. 1989. Specificity of mismatch repair following transformation of *Saccharomyces cerevisiae* with heteroduplex plasmid DNA. *Proc. Natl. Acad. Sci.* **86**: 3713.

Bishop, D.K., M.S. Williamson, S. Fogel, and R.D. Kolodner. 1987. The role of heteroduplex correction in gene conversion in *Saccharomyces cerevisiae*. *Nature* **328**: 362.

Boeke, J.D., F. Lacroute, and G.R. Fink. 1984. A positive selection for mutants lacking orotidine-5′-phosphate decarboxylase activity in yeast: 5-fluoro-orotic acid resistance. *Mol. Gen. Genet.* **197**: 345.

Boram, W.R. and H. Roman. 1976. Recombination in *Saccharomyces cerevisiae*: A DNA repair mutation associated with elevated mitotic gene conversion. *Proc. Natl. Acad. Sci.* **73**: 2828.

Borts, R.H. and J.E. Haber. 1987. Meiotic recombination in yeast: Alteration by multiple heterozygosities. *Science* **237**: 1459.

―――. 1989. Length and distribution of meiotic gene conversion tracts and crossovers in *Saccharomyces cerevisiae*. *Genetics* **123**: 69.

Borts, R.H., M. Lichten, and J.E. Haber. 1986. Analysis of meiotic-defective mutations in yeast by physical monitoring of recombination. *Genetics* **113**: 531.

Borts, R.H., M. Lichten, M. Hearn, L.S. Davidow, and J.E. Haber. 1984. Physical monitoring of meiotic recombination in *Saccharomyces cerevisiae*. *Cold Spring Harbor Symp. Quant. Biol.* **49**: 67.

Brent, R. and M. Ptashne. 1984. A bacterial repressor protein or a yeast transcriptional terminator can block upstream activation of a yeast gene. *Nature* **312**: 612.

Broach, J.R., V.R. Guarascio, and M. Jayaram. 1982. Recombination within the yeast plasmid two micron circle is site-specific. *Cell* **29**: 227.

Brunborg, G., M.A. Resnick, and D.H. Williamson. 1980. Cell cycle specific repair of DNA double strand breaks in *Saccharomyces cerevisiae*. *Radiat. Res.* **82**: 547.

Burgers, P.M.J., G.A. Bauer, and L. Tam. 1988. Exonuclease V from *Saccharomyces cerevisiae*. *J. Biol. Chem.* **263**: 8099.

Byers, B. and L. Goetsch. 1975. Electron microscopic observations on the meiotic karyotype of diploid and tetraploid *Saccharomyces cerevisiae*. *Proc. Natl. Acad. Sci.* **72**: 5056.

Calderon, I.L., C.R. Contopoulou, and R.K. Mortimer. 1983. Isolation and characterization of yeast DNA repair genes. II. Isolation of plasmids that complement the mutations *rad50-1*, *rad51-1*, *rad54-3* and *rad55-3*. *Curr. Genet.* **7**: 93.

Cao, L., E. Alani, and N. Kleckner. 1990. A pathway for generation and processing of double-strand breaks during meiotic recombination in *S. cerevisiae*. *Cell* **61**: 1089.

Carle, G.F. and M.V. Olson. 1985. An electrophoretic karyotype for yeast. *Proc. Natl. Acad. Sci.* **82**: 3756.

Carpenter, A.T.C. 1975. Electron microscopy of meiosis in *Drosophila melanogaster* females. II. The recombination nodule-a recombination-associated structure at pachytene? *Proc. Natl. Acad. Sci.* **72**: 3186.

―――. 1979. Synaptonemal complex and recombination nodules in wild-type *Drosophila melanogaster* females. *Genetics* **92**: 511.

―――. 1987. Gene conversion, recombination nodules and the initiation of meiotic synapsis. *Bioessays* **6**: 232.

Carson, M.J. 1987. "*CDC17*, the structural gene for DNA polymerase I of yeast: Mitotic hyper-recombination and effects on telomere metabolism." Ph.D. thesis, University of Washington.

Chang, L.M.S. 1977. DNA polymerases from baker's yeast. *J. Biol. Chem.* **252**: 1873.

Chen, D.S. and H. Bernstein. 1988. Yeast gene *RAD52* can substitute for phage T4 gene *46* or *47* in carrying out recombination and DNA repair. *Proc. Natl. Acad. Sci.* **85**: 6821.

Chernoff, Y.O., O.V. Kidgotko, O. Demberelijn, I.L. Luchnikova, S.P. Soldatov, V.M. Glazer, and D.A. Gordenin. 1984. Mitotic intragenic recombination in the yeast *Saccharomyces*: Marker effects on conversion and reciprocity of recombination. *Curr. Genet.* **9**: 31.

Chow, T.Y.-K. and M.A. Resnick. 1987. Purification and characterization of an endo-exonuclease from *Saccharomyces cerevisiae* that is influenced by the *RAD52* gene. *J. Biol. Chem.* **262**: 17659.

―――. 1988. An endo-exonuclease activity of yeast that requires a functional *RAD52* gene. *Mol. Gen. Genet.* **211**: 41.

Christman, M.F., F.S. Dietrich, and G.R. Fink. 1988. Mitotic recombination in the rDNA of *S. cerevisiae* is suppressed by the combined action of DNA topoisomerases I and II. *Cell* **55**: 413.

Coleman, K.G., H.Y. Steensma, D.B. Kaback, and J.R. Pringle. 1986. Molecular cloning of chromosome I DNA from *Saccharomyces cerevisiae*: Isolation and characterization of the *CDC24* gene and adjacent regions of the chromosome. *Mol. Cell. Biol.* **6**: 4516.

Davidow, L.S. and B. Byers. 1984. Enhanced gene conversion and post-meiotic segregation in pachytene-arrested *Saccharomyces cerevisiae*. *Genetics* **106**: 165.

Davies, P.J., W.E. Evans, and J.M. Parry. 1975. Mitotic recombination induced by chemical and physical agents in the yeast *Saccharomyces cerevisiae*. *Mutat. Res.* **51**: 327.

Dawson, D.S., A.W. Murray, and J.W. Szostak. 1986. An alternative pathway for meiotic chromosome segregation in yeast. *Science* **234**: 713.

Detloff, P., J. Sieber, and T.D. Petes. 1991. Repair of specific of specific base pair mismatches formed during meiotic recombination in the yeast *Saccharomyces cerevisiae*. *Mol. Cell. Biol.* **11**: 737.

DiCaprio, L. and P.J. Hastings. 1976. Gene conversion and intragenic recombination at the *SUP6* locus and the surrounding region in *Saccharomyces cerevisiae*. *Genetics* **84**: 697.

Dobzhansky, T. 1931. Translocations involving the second and fourth chromosomes of *Drosophila melanogaster*. *Genetics* **16**: 629.

Dowling, E.L., D.H. Maloney, and S. Fogel. 1985. Meiotic recombination and sporulation in repair deficient strains of yeast. *Genetics* **109**: 283.

Dresser, M. and C. Giroux. 1988. Meiotic chromosome behavior in spread preparations of yeast. *J. Cell Biol.* **106**: 567.

Egel, R. 1984. Two tightly linked silent cassettes in the mating-type region of *Schizosaccharomyces pombe*. *Curr. Genet.* **8**: 199.

Elledge, S.J. and R.W. Davis. 1987. Identification and isolation of the gene encoding the small subunit of ribonucleotide reductase from *Saccharomyces cerevisiae*: DNA damage-inducible gene required for mitotic viability. *Mol. Cell. Biol.* **7**: 2783.

Engebrecht, J. and G.S. Roeder. 1989. Yeast *mer1* mutants display reduced levels of meiotic recombination. *Genetics* **121**: 237.

Esposito, M.S. 1971. Post-meiotic segregation in *Saccharomyces*. *Mol. Gen. Genet.* **111**: 297.

―――. 1978. Evidence that spontaneous mitotic recombination occurs at the two-strand stage. *Proc. Natl. Acad. Sci.* **75**: 4436.

Esposito, M.S. and J.E. Wagstaff. 1981. Mechanisms of mitotic recombination. In *The molecular biology of the yeast* Saccharomyces: *Life cycle and inheritance* (ed. J.N. Strathern et al.), p. 341. Cold Spring Harbor Laboratory, Cold Spring Harbor, New York.

Esposito, M.S., D.T. Maleas, K.A. Bjornstad, and L.L. Holbrook. 1986. The *REC46* gene of *Saccharomyces cerevisiae* controls mitotic chromosomal stability, recombination and sporulation: Cell-type and life cycle stage-specific expression of the *rec46-1* mutation. *Curr. Genet.* **10**: 425.

Esposito, M.S., J. Hosoda, J. Golin, H. Moise, K. Bjornstad, and D. Maleas. 1984. Recombination in *Saccharomyces cerevisiae*: *REC*-gene mutants and DNA-binding proteins. *Cold Spring Harbor Symp. Quant. Biol.* **49**: 41.

Esposito, R.E. and M.S. Esposito. 1969. The genetic control of sporulation in *Saccharomyces*. I. The isolation of temperature-sensitive sporulation deficient mutants. *Genetics* **69**: 76.

———. 1974. Genetic recombination and commitment to meiosis in *Saccharomyces*. *Proc. Natl. Acad. Sci.* **71**: 3172.

Esposito, R.E. and S. Klapholz. 1981. Meiosis and ascospore development. In *The molecular biology of the yeast* Saccharomyces: *Life cycle and inheritance* (ed. J.N. Strathern et al.), p. 211. Cold Spring Harbor Laboratory, Cold Spring Harbor, New York.

Evans, D. and R.D. Kolodner. 1987. Construction of a synthetic Holliday junction analog and characterization of its interaction with a *Saccharomyces cerevisiae* endonuclease that cleaves Holliday junctions. *J. Biol. Chem.* **262**: 9160.

———. 1988. Effect of DNA structure and nucleotide sequence on Holliday junction resolution by a *Saccharomyces cerevisiae* endonuclease. *J. Mol. Biol.* **201**: 69.

Fabre, F. 1978. Induced intragenic recombination in yeast can occur during the G1 mitotic phase. *Nature* **272**: 795.

———. 1981. Mitotic recombination and repair in relation to the cell cycle in yeast. *Alfred Benzon Symp.* **16**: 399.

Fabre, F. and H. Roman. 1977. Genetic evidence for the inducibility of recombination competence in yeast. *Proc. Natl. Acad. Sci.* **74**: 1667.

———. 1979. Evidence that a single DNA ligase is involved in replication and recombination in yeast. *Proc. Natl. Acad. Sci.* **76**: 4586.

Farnet, C., R. Padmore, L. Cao, W. Raymond, E. Alani, and N. Kleckner. 1988. The *RAD50* gene of *S. cerevisiae*. *UCLA Symp. Mol. Cell. Biol.* **83**: 201.

Fasullo, M.T. and R.W. Davis. 1987. Recombinational substrates designed to study recombination between unique and repetitive sequences *in vivo*. *Proc. Natl. Acad. Sci.* **84**: 6215.

Fink, G.R. and T.D. Petes. 1984. Gene conversion in the absence of recombination. *Nature* **310**: 728.

Fink, G.R. and C.A. Styles. 1974. Gene conversion of deletions in the *HIS4* region of yeast. *Genetics* **77**: 231.

Fitzgerald-Hayes, M. 1987. Yeast centromeres. *Yeast* **3**: 187.

Fogel, S. and D.D. Hurst. 1963. Coincidence relations between gene conversion and mitotic recombination in *Saccharomyces cerevisiae*. *Genetics* **48**: 321.

———. 1967. Meiotic gene conversion in yeast tetrads and the theory of recombination. *Genetics* **57**: 455.

Fogel, S. and R.K. Mortimer. 1969. Informational transfer in meiotic gene conversion. *Proc. Natl. Acad. Sci.* **62**: 96.

———. 1970. Fidelity of meiotic gene conversion in yeast. *Mol. Gen. Genet.* **109**: 177.

Fogel, S., R.K. Mortimer, and K. Lusnak. 1981. Mechanisms of meiotic gene conversion, or "Wanderings on a foreign strand." In *The molecular biology of the yeast* Saccharomyces: *Life cycle and inheritance* (ed. J.N. Strathern et al.), p. 289. Cold Spring Harbor Laboratory, Cold Spring Harbor, New York.

Fogel, S., R. Mortimer, K. Lusnak, and F. Tavares. 1979. Meiotic gene conversion—A signal of the basic recombination event in yeast. *Cold Spring Harbor Symp. Quant. Biol.* **43**: 1325.

Friedberg, E.C. 1987. Deoxyribonucleic acid repair in the yeast *Saccharomyces cerevisiae. Microbiol. Rev.* **52**: 70.

Game, J.C. 1983. Radiation sensitive mutants and repair in yeast. In *Yeast genetics: Fundamental and applied aspects* (ed. J.F.T. Spencer et al.), p. 109. Springer-Verlag, Berlin.

Game, J.C., L.H. Johnston, and R.C. von Borstel. 1979. Enhanced mitotic recombination in a ligase-defective mutant of the yeast *Saccharomyces cerevisiae. Proc. Natl. Acad. Sci.* **76**: 4589.

Game, J.C., K.C. Sitney, V.E. Cook, and R.K. Mortimer. 1989. Use of a ring chromosome and pulsed-field gels to study interhomolog recombination, double-strand DNA breaks and sister-chromatid exchange in yeast. *Genetics* **123**: 695.

Game, J.C., T.J. Zamb, R.J. Braun, M.A. Resnick, and R.M. Roth. 1980. The role of radiation (*rad*) genes in meiotic recombination in yeast. *Genetics* **94**: 51.

Gasser, S.M. and U.K. Laemmli. 1987. A glimpse at chromosomal order. *Trends Genet.* **3**: 16.

Gaudet, A. and M. Fitzgerald-Hayes. 1987. Alterations in the adenine-plus-thymine-rich region of *CEN3* affect centromere function. *Mol. Cell. Biol.* **7**: 68.

Giroux, C.N. 1988. Chromosome synapsis and meiotic recombination. In *Genetic recombination* (ed. R. Kucherlapati and G. Smith), p. 465. American Society for Microbiology, Washington, D.C.

Giroux, C.N., M.E. Dresser, and H.F. Tiano. 1989. Genetic control of chromosome synapsis in yeast meiosis. *Genome* **31**: 88.

Goldman, S.L. and S. Smallets. 1979. Site specific induction of gene conversion: The effects of homozygosity of the *ade6* mutant *M26* of *Schizosaccharomyces pombe. Mol. Gen. Genet.* **173**: 221.

Golin, J.E. and M.S. Esposito. 1977. Evidence for joint genic control of spontaneous mutation and genetic recombination during mitosis in *Saccharomyces. Mol. Gen. Genet.* **150**: 127.

———. 1981. Mitotic recombination: Mismatch correction and replicational resolution of Holliday structures formed at the two-strand stage in *Saccharomyces. Mol. Gen. Genet.* **183**: 252.

———. 1984. Coincident gene conversion during mitosis in *Saccharomyces. Genetics* **107**: 355.

Golin, J. and S.C. Falco. 1988. The behavior of insertions near a site of mitotic gene conversion in yeast. *Genetics* **119**: 535.

Golin, J.E. and H. Tampe. 1988. Coincident recombination during mitosis in *Saccharomyces*: Distance-dependent and -independent components. *Genetics* **119**: 541.

Golin, J., S.C. Falco, and J.P. Margolskee. 1986. Coincident gene conversion events in yeast that involve a large insertion. *Genetics* **114**: 1081.

Gottlieb, S. and R.E. Esposito. 1989. A new role for a yeast transcriptional silencer gene, *SIR2*, in regulation of recombination in ribosomal DNA. *Cell* **56**: 771.

Gottlieb, S., J. Wagstaff, and R.E. Esposito. 1989. Evidence for two pathways of meiotic intrachromosomal recombination in yeast. *Proc. Natl. Acad. Sci.* **86**: 7072.

Grossenbacher-Grunder, A.M. and P. Thuriaux. 1981. Spontaneous and UV-induced recombination in radiation-sensitive mutants of *Schizosaccharomyces pombe. Mutat. Res.* **81**: 37.

Gutz, H. 1971. Site specific induction of gene conversion in *Schizosaccharomyces pombe. Genetics* **69**: 317.

Haber, J.E. and M. Hearn. 1985. *rad52*-independent mitotic gene conversion in *Saccharomyces cerevisiae* frequently results in chromosome loss. *Genetics* **111:** 7.

Haber, J.E., P.C. Thorburn, and D. Rogers. 1984. Meiotic and mitotic behavior of dicentric chromosomes in *Saccharomyces cerevisiae*. *Genetics* **106:** 185.

Haber, J.E., R.H. Borts, B. Connolly, M. Lichten, N. Rudin, and C.I. White. 1988. Physical monitoring of meiotic and mitotic recombination in yeast. *Prog. Nucleic Acid Res. Mol. Biol.* **35:** 209.

Hartwell, L. and D. Smith. 1985. Altered fidelity of mitotic chromosome transmission in cell cycle mutants of *Saccharomyces cerevisiae*. *Genetics* **110:** 381.

Hastings, P.J. 1984. Measurement of restoration and conversion: Its meaning for the mismatch repair hypothesis of conversion. *Cold Spring Harbor Symp. Quant. Biol.* **49:** 49.

———. 1987a. Meiotic recombination interpreted as heteroduplex correction. In *Meiosis* (ed. P.B. Moens), p. 107. Academic Press, New York.

———. 1987b. Models of heteroduplex formation. In *Meiosis* (ed. P.B. Moens), p. 139. Academic Press, New York.

———. 1988. Conversion events in fungi. In *Genetic recombination* (ed. R. Kucherlapati and G. Smith), p. 397. American Society for Microbiology, Washington, D.C.

Hastings, P.J. and E.A. Savage. 1983. Further evidence of a disparity between conversion and restoration in the *his1* locus of *Saccharomyces cerevisiae*. *Curr. Genet.* **8:** 23.

Hawley, R.S. 1988. Exchange and chromosomal segregation in eucaryotes. In *Genetic recombination* (ed. R. Kucherlapati and G. Smith), p. 497. American Society for Microbiology, Washington, D.C.

Hawthorne, D.C. and R.K. Mortimer. 1960. Chromosome mapping in *Saccharomyces*: Centromere-linked genes. *Genetics* **45:** 1085.

Haynes, R.H. and B.A. Kunz. 1981. DNA repair and mutagenesis in yeast. In *The molecular biology of the yeast* Saccharomyces: *Life cycle and inheritance* (ed. J.N. Strathern et al.), p. 371. Cold Spring Harbor Laboratory, Cold Spring Harbor, New York.

———. 1986. The influence of thymine nucleotide depletion on genetic stability and change in eukaryotic cells. *Curr. Science* **55:** 1.

Heyer, W.-D. and R. Kolodner. 1989. Purification and characterization of a protein from *Saccharomyces cerevisiae* that binds tightly to single-stranded DNA and stimulates a cognate strand exchange protein. *Biochemistry* **28:** 2856.

Heyer, W.-D., D. Evans, and R.D. Kolodner. 1988. Renaturation of DNA by a *Saccharomyces cerevisiae* protein that catalyzes homologous pairing and strand exchange. *J. Biol. Chem.* **263:** 15189.

Heyer, W.-D., P. Munz, H. Amstutz, R. Aebi, C. Gysler, P. Schuchert, P. Szankasi, U. Leupold, J. Kohli, V. Gamulin, and D. Soll. 1986. Inactivation of nonsense suppressor transfer RNA genes in *Schizosaccharomyces pombe*: Intergenic conversion and hot spots of mutations. *J. Mol. Biol.* **180:** 343.

Hicks, J.B., A. Hinnen, and G.R. Fink. 1979. Properties of yeast transformation. *Cold Spring Harbor Symp. Quant. Biol.* **43:** 1305.

Higgins, D.R., S. Prakash, P. Reynolds, R. Polakowska, S. Weber, and L. Prakash. 1983. Isolation and characterization of the *RAD3* gene of *Saccharomyces cerevisiae* and inviability of *rad3* deletion mutants. *Proc. Natl. Acad. Sci.* **80:** 5680.

Hinnen, A., J.R. Hicks, and G.R. Fink. 1978. Transformation of yeast. *Proc. Natl. Acad. Sci.* **75:** 1929.

Ho, K.S.Y. 1975. Induction of double-strand breaks by X-rays in a radiosensitive strain of yeast. *Mutat. Res.* **20:** 45.

Hoekstra, M.F., T.M. Naughton, and R.E. Malone. 1986. Properties of spontaneous mitotic recombination occurring in the presence of the *rad52-1* mutation of *Saccharomyces cerevisiae. Genet. Res.* **48:** 9.

Hofer, F., H. Hollenstein, F. Janner, M. Minet, P. Thuriaux, and U. Leupold. 1979. The genetic fine structure of nonsense suppressors in *Schizosaccharomyces pombe.* I. *sup3* and *sup9. Curr. Genet.* **1:** 45.

Hogset, A. and T. Oyen. 1984. Correlation between suppressed meiotic recombination and the lack of DNA strand breaks in the rDNA genes of *Saccharomyces cerevisiae. Nucleic Acids Res.* **12:** 7199.

Holliday, R. 1964. A mechanism for gene conversion in fungi. *Genet. Res.* **5:** 282.

———. 1977. Recombination and meiosis. *Philos. Trans. R. Soc. Lond. B Biol. Sci.* **277:** 359.

Hollingsworth, N.M. and B. Byers. 1989. *HOP1:* A yeast meiotic pairing gene. *Genetics* **121:** 445.

Hollingsworth, N.M., L. Goetsch, and B. Byers. 1990. The *HOP1* gene encodes a meiosis-specific component of yeast chromosomes. *Cell* **61:** 73.

Hotta, Y., S. Tabata, R.A. Bouchard, R. Piñon, and H. Stern. 1985. General recombination mechanisms in extracts of meiotic cells. *Chromosoma* **93:** 140.

Hurd, H.K., C.W. Roberts, and J.W. Roberts. 1987. Identification of the gene for the yeast ribonucleotide reductase small subunit and its inducibility by methyl methanesulfonate. *Mol. Cell. Biol.* **7:** 3673.

Hurst, D.D. and S. Fogel. 1964. Mitotic recombination and heteroallelic repair in *Saccharomyces cerevisiae. Genetics* **50:** 435.

Hurst, D.D., S. Fogel, and R.K. Mortimer. 1972. Conversion-associated recombination in yeast. *Proc. Natl. Acad. Sci.* **69:** 101.

Jackson, J.A. and G.R. Fink. 1981. Gene conversion between duplicated genetic elements in yeast. *Nature* **292:** 306.

———. 1985. Meiotic recombination between duplicated genetic elements in *Saccharomyces cerevisiae. Genetics* **109:** 303.

Jarvik, J. and D. Botstein. 1973. A genetic method for determining the order of events in a biological pathway. *Proc. Natl. Acad. Sci.* **70:** 2046.

Jensch, F., H. Kozak, N.C. Seeman, and B. Kemper. 1989. Cruciform cutting endonucleases from *Saccharomyces cerevisiae* and phage T4 show conserved reactions with branched DNAs. *EMBO J.* **8:** 4325.

Jentsch, S., J.P. McGrath, and A. Varshavsky. 1987. The yeast *RAD6* gene encodes a ubiquitin conjugating enzyme. *Nature* **329:** 131.

Jinks-Robertson, S. and T.D. Petes. 1985. High-frequency meiotic gene conversion between repeated genes on non-homologous chromosomes in yeast. *Proc. Natl. Acad. Sci.* **82:** 3350.

———. 1986. Chromosomal translocations generated by high-frequency meiotic recombination between repeated yeast genes. *Genetics* **114:** 731.

Johnston, L.H. and K.A. Nasmyth. 1978. *Saccharomyces cerevisiae* cell cycle mutant *cdc9* is defective in DNA ligase. *Nature* **274:** 891.

Judd, S.R. and T.D. Petes. 1988. Physical lengths of meiotic and mitotic gene conversion tracts in *Saccharomyces cerevisiae. Genetics* **118:** 401.

Kaback, D.B. 1989. Meiotic segregation of circular plasmid-minichromosomes from intact chromosomes in *Saccharomyces cerevisiae. Curr. Genet.* **15:** 385.

Kaback, D.B., H.Y. Steensma, and P. de Jonge. 1989. Enhanced meiotic recombination on the smallest chromosome of *Saccharomyces cerevisiae. Proc. Natl. Acad. Sci.* **86:** 3694.

Kakar, S.N. 1963. Allelic recombination and its relation to recombination of outside markers. *Genetics* **48**: 957.

Kassir, Y. and G. Simchen. 1978. Meiotic recombination and DNA synthesis in a new cell cycle mutant of *Saccharomyces cerevisiae*. *Genetics* **90**: 49.

Keil, R. and G.S. Roeder. 1984. *Cis*-acting recombination-stimulating activity in a fragment of the ribosomal DNA of *S. cerevisiae*. *Cell* **39**: 377.

Kern, R. and F. Zimmermann. 1978. The influence of defects in excision and error prone repair on spontaneous and induced mitotic recombination and mutation in *Saccharomyces cerevisiae*. *Mol. Gen. Genet.* **161**: 81.

Kim, R. and J.C. Wang. 1989. A subthreshold level of DNA topoisomerases leads to the excision of yeast rDNA as extrachromosomal rings. *Cell* **57**: 975.

Kitani, Y., L.S. Olive, and A.S. El-Ani. 1961. Transreplication and crossing-over in *Sordaria fimicola*. *Science* **134**: 668.

Klapholz, S. and R.E. Esposito. 1980a. Isolation of *spo12-1* and *spo13-1* from a natural variant of yeast that undergoes a single meiotic division. *Genetics* **96**: 567.

————. 1980b. Recombination and chromosome segregation during the single division meiosis in *spo12-1* and *spo13-1* diploids. *Genetics* **96**: 589.

Klapholz, S., C.S. Waddell, and R.E. Esposito. 1985. The role of the *SPO11* gene in meiotic recombination in yeast. *Genetics* **110**: 187.

Klar, A.J.S. and L.M. Miglio. 1986. Initiation of meiotic recombination by double-strand DNA breaks in *S. pombe*. *Cell* **46**: 725.

Klar, A.J.S. and J.N. Strathern. 1984. Resolution of recombination intermediates generated during yeast mating type switching. *Nature* **310**: 744.

Klein, H.L. 1984. Lack of association between intrachromosomal gene conversion and reciprocal exchange. *Nature* **310**: 748.

————. 1988a. Genetic analysis of repeated yeast genes. In *Recombination of genetic material* (ed. K.B. Low), p. 385. Academic Press, New York.

————. 1988b. Different types of recombination events are controlled by the *RAD1* and *RAD52* genes of *Saccharomyces cerevisiae*. *Genetics* **120**: 367.

Klein, H.L. and T.D. Petes. 1981. Intrachromosomal gene conversion in yeast. *Nature* **289**: 144.

Kolodkin, A.L., A.J.S. Klar, and F.W. Stahl. 1986. Double-strand breaks can initiate meiotic recombination in *S. cerevisiae*. *Cell* **46**: 733.

Kolodner, R.D., D. Evans, and P.T. Morrison. 1987. Purification and characterization of an activity from *Saccharomyces cerevisiae* that catalyzes homologous pairing and strand exchange. *Proc. Natl. Acad. Sci.* **84**: 5560.

Kostriken, R., J.N. Strathern, A.J.S. Klar, J.B. Hicks, and F. Heffron. 1983. A site-specific endonuclease essential for mating-type switching in *Saccharomyces cerevisiae*. *Cell* **35**: 167.

Kramer, B., W. Kramer, and H.-J. Fritz. 1984. Different base/base mismatches are corrected with different efficiencies by the methyl-directed DNA mismatch-repair system of *E. coli. Cell* **38**: 879.

Kramer, B., W. Kramer, M.S. Williamson, and S. Fogel. 1989. Heteroduplex DNA correction in yeast is mismatch-specific and requires functional *PMS* genes. *Mol. Cell. Biol.* **9**: 4432.

Kramer, W., B. Kramer, M.S. Williamson, and S. Fogel. 1989. Cloning and nucleotide sequence of DNA mismatch repair gene *PMS1* from *Saccharomyces cerevisiae*: Homology of PMS1 to procaryotic MutL and HexB. *J. Bacteriol.* **171**: 5339.

Kunz, B.A., B.J. Barclay, J.C. Game, J.G. Little, and R.H. Haynes. 1980. Induction of mitotic recombination in yeast by starvation for thymine nucleotides. *Proc. Natl. Acad. Sci.* **77**: 6057.

Kupiec, M. 1986. The *RAD50* gene of *Saccharomyces cerevisiae* is not essential for vegetative growth. *Curr. Genet.* **10:** 487.

Kupiec, M. and T.D. Petes. 1988a. Allelic and ectopic recombination between Ty elements in yeast. *Genetics* **119:** 549.

──────. 1988b. Meiotic recombination between repeated transposable elements in *Saccharomyces cerevisiae. Mol. Cell. Biol.* **8:** 2942.

Kupiec, M. and G. Simchen. 1984. Cloning and mapping the *RAD50* gene of *Saccharomyces cerevisiae. Mol. Gen. Genet.* **193:** 525.

Lacks, S.A., J.J. Dunn, and B. Greenberg. 1982. Identification of base mismatches recognized by the heteroduplex DNA repair system of *Streptococcus pneumoniae. Cell* **31:** 327.

Lambie, E.J. and G.S. Roeder. 1986. Repression of meiotic crossing-over by a centromere (*CEN3*) in *Saccharomyces cerevisiae. Genetics* **114:** 769.

──────. 1988. A yeast centromere acts in *cis* to inhibit meiotic conversion of adjacent sequences. *Cell* **52:** 863.

Larkin, J.C. and J.L. Woolford, Jr. 1983. Molecular cloning and analysis of the *CRY1* gene: A yeast ribosomal protein gene. *Nucleic Acids Res.* **11:** 403.

Lawrence, C.W. and R.B. Christensen. 1979. Metabolic suppressors of trimethoprin and ultraviolet light sensitivities of *Saccharomyces cerevisiae sarad6* mutants. *J. stBacteriol.* **139:** 866.

Lea, D.E. and C.A. Coulson. 1949. The distribution of the numbers of mutants in bacterial populations. *J. Genet.* **49:** 264.

Leblon, G. 1972a. Mechanism of gene conversion in *Ascobolus immersus.* I. Existence of a correlation between the origin of mutants induced by different mutagens and their conversion spectrum. *Mol. Gen. Genet.* **115:** 36.

──────. 1972b. Mechanism of gene conversion in *Ascobolus immersus.* II. The relationships between the genetic alterations in b_1 or b_2 mutants and their conversion spectrum. *Mol. Gen. Genet.* **116:** 322.

Lichten, M. and J.E. Haber. 1989. Position effects in ectopic and allelic mitotic recombination. *Genetics* **123:** 261.

Lichten, M., R.H. Borts, and J.E. Haber. 1987. Meiotic gene conversion and crossing-over between dispersed homologous sequences occurs frequently in *Saccharomyces cerevisiae. Genetics* **115:** 233.

Lichten, M., C. Goyon, N.P. Schultes, D. Treco, J.W. Szostak, J.E. Haber, and A. Nicolas. 1990. Detection of heteroduplex DNA molecules among the products of *Saccharomyces cerevisiae* meiosis. *Proc. Natl. Acad. Sci.* **87:** 7653.

Liebman, S. and S. Picologlu. 1988. Recombination associated with yeast retrotransposons. In *Viruses of fungi and simple eukaryotes* (ed. Y. Koltin and M.J. Leibowitz), p. 63. Dekker, New York.

Liebman, S., P. Shalit, and S. Picologlu. 1981. Ty elements are involved in the formation of deletions in *DEL1* strains of *Saccharomyces cerevisiae. Cell* **26:** 401.

Liebman, S.W., L.S. Symington, and T.D. Petes. 1988. Mitotic recombination within the centromere of a yeast chromosome. *Science* **241:** 1074.

Louis, E. and J.E. Haber. 1990. Mitotic recombination among subtelomeric Y′ repeats in *Saccharomyces cerevisiae. Genetics* **124:** 547.

Lovett, S.T. and R.K. Mortimer. 1987. Characterization of null mutants of the *RAD55* gene of *Saccharomyces cerevisiae*: Effects of temperature, osmotic strength and mating type. *Genetics* **116:** 547.

Luria, S.E. and M. Delbrück. 1943. Mutations of bacteria from virus sensitivity to virus resistance. *Genetics* **28:** 491.

Malone, R.E. 1983. Multiple mutant analysis of recombination in yeast. *Mol. Gen. Genet.* **189**: 405.

Malone, R.E. and R.E. Esposito. 1980. The *RAD52* gene is required for homothallic inter-conversion of mating type and spontaneous mitotic recombination in yeast. *Proc. Natl. Acad. Sci.* **77**: 503.

————. 1981 . Recombinationless meiosis in *Saccharomyces cerevisiae. Mol. Cell. Biol.* **1**: 891.

Malone, R.E. and M.F. Hoekstra. 1984. Relationships between a hyper-rec mutation (*rem1*) and other recombination and repair genes in yeast. *Genetics* **107**: 33.

Malone, R.E., J.E. Golin, and M.S. Esposito. 1980. Mitotic versus meiotic recombination in *Saccharomyces cerevisiae. Curr. Genet.* **1**: 241.

Malone, R.E., K. Jordan, and W. Wardman. 1985. Extragenic revertants of *rad50*, a mutation causing defects in recombination and repair. *Curr. Genet.* **9**: 453.

Malone, R.E., T. Ward, S. Lin, and J. Waring. 1990. The *RAD50* gene, a member of the double-strand-break repair epistasis group, is not required for spontaneous mitotic recombination in yeast. *Curr. Genet.* **18**: 111.

Malone, R.E., B. Montelone, C. Edwards, K. Carney, and M.F. Hoekstra. 1988. A re-examination of the role of the *RAD52* gene in spontaneous mitotic recombination. *Curr. Genet.* **14**: 211.

Maloney, D. and S. Fogel. 1980. Mitotic recombination in yeast: Isolation and character-ization of mutants with enhanced spontaneous mitotic gene conversion rates. *Genetics* **94**: 825.

————. 1987. Gene conversion, unequal crossing-over and mispairing at a non-tandem duplication during meiosis of *Saccharomyces cerevisiae. Curr. Genet.* **12**: 1.

Mann, C. and R. W. Davis. 1986. Meiotic disjunction of circular mini-chromosomes in yeast does not require DNA homology. *Proc. Natl. Acad. Sci.* **83**: 6017.

McIntosh, E.M., B. Kunz, and R. Haynes. 1986. Inhibition of DNA replication in *Saccharomyces cerevisiae* by araCMP. *Curr. Genet.* **10**: 579.

McKnight, G., T. Cardillo, and F. Sherman. 1981. An extensive deletion causing over-production of yeast iso-2-cytochrome *c. Cell* **25**: 409.

McLeod, M., F. Volkert, and J. Broach. 1984. Components of the site-specific recom-bination system encoded by the yeast plasmid 2-micron circle. *Cold Spring Harbor Symp. Quant. Biol.* **49**: 779.

Meddle, C.C., P. Kumar, J. Ham, D.A. Hughes, and I.R. Johnston. 1984. Cloning of the *CDC7* gene of *Saccharomyces cerevisiae* in association with centromeric DNA. *Gene* **34**: 179.

Menees, T. and G.S. Roeder. 1989. *MEI4*, a yeast gene required for meiotic recombina-tion. *Genetics* **123**: 675.

Meselson, M. and C. Radding. 1975. A general model for genetic recombination. *Proc. Natl. Acad. Sci.* **72**: 358.

Mikus, M. and T.D. Petes. 1982. Recombination between genes located on non-homologous chromosomes in the yeast *Saccharomyces cerevisiae. Genetics* **101**: 369.

Minet, M., A.-M. Grossenbacher-Grunder, and P. Thuriaux. 1980. The origin of a centromere effect in mitotic recombination. *Curr. Genet.* **2**: 53.

Mizuuchi, K., B. Kemper, J. Hays, and R.A. Weisberg. 1982. T4 endonuclease VII cleaves Holliday structures. *Cell* **29**: 357.

Modrich, P. 1987. DNA mismatch correction. *Annu. Rev. Biochem.* **56**: 435.

Moens, P.B. and E. Rapport. 1971a. Spindles, spindle plaques, and meiosis in the yeast *Saccharomyces cerevisiae. J. Cell Biol.* **50**: 344.

————. 1971b. Synaptic structures in the nuclei of sporulating yeast, *Saccharomyces cerevisiae. J. Cell Sci.* **9:** 665.

Moerschell, R.P., S. Tsunasawa, and F. Sherman. 1988. Transformation of yeast with synthetic oligonucleotides. *Proc. Natl. Acad. Sci.* **85:** 524.

Montelone, B.A., M.F. Hoekstra, and R.E. Malone. 1988. Spontaneous mitotic recombination in yeast: The hyper-recombinational *rem1* mutations are alleles of the *RAD3* gene. *Genetics* **119:** 289.

Montelone, B., S. Prakash, and L. Prakash. 1981a. Recombination and mutagenesis in *rad6* mutants of *Saccharomyces cerevisiae*: Evidence for multiple functions of the *RAD6* gene. *Mol. Gen. Genet.* **184:** 410.

————. 1981b. Hyperrecombination and mutator effects of the *mms9-1*, *mms13-1* and *mms21-1* mutations in *Saccharomyces cerevisiae. Curr. Genet.* **4:** 223.

————. 1981c. Spontaneous mitotic recombination in *mms8-1*, an allele of the *CDC9* gene of *Saccharomyces cerevisiae. J. Bacteriol.* **147:** 517.

Moore, C.W., D.M. Hampsey, J.F. Ernst, and F. Sherman. 1988. Differential mismatch repair can explain the disproportionalities between physical distances and recombination frequencies of *cyc1* mutations in yeast. *Genetics* **119:** 21.

Mortimer, R.K. and S. Fogel. 1974. Genetical interference and gene conversion. In *Mechanisms in recombination* (ed. R. Grell), p. 236. Plenum Press, New York.

Mortimer, R.K. and D. Schild. 1981. Genetic mapping in *Saccharomyces cerevisiae.* In *The molecular biology of the yeast* Saccharomyces: *Life cycle and inheritance* (ed. J.N. Strathern et al.), p. 11. Cold Spring Harbor Laboratory, Cold Spring Harbor, New York.

Muller, H.J. 1916. The mechanism of crossing-over. *Am. Nat.* **50:** 193.

Munz, P. and U. Leupold. 1981. Heterologous recombination between redundant tRNA genes in *Schizosaccharomyces pombe. Alfred Benzon Symp.* **16:** 264.

Muster-Nassal, C. and R. Kolodner. 1986. Mismatch correction catalyzed by cell-free extracts of *Saccharomyces cerevisiae. Proc. Natl. Acad. Sci.* **83:** 7618.

Nag, D.K., M.A. White, and T.D. Petes. 1989. Palindromic sequences in heteroduplex DNA inhibit mismatch repair in yeast. *Nature* **340:** 318.

Nakai, S. and R.K. Mortimer. 1969. Studies of the mechanism of radiation-induced mitotic segregation in yeast. *Mol. Gen. Genet.* **103:** 329.

Nakaseko, Y., Y. Adachi, S. Funahashi, O. Niwa, and M. Yanagida. 1986. Chromosome walking shows a highly homologous repetitive sequence present in all the centromeric regions of fission yeast. *EMBO J.* **5:** 1011.

Naumovski, L. and E.C. Friedberg. 1983. A DNA repair gene required for the incision of damaged DNA is essential for viability in *Saccharomyces cerevisiae. Proc. Natl. Acad. Sci.* **80:** 4818.

Neitz, M. and J. Carbon. 1987. Characterization of a centromere-linked recombination hotspot in *Saccharomyces cerevisiae. Mol. Cell. Biol.* **7:** 3871.

Newlon, C.S. 1988. Yeast chromosome structure and replication. *Microbiol. Rev.* **52:** 568.

Newlon, C.S., R.P. Green, K.J. Hardeman, K.E. Kim, L.R. Lipchitz, T.G. Palzkill, S. Synn, and S.T. Woody. 1986. Structure and organization of yeast chromosome III. *ICN-UCLA Symp. Mol. Cell. Biol.* **33:** 211.

Nickoloff, J.A., E.Y. Chen, and F. Heffron. 1986. A 24-base-pair DNA sequence from the *MAT* locus stimulates intergenic recombination in yeast. *Proc. Natl. Acad. Sci.* **83:** 7831.

Nicolas, A.D. and J.-L. Rossignol. 1989. Intermediates in homologous recombination revealed by marker-effects in *Ascobolus. Genome* **31:** 528.

Nicolas, A., D. Treco, N.P. Schultes, and J.W. Szostak. 1989. Identification of an initiation site for meiotic gene conversion in the yeast *Saccharomyces cerevisiae. Nature* **338:** 35.

Nilsson-Tillgren, T., C. Gjermannen, S. Holmberg, J.G.L. Petersen, and M.C. Kielland-Brandt. 1986. Analysis of chromosome V and the gene *ILV1* from *Saccharomyces carlsbergensis. Carlsberg Res. Commun.* **51:** 309.

Olson, L. and F.K. Zimmermann. 1978. Mitotic recombination in the absence of synaptonemal complexes in *Saccharomyces cerevisiae. Mol. Gen. Genet.* **166:** 161.

Olson, L.W., V. Eden, M. Egel-Mitani, and R. Egel. 1978. Asynaptic meiosis in fission yeast? *Hereditas* **89:** 189.

Orr-Weaver, T.L. and J.W. Szostak. 1983. Yeast recombination: The association between double-strand gap repair and crossing-over. *Proc. Natl. Acad. Sci.* **80:** 4417.

―――. 1985. Fungal recombination. *Microbiol. Rev.* **49:** 33.

Orr-Weaver, T.L., J.W. Szostak, and R. Rothstein. 1981. Yeast transformation: A model system for the study of recombination. *Proc. Natl. Acad. Sci.* **78:** 6354.

Paquette, N. and J.-L. Rossignol. 1978. Gene conversion spectrum of 15 mutants giving post-meiotic segregation in the *b2* locus of *Ascobolus immersus. Mol. Gen. Genet.* **163:** 313.

Parsons, C.A. and S.C. West. 1988. Resolution of model Holliday junctions by yeast endonuclease is dependent upon homologous DNA sequences. *Cell* **52:** 621.

Parsons, C.A., A.I.H. Murchie, D.M.J. Lilley, and S.C. West. 1989. Resolution of model Holliday junctions by yeast endonuclease: Effect of DNA structure and sequence. *EMBO J.* **8:** 239.

Perkins, D.D. 1949. Biochemical mutants in the smut fungus *Ustilago maydis. Genetics* **34:** 607.

Petes, T.D. 1980. Unequal meiotic recombination within tandem arrays of yeast ribosomal DNA genes. *Cell* **19:** 765.

Petes, T.D. and D. Botstein. 1977. Simple Mendelian inheritance of the reiterated ribosomal DNA of yeast. *Proc. Natl. Acad. Sci.* **74:** 5091.

Petes, T.D. and C.W. Hill. 1988. Recombination between repeated genes in microorganisms. *Annu. Rev. Genet.* **22:** 147.

Petes, T.D., P. Detloff, S. Jinks-Robertson, S.R. Judd, M. Kupiec, D. Nag, A. Stapleton, L.S. Symington, A. Vincent, and M. White. 1989. Recombination in yeast and the recombinant DNA technology. *Genome* **31:** 536.

Ponticelli, A.S. and G.R. Smith. 1989. Meiotic recombination-deficient mutants of *Schizosaccharomyces pombe. Genetics* **123:** 45.

Ponticelli, A.S., E.P. Sena, and G.R. Smith. 1988. Genetic and physical analysis of the M26 recombination hotspot of *Schizosaccharomyces pombe. Genetics* **119:** 491.

Powers, P.A. and O. Smithies. 1986. Short gene conversions in the human fetal globin region: A by-product of chromosome pairing during meiosis? *Genetics* **112:** 343.

Prakash, L. and S. Prakash. 1977. Isolation and characterization of MMS sensitive mutants of *Saccharomyces cerevisiae. Genetics* **86:** 33.

Prakash, L. and P. Taillon-Miller. 1981. Effects of the *rad52* gene on sister chromatid recombination in *Saccharomyces cerevisiae. Curr. Genet.* **3:** 247.

Prakash, S., L. Prakash, W. Burke, and B. Montelone. 1980. Effects of the *RAD52* gene on recombination in *Saccharomyces cerevisiae. Genetics* **94:** 31.

Pukkila, P.J., M.D. Stephens, D.M. Binninger, and B. Errede. 1986. Frequency and directionality of gene conversion events involving the *CYC7-H3* mutation in *Saccharomyces cerevisiae. Genetics* **114:** 347.

Radding, C.M. 1982. Homologous pairing and strand exchange in genetic recombination. *Annu. Rev. Genet.* **16:** 405.

Ray, A., N. Machin, and F.W. Stahl. 1989. A DNA double chain break stimulates triparental recombination in *Saccharomyces cerevisiae. Proc. Natl. Acad. Sci.* **86:** 6225.

Ray, A., I. Siddiqi, A.L. Kolodkin, and F.W. Stahl. 1988. Intra-chromosomal gene conversion induced by a DNA double-strand break in *Saccharomyces cerevisiae. J. Mol. Biol.* **201:** 247.

Resnick, M.A. 1976. The repair of double-strand breaks in DNA: A model involving recombination. *J. Theor. Biol.* **59:** 97.

————. 1987. Investigating the genetic control of biochemical events in meiotic recombination. In *Meiosis* (ed. P.B. Moens), p. 157. Academic Press, New York.

Resnick, M.A. and P. Martin. 1976. The repair of double-stranded breaks in the nuclear DNA of *Saccharomyces cerevisiae* and its genetic control. *Mol. Gen. Genet.* **143:** 119.

Resnick, M.A., M. Skaanild, and T. Nilsson-Tillgren. 1989. Lack of DNA homology in a pair of divergent chromosomes greatly sensitizes them to loss by DNA damage. *Proc. Natl. Acad. Sci.* **86:** 2276.

Resnick, M.A., T. Chow, J. Nitiss, and J. Game. 1984. Changes in the chromosomal DNA of yeast during meiosis in repair mutants and the possible role of a deoxyribonuclease. *Cold Spring Harbor Symp. Quant. Biol.* **49:** 639.

Resnick, M.A., J. Nitiss, C. Edwards, and R.E. Malone. 1986. Meiosis can induce recombination in *rad52* mutants of *Saccharomyces cerevisiae. Genetics* **113:** 531.

Rockmill, B. and G.S. Roeder. 1988. *RED1*: A yeast gene required for the segregation of chromosomes during the reductional division of meiosis. *Proc. Natl. Acad. Sci.* **85:** 6057.

————. 1990. Meiosis in asynaptic yeast. *Genetics* **126:** 563.

Rodarte-Ramon, U. and R.K. Mortimer. 1972. Radiation induced recombination in *Saccharomyces*: Isolation and genetic study of recombination deficient mutants. *Radiat. Res.* **49:** 133.

Roeder, G.S. 1983. Unequal crossing-over between yeast transposable elements. *Mol. Gen. Genet.* **190:** 117.

Roeder, G.S. and S.E. Stewart. 1988. Mitotic recombination in yeast. *Trends Genet.* **4:** 263.

Roeder, G.S., M. Smith, and E.J. Lambie. 1984. Intrachromosomal movement of genetically marked *Saccharomyces cerevisiae* transposons by gene conversion. *Mol. Cell. Biol.* **4:** 703.

Roman, H.L. 1957. Studies of recombination in yeast. *Cold Spring Harbor Symp. Quant. Biol.* **21:** 175.

Roman, H. and F. Fabre. 1983. Gene conversion and reciprocal recombination are separable events in vegetative cells of *Saccharomyces cerevisiae. Proc. Natl. Acad. Sci.* **80:** 6912.

Ronne, H. and R. Rothstein. 1988. Mitotic sectored colonies: Evidence of heteroduplex formation during direct repeat recombination. *Proc. Natl. Acad. Sci.* **85:** 2696.

Rossignol, J.-L., A. Nicolas, H. Hamza, and A. Kalogeropoulos. 1988. Recombination and gene conversion in *Ascobolus.* In *The recombination of genetic material* (ed. K.B. Low), p. 23. Academic Press, New York.

Roth, R. and S. Fogel. 1971. A selective system for yeast mutants deficient in meiotic recombination. *Mol. Gen. Genet.* **112:** 295.

Rothstein, R.J. 1983. One-step gene disruption in yeast. *Methods Enzymol.* **101:** 202.

————. 1984. Double-strand-break repair, gene conversion and postdivision segregation. *Cold Spring Harbor Symp. Quant. Biol.* **49:** 629.

Rudin, N. and J.E. Haber. 1988. Efficient repair of *HO*-induced chromosomal breaks in

Saccharomyces cerevisiae by recombination between flanking homologous sequences. *Mol. Cell. Biol.* **8**: 3918.

Sadowski, P.D., D.D. Lee, B.J. Andrews, D. Babineau, L. Beatty, M.J. Morse, G. Proteau, and D. Vetter. 1984. *In vitro* systems for genetic recombination of the DNAs of bacteriophage T7 and yeast 2-micron circle. *Cold Spring Harbor Symp. Quant. Biol.* **49**: 789.

Saeki, T., I. Machida, and S. Nakai. 1980. Genetic control of diploid recovery after gamma-irradiation in the yeast *Saccharomyces cerevisiae. Mutat. Res.* **73**: 251.

Savage, E.A. and P.J. Hastings. 1981. Marker effects and the nature of the recombination event at the *his1* locus of *Saccharomyces cerevisiae. Curr. Genet.* **3**: 37.

Scherer, S. and R.W. Davis. 1979. Replacement of chromosome segments with altered DNA sequences constructed *in vitro. Proc. Natl. Acad. Sci.* **76**: 4951.

―――. 1980. Recombination of dispersed repeated DNA sequences in yeast. *Science* **209**: 1380.

Schiestl, R.H. 1989. Nonmutagenic carcinogens induce intrachromosomal recombination in yeast. *Nature* **337**: 285.

Schiestl, R.H. and S. Prakash. 1988. *RAD1*, an excision repair gene of *Saccharomyces cerevisiae*, is also involved in recombination. *Mol. Cell. Biol.* **8**: 3619.

Schiestl, R.H., S. Igarashi, and P.J. Hastings. 1988. Analysis of the mechanism for reversion of a disrupted gene. *Genetics* **119**: 237.

Schiestl, R.H., R.D. Gietz, P.J. Hastings, and U. Wintersberger. 1990. Interchromosomal and intrachromosomal recombination in *rad18* mutants of *Saccharomyces cerevisiae. Mol. Gen. Genet.* **222**: 25.

Schild, D., I.L. Calderon, R. Contopoulou, and R.K. Mortimer. 1983a. Cloning of yeast recombination repair genes and evidence that several are non-essential genes. In *Cellular responses to DNA damage* (ed. E.C. Friedberg and B.A. Bridges), p. 417. A.R. Liss, New York.

Schild, D., B. Konforti, C. Perez, W. Gish, and R.K. Mortimer. 1983b. Isolation and characterization of yeast DNA repair genes. I. Cloning of the *RAD52* gene. *Curr. Genet.* **7**: 85.

Schmidt, H., P. Kapitza, and H. Gutz. 1987. Switching genes in *Schizosaccharomyces pombe*: Their influence on cell viability and recombination. *Curr. Genet.* **11**: 303.

Schuchert, P. and J. Kohli. 1988. The *ade6-M26* mutation of *Schizosaccharomyces pombe* increases the frequency of crossing over. *Genetics* **119**: 507.

Sclafani, R.A. and W.L. Fangman. 1984. Yeast gene *CDC8* encodes thymidylate kinase and is complemented by herpes thymidine kinase gene TK. *Proc. Natl. Acad. Sci.* **81**: 5821.

Seifert, H.S., E.Y. Chen, M. So, and F. Heffron. 1986. Shuttle mutagenesis: A method of transposon mutagenesis for *Saccharomyces cerevisiae. Proc. Natl. Acad. Sci.* **83**: 735.

Sherman, F. and H. Roman. 1963. Evidence for two types of allelic recombination in yeast. *Genetics* **48**: 253.

Simchen, G., Y. Kassir, O. Horesh-Cabilly, and A. Friedman. 1981. Elevated recombination and pairing structures during meiotic arrest in yeast of the nuclear division mutant *cdc5. Mol. Gen. Genet.* **184**: 46.

Simon, J.R. and P.D. Moore. 1987. Homologous recombination between single-stranded DNA and chromosomal genes in *Saccharomyces cerevisiae. Mol. Cell. Biol.* **7**: 2329.

Singh, H., J.J. Bieker, and L.B. Dumas. 1982. Genetic transformation of *Saccharomyces cerevisiae* with single-stranded circular DNA vectors. *Gene* **20**: 441.

Sitney, K.C., M.E. Budd, and J.L. Campbell. 1989. DNA polymerase III, a second essential DNA polymerase, is encoded by the *S. cerevisiae CDC2* gene. *Cell* **56**: 599.

Smith, G.R. 1987. Mechanism and control of homologous recombination in *Escherichia coli. Annu. Rev. Genet.* **21:** 179.

Smolik-Utlaut, S. 1982. "Spontaneous mitotic recombination in the rRNA genes of the yeast *Saccharomyces cerevisiae.*" Ph.D. thesis, University of Chicago.

Snow, R. 1968. Recombination in ultraviolet-sensitive strains of *Saccharomyces cerevisiae. Mutat. Res.* **6:** 409.

————. 1979. Maximum likelihood estimation of linkage and interference from tetrad data. *Genetics* **92:** 231.

Stapleton, A. and T.D. Petes. 1991. The Tn3 β-lactamase gene acts as a hotspot for meiotic recombination in yeast. *Genetics* **127:** 39.

Stewart, S.E. and G.S. Roeder. 1989. Transcription by RNA polymerase I stimulates mitotic recombination in *Saccharomyces cerevisiae. Mol. Cell. Biol.* **9:** 3464.

Stinchcomb, D.T., C. Mann, and R.W. Davis. 1982. Centromeric DNA from *Saccharomyces cerevisiae. J. Mol. Biol.* **158:** 157.

Strathern, J.N., C. Newlon, I. Herskowitz, and J.B. Hicks. 1979. Isolation of a circular derivative of yeast chromosome III: Implications for the mechanism of mating type interconversion. *Cell* **18:** 309.

Strathern, J.N., K.G. Weinstock, D.R. Higgins, and C.B. McGill. 1991. A novel recombinator in yeast based on gene *II* protein from bacteriophage fl. *Genetics* **127:** 61.

Strathern, J.N., A.J.S. Klar, J.B. Hicks, J.A. Abraham, J.M. Ivy, K.A. Nasmyth, and C. McGill. 1982. Homothallic switching of yeast mating type cassettes is initiated by a double-stranded cut in the *MAT* locus. *Cell* **31:** 183.

Strickland, W.N. 1958. An analysis of interference in *Aspergillus nidulans. Proc. R. Soc. Lond. B Biol. Sci.* **148:** 533.

Sugino, A., J. Nitiss, and M.A. Resnick. 1988. ATP-dependent DNA strand transfer catalyzed by protein(s) from meiotic cells of the yeast *Saccharomyces cerevisiae. Proc. Natl. Acad. Sci.* **85:** 36.

Sugino, A., B.H. Ryu, T. Sugino, L. Naumovski, and E.C. Friedberg. 1986. A new DNA-dependent ATPase which stimulates yeast DNA polymerase I and has DNA-unwinding activity. *J. Biol. Chem.* **261:** 11744.

Sun, H., D. Treco, N.P. Schultes, and J.W. Szostak. 1989. Double-strand breaks at an initiation site for meiotic gene conversion. *Nature* **338:** 87.

Sung, P., L. Prakash, S.W. Matson, and S. Prakash. 1987a. *RAD3* protein of *Saccharomyces cerevisiae* is a DNA helicase. *Proc. Natl. Acad. Sci.* **84:** 8951.

Sung, P., L. Prakash, S. Weber, and S. Prakash. 1987b. The *RAD3* gene of *Saccharomyces cerevisiae* encodes a DNA-dependent ATPase. *Proc. Natl. Acad. Sci.* **84:** 6045.

Symington, L.S. and R.D. Kolodner. 1985. Partial purification of an enzyme from *Saccharomyces cerevisiae* that cleaves Holliday junctions. *Proc. Natl. Acad. Sci.* **82:** 7247.

Symington, L.S. and T.D. Petes. 1988a. Expansions and contractions of the genetic map relative to the physical map of yeast chromosome III. *Mol. Cell. Biol.* **8:** 595.

————. 1988b. Meiotic recombination within the centromere of a yeast chromosome. *Cell* **52:** 237.

Symington, L.S., L.M. Fogarty, and R.D. Kolodner. 1983. Genetic recombination of homologous plasmids catalyzed by cell-free extracts of *Saccharomyces cerevisiae. Cell* **35:** 805.

Symington, L.S., P.T. Morrison, and R.D. Kolodner. 1984. Genetic recombination catalyzed by cell-free extracts of *Saccharomyces cerevisiae. Cold Spring Harbor Symp. Quant. Biol.* **49:** 805.

————. 1985. Plasmid recombination intermediates generated in a *Saccharomyces cerevisiae* cell-free recombination system. *Mol. Cell. Biol.* **5**: 2361.

Szankasi, P., W.D. Heyer, P. Schuchert, and J. Kohli. 1989. DNA sequence analysis of the *ade6* gene of *Schizosaccharomyces pombe. J. Mol. Biol.* **204**: 917.

Szankasi, P., C. Gysler, U. Zehntner, U. Leupold, J. Kohli, and P. Munz. 1986. Mitotic recombination between dispersed but related tRNA genes of *S. pombe* generates a reciprocal translocation. *Mol. Gen. Genet.* **202**: 394.

Szostak, J.W. and R. Wu. 1980. Unequal crossing-over in the ribosomal DNA of *Saccharomyces cerevisiae. Nature* **284**: 426.

Szostak, J.W., T.L. Orr-Weaver, R.J. Rothstein, and F.W. Stahl. 1983. The double strand-break model for recombination. *Cell* **33**: 25.

Thaler, D.S. and F.W. Stahl. 1988. DNA double-chain breaks in recombination of phage lambda and of yeast. *Annu. Rev. Genet.* **22**: 169.

Thomas, B.J. and R. Rothstein. 1989. Elevated recombination rates in transcriptionally active DNA. *Cell* **56**: 619.

Thompson, E.A. and G.S. Roeder. 1989. Expression and DNA sequence of *RED1*, a gene required for meiosis I chromosome segregation in yeast. *Mol. Gen. Genet.* **218**: 293.

Thuriaux, P. 1985. Direct selection of mutants influencing gene conversion in the yeast *Schizosaccharomyces pombe. Mol. Gen. Genet.* **199**: 365.

Thuriaux, P., M. Minet, P. Munz, A. Ahmad, D. Zbaeren, and U. Leupold. 1980. Gene conversion in nonsense suppressors of *Schizosaccharomyces pombe*. II. Specific marker effects in *sup3. Curr. Genet.* **1**: 89.

Treco, D. and N. Arnheim. 1986. The evolutionarily conserved repetitive sequence d(TG·AC)$_n$ promotes reciprocal exchange and generates unusual recombinant tetrads during yeast meiosis. *Mol. Cell. Biol.* **6**: 3934.

Villadsen, I.S., S.E. Bjorn, and A. Vrang. 1982. Exonuclease II from *Saccharomyces cerevisiae. J. Biol. Chem.* **258**: 8177.

Vincent, A. and T.D. Petes. 1989. Mitotic and meiotic gene conversion of Ty elements and other insertions in *Saccharomyces cerevisiae. Genetics* **122**: 759.

Voelkel-Meiman, K. and G.S. Roeder. 1990. A chromosome containing *HOT1* preferentially receives information during mitotic intrachromosomal gene conversion. *Genetics* **124**: 561.

Voelkel-Meiman, K., R.L. Keil, and G.S. Roeder. 1987. Recombination-stimulating sequences in yeast ribosomal DNA correspond to sequences regulating transcription by RNA polymerase. *Cell* **48**: 1071.

Von Wettstein, D., S.W. Rasmussen, and P.B. Holm. 1984. The synaptonemal complex in genetic segregation. *Annu. Rev. Genet.* **18**: 331.

Wagstaff, J.E., S. Klapholz, and R.E. Esposito. 1982. Meiosis in haploid yeast. *Proc. Natl. Acad. Sci.* **79**: 2986.

Wagstaff, J.E., S. Klapholz, C.S. Waddell, L. Jensen, and R.E. Esposito. 1985. Meiotic exchange within and between chromosomes requires a common Rec function in *Saccharomyces cerevisiae. Mol. Cell. Biol.* **5**: 3532.

Wallis, J.W., G. Chrebet, G. Brodsky, M. Rolfe, and R. Rothstein. 1989. A hyper-recombination mutation in *Saccharomyces cerevisiae* identifies a novel eukaryotic topoisomerase. *Cell* **58**: 409.

Wang, J.C. 1985. DNA topoisomerases. *Annu. Rev. Biochem.* **54**: 665.

Watabe, H.O., T. Iino, T. Kaneko, T. Shibata, and T. Ando. 1983. A new class of site-specific endo-deoxyribonucleases. *J. Biol. Chem.* **258**: 4663.

West, S.C. and A. Korner. 1985. Cleavage of cruciform DNA structures by an activity from *Saccharomyces cerevisiae. Proc. Natl. Acad. Sci.* **82**: 6445.

West, S.C., J.K. Countryman, and P. Howard-Flanders. 1983. Enzymatic formation of biparental figure-eight molecules from plasmid DNA and their resolution in *E. coli*. *Cell* **32**: 817.

West, S.C., C.A. Parsons, and S.M. Picksley. 1987. Purification and properties of a nuclease from *Saccharomyces cerevisiae* that cleaves DNA at cruciform junctions. *J. Biol. Chem.* **262**: 12752.

White, J.H., K. Lusnak, and S. Fogel. 1985. Mismatch-specific post-meiotic segregation frequency in yeast suggests a heteroduplex recombination intermediate. *Nature* **315**: 350.

White, J.H., J.F. DiMartino, R.W. Anderson, K. Lusnak, D. Hilbert, and S. Fogel. 1988. A DNA sequence conferring high postmeiotic segregation frequency to heterozygous deletions in *Saccharomyces cerevisiae* is related to sequences associated with eucaryotic recombination hotspots. *Mol. Cell. Biol.* **8**: 1253.

Whitehouse, H.L.K. 1982. *Genetic recombination*. Wiley, London.

Wildenberg, J. 1970. The relation of mitotic recombination to DNA replication in yeast pedigrees. *Genetics* **66**: 291.

Williamson, M.S., J.C. Game, and S. Fogel. 1985. Meiotic gene conversion mutants in *Saccharomyces cerevisiae*. I. Isolation and characterization of *pms1-1* and *pms1-2*. *Genetics* **110**: 609.

Willis, K.K. and H.L. Klein. 1987. Intrachromosomal recombination in *Saccharomyces cerevisiae*: Reciprocal exchange in an inverted repeat and associated gene conversion. *Genetics* **117**: 1.

Wintersberger, E. 1978. Yeast DNA polymerase: Antigenic relationships, use of RNA primer and associated exonuclease activity. *Eur. J. Biochem.* **86**: 167.

Wright, A.P.H., K. Maundrell, and S. Shall. 1986. Transformation of *Schizosaccharomyces pombe* by non-homologous, unstable integration of plasmids in the genome. *Curr. Genet.* **10**: 503.

Yanagida, M. and J.C. Wang. 1987. Yeast DNA topoisomerases and their structural genes. In *Nucleic acids and molecular biology* (ed. F. Eckstein and D.M.J. Lilley), vol. 1, p. 196. Springer-Verlag, Berlin.

Zamb, T.J. and T.D. Petes. 1981. Unequal sister-strand recombination within yeast ribosomal DNA does not require the *RAD52* gene product. *Curr. Genet.* **3**: 125.

Zickler, D. and L.W. Olson. 1975. The synaptonemal complex and the spindle plaque during meiosis in yeast. *Chromosoma* **50**: 1.

Zimmermann, F.K. 1968. Enzyme studies on the products of mitotic gene conversion in *Saccharomyces cerevisiae*. *Mol. Gen. Genet.* **101**: 171.

9

Transport across Yeast Vacuolar and Plasma Membranes

Ramon Serrano
European Molecular Biology Laboratory
6900 Heidelberg, Federal Republic of Germany

I. INTRODUCTION

The traditional approaches to transport in yeast were based on kinetic studies and on the selection of transport mutants (Cooper 1982). A later development was the application of Mitchell's chemiosmotic hypothesis, which clarified the bioenergetic aspects of transport (Eddy 1982; Serrano 1985). During recent years, several molecular approaches have complemented the knowledge previously obtained at the cellular level. One approach is the purification of transport components. This has been most successful in the case of plasma membrane (Serrano 1988b) and vacuolar (Uchida et al. 1988a) H^+-ATPases because of their abundance and the availability of an easy assay to follow purification. Permeases, even if abundant, require reconstitution into liposomes before their activities can be measured. Second, the introduction of recombinant DNA methodologies has provided an explosion of information on the primary structures of transport proteins. This has allowed their grouping into protein families and the development of models for their structures and mechanisms. A mutational analysis has been performed with the cloned plasma membrane H^+-ATPase, and it has provided insights about functional domains and physiological roles of the enzyme. A similar analysis has been started with the vacuolar H^+-ATPase. Finally, electrophysiological methodologies have provided a physical approach to ion transport at the molecular level. In particular, patch clamping has allowed the identification for the first time of ion channels in yeast.

These novel developments constitute the basis for the present chapter, which attempts to complement the review by Cooper (1982). Some classic aspects not previously covered are also briefly discussed, so that the 1982 edition and this monograph together provide a comprehensive picture of transport across yeast vacuolar and plasma membranes.

II. AN OVERVIEW OF TRAFFIC ACROSS VACUOLAR AND PLASMA MEMBRANES

A. Major and Minor Transport Systems

It is important to appreciate the differences in fluxes between transport systems. Yeast strains can grow on synthetic glucose media with generation times of 2–4 hours (specific growth rates of 0.17–0.34 h^{-1}). This implies a biosynthetic rate (amounts of yeast always refer to dry weight of cells) of 3–6 μg yeast $\times$ min^{-1} $\times$ mg yeast^{-1}. From the known composition of yeast cells and the nutrient requirements for biosynthesis and energy production (Gancedo and Serrano 1988), we can estimate the following fluxes across the yeast plasma membrane during growth (in nmole $\times$ min^{-1} $\times$ mg yeast^{-1}): glucose (120–240), NH_4^+ or other nitrogen sources (18–36), K^+ (1.5–3), P_i (1.2–2.4), purines and pyrimidines in auxotrophs (0.3–0.6 each), SO_4^{-2} (0.2–0.4), and Mg^{++} (0.1–0.2). Fluxes of other nutrients such as oligoelements and vitamins are less than 0.1 nmole $\times$ min^{-1} $\times$ mg yeast^{-1}. The effluxes of CO_2 and ethanol (150–300) are very important but seem to occur by passive diffusion (see below). The effluxes of succinate (about 1) and acetate (about 5) occur at intermediate rates.

Although the capacities of transport systems are usually in excess of the fluxes required for growth, these estimates provide a basis for the distinction between major and minor transport systems. The former are those involved in uptake of carbon and nitrogen sources and have fluxes on the order of 10–10^2 nmoles $\times$ min^{-1} $\times$ mg yeast^{-1}. Other transport systems exhibit fluxes of 10^{-2} to 1 in the same units. As the turnover numbers of these transport systems are on the order of 10^3 min^{-1} (Stein 1986) and 1 mg of yeast contains about 3×10^7 cells, the major transport systems have abundances of 10^5 to 10^6 molecules per cell. This is equivalent to 0.1–1% of the total cell protein and 5–50% of the plasma membrane protein (Serrano 1988b). Therefore, the biochemistry of these transport proteins should be relatively easy. On the other hand, minor transport systems would be expected to have much lower abundances.

One major transport system that was not taken into account in these calculations is the plasma membrane H^+-ATPase. The activity of this enzyme during growth can be calculated from its known stoichiometry (H^+/ATP = 1; Serrano 1988a) and from the requirements for active nutrient uptake by proton cotransport and for efflux of metabolically generated acids (Gancedo and Serrano 1988). In addition to the acetic and succinic acids generated during fermentation, there is a significant conversion of sugars into oxoacids during the biosynthesis of amino acids. The protons dissociated from these acids need to be expelled by the ATPase (18–36 nmoles $\times$ min^{-1} $\times$ mg yeast^{-1}) as do the protons diffusing into the cells (1–10 nmoles $\times$ min^{-1} $\times$ mg yeast^{-1}). A proton flux of 60–100 nmoles $\times$ min^{-1} $\times$ mg yeast^{-1} during growth can be estimated on the basis of these considerations. Therefore, the plasma membrane H^+-ATPase is also a major transport system. In fact, it accounts for about 50% of the plasma membrane protein of exponentially growing cells and about 25% of the plasma membrane protein of stationary-phase cells (R.

Serrano, unpubl.). The plasma membrane H⁺-ATPase is estimated to consume 10–15% of the ATP produced during yeast growth (Gancedo and Serrano 1988). During glucose fermentation in the absence of growth, the enzyme may consume up to 50% of the ATP produced because up to one H⁺ is extruded per glucose molecule fermented (Serrano 1980).

Taking into account the concentration of solutes accumulated into vacuoles (10^{-2} to 10^{-1} M) and that the vacuoles represent approximately 25% of the cell volume (Wiemken and Durr 1974), vacuolar transport systems are expected to operate with fluxes of 0.05–0.5 nmole $\times$ min^{-1} $\times$ mg yeast^{-1}. Therefore, they are less active than the major transport systems of the plasma membrane discussed above and are presumably present in low abundance in the cells.

B. Types of Transport Mechanisms and Their Distinguishing Characteristics

The transport mechanisms identified to date in yeast vacuolar and plasma membranes are diagramed in Figure 1. Number 1 indicates the two pri-

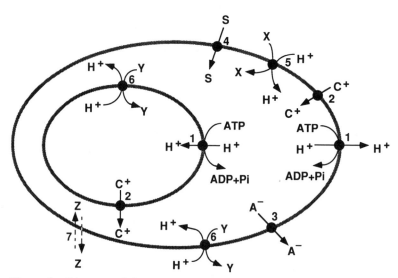

Figure 1   Diagram of the transport mechanisms known to operate in yeast. (*1*) H⁺-ATPases; (*2*) cation (C⁺) channels; (*3*) anion (A⁻) channels; (*4*) facilitated diffusion (equilibrative permeases) of sugars (S); (*5*) H⁺-symports (X = sugars, amino acids, P$_i$, SO$_4^{2-}$, purines, pyrimidines, vitamins, carboxylates); (*6*) H⁺-antiports (Y = Ca⁺⁺, S-adenosylmethionine, P$_i$ or amino acids in the vacuole, and Na⁺ or Ca⁺⁺ in the plasma membrane); (*7*) passive diffusion (Z = O₂, CO₂, ethanol, or undissociated carboxylic acids).

Table 1 Biochemical properties of the two types of H⁺-ATPases

	(E-P) H⁺-ATPases[a]	(F$_0$F$_1$) H⁺-ATPases[b]
Membrane	plasma membranes of fungi, plants, and protozoa	eubacteria, mitochondria, and chloroplasts archaebacteria and vacuolar compartment of eukaryotes
Subunits	catalytic of 100 kD; probably oligomer; may have accessory subunits	F_1 *part*: 3 copies of catalytic (50–80 kD) and 3 copies of interacting (55–60 kD) subunits F_0 *part*: 6–12 copies of proteolipid (8–16 kD) both parts may have other accessory subunits
Inhibitors	vanadate (10^{-5} M)	dicyclohexylcarbodiimide (10^{-5} M)
pH optimum	5.5–6.5	7–9
Mechanism	phosphorylated intermediate; monomer is active	no phosphorylated intermediate; obligatory cooperativity between active sites
H⁺/ATP	1–2	2–4

[a] (E-P) ATPases, ATPases with a phosphorylated intermediate. Also called P-ATPases (Pedersen and Carafoli 1987a,b), but the (E-P) nomenclature is more informative.

[b] (F$_0$F$_1$) ATPases, ATPases with distinct F$_0$ (hydrophobic) and F$_1$ (polar) parts. Also called F-ATPases (Pedersen and Carafoli 1987a,b), but the (F$_0$F$_1$) nomenclature is more informative. Vacuolar ATPases were previously considered to be a separate group (V-ATPases; Pedersen and Carafoli 1987a,b), but recent evidence from amino acid sequences suggests that they belong to the same group as the coupling factors from bacteria, mitochondria, and chloroplasts (see Section V.B).

mary pumps or chemiosmotic systems (Mitchell 1966): the plasma membrane H⁺-ATPase and the vacuolar H⁺-ATPase. The basic biochemical properties of these two types of proton pumps are summarized in Table 1. These enzymes convert the chemical energy of ATP into the physical energy of the proton gradient. This gradient then drives the movement of other molecules catalyzed by secondary active transport systems. The simplest of these are ion channels (numbers 2 and 3), which allow the passage of cations and anions driven by the membrane potential. Transmembrane channels result from the transient opening of polar pores across transmembrane proteins. They contain a fixed binding site that determines substrate specificity (Latorre and Miller 1983). K⁺ channels have been identified in both vacuolar and plasma membranes, and there is evidence for an anion channel and for channels specific for divalent cations in the plasma membrane (see below).

The fixed nature of their binding sites differentiates channels from permeases. The latter contain a binding site that alternates between two conformations in which it is exposed to one or the other side of the membrane. Therefore, the permeases, but not the channels, exhibit the phenomenon of counterflow. This consists of the acceleration of the unidirectional flux of labeled substrate by a flux of nonlabeled substrate in the opposite direction (Stein 1986). This acceleration results from the fact that the conformational change needed for changing the exposure of the binding site is rate-limiting and accelerated when the binding site is occupied. Therefore, the "return" of the binding site to the side of the labeled substrate is accelerated by the presence of nonlabeled substrate on the opposite side of the membrane. The requirement for slow conformational changes explains the low turnover numbers of permeases (10^3 min^{-1}) and their high-temperature dependence (Q_{10} = 2–3). In contrast, channels exhibit no counterflow; indeed, flux in one direction would be inhibited by flux in the opposite direction because substrate molecules on both sides of the membranes would compete for the same binding site.

The lack of gross conformational changes during transport explains the high turnover numbers of channels (10^7 to 10^{10} min^{-1}) and their low-temperature dependency (Q_{10} = 1.2–1.4). Pumps like the H$^+$-ATPases suffer from the same limitations as permeases because conformational changes determining the alternate access of binding sites at opposite sides of the membrane are also needed (Hammes 1982; Tanford 1982). Accordingly, pumps exhibit low turnover numbers similar to those of the permeases.

Facilitated diffusion (number 4) refers to a permease without energy coupling and which therefore cannot operate against the concentration gradient of the substrate. This implies the transport of neutral molecules (to avoid coupling to the electrical membrane potential) without concomitant transport of other molecules (to avoid coupling to other gradients). The permeases for glucose and galactose that have been identified in *Saccharomyces cerevisiae* are not coupled to any form of energy and result in equilibrative, passive transport. Proton symports (number 5) are permeases that catalyze the cotransport of protons and other molecules such as maltose, amino acids, phosphate, sulfate, purines, pyrimidines, lactate, and acetate. Proton antiports (number 6) are permeases that catalyze the exchange of protons for Na$^+$ or Ca^{++} in the plasma membrane and for Ca^{++}, phosphate, *S*-adenosylmethionine, or amino acids in the vacuolar membrane. Both kinds of systems coupled to protons can operate against the concentration gradient of the substrate.

One transport mechanism not identified in yeast or other eukaryotes is the vectorial phosphorylation by phosphoenolpyruvate of transported sugars, as found widely in bacteria (Saier et al. 1988). The ouabain-sensitive Na^+, K^+-ATPase typical of animal cells (Skou 1988) has not been identified in either fungi or plants. Ca^{++}-ATPases are present both in plasma membranes and in internal membranes of animal cells (Carafoli 1988). Although they have not yet been identified in yeast, their presence in plant cells (Briars et al. 1988) and the fact that fungi share most bioenergetic processes with plants (Serrano 1985) suggest that Ca^{++}-ATPases could be present also in yeast. This suggestion is supported by the recent identification (Rudolph et al. 1989) of two genes in yeast, *PMR1* and *PMR2*, that encode proteins that have some homology with animal Ca^{++}-ATPases from the sarcoplasmic reticulum and plasma membranes, respectively (see below).

Finally, small molecules, such as H_2O, O_2, CO_2, and ethanol, and hydrophobic molecules, such as undissociated carboxylic acids, can cross the membranes by passive diffusion, with the participation of neither permeases nor channels. The permeability coefficient of the former small molecules across lipid bilayers is greater than 10^{-2} cm^2 x min^{-1} (Stein 1986). As 1 mg of yeast contains about 30 cm^2 of plasma membrane surface (3 x 10^7 cells/mg; about 100 μm^2 surface/cell), passive fluxes greater than 300 nmoles x min^{-1} x mg yeast^{-1} are expected with a concentration gradient of only 1 mM. This permeability is more than enough to explain the observed fluxes of these kinds of molecules with moderate (in the millimolar range) concentration gradients.

III. SOME MISCELLANEOUS FEATURES OF TRANSPORT PHENOMENA IN YEAST

A. Reconstitution of Plasma Membrane Vesicles with Liposomes for Transport Studies

Plasma membrane vesicles from many organisms can be obtained in a sealed form suitable for transport studies. Unfortunately, yeast plasma membrane vesicles are leaky to ions and sugars (Fuhrman et al. 1976). Franzusoff and Cirillo (1983) solved this problem by the addition of soybean phospholipid liposomes followed by freezing, thawing, and brief sonication. This procedure induces the fusion of the vesicles with the liposomes and somehow reduces the passive permeability of the former. The procedure has been applied successfully to the study of the facilitated diffusion of sugars (Franzusoff and Cirillo 1983), of ATP-driven

H+ transport and membrane-potential-driven K+ transport (Calahorra et al. 1987), and of membrane-potential-driven leucine transport (Opekarova et al. 1987). These vesicle studies have confirmed the mechanisms of energy coupling for H+, K+, and amino acid transport deduced from whole-cell studies (Serrano 1985). In contrast, vacuolar membrane vesicles can be obtained in a form directly suitable for transport studies (Uchida et al. 1988a).

B. Estimates of Membrane Potential

A major hindrance to definition of the bioenergetics of active transport in yeast is the lack of a reliable method for quantitating membrane potential. Microelectrodes cannot be used because impalement of yeast cells introduces an irreversible leak that discharges the membrane within milliseconds (Bakker et al. 1986). On the other hand, the accumulation of tetraphenylphosphonium by cells of *S. cerevisiae* does not seem to be in equilibrium with the membrane potential, so that it provides only a qualitative estimate of this parameter (Eraso et al. 1984; Bakker et al. 1986; Vallejo and Serrano 1989). The claims that microelectrodes (Lichtenberg et al. 1988) and tetraphenylphosphonium (Prasad and Hofer 1986) provide valid measurements of membrane potential in *Pichia* and *Candida*, respectively, must be taken with caution because of the problems discussed above and also because the reported values (around 100 mV) are much lower than those expected from the electrogenic properties of the plasma membrane H+-ATPase. Indeed, values close to 200 mV have been determined in *Neurospora crassa*, plants, and algae (Serrano 1985), and extrapolation of the depolarization time course of impaled yeast cells also suggests a value on this order (Bakker et al. 1986).

Studies with the fluorescent probe 3,3'-dipropylthiodicarbocyanine (Kovac and Varecka 1981) revealed an additional problem: A major part of the composite membrane potential monitored in intact yeast cells appears to be represented by the membrane potential of mitochondria. It is likely that this complication also applies to measurements made with tetraphenylphosphonium.

C. Effect of Ethanol on the Proton Permeability of Membranes

S. cerevisiae tolerates much higher concentrations of ethanol than most other microorganisms, but this compound also limits yeast growth under many circumstances (D'Amore and Stewart 1987). Ethanol increases the passive permeability of the yeast plasma membrane to H^+ and thus collapses the electrochemical proton gradient and inhibits active transport processes (Leão and Van Uden 1984). This effect of ethanol at the plasma membrane seems to be more important for its toxicity than are its effects on intracellular enzymes (Jimenez and Van Uden 1985; Pascual et al. 1988). It seems that ethanol, like many anesthetics, increases the number of transient, ion-conducting defects normally present in the bilayer structure of membranes (Barchfeld and Deamer 1988). In agreement with this interpretation, increases in plasma membrane unsaturated fatty acids result in increased yeast ethanol tolerance (D'Amore and Stewart 1987).

D. Nonspecific Effects of ATPase Inhibitors on Whole Cells

In animal cells, cardiac glycosides like ouabain are fully specific inhibitors of the Na^+,K^+-ATPase, and their use has been crucial in clarifying the physiological role of this enzyme (Skou 1988). Unfortunately, there are no known inhibitors of yeast vacuolar or plasma membrane H^+-ATPases specific enough for in vivo studies. Some ATPase inhibitors, such as Dio-9, miconazole, dicyclohexylcarbodiimide, and diethylstilbestrol, may also disrupt the permeability barrier of the cells (Borst-Pauwels et al. 1983). In addition, most of them inhibit indiscriminately mitochondrial, vacuolar, and plasma membrane ATPases so that the interpretation of their in vivo effects is complicated (Goffeau and Slayman 1981). The utilization of respiratory-deficient mutants and of low concentrations of inhibitors may partially obviate these problems (Serrano 1980). However, in general, the results of whole-cell studies employing ATPase inhibitors must be viewed with caution, particularly if nonspecific effects on membrane permeability and energy metabolism are not controlled. This limitation is one of the main reasons for performing mutational analysis of ATPases (Serrano 1989). The genetic approach can provide the specific manipulation of enzyme activity required for in vivo studies (see below).

Bafilomycin has recently been proposed to be a specific inhibitor of vacuolar ATPases (E.J. Bowman et al. 1988a), but its suitability for whole-cell studies has not been demonstrated.

E. Low Activity of Ionophores in Yeast Plasma Membranes

For unknown reasons, ionophores are not very active in yeast plasma membranes, and high concentrations are required to dissipate ion gradients (Kovac et al. 1982). Polyene antibiotics at low concentrations (0.1

µg/mg cells) may be the most active and specific ionophores for deenergizing the yeast plasma membrane (Palacios and Serrano 1978). The high concentrations (10^{-4} to 10^{-3} M) of proton ionophores (dinitrophenol, hydrazone derivatives) needed to inhibit active transport in yeast produce a massive uptake of protons in exchange for intracellular K^+. Therefore, part of the observed inhibition can be explained by intracellular acidification and not by uncoupling. This secondary effect can be avoided by working at high external pH (about 6) and at high external K^+ (Palacios and Serrano 1978).

F. Irreversibility of H^+-symports and Requirement for an H^+ Gradient Even for Downhill Transport

One unusual feature of the bioenergetics of the yeast plasma membrane is that H^+-symports are irreversible, operating only in the direction of uptake and showing no significant efflux after uncoupling or removal of the external substrate (Eddy 1982; Horak 1986). This seems to be caused both by compartmentation into the vacuole (Cooper 1982) and by the kinetic properties of the permeases (Horak 1986). A second unusual feature is that the H^+-symports require an H^+ gradient to operate even under conditions of "downhill" transport (Eddy 1982). This is very apparent in the case of maltose transport, because a large excess of intracellular maltase prevents the accumulation of intracellular maltose, but the uptake of this sugar is nevertheless blocked by ionophores (Palacios and Serrano 1978). Therefore, there is a kinetic requirement for the proton gradient for downhill transport, in addition to the thermodynamic requirement during uphill transport. This phenomenon is not observed in either bacteria or animal cells, where active transport is reversible and only dependent on ion gradients for uphill transport (Stein 1986). The differences in energy requirements for downhill transport and for accumulation against a concentration gradient have recently been illustrated with yeast mutants expressing reduced levels of plasma membrane H^+-ATPase (Vallejo and Serrano 1989): The initial rates of transport are much more sensitive to membrane energization than are the final accumulation levels.

G. Utilization of the Secretory Pathway by Transport Proteins

Experiments using secretory mutants have demonstrated that several permeases (Tschopp et al. 1984) and the plasma membrane H^+-ATPase

(Holcomb et al. 1988) are externalized by the same secretory vesicles that transport periplasmic enzymes. However, the amino acid sequences of the permeases and the ATPase do not contain amino-terminal signal sequences (see below). Therefore, membrane insertion is probably mediated by internal sequences. In the case of the *Neurospora* plasma membrane H^+-ATPase, the only processing occurring during biosynthesis is the removal of the initiator methionine and the acetylation of the resulting amino-terminal alanine (Aaronson et al. 1988).

H. Specific ATPases as Markers for the Vacuolar and Plasma Membranes

In the first attempts to purify yeast plasma membranes, a morphological marker was employed (Matile et al. 1967; Wehrli et al. 1975). This was the presence of volcano-like particles of about 13.5 nm in the cytoplasmic fracture face of the yeast plasma membrane (Kubler et al. 1978). At present, the plasma membrane H^+-ATPase seems to provide the most convenient marker, because this activity can be assayed specifically even in crude fractions (Serrano 1988b). Although chitin synthase has also been proposed as a marker for the plasma membrane (Duran et al. 1975), more recent evidence indicates that most of this activity is actually present in internal membranes (Martinez and Schwencke 1988).

α-Mannosidase has been used as a marker for the vacuolar membrane (Uchida et al. 1988a), but its enrichment during vacuolar purification is much less than expected for an activity not present in other compartments of the cell. The vacuolar H^+-ATPase could provide a more specific marker, although an immunological assay would probably be needed because of its low abundance (Uchida et al. 1988a). Other possibilities are the repressible alkaline phosphatase (Klionsky and Emr 1989) and dipeptidyl aminopeptidase B (Roberts et al. 1989).

IV. PLASMA MEMBRANE H⁺-ATPASE

A. Isolation of the *PMA1* Gene

Screening with antibodies of λgt11 expression libraries (Young and Davis 1985) has allowed the cloning of the plasma membrane H^+-ATPase genes (*PMA1*) in *S. cerevisiae* (Serrano et al. 1986a) and *N. crassa* (Addison 1986; Hager et al. 1986). The expression libraries provided only fragments of the genes; these were utilized to obtain the complete genes by hybridization screening of plasmid libraries containing large inserts. In *S. cerevisiae*, the relationship of the cloned gene to the biochemically and immunologically defined enzyme could be established by the fact that a promoter deletion greatly reduced the level of antigenic protein detected with affinity-purified ATPase antibodies and the rate of ATP hydrolysis measured by a specific assay for plasma membrane H^+-ATPase (see Serrano et al. 1986a). The same *S. cerevisiae* gene was later isolated by complementation of a recessive mutation causing resistance to the ATPase inhibitor Dio-9 (see below) (Ulaszewski et al. 1987a). Although this mutation had been proposed to affect the structural gene of the ATPase (Ulaszewski et al. 1983), the possibility that it affected another regulatory component could only be discarded after the cloning of the *PMA1* gene. The corresponding gene from *Schizosaccharomyces pombe* was isolated by homology with the *S. cerevisiae* gene (Ghislain et al. 1987).

A second ATPase gene (*PMA2*) 90% homologous to *PMA1* at the amino acid level has recently been described in *S. cerevisiae* (Schlesser et al. 1988). This gene probably corresponds to a second isoform of the plasma membrane H^+-ATPase. Although *PMA1* is essential for growth (see below) and expressed at high levels, *PMA2* is not required for yeast growth and expression cannot be detected under normal growth conditions. It remains possible that *PMA2* is expressed and is significant under some special conditions.

PMA1 is located on chromosome VII, between the centromere and *LEU1*, to which it is tightly linked (<0.1 cM, about 1.2 kb between coding regions) (Serrano et al. 1986a; Ulaszewski et al. 1987a; Mortimer et al. 1989). *PMA2* is localized on chromosome XVI, about 17 cM from *GAL4* (Schlesser et al. 1988).

Two ATPase genes distantly related to *PMA1* (only 15–20% homology at the amino acid level) have recently been isolated by Rudolph et al. (1989). *PMR1* is more related to sarcoplasmic reticulum Ca^{++}-ATPases (30% homology) and contains a segment at its amino terminus that has homology with Ca^{++}-binding sites in calmodulin ("EF hands"). In addition, a strain with a null allele exhibits a requirement for high calcium concentrations in the growth medium, and its secretory pathway is perturbed (Rudolph et al. 1989). Therefore, *PMR1* probably corresponds to the endoplasmic reticulum calcium pump of *S. cerevisiae*. *PMR2* has been proposed to correspond to the plasma membrane calcium pump of *S. cerevisiae* because of the presence of a putative calmodulin-binding site at the carboxyl terminus (Rudolph et al. 1989). However, it does not show more homology with animal plasma membrane Ca^{++}-ATPases than to other (E-P) ATPases (see below). Definitive conclusions must wait the results of mutational and biochemical analyses.

```
            I                 II                III               IV
Hsc  225 IDQSAITGES    374 ILCSDKTGTLT   472 CVKGAP     534 DPPR
Hat  176 VDQSALTGES    326 VLCSDKTGTLT   422 VSKGAP     488 DPPR
Hld  197 VDEAALTGES    347 MLCSDKTGTLT   443 VTKGAP     501 DPPR
PMR1 197 IDESNLTGEN    367 VICSDKTGTLT   499 YVKGAF     580 DPPR
PMR2 174 TDESLLTGES    365 DICSDKTGTLT   559 YGKGAF     649 DPPR
Casr 175 VDQSILTGES    347 VICSDKTGTLT   513 FVKGAP     600 DPPR
Capm 238 IDESSLTGES    471 AICSDKTGTLT   599 FSKGAS     684 DPVR
NaK  211 VDNSSLTGES    370 TICSDKTGTLT   504 VMKGAP     591 DPPR
Kec  153 VDESAITGES    303 VLLLDKTGTIT   393 IRKGSV     447 DIVK

            V                 VI
Hsc  555 KMLTGDA    630 AMTGDGVNDAPSLKKADTGIA
Hat  509 KMITGDQ    585 GMTGDGVNDAPALKKADIGIA
Hld  522 KMITGDH    601 AMTGDGVNDAPALKRADVGIA
PMR1 591 IMITGDS    676 AMTGDGVNDAPALKLSDIGVS
PMR2 670 HMLTGDF    752 TMTGDGVNDSPSLKMANVGIA
Casr 621 IMITGDN    698 AMTGDGVNDAPALKKAEIGIA
Capm 705 RMVTGDN    793 AVTGDGTNDGPALKKADVGFA
NaK  612 IMVTGDH    711 AVTGDGVNDSPALKKADIGVA
Kec  468 VMITGDN    514 AMTGDGTNDAPALAQADVAVA
```

Figure 2 Most conserved motifs of (E-P) ATPases. (Hsc) H^+-ATPase from *S. cerevisiae (PMA1*; Serrano et al. 1986a); (Hat) H^+-ATPase from *Arabidopsis thaliana* (Pardo and Serrano 1989); (Hld) H^+-ATPase from *Leishmania donovani* (Meade et al. 1987); (PMR1 and PMR2) ATPase genes related to *PMA1* from *S. cerevisiae* (Rudolph et al. 1989); (Casr) Ca^{++}-ATPase from rabbit muscle sarcoplasmic reticulum (MacLennan et al. 1985); (Capm) Ca^{++}-ATPase from human plasma membrane (Verma et al. 1988); (NaK) Na^+,K^+-ATPase from sheep kidney (subunit α; Shull et al. 1985); (Kec) K^+-ATPase from *E. coli* (subunit B; Hesse et al. 1984).

B. Homology of the Plasma Membrane H^+-ATPase with a Large Family of Cation Pumps

The predicted amino acid sequence of the yeast plasma membrane H^+-ATPase has been compared with those of other cation-pumping ATPases that form phosphorylated intermediates ([E-P] ATPases). Amino acid identity between distant members of this enzyme family is only 15–20%. This is only slightly higher than the "noise" of about 5–10% identity that can be obtained by comparing unrelated proteins. However, the highly conserved motifs depicted in Figure 2 indicate that all of these enzymes are homologous and evolved from an ancestral pump. These regions represent about 6% of the entire sequences and could correspond to the basic catalytic machinery (see below).

A tentative evolutionary tree, constructed on the basis of the percentages of similarity between different members of the family, is depicted in Figure 3. Fungal plasma membrane H^+-ATPases are closely related (75% homology), as expected from the antigenic relationships observed previously (Blasco et al. 1983; Vai et al. 1986). Other plasma membrane H^+-ATPases from plants and protozoa are closer to these fungal ATPases

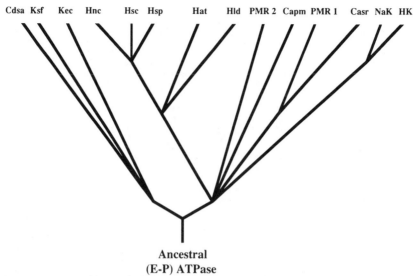

Ancestral
(E-P) ATPase

Figure 3 Tentative evolutionary tree of (E-P) ATPases based on percentages of amino acid sequence identity. (Cdsa) Cd^{++}-ATPase from *Staphylococcus aureus* (Silver et al. 1989); (Ksf) K$^+$-ATPase from *Streptococcus faecalis* (Solioz et al. 1987); (Hnc) H$^+$-ATPase from *N. crassa* (Addison 1986; Hager et al. 1986); (Hsp) H$^+$-ATPase from *S. pombe* (Ghislain et al. 1987); (HK) H$^+$,K$^+$-ATPase from rat stomach (Shull and Lingrel 1986). Other symbols are the same as those in Fig. 2. Sequence comparisons were made by the GAP program (Devereux et al. 1984). The first branching point between bacterial and eukaryotic ATPases reflects the lower level of homology between distant members of the family (15–20%). The bacterial ATPases have as the only common feature the presence of six hydrophobic stretches, whereas the eukaryotic ATPases have eight to ten stretches (Serrano and Portillo 1990). This fact justifies an ancient branching point. H$^+$-ATPases diverged early from the remainder of the eukaryotic group (at least 30% homology, 75% between fungal enzymes). The product of the *PMR1* gene of *S. cerevisiae* is slightly more related to the sarcoplasmic reticulum Ca^{++}-ATPase than to other members of the family, justifying a group of calcium pumps from internal membranes. The two K$^+$ pumps of animal cells (Na$^+$,K$^+$-ATPase and H$^+$,K$^+$-ATPase) are closely related (70% identity) and must haved diverged relatively recently. Finally, the product of the *PMR2* gene of *S. cerevisiae* and the animal plasma membrane Ca^{++}-ATPase represent up to now isolated examples of enzymes of the eukaryotic branch.

(30% homology) than to bacterial or animal (E-P) ATPases. Accordingly, there is some immunological cross-reactivity between the fungal and plant ATPases (Clement et al. 1986; Surowy and Sussman 1986) but not between fungal and animal ATPases. Therefore, plasma membrane H$^+$-

ATPases constitute an independent branch of the tree. The two types of potassium pumps of animal cells (Na+,K+-ATPases and H+,K+-ATPases) are 70% homologous and constitute another defined branch. The 30% homology between the ATPase encoded by the *PMR1* gene of *S. cerevisiae* and the sarcoplasmic reticulum Ca++-ATPases (see above) defines another branch of calcium pumps from internal membranes. On the other hand, the animal plasma membrane Ca++-ATPases and the product of the *PMR2* gene from *S. cerevisiae* are isolated examples of (E-P) ATPases (but see above for a proposed relationship between them). Finally, bacterial ATPases of this family seem to have diverged very early in the course of evolution, and the sequences available do not show any common feature, with the exception of a lower number of hydrophobic stretches (see below).

C. Topology of (E-P) ATPases

A comparison of the hydrophobicity plots of different (E-P) ATPases indicates that all of these enzymes use a similar strategy for arranging their hydrophobic regions (Serrano 1988a). Bacterial ATPases are the simplest, containing six hydrophobic stretches. Therefore, six transmembrane helices may constitute the basic pathway for cation transport. Eukaryotic ATPases seem to have from two to four additional hydrophobic stretches; the exact number is not clear in many of the sequences and will need to be determined experimentally. Figure 4 reflects some kind of consensus by assuming a total number of ten hydrophobic stretches in all eukaryotic (E-P) ATPases.

Several charged amino acids are buried in the proposed transmembrane helices of yeast ATPase and some of these are conserved in all H+-ATPases (Serrano 1989). They could form part of a polar channel across the membrane stabilized by ion pairs. Site-directed mutagenesis has demonstrated the essential role of some of these polar groups predicted to be buried inside the membrane (see below).

The amino termini of sarcoplasmic reticulum Ca++-ATPase (Reithmeier and MacLennan 1981), plasma membrane Na+,K+-ATPase (Jorgensen et al. 1982), and plasma membrane Ca++-ATPase (Sarkadi et al. 1986) seem to be located on the cytoplasmic side of the membrane. Therefore, the two central polar domains, which include the most conserved blocks of sequence, would also be located on the cytoplasmic side (Fig. 4). This is in agreement with their suggested participation in the active site (see below). The location of the carboxyl terminus is more controversial. In the plasma membrane Ca++-ATPase, it seems to be cyto-

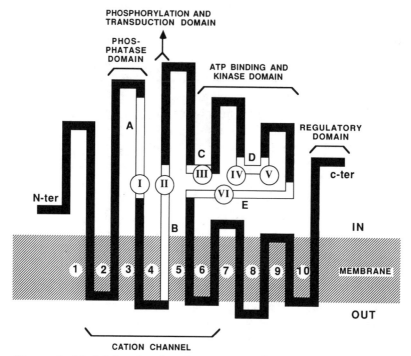

Figure 4 Model for the domain structure of yeast plasma membrane H+-ATPase. The length of the bar representing the polypeptide chain is approximately proportional to the number of amino acids in each domain. The open parts of the bar correspond to the most conserved regions (A to E) of (E-P) ATPases, representing about 200 amino acids (Serrano and Portillo 1990). The conserved motifs I–VI of Fig. 2 are also indicated. The putative transmembrane helices correspond to the ten hydrophobicity peaks of Kyte-Doolittle plots (Serrano and Portillo 1990). The first six helices are proposed to constitute the proton channel. The function of the amino-terminal domain is unknown, but the proposed functions for the other polar domains are indicated (see text). (Reprinted, with permission, from Serrano and Portillo 1990.)

plasmic and to contain a calmodulin-binding regulatory domain (Sarkadi et al. 1986; Verma et al. 1988). On the other hand, in the Na+,K+-ATPase, the carboxyl terminus seems to be extracellular (Ovchinnikov et al. 1987b, 1988). In the yeast H+-ATPase, experiments with carboxypeptidase suggest an extracellular location (Davis and Hammes 1989). However, the identification of a domain at the carboxyl terminus that regulates activity, probably by interacting with the active site (Portillo et al. 1989), suggests a cytoplasmic orientation. Recent experimental evidence with antibodies specific for the carboxyl terminus in *N. crassa* (Mandala

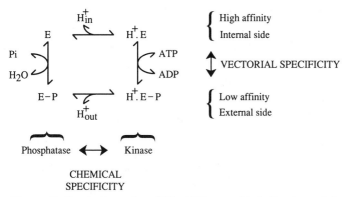

Figure 5 Catalytic cycle of H⁺-ATPase, with indication of the four basic conformational states differing in chemical and vectorial specificities.

and Slayman 1989) and *S. cerevisiae* (B.C. Monk et al., in prep.) suggests a cytoplasmic location, and this is indicated in the model of Figure 4.

D. Reaction Mechanism

As discussed recently by Jencks (1989), energy coupling between ATP hydrolysis and cation transport is effected by the existence of four basic states of (E-P) ATPases with different chemical and vectorial specificities. These states are depicted for H⁺-ATPases in Figure 5. Concerning vectorial specificity, in the unphosphorylated state, the enzyme exhibits a proton-binding site of high affinity facing the internal side of the membrane. After phosphorylation, the proton-binding site exhibits low affinity and faces the external side. With respect to chemical specificity, the unprotonated enzyme has phosphatase activity and the protonated enzyme has kinase activity. According to Hammes (1982), kinetic blocks prevent either of the two chemical conformations (unprotonated or protonated) from completing the catalytic cycle and either of the two vectorial conformations (unphosphorylated or phosphorylated) from completing the transport cycle. These blocks force the coupling between ATP hydrolysis and H⁺ transport. Although many more intermediate states are expected to occur during the catalytic cycle, these four basic conformations are enough to define energy coupling (Jencks 1989). The two basic conformations, E_1 and E_2, of previous models (Jorgensen and Andersen 1988) are equivalent to the kinase and phosphatase conformations of Jencks (1989).

The binding affinity for the proton or, more likely, hydronium ion

(Boyer 1988) would change in synchrony with the change in membrane sidedness during phosphorylation of the enzyme (Tanford 1982). Membrane-buried prolines may participate in these conformational changes because of the possibility of *cis-trans* isomerization (Brandl and Deber 1986). Accordingly, a conserved proline in transmembrane helix 4 (Pro-335) seems to be essential for activity (Portillo and Serrano 1988). The proton transport pathway across the membrane could include the essential (Portillo and Serrano 1989) Asp-730 of transmembrane helix 6 and the essential (F. Portillo and R. Serrano, unpubl.) Arg-695 of transmembrane helix 5. Other polar residues predicted to be in the membrane domain, such as Glu-703 (transmembrane helix 5) and Glu-803 (transmembrane helix 8), proved not to be essential (F. Portillo and R. Serrano, unpubl.). In the *Neurospora* ATPase, a glutamate residue in the first transmembrane helix binds the ATPase inhibitor dicyclohexylcarbodiimide (Sussman et al. 1987). However, mutagenesis of this residue in the yeast ATPase (Glu-129) indicates that it is not required for activity and that dicyclohexylcarbodiimide can bind to other sites (Portillo and Serrano 1988). It remains for future experiments to clarify the discrepancy between biochemistry and genetics in this particular case.

As previously demonstrated for animal ATPases (Jorgensen and Andersen 1988), the *Neurospora* (Addison and Scarborough 1982) and *Saccharomyces* (Perlin and Brown 1987) ATPases exhibit differential sensitivity to trypsin after ligand binding. In the *Neurospora* enzyme (Mandala and Slayman 1988), trypsin cleaves the unliganded enzyme (likely to represent the "kinase" or E_1 conformation) at three sites near the amino terminus. In the presence of vanadate and magnesium, which stabilize the "phosphatase" or E_2 conformation of the enzyme, one of these sites is protected, and a new cleavage site near the carboxyl terminus is exposed. Therefore, the ATPase undergoes significant conformational changes at both termini of the polypeptide chain during its catalytic cycle. Circular dichroism studies indicate that, like animal (E-P) ATPases, the *Neurospora* ATPase consists of 36% helix and 20% β-sheet and that there is no significant change in secondary structure during the E_1-E_2 transition (Hennessey and Scarborough 1988).

The rate constant of phosphoenzyme formation has been investigated with the quenched-flow method and is higher than 10^4 min^{-1} (Smith and Hammes 1988). As the turnover number is on the order of 10^3 min^{-1} (Serrano 1988a), phosphorylation is clearly not rate-limiting. The rate constant of phosphoenzyme hydrolysis has been estimated from the P_i-[^{18}O]H_2O exchange catalyzed by the *S. pombe* enzyme (Amory et al. 1982) and it is about 5×10^5 min^{-1}. As phosphorylation occurs in the kinase conformation and hydrolysis occurs in the phosphatase conformation, neither the conformational change nor the hydrolysis of the intermediate is rate-limiting. On the other hand, the release of bound P_i after hydrolysis is much slower (4×10^3 min^{-1}) and could be kinetically significant (Amory et al. 1982). As in animal ATPases (Jorgensen and Andersen 1988), it is likely that conformational changes in unphosphorylated enzyme ($E_2 \rightarrow E_1$ transition in the classical model or $E \rightarrow H^+.E$ in Fig. 5) are also rate limiting.

With an enzyme preparation that shows optimal ATP hydrolysis at pH 6 and 1 mM free Mg^{++}, the steady-state level of the phosphorylated intermediate is only 6% under these conditions. The intermediate accumulates to about 70% of total enzyme at pH 5.5 and 0.5 mM free Mg^{++}. This suggests that phosphoenzyme hydrolysis is optimal at higher pH and higher free Mg^{++} than required for the optimal catalytic cycle (Smith and Hammes 1988).

Phosphorylation of the enzyme by P_i, by reversal of the phosphatase reaction, is very low, amounting to only 0.02% of the enzyme molecules. This is increased about fivefold by ATP. Presumably, the enzyme in the absence of ligands exists in the kinase conformation, and only the phosphatase conformation can be phosphorylated by P_i. Phosphorylation by ATP drives the complete cycle and increases the amount of phosphatase conformation (Amory et al. 1982). The low level of phosphorylation by P_i, even in the presence of ATP, explains the very low rates of $^{32}P_i$-ATP exchange ($\sim 10^{-5}$ of the ATPase activity; Malpatida and Serrano 1981a; De Meis et al. 1987). This exchange is increased about 100-fold (to 20 nmoles $\times$ min^{-1} $\times$ mg protein^{-1}) in reconstituted vesicles energized by a proton gradient (Malpartida and Serrano 1981a), suggesting that the level of phosphatase conformation and/or the affinity for P_i increases under these conditions. A more modest increase in exchange rate (to about 2 nmoles $\times$ min^{-1} $\times$ mg protein^{-1}) can be obtained without a proton gradient at high pH and in the presence of dimethylsulfoxide (De Meis et al. 1987).

The existence of an essential binding site for free Mg^{++}, as deduced from kinetic studies (Ahlers 1981), has recently been confirmed by measuring terbium binding to the enzyme from *S. pombe* (Ronjat et al. 1987).

E. A Preliminary Overview of the Active Site of (E-P) ATPases

Chemical modification of animal (E-P) ATPases with ATP analogs and site-directed mutagenesis of yeast H^+-ATPase and the sarcoplasmic reticulum Ca^{++}-ATPase (Maruyama and MacLennan 1988) have identified several residues as likely components of the active site. Table 2

Table 2 Conserved motifs of (E-P) ATPases and their proposed functions

Motif	Proposed functions
I. TGES	hydrolysis of intermediate (mainly E, also T); preventing uncoupled ATP hydrolysis (E and S); vanadate binding (T)
II. CSDKTGTLT (CS only in eukaryotes)	phosphorylated intermediate (D); couping ATP hydrolysis to H^+ transport (D); vanadate binding (C, D, K)
III. KGAP (AP only in eukaryotes)	phosphorylation activity (K); ATP binding (phosphate part, K); FITC[a] binding (K)
IV. DPPR (PP only in eukaryotes)	ATP binding (adenine part, D)
V. M(L,I,V)TGD	ATP binding (adenine part, D)
VI. GDGXNDXP(A,S)LK (K only in eukaryotes)	ATP binding (adenine part, second D); ATP binding (phosphate part, first D); vanadate binding (first D); FSBA[b] binding (K); CIRATP[c] binding (first D)

See text for references. *X* indicates any amino acid.
[a] FITC is fluorescein 5 ′ -isothiocyanate.
[b] FSBA is 5 ′ -(*p*-fluorosulfonyl)benzoyladenosine.
[c] CIRATP is γ-[4-(*N*-2-chloroethyl-*N*-methylamino)]benzylamide ATP.

summarizes the proposed functions of these residues, and Figure 6 depicts a model for the active site and mechanism of (E-P) ATPases. The models of Figures 4 and 6 are complementary to models developed previously by Brandl et al. (1986) and Taylor and Green (1989). The present model incorporates recent results of chemical modification and mutagenesis studies that suggest a revision of the functions previously ascribed to the different domains.

In the E_1 or kinase conformation, the enzymes exhibit high affinity for ATP and a kinase activy that transfers the γ-P of ATP to the aspartate residue of motif II. Aspartate groups of motifs IV and V and the second aspartate of motif VI seem to participate in adenine binding, because mutations at these residues greatly decrease phosphorylation activity and result in altered nucleotide specificity (Portillo and Serrano 1988). In addition, the adenosine derivative FSBA (5 ′ -[*p*-fluorosulfonyl]benzoyladenosine) labels a lysine close to the second aspartate of motif VI (Ohta et al. 1986). The first aspartate of motif VI seems to bind to the γ-P of ATP because it is labeled by CIRATP (Ovchinnikov et al. 1987a). This

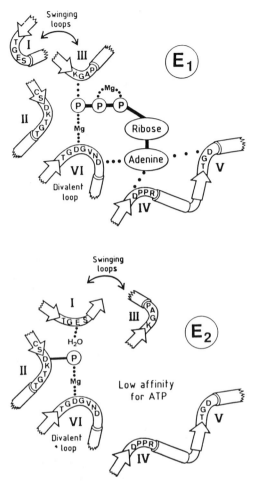

Figure 6 Model for the active site and mechanism of (E-P) ATPases. Motifs I–VI are the conserved motifs of Fig. 2 and Table 2. E_1 is the conformation catalyzing the formation of the phosphorylated intermediate (kinase reaction), and E_2 is the conformation catalyzing the hydrolysis of the phosphorylated intermediate (phosphatase reaction). Mutual replacement of the "swinging loops" I and III is proposed to mediate the change in catalytic specificity between the two conformations. Loop VI is called "divalent" because it seems to participate in both parts of the catalytic cycle. (Reprinted, with permission, from Serrano and Portillo 1990.)

binding could involve a Mg^{++} bridge. The lysine of motif III is also part of the active site because it is labeled by FITC (fluorescein isothiocyanate) and protected from this labeling by ATP (Mitchinson et al. 1982; Pardo and Slayman 1988). Mutagenesis of this residue in both the yeast (Portillo and Serrano 1988, 1989) and sarcoplasmic reticulum

(Maruyama and MacLennan 1988) ATPases greatly decreases phosphorylation activity, but no change in nucleotide specificity of the yeast ATPase is observed (Portillo and Serrano 1989). Therefore, this residue may also bind to the phosphate groups of ATP. It could function as an electrostatic catalyst to stabilize the negative charge developed during the formation of a pentacovalent intermediate (Fersht 1985).

In the E_2 or phosphatase conformation, the enzymes have low affinity for ATP and exhibit phosphatase activity. In the animal Na^+,K^+-ATPase and Ca^{++}-ATPase, in addition to the phosphorylated intermediate, exogenous p-nitrophenylphosphate is also hydrolyzed (Jorgensen and Andersen 1988). No such activity has yet been detected with the yeast plasma membrane H^+-ATPase (R. Serrano, unpubl.). Replacement of motif III by motif I could explain this change in catalytic specificity, and therefore these loops are designated as "swinging loops." Site-directed mutagenesis has demonstrated that the glutamate of motif I is essential for the hydrolysis of the phosphorylated intermediate (Portillo and Serrano 1988). It could function as a general base catalyst to promote the attack of water on the phosphate bond. The first aspartate of motif VI is also important for hydrolysis of the phosphorylated intermediate (Portillo and Serrano 1989), and it could play an accessory role by binding Mg^{++} and stabilizing the negative charge developed during the formation of a pentacovalent intermediate. Loop VI seems to participate in both kinase and phosphatase activities, and it is therefore designated as a "divalent loop."

The conserved motifs of Table 2, which probably form part of the nucleotide-binding site of (E-P) ATPases, are not related to the consensus sequences of a large family of nucleotide-binding proteins that include (F_0F_1) ATPases (Pain 1986; see below). Therefore, the nucleotide-binding site of (E-P) ATPases probably evolved independently from that of this large family.

F. Oligomeric Structure and Interacting Proteins

Despite the fact that monomers of the *Neurospora* plasma membrane H^+-ATPase seem to be active (Goormaghtigh et al. 1986), both this and the *S. cerevisiae* enzyme exhibit sigmoidal kinetics (Bowman 1983; Koland and Hammes 1986), suggesting cooperative interactions between active sites. The observation of intragenic complementation in partially defective mutants (McCusker et al. 1987) also suggests that the yeast enzyme is multimeric under physiological conditions. A stable, hexameric form of the solubilized *Neurospora* ATPase has been demonstrated (Chadwick et al. 1987), and the solubilized ATPase of *S. pombe* also exists as an

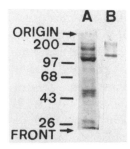

Figure 7 Polypeptide composition of yeast plasma membranes. Plasma membranes were purified from exponentially growing cells as described by Serrano (1988b), and their polypeptides were separated by SDS-PAGE (8% separating gel). (*A*) Stained for protein with Coomassie blue R-250; (*B*) stained for carbohydrate with concanavalin A and horseradish peroxidase. The position and molecular masses (in kD) of standards are indicated (R. Serrano and B. Monk, unpubl.).

oligomer of eight to ten subunits (Dufour and Goffeau 1980). The problem with these studies is, of course, that the oligomeric state could be induced by the solubilization procedure.

In *S. cerevisiae*, there are several additional observations that could be rationalized by the existence of an oligomeric state of the ATPase with a tendency to form higher-order structures. First, the H^+-ATPase is a major component of the yeast plasma membrane (100 kD band in Fig. 7A) and thus may be related to the intramembrane particles of 13.5 nm visualized by freeze-fracture electron microscopy on the cytoplasmic fracture face (Kubler et al. 1978). These particles have an abundance of 10^3 to 10^4 per μm^2 of yeast surface (Takeo et al. 1976). This is in the range estimated for ATPase molecules (10^6 per cell or 10^4 per μm^2 of cell surface). The large size of the intramembrane particles and their central hole (Kubler et al. 1978) suggest that they could be hexamers of the plasma membrane ATPase, as identified with solubilized enzyme (Chadwick et al. 1987). Indeed, recent digital image processing of the electron micrographs clearly demonstrate a hexameric structure (H. Gross, pers. comm.). These intramembrane particles have a tendency to form hexagonal arrangements, mostly in stationary-phase cells, and cover most of the surface of the yeast cells (Takeo et al. 1976; Kubler et al. 1978).

From the biochemical point of view, the yeast plasma membrane is unusual because of its refractoriness to detergent solubilization (Navarrete and Serrano 1983). Standard detergents such as Triton X-100 and bile salts can only solubilize up to 20% of the protein and up to 50% of the phospholipids of this membrane. Only zwitterionic detergents with long hydrophobic tails can effectively disperse yeast plasma membranes;

apparently, strong protein-protein interactions stabilize this membrane. After dispersion with zwitterionic detergents, the H$^+$-ATPase is in the form of large aggregates (relative molecular mass >10^6 kD) that are mostly pelleted during glycerol gradient centrifugation (Serrano 1988b). Interestingly, in stationary-phase cells, this phenomenon is exaggerated, the ATPase being more difficult to solubilize and existing as even larger aggregates after detergent treatment (R. Serrano, unpubl.). This behavior is very reminiscent of that of the intramembrane particles visualized by freeze-fracture electron microscopy and suggests that the ATPase may have an important structural role in the yeast plasma membrane. Accordingly, mutants with reduced expression of the enzyme exhibit abnormal morphologies (Vallejo and Serrano 1989).

A rigorous analysis of the *Neurospora* H$^+$-ATPase has demonstrated that purified preparations containing only the catalytic subunit of 100 kD are active as ATP-driven proton pumps after reconstitution into liposomes (Scarborough and Addison 1984). Therefore, fungal H$^+$-ATPases do not seem to require other subunits for activity. Other members of the family of (E-P) ATPases, such as the K$^+$-ATPase of *Escherichia coli* (Hesse et al. 1984) and the Na$^+$,K$^+$-ATPase of animal cells (Jorgensen and Andersen 1988), do require additional subunits. The single accessory subunit of the Na$^+$,K$^+$-ATPase (the so-called β subunit) has partial homology with one of the accessory subunits of the bacterial enzyme (the KdpC subunit; Shull et al. 1986). It remains possible that fungal H$^+$-ATPases, although active without accessory subunits, may exist in vivo in association with other membrane proteins. For example, the yeast ATPase is strongly associated with the major glycoprotein of the plasma membrane (Fig. 7). This is only weakly stained by Coomassie blue and cannot be separated from the ATPase during solubilization and gradient centrifugation (Vai et al. 1986). The intramembrane particles discussed above are also associated with the major plasma membrane glycoprotein (Maurer and Muhlethaler 1981). It is tempting to speculate that this complex is made of both the ATPase and the glycoprotein. Recent experimental results indicate that the activity of the H$^+$-ATPase in a mutant devoid of the glycoprotein is similar to that of wild type (Serrano et al. 1991). Therefore, the physiological significance of the association between the ATPase and the glycoprotein remains unknown.

G. Plasma Membrane H$^+$-ATPase Mutants

Pleiotropic mutations causing resistance to cationic drugs were first obtained by Roon et al. (1978) and more recently by Ulaszewski et al. (1983, 1986, 1987a,b) and by McCusker et al. (1987). The growth of these mutants is resistant to Dio-9, protamine, deacylated chitin, ethidium bromide, hygromycin B, and geneticin (G418). The reduced activity of the plasma membrane H^+-ATPase in these mutants suggested some relationship to either the ATPase itself or some regulatory protein. After the cloning of the plasma membrane ATPase gene *PMA1* (Serrano et al. 1986a), it became clear that these pleiotropic mutations are located in this gene.

All of these mutations cause slow growth and a decrease in active nutrient transport. Therefore, it has been suggested (McCusker et al. 1987) that the uptake of cationic drugs is reduced in mutants with reduced plasma membrane H^+-ATPase activity. Indeed, a defect in membrane potential has been measured in some of these mutants (Perlin et al. 1988), and the requirement for membrane potential for uptake of hygromycin B has been demonstrated (Vallejo and Serrano 1989). Many of these ATPase-deficient mutants also display alterations either in the kinetic properties of the enzyme (Ulaszewski et al. 1983) or in its transport activity (Perlin et al. 1988).

Some conclusions reached with these mutants have recently been confirmed using genetically engineered strains expressing either reduced amounts of wild-type ATPase (Vallejo and Serrano 1989) or normal amounts of low-activity mutant forms (Portillo and Serrano 1989). The increased sensitivity of ATPase-deficient mutants to acidification of the medium suggests that the ATPase is the major cellular system for pH regulation and that proton diffusion into the cells in acidic media may limit growth. The inhibition of active transport in these mutants also demonstrates that the ATPase is the major driving force for nutrient uptake.

These two important functions, pH regulation and nutrient uptake, may explain the correlation between ATPase activity and growth rate observed in several studies (Serrano et al. 1986a; Portillo and Serrano 1989; Vallejo and Serrano 1989). At least at low environmental pH, the plasma membrane H^+-ATPase is rate-limiting for yeast growth; any decrease in ATPase activity is associated with a decrease in overall growth rate. Intracellular pH has been measured in such cells; in media buffered at pH 4, decreased ATPase activity correlates with intracellular acidification to pH values lower than 6, which are known to be detrimental for growth (yeast growth is optimal at intracellular pH values in the range 6.8–7.3, decreasing about 50% at a value of 6.5; Gillies 1982). In contrast, when the medium is buffered at pH 6, even ATPase-deficient strains maintain an intracellular pH higher than 6.8, so that either defects in nutrient uptake or other mechanisms must explain the growth defects observed with certain mutant strains. One important clue is that at pH 6, a reduction in

ATPase activity caused by decreased expression of wild-type enzyme is much less detrimental than the same decrease in activity caused by the normal expression of a mutated ATPase. One plausible explanation for this discrepancy is that some important membrane system could be tightly associated with the ATPase and locally energized by it, instead of being energized by the overall energization of the membrane. Therefore, after reduction of the number of ATPase molecules, the overall energization of the membrane will decrease, but the tightly associated system would still be functional. It is hoped that suppressor analyses will help to clarify the cellular processes involved in growth control by the plasma membrane ATPase. As discussed below, the K^+ channel is a good candidate for such an ATPase-associated system.

Temperature-sensitive mutations of the yeast ATPase have been obtained by random mutagenesis of the cloned *PMA1* gene (Serrano et al. 1986b; Cid and Serrano 1988). ATP hydrolysis in vitro and proton efflux and amino acid uptake in vivo are all thermolabile, confirming the crucial participation of the *PMA1* product in all of these phenomena. One problem with these mutants is that growth arrest at the nonpermissive temperature (37°C) takes a very long time (7–10 hr) and does not result in any homogeneous morphology. Only a slight accumulation of unbudded cells and of cells with abortive buds has been noticed. Presumably, this means that the plasma membrane H^+-ATPase is required throughout the cell cycle. However, it is also possible that the mutations are too leaky to be informative, so that the observations should be repeated using tighter temperature-sensitive *pma1* mutations.

A yeast strain has also been constructed in which the constitutive *PMA1* promoter has been replaced by a galactose-dependent promoter (Cid et al. 1987). These cells can grow only in galactose medium; after turning off *PMA1* expression in glucose medium, there is residual growth of about three cell doublings until the preformed ATPase has been diluted to 10–15% of wild-type levels. Growth arrest under these conditions has a characteristic phenotype: Most of the culture appears as large mother cells surrounded by multiple, small, nucleated buds or daughter cells. Therefore, it seems that turning off ATPase expression impairs cytokinesis or separation and growth of daughter cells but that new rounds of budding can be started by the mother cell. One plausible explanation for this bizarre phenomenon is that the ATPase-containing plasma membrane is preferentially retained by the mother cell during budding. As the buds (and eventually daughter cells) are deficient in ATPase, they cannot grow and remain attached to the mother cell, which can itself initiate a new budding cycle because it has conserved most of its initial ATPase.

H. Regulation of ATPase Activity

The amounts of plasma membrane H^+-ATPase protein and *PMA1*
mRNA are relatively constant during the growth of yeast cells under various conditions. Only at late stationary phase are small decreases
(20–40%) of both mRNA and protein observed (Eraso et al. 1987). As
the same reduction is observed in galactose-grown cells, these changes
may be due to exhaustion of the glucose. These small effects do not
preclude considering the *PMA1* promoter as constitutive, but they may
be explained by the fact that the *PMA1* promoter contains two upstream
activating sequences (ACCCATACA) recognized by the glucose-
modulated transcription factor TUF (Capieaux et al. 1989). The constitutive promoter of *PMA1* is very strong, of strength similar to that of the
GAL1 promoter in the presence of galactose (C.G. Vallejo and R. Serrano, unpubl.).

In contrast, the activity of the plasma membrane ATPase undergoes
drastic changes depending on the physiological state of the cells. Two
factors are known to modulate the activity of the enzyme: glucose and
acid pH. Glucose metabolism produces the following reversible changes.
It increases V_{max} about threefold, reduces K_m (from 1.5–2 to 0.3–0.6
mM), increases the pH optimum (from 5.6 to 6–7), and reduces K_i for
vanadate (from 10–20 to 1–2 μM) (Serrano 1983). Therefore, common
practices such as washing the cells with water or adding sucrose to the
homogenization medium in cells with invertase activity could dramatically change the properties of the enzyme. Exposure of yeast cells to a
pH of less than 4 activates the ATPase by reducing its K_m about twofold,
other kinetic parameters remaining unaltered (Eraso and Gancedo 1987).
These two regulatory mechanisms appear to explain the changes in ATP-
ase activity during the growth of yeast in different media (Serrano 1983;
Eraso and Gancedo 1987; Eraso et al. 1987; Francois et al. 1987). The
activity decreases when the glucose is exhausted from glucose-
containing media, and it is always low in media with galactose or
gluconeogenic carbon sources. In addition, the activity increases when
the media become acidic, as during the late exponential phase in non-
buffered media with ammonia as nitrogen source. As glucose metabolism
produces intracellular acidification (Caspani et al. 1985; Purwin et al.
1986), regulation by glucose and by acid pH can both be rationalized on
the basis of the need for intracellular pH homeostasis, and the two mechanisms could be related. However, several observations indicate that
glucose and acid pH operate by different mechanisms. As indicated
above, glucose produces pleiotropic changes in kinetic properties,
whereas acid pH only affects the K_m. In addition, the activation by
glucose is greater in glucose-repressed (exponential) cells, whereas the

activation by acid pH is greater in derepressed (stationary) cells (Eraso and Gancedo 1987).

Although nothing is known about the mechanism of activation by acid pH, the fact that a phorbol ester that activates protein kinase C in animal cells produces similar K_m changes (Portillo and Mazon 1985) suggests that protein kinase C may be involved. This enzyme has recently been identified in yeast at the genetic level (Levin et al. 1990). In contrast, there is evidence for the participation of cAMP in the mechanism of glucose activation. Glucose is known to raise cAMP levels in yeast (Caspani et al. 1985; Purwin et al. 1986). Both in *S. pombe* (Foury and Goffeau 1975) and in *Dictyostelium* (Aerts et al. 1987), external cAMP activates proton pumping in whole cells. In *S. cerevisiae*, temperature-sensitive *cdc25* mutants are defective in activation of the ATPase by glucose at their nonpermissive temperature (Portillo and Mazon 1986). Although the *cdc25-1* allele utilized by Portillo and Mazon (1986) mapped to chromosome II and *CDC25* has later been mapped to chromosome XII (Johnson et al. 1987), this probably reflects a translocation of the locus in the former strain; similar defects in ATPase activation were observed with other temperature-sensitive alleles of *CDC25* that mapped to chromosome XII (F. Portillo, pers. comm.). The *CDC25* product is known to control adenylate cyclase in yeast (Broek et al. 1987). More recently (Ulaszewski et al. 1989), it has been shown that growing cultures of temperature-sensitive *cdc25* or *cdc35* (adenylate cyclase) mutants experience a decrease in plasma membrane ATPase activity when shifted to the nonpermissive temperature. This decrease in activity was prevented by cAMP in strains permeable to this nucleotide. There are observations, however, that are difficult to explain by a cAMP-mediated mechanism. Mutants with constitutively high levels of cAMP still need glucose metabolism for ATPase activation (Mazon et al. 1989), and *cdc25* mutants defective in glucose-induced cAMP increase (Munder and Küntzel 1989) still show normal activation of the ATPase by glucose metabolism (R. Serrano, unpubl.). Therefore, cAMP does not seem to mediate all nutritional responses triggered by *CDC25*. More work at the genetic and biochemical levels is needed to clarify the role of cAMP and *CDC25* in ATPase regulation.

The mechanism of activation of the plasma membrane H^+-ATPase has not been clarified. The facts that the ATPase is a phosphoprotein (McDonough and Mahler 1982; Serrano 1983), that dephosphorylation by acid phosphatase inhibits activity (Kolarov et al. 1988), and that different protein kinases are present in yeast plasma membranes (McDonough and Mahler 1982; Yanagita et al. 1987; Kolarov et al. 1988; Behrens and Mazon 1988) suggest, as the most obvious possibility, phosphorylation mediated by the cAMP-dependent protein kinase or by some other protein kinase modulated by it. However, no change in phosphorylation could be detected during activation of the ATPase by glucose (Serrano 1983, 1985). Although such negative results must always be interpreted with caution, and phosphorylation at individual sequence sites should be determined, the possibility that glucose triggers phenomena other than phosphorylation should be considered.

During solubilization of the yeast plasma membrane H^+-ATPase with strong zwitterionic detergents and purification by glycerol gradient centrifugation, the enzyme is delipidated and then requires exogenous phospholipids with negatively charged polar heads for optimal activity (Navarrete and Serrano 1983; Serrano 1988b; Serrano et al. 1988). In the presence of soybean phospholipids, which produce optimal reactivation, the difference in activity between the glucose-activated and non-activated forms of the ATPase is greatly reduced (Monk et al. 1989). In addition, a physiological lipid, oleoyl-lysophosphatidic acid, greatly increases the activity of the ATPase from starved cells but not of that from glucose-fermenting cells (Monk et al. 1989). Some rapid effects of glucose on lipid metabolism have been described (Talwalkar and Lester 1973; Kaibuchi et al. 1986), and the possibility that lipid changes mediate the regulation of the ATPase by glucose deserves further investigation.

In relation to the molecular mechanism of the regulation, two different types of mutations have been found to result in an enzyme that exhibits in glucose-starved cells the kinetic parameters of the wild-type ATPase as activated by glucose fermentation: (1) a point mutation (Ala-547→Val) affecting a residue conserved in fungal ATPases between motifs IV and V (see Fig. 2) (Cid and Serrano 1988), and (2) carboxy-terminal deletions affecting at least the last 11 amino acids of the enzyme (Portillo et al. 1989). Therefore, it can be envisaged that the carboxyl terminus of the ATPase is a regulatory domain interacting with the active site, specifically with a region between conserved motifs IV and V. This interaction may result in inhibition of ATPase activity and could perhaps be released either by protein-kinase-mediated phosphorylation or by interaction with specific lipids such as oleoyl-lysophosphatidic acid. There are no consensus phosphorylation sites (R/K R/K X S/T) for cAMP-dependent protein kinases (Cohen 1988) in the yeast ATPase, but there is a potential phosphorylation site (R X X S/T) for the calmodulin-dependent multiprotein kinase (Cohen 1988) within the last 11 amino acids of the ATPase sequence (Serrano et al. 1986a). It is interesting to note that many regulated enzymes contain inhibitory domains at their amino and carboxyl termini and that the mechanism of regulation seems to consist of displacement of the inhibitory domain either by phosphorylation or by interaction with some other molecule (Hardie 1988). For example, the Ca^{++}-ATPase of animal plasma membranes contains a calmodulin-binding regulatory domain at its carboxyl terminus (Verma et al. 1988).

Utilizing synchronized cells, a transient intracellular alkalinization has been observed at the beginning ("start") of the cell cycle of both *S. cerevisiae* (Gillies et al. 1981) and *Dictyostelium* (Aerts et al. 1985). Although this effect could be interpreted as a result of activation of the plasma membrane H$^+$-ATPase at "start," there are as yet no biochemical measurements to support this view.

V. VACUOLAR H$^+$-ATPASE

A. Isolation of Genes Encoding Subunits of the Vacuolar H$^+$-ATPase

Several different approaches have been followed to clone the genes encoding subunits of the vacuolar H$^+$-ATPase in yeast (see Table 1). The genes for the proteolipid (*VMA3*; Nelson and Nelson 1989) and interacting subunit (*VMA2* or *VAT2*; Nelson et al. 1989; Yamashiro et al. 1990) were isolated by homology with the corresponding subunits from animals and plants, respectively, or by synthesizing an oligonucleotide probe based on the amino-terminal sequence from the purified yeast polypeptide. The gene for the catalytic subunit (*VMA1*; Hirata et al. 1990) was isolated by sequencing tryptic fragments from the purified yeast polypeptide and synthesizing the corresponding oligonucleotide probes. The *VMA1* gene is identical to a previously isolated gene (*TFP1*) that by mutation confers trifluoperazine resistance (Shih et al. 1988). This coincidence suggests that the vacuolar ATPase may constitute a target for phenothiazine tranquilizers. Previously, only calmodulin (Weiss and Levin 1978) and protein kinase C (Mori et al. 1980) were considered to be targets for these drugs.

B. Homology of Vacuolar H$^+$-ATPases with a Large Family of H$^+$-ATPases

Vacuolar H$^+$-ATPases have been purified from *S. cerevisiae* (Lichko and Okorokov 1985; Uchida et al. 1988a; Kane et al. 1989), *N. crassa* (Bowman and Bowman 1988), and higher plants (Bennet et al. 1988). Their biochemical properties (see Table 1) indicate that these enzymes belong to the family of (F$_0$F$_1$) H$^+$-ATPases previously identified in bacteria, mitochondria, and chloroplasts (Senior and Wise 1983; Pedersen and Carafoli 1987a,b). Although many different accessory subunits may be present in different members of this family, the basic components of these enzymes are well defined. In the F_1 part, there are three copies of the catalytic subunit and three copies of the interacting subunit. The latter do not contain active sites but are involved in some essential cooperativity between the active sites of the three other subunits during the catalytic cycle (Senior and Wise 1983; Noumi et al. 1984; Duncan and Senior 1985). In the F_0 part, there are 6–12 copies of a low-molecular-weight proteolipid that contains a transmembrane carboxylic group involved in proton transport and modified by low concentrations of dicyclohexylcarbodiimide (Senior and Wise 1983; Walker et al. 1984).

Vacuolar H^+-ATPases do differ from "classical" (F_0F_1) H^+-ATPases in some minor properties. In vacuolar ATPases, the catalytic subunits (called A or 1) have molecular masses of 65–85 kD, greater than those of the interacting subunits (called B or 2). In mitochondrial, chloroplast, and eubacterial ATPases, the catalytic subunits (called β) have molecular masses of only 50–55 kD and are therefore smaller than the interacting subunits (called α). Alignment of the two types of catalytic subunits revealed an insertion of about 90 amino acids in vacuolar ATPases (Zimniak et al. 1988). Similarly, the proteolipid has a molecular mass of 16 kD in the first group but only 8 kD in the second group; in this case, gene duplication seems to be the explanation (Nelson and Nelson 1989). Vacuolar ATPases are more sensitive to nitrate and N-ethylmaleimide but less sensitive to azide. In the past, these differences have obscured the similarities between the two groups of enzymes, to the point that vacuolar ATPases were defined as an independent type (Pederson and Carafoli 1987a,b). The biochemical similarities described above were in conflict with such a proposition, and after the cloning and sequencing of several vacuolar ATPases, it was clear that they belong to the same large family as other (F_0F_1) H^+-ATPases (Nelson and Taiz 1989).

The amino acid sequences of the catalytic subunits of vacuolar ATPases from *S. cerevisiae* (Shih et al. 1988; Hirata et al. 1990), *Neurospora* (E.J. Bowman et al. 1988b), and carrot (Zimniak et al. 1988); of the interacting subunits from *S. cerevisiae* (Nelson et al. 1989; Yamashiro et al. 1990), *Neurospora* (B.J. Bowman et al. 1988), and *Arabidopsis* (Manolson et al. 1988); and of the proteolipid from *S. cerevisiae* (Nelson and Nelson 1989) have been deduced from the nucleotide sequences of the corresponding structural genes. The most conserved motifs of the catalytic and interacting subunits are depicted in Figure 8. In addition to the great similarities among all bacterial, mitochondrial, and vacuolar ATPases, it is important to note that in agreement with the biochemical differences discussed above, two sub-

```
A. Catalytic subunits
            I                 II               III               IV                V
YV    257 GAFGCGKTV     283 GCGERGNE      795 AEYFRD       807 MMADSSSR      861 RTGSVS
NV    246 GAFGCGKTV     272 GCGERGNE      330 AEYFRD       342 MMADSSSR      396 REGSVS
CV    252 GAFGCGKTV     278 GCGERGNE      337 AEYFRD       349 MMADSTSR      403 RNGSVT
SA    233 GPFGSGKTV     259 GCGERGNE      315 AEYFRD       326 LVADSTSR      381 RYGSVT
EC    149 GGAGVGKTV     178 GVGERTRE      228 AEKFRD       240 LFVDNIYR      287 KTGSIT
YM    188 GGAGVGKTV     217 GVGERTRE      271 AEYFRD       283 LLYDNIFR      331 KKGSVT

B. Interacting subunits
            I                 II               III               IV                V
YV    183 IAAQICRQA     214 AMGVNLET      259 AEYLAY       273 ILTDMSSY      321 RNGSIT
NV    173 IAAQICRQA     198 AMGVNLET      247 AEYYAY       271 ILTDLSSY      319 RNGSIT
AV    180 IAAQICRQA     215 AMGVNMET      264 AEYLAY       278 ILTDMSSY      326 RKGSIT
SA    166 LAAQIAKQA     187 AIGVRYDE      236 AEYLAF       250 ILIDMTNY      299 KKGSIT
EC    169 GDRQTGKTA     204 AIGQKAST      254 GEYFRD       266 IIYDDLSK      328 KTGSLT
YM    206 GDRQTGKTA     242 AVGQKRST      291 GEWFRD       303 IVYDDLSK      352 KEGSGS
```

Figure 8 Most conserved motifs in catalytic (*A*) and interacting (*B*) subunits of (F_0F_1) H$^+$-ATPases. (YV) Vacuolar ATPase from *S. cerevisiae* (A, Shih et al. 1988; Hirata et al. 1990; B, Nelson et al. 1989); (NV) vacuolar ATPase from *N. crassa* (A, E.J. Bowman et al. 1988b; B, B.J. Bowman et al. 1988); (CV) vacuolar ATPase from carrot (Zimniak et al. 1988); (AV) vacuolar ATPase from *A. thaliana* (Manolson et al. 1988); (SA) ATPase from the archaebacterium *Sulfolobus acidocalcarius* (Denda et al. 1988); (EC) ATPase from *E. coli* (Walker et al. 1984); (YM) ATPase from yeast mitochondria (A, Walker et al. 1984; B, Takeda et al. 1986).

families can be established, with vacuolar and archaebacterial ATPases in one subfamily and eubacterial and mitochondrial ATPases in the other. This is also true when the whole sequences of the subunits are considered, and cross-reaction with antibodies is only observed within each subgroup (Bowman et al. 1986). Therefore, just as mitochondria may have evolved from eubacteria by phagocytosis, vacuoles may have originated from archaebacteria by fusion and invagination (Al-Awqati 1986; Nelson and Taiz 1989).

C. Active Sites and Cooperativity of (F_0F_1) ATPases

Some years ago, Walker et al. (1982) identified two consensus sequences, A and B, present in many different nucleotide-binding proteins. Sequence A is "GXXXXGK" and sequence B is "hhhhD," where "X" means any amino acid and "h" means any hydrophobic amino acid. These two motifs are present in GTP-binding proteins such as tubulin, translation factors, G proteins and phosphoenolpyruvate carboxykinase, and in ATP-binding proteins, such as (F_0F_1) ATPases, adenylate kinase, thymidine kinase, protein kinases, myosin, and the RecA protein (Walker et al. 1982; Möller and Amons 1985; Fry et al. 1986). Recently, a family of "multipurpose" pumps involved in active transport of sugars, amino acids, phosphate, arsenite, peptides, proteins, and alkaloid drugs have been shown also to contain these two consensus sequences (Pain 1986; Higgins et al. 1988; Silver et al. 1989). The role of these sequences has been clarified in the known three-dimensional structures of adenylate kinase (Fry et al. 1986; Dreusicke and Schulz 1986), elongation factor Tu (La Cour et al. 1985), and c-Ha-*ras* p21 (De Vos et al. 1988; Pai et al. 1989; Tong et al. 1989). The glycine-rich loop (consensus sequence A) seems to form a "giant anion hole" (Dreusicke and Schulz 1986) in which the terminal phosphates of the nucleotide bind. The conserved lysine is also close to the phosphates, and it is at the beginning of a helix whose positive dipole also contributes to phosphate binding. Sequence B consists of a strand of β-sheet (the hydrophobic residues), followed by a loop in which the aspartate also binds to the terminal phosphates through a magnesium bridge.

Although the three-dimensional structures of (F_0F_1) ATPases are not known, it may be speculated that they will have similar active sites. Sequences A and B of Walker et al. (1982) correspond to motifs I and IV (see Fig. 8), respectively. It is interesting that the glycine-rich loop is present in the catalytic subunits but not in the interacting subunits. Site-directed mutagenesis of bacterial ATPases has demonstrated that the lysine of motif I and the aspartate of motif IV are involved in binding the terminal phosphate of ATP (Parsonage et al. 1988a; Yohda et al. 1988). Motifs II (Parsonage et al. 1988b) and V (Noumi et al. 1988) are also essential for catalysis, whereas the role of motif III is not known.

The cooperativity observed between active sites of (F_0F_1) ATPases (Senior and Wise 1983) has also been demonstrated with the ATPase of yeast vacuoles (Uchida et al. 1988b). It seems that product release from one active site is coupled to substrate binding to another site. It is interesting that in the family of multipurpose pumps mentioned above, there is evidence for a dimeric structure of catalytic subunits or domains (Silver et al. 1989), suggesting that two active sites must also cooperate during the catalytic cycle of these enzymes.

D. Novel Splicing Mechanism

The nucleotide sequence of the *VMA1* gene predicts a polypeptide of 119 kD, which is 1.8-fold higher than the value of 67 kD estimated from SDS-polyacrylamide gels (Shih et al. 1988; Hirata et al. 1990). Some vacuolar products are proteolytically processed during biogenesis. But proteolytic processing at amino, carboxyl, or both termini of the predicted product of *VMA1* cannot explain the large difference in the molecular mass, since tryptic peptides have been sequenced from both ends of the deduced sequence. Proteolysis coupled with a disulfide bond formation was also excluded. The lack of the typical splicing consensus sequences suggests that the *VMA1* gene does not contain introns (Hirata et al. 1990).

Alignment of the deduced sequence of the yeast catalytic subunit with the catalytic subunits of *Neurospora* and carrot vacuolar ATPases revealed the presence of a nonhomologous insert of 454 amino acids in the yeast sequence. It has been proposed that this insert is excised during biosynthesis by a novel splicing mechanism operating at the protein level (Hirata et al. 1990), and considerable support for the existence of such a mechanism has now been reported (Kane et al. 1990).

E. Vacuolar H$^+$-ATPase Mutants

Until very recently, the only evidence for the physiological role of the vacuolar ATPase was derived from in vitro experiments. The enzyme seems to operate as an electrogenic proton pump that acidifies the vacuolar compartment and drives the accumulation of other molecules into the vacuole by H$^+$-antiport systems (Uchida et al. 1988a; Klionsky et al. 1990). It was therefore of great interest to determine the physiological consequences of impairment in vacuolar ATPase activity.

Disruptions of the vacuolar ATPase subunit genes have been performed: catalytic subunit by Shih et al. (1988), Hirata et al. (1990), and Kane et al. (1990); interacting subunit by Nelson and Nelson (1990) and Yamashiro et al. (1990); and proteolipid by Nelson and Nelson (1990). In all cases, viable cells with no functional vacuolar ATPase were obtained, suggesting that the vacuolar ATPase is not essential for yeast growth. However, cells without vacuolar ATPase exhibit conditional lethality, depending on external pH and calcium concentrations. They grow well only in media buffered within a narrow pH range around 5.5. As this corresponds to the normal pH of the yeast vacuolar system, it has been proposed (Nelson and Nelson 1990) that the mutants may survive only if a proper external pH allows for the acidification of the vacuolar system by fluid-phase endocytosis. The essential cellular functions requiring vacuolar acidification have not been clarified; it has been suggested that acidification by vacuolar ATPase is essential for the proper operation of the secretory pathway, including the biogenesis of the vacuole (Klionsky et al. 1990). The growth of the ATPase-deficient mutants is also very sensitive to high calcium concentrations (50–100 mM), conditions that do not affect the growth of wild-type yeast (Shih et al. 1988; Hirata et al. 1990). Yeast vacuoles accumulate Ca^{++} using a

Ca^{++}/H^+ antiport system driven by the proton gradient generated by vacuolar ATPase (Uchida et al. 1988a). Impairment in the ATPase may thus affect the homeostasis of Ca^{++} in the cell.

Other phenotypes expected from these mutants refer to the physiological role of the accumulation of nitrogen and phosphorus stores by the vacuole. It is expected that survival of the ATPase mutants under starvation conditions should be severely compromised. Decreased growth rate under starvation conditions has already been observed in vacuolar storage mutants affecting genes not yet characterized (Klionsky et al. 1990).

VI. ION CHANNELS IN THE PLASMA MEMBRANE

A. Membrane-potential-gated K^+ Channels in Fungal and Plant Plasma Membranes

The patch-clamp technique introduced by Neher and Sakman (1976) has allowed the study of ion flow across membranes at high resolution. Application of this technique to protoplasts from plants (Hedrich et al. 1987; Schroeder et al. 1987) and yeast (Gustin et al. 1986; Saimi et al. 1988; Ramirez et al. 1989) has demonstrated the presence of similar K^+ channels in both kinds of organisms. These channels have conductivities of about 20 pS at 0.1 M K^+. At typical external K^+ concentrations of 1 mM and membrane potentials of about 0.1 V, an average opening period of 20% of total time would allow fluxes on the order of 10^6 ions per minute. Therefore, only 100–1000 channels per cell can explain the observed fluxes of K^+ in whole cells (see Section II.A). This can be compared with the estimate of about one million plasma membrane H^+-ATPase molecules per cell (see Section II.A). These channels may mediate both the uptake of K^+ in energized, hyperpolarized cells and the efflux of K^+ in depolarized cells (Borst-Pauwels 1981).

The K^+ channels of plants and yeast resemble the "delayed rectifier" of axons, responsible for the repolarization phase of the action potential (Latorre and Miller 1983). This type of channel is specific for K^+ (Na^+ conductance <5% of K^+ conductance) and is reversibly blocked by tetraethylammonium and barium. The probability of opening is voltage-dependent, depolarizing voltages being more effective than hyperpolarizing voltages (therefore, the "rectifier" character). Figure 9 illustrates the voltage dependence of the yeast channels. These channels are closed in the range from –70 mV to +30 mV. The probability of opening increases outside this range, being 50% of maximum at –120 and +80 mV (Ramirez et al. 1989).

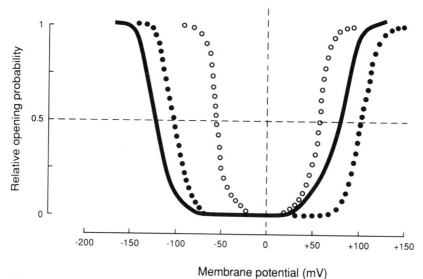

Figure 9 Voltage dependence of K[+] channels in yeast cells. The opening probability is relative to that observed at saturating voltages and is deduced from the current-voltage curves. The continuous line refers to wild-type yeast, and the circles refer to cells carrying the ATPase mutation *pma1-105* in the presence (*open circles*) or absence (*closed circles*) of added ATP (1 mM). ATP has no effect on the wild type. (Data from Ramirez et al. 1989.)

B. Relationship between the Plasma Membrane H[+]-ATPase and K[+] Channels

Although H[+] efflux in whole cells is usually observed in exchange for K[+] influx, evidence indicates that the coupling between the movement of the two cations is an indirect, electric coupling (Serrano 1985). Protons are pumped by the ATPase, and the membrane potential (negative inside) generated by this electrogenic transport drives the electrophoretic uptake of K[+], probably through the channels described above. Both in whole cells (Peña 1987) and in vesicles reconstituted with the purified yeast plasma membrane ATPase (Villalobo 1982), ATP-driven K[+] transport is blocked by H[+] ionophores. If K[+] transport were directly catalyzed by the ATPase, it should not be inhibited by H[+] ionophores. In addition, the ATPase can pump H[+] in the absence of K[+] (Malpartida and Serrano 1981b), suggesting that the enzyme does not catalyze an H[+]/K[+] exchange. Under certain circumstances, the ATPase is slightly activated by K[+], but this small effect probably does not reflect a K[+] transport activity (Serrano 1985).

There are also certain indications, however, that the ATPase is somehow connected to the K$^+$ channels. Villalobo (1984) made the interesting observation that the purified yeast ATPase reconstituted into artificial vesicles creates a K$^+$ conductance, with voltage-dependence similar to that of the K$^+$ channel described above. In addition, the ATPase ligand vanadate was found to inhibit K$^+$ transport driven by a membrane potential in the absence of ATP. More recently, Ramirez et al. (1989) have reported that the voltage dependence of the plasma membrane K$^+$ channel is altered in vesicles containing a mutant plasma membrane ATPase (allele *pma1-105*).

The opening/voltage curve of this mutant preparation is symmetrical, with a closed region from −60 to +60 mV and 50% opening probability at −100 and +100 mV (see Fig. 9) The addition of ATP increases the opening probability at low voltages, resulting in 50% probability at −50 and +50 mV; ATP has no such effect on the corresponding preparation from wild-type cells. These results raise the interesting possibility that the K$^+$ channel is somehow associated with the plasma membrane ATPase. It could be a minor membrane protein tightly bound to the ATPase and responding to conformational changes of this enzyme. Alternatively, the K$^+$ channel might actually be some transiently induced pathway within the ATPase itself. Patch-clamping experiments with mutants exhibiting reduced numbers of ATPase molecules (Vallejo and Serrano 1989) could clarify this question.

C. Regulation and Mutants of K$^+$ Transport

In growth medium with ammonia as the nitrogen source, *S. cerevisiae* grows maximally at 0.5 mM K$^+$, but in medium lacking ammonia (with arginine as the nitrogen source), a maximum growth rate is observed with only 5 μM K$^+$ (Rodriguez-Navarro and Ramos 1984). Cells growing under these conditions transport K$^+$ with an apparent K_m on the order of 10^{-5} M, whereas cells grown at the customary millimolar levels of K$^+$ display a significantly higher K_m (on the order of 10^{-3} M). The high-affinity system also has a fivefold greater V_{max}, and it is much more sensitive to uncouplers and ATP depletion. The two systems also exhibit different temperature dependencies. Part of the difference between the two systems can be explained by feedback inhibition (*trans*-inhibition) by high intracellular K$^+$. Accordingly, K$^+$ loss induced by the energy poison azide decreases the K_m and increases the V_{max} for K$^+$ transport. However, K$^+$ starvation in the absence of azide is much more effective, suggesting that in addition to the release from *trans*-inhibition, during K$^+$

starvation, there is an energy-requiring conversion of the K^+ transport system into a more efficient state (Ramos and Rodriguez-Navarro 1986). This conversion does not require protein synthesis and therefore could be a covalent modification of the transporter.

A recessive mutation of *S. cerevisiae* causing inability to grow in low-K^+ medium (0.2 mM) has been isolated (Ramos et al. 1985). The mutant displays low-affinity K^+ transport but lacks the two responses to the decrease of internal K^+ described above; thus, the two types of regulations must be somehow connected. The gene (*TRK1*) defined by this mutation has been cloned by complementation and shown to encode a plasma membrane protein that contains 12 potential membrane-spanning domains (Gaber et al. 1988). Unfortunately, this protein has no significant homology with any known protein. It is not clear whether the *TRK1* product is a K^+ channel or a membrane protein involved in the modulation of the K^+ channel. In *Neurospora*, it has been suggested that the high-affinity form of the K^+ transport system is a potassium-proton symport (Rodriguez-Navarro et al. 1986). However, it is not clear how a cation channel could exhibit a fixed stoichiometry of two transported ions; a permease-type system could explain this cotransport more easily.

The gene encoding the low-affinity K^+ transport system of *S. cerevisiae (TRK2)* has been identified recently (Ko et al. 1990). Strains with *trk1* deletions require 5 mM K^+ for growth, whereas strains with *trk1,trk2* double deletions require 100 mM K^+. In strains with *trk1* deletions, two recessive mutations, *rpd1* and *rpd3*, and one dominant mutation, *RPD2*, confer the ability to grow on 0.2 mM K^+ (Vidal et al. 1990). The three mutations increase the activity of the low-affinity transporter, suggesting that the ability of *S. cerevisiae* to grow on low-K^+ media depends on either high-affinity transport or an increased rate of low-affinity transport. *RPD2* is allelic to *TRK2*.

In plants, the K^+ channel is inhibited by increased cytosolic Ca^{++} (Schroeder and Hagiwara 1989), but this type of regulation has not been described in yeast.

D. Turgor-gated Channels

In addition to the voltage-gated K^+ channel described above, a non-specific ion channel activated by stretching of the membrane has been identified in *S. cerevisiae* (Gustin et al. 1988). This channel may play a role in turgor regulation, because osmotic stress on the cell could be relieved by the efflux of anions and cations. This mechanosensitive channel exhibited adaptation (desensitization) to mechanical stimuli.

	I	II	III	IV	V
GLsc	112 GYDTG	166 DSYGRK	285 PESPRYY	413 EFFGRR	542 YETKGLTLEEID
GAsc	86 GWDTS	144 DMYGRK	263 PESPRYL	388 ENLGRR	520 PETKGLSLEEIQ
MAsc	114 GYDTA	173 DYMGNR	293 PESPWWL	421 KYCGRF	547 PETAGRTFIEIN
LAkl	87 GYDGA	138 DWKGRK	258 PESPRWL	389 DKIGRR	515 VETKGRSLEELE
GLhu	27 GYNTG	88 NRFGRR	208 PESPRFL	329 ERAGRR	453 PETKGRTFDEIA
XYec	25 GYDTA	80 NRFGRR	221 PESPRWL	337 DKFGRK	464 PETKGKTLEELE

Figure 10 Conserved motifs in sugar permeases. (GLsc) Glucose permease from *S. cerevisiae* (Snf3p; Celenza et al. 1988); (GAsc) galactose permease from *S. cerevisiae* (Gal2p; Szkutnicka et al. 1989); (MAsc) H^+-maltose permease from *S. cerevisiae* (Mal61p; Cheng and Michels 1989); (LAkl) H^+-lactose permease from *K. lactis* (Lac12p; Chang and Dickson 1988); (GLhu) human glucose permease (Baldwin and Henderson 1989); (XYec) H^+-xylose permease from *E. coli* (Baldwin and Henderson 1989).

VII. PLASMA MEMBRANE PERMEASES

A. Homology of Yeast Sugar Permeases with a Large Family of Transport Proteins

Four yeast sugar permease genes have been cloned and sequenced: The genes for the lactose permease from *Kluyveromyces lactis* (Chang and Dickson 1988), one of the multiple (see below) glucose permeases of *S. cerevisiae* (Celenza et al. 1988), and the galactose (Szkutnicka et al. 1989) and maltose (Cheng and Michels 1989) permeases of *S. cerevisiae*. These permeases show homology with some sugar permeases from bacteria and animal cells that form a large family of homologous transport proteins (Maiden et al. 1987; Baldwin and Henderson 1989). Five conserved motifs within this family are indicated in Figure 10, and the similarity of the proteins' hydrophobicity plots is illustrated in Figure 11. All of these permeases have 12 hydrophobic stretches that could span the membrane as helices. The hydrophobic stretches adjacent in the sequence are connected by short loops and therefore are likely to be adjacent in the tertiary structure. There is a central polar domain of variable length (50–100 amino acids). These transport proteins seem to have arisen by internal gene duplication, as several regions are repeated in the amino- and carboxy-terminal halves. Motifs II and III, at the ends of hydrophobic stretches 2 and 6, are related to motifs IV and V, at the ends of hydrophobic stretches 8 and 12.

B. Derepression of High-affinity H^+ Sugar Symports by Sugar Starvation

Most yeast species have a low-affinity (K_m 10^{-3} to 10^{-2} M) constitutive sugar permease that catalyzes the facilitated diffusion of glucose, fruc-

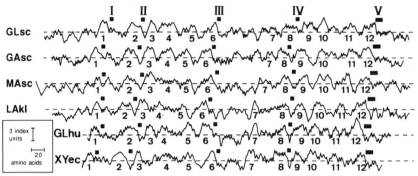

Figure 11 Hydrophobicity plots of sugar permeases. The curves show the averages of a residue-specific hydrophobicity index over a window of nine residues (Kyte and Doolittle 1982). The discontinuous lines correspond to an index value of 0, positive values above it being hydrophobic. Symbols and references for the different sequences are the same as those in Fig. 10. (1–12) Hydrophobic stretches proposed to span the membrane as helices; (I–V) conserved motifs of Fig. 10.

tose, and mannose during growth in media containing high concentrations of these sugars (Cooper 1982). Sugar starvation can derepress high-affinity (K_m 10^{-4} to 10^{-3} M) H^+-symports specific for different sugars, such as glucose in *Pichia ohmeri* (Verma et al. 1987), *Kluyveromyces marxianus* (Van den Broek et al. 1986; Gasnier 1987), *Rhodotorula* (Loureiro-Dias 1988), *Hansenula* (Loureiro-Dias 1988), and *Candida wickerhamii* (Spencer-Martins and Van Uden 1985; Loureiro-Dias 1988); xylose and L-arabinose in *Candida shehatae* (Lucas and Van Uden 1986); and lactose in *K. marxianus* (De Bruijne et al. 1988). The substrates for the H^+-galactose permease of *K. marxianus* (De Bruijne et al. 1988) and for the H^+-maltose permease of *S. cerevisiae* (Peinado and Loureiro-Dias 1986) and *Candida utilis* (Peinado et al. 1987) are also required for production of the respective permeases. Therefore, in these cases, a specific induction is required, in addition to a general derepression.

High glucose concentrations both repress the synthesis of (catabolite repression) and inactivate (catabolite inactivation) these different H^+-symports (see references above). Inactivation is irreversible because protein synthesis is required to recover activity; therefore, it probably represents proteolysis of the permeases. Using mutants affecting cAMP-dependent protein kinases in *S. cerevisiae*, Ramos and Cirillo (1989) have shown that the catabolite inactivation of sugar transporters is regulated by the cAMP cascade triggered by glucose. It is possible that phosphorylation of the permeases themselves induces their proteolytic degradation. Both in *C. wickerhamii* (Spencer-Martins and Van Uden 1985) and in *P. ohmeri* (Verma et al. 1987), derepression of high-affinity H^+-symport of glucose is accompanied by the disappearance of the low-affinity facilitated diffusion system. Therefore, it has been suggested (Spencer-Martins and Van Uden 1985) that H^+-symport may result from the combination of the permease catalyzing facilitated diffusion with an additional protein that is repressed and inactivated by high glucose concentrations.

S. cerevisiae does not develop high-affinity H^+-symports for glucose during glucose starvation (Loureiro-Dias 1988). The transport of glucose and galactose in this species always occurs by low-affinity facilitated diffusion, and this may reflect an adaptation to environments rich in sugars. One confusing feature, however, is that these two facilitated diffusion systems exhibit changes in affinity within the low-affinity range (see below). However, it must be emphasized that in *S. cerevisiae*, high-affinity glucose or galactose transport does not correspond to H^+-symport but to some form of the facilitated diffusion systems.

C. Effect of Sugar Phosphorylation on the Affinity of Facilitated Diffusion

Long ago, Van Stevenink and Rothstein (1966) made the observation that the affinity of glucose uptake in *S. cerevisiae* is much higher when the cells are metabolizing the sugar (K_m ~10^{-2} M) than when glycolysis is poisoned by iodoacetate (K_m ~10^{-1} M). As iodoacetate causes ATP depletion and the sugar is thus not phosphorylated under these conditions, it was proposed that low-affinity uptake represents facilitated diffusion and that high-affinity uptake during glycolysis occurs by transport-associated phosphorylation. Such a mechanism could be related to the phosphoenolpyruvate-dependent transport-associated phosphorylation of sugars in bacteria (Saier et al. 1988). One problem with this proposal is that if two different systems were in operation, biphasic kinetics would be observed. However, both in the absence and in the presence of iodoacetate, only one type of affinity was observed. Thus, instead of two parallel systems, a change in the affinity of a single system seemed to be in operation. A second problem is that glucose phosphorylation is catalyzed by soluble hexokinases, suggesting that transport-associated phosphorylation is unlikely and that the effect of sugar metabolism on transport must be indirect. Serrano and De la Fuente (1974) made several additional relevant observations. First, sugar metabolism decreased the K_m for uptake of nonmetabolizable sugars such as xylose, suggesting that transport-associated phosphorylation cannot be the sole explanation for the affinity changes. Second, during steady-state glycolysis, only high-affinity uptake was observed. In contrast, during short incubations (seconds) of previously starved cells in the presence of glucose, uptake of the sugar exhibited biphasic kinetics, with low- and high-affinity components, as if under these conditions, the metabolism-triggered interconversion between the two states were partial. Finally, steady-state glycolysis under the conditions of the Pasteur effect (high respiratory rate relative to fermentative rate, as in stationary-phase cells) failed to improve the affinity of glucose uptake, although sugar phosphorylation clearly occurred. It was proposed that the constitutive hexose permease may exist in two different affinity states and that the interconversion is modulated by sugar metabolism.

More recently, the effects of iodoacetate were reproduced by blocking glucose metabolism using mutants that lack hexose kinases (Bisson and Fraenkel 1983a; Lang and Cirillo 1987). Glucose uptake in a triple mutant defective in the three kinases phosphorylating glucose and fructose uptake in a double mutant defective in the two kinases phosphorylating fructose both exhibit only low-affinity uptake. In addition, this work demonstrated that phosphorylation was the only requirement for high-affinity glucose uptake, because this was observed in a mutant defective in phosphoglucose isomerase. Bisson and Fraenkel (1983a) proposed that the hexose kinases form a functional part of the hexose uptake system. This revived the idea of transport-associated phosphorylation in yeast, with the modification that the soluble kinases were proposed to interact with other membrane-bound components of the permease. Bisson and Frankel (1983b) also reported that the affinity of uptake of the non-phosphorylated sugar 6-deoxyglucose is also increased by the presence of functional kinases. This supported the hypothesis that it is not the phosphorylation of the sugar but the interaction of the kinases (and ATP) with the permeases that modulates the affinity of sugar uptake. However, no high-affinity uptake of 6-deoxyglucose could be detected in ATP- and kinase-containing cells by Schuddemat et al. (1988). These authors also observed that ATP depletion and reduction of kinase activity (by incubation of the cells with xylose) inhibited the uptake of phosphorylable sugars, such as glucose, fructose, mannose, and 2-deoxyglucose, but had no effect on the uptake of 6-deoxyglucose. The inhibition of transport observed under conditions of poor phosphorylation was the result of an increase in the K_m (see Schuddemat et al. 1986). These results appeared to support the old model of transport-associated phosphorylation.

Attempts have also been made to determine the order of labeling of the intracellular pools of free and phosphorylated hexose. After preincubation of cells with nonlabeled 2-deoxyglucose, a pulse of radioactive 2-deoxyglucose resulted in higher specific radioactivity of intracellular 2-deoxyglucose-6-phosphate than of intracellular 2-deoxyglucose (Franzusoff and Cirillo 1982). These results are consistent with a transport-associated phosphorylation mechanism in which intracellular free hexose would be derived from phosphorylated hexose released by the permease. However, as Franzusoff and Cirillo (1982) pointed out, it is crucial to demonstrate that the intracellular free 2-deoxyglucose is in the same compartment as the 2-deoxyglucose-6-phosphate; thus, caution is necessary in interpreting the results of experiments with whole cells, where problems of compartmentation and multiple pools are difficult to assess. For this reason, glucose transport has also been studied in plasma membrane vesicles after reconstitution with liposomes (Ongjoco et al. 1987). The K_m for glucose uptake by the vesicles (~8 mM) was intermediate between the values exhibited by the high-affinity (1–2 mM) and low-affinity (40–60 mM) systems in whole cells (Lang and Cirillo 1987). The authors concluded that the affinity observed in the vesicles corresponded to the low-affinity system of whole cells, but this interpretation must be taken with caution. The K_m value of the vesicles is in the range of the high-affinity transport observed with some strains (Serrano and De la Fuente 1974), and the change in phospholipid composition introduced by the reconstitution procedure and other possible changes introduced during vesicle preparation make it difficult to compare the in vitro and in vivo conditions. It was hoped that the introduction of hexokinase and ATP in the vesicles could modulate the affinity of glucose uptake, but no effects could be detected (Lang and Cirillo 1987).

The facilitated diffusion of galactose in induced cells exhibited the same change in affinity dependent on phosphorylation of the sugar as observed previously for the glucose permease (Ramos et al. 1989). In galactokinase-free cells, the K_m for galactose uptake was very high (on the order of 10^{-1} M), whereas in normal cells, biphasic kinetics were observed with K_m values on the order of 10^{-1} and 10^{-3} M. In plasma membrane vesicles reconstituted with soybean phospholipids, the K_m (10^{-2} M) exhibited an intermediate value. As in the case of glucose uptake with the same kind of vesicles, it is not clear whether this value corresponds to that of the low- or high-affinity components observed in whole cells. A further caveat in the case of galactose uptake is that it is not clear whether the galactose permease is rate-limiting for uptake. This has only been demonstrated for glucose uptake (Serrano and De la Fuente 1974) and does not seem to be the case for galactose uptake (R. Serrano, unpubl.). If the galactose permease is not rate-limiting, the measured values of K_m in whole cells are difficult to interpret.

As indicated above, one interpretation of the biphasic kinetics of glucose uptake observed during short incubations of starved cells with glucose is based on the coexistence of two states of the same permease perhaps differing by some form of covalent modification. However, some recent observations have prompted a different interpretation, namely, that low- and high-affinity uptakes are mediated by two different permeases. First, Bisson and Fraenkel (1984) observed that the high-affinity component of the kinetics of glucose uptake is greater in glucose-derepressed cells. Mutations like snf1, snf4, and snf6, which affect derepression of glucose-repressible functions, were also defective in derepression of high-affinity glucose uptake, whereas mutations like cid1 and reg1, which result in constitutive expression of glucose-repressible functions, resulted in constitutive expression of high-affinity glucose uptake (Bisson 1988a). Derepression of high-affinity glucose uptake requires a functional secretory system (Bisson 1988b) and a functional SNF3 gene (Bisson et al. 1987); SNF3 encodes a plasma membrane protein that has homology with glucose permeases (Celenza et al. 1988). Therefore, it has been proposed that the SNF3 gene product is a glucose permease of high affinity that is repressed by glucose. In addition, there are indications of the existence of several glucose permease genes in yeast; in particular, no glucose-negative mutants affected in glucose uptake have been found, suggesting that there may be multiple permease genes.

On the other hand, the low-affinity and high-affinity systems only appear to coexist in short-term experiments (Serrano and De la Fuente 1974), and the activity of low-affinity uptake is inhibited in proportion to the extent of derepression of the high-affinity process (Ramos et al. 1988). In addition, inactivation of glucose transport when protein synthesis is inhibited but glucose is available (Busturia and Lagunas 1986) results in similar inactivation of high- and low-affinity components. All of these observations seem to be most consistent with the interconversion of a single permease. One way to reconcile the two sets of observations is to postulate that SNF3 encodes a glucose-repressible glucose permease that is rapidly (<5 sec) transformed to a high-affinity state during incubation of the cells with glucose. Other glucose permease(s) may require longer incubations and therefore only exhibit the high-affinity state during steady-state glucose metabolism. It appears that measurements of the kinetics of steady-state glucose metabolism might resolve this controversy. Basically, the prediction is that snf3 null mutants would not exhibit high-affinity glucose uptake during short-time incubations but would exhibit high-affinity uptake in the steady state.

D. Recent Advances with Other Permeases

The genes for the histidine permease (*HIP1*; Tanaka and Fink 1985), arginine permease (*CAN1*; Hoffman 1985; Ahmad and Bussey 1986), uracil permease (*FUR4*; Jund et al. 1988), purine-cytosine permease (*FCY2*; Weber et al. 1988), allantoate permease (*DAL5*; Rai et al. 1988), proline permease (*PUT4*; Vandenbol et al. 1989), and general amino acid permease (*GAP1*; Jauniaux and Grenson 1990) have been cloned and sequenced. The amino acid permeases are H^+-symports (Eddy 1982) and show significant sequence homology, suggesting a common evolutionary origin (Weber et al. 1988; Jauniaux and Grenson 1990). The other permeases do not show significant homology either with each other or with any known protein. The only common feature of all these permeases is the presence of many hydrophobic stretches (from 9 to 12) that could span the membrane as helices. The purine-cytosine permease has been identified as a glycoprotein of 120 kD by photoaffinity labeling (Schmidt et al. 1984).

Overexpression of the uracil and purine-cytosine permease genes from multicopy plasmids dramatically increases the level of accumulation by the cells of uracil and cytosine, respectively (Hopkins et al. 1988). This demonstrates that accumulation does not represent thermodynamic equilibrium but rather a steady state in which uptake by the permeases operates in parallel to efflux by some other pathway (the cells employed did not metabolize the transported substrates). These experiments with overproducing strains also demonstrated proton cotransport during uracil and cytosine uptake, something very difficult to observe in normal cells because of the low rates involved.

Permeases for nitrogen sources are both inactivated and repressed by ammonia (Cooper 1982). Northern analysis has demonstrated that repression by ammonia of the general amino acid permease occurs at the RNA level (Jauniaux and Grenson 1990). Some progress has been made in clarification of the underlying mechanisms (Horak 1986; Vandenbol et al. 1987). Ammonia seems to require conversion into glutamine to exert these effects. The repression effect is mediated by the product of the *GDHCR* or *URE2* gene, which could be an aporepressor binding glutamine. The products of the *NPI1* and *NPI2* genes seem to inactivate the permeases in a continuous and stoichiometric way, possibly by forming a complex with the permeases. On the other hand, the product of the *NPR1* gene seems to reactivate the permeases in a catalytic way, whereas its own activity is inhibited by ammonia. One plausible mechanism could be that the *NPR1* product is a protein kinase that disrupts the inhibitory interaction between the permeases and the *NPI* gene products by phosphorylating one of the components.

Yeast cells extrude carbonic acid, succinate, and acetate during glucose fermentation, and they can also take up anions such as acetate and lactate, which are important gluconeogenic carbon sources. Succinate efflux seems to be driven by the negative-inside membrane potential generated by the plasma membrane H^+-ATPase (Duro and Serrano 1981). Although the detailed mechanism is unknown, it could involve a voltage-gated anion channel. Acetate efflux probably occurs by a similar mechanism. Carbonic acid efflux could also be mediated by an anion channel. However, depolarization of the plasma membrane by either inhibition of the ATPase (Serrano 1980) or addition of polyenes (Palacios and Serrano 1978) only slightly inhibits glucose fermentation, suggesting that passive diffusion of CO_2 may remove most of the carbonic acid from the cells. Acetate and lactate seem to be transported into yeast cells by two different mechanisms (Eddy and Hopkins 1985; Leão and Van Uden 1986; Cássio et al. 1987). In cells growing on glucose, there is only passive diffusion of the undissociated acids. In the case of lactic acid, this mechanism accounts for uptake rates on the order of 0.2 nmole x min^{-1} x mg $cells^{-1}$ at an external total concentration of 1 mM and pH of 3–6. It is intriguing that the permeability coefficient of undissociated acid increases when the external pH increases, compensating for the lower concentration of undissociated acid. The diffusion of acetic acid is probably much higher but has not been determined. In cells growing in lactate or acetate, there is also an electroneutral monocarboxylate-proton symport that is induced by lactate or acetate and repressed by glucose. This system exhibits a K_m for lactate (D- or L-) or acetate of 0.05 or 0.1 mM, respectively, and a V_{max} of 25 or 50 nmoles x min^{-1} x mg $cells^{-1}$, respectively. Therefore, when the concentration of monocarboxylates is in the millimolar range, the proton symport allows uptake rates two orders of magnitude greater than that of diffusion. Only at concentrations higher than 10–100 mM does diffusion become important.

Calcium uptake in yeast probably occurs through a channel of wide specificity for divalent cations, although electrophysiological evidence is not yet available. The uptake system has very low affinity (Kovac 1985), and the apparent saturation kinetics observed at millimolar Ca^{++} concentrations are caused by a reduction in the surface potential and not by saturation of a transport site (Borst-Pauwels and Thevenet 1984). A temperature-sensitive mutant (*cal1-1*) corrected by high calcium concentrations has been described by Ohya et al. (1984). It is plausible that this mutation results in thermosensitive calcium uptake because calcium is required for yeast growth (Kovac 1985). Recent evidence, however, suggests that the rescue of this mutant by high Ca^{++} may be an indirect effect (Finegold et al. 1991; Ohya et al. 1991).

Calcium efflux in *N. crassa* seems to occur by a H^+/Ca^{++} antiport (Stroobant and Scarborough 1979), but both in higher plants (Dieter and Marme 1981; Rasi-Caldogno et al. 1987; Briars et al. 1988) and in the yeast *Rhodospiridium toruloides* (Miyakawa et al. 1987), a calmodulin-dependent Ca^{++}-ATPase seems to be involved. Calcium efflux in *S. cerevisiae* is dependent on energy and stimulated by external potassium (Eilam 1982), but the exact mechanism has not been clarified. Mating pheromones increase Ca^{++} uptake in both *R. toruloides* (Miyakawa et al. 1985) and *S. cerevisiae* (Ohsumi and Anraku 1985), and this could play an essential role in the control of differentiation. In vitro experiments indicate that mating pheromone inhibits the calmodulin-dependent Ca^{++}-ATPase in *R. toruloides* (Miyakawa et al. 1987). Therefore, it seems that the observed increase in Ca^{++} uptake may reflect inhibition of efflux, rather than activation of influx. It has been shown that the intracellular level of divalent cations results from a steady-state balance between separate influx and efflux systems (Theuvenet et al. 1986).

VIII. VACUOLAR TRANSPORT SYSTEMS

Experiments with isolated vacuoles or vacuolar membrane vesicles have identified several H^+-antiport systems that allow the accumulation of *S*-adenosylmethionine (Schwencke and de Robichon-Szulmajster 1976), amino acids (Ohsumi and Anraku 1981; Sato et al. 1984; Okorokov et al. 1985), calcium (Ohsumi and Anraku 1983; Okorokov et al. 1985), and phosphate (Okorokov et al. 1985) driven by the proton gradient. The specificity of the vacuolar transport systems correlates with the specificity of amino acids and ions accumulated into the vacuole. Efflux of the accumulated compounds occurs during starvation (Sumrada and Cooper 1978; Kitamoto et al. 1988), but it is not clear whether this is mediated by the same systems involved in uptake or by separate efflux systems. A novel method for differential extraction of vacuolar and cytosolic pools has been described by Ohsumi et al. (1988). It is based on the observation that Cu^{++} causes selective lesions of the permeability barrier of the plasma membrane but that it did not affect the permeability of the vacuolar membrane. Because of its simplicity, this method may greatly facilitate future studies on the regulation of vacuolar pools.

Yeast vacuolar membranes contain a cation channel that can be differentiated from that in the plasma membrane by its large conductance (435 pS in 0.3 M KCl) and broad specificity, with little discrimination among monovalent cations (Wada et al. 1987). This channel requires a positive-inside membrane potential and cytosolic Ca^{++} for activity. It may function to discharge membrane potential across the vacuolar membrane, thus facilitating the acidification of vacuolar content by the H^+-ATPase. The properties of this channel differ from those of a vacuolar channel identified in plants, which carries both cations and anions and is activated by negative-inside membrane potentials (Hedrich et al. 1986).

IX. CONCLUDING REMARKS

The two basic problems discussed above for the different transport systems of yeast are their structure-function relationships and their physiological roles. The application of methodologies from both membrane biochemistry and molecular genetics has allowed considerable progress in the case of the plasma membrane H^+-ATPase, which remains the best-characterized transport system at the molecular level. However, much remains to be learned about this crucial enzyme, such as the detailed mechanism of proton pumping, the basis for its pace-maker control of yeast growth, and the nature of the regulatory system that modulates ATPase activity. The vacuolar H^+-ATPase has now reached the state at which both the proteins and the genes are available, and a mutational analysis is under way. Other transport systems such as the permeases, for which cloned genes are now available, will still require considerable biochemical effort to characterize the corresponding proteins. The mechanisms of regulation of the permeases present several puzzles for the future. The connection between sugar phosphorylation and transport and the mechanisms of catabolite inactivation (of sugar permeases by glucose and of nitrogen-source permeases by ammonia) are some examples. The regulation of vacuolar loading and unloading remains as a fundamental gap in knowledge. In the case of ion channels, we still await the identification of both the genes and the proteins involved, although some candidates are already available. The connection between the plasma membrane H^+-ATPase and the K^+ channel is one further challenge for future research.

ACKNOWLEDGMENTS

I express my gratitude to Professor Gerry Fink, from whom I learned that Molecular Biology can be something more than a further level of description. I also want to thank my former co-workers A. Cid, P. Eraso, B.C. Monk, F. Portillo, and C.G. Vallejo for their partnership in this lonely enterprise of scientific research.

REFERENCES

Aaronson, L.R., K.M. Hager, J.W. Davenport, S.M. Mandala, A. Chang, D.W. Speicher, and C.W. Slayman. 1988. Biosynthesis of the plasma membrane H$^+$-ATPase of *Neurospora crassa. J. Biol. Chem.* **263:** 14552.

Addison, R. 1986. Primary structure of the *Neurospora* plasma membrane H$^+$-ATPase deduced from the gene sequence. Homology to Na$^+$/K$^+$, Ca^{2+}, and K$^+$-ATPases. *J. Biol. Chem.* **261:** 14896.

Addison, R. and G.A. Scarborough. 1982. Conformational changes of the *Neurospora* plasma membrane H$^+$-ATPase during its catalytic cycle. *J. Biol. Chem.* **257:** 10421.

Aerts, R.J., R.J.W. De Wit, and M.M. Van Lookeren Campagne. 1987. Cyclic AMP induces a transient alkalinization in *Dictyostelium. FEBS Lett.* **220:** 366.

Aerts, R.J., A.J. Durston, and W.H. Moolenaar. 1985. Cytoplasmic pH and the regulation of the *Dictyostelium* cell cycle. *Cell* **43:** 653.

Ahlers, J. 1981. Temperature effects on kinetic properties of plasma membrane ATPase from the yeast *Saccharomyces cerevisiae. Biochim. Biophys. Acta* **649:** 550.

Ahmad, M. and H. Bussey. 1986. Yeast arginine permease: Nucleotide sequence of the *CAN1* gene. *Curr. Genet.* **10:** 587.

Al-Awqati, Q. 1986. Proton-translocating ATPases. *Annu. Rev. Cell Biol.* **2:** 179.

Amory, A., A. Goffeau, D.B. McIntosh, and P.D. Boyer. 1982. Exchange of oxygen between phosphate and water catalyzed by the plasma membrane ATPase from the yeast *Schizosaccharomyces pombe. J. Biol. Chem.* **257:** 12509.

Bakker, R., J. Dobbelmann, and G.W.F.H. Borst-Pauwels. 1986. Membrane potential in the yeast *Endomyces magnusii* measured by microelectrodes and TPP$^+$ distribution. *Biochim. Biophys. Acta* **861:** 205.

Baldwin, S.A. and P.J.F. Henderson. 1989. Homologies between sugar transporters from eukaryotes and prokaryotes. *Annu. Rev. Physiol.* **51:** 459.

Barchfeld, G.L. and D.W. Deamer. 1988. Alcohol effects on lipid bilayer permeability to protons and potassium: Relation to the action of general anesthetics. *Biochim. Biophys. Acta* **944:** 40.

Behrens, M.M. and M.J. Mazon. 1988. Yeast cAMP-dependent protein kinase can be associated with the plasma membrane. *Biochem. Biophys. Res. Commun.* **151:** 561.

Bennet, A.B., R.A. Leigh, and R.M. Spanswick. 1988. H$^+$-ATPase from vacuolar membranes of higher plants. *Methods Enzymol.* **157:** 579.

Bisson, L.F. 1988a. High-affinity glucose transport in *Saccharomyces cerevisiae* is under general glucose repression control. *J. Bacteriol.* **170:** 4838.

―――. 1988b. Derepression of high-affinity glucose uptake requires a functional secretory system in *Saccharomyces cerevisiae. J. Bacteriol.* **170:** 2654.

Bisson, L.F. and D.G. Fraenkel. 1983a. Involvement of kinases in glucose and fructose uptake by *Saccharomyces cerevisiae. Proc. Natl. Acad. Sci.* **80:** 1730.

―――. 1983b. Transport of 6-deoxyglucose in *Saccharomyces cerevisiae. J. Bacteriol.* **155:** 995.

―――. 1984. Expression of kinase-dependent glucose uptake in *Saccharomyces cerevisiae. J. Bacteriol.* **159:** 1013.

Bisson, L.F., L. Neigeborn, M. Carlson, and D.G. Fraenkel. 1987. The *SNF3* gene is required for high-affinity glucose transport in *Saccharomyces cerevisiae*. *J. Bacteriol.* **169:** 1656.

Blasco, F., R. Jeanjean, M. Hirn, and P. Ritz. 1983. Antigenic relationships between plasma membrane ATPases of two different yeasts, *Candida tropicalis* and *Schizosaccharomyces pombe*. *Biochem. Biophys. Res. Commun.* **115:** 1114.

Borst-Pauwels, G.W.F.H. 1981. Ion transport in yeast. *Biochim. Biophys. Acta* **650:** 88.

Borst-Pauwels, G.W.F.H. and A.P.R. Theuvenet. 1984. Apparent saturation kinetics of divalent cation uptake in yeast caused by a reduction in the surface potential. *Biochim. Biophys. Acta* **771:** 171.

Borst-Pauwels, G.W.F.H., A.P.R. Theuvenet, and A.L.H. Stols. 1983. All- or-none interactions of inhibitors of the plasma membrane ATPase with *Saccharomyces cerevisiae*. *Biochim. Biophys. Acta* **732:** 186.

Boyer, P.D. 1988. Bioenergetic coupling to protonmotive force: Should we be considering hydronium ion coordination and not group protonation? *Trends Biochem. Sci.* **13:** 5.

Bowman, B.J. 1983. Kinetic evidence for interacting active sites in the *Neurospora crassa* plasma membrane ATPase. *J. Biol. Chem.* **258:** 13002.

Bowman, B.J., R. Allen, M.A. Wechser, and E.J. Bowman. 1988. Isolation of genes encoding the *Neurospora* vacuolar ATPase. Analysis of *vma-2* encoding the 57-kDa polypeptide and comparison to *vma-1*. *J. Biol. Chem.* **263:** 14002.

Bowman, E.J. and B.J. Bowman. 1988. Purification of vacuolar membranes, mitochondria and plasma membranes from *Neurospora crassa* and modes of discriminating among the different H$^+$-ATPases. *Methods Enzymol.* **157:** 562.

Bowman, E.J., A. Sieber, and K. Altendorf. 1988a. Bafilomycins: A class of inhibitors of membrane ATPases from microorganisms, animal cells and plant cells. *Proc. Natl. Acad. Sci.* **85:** 7972.

Bowman, E.J., K. Tenney, and B.J. Bowman. 1988b. Isolation of genes encoding the *Neurospora* vacuolar ATPase. Analysis of *vma-1* encoding the 67-kDa subunit reveals homology to other ATPases. *J. Biol. Chem.* **263:** 13994.

Bowman, E.J., S. Mandala, L. Taiz, and B.J. Bowman. 1986. Structural studies of the vacuolar membrane ATPase from *Neurospora crassa* and comparison with the tonoplast membrane ATPase from *Zea mays*. *Proc. Natl. Acad. Sci.* **83:** 48.

Brandl, C.J. and C.M. Deber. 1986. Hypothesis about the function of membrane-buried proline residues in transport proteins. *Proc. Natl. Acad. Sci.* **83:** 917.

Brandl, C.J., N.M. Green, B. Korczak, and D.H. MacLennan. 1986. Two Ca^{2+} ATPases: Homologies and mechanistic implications of deduced amino acid sequences. *Cell* **44:** 597.

Briars, S.A., F. Kessler, and D.E. Evans. 1988. The calmodulin-stimulated ATPase of maize coleoptiles is a 140,000 M_r polypeptide. *Planta* **176:** 283.

Broek, D., T. Toda, T. Michaeli, L. Levin, C. Birchmeier, M. Zoller, S. Powers, and M. Wigler. 1987. The *S. cerevisiae CDC25* gene product regulates the RAS/adenylate cyclase pathway. *Cell* **48:** 789.

Busturia, A. and R. Lagunas. 1986. Catabolite inactivation of the glucose transport system in *Saccharomyces cerevisiae*. *J. Gen. Microbiol.* **132:** 379.

Calahorra, M., J. Ramirez, S. M. Clemente, and A. Peña. 1987. Electrochemical potential and ion transport in vesicles of yeast plasma membrane. *Biochim. Biophys. Acta* **899:** 229.

Capieaux, E., M.L. Vignais, A. Sentenac, and A. Goffeau. 1989. The yeast H$^+$-ATPase gene is controlled by the promoter binding factor TUF. *J. Biol. Chem.* **264:** 7437.

Carafoli, E. 1988. Membrane transport of calcium: An overview. *Methods En-*

zymol. **157**: 3.

Caspani, G., P. Tortora, G.M. Hanozet, and A. Guerritore. 1985. Glucose-stimulated cAMP increase may be mediated by intracellular acidification in *Saccharomyces cerevisiae. FEBS Lett.* **186**: 75.

Cássio, F., C. Leão, and N. Van Uden. 1987. Transport of lactate and other short-chain monocarboxylates in the yeast *Saccharomyces cerevisiae. Appl. Environ. Microbiol.* **53**: 509.

Celenza, J.L., L. Marshall-Carlson, and M. Carlson. 1988. The yeast *SNF3* gene encodes a glucose transporter homologous to the mammalian protein. *Proc. Natl. Acad. Sci.* **85**: 2130.

Chadwick, C.C., E. Goormaghtigh, and G.A. Scarborough. 1987. A hexameric form of the *Neurospora crassa* plasma membrane H^+-ATPase. *Arch. Biochem. Biophys.* **252**: 348.

Chang, Y.D. and R.C. Dickson. 1988. Primary structure of the lactose permease gene from the yeast *Kluyveromyces lactis.* Presence of an unusual transcript structure. *J. Biol. Chem.* **263**: 16696.

Cheng, Q. and C.A. Michels. 1989. The maltose permease encoded by the *MAL61* gene of *Saccharomyces cerevisiae* exhibits both sequence and structural homology to other sugar transporters. *Genetics* **123**: 477

Cid, A. and R. Serrano. 1988. Mutations of the yeast plasma membrane H^+-ATPase which cause thermosensitivity and altered regulation of the enzyme. *J. Biol. Chem.* **263**: 14134.

Cid, A., R. Perona, and R. Serrano. 1987. Replacement of the promoter of the yeast plasma membrane ATPase gene by a galactose-dependent promoter and its physiological consequences. *Curr. Genet.* **12**: 105.

Clement, J.D., M. Ghislain, J.P. Dufour, and R. Scalla. 1986. Immunodetection of a 90,000-M_r polypeptide related to yeast plasma membrane ATPase in plasma membrane from maize shoots. *Plant Sci.* **45**: 43.

Cohen, P. 1988. Protein phosphorylation and hormone action. *Proc. R. Soc. Lond. B Biol. Sci.* **234**: 115.

Cooper, T.G. 1982. Transport in *Saccharomyces cerevisiae.* In *The molecular biology of the yeast* Saccharomyces: *Metabolism and gene expression* (ed. J.N. Strathern et al.), p. 399. Cold Spring Harbor Laboratory, Cold Spring Harbor, New York.

D'Amore, T. and G.G. Stewart. 1987. Ethanol tolerance of yeast. *Enzyme Microb. Technol.* **9**: 322.

Davis, C.B. and G.G. Hammes. 1989. Topology of the yeast plasma membrane proton-translocating ATPase. *J. Biol. Chem.* **264**: 370.

De Bruijne, A.W., J. Schuddemat, P.J.A. Van den Broek, and J. Van Stevenink. 1988. Regulation of sugar transport systems of *Kluyveromyces marxianus*: The role of carbohydrates and their catabolism. *Biochim. Biophys. Acta* **939**: 569.

De Meis, L., J.P. Blanpain, and A. Goffeau. 1987. P_i-ATP exchange in the absence of proton gradient by the H^+-ATPase from yeast plasma membranes. *FEBS Lett.* **212**: 323.

Denda, K., J. Konishi, T. Oshima, T. Date, and M. Yoshida. 1988. Molecular cloning of the β-subunit of a possible non-F_0F_1 type ATP synthase from the acidothermophilic archaebacterium *Sulfolobus acidocalcarius. J. Biol. Chem.* **263**: 17251.

Devereux, J., P. Haeberli, and O. Smithies. 1984. A comprehensive set of sequence analysis programs for the VAX. *Nucleic Acids Res.* **12**: 387.

De Vos, A.M., L. Tong, M.V. Milburn, P.M. Matias, J. Jancarik, S. Noguchi, S. Nishimura, K. Miura, E. Ohtsuka, and S.H. Kim. 1988. Three-dimensional structure of an oncogene protein: Catalytic domain of human c-H-*ras* p21. *Science* **239**: 888.

Dieter, P. and D. Marme. 1981. A calmodulin-dependent, microsomal ATPase from corn (*Zea mays* L.). *FEBS Lett.* **125**: 245.

Dreusicke, D. and G.E. Schulz. 1986. The glycine-rich loop of adenylate kinase forms a giant anion hole. *FEBS Lett.* **208**: 301.

Dufour, J.P. and A. Goffeau. 1980. Molecular and kinetic properties of the purified plasma membrane ATPase of the yeast *Schizosaccharomyces pombe*. *Eur. J. Biochem.* **105**: 145.

Duncan, T.M. and A.E. Senior. 1985. The defective proton-ATPase of *uncD* mutants of *Escherichia coli*. Two mutations which affect the catalytic mechanism. *J. Biol. Chem.* **260**: 4901.

Duran, A., B. Bowers, and E. Cabib. 1975. Chitin synthetase zymogen is attached to the yeast plasma membrane. *Proc. Natl. Acad. Sci.* **72**: 3952.

Duro, A.F. and R. Serrano. 1981. Inhibiton of succinate production during yeast fermentation by deenergization of the plasma membrane. *Curr. Microbiol.* **6**: 111.

Eddy, A.A. 1982. Mechanisms of solute transport in selected eukaryotic microorganisms. *Adv. Microb. Physiol.* **23**: 1.

Eddy, A.A. and P.G. Hopkins. 1985. The putative electrogenic nitrate-proton symport of the yeast *Candida utilis*. *Biochem. J.* **231**: 291.

Eilam, Y. 1982. The effect of monovalent cations on calcium efflux in yeast. *Biochim. Biophys. Acta* **687**: 8.

Eraso, P. and C. Gancedo. 1987. Activation of yeast plasma membrane ATPase by acid pH during growth. *FEBS Lett.* **224**: 187.

Eraso, P., A. Cid, and R. Serrano. 1987. Tight control of yeast plasma membrane ATPase during changes in growth conditions and gene dosage. *FEBS Lett.* **224**: 193.

Eraso, P., M.J. Mazon, and J.M. Gancedo. 1984. Pitfalls in the measurements of membrane potential in yeast cells using tetraphenylphosphonium. *Biochim. Biophys. Acta* **778**: 516.

Fersht, A. 1985. *Enzyme structure and mechanism*. Freeman, New York.

Finegold, A.A., D.I. Johnson, C.C. Farnsworth, M.H. Gelb, S.R. Judd, J.A. Glomset, and F. Tamanoi. 1991. Protein geranylgeranyltransferase of *Saccharomyces cerevisiae* is specific for Cys-Xaa-Xaa-Leu motif proteins and requires the *CDC43* gene product but not the *DPR1* gene product. *Proc. Natl. Acad. Sci.* **88**: 4448.

Foury, F. and A. Goffeau. 1975. Stimulation of active uptake of nucleosides and amino acids by cAMP in the yeast *Schizosaccharomyces pombe*. *J. Biol. Chem.* **250**: 2354.

Francois, J., P. Eraso, and C. Gancedo. 1987. Changes in the concentration of cAMP, fructose 2,6-bisphosphate and related metabolites and enzymes in *Saccharomyces cerevisiae* during growth on glucose. *Eur. J. Biochem.* **164**: 369.

Franzusoff, A. and V.P. Cirillo. 1982. Uptake and phosphorylation of 2-deoxy-D-glucose by wild-type and single-kinase strains of *Saccharomyces cerevisiae*. *Biochim. Biophys. Acta* **688**: 295.

——. 1983. Glucose transport activity in isolated plasma membrane vesicles from *Saccharomyces cerevisiae*. *J. Biol. Chem.* **258**: 3608.

Fry, D.C., S.A. Kuby, and A.S. Mildvan. 1986. ATP-binding site of adenylate kinase: Mechanistic implications of its homology with *ras*-encoded p21, F_1-ATPase and other nucleotide-binding proteins. *Proc. Natl. Acad. Sci.* **83**: 907.

Fuhrman, G.F., C. Boehm, and A.P.R. Theuvenet. 1976. Sugar transport and potassium permeability in yeast plasma membrane vesicles. *Biochim. Biophys. Acta* **433**: 583.

Gaber, R.F., C.A. Styles, and G.R. Fink. 1988. *TRK1* encodes a plasma membrane protein required for high-affinity potassium transport in *Saccharomyces cerevisiae*. *Mol. Cell. Biol.* **8**: 2848.

Gancedo, C. and R. Serrano. 1988. Energy yielding metabolism. In *The yeasts*, edition (ed. A.H. Rose and J.S. Harrison), vol. 3, p. 205. Academic Press, New

Gasnier, B. 1987. Characterization of low- and high-affinity glucose transporters in the yeast *Kluyveromyces marxianus*. *Biochim. Biophys. Acta* **903**: 425.

Ghislain, M., A. Schlesser, and A. Goffeau. 1987. Mutation of a conserved glycine residue modifies the vanadate sensitivity of the plasma membrane H^+-ATPase from *Schizosaccharomyces pombe. J. Biol. Chem.* **262**: 17549.

Gillies, R.J. 1982. Intracellular pH and proliferation in yeast, *Tetrahymena* and sea urchin eggs. In *Intracellular pH: Its measurement, regulation and utilization in cellular functions* (ed. R. Nuccitelli and D.W. Deamer), p. 341. A.R. Liss, New York.

Gillies, R.J., K. Ugurbil, J.A. den Hollander, and R.G. Shulman. 1981. ^{31}P NMR studies of intracellular pH and phosphate metabolism during cell division cycle of *Saccharomyces cerevisiae. Proc. Natl. Acad. Sci.* **78**: 2125.

Goffeau, A. and C.W. Slayman. 1981. The proton-translocating ATPase of the fungal plasma membrane. *Biochim. Biophys. Acta* **639**: 197.

Goormaghtigh, E., C. Chadwick, and G.A. Scarborough. 1986. Monomers of the *Neurospora* plasma membrane H^+-ATPase catalyze efficient proton translocation. *J. Biol. Chem.* **261**: 7466.

Gustin, M.C., X.-L. Zhou, B. Martinac, and C. Kung. 1988. A mechanosensitive ion channel in the yeast plasma membrane. *Science* **242**: 762.

Gustin, M.C., B. Martinac, Y. Saimi, M.R. Culbertson, and C. Kung. 1986. Ion channels in yeast. *Science* **233**: 1195.

Hager, K.M., S.M. Mandala, J.W. Davenport, D.W. Speicher, E.J. Benz, and C.W. Slayman. 1986. Amino-acid sequence of the plasma membrane H^+-ATPase of *Neurospora crassa*, deduced from its genomic and cDNA sequences. *Proc. Natl. Acad. Sci.* **83**: 7693.

Hammes, G.G. 1982. Unifying concept for the coupling between ion pumping and ATP hydrolysis or synthesis. *Proc. Natl. Acad. Sci.* **79**: 6881.

Hardie, G. 1988. Pseudosubstrates turn off protein kinases. *Nature* **335**: 592.

Hedrich, R., U.I. Flugge, and J.M. Fernandez. 1986. Patch-clamp studies of ion transport in isolated plant vacuoles. *FEBS Lett.* **204**: 228.

Hedrich, R., J.I. Schroeder, and J.M. Fernandez. 1987. Patch-clamp studies on higher plant cells: A perspective. *Trends Biochem. Sci.* **12**: 49.

Hennessey, J.P. and G.A. Scarborough. 1988. Secondary structure of the *Neurospora crassa* plasma membrane H^+-ATPase as estimated by circular dichroism. *J. Biol. Chem.* **263**: 3123.

Hesse, J.E., L. Wieczorek, K. Altendorf, A.S. Reicin, E. Dorus, and W. Epstein. 1984. Sequence homology between two membrane transport ATPases, the Kdp-ATPase of *Escherichia coli* and the Ca^{++}-ATPase of sarcoplasmic reticulum. *Proc. Natl. Acad. Sci.* **81**: 4746.

Higgins, C.F., M.P. Gallagher, M.L. Mimmack, and S.R. Pearce. 1988. A family of closely related ATP-binding subunits from prokaryotic and eukaryotic cells. *BioEssays* **8**: 11.

Hirata, R., Y. Ohsumi, A. Nakano, H. Kawasaki, K. Suzuki, and Y. Anraku. 1990. Molecular structure of a gene, *VMA1*, encoding the catalytic subunit of H^+-translocating adenosine triphosphatase from vacuolar membranes of *Saccharomyces cerevisiae. J. Biol. Chem.* **265**: 6726.

Hoffman, W. 1985. Molecular characterization of the *CAN1* locus in *Saccharomyces cerevisiae*. A transmembrane protein without *N*-terminal hydrophobic signal sequence. *J. Biol. Chem.* **260**: 11831.

Holcomb, C.L., W.J. Hansen, T. Etcheverry, and R. Schekman. 1988. Secretory vesicles externalize the major plasma membrane ATPase in yeast. *J. Cell Biol.* **106**: 641.

Hopkins, P., M.R. Chevallier, R. Jund, and A.A. Eddy. 1988. Use of plasmid vectors to show that the uracil and cytosine permeases of the yeast *Saccharomyces cerevisiae* are electrogenic proton symports. *FEMS Microbiol. Lett.* **49**: 173.

Horak, J. 1986. Amino acid transport in eukaryotic microorganisms. *Biochim. Biophys. Acta* **864**: 223.

Jauniaux, J.C. and M. Grenson. 1990. *GAP1*, the general amino acid permease gene of *Saccharomyces cerevisiae*. Nucleotide sequence, protein similarity with other bakers' yeast amino acid permeases, and nitrogen catabolite repression. *Eur. J. Biochem.* **190**: 39.

Jencks, W.P. 1989. How does a calcium pump pump calcium? *J. Biol. Chem.* **264**: 18855.

Jimenez, J. and N. Van Uden. 1985. Use of extracellular acidification for the rapid testing of ethanol tolerance in yeast. *Biotechnol. Bioeng.* **27**: 1596.

Johnson, D.I., C.W. Jacobs, J.R. Pringle, L.C. Robinson, G.F. Carle, and M.V. Olson. 1987. Mapping of the *Saccharomyces cerevisiae CDC3, CDC25*, and *CDC42* genes to chromosome XII by chromosome blotting and tetrad analysis. *Yeast* **3**: 243.

Jorgensen, P.L. and J.P. Andersen. 1988. Structural basis for E_1-E_2 conformational transition in Na,K-pump and Ca-pump proteins. *J. Membr. Biol.* **103**: 95.

Jorgensen, P.L., S.J.D. Karlish, and C. Gitler. 1982. Evidence for the organization of the transmembrane segments of (Na,K)-ATPase based on labeling lipid-embedded and surface domains of the α-subunit. *J. Biol. Chem.* **257**: 7435.

Jund, R., E. Weber, and M.R. Chevallier. 1988. Primary structure of the uracil transport protein of *Saccharomyces cerevisiae*. *Eur. J. Biochem.* **171**: 417.

Kaibuchi, K., A. Miyajima, K.-I. Arai, and K. Matsumoto. 1986. Possible involvement of *RAS*-encoded proteins in glucose-induced inositolphospholipid turnover in *Saccharomyces cerevisiae*. *Proc. Natl. Acad. Sci.* **83**: 8172.

Kane, P.M., C.T. Yamashiro, and T.H. Stevens. 1989. Biochemical characterization of the yeast vacuolar H^+-ATPase. *J. Biol. Chem.* **264**: 19236.

Kane, P.M., C.T. Yamashiro, D.F. Wolczyk, N. Neff, M. Goebl, and T.H. Stevens. 1990. Protein splicing converts the yeast *TFP1* gene product to the 69-kD subunit of the vacuolar H^+-adenosine triphosphatase. *Science* **250**: 651.

Kitamoto, K., K. Yoshizawa, Y. Ohsumi, and Y. Anraku. 1988. Dynamic aspects of vacuolar and cytosolic amino acid pools of *Saccharomyces cerevisiae*. *J. Bacteriol.* **170**: 2683.

Klionsky, D.J. and S.D. Emr. 1989. Membrane protein sorting: Biosynthesis, transport and processing of yeast vacuolar alkaline phosphatase. *EMBO J.* **8**: 2241.

Klionsky, D.J., P.K. Herman, and S.D. Emr. 1990. The fungal vacuole: Composition, function and biogenesis. *Microbiol. Rev.* **54**: 266.

Ko, C.H., A.M. Buckley, and R.F. Gaber. 1990. *TRK2* is required for low affinity K^+ transport in *Saccharomyces cerevisiae*. *Genetics* **125**: 305.

Koland, J.G. and G.G. Hammes. 1986. Steady-state kinetic studies of purified yeast plasma membrane proton-translocating ATPase. *J. Biol. Chem.* **261**: 5936.

Kolarov, J., J. Kulpa, M. Baijot, and A. Goffeau. 1988. Characterization of a protein serine kinase from yeast plasma membrane. *J. Biol. Chem.* **263**: 10613.

Kovac, L. 1985. Calcium and *Saccharomyces cerevisiae*. *Biochim. Biophys. Acta* **840**: 317.

Kovac, L. and L. Varecka. 1981. Membrane potentials in respiring and respiratory-deficient yeast monitored by a fluorescent dye. *Biochim. Biophys. Acta* **637**: 209.

Kovac, L., E. Bohmerova, and P. Butko. 1982. Ionophores and intact cells. I.

Valinomycin and nigericin act preferentially on mitochondria and not on the plasma membrane of *Saccharomyces cerevisiae*. *Biochim. Biophys. Acta* **721**: 341.

Kubler, O., H. Gross, and H. Moor. 1978. Complementary structures of membrane fracture faces obtained by ultrahigh vacuum freeze-fracturing at −196°C and digital image processing. *Ultramicroscopy* **3**: 161.

Kyte, J. and R.F. Doolittle. 1982. A simple method for displaying the hydropathic character of a protein. *J. Mol. Biol.* **157**: 105.

La Cour, T.F.M., J. Nyborg, S. Thirup, and B.F.C. Clark. 1985. Structural details of the binding of guanosine diphosphate to elongation factor Tu from *E. coli* as studied by X-ray crystallography. *EMBO J.* **4**: 2385.

Lang, J.M. and V.P. Cirillo. 1987. Glucose transport in a kinaseless *Saccharomyces cerevisiae* mutant. *J. Bacteriol.* **169**: 2932.

Latorre, R. and C. Miller. 1983. Conduction and selectivity in potassium channels. *J. Membr. Biol.* **71**: 11.

Leão, C. and N. Van Uden. 1984. Effects of ethanol and other alkanols on passive proton influx in the yeast *Saccharomyces cerevisiae*. *Biochim. Biophys. Acta* **774**: 43.

⸻. 1986. Transport of lactate and other short-chain monocarboxylates in the yeast *Candida utilis*. *Appl. Microbiol. Biotechnol.* **23**: 389.

Levin, D.E., F.O. Fields, R. Kunisawa, J.M. Bishop, and J. Thorner. 1990. A candidate protein kinase C gene, *PKC1*, is required for the *S. cerevisiae* cell cycle. *Cell* **62**: 213.

Lichko, L.P. and L.A. Okorokov. 1985. What family of ATPases does the vacuolar H$^+$-ATPase belong to? *FEBS Lett.* **187**: 349.

Lichtenberg, H.C., H. Giebeler, and M. Hofer. 1988. Measurements of electrical potential differences across yeast plasma membranes with microelectrodes are consistent with values from steady-state distribution of tetraphenylphosphonium in *Pichia humboldtii*. *J. Membr. Biol.* **103**: 255.

Loureiro-Dias, M.C. 1988. Movements of protons coupled to glucose transport in yeast. A comparative study among 248 yeast strains. *Antonie Leeuwenhoek J. Microbiol.* **54**: 331.

Lucas, C. and N. Van Uden. 1986. Transport of hemicellulose monomers in the xylose-fermenting yeast *Candida shehatae*. *Appl. Microbiol. Biotechnol.* **23**: 491.

MacLennan, D.H., C.J. Brandl, B. Korczak, and N.M. Green. 1985. Amino-acid sequence of a Ca^{2+} + Mg^{2+}-dependent ATPase from rabbit muscle sarcoplasmic reticulum, deduced from its complementary DNA sequence. *Nature* **316**: 696.

Maiden, M.C.J., E.O. Davis, S.A. Baldwin, D.C.M. Moore, and P.J.F. Henderson. 1987. Mammalian and bacterial sugar transport proteins are homologous. *Nature* **325**: 641.

Malpartida, F. and R. Serrano. 1981a. Reconstitution of the proton-translocating ATPase of yeast plasma membranes. *J. Biol. Chem.* **256**: 4175.

⸻. 1981b. Proton translocation catalyzed by the purified plasma membrane ATPase reconstituted in liposomes. *FEBS Lett.* **131**: 351.

Mandala, S.M. and C.W. Slayman. 1988. Identification of tryptic cleavage sites for two conformational states of the *Neurospora* plasma membrane H$^+$-ATPase. *J. Biol. Chem.* **263**: 15122.

⸻. 1989. The amino and carboxyl termini of the *Neurospora* plasma membrane H$^+$-ATPase are cytoplasmically located. *J. Biol. Chem.* **264**: 16276.

Manolson, M.F., B.F.F. Ouellette, M. Filion, and R.J. Poole. 1988. cDNA sequence and homologies of the "57-kDa" nucleotide-binding subunit of the vacuolar ATPase from *Arabidopsis*. *J. Biol. Chem.* **263**: 17987.

Martinez, A.F. and J. Schwencke. 1988. Chitin synthetase activity is bound to the chitosomes and to the plasma membrane in protoplasts of *Saccharomyces cerevisiae*.

Biochim. Biophys. Acta **946**: 328.

Maruyama, K. and D.H. MacLennan. 1988. Mutation of aspartic acid-351, lysine-352 and lysine-515 alters the Ca^{2+} transport activity of the Ca^{2+}-ATPase expressed in COS-1 cells. *Proc. Natl. Acad. Sci.* **85**: 3314.

Matile, P., H. Moor, and K. Muhlethaler. 1967. Isolation and properties of the plasmalemma in yeast. *Arch. Mikrobiol.* **58**: 201.

Maurer, A. and K. Muhlethaler. 1981. Specific labeling of glycoproteins in yeast plasma membrane with Concanavalin A. *Eur. J. Cell Biol.* **25**: 58.

Mazon, M.J., M. Behrens, F. Portillo, and R. Piñon. 1989. cAMP- and *RAS*-independent nutritional regulation of plasma membrane H^+-ATPase in *Saccharomyces cerevisiae. J. Gen. Microbiol.* **135**: 1453.

McCusker, J.H., D.S. Perlin, and J.E. Haber. 1987. Pleiotropic plasma membrane ATPase mutations of *Saccharomyces cerevisiae. Mol. Cell. Biol.* **7**: 4082.

McDonough, J.P. and H.P. Mahler. 1982. Covalent phosphorylation of the Mg^{2+}-dependent ATPase of yeast plasma membranes. *J. Biol. Chem.* **257**: 14579.

Meade, J.C., J. Shaw, G. Gallagher, S. Lemaster, and J.R. Stringer. 1987. Structure and expression of a tandem gene pair in *Leishmania donovani* that encodes a protein structurally homologous to eukaryotic cation-transporting ATPases. *Mol. Cell. Biol.* **7**: 3937.

Mitchell, P. 1966. Chemiosmotic coupling in oxidative and photosynthetic phosphorylation. *Biol. Rev.* **41**: 445.

Mitchinson, C., A.F. Wilderspin, B.J. Trinnaman, and N.M. Green. 1982. Identification of a labeled peptide after stoichiometric reaction of fluorescein isothiocyanate with the Ca^{2+}-dependent ATPase of sarcoplasmic reticulum. *FEBS Lett.* **146**: 87.

Miyakawa, T., T. Tachikawa, Y.K. Jeong, E. Tsuchiya, and S. Fukui. 1985. Transient increase of Ca^{2+} uptake as a signal for mating-pheromone-induced differentiation in the heterobasidiomycetous yeast *Rhodosporidium toruloides. J. Bacteriol.* **162**: 1304.

————. 1987. Inhibition of membrane Ca^{2+}-ATPase in vitro by mating pheromone in *Rhodosporidium toruloides*, a heterobasidiomycetous yeast. *Biochem. Biophys. Res. Commun.* **143**: 893.

Möller, W. and R. Amons. 1985. Phosphate-binding sequences in nucleotide-binding proteins. *FEBS Lett.* **186**: 1.

Monk, B.C., C. Montesinos, and R. Serrano. 1989. Sidedness of yeast plasma membrane vesicles and mechanisms of activation of the ATPase by detergents. *Biochim. Biophys. Acta* **981**: 226.

Mori, T., Y. Takai, R. Minakuchi, B. Yu, and Y. Nishizuka. 1980. Inhibitory action of chlorpromazine, dibucaine and other phospholipid-interacting drugs on calcium-activated, phospholipid-dependent protein kinase C. *J. Biol. Chem.* **255**: 8378.

Mortimer, R.K., D. Schild, C.R. Contopoulou, and J.A. Kans. 1989. Genetic map of *Saccharomyces cerevisiae*, edition 10. *Yeast* **5**: 321.

Munder, T. and H. Küntzel. 1989. Glucose-induced cAMP signaling in *Saccharomyces cerevisiae* is mediated by the CDC25 protein. *FEBS Lett.* **242**: 341.

Navarrete, R. and R. Serrano. 1983. Solubilization of yeast plasma membranes and mitochondria by different types of non-denaturing detergents. *Biochim. Biophys. Acta* **728**: 403.

Neher, E. and B. Sakman. 1976. Single-channel currents recorded from membrane of denervated frog muscle fibres. *Nature* **260**: 779.

Nelson, H. and N. Nelson. 1989. The progenitor of ATP synthases was closely related to the current vacuolar H^+-ATPase. *FEBS Lett.* **247**: 147.

————. 1990. Disruption of genes encoding subunits of yeast vacuolar H^+-ATPase causes conditional lethality. *Proc. Natl. Acad. Sci.* **87:** 3503.

Nelson, H., S. Mandiyan, and N. Nelson. 1989. A conserved gene encoding the 57-kDa subunit of the yeast vacuolar H+-ATPase. *J. Biol. Chem.* **264:** 1775.

Nelson, N. and L. Taiz. 1989. The evolution of H+-ATPases. *Trends Biochem. Sci.* **14:** 113.

Noumi, T., M. Futai, and H. Kanazawa. 1984. Replacement of serine 373 by phenylalanine in the α subunit of *Escherichia coli* F_1-ATPase results in loss of steady-state catalysis by the enzyme. *J. Biol. Chem.* **259:** 10076.

Noumi, T., M. Maeda, and M. Futai. 1988. A homologous sequence between H+- ATPase (F_0F_1) and cation-transporting ATPases. Thr285–Asp replacement in the β subunit of *Escherichia coli* F_1 changes its catalytic properties. *J. Biol. Chem.* **263:** 8765.

Ohsumi, Y. and Y. Anraku. 1981. Active transport of basic amino-acids driven by a proton-motive force in vacuolar membrane vesicles of *Saccharomyces cerevisiae. J. Biol. Chem.* **256:** 2079.

————. 1983. Calcium transport driven by a proton-motive force in vacuolar membrane vesicles of *Saccharomyces cerevisiae. J. Biol. Chem.* **258:** 5614.

————. 1985. Specific induction of Ca^{2+} transport activity in *MATa* cells of *Saccharomyces cerevisiae* by a mating pheromone, α factor. *J. Biol. Chem.* **260:** 10482.

Ohsumi, Y., K. Kitamoto, and Y. Anraku. 1988. Changes in the permeability barrier of the yeast plasma membrane by cupric ion. *J. Bacteriol.* **170:** 2676.

Ohta, T., K. Nagano, and M. Yoshida. 1986. The active site structure of Na+/K+-transporting ATPase: Location of the 5′-(p-fluorosulfonyl) benzoyladenosine binding site and soluble peptides released by trypsin. *Proc. Natl. Acad. Sci.* **83:** 2071.

Ohya, Y., Y. Ohsumi, and Y. Anraku. 1984. Genetic study of the role of calcium ions in the cell division cycle of *Saccharomyces cerevisiae*: A calcium-dependent mutant and its trifluoperazine dependent pseudorevertant. *Mol. Gen. Genet.* **193:** 389.

Ohya, Y., M. Goebl, L.E. Goodman, S. Petersen-Bjorn, J.D. Friesen, F. Tamanoi, and Y. Anraku. 1991. Yeast *CAL1* is a structural and functional homologue to the *DPR1* (*RAM*) gene involved in Ras processing. *J. Biol. Chem.* (in press).

Okorokov, L.A., T.V. Kulakovskaya, L.P. Lichko, and E.V. Polorotova. 1985. H+/ion antiport as the principal mechanism of transport in the vacuolar membrane of the yeast *Saccharomyces carlsbergensis. FEBS Lett.* **192:** 303.

Ongjoco, R., K. Szkutnicka, and V. P. Cirillo. 1987. Glucose transport in vesicles reconstituted from *Saccharomyces cerevisiae* membranes and liposomes. *J. Bacteriol.* **169:** 2926.

Opekarova, M., A.J.M. Driessen, and W.N. Konings. 1987. Protonmotive-force-driven leucine uptake in yeast plasma membrane vesicles. *FEBS Lett.* **213:** 45.

Ovchinnikov, Y.A., K.N. Dzhandzugazyan, S.V. Lutsenko, A.A. Mustayev, and N.N. Modyanov. 1987a. Affinity modification of E_1-form of Na+, K+- ATPase revealed Asp-710 in the catalytic site. *FEBS Lett.* **217:** 111.

Ovchinnikov, Y.A., N.M. Arzamazova, E.A. Arystarkhova, N.M. Gevondyan, N.A. Aldanova, and N.N. Modyanov. 1987b. Detailed structural analysis of exposed domains of membrane-bound Na+,K+-ATPase. A model of transmembrane arrangement. *FEBS Lett.* **217:** 269.

Ovchinnikov, Y.A., N.M. Luneva, E.A. Arystarkhova, N.M. Gevondyan, N.M. Arzamazova, A.T. Kozhich, V.A. Nesmeyanov, and N.N. Modyanov. 1988. Topology of Na+,K+-ATPase. Identification of the extra- and intracellular hydrophilic loops of the catalytic subunit by specific antibodies. *FEBS Lett.* **227:** 230.

Pai, E.F., W. Kabsch, U. Krengel, K.C. Holmes, J. John, and A. Wittinghofer. 1989. Structure of the guanine-nucleotide-binding domain of the Ha-*ras* oncogene product p21 in the triphosphate conformation. *Nature* **341**: 209.

Pain, R.H. 1986. Evolutionary economics. *Nature* **323**: 393.

Palacios, J. and R. Serrano. 1978. Proton permeability induced by polyene antibiotics. A plausible mechanism for their inhibition of maltose fermentation in yeast. *FEBS Lett.* **91**: 198.

Pardo, J.M. and R. Serrano. 1989. Structure of a plasma membrane H$^+$-ATPase gene from the plant *Arabidopsis thaliana*. *J. Biol. Chem.* **264**: 8557.

Pardo, J.P. and C.W. Slayman. 1988. The fluorescein isothiocyanate-binding site of the plasma membrane H$^+$-ATPase of *Neurospora crassa*. *J. Biol. Chem.* **263**: 18664.

Parsonage, D., M.K. Al-Shawi, and A.E. Senior. 1988a. Directed mutations of the strongly conserved lysine 155 in the catalytic nucleotide-binding domain of β-subunit of F$_1$-ATPase from *Escherichia coli*. *J. Biol. Chem.* **263**: 4740.

Parsonage, D., S. Wilke-Mounts, and A.E. Senior. 1988b. *E. coli* F$_1$-ATPase: Site-directed mutagenesis of the β subunit. *FEBS Lett.* **232**: 111.

Pascual, C., A. Alonso, I. Garcia, C. Romay, and A. Kotyk. 1988. Effect of ethanol on glucose transport, key glycolytic enzymes and proton extrusion in *Saccharomyces cerevisiae*. *Biotechnol. Bioeng.* **32**: 374.

Pedersen, P. L. and E. Carafoli. 1987a. Ion motive ATPases. I. Ubiquity, properties and significance to cell function. *Trends Biochem. Sci.* **12**: 146.

––––––. 1987b. Ion motive ATPases. II. Energy coupling and work output. *Trends Biochem. Sci.* **12**: 186.

Peinado, J.M. and M.C. Loureiro-Dias. 1986. Reversible loss of affinity induced by glucose in the maltose-H$^+$ symport of *Saccharomyces cerevisiae*. *Biochim. Biophys. Acta* **856**: 189.

Peinado, J.M., A. Barbero, and N. Van Uden. 1987. Repression and inactivation by glucose of the maltose transport system of *Candida utilis*. *Appl. Microbiol. Biotechnol.* **26**: 154.

Peña, A. 1975. Studies on the mechanism of K$^+$ transport in yeast. *Arch. Biochem. Biophys.* **167**: 397.

Perlin, D.S. and C.L. Brown. 1987. Identification of structurally distinct catalytic intermediates of the H$^+$-ATPase from yeast plasma membranes. *J. Biol. Chem.* **262**: 6788.

Perlin, D.S., C.L. Brown, and J.E. Haber. 1988. Membrane potential defect in hygromycin B-resistant *pma1* mutants of *Saccharomyces cerevisiae*. *J. Biol. Chem.* **263**: 18118.

Portillo, F. and M.J. Mazon. 1985. Activation of yeast plasma membrane ATPase by phorbol ester. *FEBS Lett.* **192**: 95.

––––––. 1986. The *Saccharomyces cerevisiae* start mutant carrying the *cdc25* mutation is defective in activation of plasma membrane ATPase by glucose. *J. Bacteriol.* **168**: 1254.

Portillo, F. and R. Serrano. 1988. Dissection of functional domains of the yeast proton-pumping ATPase by directed mutagenesis. *EMBO J.* **7**: 1793.

––––––. 1989. Growth control strength and active site of yeast plasma membrane ATPase studied by directed mutagenesis. *Eur. J. Biochem.* **186**: 501.

Portillo, F., I.F. de Larrinoa, and R. Serrano. 1989. Deletion analysis of yeast plasma membrane H$^+$-ATPase and identification of a regulatory domain at the carboxyl-terminus. *FEBS Lett.* **247**: 381.

Prasad, R. and M. Hofer. 1986. Tetraphenylphosphonium is an indicator of negative membrane potential in *Candida albicans*. *Biochim. Biophys. Acta* **861**: 377.

Purwin, C., K. Nicolay, W.A. Scheffers, and H. Holzer. 1986. Mechanism of control of adenylate cyclase activity in yeast by fermentable sugars and carbonyl cyanide m-chlorophenylhydrazone. *J. Biol. Chem.* **261:** 8744.

Rai, R., F.S. Genbauffe, and T.G. Cooper. 1988. Structure and transcription of the allantoate permease gene (*DAL5*) from *Saccharomyces cerevisiae. J. Bacteriol.* **170:** 266.

Ramirez, J.A., V. Vacata, J.H. McCusker, J.E. Haber, R.K. Mortimer, W.G. Owen, and H. Lecar. 1989. ATP-sensitive K^+ channels in a plasma membrane H^+-ATPase mutant of the yeast *Saccharomyces cerevisiae. Proc. Natl. Acad. Sci.* **86:** 7866.

Ramos, J. and V.P. Cirillo. 1989. Role of cyclic AMP-dependent protein kinase in catabolite inactivation of the glucose and galactose transporters in *Saccharomyces cerevisiae. J. Bacteriol.* **171:** 3545.

Ramos, J. and A. Rodriguez-Navarro. 1986. Regulation and interconversion of the potassium transport systems of *Saccharomyces cerevisiae* as revealed by rubidium transport. *Eur. J. Biochem.* **154:** 307.

Ramos, J., P. Contreras, and A. Rodriguez-Navarro. 1985. A potassium transport mutant of *Saccharomyces cerevisiae. Arch. Microbiol.* **143:** 88.

Ramos, J., K. Szkutnicka, and V.P. Cirillo. 1988. Relationship between low- and high-affinity glucose transport systems of *Saccharomyces cerevisiae. J. Bacteriol.* **170:** 5375.

―――. 1989. Characteristics of galactose transport in *Saccharomyces cerevisiae* cells and reconstituted lipid vesicles. *J. Bacteriol.* **171:** 3539.

Rasi-Caldogno, F., M.C. Pugliarello, and M.I. De Michelis. 1987. The Ca^{2+} transport ATPase of plant plasma membrane catalyzes an H^+/Ca^{2+} exchange. *Plant Physiol.* **83:** 994.

Reithmeier, R.A.F. and D.H. MacLennan. 1981. The NH_2 terminus of the $(Ca^{2+} + Mg^{2+})$-ATPase is located on the cytoplasmic surface of the sarcoplasmic reticulum membrane. *J. Biol. Chem.* **256:** 5957.

Roberts, C.J., G. Pohlig, J.H. Rothman, and T.H. Stevens. 1989. Structure, biosynthesis, and localization of dipeptidyl aminopeptidase B, an integral membrane glycoprotein of the yeast vacuole. *J. Cell Biol.* **108:** 1363.

Rodriguez-Navarro, A. and J. Ramos. 1984. Dual system for potassium transport in *Saccharomyces cerevisiae. J. Bacteriol.* **159:** 940.

Rodriguez-Navarro, A., M.R. Blatt, and C.L. Slayman. 1986. A potassium-proton symport in *Neurospora crassa. J. Gen. Physiol.* **87:** 649.

Ronjat, M., J.J. Lacapere, J.P. Dufour, and Y. Dupont. 1987. Study of the nucleotide binding site of the yeast *Schizosaccharomyces pombe* plasma membrane H^+-ATPase using formycin triphosphate-terbium complex. *J. Biol. Chem.* **262:** 3146.

Roon, R.J., F.S. Larimore, G.M. Meyer, and R.A. Kreisle. 1978. Characterization of a Dio-9-resistant strain of *Saccharomyces cerevisiae. Arch. Biochem. Biophys.* **185:** 142.

Rudolph, H.K., A. Antebi, G.R. Fink, C.M. Buckley, T.E. Dorman, J. LeVitre, L.S. Davidow, J. Mao, and D.T. Moir. 1989. The yeast secretory pathway is perturbed by mutations in *PMR1*, a member of a Ca^{2+} ATPase family. *Cell* **58:** 133.

Saier, M.H., M. Yamada, B. Erni, K. Suda, J. Lengeler, R. Ebner, P. Argos, B. Rak, K. Schnetz, C.A. Lee, G.C. Stewart, F. Breidt, E.B. Waygood, K.G. Peri, and R.F. Doolittle. 1988. Sugar permeases of the bacterial phosphoenolpyruvate-dependent phosphotransferase system: Sequence comparisons. *FASEB J.* **2:** 199.

Saimi, Y., B. Martinac, M.C. Gustin, M.R. Culbertson, J. Adler, and C. Kung. 1988. Ion channels in *Paramecium*, yeast and *Escherichia coli. Trends Biochem. Sci.* **13:** 304.

Sarkadi, B., A. Enyedi, Z. Foldes-Papp, and G. Gardos. 1986. Molecular characterization of the *in situ* red cell membrane calcium pump by limited proteolysis. *J. Biol. Chem.* **261:** 9552.

Sato, T., Y. Ohsumi, and Y. Anraku. 1984. Substrate specificities of active transport systems for amino acids in vacuolar membrane vesicles of *Saccharomyces cerevisiae*: Evidence of seven independent proton/amino acid antiport systems. *J. Biol. Chem.* **259**: 11509.

Scarborough, G.A. and R. Addison. 1984. On the subunit composition of the *Neurospora* plasma membrane H⁺-ATPase. *J. Biol. Chem.* **259**: 9109.

Schlesser, A., S. Ulaszewski, M. Ghislain, and A. Goffeau. 1988. A second transport ATPase gene in *Saccharomyces cerevisiae*. *J. Biol. Chem.* **263**: 19480.

Schmidt, R., M.F. Manolson, and M.R. Chevallier. 1984. Photoaffinity labeling and characterization of the cloned purine-cytosine transport system in *Saccharomyces cerevisiae*. *Proc. Natl. Acad. Sci.* **81**: 6276.

Schroeder, J.I. and S. Hagiwara. 1989. Cytosolic calcium regulates ion channels in the plasma membrane of *Vicia faba* guard cells. *Nature* **338**: 427.

Schroeder, J. I., K. Raschke, and E. Neher. 1987. Voltage dependence of K⁺ channels in guard-cell protoplasts. *Proc. Natl. Acad. Sci.* **84**: 4108.

Schuddemat, J., P.J.A. Van den Broek, and J. Van Steveninck. 1986. Effect of xylose incubation on the glucose transport system in *Saccharomyces cerevisiae*. *Biochim. Biophys. Acta* **861**: 489.

————. 1988. The influence of ATP on sugar uptake mediated by the constitutive glucose carrier of *Saccharomyces cerevisiae*. *Biochim. Biophys. Acta* **937**: 81.

Schwencke, J. and H. de Robichon-Szulmajster. 1976. The transport of S-adenosyl-L-methionine in isolated yeast vacuoles and spheroplasts. *Eur. J. Biochem.* **65**: 49.

Senior, A.E. and J.G. Wise. 1983. The proton-ATPase of bacteria and mitochondria. *J. Membr. Biol.* **73**: 105.

Serrano, R. 1980. Effect of ATPase inhibitors on the proton pump of respiratory-deficient yeast. *Eur. J. Biochem.* **105**: 419.

————. 1983. In vivo glucose activation of the yeast plasma membrane ATPase. *FEBS Lett.* **156**: 11.

————. 1985. *Plasma membrane ATPase of plants and fungi.* CRC Press, Boca Raton, Florida.

————. 1988a. Structure and function of proton translocating ATPase in plasma membranes of plants and fungi. *Biochim. Biophys. Acta* **947**: 1.

————. 1988b. H⁺-ATPase from plasma membranes of *Saccharomyces cerevisiae* and *Avena sativa* roots: Purification and reconstitution. *Methods Enzymol.* **157**: 533.

————. 1989. Structure and function of plasma membrane ATPase. *Annu. Rev. Plant Physiol. Plant Mol. Biol.* **40**: 61.

Serrano, R. and G. De la Fuente. 1974. Regulatory properties of the constitutive hexose transport in *Saccharomyces cerevisiae*. *Mol. Cell. Biochem.* **5**: 161.

Serrano, R. and F. Portillo. 1990. Catalytic and regulatory sites of yeast plasma membrane H⁺-ATPase studied by directed mutagenesis. *Biochim. Biophys. Acta* **1018**: 195.

Serrano, R., M.C. Kielland-Brandt, and G.R. Fink. 1986a. Yeast plasma membrane ATPase is essential for growth and has homology with (Na⁺+K⁺), K⁺ and Ca²⁺-ATPases. *Nature* **319**: 689.

Serrano, R., C. Montesinos, and A. Cid. 1986b. A temperature-sensitive mutant of the yeast plasma membrane ATPase obtained by in vitro mutagenesis. *FEBS Lett.* **208**: 143.

Serrano, R., C. Montesinos, and J. Sanchez. 1988. Lipid requirements of the plasma membrane ATPase from oat roots and yeast. *Plant Sci.* **56**: 117.

Serrano, R., C. Montesinos, M. Roldan, G. Garrido, C. Ferguson, K. Leonard, B.C. Monk, D.S. Perlin, and E.W. Weiler. 1991. Domains of yeast plasma membrane and ATPase-associated glycoprotein. *Biochim. Biophys. Acta* (in press).

Shih, C.-K., R. Wagner, S. Feinstein, C. Kanik-Ennulat, and N. Neff. 1988. A dominant trifluoperazine resistance gene from *Saccharomyces cerevisiae* has homology with F_0F_1 ATP synthase and confers calcium-sensitive growth. *Mol. Cell. Biol.* **8**: 3094.

Shull, G.E. and J.B. Lingrel. 1986. Molecular cloning of the rat stomach (H^++K^+) ATPase. *J. Biol. Chem.* **261**: 16788.

Shull, G.E., L.K. Lane, and J.B. Lingrel. 1986. Amino-acid sequence of the β-subunit of the (Na^++K^+)ATPase deduced from a cDNA. *Nature* **321**: 429.

Shull, G.E., A. Schwartz, and J.B. Lingrel. 1985. Amino acid sequence of the catalytic subunit of the sodium-potassium ATPase deduced from a complementary DNA. *Nature* **316**: 691.

Silver, S., G. Nucifora. L. Chu, and T. K. Misra. 1989. Bacterial resistance ATPases: Primary pumps for exporting toxic cations and anions. *Trends Biochem. Sci.* **14**: 76.

Skou, J.C. 1988. Overview: The Na,K-pump. *Methods Enzymol.* **156**: 1.

Smith, K.E. and G.G. Hammes. 1988. Studies of the phosphoenzyme intermediate of the yeast plasma membrane proton-translocating ATPase. *J. Biol. Chem.* **263**: 13774.

Solioz, M., S. Mathews, and P. Furst. 1987. Cloning of the K^+-ATPase of *Streptococcus faecalis*. Structural and evolutionary implications of its homology to the KdpB-protein of *Escherichia coli*. *J. Biol. Chem.* **262**: 7358.

Spencer-Martins, I. and N. Van Uden. 1985. Catabolite interconversion of glucose transport systems in the yeast *Candida wickerhamii*. *Biochim. Biophys. Acta* **812**: 168.

Stein, W.D. 1986. *Transport and diffusion across cell membranes.* Academic Press, New York.

Stroobant, P. and G.A. Scarborough. 1979. Active transport of calcium in *Neurospora* plasma membrane vesicles. *Proc. Natl. Acad. Sci.* **76**: 3102.

Sumrada, R. and T.G. Cooper. 1978. Control of vacuole permeability and protein degradation by the cell cycle arrest signal in *Saccharomyces cerevisiae*. *J. Bacteriol.* **136**: 234.

Surowy, T.K. and M.R. Sussman. 1986. Immunological cross-reactivity and inhibitor sensitivities of the plasma membrane H^+-ATPase from plants and fungi. *Biochim. Biophys. Acta* **848**: 24.

Sussman, M.R., J.E. Strickler, K.M. Hager, and C.W. Slayman. 1987. Location of a dicyclohexylcarbodiimide-reactive glutamate residue in the *Neurospora crassa* plasma membrane H^+-ATPase. *J. Biol. Chem.* **262**: 4569.

Szkutnicka, K., J.F. Tschopp, L. Andrews, and V.P. Cirillo. 1989. Sequence and structure of the yeast galactose transporter. *J. Bacteriol.* **171**: 4486.

Takeda, M., W.-J. Chen, J. Saltzgaber, and M.G. Douglas. 1986. Nuclear genes encoding the yeast mitochondrial ATPase complex: Analysis of *ATP1* coding the F_1-ATPase α subunit and its assembly. *J. Biol. Chem.* **261**: 15126.

Takeo, K., M. Shigeta, and Y. Takagi. 1976. Plasma membrane ultrastructural differences between the exponential and the stationary phases of *Saccharomyces cerevisiae* as revealed by freeze-etching. *J. Gen. Microbiol.* **97**: 323.

Talwalkar, R.T. and R.L. Lester. 1973. The response of diphosphoinositide and triphosphoinositide to perturbations of the adenylate energy charge in cells of *Saccharomyces cerevisiae*. *Biochim. Biophys. Acta* **306**: 412.

Tanaka, J. and G.R. Fink. 1985. The histidine permease gene (*HIP1*) of *Saccharomyces cerevisiae*. *Gene* **38**: 205.

Tanford, C. 1982. Simple model for the chemical potential change of a transported ion in active transport. *Proc. Natl. Acad. Sci.* **79**: 2882.

Taylor, W.R. and N.M. Green. 1989. The predicted secondary structures of the nucleotide-binding sites of six cation-translocating ATPases lead to a probable tertiary fold. *Eur. J. Biochem.* **179:** 241.

Theuvenet, A.P.R., B.J.W.M. Niewenhuis, J. Van de Mortel, and G.W.F.H. Borst-Pauwels. 1986. Effect of ethidium bromide and DEAE-dextran on divalent cation accumulation in yeast. Evidence for an ion-selective extrusion pump. *Biochim. Biophys. Acta* **855:** 383.

Tschopp, J., P.C. Esmon, and R. Schekman. 1984. Defective plasma membrane assembly in yeast secretory mutants. *J. Bacteriol.* **160:** 966.

Tong, L., A.M. de Vos, M.V. Milburn, J. Jancarik, S. Noguchi, S. Nishimura, K. Miura, E. Ohtsuka, and S.H. Kim. 1989. Structural differences between a *ras* oncogene protein and the normal protein. *Nature* **337:** 90.

Uchida, E., Y. Ohsumi, and Y. Anraku. 1988a. Purification of yeast vacuolar membrane H^+-ATPase and enzymological discrimination of three ATP-driven proton pumps in *Saccharomyces cerevisiae. Methods Enzymol.* **157:** 544.

―――――. 1988b. Characterization and function of catalytic subunit *a* of H^+-translocating ATPase from vacuolar membranes of *Saccharomyces cerevisiae.* A study with 7-chloro-4-nitrobenzo-2-oxa-1,3-diazole. *J. Biol. Chem.* **263:** 45.

Ulaszewski, S., E. Balzi, and A. Goffeau. 1987a. Genetic and molecular mapping of the *pma1* mutation conferring vanadate resistance to the plasma membrane ATPase from *Saccharomyces cerevisiae. Mol. Gen. Genet.* **207:** 38.

Ulaszewski, S., A. Coddington, and A. Goffeau. 1986. A new mutation for multiple drug resistance and modified plasma membrane ATPase activity in *Schizosaccharomyces pombe. Curr. Genet.* **10:** 359.

Ulaszewski, S., M. Grenson, and A. Goffeau. 1983. Modified plasma-membrane ATPase in mutants of *Saccharomyces cerevisiae. Eur. J. Biochem.* **130:** 235.

Ulaszewski, S., F. Hilger, and A. Goffeau. 1989. Cyclic AMP controls the plasma membrane H^+-ATPase activity from *Saccharomyces cerevisiae. FEBS Lett.* **245:** 131.

Ulazewski, S,. J.C. Van Herck, J.P. Dufour, J. Kulpa, B. Nieuwenhuis, and A. Goffeau. 1987b. A single mutation confers vanadate resistance to the plasma membrane H^+-ATPase from the yeast *Schizosaccharomyces pombe. J. Biol. Chem.* **262:** 223.

Vai, M., L. Popolo, and L. Alberghina. 1986. Immunological cross-reactivity of fungal and yeast plasma membrane H^+-ATPase. *FEBS Lett.* **206:** 135.

Vallejo, C.G. and R. Serrano. 1989. Physiology of mutants with reduced expression of plasma membrane H^+-ATPase. *Yeast* **5:** 307.

Vandenbol, M., J.C. Jauniaux, and M. Grenson. 1989. Nucleotide sequence of the *Saccharomyces cerevisiae PUT4* proline-permease-encoding gene: Similarities between *CAN1, HIP1* and *PUT4* permeases. *Gene* **83:** 153.

Vandenbol, M., J.C. Jauniaux, S. Vissers, and M. Grenson. 1987. Isolation of the *NPR1* gene responsible for the reactivation of ammonia-sensitive amino-acid permeases in *Saccharomyces cerevisiae.* RNA analysis and gene dosage effects. *Eur. J. Biochem.* **164:** 607.

Van den Broek, P.J.A., J. Schuddemat, C.C.M. Van Leeuwen, and J. Van Steveninck. 1986. Characterization of 2-deoxyglucose and 6-deoxyglucose transport in *Kluyveromyces marxianus*: Evidence for two different transport mechanisms. *Biochim. Biophys. Acta* **860:** 626.

Van Steveninck, J. and A. Rothstein. 1966. Sugar transport and metal binding in yeast. *J. Gen. Physiol.* **49:** 235.

Verma, A.K., A.G. Filoteo, D.R. Stanford, E.D. Wieben, J.T. Penniston, E.E. Strehler, R. Fischer, R. Heim, G. Vogel, S. Mathews, M.A. Strehler-Page, P. James, T. Vorherr, J. Krebs, and E. Carafoli. 1988. Complete primary structure of a human plasma membrane Ca^{2+} pump. *J. Biol. Chem.* **263:** 14152.

Verma, R.S., I. Spencer-Martins, and N. Van Uden. 1987. Role of de novo protein synthesis in the interconversion of glucose transport systems in the yeast *Pichia ohmeri. Biochim. Biophys. Acta* **900:** 139.

Vidal, M., A.M. Buckley, F. Hilger, and R.F. Gaber. 1990. Direct selection for mutants with increased K$^+$ transport in *Saccharomyces cerevisiae. Genetics* **125:** 313.

Villalobo, A. 1982. Potassium transport coupled to ATP hydrolysis in reconstituted proteoliposomes of yeast plasma membrane ATPase. *J. Biol. Chem.* **257:** 1824.

———. 1984. Energy-dependent H$^+$ and K$^+$ translocation by reconstituted yeast plasma membrane ATPase. *Can. J. Biochem. Cell Biol.* **62:** 865.

Wada, Y., Y. Ohsumi, M. Tanifuji, M. Kasai, and Y. Anraku. 1987. Vacuolar ion channel of the yeast *Saccharomyces cerevisiae. J. Biol. Chem.* **262:** 17260.

Walker, J.E., M. Saraste, and N.J. Gay. 1984. The *unc* operon. Nucleotide sequence, regulation and structure of ATP-synthase. *Biochim. Biophys. Acta* **768:** 164.

Walker, J.E., M. Saraste, M.J. Runswick, and N.J. Gay. 1982. Distantly related sequences in the α- and β-subunits of ATP synthase, myosin, kinases and other ATP-requiring enzymes and a common nucleotide binding fold. *EMBO J.* **1:** 945.

Weber, E., M.R. Chevallier, and R. Jund. 1988. Evolutionary relationship and secondary structure prediction in four transport proteins of *Saccharomyces cerevisiae. J. Mol. Evol.* **27:** 341.

Wehrli, E., C. Boehm, and G.F. Fuhrmann. 1975. Yeast plasma membrane vesicles suitable for transport studies. *J. Bacteriol.* **124:** 1594.

Weiss, B. and R. Levin. 1978. Mechanism for selectively inhibiting the activation of cyclic nucleotide phosphodiesterase and adenylate cyclase by anti-psychotic agents. *Adv. Cyclic Nucleotide Res.* **9:** 285.

Wiemken, A. and M. Durr. 1974. Characterization of amino acid pools in the vacuolar compartment of *Saccharomyces cerevisiae. Arch. Microbiol.* **101:** 45.

Yamashiro, C.T., P.M. Kane, D.F. Wolczyk, R.A. Preston, and T.H. Stevens. 1990. Role of vacuolar acidification in protein sorting and zymogen activation: A genetic analysis of the yeast vacuolar proton-translocating ATPase. *Mol. Cell. Biol.* **10:** 3737.

Yanagita, Y., M. Abdel-Ghany, D. Raden, N. Nelson, and E. Racker. 1987. Polypeptide-dependent protein kinase from bakers' yeast. *Proc. Natl. Acad. Sci.* **84:** 925.

Yohda, M., S. Ohta, T. Hisabori, and Y. Kagawa. 1988. Site-directed mutagenesis of stable adenosine triphosphate synthase. *Biochim. Biophys. Acta* **933:** 156.

Young, R.A. and R.W. Davis. 1985. Immunoscreening λ-gt11 recombinant DNA expression libraries. In *Genetic engineering* (ed. J.K. Setlow and A. Hollaender), vol. 8, p. 29. Plenum Press, New York.

Zimniak, L., P. Dittrich, J.P. Gogarten, H. Kibak, and L. Taiz. 1988. The cDNA sequence of the 69-kDa subunit of the carrot vacuolar H$^+$-ATPase. Homology to the beta-chain of F$_0$F$_1$-ATPases. *J. Biol. Chem.* **263:** 9102.

10
The Ribosome and Its Synthesis

John L. Woolford, Jr.
Department of Biological Sciences
Carnegie Mellon University
Pittsburgh, Pennsylvania 15213

Jonathan R. Warner
Department of Cell Biology
Albert Einstein College of Medicine
Bronx, New York 10461

I. INTRODUCTION

The ribosome can be considered the blue-collar worker of the cell, faithfully carrying out the designs provided by the structural genes. In this role, the ribosome is central to the growth and maintenance of the cell, and, in turn, many of the functions of the cell are involved in ensuring a steady and regulated supply of ribosomes. Studies of the biosynthesis of ribosomes and its regulation provide us with an entrée into many of the key questions of molecular and cellular biology. Expression of ribosomal RNA (rRNA) and protein genes involves most of the known machinery and processes of eukaryotic gene expression: transcription by RNA polymerases I, II, and III; posttranscriptional modification and process-ing of precursor RNAs; and translation of mRNAs. Efficient biosynthesis of ribosomes must involve regulation of these processes to produce equal amounts of rRNAs and ribosomal proteins in proportion to cellular growth rates. Consequently, studies of ribosome biosynthesis are closely intertwined with those of growth rate regulation. rRNA transcription and processing as well as ribosome assembly take place within a special organelle, the nucleolus; little is known about its structure and operation. The ribosomal proteins are synthesized in the cytoplasm and transported to the nucleus. The completed ribosomal subunits are transported from the nucleolus to the cytoplasm. Almost nothing is known about either of these processes. The processing of rRNA and the role of ribosomal proteins, nonribosomal proteins, and small nucleolar RNAs in this pro-cessing are similarly hazy. These are questions of fundamental impor-tance for all eukaryotic cells. The genetic, molecular biological, and biochemical techniques available in *Saccharomyces* provide a unique op-portunity to approach them.

The ribosome was considered in the previous edition of this monograph (Warner 1982) to which the reader is referred for a summary of the earlier literature. In this paper, we emphasize the progress during the past 8 years primarily in our understanding of yeast ribosome biogenesis, largely because the structure of ribosomes has been relatively neglected during that time. For other recent reviews of the yeast ribosome, see Warner (1989) and Raué and Planta (1990), and for a review of the structure of yeast ribosomes, see Lee (1990). The function of yeast ribosomes is dealt with primarily in the chapter on protein synthesis elsewhere in this volume (Hinnebusch and Liebman).

II. RIBOSOMAL RNA

Each ribosome contains four RNA molecules. The 18S RNA (1798 nucleotides) of the 40S subunit and the 25S RNA (3392 nucleotides) and associated 5.8S RNA (158 nucleotides) of the 60S subunit all derive from a common 35S RNA precursor transcribed by RNA polymerase I. The 5S RNA (121 nucleotides) of the 60S subunit is transcribed by RNA polymerase III. The structure of these RNAs has been described in detail (Hogan et al. 1984a,b; Raué et al. 1990). Although the primary structures of the yeast molecules are unique, their secondary, and presumably tertiary, structures are essentially identical to those of their counterparts in bacteria and in higher eukaryotes (Dams et al. 1988; Gutell et al. 1990).

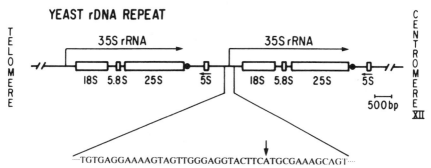

Figure 1 rRNA genes of *S. cerevisiae*. Two repeats are shown. The regions corresponding to mature RNA and to the transcripts are indicated. Closed circles indicate the location of the enhancer. Below is shown the sequence around the transcription initiation site for the 35S RNA (↓).

III. RIBOSOMAL RNA GENES

The rRNA genes of *Saccharomyces cerevisiae* are organized in a single tandem array of 9.1-kb repeating units on chromosome XII (Petes and Botstein 1977). The number of repeats is strain-dependent, ranging from about 100 to 200. Even within a single strain, there is a fluctuation of copy number, presumably due to unequal sister-chromatid exchange (Olson, this volume). Recombination within the rDNA genes shows a number of unusual properties (for a brief review, see Warner 1989) that are discussed in this volume (Petes et al.).

The repeating unit is arranged as described in Figure 1 (Philippsen et al. 1978). It is unusual in that it contains the genes both for the 35S ribosomal precursor RNA and for 5S RNA, which are transcribed in opposite directions. These sequence data should be used with caution because there are not only restriction site polymorphisms between the repeats (Petes 1980), but also substantial insertions or deletions (Jemtland et al. 1986). The latter appear to be confined to certain regions of the nontranscribed spacer.

IV. TRANSCRIPTION OF 35S RIBOSOMAL RNA

The transcription of eukaryotic rRNA genes has been reviewed recently (Sollner-Webb and Tower 1986). In yeast cells, the major portion of the rDNA repeat consists of a single transcription unit, leading to the synthesis of a 6.6-kb 35S rRNA precursor that is processed to form the 18S RNA found in the 40S ribosomal subunit and the 25S and 5.8S RNAs found in the 60S subunit (Fig. 2). As in all eukaryotes, rRNA is

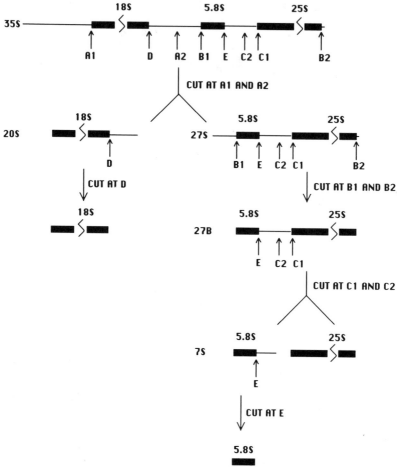

Figure 2 Proposed sequence of events in the processing of ribosomal precursor RNA. (Adapted from Veldman et al. 1981.)

transcribed by the multisubunit enzyme, RNA polymerase I (for review, see Sollner-Webb and Tower 1986). The yeast RNA polymerases are discussed in detail (Sentenac, these volumes) and have been reviewed previously by Sentenac (1985). The gene for the largest subunit, *RPA190*, has been cloned and sequenced (Memet et al. 1988). It resembles both the β′ subunit of *Escherichia coli* RNA polymerase and the largest subunits of pol II and pol III. Temperature-sensitive mutations in *RPA190* have been isolated (Wittekind et al. 1988). Because they shut off rRNA synthesis only slowly after shift to the nonpermissive temperature, these mutations are probably defective in the assembly rather than the function of the enzyme. The gene *RPC40* codes for a 40-kD polypeptide that is common to pol I and pol III (Mann et al. 1987).

Despite considerable effort, there is still little understanding of the details of the transcription of the ribosomal precursor RNA, although it makes up more than 60% of the total transcription in a rapidly growing yeast cell. The initiation site (the origin in the numbering systems used below) has been identified because it can be found bearing a triphosphate terminus (Fig. 2) (Klootwijk et al. 1979; Nikolaev et al. 1979; Klemenz and Geiduschek 1980; Bayev et al. 1981). There is substantial homology at the initiation sites of several fungal rRNA genes (Verbeet et al. 1984).

Effective in vitro systems from yeast in which transcription of rRNA is dependent on added templates are only now being developed (Riggs and Nomura 1990; R. Kornberg and R. Reeder, pers. comm.). With these systems in hand, it may be possible to determine directly the sequences responsible for rRNA transcription. Until now, such analyses have been conducted in vivo using plasmids carrying mini-rRNA genes marked with reporter sequences. A study of linker-scanning mutations in the region immediately upstream of the initiation site revealed that nucleotides −75 to −50 and −25 to +8 were particularly important for transcription (Musters et al. 1989a). A third domain, beyond −140, plays a lesser role. These regions are efficient as promoters of transcription, but only in the presence of the rDNA enhancer, a 190-bp region in the spacer region that lies 2 kb upstream of the initiation site and 100 nucleotides downstream from the 3′ end of the 35S transcript (Elion and Warner 1984, 1986). This enhancer is unusual for yeast in that it is effective both upstream and downstream from the transcription unit and can be effective from a distance of as much as 6 kb (Elion and Warner 1986; Johnson and Warner 1989). Deletion analysis of sequences within the enhancer suggests that no single region is responsible for the enhancer activity (Mestel et al. 1989).

The site of termination of transcription of rRNA has been harder to establish. It is now evident that the terminus of 35S RNA is a site for processing (Kempers-Veenstra et al. 1986; Yip and Holland 1989), probably carried out by a protein encoded by the *RNA82* gene. Both run-on transcription and transcription in isolated nuclei suggest that transcription continues into the enhancer region. A number of termination sites have been identified between the end of 35S RNA and the end of 5S RNA (van der Sande et al. 1989). Finally, as found in many other organisms, there is a strong termination site just upstream of the promoter that may serve to protect the initiation complex from runaway RNA polymerase molecules (van der Sande et al. 1989).

Although neither the precise site nor the mechanism of termination of transcription is clear, it is intriguing that transcription continues into the enhancer. The tandem nature of the rRNA genes has led a number of workers to suggest the possible coupling of termination and initiation of transcription (for review, see Sollner-Webb and Tower 1986). In the case of *S. cerevisiae*, it seems likely that some sort of loop is formed to bring the termination region near the initiation region (Kempers-Veenstra et al. 1986). However, results with a tandemly repeated test gene show that the enhancer itself need not be transcribed nor be adjacent to the target gene, suggesting that the looping be considered in three dimensions, where many enhancers and many promoters come together to form an array (Johnson and Warner 1989).

A protein, termed Reb1p, that binds to specific sequences in both the enhancer and the promoter regions of the spacer has been identified (Morrow et al. 1989, 1990), and its gene, *REB1*, has been cloned (Ju et al. 1990). Despite the potential for this protein to be involved in such an array, experimental analyses thus far suggest that it may not be a regulator (Kulkens et al. 1989), but instead may contribute to the arrangement of the rDNA chromatin (Chasman et al. 1990).

A recent interesting observation is that inactivation of both topoisomerases I and II, but not either separately, leads to relatively specific inhibition of rRNA transcription in *S. cerevisiae* (Brill et al. 1987). This finding suggests that DNA conformation must play some, as yet unclear, role in rRNA transcription.

V. TRANSCRIPTION OF 5S RIBOSOMAL RNA

5S RNA is transcribed by RNA polymerase III (for review, see Geiduschek and Tocchini-Valentini 1988), after the prior association of the gene with the factors TFIIIA, TFIIIB, and TFIIIC (Segall 1986). 5S RNA is used in equimolar amounts with the other rRNAs. Indeed, there appears to be roughly equimolar synthesis of 5S RNA and the 5.8S RNA derived from the 35S RNA (Neigeborn and Warner 1990). However, the mechanism that normally balances the transcription of 35S and 5S RNAs is unknown. A strain defective in 5S RNA transcription continues to transcribe 35S RNA for many hours (Gudenus et al. 1988). A strain defective in 35S RNA transcription continues to transcribe 5S RNA for many hours (Wittekind et al. 1990). Although the 5S gene lies between the rRNA enhancer and the initiation site for 35S RNA synthesis, the enhancer has no effect on 5S transcription (Neigeborn and Warner 1990). Indeed, sequences in the rDNA repeat may inhibit the transcription of the 5S gene, since a 5S gene within a fragment of the rDNA repeat that contains only 424 bp is transcribed much more efficiently than is the 5S gene within the complete rDNA repeat (Van Ryk et al. 1990). Perhaps the transcription of 5S and 35S RNAs is generally kept in balance indirectly, through the availability of the ribosomal protein that binds 5S RNA, as suggested by the in vitro experiments of Brow and Geiduschek (1987).

VI. THE YEAST NUCLEOLUS

Ribosome assembly in eukaryotic cells takes place in a specialized region of the nucleus termed the nucleolus (for review, see Scheer and Benavente 1990; Warner 1990). Sillevis-Smitt et al. (1973) showed many years ago that, in *Saccharomyces*, a region of the nucleus termed the "gray" or "dense" crescent contains most of the ribosomal precursor RNA, suggesting that it is the yeast equivalent of the nucleolus, although it lacks the nucleolar fine structure observed in larger cells. Recently, N. Dvorkin et al. (pers. comm.) have demonstrated by in situ hybridization that the dense crescent is indeed the site of the rDNA. Furthermore, it is the location both of the yeast equivalent of fibrillarin, encoded by the essential *NOP1* gene (Aris and Blobel 1988; Hurt et al. 1988), and of the product of the *SSB1* gene (Jong et al. 1987). Fibrillarin has been identified as a major component of the nucleolus in many higher organisms. The fibrillarin of *S. cerevisiae* is associated with small nucleolar RNA molecules (Table 1) (Schimmang et al. 1989; Henriquez et al. 1990), some of which are homologous to small RNA molecules isolated from nucleoli of higher organisms. One of the latter, U3, has been demonstrated to bind to ribosomal precursor RNA and to participate in an early processing step (Kass et al. 1990). The small nucleolar RNAs of *Saccharomyces* are identified by their retention in the nucleus after the RNAs involved in splicing have been extracted, by their binding to ribosomal precursor RNAs (Tollervey 1987), and by their coprecipitation by antifibrillarin antibodies (Table 1) (Schimmang et al. 1989).

The function of each of the nucleolar RNAs has been examined by gene disruption; many of them are dispensable (Table 1). At least one of the two genes that encode U3 (snR17) must be present for growth (Hughes et al. 1987). Disruption of the gene for snR10 leads to slow growth and cold sensitivity, apparently due to slow and inaccurate processing of 35S precursor RNA (Tollervey 1987). Examination of the closely linked genes for snR128 (U14) and snR190 showed that the latter is dispensable, but that deletion of the gene for U14 leads to inefficient production of 18S RNA, resulting eventually in a severe depletion of 40S subunits (Li et al. 1990). Surprisingly, U14 is associated not with a precursor to 18S RNA but with the 27S precursor to 25S RNA (Zagorski

Table 1 Small nucleolar RNAs of *Saccharomyces cerevisiae*

RNA	Size	Essential	Bind to	Precipitated by anti-yeast fibrillarin	Notes
snR3	194	no	35S	++	
snR4	192	no	35S	++	
snR5	205	no	20S	++	
snR8	189	no	20S	++	
snR9	188	no	20S–35S	++	
snR10	245	yes/no	35S	++	deletion leads to cold sensitivity; poor rRNA processing
snR17(U3)	328	yes	25S	+	essential for growth
snR128(U14)	128	yes	27S	+	required for 18S rRNA formation
snR190	190	no	20S	++	

Data from Reidel et al. (1986); Hughes et al. (1987); Tollervey (1987); Parker et al. (1988); Zagorski et al. (1988).

et al. 1988). Applying the techniques of yeast molecular genetics will undoubtedly lead to new insights into the structure and function of the nucleolus.

VII. PROCESSING OF RIBOSOMAL PRECURSOR RNA

rRNA is transcribed as a single polynucleotide of 6587 nucleotides that is subjected to methylation on both ribose and bases, to pseudouridylation, and to cleavage by both exo- and endonucleases to yield the 18S RNA that is in the 40S subunit and the 25S RNA and associated 5.8S RNA that are in the 60S subunit. Methylation occurs largely on completed 35S molecules (Udem and Warner 1972). Because methylation occurs only in the conserved regions of the 35S molecule, one can analyze the synthesis, processing, and, in some cases, degradation of rRNA by pulse labeling newly formed molecules with the methyl group of methionine and observing the flow of label into intermediate and mature species during a chase (Udem and Warner 1972; Gorenstein and Warner 1977; Sachs and Davis 1990).

The cleavage steps involved in the processing of the 35S precursor are outlined in Figure 2. It is not clear whether the order and arrangement of steps as shown hold for every molecule or whether there may be some flexibility. Veinot-Drebot et al. (1988) suggest that cleavage A1 occurs before completion of the transcript in most cases. Analysis of pulse-labeled rRNA on high-resolution gels now suggests that there is an early step, not shown in Figure 2, that generates a molecule migrating slightly faster than 35S RNA (Sachs and Davis 1990; K. Shuai and J.R. Warner, in prep.). Possibly, this molecule corresponds to that produced in an early U3 snRNA-dependent cleavage in vertebrate cells (Kass et al. 1990).

The processing pattern shown in Figure 2 has been derived entirely from experiments carried out in vivo. No system that carries out processing in vitro has yet been developed in yeast, except for the formation of the 3' end (Yip and Holland 1989). However, four experimental approaches to the processing of ribosomal precursor RNA are now beginning to show promise.

1. Musters et al. (1989b) have marked individual rRNA genes by inserting an oligonucleotide in either the 18S or the 25S sequences. When located appropriately, the insertion has no effect on the processing or the function of the transcripts. These authors have used these marked genes to study the effect on processing of the deletion or manipulation of other sequences within the transcript. Surprisingly, deletion of most of the 18S RNA sequences has little effect on the processing

events that yield 25S and 5.8S RNAs. On the other hand, almost any manipulation of the sequences between 5.8S and 25S prevents the 27S precursor from being processed to 5.8S and 25S (Fig. 2) (Veldman et al. 1981; Musters et al. 1990).

Strains have been constructed that contain no chromosomal rDNA repeats and express all their functional rRNA from plasmids (E. Morgan, pers. comm.). This system, in permitting mutations to be introduced into all copies of rDNA, should facilitate studies of rRNA synthesis, structure, and function.

2. As components of the nucleolus are identified, e.g., the small nucleolar RNAs and proteins such as fibrillarin, one can manipulate their structures and begin to probe their roles in ribosome formation. It is interesting that either deletion of ribosomal protein L46 or mutation of a putative RNA helicase leads to cold sensitivity and poor rRNA processing (Sachs and Davis 1990).

3. A number of mutants that appear to be defective in the processing of rRNA have been isolated over the years (Bayliss and Ingraham 1974; Andrew et al. 1976; Gritz et al. 1982; Mitlin and Cannon 1984). Of these, only one has been analyzed in detail. In the mutant carrying a temperature-sensitive allele of *RRP1*, the 27S RNA is rapidly degraded, rather than being processed to form 25S and 5.8S RNAs (Andrew et al. 1976; Gorenstein and Warner 1977). *RRP1* has been cloned (Fabian and Hopper 1987) and sequenced (R. Crouch, pers. comm.) without revealing anything beyond the fact that its product is not likely to be a ribosomal protein. The isolation of suppressors of *rrp1*, such as *srd1*, may in time reveal its function (Fabian et al. 1990).

Recent screens of temperature-sensitive mutants have yielded new candidates for rRNA-processing mutants, with defects in the 5′ end of 5.8S RNA as well as in the handling of the 3′ portion of 20S RNA and of the sequences downstream from the 35S transcript (K. Shuai and J. Warner; L. Lindahl; both unpubl.).

4. It has long been clear that ribosomal proteins are essential for the processing of ribosomal precursor RNA (Warner and Udem 1972; Warner and Gorenstein 1977). The ribosomal proteins bind to the newly formed transcript (Trapman et al. 1975); the complex is the substrate for processing. An alternative approach to studying processing, then, is to deplete the cell of an individual ribosomal protein, either by deletion of one or both of its genes (Abovich et al. 1985; Lucioli et al. 1988; Tsay et al. 1988; Finley et al. 1989) or by manipulation of the promoter (Nam and Fried 1986; Moritz et al. 1990). Although this frequently leads to inefficient processing of the

RNA associated with that protein, and a marginal imbalance in the ratio of 60S to 40S subunits, none of the mutants or conditions generated in this way have led to a crisp phenotype, i.e., of the blocking of a particular step in the processing.

VIII. RIBOSOME ASSEMBLY

Stepwise reconstitution of *E. coli* 30S and 50S subunits from purified components and analysis of the topography of rRNAs and ribosomal proteins revealed that bacterial ribosome assembly proceeds in vitro through a complex pathway of interdependent and independent protein-protein and protein-RNA interactions (for review, see Stern et al. 1989). However, only partially dissociated eukaryotic ribosomes have been reconstituted (Cox and Greenwell 1980; Vioque and Palacian 1985; Lee and Anderson 1986). The failure to reconstitute eukaryotic ribosomes from totally separate components successfully has hindered efforts to determine a high-resolution assembly map in vitro.

In-vivo-labeling experiments in yeast showed that a 90S preribosomal particle containing 35S RNA is processed to form a nuclear 66S particle containing 27S RNA and a cytoplasmic 43S particle containing 20S RNA (Trapman et al. 1975). Final steps in the maturation of the 40S subunit, including processing of 20S RNA to 18S RNA, occur in the cytoplasm (Udem and Warner 1973). Although the precise protein components of these ribosomal precursors are undefined, they apparently contain a number of nonribosomal and ribosomal proteins (Warner 1971). A low-resolution picture of yeast ribosome assembly and structure was inferred by following in vivo the incorporation into subunits of individual ribosomal proteins (Kruiswijk et al. 1978) and by assaying in vitro the order with which individual ribosomal proteins can be dissociated from ribosomes by increasingly stringent washes (El-Baradi et al. 1984; Lee et al. 1985). Proteins were distinguished as assembling relatively early or late. Among the early assembling proteins are some that are tightly associated with 25S or 5.8S RNA in high-salt-washed ribosomal core particles. One example is L25, which binds tightly and reversibly to a region of the 25S RNA that is homologous to the binding site on *E. coli* 23S RNA of L23, one of the primary binding proteins in the assembly of *E. coli* 50S particles (El-Baradi et al. 1984). Similarly, L15 of *S. cerevisiae* binds to a site analogous to that of L11 of *E. coli* (El-Baradi et al. 1987).

Assembled ribosomal subunits are exported from the nucleus to the cytoplasm, presumably through the nuclear pores. An intriguing double mutation in the gene *RNA1* appears to block such export (Hutchison et al.

1969; Traglia et al. 1989), but since it also blocks mRNA transport and tRNA processing (Hopper et al. 1978), it has been of limited value in probing this key step in ribosome biosynthesis.

IX. NUCLEOCYTOPLASMIC INTERACTION

An important aspect of ribosome biosynthesis is the transport of materials across the nuclear envelope. The 75 ribosomal proteins must be imported from the cytoplasm, where they are synthesized. The 60S and 40S subunits, of not inconsiderable size, must be exported to the cytoplasm. Nuclear localization signals have been identified for ribosomal proteins L3 (Moreland et al. 1985), L25 (H.A. Raué, pers. comm.), and L29 (Underwood and Fried 1990). L29 has two similar signals –KTRKHRG– and –KHRKHPG–. Neither is identical to any of the numerous other nuclear localization signals identified, but they do share their motif of basic amino acids. For L25, H. Raué et al. (pers. comm.) have determined that the nuclear localization signal is distinct from the RNA-binding domain. Thus, the localization of L25 to the nucleus is *not* due to its binding to ribosomal precursor RNA.

Alteration of the nuclear localization signal on L29 led to reduced ribosome synthesis and slow cell growth. It is clear, then, that even for this rather small ribosomal protein of 149 amino acids, a nuclear localization signal is necessary for effective targeting. Although experiments based on cytoplasmic injection of foreign proteins have suggested that the nucleus is accessible to proteins with a relative molecular weight of less than 50,000 (Paine and Feldherr 1972), it seems likely that each ribosomal protein will have its own nuclear localization signal or will be transported in association with another protein that has one.

X. RIBOSOMAL PROTEINS

Two-dimensional gel electrophoresis has identified up to 45 different proteins from the 60S ribosomal subunit and 32 proteins from the 40S ribosomal subunit. Although multiple nomenclatures have been used for yeast ribosomal proteins, a uniform nomenclature (Bollen et al. 1981a; Michel et al. 1983) has been adopted by most laboratories. Ultimately, a comprehensible nomenclature for eukaryotic ribosomal proteins may result from a thorough cataloging of ribosomal protein homologs from several different organisms.

More than half of the yeast ribosomal proteins have been purified and subjected to at least partial amino acid sequence analysis (for review, see Lee 1990). The complete amino acid sequences of 33 ribosomal proteins have been inferred from the nucleotide sequences of the cloned genes. Tables 2 and 3 list those proteins for which either complete amino acid sequences are known or the genes have been cloned. Some yeast ribosomal proteins are unusual. For example, L47 is translated as a protein of 25 amino acids, of which 17 are arginine or lysine (Suzuki et al. 1990).

Immunological assays and comparisons of nucleotide sequences of ribosomal protein genes led to identification of homologs for 29 of the yeast ribosomal proteins among ribosomal proteins of other eukaryotes, eubacteria, and archaebacteria (see Tables 2 and 3). There has been extraordinary conservation of the ribosomal proteins during the evolution of eukaryotes. Generally, the ribosomal proteins of yeast and mammals are more than 75% identical. On the other hand, in only a few cases has it been possible to identify bacterial homologs of yeast ribosomal proteins. Nevertheless, comparisons with the well-studied ribosomal proteins, such as those in bacteria, may provide clues for the location and function of the homologous proteins in yeast ribosomes. For example, the omnisuppressor genes *SUP44* and *SUP46* encode proteins nearly identical to the *E. coli* proteins *S5* and *S4*, respectively, that are also involved in the accuracy of translation (All-Robyn et al. 1990; S. Liebman, pers. comm.). In some cases, the correspondence of proteins is more apparent through function than through sequence. An example is L25, which binds to the same location on the rRNA as does L23 of *E. coli*, in spite of their having very limited sequence similarity (see El-Baradi et al. 1985).

The yeast acidic ribosomal proteins A0, A1, L44, L44′, and L45 are similar to each other and related to the human P0, P1, and P2 proteins (Mitsui and Tsurugi 1988a,b,c; Remacha et al. 1988; Newton et al. 1990). These proteins cycle on and off the ribosome in the cytoplasm and are needed for interactions of soluble factors with the ribosome. The related proteins in *E. coli*, L7/L12 and L10, are present in the evolutionarily conserved stalk of the large subunit and are involved in factor-dependent GTP hydrolysis during protein synthesis (for review, see Traut et al. 1985). Yeast L15(YL23) is immunologically and functionally related to *E. coli* L11, a protein known to interact with the L7/L12 acidic proteins and to function in GTP hydrolysis, peptidyl transfer, and antibiotic binding (Juan-Vidales et al. 1983). *E. coli* L11 protein can reconstitute the activity of yeast ribosome core particles deficient in yeast L15.

Antibiotic resistance or sensitivity provides a convenient means to as-

Table 2 60S Ribosomal subunit proteins and their genes

Protein[a]	Gene copy number	Intron	Essential	Homologs[b]	References[c]
L1 YL3 rp3,4	1	no	yes	L5(X,R)	1
L2 YL2 rp2	2	no	yes	L1(X,D)	2, 3
L3 YL1 rp1	1 (*TCM1*, XVR)	no	yes	L1(T,A)	4–6
L4 YL5 rp6	2	no	n.d.	L7a(H,M)	7, 8
L16 YL22 rp39	2 (*RPL16A*, VIIR; *RPL16B*, XVIR)	no	yes	L5(E); HL19(H.c.); L21(T)	9–13
L17a YL32	2	yes	n.d.	—	10, 14
L25 YL25	1	yes	n.d.	L23(E)	10, 14, 15
L29 YL24 rp44	1 (*CYH2*, VIIL)	yes	yes	L29(N); L27a(M)	6, 16, 17
L30 YL21 rp29	2	yes	no	—	7, 18, 19
L32 YL38 rp73	1	yes	yes	L30(M,R)	7, 20
L34 YL36	2	yes	n.d.	L31(R)	10, 21
L37 YL37 rp47	1	n.d.	n.d.	L35a(R); L32(X)	7, 22
L41[d] YL27	n.d.[e]	n.d.	n.d.	L36a(H,R)	23
L43[d] YL35	n.d.	n.d.	n.d.	L37(R)	24
L46 YL40	1 (*RPL46*, X)	yes	no	L39(R); L36(Sch)	10, 25, 26
L47 YL41	2	no	n.d.	—	27
L?	2 (*UBI1*, *UBI2*)	yes	yes	—	28, 29

				L18(R); L14(X)	
rp24	n.d.	n.d.	n.d.	—	7
rp28	2 (*RP28A*[f], XVL; *RP28B*, XIVL)	yes	yes	L18(R); L14(X)	7, 30, 31
rp58	n.d.	n.d.	n.d.	—	7
A0 A0	1	no	n.d.	P0(H)	32, 33
A1 A1	1	no	n.d.	P1(H)	33, 34
L44'	1	yes	yes	P1(H)	33, 35
L44 A2	1	no	no	P2(H)	33, 35–37
L45 YPA1	1	no	no	P2(H)	33, 35, 36

[a] Three different general nomenclatures used, where known, are L or S (Michel et al. 1983), YL or YS (Otaka and Osawa 1983), and rp (Warner and Gorenstein 1978b). The acidic proteins are designated by A0, A1, and A2 (Mitsui and Tsurugi 1988a,b,c) as well as by the L (Remacha et al. 1988, 1990) or YP (Itoh 1980) nomenclatures.

[b] Homologies were identified in the original references describing protein or gene sequences or were from a collection of ribosomal protein sequences assembled by Tatsuo Tanaka (Department of Biochemistry, Yamagata University School of Medicine, Yamagata, 990-23, Japan). (A) *Arabidopsis thaliana*; (C), *Chlamydomonas reinhardtii*; (Ch) chicken; (D) *Drosophila melanogaster*; (E) *E. coli*; (H) *Homo*; (Ha) hamster; (H.m.) *Halobacterium morismortui*; (H.c.) *Halobacterium cutirubrum*; (M) *Mus*; (N) *Neurospora crassa*; (R) *Rattus*; (Sch) *Schizosaccharomyces pombe*; (T) *Tetrahymena thermophila*; (Tr) *Trypanosome*; (X) *Xenopus laevis*.

[c] References: (1) Y.-F. Tsay et al., in prep.; (2) Lucioli et al. 1988; (3) Presutti et al. 1988; (4) Fried and Warner 1981; (5) Schultz and Friesen 1983; (6) Fried et al. 1985; (7) Fried et al. 1981; (8) Arevalo and Warner 1990; (9) Woolford et al. 1979; (10) Bollen et al. 1981b; (11) Teem et al. 1984; (12) Rotenberg et al. 1988; (13) J. Woolford, in prep.; (14) Leer et al. 1984; (15) El-Baradi et al. 1984; (16) Fried and Warner 1982; (17) Kaufer et al. 1983; (18) Mitra and Warner 1984; (19) Baronas-Lowell and Warner 1990; (20) Dabeva and Warner 1987; (21) Schaap et al. 1984; (22) G. Santangelo, pers. comm.; (23) Itoh and Wittman-Liebold 1978; (24) Otaka et al. 1984; (25) Leer et al. 1985b; (26) Sachs and Davis 1990; (27) Suzuki et al. 1990; (28) Ozkaynak et al. 1987; (29) Finley et al. 1989; (30) Molenaar et al. 1984; (31) Donovan et al. 1990; (32) Mitsui and Tsurugi 1988a; (33) Newton et al. 1990; (34) Mitsui and Tsurugi 1988b; (35) Remacha et al. 1990; (37) Mitsui and Tsurugi 1988c.

[d] The genes encoding L41 and L43 have not been cloned; the complete amino acid sequences of these proteins were determined. All other data in Tables 2 and 3 are derived from cloned genes.

[e] n.d. indicates not determined.

[f] The names of these genes are provisional until the protein has been identified in gel systems using the L or S nomenclature (Michel et al. 1983). Similar rules for nomenclature apply for other ribosomal protein genes whose proteins are not yet identified in this way.

Table 3 40S Ribosomal subunit proteins and their genes

Protein[a]	Gene copy number	Intron	Essential	Homologs[b]	References[c]
S4 YS5 rp12	1 (*SUP44*)	no	yes	S5(E)	1
S7 YS6 rp5	2	yes	n.d.	S4(R)	2, 3
S10 YS10 rp9	2	yes	yes	S6(H,M,R,Sch)	2, 4–6
S11	n.d.[d]	n.d.	n.d.	—	2
S13 YS11	2 (*SUP46*)	yes	n.d.	S4(E)	1a
S16 YS16 rp55	2 (*RPS16A*, XVL; *RPS16B*, XIVL)	yes	yes	S12(H.m.)	2, 7–9
S21 rp52	1	yes	n.d.	—	10, 11
S24 YS22 rp50	2 (*RPS24A*, X; *RPS24B*, ?)	no	n.d.	S16(H.m.) S20(H.c.)	7, 12
S26 YS25	n.d.	yes	n.d.	S21(R) S28 (Sch)	13, 14
S31 YS23	2	yes	n.d.	S25(R)	7, 15
S33 YS27	1	no	n.d.	—	7, 16
S37 YS24	1 (*UBI3*)	no	no	S27a(R)	17–19
rp10	n.d.	n.d.	n.d.	—	2
rp51	2	yes	yes	S17(H,R,Ha,Ch)	10, 20
rp59	2 (*CRY1*, IIIR; *CRY2*,?)	yes	yes	S14(H,Ha,D,N,Tr,C) S11(E)	21–24
rp63	n.d.	n.d.	n.d.	—	10

[a] Nomenclatures are the same as those in Table 2.
[b] Homologs were identified and abbreviations of species are the same as those described in footnotes to Table 2.
[c] References: (1) All-Robyn et al. 1990; (1a) S.W. Liebman, pers. comm.; (2) Fried et al. 1981; (3) D. Synetos and J.R. Warner, in prep.; (4) Leer et al. 1982; (5) Leer et al. 1985c; (6) Kruse et al. 1985; (7) Bollen et al. 1981b; (8) Molenaar et al. 1984; (9) Donovan et al. 1990; (10) Woolford et al. 1979; (11) J. Woolford, unpubl.; (12) Leer et al. 1985b; (13) Itoh et al. 1985; (14) Suzuki and Otaka 1988; (15) Nieuwint et al. 1985; (16) Leer et al. 1983; (17) Ozkaynak et al. 1987; (18) Finley et al. 1989; (19) Otaka et al. 1984; (20) Abovich et al. 1985; (21) Larkin and Woolford 1983; (22) Himmelfarb et al. 1984; (23) Larkin et al. 1987; (24) A. Paulovich et al., in prep.
[d] n.d. indicates not determined.

say expression and function of certain ribosomal proteins and to define specific functional domains of the ribosome. Mutations in four genes, *CRY1*, *CRY2*, *CYH2*, and *TCM1*, confer resistance to antibiotics that inhibit specific steps in protein synthesis (Cooper et al. 1967; Jiménez et al. 1975; Skogerson et al. 1973; Grant et al. 1974, 1976; A. Paulovich and J. Woolford, in prep.). Cloning of these genes unambiguously identified ribosomal proteins responsible for these phenotypes (Fried and Warner 1981, 1982; Larkin and Woolford 1983; Himmelfarb et al. 1984; A. Paulovich and J. Woolford, in prep.).

Cryptopleurine binds with high affinity to the 40S subunit and blocks the EF-2 and GTP-dependent step of translocation (Bucher and Skogerson 1976; Dölz et al. 1982). Cryptopleurine resistance can result from mutations in either of the duplicated, unlinked *CRY1* and *CRY2* genes (Skogerson et al. 1973; Grant et al. 1974; A. Paulovich and J. Woolford, unpubl.) that encode rp59 (Larkin and Woolford 1983; Himmelfarb et al. 1984; Larkin et al. 1987; A. Paulovich and J. Woolford, in prep.). Ribosomal protein 59 is 70–88% homologous to ribosomal proteins identified in humans, hamsters, maize, *Drosophila melanogaster*, *Chlamydomonas reinhardtii*, trypanosomes, and *Neurospora crassa*; rp59 is 45% homologous to *E. coli* ribosomal protein S11 (for review, see A. Paulovich and J. Woolford, in prep.).

Mutations in *CYH2* confer resistance to high concentrations of cycloheximide, an inhibitor of peptidyltransferase (Cooper et al. 1967); *CYH2* encodes L29 (Stocklein and Piepersberg 1980; Fried and Warner 1982). The mouse ribosomal protein L27 is both a structural and functional homolog of L29, as a mouse L27 cDNA clone can complement a lethal *cyh2* null allele (Fleming et al. 1989).

Trichodermin binds to the 60S subunit and blocks termination and elongation by interfering with the peptidyltransferase center (Schindler et al. 1974; Jiménez et al. 1975). Trichodermin resistance maps to the *TCM1* gene (Jiménez et al. 1975; Grant et al. 1976) cloned by Fried and Warner (1981) and Schultz and Friesen (1983).

Ribosomal proteins are posttranslationally modified by phosphorylation, methylation, or acetylation, but the significance of these modifications remains unknown (for review, see Lee 1990). For example, S10, one of the major phosphorylated proteins, and its mammalian homolog, S6, are phosphorylated in response to cell proliferation. Nevertheless, yeast containing mutant S10 that cannot be phosphorylated grow identically to wild-type cells (Johnson and Warner 1987). Mutations in either of two genes, *NAT1* or *ARD1*, block amino-terminal acetylation of proteins (Mullen et al. 1989). In cells carrying such mutations, at least 11 ribosomal proteins have altered electrophoretic mobility, suggesting that, in wild-type cells, these proteins are acetylated (J.R. Warner, unpubl.). Among these are rp39, rp52, rp56, and rp59. The lack of acetylation leads to no apparent change in growth rate.

Two ribosomal proteins are derived by cleavage from fusion proteins containing ubiquitin at their amino termini. A 60S ribosomal subunit protein is derived from the product of *UBI1* or *UBI2*; the 40S ribosomal subunit protein S37 is encoded by *UBI3* (Finley et al. 1989; Finley, these volumes). The mammalian homolog of S37, S27a, is also synthesized as a ubiquitin fusion protein (Redman and Rechsteiner 1989). An apparent function of the ubiquitin moiety expressed from *UBI3* is to facilitate assembly of S37 into the 40S subunit; mutants lacking the ubiquitin-coding portion of *UBI3* are deficient in small ribosomal subunits and in conversion of 20S pre-rRNA to 18S RNA. It is not clear whether ubiquitin is directly used for assembly of S37 into ribosomes or whether it protects S37 from turnover prior to assembly. *UBI1–UBI3* normally are the cell's primary sources of ubiquitin; *UBI4*, encoding polyubiquitin, is derepressed under conditions where ubiquitin is not expressed from *UBI1–UBI3* (Finley et al. 1989; Finley, these volumes).

XI. RIBOSOMAL PROTEIN GENES

Genes for 38 different ribosomal proteins have been cloned (Tables 2 and 3). Where possible, *RPL* or *RPS* are used to designate genes encoding proteins from the large or small ribosomal subunits, respectively, on the basis of the protein nomenclature of Bollen et al. (1981a) and Michel et al. (1983). Other gene names are provisional until the proteins are located in a gel system using the L or S nomenclature.

Unlike the situation in *E. coli*, yeast ribosomal protein genes are generally unlinked. However, five pairs of ribosomal protein genes are linked. The duplicated genes encoding rp28 and rpS16 are linked to each other in a head-to-tail configuration; one pair maps to chromosome XVL and the other pair maps to chromosome XIVL (Molenaar et al. 1984; Papciak and Pearson 1987). *RPH16A* and *MRP13* (a mitochondrial ribosomal protein gene) (Partaledis and Mason 1988; J. Woolford, unpubl.) are oriented head to tail on VIIR. *RPL46* and *RPS24A* are adjacent head to head (Leer et al. 1985a), as is the pair *RPL30A* and *RPL32* (Mitra and Warner 1984). In all such cases, each gene is transcribed independently.

Slightly more than half of the cloned ribosomal protein genes are duplicated (see Tables 2 and 3). In each case tested, both copies are expressed with ratios of expression ranging from 1:1 to 10:1. It is not clear why yeast ribosomal protein genes, unlike most other yeast genes, are duplicated. Other lower eukaryotes typically contain one copy of each ribosomal protein gene, whereas mammals have one functional copy of each ribosomal protein gene, and numerous pseudogenes per haploid genome (for review, see Jacobs-Lorena and Fried 1987). There is no evidence for differential expression of duplicate ribosomal protein genes under particular conditions, such as sporulation, or for qualitative differences in function of the two copies. The proteins encoded by duplicate ribosomal protein genes usually differ by no more than one or two amino acids. The slow-growth phenotype that often results from deletion of one copy can be suppressed by introduction of extra copies of the remaining functional gene. The noncoding regions of most duplicated ribosomal protein genes, including the introns, have diverged almost completely.

More than half of the ribosomal protein genes contain an intron (Tables 2 and 3), whereas only a dozen of the hundreds of other yeast genes thus far cloned do. Unlike genes of metazoans, which typically contain several introns from 50 nucleotides to 100 kb long, yeast mosaic genes typically contain a single intron, usually 300–500 nucleotides long, located near the 5' end.

Ribosomal protein genes have been used extensively in the study of *cis*-acting sequences and *trans*-acting factors necessary for yeast mRNA splicing in vivo and in vitro (for review, see Woolford 1989). The genes *RNA2–RNA9* and *RNA11*, originally thought to regulate ribosome biosynthesis (Hartwell et al. 1970; Gorenstein and Warner 1976; Warner and Gorenstein 1977), subsequently were found to be defective in splicing of all nuclear pre-mRNAs (Rosbash et al. 1981; Larkin and Woolford 1983). Therefore, the Rna$^-$ phenotype is due to the rapid turnover of rRNA that occurs in the absence of ribosomal proteins. These genes have been renamed *PRP2–PRP11*, for *pre-RNA processing*. Twenty new *PRP* genes have now been identified (Vijayraghavan et al. 1989; Burgess et al. 1990; J. Maddock et al., in prep.; E. Strauss and C. Guthrie, pers. comm.).

Gene disruption experiments have proven that 12 yeast ribosomal proteins are essential (see Tables 2 and 3), most likely because they are necessary for assembly of their respective ribosomal subunits. This has been shown directly for L3, L16, L29, and rp59 (Nam and Fried 1986; Moritz et al. 1990). Five proteins (S37, L30, L44, L45, and L46) are not essential. Strains lacking L46 are cold sensitive and have diminished levels of 60S ribosomal subunits (Sachs and Davis 1989, 1990); those lacking S37 are deficient in small ribosomal subunits (Finley et al. 1989). Mutants lacking L30 grow slowly and apparently contain altered 60S subunits (Baronas-Lowell and Warner 1990). Although mutants lacking either L44 or L45 are viable, those lacking both proteins are not (Remacha et al. 1990). These two acidic proteins are quite similar (80%

conserved or identical amino acids), but it is thought that they function cooperatively in ribosome assembly or activity, rather than carrying out identical functions. Further analysis of the function of individual ribosomal proteins in ribosome assembly and in translation will be facilitated by mutagenesis of cloned ribosomal protein genes.

XII. REGULATION OF RIBOSOMAL PROTEIN SYNTHESIS IN LOGARITHMICALLY GROWING CELLS

A. Transcriptional Mechanisms

Equimolar quantities of ribosomal proteins and rRNAs accumulate in logarithmically growing yeast cells. The paradigm from studies with *E. coli* is that balanced accumulation of ribosomal proteins and rRNAs results from posttranscriptional feedback mechanisms (Nomura et al. 1984). In yeast, on the other hand, synthesis of ribosomal proteins is controlled primarily at the level of transcription of their genes.

Yeast ribosomal protein mRNAs are approximately equimolar and are moderately abundant. The equimolarity of four different ribosomal protein mRNAs results from their identical rates of transcription and similar rates of turnover (Kim and Warner 1983a).

Several conserved sequence motifs are present upstream of ribosomal protein genes (Teem et al. 1984; Leer et al. 1985b). Deletion and linker insertion mutagenesis as well as promoter substitutions showed that two of these conserved sequences, HOMOL1 (AACATC C/T G/A T A/G CA) and RPG (ACCATACAT T/C T/A) are, in fact, promoter elements for the *RPL16A*, *RPL25*, *CYH2*, *CRY1*, *RPL46*, and *RPS24A* ribosomal protein genes (Rotenberg and Woolford 1986; Woudt et al. 1986, 1987; Larkin et al. 1987; Schwindinger and Warner 1987; Kraakman et al. 1989). Two copies of these promoter elements are present in either orientation and at varying distances from each other 250–450 nucleotides upstream of the ATG initiator codon of most ribosomal protein genes. A T-rich region present between HOMOL1 or RPG and the 5′ ends of ribosomal protein genes is also necessary for transcription of ribosomal protein genes (Rotenberg and Woolford 1986). A single protein, Rap1p (repressor/activator-binding protein), binds to both RPG and HOMOL1 (Huet et al. 1985; Huet and Sentenac 1987; Vignais et al. 1987). Planta and Raué (1988) have suggested that there is one general consensus promoter and Rap1p-binding site, UAS_{RPG} (5′-ACACCCATACATTT-3′). Single substitutions of all nucleotides except those underlined have little effect on upstream activating sequence (UAS_{RPG}) activity, consistent with the observed natural variation in UAS_{RPG} sequences (Nieuwint et al. 1989).

UAS$_{RPG}$ sequences are not present 5' of *TCM1*, *RPS33*, *RPL2A*, and *RPL2B*, although these genes are expressed similarly to the other ribosomal protein genes that contain UAS$_{RPG}$. A protein different from Rap1p, designated Taf (*TCM1* activation factor) or Suf (*S33* upstream factor), binds to a sequence, UAS$_T$, with the consensus RTCRYN$_5$ACG, that is upstream of *TCM1*, *RPS33*, *RPL2A*, and *RPL2B* and activates the transcription of these genes (Vignais et al. 1987; Hamil et al. 1988; Della-Seta et al. 1990).

Evidence to date suggests that Rap1p is identical to the proteins Grf1p (*general regulatory factor 1*) and Tuf (*translational upstream factor*) and performs several different roles (Shore and Nasmyth 1987; Buchman et al. 1988). Rap1p/Grf1p/Tuf, hereafter designated Rap1p, is a general transcription factor similar to the mammalian SP1 protein, CCAAT box binding factors, or octamer sequence binding factors. The UAS$_{RPG}$ to which Rap1p binds functions as a promoter for a wide variety of yeast genes, including those involved in translation, transcription, mating-type regulation, glycolysis, and nutrient transport (for review, see Capieaux et al. 1989). Although Rap1p is involved in activating the transcription of most of the genes containing UAS$_{RPG}$, binding of Rap1p to the silencer (E) regions flanking the silent mating-type loci, *HMR* and *HML*, is necessary for repression of transcription of these loci (Shore and Nasmyth 1987; Buchman et al. 1988). Rap1p also may play a structural role; it binds to C$_{1-3}$A sequences (resembling UAS$_{RPG}$) present in yeast telomeres (Buchman et al. 1988; Longtine et al. 1989; Biswas and Biswas 1990). Apparently, Rap1p performs different functions depending both on the sequence context flanking the protein-binding site and on the spectrum of interacting proteins present.

Taf/Suf is also a multifunctional protein. DNA-binding competition assays (Dorsman et al. 1989) suggested that Taf/Suf is identical to GFI, an activator of nucleus-encoded mitochondrial proteins (Dorsman et al. 1988); Abf1p, an autonomously replicating sequence (*ARS*)-binding protein necessary for *ARS* function (Shore et al. 1987; Buchman et al. 1988; Diffley and Stillman 1988; Sweder et al. 1988); and SBF-B, a protein that either activates transcription at *HMR* when it binds the I site or represses *HMR* transcription when it binds the E site in concert with Rap1p (Shore et al. 1987; Buchman et al. 1988; Kimmerly et al. 1988). This protein will be referred to as Abf1p, since the gene encoding it has been cloned and named *ABF1* (Diffley and Stillman 1989; Rhode et al. 1989). Like Rap1p, Abf1p may function as a general transcription factor. The UAS$_T$ sequence to which Abf1p binds is found upstream of a number of yeast genes encoding proteins involved in translation, transcription, general metabolism (including glycolysis), and cell differentiation (for review, see Della-Seta et al. 1990; Campbell and Newlon, this volume).

Four different results suggest that the constitutive balanced transcription of ribosomal protein genes does not result simply from binding of Rap1p or Abf1p to UAS$_{RPG}$ or UAS$_T$ sequences, but may involve additional *cis*-acting nucleotide sequences and *trans*-acting molecules: (1) The *GCR1* gene product, distinct from Rap1p or Abf1p, is necessary for transcription of *TEF1*, *TEF2*, and *CRY1*, as well as glycolytic enzyme genes, all of which contain UAS$_{RPG}$, and is required for UAS$_{RPG}$-dependent expression from a *LAC4* reporter gene (Santangelo and Tornow 1990). Gcr1p is not required for either Rap1p expression or binding of Rap1p to UAS$_{RPG}$ in vitro. (2) An 82-kD polypeptide distinct from Rap1p or Abf1p can be photochemically cross-linked in vitro to DNA containing either the UAS$_{RPG}$ or the UAS$_T$ sequence (Hamil et al. 1988). (3) Deletion mutations of ribosomal protein genes that lack the UAS$_{RPG}$ sequence but contain the upstream T-rich sequences are still transcribed, although at somewhat lower rates (Larkin et al. 1987). (4) Steady-state levels of ribosomal protein mRNAs are not measurably affected in *rap1*^{ts} strains, although genes that require only UAS$_{RPG}$ for activity are no longer expressed (D. Shore, pers. comm.). *rap1*^{ts} mutants are also altered in telomere structure, activation of *MATα1*, and repression of *HMR* (Lustig et al. 1990; D. Shore, unpubl.). More detailed analysis using mutant *RAP1* and *ABF1* genes (Shore and Nasmyth 1987; Chambers et al. 1989; Diffley and Stillman 1989; Halfter et al. 1989; Rhode et al. 1989; Henry et al. 1990) will be necessary to characterize the many functions of these proteins and to identify other molecules with which they interact, including some that may be necessary for transcription of ribosomal protein genes.

How can one account for the balance of constitutive expression of single copy and duplicated ribosomal protein genes, keeping in mind that the relative expression of the duplicate genes varies as much as tenfold? There is some correlation between the relative affinity for binding of Rap1p or Abf1p to the UASs in vitro and the relative levels of mRNAs in vivo for both the pair of genes encoding rp28 and the *RPL2* gene pair (Vignais et al. 1987; Della-Seta et al. 1990). However, there is no obvious correlation between the relative expression of a duplicate pair of ribosomal protein genes and the number of UAS$_{RPG}$ elements 5′ of each gene, the orientation or spacing of UAS$_{RPG}$ sequences with respect to each other or the respective genes, or the precise UAS$_{RPG}$ sequence for each gene. In summary, although it is clear that, for the most part, the synthesis of ribosomal proteins is controlled at the level of the transcription of their genes and that the binding sites for Rap1p and Abf1p are important for this transcription, a quantitative understanding of regulation of the amounts of the mRNAs for the various ribosomal proteins remains elusive.

B. Posttranscriptional Mechanisms

Accumulation of equimolar quantities of ribosomal proteins probably does not result from *precisely* equal rates of transcription (and translation) of mRNAs for each of the approximately 77 ribosomal proteins. The absence of ribosomal protein operons in yeast and the assembly of ribosomes, a compartment different from that in which ribosomal proteins are synthesized, make it unlikely that yeast ribosome biosynthesis is regulated by a mechanism similar to the autogenous feedback inhibition of translation observed in bacteria (Nomura et al. 1984). If modest differences in rates of transcription of ribosomal protein genes do occur, they may be compensated by a combination of posttranscriptional mechanisms.

Posttranscriptional control has been investigated by assaying expression of ribosomal constituents in response to two perturbations: (1) increasing the copy number of individual ribosomal protein genes in wild-type cells and (2) inactivating expression of rRNA or a single ribosomal protein.

Rates of transcription, steady-state levels of pre-mRNA and mRNA, and rates of ribosomal protein synthesis and turnover were measured in transformants containing 5–20 extra copies of different ribosomal protein genes (Pearson et al. 1982; Himmelfarb et al. 1984; Abovich et al. 1985; Warner et al. 1985; El-Baradi et al. 1986; Maicas et al. 1988; Tsay et al. 1988). Excess ribosomal protein mRNA is synthesized and accumulates, but not always in proportion to the ribosomal protein gene copy number. The excess mRNAs are associated with a normal number of ribosomes in polyribosomes and are translated at wild-type rates to yield a proportionate excess of proteins. However, excess ribosomal proteins do not accumulate, but turn over with half-lives ranging from 30 seconds to 24 minutes. Expression of an excess ribosomal protein does not affect the synthesis of other ribosomal proteins, as it frequently does in *E. coli* (for review, see Nomura et al. 1984).

An approach complementary to gene-dosage experiments is to examine ribosomal protein synthesis upon depletion of an individual ribosomal component. This method has the advantage that regulation of all ribosomal proteins can be assessed simultaneously at more physiological concentrations of ribosomal protein genes, transcripts, and proteins. Wittekind et al. (1990) assayed expression of ribosomal protein genes when transcription of 35S rRNA was limited by depletion of Rpa190p, a subunit of RNA polymerase I. A tenfold decline in 35S rRNA synthesis is paralleled by a decrease in accumulation of all ribosomal proteins analyzed. Synthesis of most ribosomal proteins is not affected, but the proteins are degraded because they have no nascent rRNA with which to assemble.

Similar results were obtained when synthesis of a component specific to one or the other ribosomal subunit was perturbed. When processing of the 27S RNA to the 25S and 5.8S species is blocked by a temperature-sensitive mutation in *RRP1*, ribosomal proteins are synthesized at normal rates (Gorenstein and Warner 1977). Those proteins that are components of the 60S subunit are degraded with half-lives from 7 to 24 minutes, but proteins in the 40S subunit are stable. When the 60S subunit proteins L3, L29, or L16, or the 40S subunit protein rp59, is depleted by shutting off transcription of *GAL* promoter fusion constructs, rRNA and other ribosomal proteins are synthesized normally (Nam and Fried 1986; Moritz et al. 1990). However, protein and RNA molecules destined for the same subunit as the depleted protein are destabilized and that subunit does not assemble.

The accumulated evidence suggests that there is little if any regulation of the translation of excess yeast ribosomal protein mRNAs. The original gene-dosage experiments suggesting that ribosomal proteins are not synthesized at rates proportional to excess mRNAs were misleading because they employed relatively long pulse-labeling times that would not have detected the rapid rate of turnover of some ribosomal proteins (Pearson et al. 1982; Himmelfarb et al. 1984; Warner et al. 1985). Nevertheless, the apparent efficiency of translation of a few ribosomal proteins can also be quite low even using very short (30–45 sec) labeling periods (Warner et al. 1985; Maicas et al. 1988; Wittekind et al. 1990). Do these proteins have exceptionally short half-lives when over-produced? Or is translation of these ribosomal protein mRNAs regulated in some fashion that has not yet been detected?

How and where are excess unassembled ribosomal proteins degraded so rapidly? Turnover of ribosomal proteins is scored by loss of each protein from its diagnostic position of migration in two-dimensional gels and therefore may mark only the first step in the degradation pathway of ribosomal proteins. Is there a specific machinery for the degradation of ribosomal proteins or are unassembled ribosomal proteins simply recognized as mislocalized and scavenged by a general pathway for protein degradation? If the latter, the pathway does not involve vacuolar proteases: Ribosomal proteins synthesized in *pep4–3* strains deficient in vacuolar proteases have half-lives identical to those in wild-type cells

(Tsay et al. 1988). Although it seems likely that the recognition of a ribosomal protein as being in excess occurs in the nucleolus, it is possible that turnover could occur in the cytoplasm, either upon failure of individual ribosomal proteins or subassembly complexes of ribosomal proteins to enter the nucleus or upon exit of these proteins from the nucleus in the absence of productive assembly.

C. Autogenous Regulation of Ribosomal Protein Expression

Although the majority of ribosomal protein genes studied do not appear to be regulated by feedback mechanisms, two distinct exceptions have been found. An imbalance in the synthesis of ribosomal precursor RNA and ribosomal protein L32, due either to the expression of extra copies of *RPL32* (Dabeva et al. 1986) or to a deficiency of RNA polymerase I (Wittekind et al. 1990), results in the accumulation of unspliced *RPL32* pre-mRNA. Regulation of *RPL32* pre-mRNA splicing has been localized to the 5′ end of exon 1 and the 5′ splice site (F.J. Eng and J.R. Warner, in prep.). L32 protein may stabilize base pairing between these two sequences in the 5′ end of *RPL32* pre-mRNA and prevent its splicing by blocking association of the U1 snRNP with the 5′ splice site. Although the frequent occurrence of introns in ribosomal protein genes suggests that splicing might be a target for coordinated regulation of these genes, splicing of none of six other ribosomal protein genes tested is similarly regulated (Abovich et al. 1985; Warner et al. 1985; El-Baradi et al. 1986; Maicas et al. 1988; Tsay et al. 1988). Splicing of the third intron of the *Xenopus laevis RPL1* gene is inhibited by excess amounts of free L1 (Bozzoni et al. 1984), but splicing of the homologous genes in yeast, *RPL2A* and *RPL2B*, is not similarly regulated (I. Bozzoni, pers. comm.).

  CRY1, encoding rp59, is expressed at tenfold higher levels than *CRY2*. Cessation of transcription of a *GAL1 CRY1* promoter fusion in a *cry1:TRP1* deletion strain results in an eightfold increase in *CRY2* mRNA within 20 minutes (A. Paulovich and J. Woolford, unpubl.). Because this effect is mediated through sequences 5′ of the *CRY2* transcription site, a reasonable hypothesis is that free rp59 present in the nucleus, in modest excess of that assembling into 40S ribosomal subunits, directly or indirectly represses transcription of *CRY2*.

  Clearly, assembly of ribosomes can be more easily coupled with feedback regulation of splicing or transcription, which are also nuclear events, than with translation, which occurs in the cytoplasm. The physiological significance of these regulatory events is unclear. It remains to be seen whether *RPL32* and *CRY2* are exceptional among ribosomal protein genes in being autogenously regulated.

XIII. COORDINATE CHANGES OF RIBOSOME BIOSYNTHESIS

A coordinated increase or decrease of ribosomal protein synthesis that occurs in proportion to different physiological requirements for protein synthesis is mediated primarily by control of transcription of ribosomal protein genes, rather than by posttranscriptional mechanisms.

Yeast grown in minimal medium using glucose as a carbon source have a growth rate three times faster than that of yeast utilizing ethanol, contain two and one-half times as many ribosomes, and synthesize rRNA and ribosomal proteins at a sevenfold higher rate (Kief and Warner 1981). Within 5 minutes after shifting cells from ethanol to glucose-containing medium, transcription of both rRNA and mRNA for ribosomal proteins increases and reaches a threefold higher level by 30–60 minutes. Expression of constructs containing only 5′ non-transcribed sequences of *RPS16A* increases at the same rate as does that of intact ribosomal protein genes in response to such a nutritional upshift (Donovan and Pearson 1986). The UAS$_{RPG}$ of *RPL25* is sufficient for the response (Herruer et al. 1987). However, it is not yet clear whether carbon-source-dependent changes in transcription of ribosomal protein genes result simply from altered amounts or activity of the Rap1p or Abf1p transcription factors (Chambers et al. 1989; Herruer et al. 1989). Other protein factors and *cis*-acting DNA sequences may be involved. For example, four different protein-binding sequences upstream of *RPS33* may function differently in ethanol versus glucose medium (Herruer et al. 1989).

Yeast cells grow at roughly the same rate at 23°C and 36°C. Yet cells shifted abruptly from 23°C to 36°C undergo a complex pattern of changes. Synthesis of more than 300 proteins is transiently repressed with varying rates and extents of decline and recovery, synthesis of more than 80 other proteins is induced, and synthesis of many other proteins is unaffected (Miller et al. 1982). The synthesis of nearly all ribosomal proteins is particularly sensitive to such a temperature shift, dropping by nearly 80% within 20 minutes, relative to synthesis of total proteins (Gorenstein and Warner 1976). By 60 minutes after the shift, synthesis is restored to normal rates. This effect results from a transient decrease in the rate of transcription of ribosomal protein genes (Kim and Warner 1983b). Ribosomal protein mRNA levels subsequently decline, as preexisting mRNAs are turned over (Lindquist 1981), possibly at an enhanced rate (Herruer et al. 1988). Upon resumption of transcription, ribosomal protein mRNA levels and synthesis of ribosomal proteins return to normal (Lindquist 1981; Rosbash et al. 1981; Kim and Warner 1983b; Larkin et al. 1987; Schwindinger and Warner 1987). Although this heat-shock effect is transcriptional, UAS$_{RPG}$ is neither necessary nor sufficient to mediate the effect of heat shock on transcription (Larkin et al. 1987; Herruer et al. 1988). Nor does deletion of substantial sequences between the UAS_{RPG} and the transcription initiation site of *CYH2* eliminate the heat-shock response of this gene (Schwindinger and Warner 1987).

Yeast exhibits a stringent response. Deprivation of amino acids causes reduced synthesis of rRNA and ribosomal proteins, whereas synthesis of tRNA and of most mRNAs is unaffected (Shulman et al. 1976; Warner and Gorenstein 1978a). This regulation occurs at the level of transcription. rRNA transcription decreases five- to sixfold within 2 hours, and the amounts of translatable ribosomal protein mRNA drop two- to fourfold within 5 hours (Warner and Gorenstein 1978a). Expression of fusions of the *RPL16A*, *CRY1*, *RPS16A*, or *RPL25* promoters to *lacZ* decreases in response to histidine limitation (C. Moehle and A. Hinnebusch, in prep.). Although UAS_{RPG} is sufficient for this response, the specific activity of Rap1p binding in cell extracts is unaffected (C. Moehle and A. Hinnebusch, pers. comm.).

Upon deprivation of nitrogen, diploid cells can sporulate. Although RNA and protein synthesis continues, ribosomal protein synthesis decreases five- to tenfold (Pearson and Haber 1980) as a result of a decrease in ribosomal protein mRNA levels (Kraig et al. 1982). It is not known whether this results from lower levels of transcription, increased rates of mRNA turnover, or both.

The results of a variety of experiments of the type described above suggest that the relative rate of ribosome biosynthesis is a sensitive barometer for the future prospects of the cell. If the prospects improve, e.g., on addition of glucose, the cell makes more ribosomes; if the prospects are foreboding, e.g., as an amino acid becomes limiting or the temperature changes abruptly, the cell makes fewer ribosomes. Clearly, the synthesis of ribosomal components is sensitive to some signal(s) generated by the cell's metabolism or milieu. But we have no idea what the signal is, or even if the same signal controls the synthesis of both rRNA and ribosomal proteins.

XIV. SUMMARY AND PROSPECTS

The following are some general conclusions that can be drawn from the decade of cloning:

1. The ribosomal proteins of all eukaryotes have been highly conserved through evolution. It is likely that most of the eukaryotic ribosomal

proteins have counterparts in the prokaryotic ribosome, although the relationship is not always obvious from the sequence alone.

2. For the most part, regulation of synthesis of the ribosomal proteins occurs at the level of transcription, which is controlled by a small number of transcription factors. The details of the regulation remain unclear. The regulation of ribosomal protein synthesis is generally global, without provision for the regulation of individual genes; exceptions occur.

3. The transcription of rRNA is supported by an enhancer that lies at the very end of the transcription unit. It is not known how rRNA transcription is controlled or whether a single signal controls the transcription of both rRNA and ribosomal protein genes.

4. In yeast, the processing of rRNA and its assembly with ribosomal proteins takes place in a nucleolus that contains many if not most of the components found in the nucleolus of higher eukaryotes.

The following are key questions that we hope to answer in the next decade: (1) What signals control the transcription of rRNA and ribosomal protein genes? How are they generated, and how do they function? (2) What components are essential for the processing of ribosomal precursor RNA and its assembly with ribosomal proteins? How do they work? (3) How is the nucleolus organized? (4) How does the nucleolus judge that a ribosomal particle is complete, and then how is the completed subunit exported to the cytoplasm?

ACKNOWLEDGMENTS

Work from the laboratory of J.L.W. has been supported by grants from NIGMS (GM-28301) and National Cancer Institute (CA-01000), and that from the laboratory of J.R.W. by grants from NIGMS (GM-25532), National Cancer Institute (CA-13330), and the American Cancer Society (MV-323T).

REFERENCES

Abovich, N., L. Gritz, L. Tung, and M. Rosbash. 1985. Effect of *RP51* gene dosage alterations on ribosome synthesis in *Saccharomyces cerevisiae*. *Mol. Cell. Biol.* **5:** 3429.

All-Robyn, J.A., N. Brown, E. Otaka, and S.W. Liebman. 1990. Sequence and functional similarity between a yeast ribosomal protein and the *E. coli* S5 *ram* protein. *Mol. Cell. Biol.* **10:** 6544.

Andrew, C., A.K. Hopper, and B.D. Hall. 1976. A yeast mutant defective in the processing of 27S rRNA precursor. *Mol. Gen. Genet.* **144:** 29.

Arevalo, S.G. and J.R. Warner. 1990. Ribosomal protein L4 of *Saccharomyces cerevisiae*: The gene and its protein. *Nucleic Acids Res.* **18:** 1447.

Aris, J.P. and G. Blobel. 1988. Identification and characterization of a yeast nucleolar protein that is similar to a rat liver nucleolar protein. *J. Cell Biol.* **107:** 17.

Baronas-Lowell, D.M. and J.R. Warner. 1990. Ribosomal protein L30 is dispensable in the yeast *Saccharomyces cerevisiae. Mol. Cell. Biol.* **10:** 5235.

Bayev, A.A., O.I. Georgiev, A.A. Hadjiolov, K.G. Skryabin, and V.M. Zakharyev. 1981. The structure of the yeast ribosomal RNA genes. 3. Precise mapping of the 18S and 25S rRNA genes and structure of the adjacent regions. *Nucleic Acids Res.* **9:** 789.

Bayliss, F.T. and J.L. Ingraham. 1974. Mutation in *Saccharomyces cerevisiae* conferring streptomycin and cold sensitivity by affecting ribosome formation and function. *J. Bacteriol.* **118:** 319.

Biswas, S.B. and E.E. Biswas. 1990. ARS binding factor 1 of the yeast *Saccharomyces cerevisiae* binds to sequences in telomeric and nontelomeric autonomously replicating sequences. *Mol. Cell. Biol.* **10:** 810.

Bollen, G.H.P.M., W.H. Mager, and R.J. Planta. 1981a. High resolution mini-two-dimensional gel electrophoresis of yeast ribosomal proteins. A standard nomenclature for yeast ribosomal proteins. *Mol. Biol. Rep.* **8:** 37.

Bollen, G.H.P.M.,, L.H. Cohen, W.H. Mager, A.W. Klaassen, and R.J. Planta. 1981b. Isolation of cloned ribosomal protein genes from the yeast *Saccharomyces carlsbergensis. Gene* **14:** 279.

Bozzoni, I., P. Fragapane, F. Annesi, P. Pierandrei-Amaldi, F. Amaldi, and E. Beccari. 1984. Expression of two *Xenopus laevis* ribosomal protein genes in injected frog oocytes. A specific splicing block interferes with the L1 RNA maturation. *J. Mol. Biol.* **180:** 987.

Brill, S.J., S. DiNardo, K. Voelkel-Meiman, and R. Sternglanz. 1987. Need for DNA topoisomerase activity as a swivel for DNA replication and for transcription of ribosomal RNA. *Nature* **326:** 414.

Brow, D.A. and E.P. Geiduschek. 1987. Modulation of yeast 5S rRNA synthesis *in vitro* by ribosomal protein YL3. *J. Biol. Chem.* **262:** 13953.

Bucher, K. and L. Skogerson. 1976. Cryptopleurine—An inhibitor of translocation. *Biochemistry* **22:** 4755.

Buchman, A.R., W.J. Kimmerly, J. Rine, and R.D. Kornberg. 1988. Two DNA-binding factors recognize specific sequences at silencers, upstream activating sequences, autonomously replicating sequences, and telomeres in *Saccharomyces cerevisiae. Mol. Cell. Biol.* **8:** 210.

Burgess, S., J. Couto, and C. Guthrie. 1990. A putative ATP binding protein influences the fidelity of branchpoint recognition in splicing. *Cell* **60:** 705.

Capieaux, E., M.-L. Vignais, A. Sentenac, and A. Goffeau. 1989. The yeast H^+-ATPase gene is controlled by the promoter binding factor TUF. *J. Biol. Chem.* **264:** 7437.

Chambers, A., J.S.H. Tsang, C. Stanway, A. Kingsman, and S.M. Kingsman. 1989. Transcriptional control of the *Saccharomyces cerevisiae PGK* gene by RAP1. *Mol. Cell. Biol.* **9:** 5516.

Chasman, D.I., N.F. Lue, A.R. Buchman, J.W. LaPointe, Y. Lorch, and R.D. Kornberg. 1990. A yeast protein that influences the chromatin structure of UAS_G and functions as a powerful auxiliary gene activator. *Genes Dev.* **4:** 503.

Cooper, D., D.D. Banthorpe, and D. Wilkie. 1967. Modified ribosomes conferring resistance to cycloheximide in mutants of *Saccharomyces cerevisiae. J. Mol. Biol.* **26:** 347.

Cox, R.A. and P. Greenwell. 1980. Protein synthesis by hybrid ribosomes reconstructed from rabbit reticulocyte ribosomal core-particles and amphibian or fungal split-proteins. *Biochem. J.* **186:** 861.

Dabeva, M.D. and J.R. Warner. 1987. The yeast ribosomal protein L32 and its gene. *J. Biol. Chem.* **262:** 16055.

Dabeva, M.D., M.A. Post-Beittenmiller, and J.R. Warner. 1986. Autogenous regulation of splicing of the transcript of a yeast ribosomal protein gene. *Proc. Natl. Acad. Sci.* **83:** 5854.

Dams, E., L. Hendriks, Y. Van de Peer, J.-M. Neefs, G. Smits, I. Vandenbempt, and R. De Wachter. 1988. Compilation of small ribosomal subunit RNA sequences. *Nucleic Acids Res.* **16:** r87.

Della-Seta, F., S.-A. Ciafré, C. Marck, B. Santoro, C. Presutti, A. Sentenac, and I. Bozzoni. 1990. The ABF1 factor is the transcriptional activator of the L2 ribosomal protein genes in *Saccharomyces cerevisiae*. *Mol. Cell. Biol.* **10:** 2437.

Diffley, J.F.X. and B. Stillman. 1988. Purification of a yeast protein that binds to origins of DNA replication and a transcriptional silencer. *Proc. Natl. Acad. Sci.* **85:** 2120.

————. 1989. Similarity between the transcriptional silencer binding proteins ABF1 and RAP1. *Science* **246:** 1034.

Dölz, H., D. Vázquez, and A. Jiménez. 1982. Quantitation of the specific interaction of [14a–^{3}H] cryptopleurine with 80S and 40S ribosomal species from the yeast *Saccharomyces cerevisiae*. *Biochemistry* **21:** 3181.

Donovan, D.M. and N.J. Pearson. 1986. Transcriptional regulation of ribosomal proteins during a nutritional upshift in *Saccharomyces cerevisiae*. *Mol. Cell. Biol.* **6:** 2429.

Donovan, D.M., M.P. Remington, D.A. Stewart, J.C. Crouse, D.J. Miles, and N.J. Pearson. 1990. Functional analysis of a duplicated linked pair of ribosomal protein genes in *Saccharomyces cerevisiae*. *Mol. Cell. Biol.* **10:** 6097.

Dorsman, J.C., W.C. Van Heeswijk, and L.A. Grivell. 1988. Identification of two factors that bind to the upstream sequences of a number of nuclear genes coding for mitochondrial proteins and to genetic elements important for cell division in yeast. *Nucleic Acids Res.* **16:** 7287.

Dorsman, J.C., M.M. Doorenbosch, C.T.C. Maurer, J.H. de Winde, W.H. Mager, R.J. Planta, and L.A. Grivell. 1989. An ARS/silencer binding factor also activates two ribosomal protein genes in yeast. *Nucleic Acids Res.* **17:** 4917.

El-Baradi, T.T., H.A. Raué, V.C.H. de Regt, and R.J. Planta. 1984. Stepwise dissociation of yeast 60S ribosomal subunits by LiCl and identification of L25 as a primary 26S rRNA binding protein. *Eur. J. Biochem.* **144:** 393.

El-Baradi, T.T., H.A. Raué, V.C.H. de Regt, E.C. Verbree, and R.J. Planta. 1985. Yeast ribosomal protein L25 binds to an evolutionary conserved site on yeast 26S and *E. coli* 23S rRNA. *EMBO J.* **4:** 2101.

El-Baradi, T.T., A.F.M. van der Sande, W.H. Mager, H.A. Raué, and R.J. Planta. 1986. The cellular level of yeast ribosomal protein L25 is controlled principally by rapid degradation of excess protein. *Curr. Genet.* **10:** 733.

El-Baradi, T.T., V.C. de Regt, S.W. Einerhand, J. Teixido, R.J. Planta, J.P. Ballesta, and H.A. Raué. 1987. Ribosomal proteins EL11 from *Escherichia coli* and L15 from *Saccharomyces cerevisiae* bind to the same site in both yeast 26S and mouse 28S rRNA. *J. Mol. Biol.* **195:** 909.

Elion, E.A. and J.R. Warner. 1984. The major promoter element of rRNA transcription in yeast lies 2 kb upstream. *Cell* **39:** 663.

————. 1986. An RNA polymerase I enhancer in *Saccharomyces cerevisiae*. *Mol. Cell. Biol.* **6:** 2089.

Fabian, G.F. and A.K. Hopper. 1987. *RRP1*, a *Saccharomyces cerevisiae* gene affecting rRNA processing and production of mature ribosomal subunits. *J. Bacteriol.* **169**: 1571.

Fabian, G.F., S.M. Hess, and A.K. Hopper. 1990. *srd1*, a *Saccharomyces cerevisiae* suppressor of the temperature-sensitive pre-rRNA processing defect of *rrp1-1*. *Genetics* **124**: 497.

Finley, D., B. Bartell, and A. Varshavsky. 1989. The tails of ubiquitin precursors are ribosomal proteins whose fusions to ubiquitin facilitates ribosome biogenesis. *Nature* **338**: 394.

Fleming, G., P. Belhumeur, D. Skup, and H.M. Fried. 1989. Functional substitution of mouse ribosomal protein L27′ for yeast ribosomal protein L29 in yeast ribosomes. *Proc. Natl. Acad. Sci.* **86**: 217.

Fried, H.M. and J.R. Warner. 1981. Cloning of yeast gene for trichodermin resistance and ribosomal protein L3. *Proc. Natl. Acad. Sci.* **78**: 238.

―――. 1982. Molecular cloning and analysis of yeast gene for cycloheximide resistance and ribosomal protein L29. *Nucleic Acids Res.* **10**: 3133.

Fried, H.M., H.G. Nam, S. Loechel, and J. Teem. 1985. Characterization of yeast strains with conditionally expressed variants of ribosomal protein genes *tcm1* and *cyh2*. *Mol. Cell. Biol.* **5**: 99.

Fried, H.M., N.J. Pearson, C.H. Kim, and J.R. Warner. 1981. The genes for fifteen ribosomal proteins of *Saccharomyces cerevisiae*. *J. Biol. Chem.* **256**: 10176.

Geiduschek, E.P. and G.P. Tocchini-Valentini. 1988. Transcription by RNA polymerase III. *Annu. Rev. Biochem.* **57**: 873.

Gorenstein, C. and J.R. Warner. 1976. Coordinate regulation of the synthesis of eukaryotic ribosomal proteins. *Proc. Natl. Acad. Sci.* **73**: 1547.

―――. 1977. Synthesis and turnover of ribosomal proteins in the absence of 60S subunit assembly in *Saccharomyces cerevisiae*. *Mol. Gen. Genet.* **157**: 327.

Grant, P., L. Sánchez, and A. Jiménez. 1974. Cryptopleurine resistance: Genetic locus for a 40S ribosomal component in *Saccharomyces cerevisiae*. *J. Bacteriol.* **120**: 1308.

Grant, P.G., D. Schindler, and J.E. Davies. 1976. Mapping of trichodermin resistance in *Saccharomyces cerevisiae*: A genetic locus for a component of the 60S ribosomal subunit. *Genetics* **83**: 667.

Gritz, L.R., J.A. Mitlin, M. Cannon, B. Littlewood, C.J. Carter, and J.E. Davies. 1982. Ribosome structure, maturation of ribosomal RNA and drug sensitivity in temperature-sensitive mutants of *Saccharomyces cerevisiae*. *Mol. Gen. Genet.* **188**: 384.

Gudenus, R., S. Mariotte, A. Moenne, A. Ruet, S. Memet, J.M. Buhler, A. Sentenac, and P. Thuriaux. 1988. Conditional mutants of *RPC160*, the gene encoding the largest subunit of RNA polymerase C in *Saccharomyces cerevisiae*. *Genetics* **119**: 517.

Gutell, R.R., M.N. Schnare, and M.W. Gray. 1990. A compilation of large subunit (23S-like) ribosomal RNA sequences presented in a secondary structure format. *Nucleic Acids Res.* (suppl.) **18**: 2319.

Halfter, H., B. Kavety, J. Vandekerckhove, F. Kiefer, and D. Gallwitz. 1989. Sequence, expression and mutational analysis of BAF1, a transcriptional activator and *ARS1*-binding protein of the yeast *Saccharomyces cerevisiae*. *EMBO J.* **8**: 4265.

Hamil, K.G., H.G. Nam, and H.M. Fried. 1988. Constitutive transcription of yeast ribosomal protein gene *TCM1* is promoted by uncommon *cis*- and *trans*-acting elements. *Mol. Cell. Biol.* **8**: 4328.

Hartwell, L., C.S. McLaughlin, and J.R. Warner. 1970. Identification of ten genes that control ribosome formation in yeast. *Mol. Gen. Genet.* **109**: 42.

Henriquez, R., G. Blobel, and J.P. Aris. 1990. Isolation and sequencing of *NOP1*. *J. Biol. Chem.* **265**: 2209.

Henry, Y.A.L., A. Chambers, J.S.H. Tsang, A.J. Kingsman, and S.M. Kingsman. 1990.

Characterization of the DNA binding domain of the yeast RAP1 protein. *Nucleic Acids Res.* **18**: 2617.

Herruer, M.H., W.H. Mager, T.M. Doorenbosch, P.L.M. Wessels, T.M. Wassenaar, and R.J. Planta. 1989. The extended promoter of the gene encoding ribosomal protein S33 in yeast consists of multiple protein binding elements. *Nucleic Acids Res.* **17**: 7427.

Herruer, M.H., W.H. Mager, H.A. Raué, P. Vreken, E. Wilms, and R.J. Planta. 1988. Mild temperature shock affects transcription of yeast ribosomal protein genes as well as the stability of their mRNAs. *Nucleic Acids Res.* **16**: 7917.

Herruer, M.H., W.H. Mager, L.P. Woudt, R.T. Nieuwint, G.M. Wassenaar, P. Groeneveld, and R.J. Planta. 1987. Transcriptional control of yeast ribosomal protein synthesis during the carbon-source upshift. *Nucleic Acids Res.* **15**: 10122.

Himmelfarb, H.J., A. Vassarotti, and J.D. Friesen. 1984. Molecular cloning and biosynthetic regulation of the *cry1* gene of *Saccharomyces cerevisiae*. *Mol. Gen. Genet.* **195**: 500.

Hogan, J.J., R.R. Gutell, and H.F. Noller. 1984a. Probing the conformation of 18S rRNA in yeast 40S ribosomal subunits with kethoxal. *Biochemistry* **23**: 3322.

————. 1984b. Probing the conformation of 26S rRNA in yeast 60S ribosomal subunits with kethoxal. *Biochemistry* **23**: 3330.

Hopper, A., F. Banks, and V. Evangelidis. 1978. A yeast mutant which accumulates precursor tRNAs. *Cell* **14**: 211.

Huet, J. and A. Sentenac. 1987. TUF, the yeast DNA-binding factor specific for UAS$_{rpg}$ upstream activating sequences: Identification of the protein and its DNA-binding domain. *Proc. Natl. Acad. Sci.* **84**: 3648.

Huet, J., P. Cottrelle, M. Cool, M.-L. Vignais, D. Thiele, C. Marck, J.-M. Buhler, A. Sentenac, and P. Fromageot. 1985. A general upstream binding factor for genes of the yeast translational apparatus. *EMBO J.* **4**: 3539.

Hughes, J.M.X., D.A.M. Konings, and G. Cesareni. 1987. The yeast homologue of U3 snRNA. *EMBO J.* **6**: 2145.

Hurt, E.C., A. McDowall, and T. Schimmang. 1988. Nucleolar and nuclear envelope proteins of the yeast *Saccharomyces cerevisiae*. *Eur. J. Cell Biol.* **46**: 554.

Hutchison, H.T., L.H. Hartwell, and C.S. McLaughlin. 1969. Temperature-sensitive yeast mutant defective in ribonucleic acid production. *J. Bacteriol.* **99**: 807.

Itoh, T. 1980. Primary structure of yeast acidic protein YPA1. *FEBS Lett.* **114**: 119.

Itoh, T. and B. Wittman-Liebold. 1978. The primary structure of protein L44 from the large subunit of yeast ribosomes. *FEBS Lett.* **96**: 399.

Itoh, T., E. Otaka, and K.A. Matsui. 1985. Primary structures of ribosomal protein YS25 from *Saccharomyces cerevisiae* and its counterparts from *Schizosaccharomyces pombe* and rat liver. *Biochemistry* **24**: 7418.

Jacobs-Lorena, M. and H.M. Fried. 1987. Translational regulation of ribosomal protein gene expression in eukaryotes. In *Translational regulation of gene expression* (ed. J. Ilan), p. 63. Plenum Press, New York.

Jemtland, R., E. Maehlum, O.S. Gabrielsen, and T.B. Oyen. 1986. Regular distribution of length heterogeneities within non-transcribed spacer regions of cloned and genomic rDNA of *Saccharomyces cerevisiae*. *Nucleic Acids Res.* **14**: 5145.

Jiménez, A., L. Sánchez, and D. Vázquez. 1975. Simultaneous ribosomal resistance to trichodermin and anisomycin in *Saccharomyces cerevisiae* mutants. *Biochim. Biophys. Acta* **383**: 427.

Johnson, S.P. and J.R. Warner. 1987. Phosphorylation of the *Saccharomyces cerevisiae* equivalent of ribosomal protein S6 has no detectable effect on growth. *Mol. Cell. Biol.* **7**: 1338.

————. 1989. Unusual enhancer function in yeast rRNA transcription. *Mol. Cell. Biol.* **9:** 4986.

Jong, A.Y.-S., M.W. Clark, M. Gilbert, A. Oehm, and J.L. Campbell. 1987. *Saccharomyces cerevisiae* SSB1 protein and its relationship to nucleolar RNA-binding proteins. *Mol. Cell. Biol.* **7:** 2947.

Ju, Q., B.E. Morrow, and J.R. Warner. 1990. *REB1*, a yeast DNA-binding protein with many targets, is essential for cell growth and bears some resemblance to the oncogene *myb*. *Mol. Cell Biol.* **10:** 5226.

Juan-Vidales, F., F. Sanchez Madrid, M.T. Saenz-Robles, and J.P.G. Ballesta. 1983. Purification and characterization of two ribosomal proteins of *Saccharomyces cerevisiae*. Homologies with proteins from eukaryotic species and with bacterial protein ECL 11. *Eur. J. Biochem.* **136:** 275.

Kass, S., K. Tyc, J.A. Steitz, and B. Sollner-Webb. 1990. The U3 small nucleolar ribonucleoprotein functions in the first step of preribosomal RNA processing. *Cell* **60:** 897.

Kaufer, N.F., H.M. Fried, W.F. Schwindinger, M. Jasin, and J.R. Warner. 1983. Cycloheximide resistance in yeast: The gene and its protein. *Nucleic Acids Res.* **11:** 3123.

Kempers-Veenstra, A.E., J. Oliemans, H. Offenberg, A.F. Dekker, P.W. Piper, R.J. Planta, and J. Klootwijk. 1986. 3′-end formation of transcripts from the yeast rRNA operon. *EMBO J.* **5:** 2703.

Kief, D.R. and J.R. Warner. 1981. Coordinate control of syntheses of ribosomal ribonucleic acid and ribosomal proteins during nutritional shift-up in *Saccharomyces cerevisiae*. *Mol. Cell. Biol.* **1:** 1007.

Kim, C.H. and J.R. Warner. 1983a. The mRNA for ribosomal proteins in yeast. *J. Mol. Biol.* **165:** 79.

————. 1983b. Mild temperature shock alters the transcription of a discrete class of *Saccharomyces cerevisiae* genes. *Mol. Cell. Biol.* **3:** 457.

Kimmerly, W., A. Buchman, R. Kornberg, and J. Rine. 1988. Roles of two DNA-binding factors in replication, segregation, and transcriptional repression mediated by a yeast silencer. *EMBO J.* **7:** 2241.

Klemenz, R. and E.P. Geiduschek. 1980. The 5′ terminus of the precursor ribosomal RNA of *Saccharomyces cerevisiae*. *Nucleic Acids Res.* **8:** 2679.

Klootwijk, J., P. de Jonge, and R.J. Planta. 1979. The primary transcript of the ribosomal repeating unit in yeast. *Nucleic Acids. Res.* **6:** 27.

Kraakman, L.S., W.H. Mager, K.T.C. Maurer, R.T.M. Nieuwint, and R.J. Planta. 1989. The divergently transcribed genes encoding yeast ribosomal proteins L46 and S24 are activated by shared RPG-boxes. *Nucleic Acids Res.* **17:** 9693.

Kraig, E., J.E. Haber, and M. Rosbash. 1982. Sporulation and *rna2* lower ribosomal protein mRNA levels by different mechanisms in *Saccharomyces cerevisiae*. *Mol. Cell. Biol.* **2:** 1199.

Kruiswijk, T., R.J. Planta, and J.M. Krop. 1978. The course of assembly of ribosomal subunits in yeast. *Biochem. Biophys. Acta* **517:** 378.

Kruse, C., S.P. Johnson, and J.R. Warner. 1985. Phosphorylation of the yeast equivalent of ribosomal protein S6 is not essential for growth. *Proc. Natl. Acad. Sci.* **82:** 7515.

Kulkens, T., H. van Heerikhuizen, J. Klootwijk, J. Oliemans, and R.J. Planta. 1989. A yeast ribosomal DNA-binding protein that binds to the rDNA enhancer and also close to the site of Pol I transcription initiation is not important for enhancer functioning. *Curr. Genet.* **16:** 351.

Larkin, J.C. and J.L. Woolford, Jr. 1983. Molecular cloning and analysis of the *CRY1* gene: A yeast ribosomal protein gene. *Nucleic Acids Res.* **11:** 403.

Larkin, J.C., J.R. Thompson, and J.L. Woolford, Jr. 1987. Structure and expression of the *Saccharomyces cerevisiae CRY1* gene: A highly conserved ribosomal protein gene. *Mol. Cell. Biol.* **7:** 1764.

Lee, J.C. 1990. Ribosomes from *Saccharomyces cerevisiae*. In *The yeasts* (ed. A.H. Rose and J.S. Harrison), vol. 4, p. 489. Academic Press, New York.

Lee, J.C. and R. Anderson. 1986. Partial reassembly of yeast 60S ribosomal subunits *in vitro* following controlled disassociation under nondenaturing conditions. *Arch. Biochem. Biophys.* **245:** 248.

Lee, J.C., R. Anderson, Y.C. Yeh, and P. Horowitz. 1985. Extraction of proteins from *Saccharomyces cerevisiae* ribosomes under nondenaturing conditions. *Arch. Biochem. Biophys.* **237:** 292.

Leer, R.J., M.M. van Raamsdonk-Duin, W.H. Mager, and R.J. Planta. 1985a. Conserved sequences upstream of yeast ribosomal protein genes. *Curr. Genet.* **9:** 273.

Leer, R.J., M.M. van Raamsdonk-Duin, M.J. Hagendoorn, W.H. Mager, and R.J. Planta. 1984. Structural comparison of yeast ribosomal protein genes. *Nucleic Acids Res.* **12:** 6685.

Leer, R.J., M.M. van Raamsdonk-Duin, P. Kraakman, W.H. Mager, and R.J. Planta. 1985b. The genes for yeast ribosomal proteins S24 and L46 are adjacent and divergently transcribed. *Nucleic Acids Res.* **13:** 701.

Leer, R.J., M.M.C. van Raamsdonk-Duin, C.M.T. Molenaar, L.H. Cohen, W.H. Mager, and R.J. Planta. 1982. The structure of the gene coding for the phosphorylated ribosomal protein S10 in yeast. *Nucleic Acids Res.* **10:** 5869.

Leer, R. J., M.M. van Raamsdonk-Duin, C.M. Molenaar, H.M. Witsenboer, W.H. Mager, and R.J. Planta. 1985c. Yeast contains two functional genes coding for ribosomal protein S10. *Nucleic Acids Res.* **13:** 5027.

Leer, R.J., M.M. van Raamsdonk-Duin, P.J. Schoppink, M.T. Cornelissen, L.H. Cohen, W.H. Mager, and R.J. Planta. 1983. Yeast ribosomal protein S33 is encoded by an unsplit gene. *Nucleic Acids Res.* **11:** 7759.

Li, H.V., J. Zagorski, and M.J. Fournier. 1990. Depletion of U14 small nuclear RNA (snR128) disrupts production of 18S rRNA in *Saccharomyces cerevisiae. Mol. Cell. Biol.* **10:** 1145

Lindquist, S. 1981. Regulation of protein synthesis during heat shock. *Nature* **293:** 311.

Longtine, M.S., N.M. Wilson, M.E. Petracek, and J. Berman. 1989. A yeast telomere binding activity binds to two related telomere sequence motifs and is indistinguishable from RAP1. *Curr. Genet.* **16:** 225.

Lucioli, A., C. Presutti, S. Ciafre, E. Caffarelli, P. Fragapane, and I. Bozzoni. 1988. Gene dosage alteration of L2 ribosomal protein genes in *Saccharomyces cerevisiae:* Effects on ribosome synthesis. *Mol. Cell. Biol.* **8:** 4792.

Lustig, A.J., S. Kurtz, and D. Shore. 1990. Involvement of the silencer and UAS binding protein RAP1 in regulation of telomere length. *Science* **250:** 549.

Maicas, E., F.G. Pluthero, and J.D. Friesen. 1988. The accumulation of three yeast ribosomal proteins under conditions of excess mRNA is determined primarily by fast protein decay. *Mol. Cell. Biol.* **8:** 169.

Mann, C., J.M. Buhler, I. Treich, and A. Sentenac. 1987. *RPC40*, a unique gene for a subunit shared between yeast RNA polymerases A and C. *Cell* **48:** 627.

Memet, S., M. Gouy, C. Marck, A. Sentenac, and J.N. Buhler. 1988. *RPA190*, the gene coding for the largest subunit of yeast RNA polymerase A. *J. Biol. Chem.* **263:** 2830.

Mestel, R., M. Yip, J.P. Holland, E. Wang, J. Kang, and M.J. Holland. 1989. Sequences within the spacer region of yeast rRNA cistrons that stimulate 35S rRNA synthesis *in vivo* mediate RNA polymerase I-dependent promoter and terminator activities. *Mol. Cell. Biol.* **9:** 1243.

Michel, S., R.R. Traut, and J.C. Lee. 1983. Yeast ribosomal proteins: Electrophoretic analysis in four two-dimensional gel systems—Correlation of nomenclatures. *Mol. Gen. Genet.* **191:** 251.

Miller, M.J., N.-H. Xuong, and E.P. Geiduschek. 1982. Quantitative analysis of the heat shock response of *Saccharomyces cerevisiae. J. Bacteriol.* **151:** 311.

Mitlin, J.A. and N. Cannon. 1984. Defective processing of ribosomal precursor RNA in *Saccharomyces cerevisiae. Biochem. J.* **220:** 461.

Mitra, G. and J.R. Warner. 1984. A yeast ribosomal protein gene whose intron is in the 5′ leader. *J. Biol. Chem.* **259:** 9218.

Mitsui, K. and K. Tsurugi. 1988a. cDNA and deduced amino acid sequence of 38 kDa-type acidic ribosomal protein AO from *Saccharomyces cerevisiae. Nucleic Acids Res.* **16:** 3573.

————. 1988b. cDNA and deduced amino acid sequence of acidic ribosomal protein A1 from *Saccharomyces cerevisiae. Nucleic Acids Res.* **16:** 3574.

————. 1988c. cDNA and deduced amino acid sequence of acidic ribosomal protein A2 from *Saccharomyces cerevisiae. Nucleic Acids Res.* **16:** 3575.

Molenaar, C.M., L.P. Woudt, A.E.M. Jansen, W.H. Mager, R.J. Planta, D.M. Donovan, and N.J. Pearson. 1984. Structure and organization of two linked ribosomal protein genes in yeast. *Nucleic Acids Res.* **12:** 7345.

Moreland, R.B., H.G. Nam, L.M. Hereford, and H.M. Fried. 1985. Identification of a nuclear localization signal of a yeast ribosomal protein. *Proc. Natl. Acad. Sci.* **82:** 6561.

Moritz, M., A.G. Paulovich, Y.-F. Tsay, and J.L. Woolford, Jr. 1990. Depletion of yeast ribosomal proteins in L16 or rp59 disrupts ribosome assembly. *J. Cell Biol.* **111:** 2261.

Morrow, B.E., S.P. Johnson, and J.R. Warner. 1989. Proteins that bind to the yeast rDNA enhancer. *J. Biol. Chem.* **264:** 9061.

Morrow, B.E., Q. Ju, and J.R. Warner. 1990. Purification and characterization of the yeast rDNA binding protein *REB1. J. Biol. Chem.* **265:** 20778.

Mullen, J.R., P.S. Kayne, R.P. Moerschell, S. Tsunasawa, M. Gribskov, M. Colavito-Shepanski, M. Grunstein, F. Sherman, and R. Sternglanz. 1989. Identification and characterization of genes and mutants for an N-terminal acetyltransferase from yeast. *EMBO J.* **8:** 2067.

Musters, W., R.J. Planta, H. van Heerikhuizen, and H.A. Raué. 1990. Functional analysis of the transcribed spacers of *Saccharomyces cerevisiae* ribosomal DNA: It takes a precursor to form a ribosome. In *The ribosome: Structure, function and evolution* (ed. W. Hill et al.), p. 435. American Society for Microbiology, Washington, D.C.

Musters, W., J. Knol, P. Maas, A.F. Dekker, H. van-Heerikhuizen, and R.J. Planta. 1989a. Linker scanning of the yeast RNA polymerase I promoter. *Nucleic Acids Res.* **17:** 9661.

Musters, W., J. Venema, G. van der Linden, H. van Heerikhuizen, J. Klootwijk, and R.J. Planta. 1989b. A system for the analysis of yeast ribosomal DNA mutations. *Mol. Cell. Biol.* **9:** 551.

Nam, H.G. and H. Fried. 1986. Effects of progressive depletion of *TCM1* or *CYH2* mRNA on *Saccharomyces cerevisiae* ribosomal protein accumulation. *Mol. Cell. Biol.* **6:** 1535.

Neigeborn, L. and J.R. Warner. 1990. Expression of yeast 5S RNA is independent of the rDNA enhancer region. *Nucleic Acids Res.* **18:** 4179.

Newton, C.H., L.C. Shimmin, J. Yee, and P.P. Dennis. 1990. A family of genes encode the multiple forms of the *Saccharomyces cerevisiae* ribosomal proteins equivalent to the *Escherichia coli* L12 protein and a single form of the L10-equivalent ribosomal protein. *J. Bacteriol.* **172:** 579.

Nieuwint, R.T.M., W.H. Mager, K.C.T. Maurer, and R.J. Planta. 1989. Mutational analysis of the upstream activation site of yeast ribosomal protein genes. *Curr. Genet.* **15:** 247.

Nieuwint, R.T.M., C.M.T. Molenaar, J.H. van Bommell, M.M.C. van Raamsdonk-Duin, W.H. Mager, and R.J. Planta. 1985. The gene for yeast ribosomal protein S31 contains an intron in the leader sequence. *Curr. Genet.* **10:** 1.

Nikolaev, N., O.I. Georgiev, P.V. Venkov, and A.A. Hadjiolov. 1979. The 37S precursor to ribosomal RNA is the primary transcript of ribosomal RNA genes in *Saccharomyces cerevisiae. J. Mol. Biol.* **127:** 297.

Nomura, M., R. Gourse, and G. Baughman. 1984. Regulation of the synthesis of ribosomes and ribosomal components. *Annu. Rev. Biochem.* **53:** 75.

Otaka, E. and S. Osawa. 1981. Yeast ribosomal proteins. V. Correlation of several nomenclatures and proposal of a standard nomenclature. *Mol. Gen. Genet.* **181:** 176.

Otaka, E., K. Higo, and T. Itoh. 1984. Yeast ribosomal proteins. VIII. Isolation of two proteins and sequence characterization of twenty-four proteins from cytoplasmic ribosomes. *Mol. Gen. Genet.* **195:** 544.

Ozkaynak, E., D. Finley, M.J. Solomon, and A. Varshavsky. 1987. The yeast ubiquitin genes: A family of natural gene fusions. *EMBO J.* **6:** 1429.

Paine, P.L. and C.M. Feldherr. 1972. Nucleocytoplasmic exchange of macromolecules. *Exp. Cell Res.* **74:** 81.

Papciak, S.N. and N.J. Pearson. 1987. Genetic mapping of two pairs of linked ribosomal protein genes in *Saccharomyces cerevisiae. Curr. Genet.* **11:** 445.

Parker, R., T. Simmons, E.O. Shuster, P.C. Siciliano, and C. Guthrie. 1988. Genetic analysis of small nuclear RNAs in *Saccharomyces cerevisiae*: Viable sextuple mutant. *Mol. Cell. Biol.* **8:** 3150.

Partaledis, J.A. and T.L. Mason. 1988. Structure and regulation of a nuclear gene in *Saccharomyces cerevisiae* that specifies MRP13, a protein of the small subunit of the mitochondrial ribosome. *Mol. Cell. Biol.* **8:** 3647.

Pearson, N.J. and J.E. Haber. 1980. Changes in regulation of ribosomal protein synthesis during vegetative growth and sporulation of *Saccharomyces cerevisiae. J. Bacteriol.* **143:** 1411.

Pearson, N.J., H.M. Fried, and J.R. Warner. 1982. Yeast use translational control to compensate for extra copies of a ribosomal protein gene. *Cell* **29:** 347.

Petes, T.D. 1980. Unequal meiotic recombination within tandem arrays of yeast ribosomal DNA genes. *Cell* **19:** 765.

Petes, T.D. and D. Botstein. 1977. Simple Mendelian inheritance of the reiterated ribosomal DNA of yeast. *Proc. Natl. Acad. Sci.* **74:** 5091.

Philippsen, P., M. Thomas, R.A. Kramer, and R.W. Davis. 1978. Unique arrangement of coding sequences for 5S, 5.8S, 18S and 25S ribosomal RNA in *Saccharomyces cerevisiae* as determined by R-loop and hybridization analysis. *J. Mol. Biol.* **123:** 387.

Planta, R.J. and H.A. Raué. 1988. Control of ribosome biogenesis in yeast. *Trends Genet.* **4:** 64.

Presutti, C., A. Lucioli, and I. Bozzoni. 1988. Ribosomal protein L2 in *Saccharomyces cerevisiae* is homologous to ribosomal protein L1 in *Xenopus laevis. J. Biol. Chem.* **263:** 6188.

Raué, H.A. and R.J. Planta. 1990. Ribosome biogenesis in yeast. *Prog. Biophys. Mol. Biol.* **41:** (in press).

Raué, H.A., W. Musters, C.A. Rutgers, J. Van't Riet, and R.J. Planta. 1990. rRNA: From structure to function. In *The ribosome: Structure, function and evolution* (ed. W. Hill et al.), p. 217. American Society for Microbiology, Washington, D.C.

Redman, K.L. and M. Rechsteiner. 1989. The long ubiquitin extension is ribosomal protein S27a. *Nature* **338:** 438.

Reidel, N., J.A. Wise, H. Swerdlow, A. Mak, and C. Guthrie. 1986. Small nuclear RNAs from *Saccharomyces cerevisiae*: Unexpected diversity in abundance, size, and molecular complexity. *Proc. Natl. Acad. Sci.* **83:** 8097.

Remacha, M., C. Santos, and J.P. Ballesta. 1990. Disruption of single-copy genes encoding acidic ribosomal proteins in *Saccharomyces cerevisiae*. *Mol. Cell. Biol.* **10:** 2182.

Remacha, M., M.T. Saenz-Robles, M.D. Vilella, and J.P. Ballesta. 1988. Independent genes coding for three acidic proteins of the large ribosomal subunit from *Saccharomyces cerevisiae*. *J. Biol. Chem.* **263:** 9094.

Rhode, P.R., K.S. Sweder, K.F. Oegema, and J.L. Campbell. 1989. The gene encoding ARS-binding factor I is essential for the viability of yeast. *Genes Dev.* **3:** 1926.

Riggs, D. and M. Nomura. 1990. Specific transcription of *Saccharomyces cerevisiae* 35S rDNA by RNA polymerase I *in vitro*. *J. Biol. Chem.* **265:** 7596.

Rosbash, M., P.K.W. Harris, J.L. Woolford, Jr., and J.L. Teem. 1981. The effect of temperature sensitive RNA mutants on the transcription products from cloned ribosomal protein genes of yeast. *Cell* **24:** 679.

Rotenberg, M.O. and J.L. Woolford, Jr. 1986. Tripartite upstream promoter element essential for expression of *Saccharomyces cerevisiae* ribosomal protein genes. *Mol. Cell. Biol.* **6:** 674.

Rotenberg, M.O., M. Moritz, and J.L. Woolford, Jr. 1988. Depletion of *Saccharomyces cerevisiae* ribosomal protein L16 causes a decrease in 60S ribosomal subunits and formation of half-mer polyribosomes. *Genes Dev.* **2:** 160.

Sachs, A.B. and R.W. Davis. 1989. The poly(A) binding protein is required for poly(A) shortening and the 60S ribosomal subunit-dependent translation initiation. *Cell* **58:** 857.

————. 1990. Translation initiation and ribosomal biogenesis: Involvement of a putative rRNA helicase and *RPL46*. *Science* **247:** 1077.

Santangelo, G.M. and J. Tornow. 1990. Efficient transcription of the glycolytic gene *ADH1* and three translational component genes requires the *GCR1* product, which can act through TUF/GRF/RAP binding sites. *Mol. Cell. Biol.* **10:** 859.

Schaap, P.J , C.M.T. Molenaar, W.H. Mager, and R.J. Planta. 1984. The primary structure of a gene encoding yeast ribosomal protein L34. *Curr. Genet.* **9:** 47.

Scheer, U. and R. Benavente. 1990. Functional and dynamic aspects of the mammalian nucleolus. *BioEssays* **12:** 14.

Schimmang, T., D. Tollervey, H. Kern, R. Frank, and E.C. Hurt. 1989. A yeast nucleolar protein related to mammalian fibrillarin is associated with small nucleolar RNA and is essential for viability. *EMBO J.* **8:** 4015.

Schindler, D., P. Grant, and J. Davies. 1974. Trichodermin resistance-mutation affecting eukaryotic ribosomes. *Nature* **248:** 535.

Schultz, L.D. and J.D. Friesen. 1983. Nucleotide sequence of the *tcm1* gene (ribosomal protein L3) of *Saccharomyces cerevisiae*. *J. Bacteriol.* **155:** 8.

Schwindinger, W.F. and J.R. Warner. 1987. Transcriptional elements of the yeast ribosomal protein gene *CYH2*. *J. Biol. Chem.* **262:** 5690.

Segall, J. 1986. Assembly of a yeast 5S RNA gene transcription complex. *J. Biol. Chem.* **261:** 11578.

Sentenac, A. 1985. Eukaryotic RNA polymerases. *Crit. Rev. Biochem.* **18:** 31.

Shore, D. and K. Nasmyth. 1987. Purification and cloning of a DNA-binding protein that binds to both silencer and activator elements. *Cell* **51:** 721.

Shore, D., D.J. Stillman, A.H. Brand, and K.A. Nasmyth. 1987. Identification of silencer binding proteins from yeast: Possible roles in *SIR* control and DNA replication. *EMBO J.* **6:** 461.

Shulman, R.W., C.E. Sripati, and J.R. Warner. 1976. Noncoordinated transcription in the absence of protein synthesis in yeast. *J. Biol. Chem.* **252:** 1344.

Sillevis-Smitt, W.W., J.M. Vlak, I. Molenaar, and T.H. Rozijn. 1973. Nucleolar function of the dense crescent in the yeast nucleus. A biochemical and ultrastructural study. *Exp. Cell Res.* **80:** 313.

Skogerson, L., C. McLaughlin, and E. Wakatama. 1973. Modification of ribosomes in cryptopleurine-resistant mutants of yeast. *J. Bacteriol.* **116:** 818.

Sollner-Webb, B. and J. Tower. 1986. Transcription of cloned eukaryotic ribosomal RNA genes. *Annu. Rev. Biochem.* **55:** 801.

Stern, S., T. Powers, L.-M. Changchien, and H.F. Noller. 1989. RNA-protein interactions in 30S ribosomal subunits: Folding and function of 16S rRNA. *Science* **244:** 783.

Stocklein, W. and W. Piepersberg. 1980. Altered ribosomal protein L29 in a cycloheximide-resistant strain of *Saccharomyces cerevisiae. Curr. Genet.* **1:** 177.

Suzuki, K. and E. Otaka. 1988. Cloning and nucleotide sequence of the gene encoding yeast ribosomal protein YS25. *Nucleic Acids Res.* **16:** 6223.

Suzuki, K., T. Hashimoto, and E. Otaka. 1990. Yeast ribosomal proteins. XI. Molecular analysis of two genes encoding YL41, an extremely small and basic ribosomal protein, from *Saccharomyces cerevisiae. Curr. Genet.* **17:** 185.

Sweder, K.S., P.R. Rhode, and J.L. Campbell. 1988. Purification and characterization of proteins that bind to yeast *ARS*s. *J. Biol. Chem.* **263:** 17270.

Teem, J.L., N. Abovich, N.F. Kaufer, W.F. Schwindinger, J.R. Warner, A. Levy, J. Woolford, R.J. Leer, M.M.C. van Raamsdonk-Duin, W.H. Mager, R.J. Planta, L. Schultz, J.D. Friesen, H. Fried, and M. Rosbash. 1984. A comparison of yeast ribosomal protein gene DNA sequences. *Nucleic Acids Res.* **12:** 8295.

Tollervey, D. 1987. A yeast small nuclear RNA is required for normal processing of pre-ribosomal RNA. *EMBO J.* **6:** 4169.

Traglia, H.M., N.S. Atkinson, and A.K. Hopper. 1989. Structural and functional analyses of *Saccharomyces cerevisiae* wild-type and mutant *RNA1* genes. *Mol. Cell. Biol.* **9:** 2989.

Trapman, J., J. Retel, and R.J. Planta. 1975. Ribosomal precursor particles from yeast. *Exp. Cell Res.* **90:** 95:

Traut, R.R., D.S. Tewari, A. Sommer, G.R. Gavino, H.M. Olson, and D.C. Glitz. 1985. Protein topography of ribosomal functional domains: Effects of monoclonal antibodies to different epitopes in *Escherichia coli* protein L7/L12 on ribosome function and structure. In *Structure, function, and genetics of ribosomes* (ed. B. Hardesty and G. Kramer), p. 286. Springer-Verlag, New York.

Tsay, Y.-F., J.R. Thompson, M.O. Rotenberg, J.C. Larkin, and J.L. Woolford, Jr. 1988. Ribosomal protein synthesis is not regulated at the translational level in *Saccharomyces cerevisiae*: Balanced accumulation of ribosomal proteins L16 and rp59 is mediated by turnover of excess protein. *Genes Dev.* **2:** 664.

Udem, S.A. and J.R. Warner. 1972. Ribosomal RNA synthesis in *Saccharomyces cerevisiae. J. Mol. Biol.* **65:** 227.

———. 1973. The cytoplasmic maturation of a ribosomal precursor ribonucleic acid in yeast. *J. Biol. Chem.* **248:** 1412.

Underwood, M.R. and H.M. Fried. 1990. Characterization of nuclear localizing sequences derived from yeast ribosomal protein L29. *EMBO J.* **9:** 91.

van der Sande, C.A., T. Kulkens, A.B. Kramer, I.J. de-Wijs, H. van Heerikhuizen, J.

Klootwijk, and R.J. Planta. 1989. Termination of transcription by yeast RNA polymerase I. *Nucleic Acids Res.* **17:** 9127.

Van Ryk, D.I., Y. Lee, and R.N. Nazar. 1990. Efficient expression and utilization of mutant 5S rRNA in *Saccharomyces cerevisiae. J. Biol. Chem.* **265:** 8377.

Veinot-Drebot, L.M., R.A. Singer, and G.C. Johnston. 1988. Rapid initial cleavage of nascent pre-rRNA transcripts in yeast. *J. Mol. Biol.* **199:** 107.

Veldman, G.M., J. Klootwijk, H. van Heerikhuizen, and R.J. Planta. 1981. The nucleotide sequence of the intergenic region between the 5.8S and 26S rRNA genes of the yeast ribosomal RNA operon. Possible implications for the interaction between 5.8S and 26S rRNA and the processing of the primary transcript. *Nucleic Acids Res.* **9:** 4847.

Verbeet, M.P., J. Klootwijk, H. van Heerikhuizen, R.D. Fontijn, E. Vreugdenhil, and R.J. Planta. 1984. A conserved sequence element is present around the transcription initiation site for RNA polymerase A in *Saccharomycetoideae. Nucleic Acids Res.* **12:** 1137.

Vignais, M.L., L.P. Woudt, G.M. Wassenaar, W.H. Mager, A. Sentenac, and R.J. Planta. 1987. Specific binding of TUF factor to upstream activation sites of yeast ribosomal protein genes. *EMBO J.* **6:** 1451.

Vijayraghavan, U., M. Company, and J. Abelson. 1989. Isolation and characterization of pre-mRNA splicing mutants of *Saccharomyces cerevisiae. Genes Dev.* **3:** 1206.

Vioque, A. and E. Palacian. 1985. Partial reconstitution of 60S ribosomal subunits from yeast. *Mol. Cell. Biochem.* **66:** 55.

Warner, J.R. 1971. The assembly of ribosomes in yeast. *J. Biol. Chem.* **246:** 447.

———. 1982. The yeast ribosome: Structure, function and synthesis. In *The molecular biology of the yeast* Saccharomyces: *Metabolism and gene expression* (ed. J. Strathern et al.), p. 529. Cold Spring Harbor Laboratory, Cold Spring Harbor, New York.

———. 1989. Synthesis of ribosomes in *Saccharomyces cerevisiae. Microbiol. Rev.* **53:** 256.

———. 1990. The nucleolus and ribosome formation. *Curr. Opin. Cell Biol.* **2:** 521.

Warner, J.R. and G. Gorenstein. 1977. The synthesis of eukaryotic ribosomal proteins *in vitro. Cell* **11:** 201.

———. 1978a. Yeast has a true stringent response. *Nature* **275:** 338.

———. 1978b. The ribosomal proteins of *Saccharomyces cerevisiae. Methods Cell Biol.* **20:** 45.

Warner, J.R. and S.A. Udem. 1972. Temperature sensitive mutations affecting ribosome synthesis in *Saccharomyces cerevisiae. J. Mol. Biol.* **65:** 243.

Warner, J.R., G. Mitra, W.F. Schwindinger, M. Studeny, and H.M. Fried. 1985. *Saccharomyces cerevisiae* coordinates the accumulation of yeast ribosomal proteins by modulating mRNA splicing, translational initiation, and protein turnover. *Mol. Cell. Biol.* **5:** 1512.

Wittekind, M., J. Dodd, L. Vu, J.M. Kolb, J.-M. Buhler, A. Sentenac, and M. Nomura. 1988. Isolation and characterization of temperature-sensitive mutations in *RPA190*, the gene encoding the largest subunit of RNA polymerase I from *Saccharomyces cerevisiae. Mol. Cell. Biol.* **8:** 3997.

Wittekind, M., J.M. Kolb, J. Dodd, M. Yamagishi, S. Memet, J.-M. Buhler, and M. Nomura. 1990. Conditional expression of *RPA190*, the gene encoding the largest subunit of yeast RNA polymerase I: Effects of decreased rRNA synthesis on ribosomal protein synthesis. *Mol. Cell. Biol.* **10:** 2049.

Woolford, J.L., Jr. 1989. Nuclear pre-mRNA splicing in yeast. *Yeast* **5:** 439.

Woolford, J.L., Jr., L.M. Hereford, and M. Rosbash. 1979. Isolation of cloned DNA sequences containing ribosomal protein genes from *Saccharomyces cerevisiae. Cell* **18:** 1247.

Woudt, L.P., A.B. Smit, W.H. Mager, and R.J. Planta. 1986. Conserved sequence elements upstream of the gene encoding yeast ribosomal protein L25 are involved in transcription activation. *EMBO J.* **5:** 1037.

Woudt, L.P., W.H. Mager, R.T.M. Nieuwint, G.M. Wassenaar, A.C. van der Kuyl, J.J. Murre, M.F.M. Murre, M.F.M. Hoekman, P.G.M. Brockhoff, and R.J. Planta. 1987. Analysis of upstream activation sites of yeast ribosomal protein genes. *Nucleic Acids Res.* **15:** 6037.

Yip, M.T. and M.J. Holland. 1989. *In vitro* RNA processing generates mature 3′ termini of yeast 35 and 25S ribosomal RNAs. *J. Biol. Chem.* **264:** 4045.

Zagorski, J., D. Tollervey, and M.J. Fournier. 1988. Characterization of an *SNR* gene locus in *Saccharomyces cerevisiae* that specifies both dispensable and essential small nuclear RNAs. *Mol. Cell. Biol.* **8:** 3282.

11

Protein Synthesis and Translational Control in *Saccharomyces cerevisiae*

Alan G. Hinnebusch

Section on Molecular Genetics of Lower
Eukaryotes, Laboratory of Molecular Genetics,
National Institute of Child Health and Human
Development, National Institutes of Health,
Bethesda, Maryland 20892

Susan W. Liebman

The Laboratory for Molecular Biology
Department of Biological Sciences
College of Liberal Arts and Sciences
The University of Illinois at Chicago
Chicago, Illinois 60680

Volume I. The Molecular and Cellular Biology of the Yeast Saccharomyces: *Genome Dynamics, Protein Synthesis, and Energetics.* Copyright 1991 Cold Spring Harbor Laboratory Press 0-87969-355-X/91 $3 + 00

I. INTRODUCTION

The process of translation in *Saccharomyces cerevisiae* does not appear to differ fundamentally from the way it occurs in other organisms. Thus, much of what follows represents a comparison of the components and reactions of the protein synthetic machinery in yeast with their counterparts in other organisms, both prokaryotic and eukaryotic. For example, aminoacylation of tRNA and polypeptide chain elongation appear to be very similar in yeast and *Escherichia coli*, and, although the mechanism of translation initiation in *Saccharomyces* differs importantly from that described for bacteria, it closely resembles the process elucidated for plants and animals.

Despite these similarities with other well-studied systems, the powerful genetic techniques developed for *Saccharomyces* make it very worthwhile to study protein synthesis in this organism. As described below, the genes encoding many components of the translational apparatus have been isolated from yeast and are being manipulated with genetic techniques to learn the complete functions of their gene products in vivo. This enterprise is particularly important for the reactions involved in translation initiation for which many differences exist between eukaryotes and prokaryotes. Mutational analysis of yeast should indicate whether the large number of proteins implicated in the process of initiation in eukaryotes from biochemical studies of in vitro systems is actually required for translation in vivo. This approach is extremely difficult for the multicellular organisms used most frequently to study eukaryotic protein synthesis. In addition, the isolation of mutations that alter the efficiency or fidelity of protein synthesis in yeast is likely to identify important factors that were undetected in studies limited to cell-free extracts. To complement the genetic approach, in vitro translation systems have been developed for *Saccharomyces* that are being used increasingly to analyze the partial reactions carried out by individual components of the protein synthesis machinery.

Another important reason to study translation in yeast is the ease with which gene-specific regulatory phenomena can be detected and analyzed. A case in point concerns the regulatory potential of upstream open reading frames (uORFs) in eukaryotic messenger RNA (mRNA). These sequences are potent inhibitors of initiation and are present in many interesting transcripts in higher eukaryotic organisms; however, it is only in *Saccharomyces* that they have been shown to modulate translational efficiency of mRNAs in response to environmental stimuli. Also unique to yeast is the demonstration that translation of mitochondrial transcripts is controlled by activator proteins that are encoded in the nucleus. In addition, although the sequence-specific frameshift events needed for translation of Ty and L-A mRNAs are related to those described previously for retroviral mRNAs, the cellular factors required for such specialized elongation reactions will probably be identified first in yeast using genetic approaches.

II. THE PROCESS OF TRANSLATION IN *S. CEREVISIAE*

A. tRNAs and Aminoacyl-tRNA Synthetases

1. Number of tRNA Species and Codon Usage

The first step in protein synthesis is the activation of amino acids by esterification to their cognate transfer RNA (tRNA) molecules. This reaction is carried out by a different aminoacyl-tRNA synthetase for each amino acid. One molecule of ATP is consumed in a reaction that occurs in two steps on the enzyme: (1) Aminoacyl adenylate is formed with the release of pyrophosphate and (2) the aminoacyl moiety is transferred to the free 3'-hydroxyl group of tRNA (Schimmel 1987). Because of degeneracy in the genetic code, many of the 20 amino acids are esterified to more than one species of tRNA. In fact, about 40 different tRNA species have been resolved by two-dimensional polyacrylamide gel electrophoresis (Ikemura 1982), and 38 *S. cerevisiae* tRNA species have been identified by isolation and sequence analysis of tRNAs or their structural genes (Table 1). On the basis of the wobble rules first proposed by Crick (1966) and the number and kind of tRNA sequences known at the time in *S. cerevisiae*, Guthrie and Abelson (1982) proposed a modified set of rules for third-position codon-anticodon pairing that would limit the number of tRNA species to about 46: (1) A single tRNA will read two codons that differ only by ending in U or C; in such cases, an I (inosine) or a G nucleotide is found in the first position of the anticodon (read 5' to 3') that can pair with either U or C in the codon. (2) Two different tRNAs are required to read two codons that differ only by ending in A or G; in this instance, U-G base pairs at the wobble position are not permitted. (3) According to rules 1 and 2, it follows that three tRNAs are required to read a set of four codons that differ only in the third position (see Section II.C. and Table 7 below). All of the new tRNA species identified since 1982 have anticodons predicted by the above rules; therefore, it seems likely that only about eight additional tRNAs remain to be identified in *Saccharomyces*.

Isoaccepting tRNAs are generally not present in cells in equal amounts (Bennetzen and Hall 1982; Ikemura 1982), and the major isoacceptor for each amino acid invariably recognizes the codon or codons that are used preferentially to specify that amino acid in highly expressed yeast genes. Codon bias in this group of genes is quite extreme, with most amino acids being encoded by only 25 of the 61 possible triplets (Table 1) (Sharp et al. 1988). In fact, the pattern of codon usage is a good indicator of whether or not a gene is highly expressed in *Saccharomyces* (Bennetzen and Hall 1982; Sharp et al. 1986; Sharp and Li 1987). These observations led to the suggestion that codon usage in highly expressed genes is driven by selection to coincide with the most abundant tRNA species and thereby optimize translation rates. Alternatively, codons recognized by minor isoaccepting species may be avoided in such genes to prevent depletion of aminoacyl-tRNA pools for minor tRNA species (Bennetzen and Hall 1982; Ikemura 1982; Sharp et al. 1986). Codon-usage bias is also observed in *Saccharomyces* when a tRNA can recognize more than one codon, i.e., when two codons differ only by ending in U or C. In some cases, this ensures that at least one stable G-C base pair is formed in the codon-anticodon interaction, a preference that can be rationalized on the basis of increased translational efficiency (Bennetzen and Hall 1982; Sharp et al. 1986). However, there are exceptions to this trend, and the codon preferences seen in *S. cerevisiae* do not apply to all organisms (Sharp et al. 1986).

The effect of replacing a large fraction of the preferred codons with those rarely used in highly expressed genes was determined for the gene that encodes phosphoglycerate kinase (*PGK1*). Introduction of 22 rare synonomous codons at the beginning of the gene had no detectable effect on expression, but introduction of 64, 99, or 164 rare codons progressively reduced gene expression, reaching a value of only 10–15% of wild type in the case of the 164-codon replacement. These alterations also reduced the abundance of (*PGK1*) mRNA, complicating the analysis somewhat; however, the translational efficiency of the most extensively substituted transcript did appear to be lower than that of wild-type *PGK1* mRNA (Hoekema et al. 1987).

2. Sequence Determinants of tRNA Identity in Aminoacylation Reactions

All tRNAs interact with some of the same components of the translational machinery, such as elongation factors and ribosomal binding sites. This fact probably explains why all tRNA sequences can be arranged in the cloverleaf two-dimensional folding scheme (Fig. 1A) and contain highly conserved residues thought to stabilize similar tertiary structures (Fig. 1B) (Sprinzl et al. 1989). However, each tRNA is involved in a specific interaction with the particular aminoacyl-tRNA synthetase (AA-RS) that attaches an amino acid to its 3′ terminus.

Because it is unique to each tRNA, the anticodon is an obvious choice for a sequence determinant of tRNA identity in aminoacylation reactions. In fact, the anticodon is critical for specific tRNA recognition by Met-RS, Gln-RS, Val-RS, and Trp-RS, in *E. coli* (Fig. 1A) (for review, see Schimmel 1987; Schulman and Abelson 1988; Yarus 1988). There is evidence that alteration of the anticodon of initiator tRNA$_i^{Met}$ reduces the efficiency of aminoacylation by Met-RS in *S. cerevisiae* (Cigan et al.

Table 1 tRNAs and their structural genes in *S. cerevisiae*

Amino acid	Codon (5'-3')	Preferred codons	Anticodons (5'-3') predicted	found in tRNA	found in DNA	Number of genes
Ala	GCU	GCU	IGC	IGC	AGC[1]	≥6[2]
	GCC	GCC	IGC			
	GCA		U*GC		TGC	
	GCG		CGC			
Arg	CGU		ICG	ICG	ACG	
	CGC		ICG			
	CGA		U*CG			
	CGG		CCG			
	AGA	AGA	U*CU	U*CU	TCT	≥8[2]
	AGG		CCU		CCT	1[3]
Asn	AAU		GUU	GUU		
	AAC	AAC				
Asp	GAU	GAU	GUC	GUC	GTC	14–16[3,10]
	GAC	GAC	GUC			
Cys	UGU	UGU	GCA	GCA	GCA	
	UGC		GCA			
Gln	CAA	CAA	U*UG		TTG	
	CAG		CUG		CTG[11,12]	
Glu	GAA	GAA	U*UC	U*UC	TTC	14[4]
	GAG		CUC		CTC[4]	1[4]
Gly	GGU	GGU	ICC			
	GGC		ICC	GCC	GCC	15[5]
	GGA		U*CC	UCC	TCC[5]	3[5]
	GGG		CCC	CCC	CCC[13-16]	≥2[13-16]
His	CAU		GUG	GUG	GTG	7[6]
	CAC	CAC	GUG			
Ile	AUU	AUU	IAU	IAU	AAT[1]	
	AUC	AUC	IAU			
	AUA		U*AU		TAT	
Leu	UUA		U*AA	ZAA	TAA[17-19]	≥5[2]
	UUG	UUG	CAA	CAA	CAA	≥9[20,21]
	CUU		IAG			
	CUC		IAG			
	CUA		U*AG	UAG		
	CUG		CAG			

632

Table 1 (Continued.)

Amino acid	Codon (5′-3′)	Preferred codons	Anticodons (5′-3′)			Number of genes
				found in		
			predicted	tRNA	DNA	
Lys	AAA		U*UU	U*UU	TTT	
	AAG	AAG	CUU	CUU	CTT	
Met$_m$	AUG	n.a.	CAU	CAU	CAT	
Met$_i$	AUG	n.a.	CAU	CAU	CAT	4[7]
Phe	UUU		GAA	G$_m$AA	GAA	10[2]
	UUC	UUC				
Pro	CCU		IGG		AGG[8]	
	CCC		IGG			
	CCA	CCA	U*GG	U*GG	TGG[9]	5[9]
	CCG		CGG			
Ser	UCU	UCU	IGA	IGA	AGA	11[2]
	UCC	UCC	IGA			
	UCA		U*GA	U*GA	TGA	3[2]
	UCG		CGA	CGA	CGA	1[2]
	AGU		GCU		GCT	
	AGC					
Thr	ACU	ACU	IGU	IGU		≥4[2]
	ACC	ACC	IGU			
	ACA		U*GU			
	ACG		CGU			
Trp	UGG	n.a.	CCA	C$_m$CA	CCA	≥4[2]
Tyr	UAU		GUG	GψA	GTA	8[2]
	UAC	UAC				
Val	GUU	GUU	IAC	IAC	AAC	
	GUC	GUC	IAC			
	GUA		U*AC	N*AC		
	GUG		CAC	CAC		≥7[2]

The preferred codons are those found in highly expressed yeast genes to the near exclusion of synonymous codons (Sharp and Li 1987). The predicted anticodons are based on the rules summarized by Guthrie and Abelson (1982). The anticodons for the most abundant isoacceptor tRNAs (Bennetzen and Hall 1982; Ikemura 1982) are in boldface type. n.a. indicates not applicable. Unless otherwise indicated, references for the nucleotide sequences of tRNAs (or the corresponding genes) with the indicated anticodons are given in Sprinzl et al. (1989).

References: (1) Sandmeyer et al. 1988; (2) Guthrie and Abelson 1982; (3) Gafner et al. 1983; (4) Stucka et al.. 1987; (5) Mendenhall et al. 1987; (6) del Rey et al. 1983; (7) Cigan and Donahue 1986; (8) Cummins et al. 1982; (9) Cummins et al. 1985; (10) Kolman et al. 1988; (11) Calderon et al. 1984; (12) Winey et al. 1989a; (13) Gaber et al. 1983; (14) Mendenhall and Culbertson 1988; (15) Ball et al. 1988; (16) Culbertson et al. 1980; (17) Hawthorne and Mortimer 1968; (18) Ono et al. 1979b; (19) Ono et al. 1985; (20) Liebman et al. 1977; (21) Liebman et al. 1984.

A.

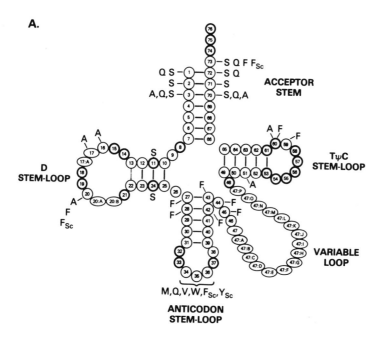

B.

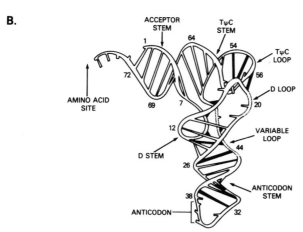

Figure 1 (*See facing page for legend.*)

1988a). In addition, mutations in the anticodons of *S. cerevisiae* tRNA^{Tyr} and tRNA^{Phe} increase the K_m or reduce the V_{max} values in reactions with their cognate synthetases by about tenfold. Some of these mutations also produce low-level misacylation by another synthetase (Bare and Uhlenbeck 1985, 1986). However, a reduced K_m/V_{max} ratio was also observed for two mutations outside the yeast tRNA^{Phe} anticodon: the G20 residue in the "D" loop and A73 located four residues from the 3'end of the molecule (the "discriminator" base) (Sampson and Uhlenbeck 1988; Sampson et al. 1989). Moreover, efficient aminoacylation of tRNA^{Met}, tRNA^{Arg}, and tRNA^{Tyr} by Phe-RS required introduction of the tRNA^{Phe} anticodon, the G20, and the A73 residues into these tRNAs (along with a few additional changes intended to promote a tertiary structure similar to that of tRNA^{Phe}). Given the sequence differences among the three tRNAs that were converted to Phe-RS substrates, it was concluded that 48 of 76 nucleotides in tRNA^{Phe} are not uniquely required for its recognition by Phe-RS. Another 16 positions are invariant among all *Saccharomyces* cytoplasmic tRNAs (Fig. 1A), leaving 12 or fewer nucleotides (including the five just mentioned) responsible for specific recognition by Phe-RS (Sampson et al. 1989). Interestingly, a rather different set of nucleotides are required for aminoacylation by *E. coli* Phe-RS (McClain and Foss 1988a), suggesting that different solutions to the tRNA identity problem exist for the same tRNA species in different organisms.

The anticodon is probably not a critical identity determinant for all tRNAs, since many amino acids are esterified to two or more tRNAs with different anticodons (Table 1); yet, so far as is known, all isoacceptors for a given amino acid are charged by the same synthetase. In addition, certain nonsense suppressors with altered anticodons can be charged

Figure 1 (*A*) tRNA aminoacylation identity determinants. The cloverleaf representation of the secondary structure of tRNA shown is that of Sprinzl et al. (1989). Residues with thicker circles are invariant or nearly so among all known tRNAs. Residues invariant in *Saccharomyces* tRNAs are A14, G18, G19, A21, U33, G53, U54, U55, C56, A58, C61, C74, C75, and A76 (Sampson et al. 1989). Nucleotide positions important in determining tRNA identity are indicated using the single-letter code to designate the amino acid (adapted from Yarus [1988] and based on reviews in Schimmel [1987], Schulman and Abelson [1988], and Yarus [1988]). With the exception of *S. cerevisiae* tRNA^{Phe} (F$_{Sc}$) and tRNA^{Tyr} (Y$_{Sc}$), all other identified determinants apply to *E. coli* tRNAs and not necessarily to the corresponding yeast tRNA species. (*B*) Representation of the three-dimensional structure of yeast tRNA^{Phe} with selected residues numbered. (Adapted from Schimmel and Soll 1979.)

by the same enzyme that activates the wild-type tRNA. In fact, an *E. coli* tRNA[Leu] amber suppressor was converted to a serine-inserting suppressor by altering only 3 bp in the acceptor stem and 1 bp in the D stem and the discriminator base (Normanly et al. 1986; Schulman and Abelson 1988). When two of these base pairs in the acceptor stem of an *E. coli* serine suppressor tRNA were altered, misacylation by Gln-RS occurred (Rogers and Soll 1988). In another striking instance of altered tRNA identity, introduction of the single-base-pair G3-U70 into the acceptor stems of tRNA[Lys], a tRNA[Cys] amber suppressor, and a tRNA[Phe] amber suppressor led to significant misacylation of these mutant tRNAs by *E. coli* Ala-RS. Moreover, elimination of this G-U base pair from an amber suppressor tRNA[Ala] destroyed its suppressor activity and the ability to be aminoacylated in vitro by Ala-RS (Prather et al. 1984; Hou and Schimmel 1988; McClain and Foss 1988b). tRNA[Ala] is the only species in *E. coli* that contains the G3-U70 base pair, and this holds true in *S. cerevisiae* as well (Sprinzl et al. 1989). Figure 1A summarizes the results of these and other experiments in which aminoacylation identity was altered by a small number of base substitutions in tRNAs from *E. coli* and *Saccharomyces*. The identity set determined for a tRNA from *E. coli* does not necessarily apply to the corresponding yeast tRNA, and vice versa.

It has been proposed that synthetases contact tRNA along the inside of the three-dimensional L-shaped structure (Fig. 1B) (Schimmel and Soll 1979). The results of experiments done to map the phosphates in *Saccharomyces* tRNA[Phe] that are protected by Phe-RS against alkylation by ethylnitrosourea were considered to be in general accord with this diagonal-side model for tRNA-synthetase interaction; however, a different binding scheme was proposed for the tRNA[Asp]/Asp-RS complex from the results of the same type of analysis (Romby et al. 1985). Recently, the first crystal structure of a synthetase complexed with its cognate tRNA was solved (Rould et al. 1989). The results indicate that Gln-RS of *E. coli* makes contact with tRNA[Gln] from the anticodon to the acceptor stem along the entire inside of the L of the tRNA, as predicted in the diagonal-side model. In accord with these findings, nucleotides in both the anticodon and the acceptor stem are known to be important for recognition of tRNA[Gln] by Gln-RS (Fig. 1A) (Perona et al. 1989).

In contrast, *E. coli* Ala-RS appears to contact primarily the acceptor and TψC helices, which are juxtaposed in the tRNA three-dimensional structure (Fig. 1B), because a truncated version of tRNA[Ala] containing only the residues in these two helices could be aminoacylated by Ala-RS in vitro with kinetics very similar to that seen for wild-type tRNA[Ala]. In fact, aminoacylation at reduced rates was observed for a helix containing only the 7 bp of the tRNA[Ala] acceptor stem (Francklyn and Schimmel 1989). These results are in accord with the primacy of the G3-U70 base pair of the acceptor stem in determining aminoacylation by Ala-RS (Fig. 1A).

3. Aminoacyl-tRNA Synthetase Structural Genes and Mutations

Table 2 summarizes the phenotypes of mutations in structural genes encoding cytoplasmic synthetases in *S. cerevisiae*. Among the point mutations generated in vivo, some only lead to auxotrophy for the cognate amino acid (*asn3-1* and *gln4-1*), whereas the *mes1-1* and *ils1-1* mutations cause lethality on rich medium at 36°C, leading to arrest in the G_1 phase of the cell cycle. All known deletions of genes encoding cytoplasmic synthetases are unconditionally lethal in haploid cells. The *hts1-150* (Natsoulis et al. 1986) and *vas1-1* (Chatton et al. 1988) mutations are of special interest because they have no effect on cell growth and division when cultures are grown on a fermentable carbon source such as glucose, but they are lethal when cells are grown on carbon sources where respiration is required (Pet⁻ phenotype). Each of these two mutations eliminates the first ATG codon present in-frame with the protein-coding sequence of the gene. For both, only a fraction of the mRNA transcripts begins upstream of the first ATG codon, with the majority initiating between the first and second in-frame start sites. Consequently, synthesis of the protein products from the predominant *HTS1* and *VAS1* mRNA species should be unaffected by the *hts1-150* and *vas1-1* mutations, respectively. The simplest explanation for these results is that proteins synthesized from the longer transcripts are localized in the mitochondria and used exclusively for amino acid activation in this organelle. Protein synthesis directed by the shorter transcripts is expected to initiate at the second AUG codon, with the enzymes thus produced lacking amino-terminal residues required for mitochondrial localization. These shorter forms of the enzymes would be required under all growth conditions for cytoplasmic protein synthesis. This interpretation is supported by the fact that cytoplasmic Val-RS contains the amino-terminal peptide predicted from translation of the shorter *VAS1* transcripts but lacks peptides expected from initiation at the 5′-proximal AUG codon on the longer transcripts (Chatton et al. 1988).

Most mitochondrial synthetases are not produced in this way, as shown for Leu-RS, Phe-RS (α subunit), Trp-RS, Thr-RS, Asp-RS, and Met-RS by the isolation of the genes that encode only the mitochondrial forms of the enzymes. Deletions of these genes eliminate aminoacylation of the cognate mitochondrial tRNAs, and thus produce a Pet⁻ phenotype;

Table 2 Phenotypes of mutations in structural genes for *S. cerevisiae* aminoacyl-tRNA synthetases

Enzyme	Gene	Mutation	Type	Phenotype	Refs.
Asn-RS	*ASN3*	*asn3-1*		recessive Asn requirement; tenfold increase in K_m for Asn; decreased V_{max} of aminoacylation	1
Gln-RS	*GLN4*	*gln4-1*		recessive Gln auxotrophy when combined with leaky *gln1-105*; 20-fold increase in K_m for Gln	2
		gln4Δ	deletion of all coding sequences	recessive unconditional lethal	3
		gln4Δ202	deletion of amino-terminal 65 codons	no phenotype when present on multicopy plasmid; reduces specific enzyme activity about 20-fold	3
His-RS	*HTS1*	*hts1-1*		recessive temperature-sensitive His auxotrophy; no enzyme activity detectable in vitro	4–6
		hts1-2	deletion of 86% of of coding region	recessive unconditional lethal	6
		hts1-150	10 bp deletion removing 5′-proximal in-frame ATG	recessive Pet⁻	6
Ile-RS	*ILS1*	*ils1-1* *ils1-2*		recessive temperature-sensitive lethal; inhibition of protein synthesis at 36°C; G_1 arrest and derepression of enzymes under general amino acid control at 34°C; temperature-sensitive aminoacylation activity in vitro	7–11
Met-RS	*MES1*	*mes1-1*	Gly-502 converted to Asp	recessive temperature-sensitive lethal; Met auxotrophy at 23°C; inhibition of protein synthesis and G_1 arrest at 36°C; derepression of Met biosynthetic enzymes at 28°C	7,12–18

Enzyme	Gene	Allele	Molecular defect	Phenotype	Ref.
		mes1-100	Asp-420 converted to Tyr	suppresses His⁻ phenotype of his4-318 (initiation codon mutation) in strains containing altered initiator-tRNA^{Met} with CCU anticodon	19
		MES1-Δ6-185	deletion of amino-terminal residues 6–185	wild-type, at least when gene product overexpressed	23
		mes1-Δ2-190	deletion of amino-terminal residues 2–190	severe reduction in aminoacylation activity in vitro	23
Thr-RS	BOR3 (THS1)	BOR3-13		dominant resistance to borrelidin; borrelidin-insensitive aminocylation activity in vitro	20
		ths1:HIS3	deletion of 52% of coding region	recessive unconditional lethal	21
Val-RS	VAS1	vas1-1	point mutation in 5′-proximal in-frame ATG codon	recessive Pet⁻	22
		vas1-2	point mutation in second in-frame ATG codon	recessive unconditional lethal	22

(1) Ramos and Wiame 1979; (2) Mitchell and Ludmerer 1984; (3) Ludmerer and Schimmel 1987b; (4) Thonart et al. 1976; (5) Hilger et al. 1982; (6) Natsoulis et al. 1986; (7) Hartwell and McLaughlin 1968a; (8) Hartwell and McLaughlin 1968b; (9) McLaughlin et al. 1969; (10) Messenguy and Delforge 1976; (11) Niederberger et al. 1983; (12) McLaughlin and Hartwell 1969; (13) Cherest et al. 1975; (14) Unger 1977; (15) Fasiolo et al. 1981; (16) Cherest et al. 1971; (17) Unger and Hartwell 1976; (18) Chatton et al. 1987; (19) Cigan et al. 1988a; (20) Nass and Poralla 1976; (21) Pape and Tzagoloff 1985; (22) Chatton et al. 1988; (23) Walter et al. 1989.

Table 3 Size and quaternary structure of *S. cerevisiae* aminoacyl-tRNA synthetases and sequence similarities with the corresponding *E. coli* enzymes

Enzyme	Quaternary structure	Length of polypeptide	Residues compared in enzymes from[a]		% Identity	Refs.
			S. cerevisiae	*E. coli*		
Asp-RS	α2	557	n.a.	n.a.		1,2
Gln-RS	unknown	809	225–786	1–551	40	3
His-RS	α2	526	1–526	1–424	28	2,4
Ile-RS	α	1073	1–959	12–939	27	5,6,13
Lys-RS	α2	591	n.a.	n.a.		7
Met-RS	α2	751	192–751	1–677	20	2,8
Ser-RS	α2	462	n.a.	n.a.		9,10
Thr-RS	unknown	734	50–725	1–642	38	11
Val-RS	α	1104	177–726	50–618	23[b]	12
Phe-RSβ	α2 β2	503	220–480	100–310	32	14

References: (1) Sellami et al. 1986; (2) Schimmel 1987; (3) Ludmerer and Schimmel 1987a; (4) Natsoulis et al. 1986; (5) Bhanot et al. 1974; (6) Englisch et al. 1987; (7) Mirande and Waller 1988; (8) Walter et al. 1983; (9) Heider et al. 1971; (10) Weygand-Durasevic et al. 1987; (11) Pape and Tzagoloff 1985; (12) Jordana et al. 1987; (13) Martindale et al. 1989; (14) Sanni et al. 1988.

[a] n.a. indicates not applicable.

[b] This alignment was done with *E. coli* Ile-RS.

however, the haploid deletion strains grow normally on medium containing glucose as the carbon source (Myers and Tzagoloff 1985; Pape et al. 1985; Koerner et al. 1987; Tzagoloff et al. 1988, 1989; Gampel and Tzagoloff 1989). These results imply the existence of separate genes encoding cytoplasmic forms of the corresponding synthetases, and indeed, the structural genes for cytoplasmic Thr-RS, Asp-RS, Met-RS, and Phe-RS have been isolated and characterized (see Table 3).

The subunit structures and polypeptide chain lengths of the cytoplasmic synthetases for which the structural genes have been cloned and sequenced are listed in Table 3. Considering that these enzymes carry out the same reaction for different amino acids, there is surprising variation in their physical characteristics. There is also little sequence similarity among enzymes that activate different amino acids in the same organism. In contrast, considerable sequence similarity exists between enzymes that activate the same amino acid in *Saccharomyces* and *E. coli* (Schimmel 1987) (Table 3), suggesting that these proteins originate from a common ancestral sequence.

Comparing the primary structures of the same enzymes from *S.*

cerevisiae and *E. coli* reveals that some yeast enzymes contain sizeable amino-terminal (Gln-RS, Met-RS, Val-RS, Thr-RS, Phe-RSβ) or carboxy-terminal (Ile-RS) segments not found in their bacterial counterparts (Table 3). Presumably, these extra amino acids are dispensable for catalysis and are required instead for regulation or cellular localization of the enzymes. *Saccharomyces* Gln-RS is a case in point, containing an amino-terminal extension of 224 amino acids relative to the *E. coli* enzyme. In-frame deletions of the *GLN4* sequences encoding this segment reduce the level of Gln-RS activity in vivo; however, enough activity remains to support growth if the mutant gene is present in many copies (see Table 2). These results are consistent with an accessory role for the amino terminus of yeast Gln-RS. The same appears to be true for most of the amino-terminal segment of *Saccharomyces* Met-RS; however, there is some evidence that a portion of this nonconserved segment affects the functioning of the conserved catalytic core of the enzyme (Walter et al. 1989).

S. cerevisiae synthetases bind to polyanionic chromatographic supports to which their bacterial counterparts do not (Cirakoglu and Waller 1985). In addition, Lys-RS and Asp-RS from *S. cerevisiae* contain amino-terminal extensions that can be removed by protease digestion without loss of enzyme activity in vitro, and removal of these segments abolishes the affinity for polyanionic carriers (Cirakoglu and Waller 1985; Lorber et al. 1988). The amino-terminal segments of both enzymes are basic in character, accounting for their binding affinity for polyanions. The same is true of Gln-RS, Thr-RS, and Val-RS from yeast (Mirande and Waller 1988). It was proposed that these basic extensions are responsible for compartmentalization of the enzymes through their interaction with negatively charged sites, for example, on the cytoskeleton. At present, there is no definitive evidence for association of the yeast enzymes with one another or with cellular structures.

4. Functional Domains of Aminoacyl-tRNA Synthetases

X-ray crystallographic data are available for the amino-terminal portions of Met-RS from *E. coli* and Tyr-RS from *Bacillus stearothermophilus* (for review, see Schimmel 1987), as well as for the entire Gln-RS of *E. coli*, as mentioned above. All three proteins were crystallized as complexes with ATP, or tyrosyl adenylate in the case of Tyr-RS, and were found to bind ATP in a domain that is structurally related to the nucleotide-binding fold identified in dehydrogenases and kinases. This domain consists of five β strands and four α helices, with a segment of variable size inserted between the third β-strand and third α-helix (Fig.

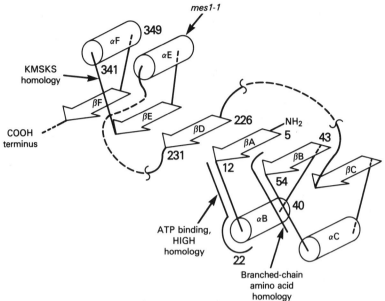

Figure 2 Schematic representation of the amino-terminal nucleotide-binding fold of Met-RS from *E. coli* that also occurs with some differences in Tyr-RS of *B. stearothermophilus* and Gln-RS of *E. coli*. α helices and β strands are lettered; the positions of selected residues are indicated and numbered from the amino terminus (NH₂). The dashed lines and the D β-strand (βD) indicate segments of the polypeptide chain that do not contribute directly to the nucleotide-binding fold, serving as connectors between different structural elements in the fold. In the case of Gln-RS of *E. coli*, these connector residues play an important role in positioning the 3′ end of tRNA near the bound aminoacyl adenylate. The locations of three different conserved sequence motifs are indicated, as well as the amino acid substitution in Met-RS associated with the *S. cerevisiae mes1-1* mutation. (Adapted from Starzyk et al. 1987.)

2). Two histidine residues in the amino-terminal half of the fold (at the beginning of the B α-helix in Fig. 2) interact with ATP. Amino acids surrounding these histidines are loosely conserved among several different bacterial synthetases in a region close to the amino terminus of each enzyme (the "HIGH" sequence or His-Ile-Gly-His; for review, see Schimmel 1987). Related sequences also occur in certain *S. cerevisiae* synthetases, in each case near the beginning of the catalytic core defined by homology with the bacterial counterpart (Fig. 3A). The HIGH sequence probably identifies the ATP-binding site in those yeast enzymes in which it occurs (Ludmerer and Schimmel 1987a; Schimmel 1987).

Ile-RS from *S. cerevisiae* and *E. coli* each contains a short stretch of

A
```
Gln-RS Sc[1]    254   T R F P P E P N G Y L H I G H
Gln-RS Ec[1]     30   T R F P P E P N G Y L H I G H

Glu-RS Ec[2]      5   T R F A P S P T G Y L H V G G

Ile-RS Sc[3]     42   F D G P P F A T G T P H Y G H
Ile-RS Ec[3]     53   H D G P P Y A N G S I H I G H

Leu-RS Sc-mt[4,5] 52  L C Q F P Y P S G A L H I G H
Leu-RS Ec[4]     41   L S M L P Y P S G R L H M G H

Met-RS Sc[6]    201   T S A L T Y V N N V P H L G N
Met-RS Ec[7]     11   T C A L P Y A N G S I H L G H

Ser-RS Sc[8]    162   D G Y D P D R - G V K I C G H

Trp-RS Sc-mt[9]  38   S M I Q P T - - G C F H L G N
Trp-RS Ec[9]      7   S G A Q P S - - G E L T I G N
Trp-RS Bs[9]      5   S G I Q P S - - G V I T I G N

Tyr-RS Ec[2]     37   C G F D P T A - D S L H L G H
Tyr-RS Bs[2]     35   C G F D P T A - D S L H I G H

Val-RS Sc[10]   184   P A P P P N V T G A L H I G H
Val-RS Ec[11]    38   M I P P P N V T G S L H M G H

CONSENSUS                   P - A - G - L H I G H
                                      V       I   L   N
```

B
```
Ile-RS Sc       79   E R R F G W D T H G V P I E
Ile-RS Ec       90   P Y V P G W D C H G L P I E

Leu-RS Sc-mt    63   I H P M G W D A F G L P A E
Leu-RS Ec       61   L Q P I G W D A F G L P A E

Met-RS Sc      238   L F I C G T D E Y G T A T E
Met-RS Ec       46   N F I C A D D A H G T P I M

Val-RS Sc      222   L F L P G F D H A G I A T Q

CONSENSUS                  G - D - - G - P - E
                                            A
```

C
```
Gln-RS Sc      492   T G T V L S K R K I A
Gln-RS Ec      264   E Y T V M S K R K L N

Ile-RS Sc      599   D G R K M S K S L K N
Ile-RS Ec      599   Q G R K M S K S I G N

Leu-RS Sc-mt   644   S Y E K M S K S K Y N
Leu-RS Ec      618   G M S K M S K S K N N

Met-RS Sc      522   E N G K F S K S R G V
Met-RS Ec      329   N G A K M S K S R G T

Trp-RS Sc-mt   240   P E K K M S K S D P N
Trp-RS Ec      192   P T K K M S K S D D N
Trp-RS Bs      189   P T K K M S K S D P D

Tyr-RS Ec      223   T V P L I T K A D G T
Tyr-RS Bs      219   T I P L V T K A D G T

Val-RS Sc      700   Q G R K M S K S L G N

CONSENSUS            K M S K S - G N
```

Figure 3 Amino acids conserved among different synthetases from *E. coli* (*Ec*), *B. stearothermophilus* (*Bs*), and *S. cerevisiae* (*Sc*); (mt) mitochondrial form. The position of the first residue in each stretch is indicated relative to the amino terminus of the protein. Sequence data were taken from the following sources: [1]Ludmerer and Schimmel (1987a); [2]Schimmel (1987); [3]Englisch et al. (1987); [4]Labouesse et al. (1987); [5]Tzagoloff et al. (1988); [6]Walter et al. (1983); [7]Dardel et al. (1984); [8]Weygand-Durasevic et al. (1987); [9]Myers and Tzagoloff (1985); [10]Jordana et al. (1987); [11]Heck and Hatfield (1988).

sequence similarity located 25 residues toward the carboxyl terminus from the HIGH sequence. In *B. stearothermophilus* Tyr-RS, these residues occur between the B β-strand and the C α-helix, in close proximity to the bound tyrosyl adenylate (see Fig. 2). Interestingly, it was shown that changing Gly-94 to Arg-94 in the first position of this stretch of amino acids in *E. coli* Ile-RS increased the K_m for isoleucine 6000-fold, while having only a minor effect on the K_m for ATP (Clarke et al. 1988). This finding suggests that these sequences form part of the binding site for isoleucine on the bacterial Ile-RS. Homology with this region of Ile-RS can also be found in Met-RS and Leu-RS from *E. coli*, and in Val-RS, Met-RS, and mitochondrial Leu-RS from *S. cerevisiae* (Fig. 3B) (Jordana et al. 1987; Herbert et al. 1988). Given that these enzymes all activate large aliphatic amino acids, the residues they have in common may be involved in specific amino acid binding.

Residues in the E and F α helices of Tyr-RS and Gln-RS are also thought to contribute to binding of the amino acid (Schimmel 1987; Rould et al. 1989). Interestingly, the *S. cerevisiae* *mes1-1* mutation causes a glycine to aspartic acid substitution at position 502, expected to occur in the beginning of the E α-helix of this enzyme (see Fig. 2). The charging defect associated with *mes1-1* can be overcome at 25°C by addition of methionine to the medium. Because the mutant enzyme is unstable in vitro even at 25°C, it was suggested that *mes1-1* lowers the K_m for methionine by destabilizing the nucleotide-binding fold (Chatton et al. 1987).

Adenylation of the amino acid can be carried out by amino-terminal fragments of two bacterial synthetases that are inactive for binding and aminoacylation of tRNA (Schimmel 1987), suggesting that some of the residues needed for tRNA binding are located closer to the carboxyl terminus of these enzymes than those involved in adenylation. A lysine residue just preceding the F α-helix in the nucleotide-binding fold of *E. coli* Met-RS was affinity-labeled by the 3′-adenosine dialdehyde derivative of tRNA[Met] (Hountondji and Blanquet 1985), and a lysine at the same position in *E. coli* Tyr-RS was similarly labeled (Hountondji et al. 1986b). These results suggested that the reactive 3′ end of tRNA binds at the carboxy-terminal portion of the nucleotide-binding fold, in close proximity to bound aminoacyl adenylate (see Fig. 2). The affinity-labeled lysines are present in a short region of sequence homology, known as the KMSKS sequence (Lys-Met-Ser-Lys-Ser), that is shared by several synthetases that also contain the HIGH homology, including four enzymes from *S. cerevisiae* (Fig. 3C) (Hountondji et al. 1986a).

The crystal structure of the Gln-RS/tRNA[Gln] complex confirms the idea that the 3′ end of tRNA comes in close contact with the amino acid and ATP in the nucleotide-binding fold; however, no direct contact was seen between the 3′ end of tRNA^{Gln} and the KMSKS-related residues of this enzyme. In addition, the 110 amino acids inserted between the two halves of the nucleotide-binding fold of Gln-RS are directly involved in positioning the reactive end of tRNA^{Gln} by melting out the U1-A72 base pair at the top of the acceptor stem and stabilizing a hairpin turn of the single-stranded 3′ end of the tRNA into the active site. The remainder of Gln-RS interacts closely with the anticodon loop of the tRNA and may therefore function in discriminating against noncognate tRNAs (Rould et al. 1989).

B. Initiation of Translation

1. mRNA Sequence Requirements for Efficient Initiation

a. AUG Is the Only Efficient Initiation Codon in S. cerevisiae. A recent compilation of 131 yeast genes suggests that all 131 use an AUG triplet as the initiation codon (Cigan and Donahue 1987). Quantitative measurements of the expression of *CYC1*, *CYC7*, and *HIS4* alleles containing initiation codon mutations show that non-AUG codons support levels of protein synthesis only a few percentage or less of that seen with an AUG codon (Table 4). The precise ranking of non-AUG codons by their efficiency as initiation sites varies somewhat among the three genes (and thus may be influenced by sequence context); however, UUG, AUU, AUA, and GUG appear to be the best substitutes for AUG, whereas AAG and AGG are the worst substitutes for AUG, with the two groups differing by at least one order of magnitude in their efficiency as start sites.

b. Nucleotides Surrounding the Initiation Codon. There is a strong preference for A nucleotides (~45%) and a scarcity of G nucleotides (~12%) in the mRNA leaders of yeast genes (Cigan and Donahue 1987; Hamilton et al. 1987; Laz et al. 1987). The preference for A is particularly striking at the −3 position relative to the ATG codon (75%) and occurs at the expense of pyrimidines, the latter accounting for only 3% of the residues found at −3. This nonrandom sequence distribution is even more striking among genes that are highly expressed, for which occurrence of A nucleotides is 100% at −3 and extends well above 50% at −1, −2, −5, −6, and −7. At the latter five positions, and also at −4, a pyrimidine always occurs when an A nucleotide is not present (Cigan and Donahue 1987; Hamilton et al. 1987). Because the strongest bias occurs among

Table 4 Efficiency of translation using non-AUG initiation codons

Triplet	% Protein expressed from			
	CYC1[a]	*CYC1-galK*[b]	*CYC7-lacZ*[c]	*HIS4-lacZ*[d]
AUG	100	100	100	100
UUG	1.0	0.3–6.9[e]	0.37	0.8
AUU	0.1	n.d.[f]	0.38	1.5
GUG	0.1	n.d.	0.50	1.3
AUA	0.1	0.4	0.29	2.1
CUG	n.d.	n.d.	0.22	0.7
ACG	n.d.	n.d.	0.39	0.2
AUC	0.1	0	0.05	0.9
AAG	0	n.d.	0.02	0.6
AGG	0	n.d.	0.04	0.4

Data from [a]Laz et al. 1988; [b]Zitomer et al. 1984; [c]Clements et al. 1988; [d]Donahue and Cigan 1988.

[e] These values correspond to two different constructs that differ at the −3 position, with the lower value observed for a T nucleotide and the higher value observed for an A nucleotide at −3.

[f] n.d. indicates not done.

highly expressed genes, the observed nucleotide preferences may be indicative of sequence requirements for efficient initiation (see below).

The distribution of nucleotides downstream from the AUG codon is also nonrandom: 28% of all *Saccharomyces* genes contain a UCU serine codon immediately following the initiation codon. Thr (10%), Leu (9.5%), Ala and Val (7.8% each), Lys and Phe (6% each), and Asn, Gly, and Pro (4% each) are the next most abundant penultimate amino-terminal residues (Cigan and Donahue 1987). On the basis of this preference for second codon usage, the amino-terminal methionine is expected to be cleaved from the majority of yeast proteins by an amino-terminal peptidase (Sherman et al. 1985; Huang et al. 1987). Thus, the non-randomness seen at the +4 to +6 positions may reflect constraints on amino-terminal protein sequences, rather than sequence requirements for efficient translation initiation (Cigan and Donahue 1987; Hamilton et al. 1987).

The consensus sequence surrounding *S. cerevisiae* start codons (5′-(A/Y) A (A/Y) A (A/Y) A A U G U C U-3′) differs from that compiled for vertebrate mRNAs (5′-G C C G C C (A/G) C C A U G G-3′) (Kozak 1987a), except for the strong preference for A at −3 (61% in vertebrate mRNAs). The functional significance of the vertebrate sequence was established primarily by showing that mutations in the initiation region of the rat preproinsulin gene that reduce its agreement with the consensus sequence lowered preproinsulin synthesis in cultured animal cells. The greatest reductions (four- to fivefold) were observed for substitutions of the A at −3 or the G at +4 (Kozak 1986a).

The influence of nucleotides surrounding an AUG start codon for translation initiation in *S. cerevisiae* was first addressed for the *CYC1* gene. Revertants of *cyc1* initiation codon mutants were isolated in which new ATG codons were present at any of six different locations in a 37-nucleotide region encompassing the normal *CYC1* translation start site. Most of the revertants express wild-type levels of Cyc1p, dismissing a strict requirement for particular sequences upstream of the start codon for efficient initiation in yeast. Among the revertants that express wild-type levels of Cyc1p are several in which only the three nucleotides following the ATG codon vary (ATA, AGG, AAG, GTG, or CTG), suggesting that many different +4 to +6 sequences are also compatible with efficient initiation on *CYC1* mRNA (Sherman and Stewart 1982).

Subsequent studies (Baim and Sherman 1988) showed that single-base substitutions at *CYC1* that replace an A or G nucleotide at −3 with a T or C result in approximately twofold reductions in gene expression (Table 5). Similar results were obtained for single-base substitutions at the −3 position at *HIS4*. Changing the A at −3 to a G or T nucleotide led to 20–40% reductions in the expression of a *HIS4-lacZ* fusion (Table 5) (Cigan et al. 1988b) . A small effect was also observed when the +4 and +5 positions at *HIS4* were altered from GT to TC. Thus, although nucleotides surrounding the start codon influence the efficiency of initiation in *S. cerevisiae*, they appear to do so to a lesser extent than in mammalian cells.

c. Influence of Secondary Structure in the mRNA Leader on Translation Initiation. It is generally accepted that, in eukaryotes, a 43S complex containing the small ribosomal subunit, Met-tRNA$_i^{Met}$, and various initiation factors associates with the capped 5′ end of the mRNA and advances in the 3′ direction until the first AUG codon in a suitable sequence context is encountered, whereupon translation begins (for review, see Kozak 1989). This process has been referred to as scanning. Little or no effect on the efficiency of translation has been seen in response to changes in the length of untranslated leader sequences for several mammalian transcripts (Johansen et al. 1984; Kozak 1987b), and the same conclusion seems to apply to *S. cerevisiae* mRNAs (Baim and Sherman 1988; Cigan et al. 1988b). These results imply that scanning is not a rate-limiting step in the initiation process. However, shortening the leader length of the yeast *PGK1* transcript to 21 nucleotides or fewer reduced its translational efficiency by 50% from that of similar transcripts containing longer leaders. To explain this effect, it was proposed that 80S ribosomes at the AUG start codon sterically hinder binding of 43S complexes to the mRNA cap when the leader length is comparable to the diameter of the ribosome (van den Heuvel et al. 1989).

In contrast to the modest effects associated with differences in leader lengths, translation of *Saccharomyces* mRNAs is greatly inhibited by inserting sequences in the leader with the potential to form stable secondary structures. This inhibitory effect is an important piece of evidence that the scanning mechanism of translation initiation operates in yeast. Introduction of stem-loop structures with 9 or 15 bp of 60% G+C content at a position either 9 or 46 nucleotides 3′ to the mRNA cap site reduced *HIS4-lacZ* expression by 95–99%, without affecting the steady-state level of *HIS4-lacZ* mRNA (Fig. 4) (Cigan et al. 1988b). The inhibitory effect of the insertion made 46 nucleotides downstream from the cap site most probably interferes with scanning rather than initial binding of 40S subunits to the 5′ end of the mRNA.

Insertion of an 8-bp (40% G+C) stem-loop in the *CYC1* mRNA leader immediately upstream of the start codon reduced gene expression 20-fold (*cyc1-728*), without affecting the steady-state amount of the transcript. It was stated that this insertion reduces the average polysome size of *CYC1* mRNA, indicative of a reduced rate of translation initiation (Baim and Sherman 1988). Revertants of *cyc1-728* were isolated in which *CYC1* expression was elevated fourfold, and two such revertants contained 1- or 2-bp substitutions that reduced the predicted length of the stem. Another class of revertants had new ATG codons, in-frame with the *CYC1*-coding sequences, that were introduced upstream of the secondary structure (Fig. 4). The stem-loop may be less inhibitory to *CYC1* expression in this second class of revertants because it is located downstream, rather than

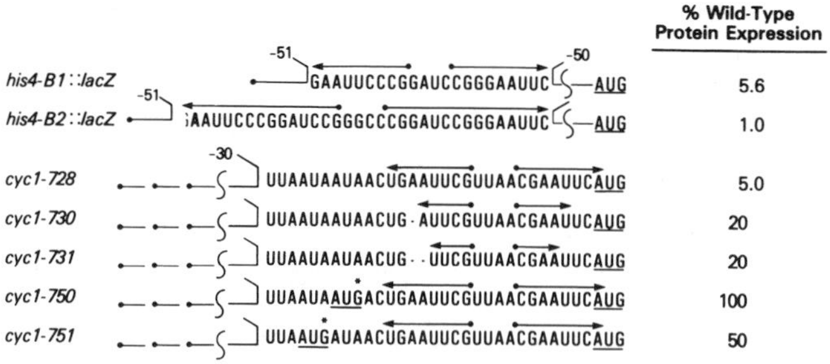

Figure 4 Inhibitory effects of sequences with the potential to form secondary structure on translation of *HIS4-lacZ* and *CYC1* mRNAs. Positions in the mRNA leader sequences are indicated relative to the AUG start codon (+1). The multiple 5′ ends of *CYC1* mRNA are indicated. Gene expression is given relative to the corresponding wild-type gene containing no insertion in the leader.

upstream, from the presumed AUG start codon (Baim and Sherman 1988). If so, this would imply that elongation is less sensitive than initiation to mRNA secondary structure in yeast, as appears to be the case for preproinsulin mRNA in mammalian cells (Kozak 1986b).

Secondary structure insertions of much greater stability than those just described are required for an equivalent inhibition of translation in mammalian cells (Kozak 1986b; Pelletier and Sonenberg 1985). The different sensitivities to increased secondary structure between *Saccharomyces* and mammalian cells are probably related to the fact that yeast mRNA leaders are intrinsically unstructured due to the high content of A nucleotides and scarcity of G nucleotides. Thus, yeast initiation complexes may be unable to scan through structured regions of even moderate stability.

The m transcript of the M1 double-stranded RNA of the *S. cerevisiae* killer virus, which encodes the killer toxin, contains a hairpin structure at its 5′ terminus that includes the initiation codon for preprotoxin synthesis (Leibowitz et al. 1983; Hannig et al. 1984). Hybridization of oligonucleotides complementary to the 5′-proximal side of this stem loop inhibits translation of m RNA in a cell-free translation system prepared from yeast, suggesting that formation of mRNA-complementary DNA (cDNA) hybrids impedes scanning by the preinitiation complex in the same manner postulated above for RNA secondary structure. In contrast, hybridization of complementary oligodeoxyribonucleotides to the 3′-proximal side of the stem stimulates m RNA translation, suggesting that destabilization of the stem increases the accessibility of the AUG start codon and that formation of limited mRNA-cDNA hybrids downstream from the initiation site has little effect on elongation. It was proposed that the inhibitory effect of the stem loop in m RNA might be destabilized in vivo by base pairing with either ribosomal RNA (rRNA) or a cRNA with a regulatory function (Leibowitz et al. 1988). This model is interesting considering that killer virus transcripts such as m may be uncapped (Leibowitz et al. 1988) and that cap-associated initiation factors are thought to be required primarily for unwinding secondary structure in the mRNA leader (see Section II.B.2 below).

d. Inhibitory Effect of Upstream Initiation Sites in the mRNA Leader. For most yeast mRNAs, the 5′-proximal AUG codon is the initiation site for protein synthesis. In a group of 131 yeast genes, only 6 have upstream ATG codons that are out of frame with the protein-coding sequences, or separated from the latter by a stop codon. These 6 upstream

ATG codons, because of the contextual constraint, should be unable to function as start sites for protein synthesis (Cigan and Donahue 1987). The rarity of upstream AUG codons in *S. cerevisiae* is consistent with the fact that these sequences invariably lower expression of downstream protein-coding sequences when introduced into yeast transcripts. This phenomenon was first demonstrated by recombining *cyc1* mutant alleles containing ATG codons at different locations near the beginning of the gene to generate new alleles containing two closely spaced initiation sites. An example is *cyc1-340a*, which contains two in-frame ATG codons separated by three codons (Fig. 5A). Analysis of the *cyc1-340a* protein product showed that essentially all initiation occurred at the 5′-proximal AUG codon. The *cyc1-341* and *cyc1-599* alleles additionally illustrate the fact that no cytochrome *c* is expressed from the second ATG codon even when the first ATG codon is followed by an in-frame stop codon located 3 or 18 nucleotides, respectively, upstream of the second ATG codon. Both of the latter were reverted to give nearly wild-type *CYC1* expression by point mutations that removed the upstream ATG codon (*CYC1-341D* and *CYC1-599B*) or by elimination of the stop codon that separates the two ATG codons, thereby fusing the Cyc1p-coding sequences in-frame to the 5′-proximal start site (*CYC1-599A*; Fig. 5A) (Sherman and Stewart 1982). Similar results were obtained from an analysis of *CYC1-galK* fusions containing a portion of the *CYC1* initiation region located upstream of the GalK protein-coding sequences (Fig. 5B) (Zitomer et al. 1984). The inhibitory effect of upstream AUG codons on *CYC1* expression is consistent with the scanning mechanism for translation initiation in that prior recognition of 5′-proximal AUG codons interferes with initiation at the authentic *CYC1* start site located downstream.

Werner et al. (1987) introduced a heterologous upstream open reading frame (uORF) into the yeast *CPA1* transcript at a position 46 nucleotides 5′ to the *CPA1* ATG codon. When the preferred A nucleotide was present at the −3 position of the uORF, *CPA1* expression was about 20% of that seen in the absence of a uORF. Changing the −3 and −2 nucleotides from AT to TA completely abolished the inhibitory effect of the uORF on *CPA1* expression (Fig. 5C). Similarly, the inhibitory effect on *CYC1* expression of a five-codon uORF present in the leader of *cyc1-362* was reduced from 99% to 80% by altering the A at the −1 position of the uORF to a T nucleotide (I. Pinto and M. Hampsey, pers. comm.). These results are reminiscent of mutations surrounding an upstream out-of-frame ATG codon in the preproinsulin transcript that led to increased initiation at the downstream start codon. Presumably, the initiation complex is able to scan past an upstream AUG codon present in an un-

Protein Expression
(% amount seen with
no upstream AUGs)

A

CYC1[1]	AUAAUGACUGAAUUCAAGGCCGGUUCUGCUAAG...	(100)
cyc1-340a	AUAAUGACUGAAUUCAUGGCCGGUUCUGCUAAG...	100
cyc1-341	AUAAUGACUUAAUUCAUGGCCGGUUCUGCUAAG...	0
CYC1-341D	AUAAUAACUUAAUUCAUGGCCGGUUCUGCUAAG...	50
cyc1-599	AUAAUGACUUAAUUCCAAGCCGGUUCUGCUAUG...	0
CYC1-599B	AUAAUAACUUAAUUCCAAGCCGGUUCUGCUAUG...	20
CYC1-599A	AUAAUGACUCAAUUCCAAGCCGGUUCUGCUAUG...	100

B

CYC1-galk-R6[2]	AUAAUGACUGAAUUCCUG...27 nt...UAAGAAAUG...	0.5
CYC1-galk-R3	AUAAUGACUGAAUUCCUA...23 nt...UAAGAAAUG...	0.6
CYC1-galk-R3-7	AUAAUUGACUGAAUUCCUA...23 nt...UAAGAAAUG...	29
CYC1-galk-R3-67	AUAAAAACUGAAUUCCUA...23 nt...UAAGAAAUG	58

C

CPA1-O-B[3]	AUUAUGUGU...78 nt...GCUUAAUAA...37 nt...CAAAUG	20
CPA1-O-E	UAUAUGUGU...78 nt...GCUUAAUAA...37 nt...CAAAUG	100

D

HIS4-lacZ[4]	AUAAUGGUUUUGCCGAUUCUACCGUUAAUUGAUG	<0.1
	CUAAUGGUUUUGCCGAUUCUACCGUUAAUUGAUG	<0.1
	UUAAUGGUUUUGCCGAUUCUACCGUUAAUUGAUG	<0.2
	AUAAUAGUUUUGCCGAUUCUACCGUUAAUUGAUG	42
	AUACUGGUUUUGCCGAUUCUACCGUUAAUUGAUG	34
	AUAACGGUUUUGCCGAUUCUACCGUUAAUUGAUG	19

E

GCN4 (uORF1)[5]	AAAAUGGCUUGCUAAACC... 353 nt ... AAUAUG	33
GCN4 (uORF1')	AAAAUGGCUUGCUAAACC... 184 nt ... AAUAUG	16
GCN4 (uORF4)	AAGAUGUUUCCGUAACGG... 146 nt ... AAUAUG	0.5
GCN4 (uORF4')	AAGAUGUUUCCGUAACGG... 353 nt ... AAUAUG	3

Figure 5 Inhibitory effect of upstream AUG codons on translation initiation. An arrow originates from the AUG codon that functions as the start site for synthesis of the protein encoded by each allele. Double underlines indicate the in-frame stop codons. Asterisks designate point mutations that distinguish a particular construct from one preceding it in the list for that gene. In the case of the *GCN4* constructs, the uORF shown is the only one present in the mRNA leader, the other three having been removed by point mutations or deletions. uORF1' and uORF4' differ from the immediately preceding constructs only in the distance between the uORF and the *GCN4* start codon. See text for details. [1]Sherman and Stewart (1982); [2]Zitomer et al. (1984); [3]Werner et al. (1987); [4]Cigan et al. (1988b) and Donahue and Cigan (1988); [5]Mueller and Hinnebusch (1986), Williams et al. (1988), and Miller and Hinnebusch (1989).

favorable sequence context and initiate at a downstream start site (Kozak 1986a).

Genetic studies on *HIS4* confirmed the strong inhibitory effect of upstream out-of-frame AUG codons on translation initiation at downstream start sites. The polar frameshift mutation *his4-100* in the *HIS4A*-coding region abolishes expression of all three enzymatic activities encoded at *HIS4* by generating an in-frame stop codon (Donahue et al. 1982). His4C+ revertants of *his4-100* were isolated that contain single-base-pair substitutions in the wild-type *HIS4* start codon. These mutations are thought to allow initiation at the second AUG codon that occurs in the reading frame that restores synthesis of the *HIS4* polypeptide following the *his4-100* frameshift. Thus, the second AUG codon is an efficient translational start site but can function only in the absence of the normal *HIS4* initiation codon located upstream. This interpretation was confirmed by fusing *lacZ*-coding sequences in-frame with the second ATG codon and measuring expression of β-galactosidase activity in the presence or absence of the normal *HIS4* start site present upstream and out of frame with the second ATG codon (Fig. 5D) (Donahue and Cigan 1988). Changing the A to a T nucleotide at the −3 position of the first ATG codon in this construct did not increase initiation at the second start site, despite the fact that this substitution reduces the initiation efficiency at the first ATG codon by about 40% (Table 5 and Fig. 5D) (Cigan et al. 1988b). However, because the 5′-proximal ORF overlaps the second initiation site in these *HIS4* constructs, translation of the uORF might interfere with initiation at the second start site by those 40S subunits that succeed in bypassing the uORF initiation site.

Studies on several mammalian transcripts have shown that the inhibitory effect of an upstream ATG codon can be reduced by inserting an in-frame stop codon that terminates translation of the uORF in the vicinity of the downstream start site (Johansen et al. 1984; Kozak 1984; Liu et al. 1984). In the case of the preproinsulin transcript, protein synthesis progressively increased as the separation between a uORF and the preproinsulin start site was increased from 2 to 79 nucleotides (Kozak 1987b). Other studies showed that insertion of a stop codon reduced the inhibitory effect of the upstream ATG codon even when the resulting uORF overlaps the beginning of the downstream protein-coding sequences (Johansen et al. 1984; Peabody and Berg 1986; Thomas and Capecchi 1986; Sedman et al. 1989). These observations were interpreted to indicate that translation can reinitiate at a downstream AUG codon following termination at a nearby stop codon. The fact that no *CYC1* expression was detected from alleles containing a uORF in which the stop codon is very close to the *CYC1* start site could mean that no

Table 5 Effect on translational efficiency of altering the sequence context of the initiation codon

Allele	Initiation region			% Protein expression
CYC1	AUA	*AUG*	ACU	100
	*		**	
CYC1-13-A	UUA	*AUG*	AUA	100
CYC1-730	UUC	*AUG*	GCC	20
	*			
CYC1-732	CUC	*AUG*	GCC	20
	*			
CYC1-735	GUC	*AUG*	GCC	50
	*			
CYC1-738	AUC	*AUG*	GCC	50
HIS4-lacZ	AUA	*AUG*	GUU	100
	*			
	CUA	*AUG*	GUU	97
	*			
	GUA	*AUG*	GUU	77
	*			
	UUA	*AUG*	GUU	61
			*	
	AUA	*AUG*	AUU	88
			*	
	AUA	*AUG*	CUU	102
			*	
	AUA	*AUG*	UUU	92

Asterisks are placed above nucleotides that differ between that allele and the first allele in each set, either *CYC1*, *CYC1-730*, or *HIS4-lacZ*.

reinitiation occurs on this yeast mRNA (Fig. 5A,B). Alternatively, efficient reinitiation may require a greater separation between the uORF and the second start codon, as suggested for the preproinsulin transcript.

GCN4 mRNA represents a rare instance in which upstream AUG codons occur naturally in the transcript (Hinnebusch 1984; Thireos et al. 1984). As expected, these sequences impose a strong translational barrier to *GCN4* expression, as removal of all four upstream ATG codons leads to an approximately 100-fold increase in expression of a *GCN4-lacZ* fusion. When present singly in the leader, uORF4 (numbered from the 5' end) lowers *GCN4* expression by 99% (Mueller and Hinnebusch 1986) and thus resembles the small uORFs present in the *cyc1-341* and *cyc1-599* alleles that exert a complete block to gene expression. In contrast, uORF1 inhibits *GCN4* expression by only about 66% (Fig. 5E), despite the fact that the sequence context of its start codon appears to be compatible with efficient initiation (e.g., an A is present at −3). In fact, a uORF1-*lacZ* fusion is expressed as efficiently as a *GCN4-lacZ* construct lacking all four uORFs (Mueller et al. 1988). Thus, *GCN4* uORF1 and the aforementioned heterologous uORF introduced at *CPA1* are two cases in which a 5'-proximal AUG codon present in a favorable sequence context is not used exclusively as the translation start site on the mRNA. In both instances, translation of the uORF would terminate well upstream of the AUG codon used to initiate synthesis of the protein, raising the possibility that reinitiation does occur in *Saccharomyces* if a sufficient distance separates the uORF from the next start codon.

e. Expression of Two Protein Products from the Same Gene Using Different Translation Start Sites. Several genes in *S. cerevisiae* express two proteins that differ by only the presence of an amino-terminal extension. In the case of the *SUC2* gene product, invertase, this extension provides a signal sequence for secretion of the enzyme. Secreted invertase is made from a transcript that begins upstream of the first ATG codon in the *SUC2* ORF. A shorter mRNA is also produced that begins downstream from the first ATG codon. Invertase made from the latter lacks a signal sequence and thus remains in the cytoplasm. Presumably, initiation at the first AUG codon on the longer transcripts precludes synthesis of the shorter cytoplasmic form of the enzyme from the same mRNAs. This translational exclusion allows differential regulation of the two forms of invertase through independent transcriptional controls over production of the transcripts: Synthesis of the larger transcript is repressed by glucose, whereas the smaller mRNA is expressed constitutively (Carlson and Botstein 1982; Perlman et al. 1982).

Other examples of this type of gene organization have been described for genes whose products function in both the cytoplasm and the mitochondria. The *HTS1* and *VAS1* genes were mentioned earlier. Two other examples are *LEU4*, which encodes the leucine biosynthetic enzyme α-isopropylmalate synthase (Beltzer et al. 1986, 1988), and *TRM1*, which encodes a tRNA methyltransferase (Ellis et al. 1987). Interestingly, the start codon for the long form of Trm1p is only two nucleotides from the 5' end of the longer transcript that contains this start site. The fact that removal of the second in-frame ATG codon does not abolish mitochondrial enzyme activity confirms that the first ATG codon is a functional start site, despite its close proximity to the 5' end of the mRNA. A related sequence organization has been reported for the *MOD5* gene, which also encodes a tRNA-modifying enzyme found in both mitochondria and the cytoplasm (Najarian et al. 1987). The two known *MOD5* transcripts begin upstream of both in-frame ATG codons; however, the major transcript contains the first AUG codon immediately adjacent to the cap site. This unusual location may preclude its recognition as a start codon, thereby causing the major transcript to produce only the short form of Mod5p. The minor *MOD5* transcript begins 11 nucleotides further upstream of the first ATG codon and should thus produce only the large form of the enzyme. It will be interesting to test these predictions by mutagenesis of the two *MOD5* potential start codons.

2. Reactions Involved in Translation Initiation

Polypeptide chain initiation involves decoding the AUG methionine codon. In all eukaryotic organisms, a specialized $tRNA_i^{Met}$ is used for this purpose. The reaction requires charged $tRNA_i^{Met}$, mRNA, ribosomal subunits, several initiation factors (eIFs), GTP, and ATP. On the basis of biochemical analyses of protein synthesis in several eukaryotes, the initiation process has been divided into several distinct steps (Fig. 6) (for review, see Moldave 1985): (1) dissociation of 80S ribosomes into large (60S) and small (40S) subunits and formation of a complex between eIF-3 and the small subunit; (2) formation of a ternary complex containing Met-tRNA$_i^{Met}$, GTP, and eIF-2; (3) binding of the ternary complex to the 40S/eIF-3 complex to form a 43S preinitiation complex; (4) association of mRNA with initiation factors eIF-4F, eIF-4A, and eIF-4B, mediated by the mRNA 5'cap; (5) binding of the 43S preinitiation complex to the mRNA near the 5'cap, followed by scanning of the complex to the initiation codon; (6) reaction of the preinitiation complex with eIF-5, catalyzing hydrolysis of GTP, release of eIF-2/GDP and eIF-3, and joining of the 60S subunit to form an 80S initiation complex at the AUG start codon. The following summarizes the information that exists regarding these steps in *Saccharomyces*.

a. Formation of the 40S/eIF-3 Preinitiation Complex. eIF-3, a multi-subunit protein of 7–10 polypeptide chains, has been implicated in two

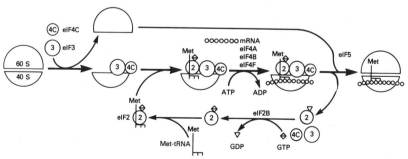

Figure 6 Hypothetical representation of the process of translation initiation in eukaryotes (from J.W.B. Hershey et al., pers. comm.). With the exception of eIF-4C and eIF-5, the functions of the various initiation factors are discussed in the text.

important activities. One is a ribosome dissociation or antiassociation function needed to maintain a pool of free 40S subunits that can participate in the initiation process; the second function is to facilitate binding of eIF-2/GTP/Met-tRNA$_i^{Met}$ to the 40S subunit (Moldave 1985). Some evidence for the latter has come from studies on the temperature-sensitive lethal mutation *ts187* isolated in *S. cerevisiae*. This mutation, now known as *prt1-1*, reduces incorporation of amino acids into protein within minutes following a temperature shift from 23°C to 37°C (Hartwell and McLaughlin 1968a, 1969). Polysomes disintegrate immediately following the temperature shift, and this event is prevented by treatment with cycloheximide, an inhibitor of chain elongation. Thus, it appears that elongation of nascent chains can be completed, but new chains are not initiated in *prt1-1* cells incubated at the restrictive temperature.

The nature of the *prt1-1* lesion was studied in a template-dependent cell-free translation system developed for *Saccharomyces* (Gasior et al. 1979a,b; Tuite et al. 1980), where it is possible to examine the individual reactions involved in protein synthesis (Gasior et al. 1979a). Feinberg et al. (1982) showed that extracts prepared from *prt1-1* cells heat treated at 39°C were defective for translation of yeast mRNA but competent for translation of poly(U), confirming that elongation reactions are intact at the restrictive temperature.

Termination and release of completed polypeptide chains were also unaffected by the heat treatment, as was the efficiency of eIF-2/GTP/Met-tRNA$_i^{Met}$ ternary complex formation. In contrast, binding of ternary complexes to 40S subunits was reduced. Because 40S ribosomes from heat-treated mutant cells were still competent for binding the ternary complexes present in extracts of heat-treated wild-type cells, it was suggested that *prt1-1* impairs a soluble factor required to bind eIF-2/GTP/Met-tRNA$_i^{Met}$ to the 40S subunit (see Table 6). In more recent work summarized by Moldave and McLaughlin (1988), an activity purified from wild-type cells on the basis of in vitro complementation of a *prt1-1* mutant extract for failure to bind ternary complexes to the wild-type 40S subunits apparently has characteristics described previously for eIF-3 from other sources. It remains to be determined whether the *PRT1* product is a subunit of eIF-3.

The *PRT1* gene has been cloned, and the predicted amino acid sequence shows no striking homology with any known protein. As would be expected, deletion of *PRT1* is lethal (Table 6) (Hanic-Joyce et al. 1987). *cdc63-1* is a temperature-sensitive allele of *PRT1* that differs from *prt1-1* in causing arrest in the G_1 phase of the cell cycle at the restrictive temperature (Bedard et al. 1981; Hanic-Joyce 1985).

b. eIF-2 and Formation of the eIF-2/GTP/Met-tRNA$_i$^Met Ternary Complex. The best-known functions of eIF-2 are formation of the ternary complex and subsequent binding of this complex to the 40S ribosomal subunit. eIF-2 is specific for binding Met-tRNA$_i$^Met and will not interact with any other aminoacylated tRNA species (Moldave 1985). eIF-2 has been purified from many organisms, including *Saccharomyces*, by assaying for GTP-dependent binding of radioactively labeled Met-tRNA$_i$^Met. The initial studies conducted by Baan et al. (1981) suggested that *Saccharomyces* eIF-2 is about 130 kD and consists of three nonidentical subunits of 31, 49.6, and 46.5 kD, presumed to be the α, β, and γ subunits, respectively. The 31-kD species could be phosphorylated in vitro by an endogenous kinase and occurs as a phosphoprotein in vivo (Romero and Dahlberg 1986), consistent with its identification as eIF-2α, as this subunit is phosphorylated in other eukaryotes (Moldave 1985). A second study yielded similar results concerning the native molecular mass of eIF-2; however, only two subunits were detected by SDS-PAGE: a 36-kD species and a 54-kD species that seemed to be present in twice the molar amount of the former (Ahmad et al. 1985a). The 54-kD species cross-reacted with antiserum specific for the γ subunit of rabbit reticulocyte eIF-2 and the 36-kD polypeptide was phosphorylated by the reticulocyte heme-regulated kinase known to phosphorylate the α subunit of rabbit eIF-2 (Ahmad et al. 1985b). These findings suggested that the 36- and 54-kD polypeptides in this preparation were the α and γ subunits of *Saccharomyces* eIF-2, respectively. The identity of the β subunit was left unresolved.

The issue of the subunit molecular weights of the α and β subunits was recently clarified by isolation of the structural genes for these two proteins. Three unlinked genes, *SUI1*, *SUI2*, and *SUI3* (*suppressor of initiation defect*), were identified by mutations that increase *HIS4* expression 10- to 50-fold when an ATT replaces the normal *HIS4* ATG start codon (Fig. 7). (Part of the increased expression results from elevated *HIS4* transcription [see Section III.A.2.f. below]; after correction for this contribution, the mutations in the *SUI* genes appear to increase translation of the mutant *HIS4* mRNA two- to tenfold [Castilho-Valavicius et al. 1990].) The *SUI* gene mutations increase *HIS4* translation regardless of which non-ATG codon is present at the initiation site. Sequence analysis of the amino termini of fusion proteins produced by a *HIS4-lacZ* construct containing an ATT start codon suggests that initiation occurs in *sui2-1* and *SUI3-2* strains at a UUG present at the third codon in the *HIS4*-coding sequence. This suggestion was confirmed by showing that a *HIS4-lacZ* fusion containing a TTG codon at the normal start site, as well as at the third codon, produced two fusion proteins with the amino

Table 6 Phenotypes of mutations affecting initiation factors in *S. cerevisiae*

Gene	Allele	Phenotype of mutation	Presumed function of gene product in translation initiation
PRT1	*prt1-1*	inhibition of polypeptide chain initiation at 37°C, leading to cell death; binding of eIF-2/GTP/Met-tRNA$_i^{Met}$ to 40S subunit defective at 37°C in vitro	formation of 43S preinitiation complex containing 40S subunit and eIF-2/GTP/Met-tRNA$_i^{Met}$
	prt1Δ	unconditionally lethal	
	cdc63-1	temperature-sensitive lethal; arrest in G$_1$ phase of cell cycle at 36°C	
SUI2	*sui2-1*	increased utilization of TTG as start codon at *HIS4*; reduced eIF-2/GTP/Met-tRNA$_i^{Met}$ formation in vitro	α subunit of eIF-2, required for association of Met-tRNA$_i^{Met}$ and GTP with 40S subunit
	sui2Δ	unconditionally lethal	
SUI3	*SUI3-2*	increased utilization of TTG as start codon at *HIS4*; reduced eIF-2/GTP/Met-tRNA$_i^{Met}$ formation in vitro	β subunit of eIF-2, required for association of Met-tRNA$_i^{Met}$ and GTP with 40S subunit
	sui3Δ	unconditionally lethal	

CDC33	*cdc33Δ*	unconditionally lethal	mRNA cap-binding protein eIF-4E, required for binding of 43S preinitiation complex at AUG start codon
	cdc33-1	temperature-sensitive lethal; arrest in G_1 phase of cell cycle at 37°C; reduces cap-binding activity in vitro	
	cdc33-4-2	temperature-sensitive lethal; severe reductions in protein synthesis at 37°C in vivo and in cap-binding activity in vitro	
TIF1	*tif1Δ*	no phenotype alone; unconditionally lethal when combined with *tif2Δ*	eIF-4A, mRNA-binding factor required for interaction of 43S preinitiation complex with AUG start codon; thought to catalyze ATP-dependent unwinding of mRNA secondary structure
TIF2	*tif2Δ*	no phenotype alone; unconditionally lethal when combined with *tif1Δ*	eIF-4A, mRNA-binding factor required for interaction of 43S preinitiation complex with AUG start codon; thought to catalyze ATP-dependent unwinding of mRNA secondary structure

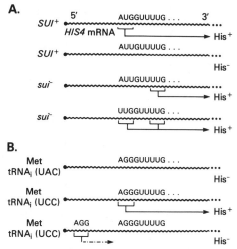

Figure 7 Efficient initiation at non-AUG start codons in *sui* mutants (*A*) and in strains expressing a tRNA$_i$^Met^ with a 3′-UCC-5′ anticodon rather than 3′-UAC-5′(*B*). The mRNA leader sequences of wild-type and mutant *HIS4* mRNAs are depicted with an arrow originating from the codon expected to function as the predominant initiation site. In the last construct in *B*, translation from the 5′-proximal AGG codon is out of frame with the *HIS4*-coding sequences.

termini expected from initiation at each of the two sites (Fig. 7) (Donahue et al. 1988; Cigan et al. 1989).

DNA sequence analysis revealed that *SUI2* and *SUI3* encode proteins with substantial similarity in size and sequence to human eIF-2α and eIF-2β, respectively: Sui3p shows 42% sequence identity with human eIF-2β; Sui2p is 58% identical to human eIF-2α (Donahue et al. 1988; Pathak et al. 1988; Ernst et al. 1987; Cigan et al. 1989). Additional support for the idea that Sui2p and Sui3p are the α and β subunits of eIF-2 was provided by the following observations: (1) Sui2p and Sui3p, detected immunologically with antisera raised against TrpE-Sui fusion proteins, copurify with eIF-2 biochemical activity; (2) ternary complex formation is defective in *sui2-1*, *SUI3-2*, and *SUI3-3* strains; (3) Sui2p is phosphorylated in vivo; and (4) Deletion of *SUI2* or *SUI3* is lethal (Donahue et al. 1988; Cigan et al. 1989). Although the mutations in the *SUI* genes reduce ternary complex formation in vitro, it is not known whether this defect is expressed in vivo, nor how it relates to the increased efficiency with which UUG can be used as a start codon in the mutants. Interestingly, the β subunits of mammalian and *Saccharomyces* eIF-2 each contain a "zinc finger" motif close to the carboxyl terminus, and all three mutations in the *SUI3* gene alter amino acids near this motif. This observation could indicate that eIF-2β makes direct contact with mRNA and plays an important role in recognition of the AUG start codon. In this model, the mutations in *SUI3* would increase the efficiency with which UUG can be chosen as a start site by the scanning initiation complex (Donahue et al. 1988; Cigan et al. 1989).

Antisera to Sui2p and Sui3p cross-react with 36-kD proteins present in partially purified preparations of *Saccharomyces* eIF-2, in agreement with the molecular masses predicted from the DNA sequences of *SUI2* and *SUI3* (Donahue et al. 1988; Cigan et al. 1989). However, apparently none of the three major polypeptides present in eIF-2 preparations made according to the method of Baan et al. (1981) cross-react with Sui2p antibodies (Cigan et al. 1989), raising the possibility that the 31-kD phosphorylated protein present in these preparations is not eIF-2α (Romero and Dahlberg 1986). Because phosphorylation of the 31-kD protein was unaffected by a variety of stress conditions in which variations in the rates of protein synthesis were expected, it was concluded that phosphorylation of eIF-2α does not alter the rate of protein synthesis in *Saccharomyces* (Romero and Dahlberg 1986). This issue should be reexamined using Sui2p antiserum to probe the phosphorylation state of eIF-2α under different conditions.

c. Regulation of eIF-2/GTP/Met-tRNA$_i$^{Met} Ternary Complex Formation by Ancillary Factors. It is generally accepted that eIF-2 is released at the end of the initiation reaction in a stable complex with GDP that is inactive for further binding of Met-tRNA$_i$^{Met}. The high affinity of mammalian eIF-2 for GDP in the presence of Mg^{++} necessitates an exchange factor to replace GDP with GTP. This factor, known as eIF-2B, is thought to play a central role in the regulation of eIF-2α activity by phosphorylation in heme-deprived rabbit reticulocytes and adenovirus-infected human cells (Safer 1983; Schneider and Shenk 1987). The prevailing model postulates that eIF-2/GDP binary complexes containing a phosphorylated α subunit form a stable inactive complex with eIF-2B, sequestering the latter and blocking the GDP-GTP exchange reaction needed to regenerate active eIF-2/GTP (Safer 1983). It has been reported (Ahmad et al. 1985b) that *Saccharomyces* eIF-2 differs from its mammalian counterpart in that (1) none of the former is found in a binary complex with GDP, (2) ternary complex formation by yeast eIF-2 is not inhibited by Mg^{++} (Mg^{++} stabilizes mammalian eIF-2/GDP complexes), and (3) preformed yeast eIF-2/GDP freely exchanges with exogenous GTP in the presence of Mg^{++}. In addition, Ahmad et al. (1985b) reported that phosphorylation of *Saccharomcyes* eIF-2α by the heme-responsive kinase from rabbit reticulocytes had no effect on GDP exchange or ternary complex formation by the yeast factor. On the basis of these findings, these workers suggested that eIF-2B activity is not required in *Saccharomyces* to regenerate eIF-2/GTP following completion of the initiation cycle. Consequently, eIF-2α phosphorylation might not inhibit protein synthesis in yeast. It is important to be certain that the eIF-2 preparations used in such studies do not contain guanine nucleotide exchange activity.

Although they found no evidence for eIF-2B in *Saccharomyces*, Ahmad et al. (1985a) detected an eIF-2 ancillary factor, called Co-eIF-2A, that stimulates ternary complex formation two- to threefold at low eIF-2 concentrations. Polyclonal antibodies raised against Co-eIF-2A had an inhibitory effect on translation in a cell-free extract from wild-type *S. cerevisiae*, consistent with an important role for this factor in protein synthesis. The antibodies inhibited binding of Met-tRNA$_i^{Met}$ to 40S subunits. *Saccharomyces* Co-eIF-2A also stimulated ternary complex formation by rabbit reticulocyte eIF-2. The degree of stimulation was greatest in the absence of Mg^{++}, a condition where GDP-GTP exchange by eIF-2B is not required, suggesting that yeast Co-eIF-2A does not mimic eIF-2B activity. It was proposed that Co-eIF-2A stimulates ternary complex formation by stabilizing the preformed complex. Several polypeptides with Co-eIF-2A activity are present in high-salt extracts of *Saccharomyces* ribosomes that are probably generated by proteolysis of an 80-kD species (Nasrin et al. 1986).

d. Binding of the Small Ribosomal Subunit to mRNA. It is generally assumed that the ternary complex binds to the 40S subunit prior to association of either species with the mRNA. The 43S preinitiation complex thus formed binds to the mRNA and selects the AUG start codon. This reaction is thought to require additional factors that interact with the mRNA at the 5' cap structure (m^{7}G(5')ppp(5')N, where N is any nucleotide). Numerous observations suggest that capped mRNAs are more efficiently translated in eukaryotic cell-free systems; moreover, cap analogs inhibit translation in these systems, as would be expected if the cap structure mediates association of important initiation factors with the mRNA (Sonenberg 1988). In fact, several mammalian protein factors that stimulate translation in vitro exhibit specific interactions with the 5' cap. One such factor, eIF-4F, is a complex of three proteins with molecular masses of 24, 50, and 220 kD. The 50-kD protein appears to be identical to eIF-4F, a factor that catalyzes ATP-dependent bidirectional unwinding of RNA secondary structure and is believed to carry out the same reaction when functioning as a component of eIF-4F (Rozen et al. 1990). The 24-kD protein, also known as eIF-4E, is the cap-binding component of eIF-4F. Another factor, eIF-4B, stimulates and may even be essential for the RNA helicase activities of eIF-4F and eIF-4A. A working model to explain the functions of these proteins suggests that a direct interaction occurs between the 24-kD component of eIF-4F and the mRNA cap. Once eIF-4F is bound, it acts with eIF-4B to unwind secondary structure in the vicinity of the cap, allowing free molecules of eIF-4A and eIF-4B to bind to the mRNA. The latter melt the remaining secondary structure in the mRNA leader in a processive reaction that requires ATP hydrolysis. The 43S preinitiation complex interacts with the mRNA at or near the cap and scans the now-unstructured leader for the initiation codon, or binds directly to the start codon in a reaction facilitated by eIF-4B (Sonenberg 1988; Rozen et al. 1990).

Yeast mRNAs are capped (Sripati et al. 1976), and it was reported that capping yeast killer virus m mRNA increases its translational efficiency in a yeast cell-free system (Leibowitz et al. 1988). In addition, cap analogs inhibit translation in *S. cerevisiae* cell-free extracts (Gasior et al. 1979b; Szczesna and Filipowicz 1980; Tuite et al. 1980; Hussain and Leibowitz 1986). The 24-kD cap-binding protein eIF-4E has been isolated from *S. cerevisiae* by its affinity for the cap structure. This protein can be specifically cross-linked to mRNA containing an oxidized cap, and a monoclonal antibody prepared against the purified protein inhibits translation in a yeast cell-free system in a dose-dependent fashion (Altmann et al. 1985). The structural gene for the cap-binding factor was isolated and shown to encode a protein of 213 amino acids (Altmann et al. 1987) that is about 30% identical to mouse and human eIF-4E (Sonenberg 1988). It was realized subsequently that this is the same gene as *CDC33* (Table 6) (Brenner et al. 1988).

The temperature-sensitive lethal mutation *cdc33-1* leads to cell-cycle arrest in the G_1 phase at 37°C (Pringle and Hartwell 1982), the same phenotype seen for conditional lethal mutations in certain aminoacyl-tRNA synthetase genes (see Table 2) and for the *cdc63-1* allele of *PRT1* (Table 6). Not surprisingly, a deletion of *CDC33* is unconditionally lethal (Table 6) (Altmann et al. 1987); however, it is remarkable that this lethality can be overcome by expression of the mouse eIF-4E cDNA from a yeast promoter (Altmann et al. 1989b). Only a small difference in growth rate was observed between yeast strains expressing wild-type *CDC33*-coding sequences versus the mouse homolog from the same promoter. These results strongly suggest that the 24-kD cap-binding protein plays an important role in translation initiation in yeast and that interactions between eIF-4E and other translation factors have been highly conserved during evolution.

Direct support for the role of Cdc33p in protein synthesis was provided by the demonstration that a cell-free translation system prepared from the temperature-sensitive *cdc33-4-2* mutant programmed with wild-type yeast mRNA had little activity and functioned even less after preincubation at the restrictive temperature. Translation in this extract could be rescued by the addition of purified Cdc33p synthesized in bacteria. As expected, addition of Cdc33p to the mutant extract stimulated translation of total yeast mRNA much more than it did for alfalfa mosaic virus RNA 4 (AMV-4), as there is evidence that AMV-4 can be translated in a cap-independent fashion. The *cdc33-4-2* product was shown to have reduced affinity for the cap analog m^{7}GDP in vitro, as do the products of *cdc33-1* and several mutant alleles containing substitutions in certain highly conserved tryptophan residues (Table 6) (Altmann et al. 1988, 1989a; Altmann and Trachsel 1989).

A 150-kD polypeptide was identified in *S. cerevisiae* that, like eIF-4E, can be specifically cross-linked to capped mRNA and partially purified by affinity chromatography using the cap analog m^{7}GDP bound to agarose. In addition, the 150-kD species cosediments with eIF-4E in sucrose density gradient centrifugation and copurifies with eIF-4E in ion-exchange chromatography, suggesting that these two proteins exist in a complex. This complex may be the yeast equivalent of eIF-4F (Goyer et al. 1989). If so, it resembles the corresponding complex found in plants in lacking the 50-kD eIF-4A, the catalytic moiety of the RNA helicase.

The lack of eIF-4A in the aforementioned complex is probably not due to the absence of eIF-4A in yeast, since two genes (*TIF1* and *TIF2*) were isolated recently whose predicted protein products are 65% identical in sequence to mouse eIF-4A (Linder and Slonimski 1988, 1989). Deletion of *TIF1* or *TIF2* singly has no apparent phenotype, but deletion of both genes is lethal, as would be expected if the products of these genes were required for translation initiation (Table 6). Strong support for this assertion was provided by showing that cell-free translation systems depleted of both Tif1p and Tif2p are inactive and can be reactivated by the addition of Tif1p purified from a wild-type strain (Blum et al. 1989).

The *TIF* genes were isolated on the basis of their ability to suppress a missense mutation in the mitochondrial gene *oxi2* (encodes subunit III of cytochrome oxidase) when present on a high-copy-number plasmid. The mechanism of this suppression remains to be elucidated (Linder and Slonimski 1989). Interestingly, it appears that Tif1p and Tif2p belong to a family of proteins with diverse functions found in both eukaryotes and prokaryotes, named the D-E-A-D box family after one of the most highly conserved blocks of amino acids present in these proteins (Asp-Glu-Ala-Asp) (Linder et al. 1989). Seven additional members of the family have been identified in *Saccharomyces*. Two of these are involved in RNA splicing, a function that can be easily rationalized for an RNA helicase (Chang et al. 1990). The roles of the other five D-E-A-D box proteins present in yeast are not known.

e. Selection of the AUG Start Codon. The fact that mutations in the eIF-2α and eIF-2β structural genes (*SUI2* and *SUI3*) affect the specificity of start-site selection implies that the eIF-2/GTP/Met-tRNA$_i^{Met}$ ternary complex plays a central role in the scanning process. Additional support for this idea has come from the fact that altering the anticodon of tRNA$_i^{Met}$ changes the codon that is used as the translation start site. Cigan et al. (1988a) isolated a tRNA$_i^{Met}$ gene from *Saccharomyces* and changed the anticodon from 3′-UAC-5′ to 3′-UCC-5′. When introduced back into cells on a high-copy-number plasmid, the mutant tRNA$_i^{Met}$ gene restored expression from a *his4* allele in which the normal ATG start codon was replaced by AGG. Expression of *his4* alleles containing other non-AUG start codons was not affected by the mutant tRNA$_i^{Met}$, showing that codon-anticodon pairing was required for recognition of the non-AUG start site. In addition, an out-of-frame AGG codon introduced into the mRNA leader upstream of the AGG start site abolished the ability of the mutant tRNA$_i^{Met}$ gene to restore gene expression (Fig. 7). Thus, the 5′-proximal AGG codon was used preferentially as the translation start site, another hallmark of normal initiation. Given that the sequence context of the start codon can be altered with little change in translational efficiency of the mRNA, it seems clear from these results that complementarity between the start codon and the anticodon of tRNA$_i^{Met}$ predominates over other interactions between mRNA and the initiation complex in selecting the start site during the scanning process. This situation stands in contrast to that described in prokaryotes, where pairing between the 3′ end of 16S mRNA and the Shine-Delgarno sequence is critical for initiation site selection. Thus, the fact that the initiation process begins at the 5′ end of the mRNA, coupled with the primacy of the AUG codon/tRNA$_i^{Met}$ interaction in selecting the start site, ensures that translation initiates predominantly at the 5′-proximal AUG codon in yeast mRNAs.

The conclusion that tRNA$_i^{Met}$ and eIF-2 play a direct role in start codon selection (Cigan et al. 1988a; Donahue et al. 1988) is in accord with the idea that eIF-2/GTP/Met-tRNA$_i^{Met}$ binds to the 40S subunit prior to association of the resulting 43S preinitiation complex with mRNA (Moldave 1985). However, a different mechanism was proposed for *Saccharomyces* by Nasrin et al. (1986) on the basis of results which suggest that an AUG codon is required for binding of ternary complex to 40S subunits. These workers postulate that the AUG codon in the mRNA is bound first by the 40S subunit, with the help of mRNA-binding factors, followed by eIF-2/GTP/Met-tRNA$_i^{Met}$. In fact, this alternative model is not inconsistent with the effects of altering eIF-2 and tRNA$_i^{Met}$ structure on the specificity of initiation if one proposes that the ternary complex functions to stabilize a transient 40S/mRNA association at the initiation codon. More experiments will be required to determine the precise sequence of events that occurs to bring together the 40S subunit and the ternary complex at the AUG start codon.

C. Elongation

1. Overview

The elongation phase of protein synthesis is quite similar in eukaryotes and bacteria and has been reviewed in detail by Moldave (1985) and Miller and Weissbach (1977). The soluble protein factors required for elongation in eukaryotes, elongation factors 1 (EF-1) and 2 (EF-2), have analogous functions to the bacterial elongation factors EF-Tu and EF-G, respectively. Yeast and other fungi also require another elongation factor, EF-3.

The process of elongation can be divided into three stages: (1) binding of aminoacylated tRNA to the ribosomal A site, (2) formation of the peptide bond, and (3) translocation. In the first stage, EF-1 is required for the GTP-dependent binding of aminoacylated tRNA to the ribosomal A site and also participates in the proofreading of the codon-anticodon match. In the second stage, the ribosome promotes the transfer of the nascent peptide bound to the tRNA in the P site to the α amino group of the aminoacylated tRNA in the A site. This transfer and concurrent peptide bond formation are catalyzed by the peptidyltransferase center of the ribosome and do not require any soluble factors. In yeast, both proteins and rRNA of the 60S subunit appear to be involved (Perez-Gosalbez et al. 1978, 1983). During translocation, the final stage, EF-2 promotes the movement of the ribosome along the mRNA, causing the transfer of the new peptidyl-tRNA to the P site while releasing the stripped tRNA.

The details of the *S. cerevisiae* translation system have been the subject of two recent reviews (Chakraburtty and Kamath 1988; Tuite 1989). Here, we briefly summarize the properties of the yeast elongation factors.

2. EF-1

a. Biochemical Components. The α subunit of yeast cytoplasmic EF-1 has been isolated and characterized by several groups (Richter and Lipmann 1970; Spremulli and Ravel 1976; Dasmahapatra et al. 1981; Thiele et al. 1985). It is composed of a single polypeptide chain with a molecular mass of about 50 kD and an isoelectric point of 8.9 (Dasmahapatra et al. 1981). Most of EF-1α is isolated as a monomer. However, as in other eukaryotes, yeast EF-1α is also found in high-molecular-mass forms associated with two other polypeptide chains called EF-1β and EF-1γ (Iwasaki and Kaziro 1979; Nolan et al. 1979; Merrick et al. 1980). The molecular masses of these subunits are 33 kD and 48 kD, respectively (Saha and Chakraburtty 1986; Chakraburtty and Kamath 1988). In addition to the cytoplasmic EF-1α, the yeast mitochondrial form, mEF-Tu, has also been isolated (Rosenthal and Bodley 1987).

b. Biological Functions. EF-1 is one of the most abundant proteins in rapidly growing cells. Its major function is to facilitate the binding of the correct aminoacyl-tRNA to the ribosomal A site. This occurs when the α subunit of EF-1 forms a binary complex with GTP, which then forms a ternary complex with the aminoacyl-tRNA. It is this ternary complex that binds reversibly to the ribosome. This ternary complex is reminiscent of the eIF-2/GTP/tRNA$_i^{Met}$ complex mentioned above. In *E. coli*, association of the ternary complex with the ribosome reveals a second tRNA-binding site in EF-Tu, the bacterial counterpart of EF-1α. Since peptidyl-tRNA can bind to this new site, EF-Tu has been proposed to be involved in positioning the incoming tRNA correctly relative to the peptidyl-tRNA (van Noort et al. 1985; Hughes et al. 1987).

Extensive studies on the bacterial EF-Tu (Thompson 1988) suggest that rejection of incorrectly bound aminoacyl-tRNA occurs in two steps. At each step, the EF-Tu (and by analogy, EF-1α) activity provides an internal standard against which the strength of the codon:anticodon bond can be measured. To pass the first test, the aminoacyl-tRNA/EF-1α/GTP/ribosome complex must remain stable long enough for the GTP to be hydrolyzed to GDP. Presumably, at this stage, noncognate tRNAs will frequently be rejected because they dissociate faster than the EF-1α-promoted GTP hydrolysis can take place. To pass the next proofreading step, the aminoacyl-tRNA must remain in the ribosome complex long enough for the EF-1α:GDP to dissociate. Here, rejection of noncognate tRNAs occurs because they will dissociate from the complex faster than EF-1α:GDP. EF-1β and its bacterial counterpart EF-Ts are able to stimulate a nucleotide exchange reaction that regenerates EF-1α:GTP from the dissociated EF-1α:GDP. The function of EF-1γ is unknown.

c. Genes. Two unlinked genes, *TEF1* (on the right arm of chromosome XVI) and *TEF2* (on the right arm of chromosome II), code for cytoplasmic EF-1α (Nagata et al. 1984; Schirmaier and Philippsen 1984; Cottrelle et al. 1985b; Nagashima et al. 1986; Sandbaken and Culbertson 1988). The ORFs of these genes differ in only two positions, although their 5′- and 3′-flanking regions differ considerably. Both genes code for identical proteins of 458 amino acids. These yeast genes show about 80% identity with EF-1α genes from *Artemia salina* (van Hemert et al. 1984) and humans (Brands et al. 1986). They are about 30% identical to *E. coli* EF-Tu (An and Friesen 1980; Yokota et al. 1980) and to yeast mitochondrial mEF-Tu (Nagata et al. 1983). Both the *TEF1* and *TEF2* genes are efficiently transcribed at about the same level (Nagashima et al. 1986). The presence of either *TEF* gene alone is sufficient to support normal growth rates, whereas disruption of both *TEF* genes is lethal (Cottrelle et al. 1985a).

TEF gene dosage has recently been shown to affect translational accuracy (Song et al. 1989; Song 1987; Liebman et al. 1988). Deletion of one of the *TEF* genes reduced the levels of misreading associated with the omnipotent suppressors *sup35* and *sup45*, the amber suppressor *SUP7*-**a**, and the drug paromomycin. In contrast, an extra copy of a *TEF* gene enhanced the misreading induced by omnipotent suppressors. One possible explanation of these results is that higher concentrations of EF-1α may better enable aminoacyl-tRNA to compete with termination factors when a nonsense codon is in the A site.

d. Mutations. Mutations in the yeast *TEF* genes also alter translational accuracy (Sandbaken and Culbertson 1988). Single mutational changes in *TEF2* were shown to cause dominant suppression of both nonsense and frameshift mutants. However, only certain nonsense and frameshift alleles were suppressed. In particular, suppressible frameshift alleles were always associated with nearby downstream nonsense codons. Bacterial EF-Tu mutations with similar sequence-specific suppressor activity have also been described (Hughes et al. 1987). How the downstream nonsense codons are involved in the suppression of frameshift mutations is unknown. Nevertheless, the fact that the same altered elongation factor can suppress both nonsense and frameshift mutations and that the suppressible frameshift alleles are always associated with nonsense codons led to the suggestion that EF-1α may use the same mechanism(s) to control amino acid misincorporation and reading frame errors (Hughes et al. 1987; Sandbaken and Culbertson 1988).

In the Sandbaken and Culbertson (1988) study, sequence analysis of *TEF2* suppressors showed that mutations were clustered in two areas: in or near the GTP-binding/hydrolysis domain or near a presumptive aminoacyl-tRNA-binding region. According to the Thompson (1988) model, alterations in either of these regions might be predicted to alter translational fidelity. It will be of interest to determine whether the EF-1α encoded by any of the *tef2* suppressors results in an increased rate of GTP hydrolysis, thereby decreasing the time period for proofreading, as the model predicts. Similarly, increased binding of aminoacyl-tRNA might be expected to permit incorporation of amino acids from non-cognate tRNA. Indeed, one of the *tef2* mutants did cause increased G-U wobble in the third codon position. Frameshifting might occur from alterations in EF-1α directly or by its acceptance of noncognate aminoacyl-tRNA, either of which might lead to effects on the positioning of incoming tRNA.

3. EF-2

a. Biochemical Components. Yeast EF-2 is a single polypeptide chain with a molecular weight of between 80,000 and 100,000 (Richter and Flink 1971; Skogerson and Wakatama 1976; Spremulli and Ravel 1976; Bodley et al. 1984). Diphtheria toxin inactivates this factor, as it does EF-2 from other eukaryotes (Collier 1975; Pappenheimer 1977). The toxin acts by catalyzing the transfer of ADP-ribose from nicotinamide adenosine diphosphate to a modified histidine residue, diphthamide, unique to EF-2 (van Ness et al. 1980; Dunlop and Bodley 1983). Diph-thamide is synthesized posttranslationally in a highly conserved region of EF2, but its biological role is unknown (Brown and Bodley 1979; Wold 1981).

b. Biological Functions. EF-2 (sometimes called G) promotes peptide translocation. This includes the transfer of the peptidyl-tRNA from the A to the P site, concurrent movement of the message, and release of the free tRNA from the P site. EF-2 binds strongly to both GTP and GDP. Translocation is associated with the hydrolysis of GTP, which brings about a conformational change in EF-2, necessary for efficient tRNA release (Robertson and Wintermeyer 1987).

c. Genes and Mutations. Diphtheria-toxin-resistant mutants in Chinese hamster ovary cells were shown to result from either codominant mutations in the EF-2 gene or recessive mutations in genes responsible for posttranslational modification of EF-2 (Moehring et al. 1980). In yeast, five complementation groups of recessive mutations that result in diphtheria-toxin-resistant EF-2 have been identified (Chen et al. 1985). By analogy with the Chinese hamster mutants, these mutants are believed to be defective in enzymatic activities required for the posttranslational synthesis of diphthamide. Indeed, the EF-2 found in one mutant, *dph5*, could be converted to an ADP-ribose-accepting form by an enzyme found in *DPH5* cells. Since the mutations have no deleterious effect on the growth of the yeast strains, it appears that diphthamide is not required for EF-2 function under normal growth conditions.

4. EF-3

a. Biochemical Components and Biological Function. Although the ribosomes from most well-studied eukaryotic systems can translate in the presence of only two elongation factors, translation with ribosomes from yeast and a variety of other fungi requires a third elongation factor (Skogerson and Wakatama 1976; Uritani and Miyazaki 1988a). This factor, called EF-3, has been purified and shown to consist of a single polypeptide chain with a molecular weight of 125,000 and an isoelectric point of 5.9 (Dasmahapatra and Chakraburtty 1981). EF-3 has ribosome-dependent GTPase and ATPase activites that are not present on salt-washed yeast ribosomes (Skogerson 1979; Dasmahapatra and Chakraburtty 1981). In contrast, well-washed ribosomes from eukaryotes that do not require EF-3 retain ATPase and GTPase activity. This suggests that EF-3 may be a solubilized form of a ribosomal component (Uritani and Miyazaki 1988a). Experiments in which the concentrations of ATP are varied, or in which the ATPase activity is inhibited, suggest that hydrolysis of ATP by EF-3 is required for the yeast peptide elongation reaction (Miyazaki et al. 1988; Uritani and Miyazaki 1988b).

The specific role that EF-3 plays in translation is unknown. Immunoinactivation studies with antibody specific for this factor indicated that it is required for every cycle of elongation, since the addition of antibody immediately inhibited any further protein synthesis (Hutchison et al. 1984). EF-3 was found to be associated with polysomes and to a lesser extent with monosomes but not with ribosomal subunits (Hutchison et al. 1984). Experiments involving partial reactions suggested that EF-3 is not necessary for peptide bond formation or translocation, nor does it stimulate the nucleotide exchange reaction required by EF-1 (Skogerson and Engelhardt 1977; Dasmahapatra and Chakraburtty 1981). A recent report, however, indicates that EF-3 does stimulate translocation in a partial reaction when a limited amount of EF-2 is used. Furthermore, although EF-3 is not required for the EF-1α-dependent aminoacyl-tRNA-binding step, it clearly stimulates this reaction (Kamath and Chakraburtty 1989) in a manner specific for cognate, but not noncognate, aminoacyl-tRNAs, suggesting a possible role for EF-3 in maintaining fidelity (Uritani and Miyazaki 1988b).

b. Genes and Mutations. The gene encoding yeast EF-3 (*YEF-3*) has been cloned by immunoscreening of a yeast genomic library in λgt11 (Qin et al. 1987). The cloned *YEF-3* gene complemented a yeast mutant impaired in the elongation phase of protein synthesis whose EF-3 activity is thermolabile in an in vitro assay (Herrera et al. 1984; Kamath and Chakraburtty 1986a,b). *YEF-3* is a single-copy essential gene located on chromosome XII (Qin et al. 1990). The ORF codes for a deduced protein of 1044 amino acids with a molecular weight of 115,860. The codon bias is 0.87, suggesting that *YEF-3* is expressed at a high level. The transcript is about 3.4 kb and there are no introns (Qin et al. 1987). The amino acid composition and hydropathy profile suggest that the polypeptide is a soluble protein.

The carboxy-terminal region of the protein is highly charged, containing three polylysine blocks that are proposed to interact directly with RNA. The protein sequence contains an internal repeat of about 200 amino acids, including an ATP-binding site that shares significant similarity with some drug-resistant and transport proteins (Qin et al. 1990). Thus, EF-3 belongs to a family of proteins that bind ATP and appear to couple ATP hydrolysis to various biological processes (Higgins et al. 1986; Qin et al. 1990). EF-3 is proposed to transduce nucleotide triphosphate energy into the mechanical energy needed for translocation during protein synthesis (Qin et al. 1990).

5. Other Elongation Factors?

The biochemical studies described above have revealed the need for only three yeast elongation factors. These factors are highly abundant proteins, and it is unclear whether nonabundant proteins could have been detected by these biochemical techniques. The possibility that there may be additional elongation factors is suggested by the recent cloning of a poorly expressed essential gene, *sup35*, that is highly homologous to EF-1α (see Section II.D.4 below). This gene was first uncovered by a genetic approach that may reveal additional factors in the future.

D. Informational Suppression

1. Overview

One approach to detecting genes important in translation has been to isolate translational suppressors and genes that alter the efficiency of these suppressors. This method has been used successfully in *E. coli* and has led to a better understanding of the protein synthetic apparatus and the genes that control translational fidelity.

In yeast, two major classes of translational suppressors have been isolated that can misread nonsense codons or cause a change in the reading frame. The first class results from alterations in tRNAs. The nonsense suppressors are generally codon-specific, suppressing only UAA, or UAG, or UGA; the frameshift suppressors are generally amino-acid-specific, suppressing +1 frameshifts of glycine or proline codons. The other class, omnipotent suppressors, can cause misreading of a variety of different codons. Mutations in unlinked genes that either reduce (antisuppressors) or enhance (allosuppressors) the efficiency of both classes of suppressors have also been isolated. These genes are presumed to code for components of the translational apparatus or for products that modify these components. By using this layered approach to isolate mutants, one identifies genes whose products are likely to interact. This material has been reviewed previously (Sherman 1982). Here, we concentrate on recent results.

2. Suppressible Markers

It is widely appreciated that suppression of markers depends on the nature of the suppressible codon, the acceptability of the amino acid replacement at the mutant site, and the efficiency of the suppressor. Often overlooked is the fact that the same suppressible codon in different sequence contexts is frequently suppressed with widely differing efficiencies. This context effect is observed even when the amino acid inserted by the suppressor leads to fully functional protein in intragenic revertants (Liebman et al. 1975). There are also rare contexts in which single codons can be suppressed by both UAA- and UAG-specific suppressors: *leu1-101* (Hawthorne and Leupold 1974), *lys2-59* (Chattoo et al. 1979), and *rad6-3* (Lawrence and Christensen 1976). The DNA sequence of the *rad6-3* allele showed that it contains a UAA lesion (Reynolds et al. 1985).

The choice of suppressible alleles dictates the type of suppressor that can be isolated. Obviously, only suppressors with matching codon specificity and that insert an acceptable amino acid can be obtained. Selection of suppressors of poorly or easily suppressed alleles will allow the isolation of efficient or less efficient suppressors, respectively.

3. tRNA Suppressors

The best-studied yeast nonsense suppressors arise from alterations in tRNA genes that change the anticodon so that it decodes a stop codon (Sherman 1982). Rules that predict the allowed pairings between the codon and anticodon at the third or "wobble" position of the codon were first proposed by Crick in 1966 (Crick 1966). Almost all of the evidence gathered since that time supports these rules in *E. coli*. However, in eukaryotes, the wobble rules appear to be more restricted. On the basis of sequence information for *S. cerevisiae* tRNA species, Guthrie and Abelson (1982) proposed a revised set of wobble rules for yeast (see Table 7). These rules are consistent with the codon specificity differences observed between prokaryotic and yeast UAA suppressors. UAA suppressors in bacteria suppress both UAA and UAG codons, presumably via wobble pairing between the U in the first position of the suppressor tRNA anticodon and the G in the third position of the UAG codon. In yeast, UAA suppressors recognize only UAA codons and not UAG codons. The inability, in yeast, of U in the anticodon to wobble pair with G is thought to result from hypermodifications of the uridine (see Sherman 1982; Gelugne and Bell 1988). The uridine in the wobble position of the anticodon in *S. cerevisiae* UAA suppressors is modified to 5-methoxycarbonyl-methyl-2-thiouridine (Waldron et al. 1981; Laten et al. 1983). The absence of the U-G wobble pairing also explains why the dominant amber suppressor *SUP61*, which causes an anticodon change in the unique UCG decoding seryl-tRNA, is a recessive-lethal, despite the presence of a UCA decoding seryl-tRNA (see Brandriss et al. 1976; Etcheverry et al. 1982).

An updated summary of known UAA and UAG suppressors and their properties is presented in Table 8. The current list of presumptive UGA

Table 7 Wobble rules of yeast tRNAs

Third position of codon	First position of anticodon	
	E. coli	*Saccharomyces*
U	A, G, or I	G or I
C	G or I	G or I
A	U or I	U*
G	C or U	C

U* indicates a modified uridine residue. (Adapted from Guthrie and Abelson 1982.)

Table 8 UAA and UAG suppressors

Mutation	Amino acid replacement or tRNA altered	Map position	Possible equivalents	Codon suppressed	Suppressor anticodon (5'-3') found in		References
					tRNA	DNA	
SUP2	Tyr	IV R		UAA, or UAG			1–4
SUP3	Tyr	XV L		UAA, or UAG			1, 3, 4
SUP4	Tyr	X R		UAA, or UAG		TTA[a]	1–4, 20
SUP5	Tyr	XIII L		UAA, or UAG	CψA[b]		1–4, 21
SUP6	Tyr	VI R		UAA, or UAG			1, 3, 4, 5
SUP7	Tyr	X L		UAA, or UAG			1, 3, 4, 5
SUP8	Tyr	XIII R		UAA, or UAG			1, 3, 4
SUP11	Tyr	VI R		UAA, or UAG			1–4
SUP16	Ser	XVI R	SUQ5, SUP15	UAA	SUA[c]	TTA	4, 6, 7, 17 24, 26
SUP17	Ser	IX L		UAA			7, 17
SUP19	Ser	V R	SUP20, SUP85	UAA, or UGA			4, 8, 17, 19
SUP22	Ser	IX L		UAA			8
SUP25	Ser	XI R		UAA			4
SUP26	Leu	XII R	SUP86[d]	UAA, or UGA[d]			9
SUP27	Leu	IV R	SUP88[d]	UAA, or UGA[d]			9
SUP28	Leu	XIV R	SUP152	UAA, or UGA			9, 13
SUP29	Leu	X C	SUP30	UAA			3, 9
SUP2	Tyr	IV R		UAA, or UAG			1–4
SUP32	Leu	unknown		UAA			9

SUP33	Leu	XI L	SUP161	UAA, or UGA			9, 13
SUP50	Leu	frag. 6		UAG			4
SUP52	Leu	X C	SUP51	UAG			11
SUP53	Leu	III L		UAG		CTA	12, 25
SUP54	Leu	VII L		UAG			12
SUP55	Leu	unknown		UAG			12
SUP56	Leu	I R		UAG			12
SUP57	Leu[e]	VI R		UAG			12
SUP58	Leu[e]	XI L		UAG			12
SUP59	Leu[e]	unknown		UAG			12
SUP60	Leu[e]	unknown		UAG			12
SUP61	Ser	III R	SUP-RL1	UAG	CUA	CTA	4, 10, 16, 18, 22, 23
SUP70[f]	Glu		SUP60[f], SUP-1A	UAG		CTA	14, 15

References: (1) Gilmore et al. 1971; (2) Liebman et al. 1976; (3) Hawthorne and Mortimer 1968; (4) Mortimer and Hawthorne 1973; (5) Sherman et al. 1973; (6) Liebman et al. 1975; (7) Ono et al. 1979a; (8) Ono et al. 1981; (9) Ono et al. 1979b; (10) Brandriss et al. 1976; (11) Liebman et al. 1977; (12) Liebman et al. 1984; (13) Ono et al. 1985; (14) Calderon et al. 1984; (15) Winey et al. 1989a; (16) Etcheverry et al. 1982; (17) Broach et al. 1981; (19) Piper 1978; (19) Hawthorne 1981; (20) Goodman et al. 1977; (21) Piper et al. 1976; (22) Etcheverry et al. 1979; (23) Olson et al. 1981; (24) Waldron et al. 1981; (25) Fischhoff et al. 1984; (26) Laten et al. 1983.

[a] The suppressor with this anticodon acts specifically on UAA codons.
[b] The suppressor with this anticodon acts specifically on UAG codons.
[c] S represents 5-methoxycarbonyl-methyl-2-thiouridine.
[d] Allelism with the UGA suppressors SUP86 and SUP88 is presumed from the identical map positions and by analogy with the documented UAA/UGA leucine-inserting allelic suppressors SUP28/SUP152 and SUP33/SUP161.
[e] The amino acid inserted or tRNA altered in these cases has been presumed. The assignments were made because these suppressors have the same specificity and efficiency as other suppressors known to be associated with the indicated amino acid insertion or altered tRNA.
[f] This suppressor has been renamed to avoid confusion with the leucine-inserting suppressor SUP60 and to conform with the standard nomenclature.

Table 9 Presumptive UGA suppressors

Class I		Class II	
mutation	map position	locus	map position
SUP28	XIV R	SUP71	V R
SUP33	XI L	SUP72	II R
SUP85	V R	SUP73	X L
SUP86	XII R	SUP74	X L
SUP87	II R	SUP75	XI L
SUP88	IV R	SUP76	VII R
SUP90	IX L	SUP77	VII R
SUP160	XV C	SUP78	XIII R
SUP165	XIV R	SUP79	XIII L
		SUP80	IV R
		SUP154	VII L
		SUP155	VII R

Class I suppressors are more efficient than class II suppressors. (Adapted from Hawthorne 1976, 1881; Sherman 1982; Ono et al. 1988.)

and frameshift suppressors are in Tables 9 and 10, respectively. Alterations in tRNA that result in frameshift suppressors have been shown to include base-pair insertions or changes that result in a larger anticodon loop (Cummins et al. 1982, 1985; Gaber and Culbertson 1982a, 1984; Winey et al. 1986; Ball et al. 1988; Mendenhall and Culbertson 1988). Examples of multiple changes in the anticodon loop have also been reported (Mendenhall et al. 1987).

In addition to the suppressors that have been selected in vivo, nonsense suppressors have also been constructed in vitro. Synthetic UAA and UAG suppressor genes were built from a tRNA^{Phe} gene (Masson et al. 1987), and a UAG suppressor was made from a tRNA^{Trp} gene (Kim and Johnson 1988).

A low level of UAG suppression has also been associated with normal levels of the wild-type glutamine tRNA$_{CAG}$ that decodes CAG codons (Weiss and Friedberg 1986). When this tRNA with the anticodon 5'-CUG-3' is present on a multicopy plasmid, suppression efficiency is increased (Calderon et al. 1984; Weiss et al. 1987). Likewise, when expressed in sufficient quantity, the glutamine tRNA$_{CAA}$, which normally decodes CAA codons, can suppress UAA (Pure et al. 1985). Suppression by each of these tRNAs requires G-U mispairing at the first codon position. Disruption of the unique tRNA$_{CAG}$ is lethal unless rescued by an excess of tRNA$_{CAA}$ to allow CAG codons to be decoded (Weiss and Friedberg 1986). This rescue requires G-U wobble in the third codon position and has been used as an assay to detect increased G-U wobble (Sandbaken and Culbertson 1988).

4. Omnipotent Suppressors

a. Isolation. Omnipotent suppressors have been obtained in many selections and screens for mutations that compromise the accuracy of translation. They have surfaced as suppressors of two or more different nonsense codons, as codon nonspecific frameshift suppressors, as enhancers of nonsense suppressors, and as modifiers that alter the codon specificity of nonsense suppressors.

Omnipotent suppressors were originally obtained by selecting for suppressors that could act simultaneously on more than one type of nonsense codon. In a variety of such studies, only two loci were revealed that could mutate to suppressors capable of acting on at least one UAA and one UAG marker (Inge-Vechtomov and Andrianova 1970; Hawthorne and Leupold 1974; Gerlach 1975). These recessive suppressors were named *sup35* (also called *sup2* and *supP*), and *sup45* (also called *sup1* and *supQ*). Three codon nonspecific suppressors, *suf12*, *suf13*, and *suf14*, were also isolated by their ability to act on more than one type of frameshift allele: +1 insertions in either glycine or proline codons (Culbertson et al. 1982). They are each recessive, and *suf12* is allelic to *sup35* (Wilson and Culbertson 1988). Another gene, *GST1*, that is identical to *SUP35* has recently been cloned and sequenced (Kikuchi et al. 1988). The *gst1* mutation was obtained from a selection scheme that was designed to isolate mutants with defective centromere function but that apparently also selected translational suppressors. Selection was for growth on −Leu medium of a strain carrying the promoter-defective *LEU2-d* allele on a centromere plasmid and an undescribed *leu2* mutation in the chromosome. Although defective centromere function could lead to an increase in plasmid copy number, resulting in growth on −Leu plates, translational suppression of the chromosomal *leu2* mutation would likewise mask the leucine requirement. In fact, the *gst1* mutant was not defective in centromere function, although it did have a temperature-sensitive defect in G_1 to S phase transition.

Recent searches for omnipotent suppressors have been done in strains known to carry translational modifiers that might alter the spectrum of omnipotent suppressors that are recovered. Thus, suppressors were isolated in the presence of the [*psi$^+$*] factor (Ono et al. 1984, 1989; Waekem and Sherman 1990), which enhances the efficiency of certain dominant codon-specific suppressors; suppressors were also isolated in the pres-

Table 10 Frameshift suppressors

Mutation	Group[a]	Altered tRNA	Wild-type anticodon (5'-3')	Suppressor anticodon (5'-3')	Map location	References
SUF1	II	Gly	UCC	CCCC	XV L	1, 2
SUF3	II	Gly	CCC	CCCC	IV R	1, 5
SUF4	II	Gly	UCC	CCCC	VII R	1, 2
SUF5	II	Gly	CCC	C/UCCC	XV R	6, 7
SUF6	II	Gly	UCC		XIV L	2
SUF17	II	Gly	GCC	CCCC	XV L	2, 11
SUF15	II	Gly[b]			VII R	11
SUF16	II	Gly	GCC	GCCC	III R	2, 11, 12
SUF18	II	Gly[b]			VI R	11
SUF19	II	Gly[b]			V L	11
SUF20	II	Gly	GCC	GCCC	VI R	2, 11
SUF21	II	Gly[b]			XVI R	11
SUF22	II	Gly[b]			XIII R	11
SUF23	II	Gly[b]			X R	11
SUF24	II	Gly[b]			IV R	11
SUF25	II	Gly[b]			IV L	11
trn2	II	Gly	GCC		XV R	2
SUF2	III	Pro	AGG	AGGG	III R	3, 4, 13
SUF10	III	Pro	AGG	AGGG	XIV L	10, 14
SUF7	III	Pro	UGG	UGGG[c]	XIII L	8–10, 14, 15
SUF8	III	Pro	UGG	UGGG[c]	VIII R	8–10, 14, 15
SUF9	III	Pro	UGG	UGGG[c]	VI L	10, 14, 15
suf11	III	Pro	UGG	UGGG[c]	XV R	1, 10, 14, 15

References: (1) Gaber et al. 1983; (2) Mendenhall et al. 1987; (3) Culbertson et al. 1977; (4) Cummins and Culbertson 1981; (5) Mendenhall and Culbertson 1988; (6) Ball et al. 1988; (7) Culbertson et al. 1980; (8) Cummins et al. 1985; (9) Winey et al. 1986; (10) Cummins et al. 1980; (11) Gaber and Culbertson 1982b; (12) Gaber and Culbertson 1982a; (13) Cummins et al. 1982; (14) Winey et al. 1989b; (15) Curran and Yarus 1987.

[a] Group II suppressors act on +1 frameshifts in glycine codons, and group III suppressors act on +1 frameshifts in proline codons.

[b] The altered tRNA in these cases has been presumed because these suppressors have the same specificity and efficiency as *SUF16*.

[c] Curran and Yarus (1987) have suggested that the anticodon in these cases may actually be UUGG.

ence of the [eta+] factor (All-Robyn et al. 1990b), which causes lethality of certain *sup35* and *sup45* alleles (see section below). In these studies, dominant or semidominant omnipotent suppressors at other loci, namely, *SUP39*, *SUP42*, *SUP43*, *SUP44*, *SUP46*, and *SUP139*, were obtained, in addition to recessive suppressors.

Another approach yielding omnipotent suppressors has been to select for mutations that enhance the efficiency of a tRNA suppressor. Ono et al. (1982) isolated mutations in three genes (*SUP111*, *SUP112*, and *SUP113*) that enhanced the efficiency of the leucine-inserting UAA suppressor *SUP29* in [*psi*$^+$] strains. Subsequently, it was shown (Ono et al. 1986) that certain alleles of these loci act as recessive omnipotent suppressors. Likewise, of five loci (*sal1* through *sal5*) identified as allosuppressors that enhance the efficiency of the serine-inserting UAA suppressor *SUQ5* (Cox 1977), two turned out to be alleles of omnipotent suppressors: *sal3* = *sup35* (Crouzet and Tuite 1987) and *sal4* = *sup45* (Crouzet et al. 1988). Finally, of four mutations (*mos1*, *mos2*, *MOS3*, and *MOS4*) that were isolated by virtue of their ability to extend the codon specificity of the tyrosine-inserting UAA suppressor *SUP4* to UAG codons (see below), two appear to be alleles of omnipotent suppressors: *MOS3* = *SUP46* and *mos1* = *sup45* (Gelugne and Bell 1988).

As described above, suppressor mutations are often isolated on the basis of different phenotypes. Detection of each phenotype may require a specific array of accessory markers. Furthermore, allelic mutations in a given suppressor locus may not have the same phenotypes. For these reasons, it is often difficult to score crosses or complementation tests between different suppressor-bearing strains. In addition, poor sporulation or spore viability sometimes makes analysis of these suppressors difficult. Another approach used to determine if suppressors are allelic is to map their chromosomal positions relative to other markers. Finally, cloning has been used to reveal the identity or distinction between suppressor loci. The known omnipotent suppressors, together with presumed allelic relationships, are listed in Table 11.

Recently, cycloheximide-resistant temperature-sensitive lethal (*crl*) mutants were isolated that defined 22 complementation groups. The mutations are proposed to affect translational fidelity (McCusker and Haber 1988a,b). The *crl* mutations cause hypersensitivity to the aminoglycoside antibiotic hygromycin B that increases misreading, suppress one nonsense mutation, and often cause osmotic or cold sensitivity. These phenotypes are also commonly associated with omnipotent suppressors. The relationship of the *crl* mutations to the defined omnipotent suppressor loci has not been examined.

b. Properties of Omnipotent Suppressors. A large number of *sup35* and *sup45* suppressor alleles have been isolated that differ in their efficiencies of suppression. Many of these suppressors have been shown to exhibit allele-specific pleiotropic phenotypes that generally appear to be unrelated to their suppressor phenotypes (for review, see Surguchov et al. 1984). Most cause poor growth, whereas some cause high- or low-tem-

Table 11 Omnipotent suppressors

Mutation	Possible equivalents	Product	Map position	References
sup35[a]	*sup2*[a], *supP*[a], *sufl2*[b], *sal3*[c], *gst1*[a]?	EF-1α-like factor	IV R	1–7
sup45[a]	*sup1*[a], *supQ*[a], *sal4*[c], *mos1*[d]	not a ribosomal protein	II R	1–3, 8, 9
sup111[c]		?	VIII R	10
sup112[c]		?	VII R	10
sup113[c]		?	XIII R	10
sufl3[b]		?	XV R	7
sufl4[b]		?	XIV L	7
SUP39[a]		?	unknown[e]	11, 12
SUP42[a]		?	IV	15
SUP43[a]		?	XV	13, 15
SUP44[a]	*SUP38*[a], *SUP138*[a]	ribosomal protein S4	VII L	11–13, 15–17
SUP46[a]	*SUP40*[a], *MOS3*[d]	ribosomal protein S13	II R	9, 11, 14, 18
SUP139[a]		?	V R	16

References: (1) Hawthorne and Leupold 1974; (2) Inge-Vechtomov and Andrianova 1970; (3) Gerlach 1975; (4) Wilson and Culbertson 1988; (5) Crouzet and Tuite 1987; (6) Kikuchi et al. 1988; (7) Culbertson et al. 1982; (8) Crouzet et al. 1988; (9) Gelugne and Bell 1988; (10) Ono et al. 1982; Ono et al 1986; (11) Liebman et al. 1988; (12) All-Robyn et al. 1990a; (13) Eustice et al. 1986; (14) Ono et al. 1984; (15) Waekem and Sherman 1990; (16) Ono et al. 1989; (17) All-Robyn et al. 1990b; (18) A. Vincent and S.W. Liebman, unpubl.
[a] Obtained as a suppressor of nonsense codons.
[b] Obtained as a codon-nonspecific frameshift suppressor.
[c] Obtained as an enhancer of a nonsense suppressor.
[d] Obtained as a modifier that alters the codon specificity of nonsense suppressors.
[e] *SUP39* is unlinked to *sup35, sup45, SUP42, SUP43, SUP44, SUP46,* and *SUP139*.

perature sensitivity for growth, osmotic sensitivity, respiratory deficiency, or reduced spore viability. In certain cases, this temperature-sensitive growth cannot be observed in a medium of high osmolarity or one that contains cycloheximide or trichodermin. Increased osmolarity is thought to obscure the temperature-sensitive phenotype by altering the conformation of the mutant protein. Cycloheximide and trichodermin are also postulated to compensate for the mutational defect by inducing a conformational change upon binding to the ribosome.

The aminoglycoside antibiotic paromomycin can induce misreading, causing phenotypic suppression of some markers (Palmer et al. 1979a; Singh et al. 1979). When strains bearing certain *sup35* and *sup45* alleles are grown in the presence of paromomycin, there can be an interaction leading to a higher level of suppression than is associated with either the drug or the suppressor alone. Numerous suppressor alleles cause hypersensitivity to the drug, presumably by causing an intolerable level of misreading (Surguchov et al. 1984).

Alleles of *sup35* and *sup45* can interact with dominant tRNA suppressors in various ways in a haploid cell (Ter-Avanesyan and Inge-Vechtomov 1980; Song and Liebman 1985). For some alleles, there is no visible interaction; for other alleles, the efficiency of the omnipotent suppressor is reduced by the presence of the tRNA suppressor. Still other *sup35* and *sup45* alleles show synthetic lethality when combined with a particular tRNA suppressor. Finally, some alleles of *sup35* and *sup45* are lethal in the presence of certain non-Mendelian factors (see below).

Allele-specific differences in suppression efficiency and paromomycin sensitivity have also been observed for the dominant and semidominant suppressors *SUP42*, *SUP43*, *SUP44*, and *SUP46* (Waekem and Sherman 1990). The single *SUP39* allele also causes paromomycin sensitivity (All-Robyn et al. 1990b). *SUP46* causes the insertion of serine at a UAA site in cytochrome *c* (Ono et al. 1981).

A variety of biochemical data suggest that the translational inaccuracy caused by omnipotent suppressors results from altered ribosomes (Smirnov et al. 1973, 1974, 1976, 1978; Surguchov et al. 1980). For example, in vitro poly(U) translation assays showed that the 40S ribosomal subunits isolated from strains carrying the suppressors *sup35*, *sup45*, *SUP44*, or *SUP46* induce translational misreading (Masurekar et al. 1981; Eustice et al. 1986). Furthermore, all *SUP46* alleles cause the same electrophoretic alteration of the S13 (formerly called S11) protein from the 40S ribosomal subunit (Ishiguro 1981; Ono et al. 1981). This suggests that *SUP46* codes either for S13 or for an enzyme that modifies S13. A different protein from the small subunit of a strain containing a semidominant allele of *SUP35* also had an altered electrophoretic mobility (Eustice et al. 1986). Various protein spots were also missing or fainter on gels of ribosomal proteins isolated from *sup45*, *SUP43*, and *SUP44* strains (Eustice et al. 1986).

c. Cloned Genes. The biochemical data described above suggested that *sup35*, *sup45*, *SUP44*, and *SUP46* were likely to code for components of the 40S ribosomal subunit or enzymes that modify the subunit. Analyses of the first cloned omnipotent suppressors, *sup35* and *sup45*, indicate they are not ribosomal proteins. In addition, the allele-specific phenotypes associated with these two omnipotent suppressors argue against ribosomal modification activities. Recent analyses of the cloned *sup44* and *sup46* genes show that they do encode ribosomal proteins.

The *SUP45* gene has been cloned and shown to be unique and essential (Breining et al. 1984; Himmelfarb et al. 1985; Crouzet et al. 1988). It encodes a 49-kD acidic protein that is larger than the largest known yeast ribosomal protein (Himmelfarb et al. 1985; Breining and Piepersberg 1986). Unlike most ribosomal protein genes, *SUP45* does not contain any introns (Breining and Piepersberg 1986; Surguchov et al. 1986), and its codon usage differs from that typical of yeast ribosomal proteins or other highly expressed genes (Breining and Piepersberg 1986). Although the steady-state level of the 1.5–1.6-kb *SUP45* transcript is less than 10% of that of ribosomal protein mRNAs, *SUP45* transcription appears to be regulated coordinately with that of other genes that encode components of the translational machinery (Himmelfarb et al. 1985). The conserved sequence, HOMOL 1, first found preceding ribosomal protein genes (Leer et al. 1985; Rotenberg and Woolford 1986; Woudt et al. 1986) and now also associated with a variety of other genes (Brand et al. 1987; Shore et al. 1987; Buchman et al. 1988; Kimmerly et al. 1988), is present in the upstream region of *SUP45* (Breining and Piepersberg 1986). Although the *SUP45* sequence contains homologies with aminoacyl-tRNA synthetase and ATPase sequences (Breining and Piepersberg 1986), its function in translation remains unknown.

The *SUP35* gene has also been cloned and shown to be unique and essential (Crouzet and Tuite 1987; Kushnirov et al. 1987; Kikuchi et al. 1988; Wilson and Culbertson 1988). The low level of transcript and the codon bias both suggest that it is not highly expressed (Kushnirov et al. 1988; Wilson and Culbertson 1988). The gene sequence revealed three distinct domains. Stretches rich in glutamine and glycine are repeated in the amino-terminal domain.

The second domain toward the carboxyl terminus contains a high content of charged amino acids. The third domain has extensive homology with the EF-1α family of elongation factors that includes the regions involved in GTP and tRNA binding. Despite the extensive similarity between EF-1α and Sup35p, the two are not interchangeable since both are essential. A major transcript of about 2.3 kb (Surguchov et al. 1986; Crouzet and Tuite 1987; Kikuchi et al. 1988; Kushnirov et al. 1988; Wilson and Culbertson 1988) and a minor transcript of 1.4 kb (Surguchov et al. 1986; Kikuchi et al. 1988) have been reported for the *SUP35* gene. In vitro translation of hybrid-selected mRNA yielded a protein of about 80 kD (Crouzet and Tuite 1987), the size expected for the 2.3-kb mRNA. Although experiments with mutants deficient in splicing indicate that there are no introns in *SUP35* (Surguchov et al. 1986), the possibility of alternative splicing has been suggested because of the presence of conserved sequences typically found in yeast introns (Kushnirov et al. 1988). In addition, the possible translation of two other proteins that might initiate at internal AUGs has been proposed (Kushnirov et al. 1988).

On the basis of gene size, nucleotide sequence, and restriction maps, it appears that the *SUP35* gene is distinct from all of the known elongation factors. Possibly, *SUP35* encodes a new soluble translation factor, present in such low abundance that it was not detected by biochemical methods. Indeed, a novel translation factor, SelBp, has recently been identified in *E. coli* that has extensive homology with EF-Tu. SelBp appears to function as an alternative elongation factor to EF-Tu, specific for the insertion of selenocysteine at certain UGA codons (Forchhammer et al. 1989). How changes in such a factor might cause a change in the electrophoretic mobility of a ribosomal protein remains to be elucidated.

The *sup44* and *sup46* genes have recently been cloned and shown to encode the yeast ribosomal proteins S4 and S13, respectively (All-Robyn et al. 1990a; A. Vincent and S.W. Liebman, unpubl.). The *sup44* gene is essential and exists as a single copy without an intron; *sup46* is one of a duplicated pair of genes and contains an intron. Computer analyses revealed that Sup44p is 59% identical to a mouse protein, encoded by *LLrep3*, whose function was previously unknown (Heller et al. 1988). In addition, the S5 ribosomal proteins from *B. stearothermophilus* (Kimura 1984) and *E. coli* (Wittmann-Liebold and Greuer 1978) are, respectively, 30% and 26% identical to Sup44p. The four proteins most similar to Sup46p are the S4 ribosomal proteins from maize chloroplast (Subramanian et al. 1983), tobacco chloroplast (Shinozaki et al. 1980), liverwort chloroplast (Ohyama et al. 1986), and *E. coli* (Schiltz and Reinbolt 1975). There is 28% identity between Sup46p and the S4 maize chloroplast protein over a 132-amino-acid overlap. The similarity between the yeast Sup44p and Sup46p proteins and the respective *E. coli* S5 and S4 proteins is particularly striking since it often has not been possible to cross-identify homologous ribosomal proteins between prokaryotes (or chloroplasts) and eukaryotes. Of even greater interest is the fact that *ram* (*r*ibosomal *am*biguity) mutations in structural genes of *E. coli* ribosomal proteins S4 and S5 have been shown to increase translational error frequencies (for review, see Piepersberg et al. 1980). In addition, a *SUP44* suppressor mutation was sequenced and found to cause an amino acid substitution of tyrosine for serine in Sup44p that corresponds to an alteration at position 126 in the *E. coli* S5 protein. This alteration is not far from the alterations at positions 103 and 111 in S5 that affect fidelity. Sup44p and Sup46p are the first examples of ribosomal proteins whose function and structure have been conserved between prokaryotes and eukaryotes.

5. Suppressor Modifiers

a. Antisuppressors.   One way to identify genes that affect translation is to obtain mutations that alter the efficiency of translational suppressors. A variety of unlinked antisuppressor mutations have been obtained that lower the efficiency of tRNA suppressors (McCready and Cox 1973; Laten et al. 1978; Hopper et al. 1981). One of these, *los1*, has been shown to cause a defect in tRNA processing. Another, *prt1* (Hopper et al. 1981), may affect an initiation factor, eIF-3, that is involved in the dissociation of 80S ribosomes (see above). The *mod5* antisuppressor is a mutation in the structural gene for Δ^2-isopentenyl pyrophosphate:tRNA isopentenyl transferase, which is required for the isopentylation of both cytoplasmic and mitochondrial tRNAs (Laten et al. 1978; Dihanich et al. 1987; Najarian et al. 1987). Likewise, antisuppressors that affect the modification of tRNAs have been identified in *Schizosaccharomyces pombe* (Janner et al. 1980; Heyer et al. 1984; Grossenbacher et al. 1986).

Antisuppressors that reduce the efficiency of omnipotent suppressors have also been obtained: *asu9*, *ASU10*, *asu11*, and *ade3* (Liebman and Cavenagh 1980; Liebman et al. 1980; Ishiguro 1981; Song and Liebman 1989). Two of these, *asu9* and *asu11*, resulted in an altered two-dimensional electrophoretic pattern of ribosomal proteins (Liebman and Cavenagh 1980; Ishiguro 1981). The mechanism by which *ade3* affects suppression efficiency is unknown.

b. Allosuppressors.   Several allosuppressor mutations that enhance the efficiency of unlinked tRNA suppressors have been isolated: *sal1-sal6* (Cox 1977; Song and Liebman 1987); *sup111*, *sup112*, and *sup113* (Ono et al. 1982, 1986); and *upf1* and *upf2* (Culbertson et al. 1980). Alleles of *sal2* and *sal6* were also shown to increase the efficiency of omnipotent suppressors (Song and Liebman 1987). In addition, the *sal2-B* allele (Song and Liebman 1987) and the *mos1*, *mos2*, *MOS3*, and *MOS4* mutations extend the action spectra of ochre-specific suppressors, enabling them to suppress amber mutations (Gelugne and Bell 1988). These mutations are candidates for identifying tRNA modification enzymes. Some of these allosuppressors (see above) also have omnipotent suppressor activity. Finally, extra copies of the wild-type *STP1* gene have been found to increase the efficiency of a defective *SUP4* suppressor. *STP1* is thought to encode a product that is involved in pre-tRNA splicing (Wang and Hopper 1988) and that is specific for certain pre-tRNAs.

c. Non-Mendelian Factors.   In addition to the Mendelian modifiers described above, non-Mendelian factors have been found to alter the activity of suppressors. The best studied of these, the [*psi*+] factor, is the subject of a recent detailed review (Cox et al. 1988) and will be discussed only briefly here. The in vivo efficiencies of various ochre, frameshift, and UGA suppressors are increased by [*psi*+], although amber suppressors are not affected (Cox 1965, 1971; Liebman et al. 1975; Culbertson et al. 1977; Liebman and Sherman 1979; Ono et al. 1979a,b, 1981, 1988; Cummins et al. 1980; E.F. Schmitt and M.V. Olson, pers. comm. cited in Cox et al. 1988). Likewise, phenotypic in vivo suppression caused by the aminoglycoside antibiotic paromomycin is also enhanced by [*psi*+] (Palmer et al. 1979a,b). In vitro, suppression of all nonsense alleles is more efficient when extracts are prepared from [*psi*+] compared with [*psi*−] strains (Tuite et al. 1981, 1983).

Another factor, [*eta*+], has many similarities to [*psi*+] (Liebman and All-Robyn 1984). Both are inherited in a non-Mendelian fashion and can be cured by guanidine hydrochloride (Lund and Cox 1981; Liebman and All-Robyn 1984). However, the two factors differ in their pattern of inheritance. Whereas [*psi*+] is inherited by all meiotic products (Cox 1965, 1971), [*eta*+] segregates irregularly in meiosis.

Some alleles of the omnipotent suppressors *sup35* and *sup45* are lethal in [*eta*+] backgrounds. The efficiencies of *sup35* and *sup45* alleles that are not lethal in combination with [*eta*+] are unaffected by loss of the factor (All-Robyn et al. 1990b). Although [*psi*+] did not affect the efficiency or viability of any tested alleles of *sup35* and *sup45* (Liebman and Cavenagh 1980; Liebman and All-Robyn 1984; Ono et al. 1984; B.S. Cox, unpubl. cited in Cox et al. 1988), all alleles of *sup35* and *sup45* isolated as *sal3* or *sal4* mutations were lethal in [*psi*+] backgrounds (Cox 1977). These differences in sensitivities to the [*psi*+] factor may represent allele-specific effects; another possible reason for this discrepancy is that the *sal* strains may also be [*eta*+].

As mentioned above, [*psi*+] increases the efficiency of ochre-specific tRNA suppressors. In contrast, [*eta*+] has no such activity (Liebman and All-Robyn 1984), although it does increase the efficiencies of the omnipotent suppressors *SUP39*, *SUP44*, and *SUP46* (All-Robyn et al. 1990b). Despite these differences and because the molecular nature of the [*psi*+] and [*eta*+] factors is unknown, it is possible that they represent different forms of the same factor.

6. Informational Suppression in Mitochondria

Mitochondria use their own translational apparatus to synthesize the polypeptides they encode. Thus, appropriate alterations in this apparatus could cause translational suppression of mitochondrial mutations. Since some translational components (factors, synthetases, ribosomal proteins) are encoded by nuclear genes, and other translational components (23S rRNA, 15S rRNA, numerous tRNAs, one ribosomal protein) are encoded by mitochondrial genes, translational suppressors might map to either the nuclear genome or the mitochrondrial genome.

Polypeptides encoded by mitochondrial DNA (e.g., subunits of cytochrome *b*, cytochrome oxidase, and ATPase) often contain introns that themselves encode proteins, called maturases, required for splicing pre-mRNA. Nonsense and frameshift mutations have been identified within the structural genes and maturases (Fox and Weiss-Brummer 1980; Fox and Staempfli 1982; Dujardin et al. 1984), and nuclearly (*nam*) and mitochondrially (*mim*) encoded suppressors of these mutations have been isolated (Dujardin et al. 1980; Groudinsky et al. 1981). Several of these conform to the definition of informational suppressors, since they are allele-specific but not locus-specific.

Nucleus-encoded informational suppressors include *nam1* (Groudinsky et al. 1981), *nam3* (Kruszewska and Slonimski 1984a,b), and *R705* (Zoladek et al. 1985). *NAM1* has recently been cloned, sequenced, and mapped to chromosome IV. A gene-disruption experiment showed that it is required for overall mitochondrial protein synthesis, as well as for splicing reactions specific for the COX I pre-mRNA (Ben Asher et al. 1989). The *nam3* and *R705* mutations are unlinked and each causes changes in the 37S mitochondrial ribosomal subunit (Boguta et al. 1988).

Mitochondrion-encoded informational suppressors include *MIM1-1* (Dujardin et al. 1980), *mim3-1* (Kruszewska and Slonimski 1984a,b), *MSU1* (Fox and Staempfli 1982), and *MF1* (Weiss-Brummer et al. 1987). *MIM1* maps to the 23S rDNA gene, and *mim3*, *MSU1*, and *MF1* all map to the 15S rDNA gene. The *MSU1* suppressor contains a substitution of an invariant nucleotide within the highly conserved "530-loop," suggesting that this region of 15S rRNA has an important influence on translational accuracy (Shen and Fox 1989).

III. GENE-SPECIFIC TRANSLATIONAL CONTROL IN *S. CEREVISIAE*

A. Translational Control by uORFs

1. CPA1 Expression Regulated by a Single uORF

CPA1 encodes an enzyme in the arginine biosynthetic pathway that is repressed about fivefold by arginine. RNA blot-hybridization studies revealed little or no change in the steady-state level of *CPA1* mRNA upon arginine addition, suggesting that enzyme repression occurs at the translational level (Messenguy et al. 1983). The *CPA1* transcript contains

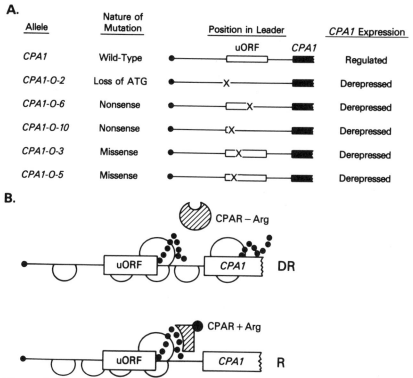

Figure 8 Translational control of *CPA1* expression. (*A*) Effects of point mutations in the uORF on expression of *CPA1* enzyme activity. The positions of point mutations in the uORF are indicated by Xs. (*B*) Hypothetical model for uORF-mediated translational repression. Most scanning 40S subunits bypass the uORF start codon but initiate efficiently at the *CPA1*-coding sequences. Translation of the uORF is without consequence in the absence of arginine (–Arg). In the presence of arginine, a *trans*-acting factor designated CPAR is altered and acts in conjunction with the leader peptide to arrest the progression of ribosomes translating the uORF, which in turn blocks the advance of scanning 40S subunits to the *CPA1* start codon. The structure of the peptide plays a critical role in stalling the progression of elongating ribosomes.

a single uORF of 25 codons (Nyunoya and Lusty 1984; Werner et al. 1985) and removal of the uORF start codon leads to derepression of *CPA1* enzyme expression in the presence of excess arginine, while apparently having little effect on the *CPA1* mRNA level. Interestingly, a variety of missense and nonsense mutations mapping throughout the uORF also lead to derepressed *CPA1* expression in a *cis*-dominant fashion (*CPA1-O* mutations) (Fig. 8A) (Werner et al. 1987).

Given their various positions within the uORF, it seems unlikely that the *CPA1-O* mutations impair regulation by interfering with recognition of the uORF AUG codon as a translational start site. An alternative explanation is that the peptide product of the 25-codon uORF functions in translational repression. Because the *CPA1-O* mutations are only expressed in *cis*, the putative leader peptide would necessarily be confined to regulating the translation of its own mRNA. Another important aspect of the *CPA1* uORF is that removal of its AUG codon does not lead to in-

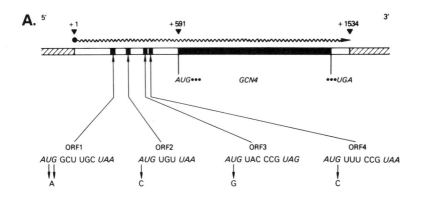

B.

GCN4 UPSTREAM ORFs	GCN4-lacZ Enzyme Activity (U)					
	wt		gcn2		gcd1	
	R	DR	R	DR	R	DR
■ ■ ■ ■	10	80	5	6	280	340
X X X X	740	470	1300	1300	1400	700
■ X X X	240	390	560	310	580	790
X X ■ X	13	30	52	60	59	104
X X X ■	5	12	6	11	15	16
■ ■ X X	110	290	150	190	370	520
■ X ■ X	23	88	26	35	290	360
■ X X ■	14	97	9	21	325	380

C.

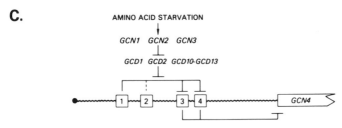

Figure 9 (See facing page for legend.)

creased *CPA1* expression under derepressing conditions, implying that during normal growth conditions in medium lacking arginine, the *CPA1* upstream AUG codon has little or no effect on initiation at the *CPA1* start site. This finding was interpreted to indicate that most scanning 40S subunits bypass the uORF AUG codon and initiate at *CPA1* downstream. To account for the regulatory function of the leader peptide, it was proposed that in the presence of arginine, the small amount of peptide produced blocks initiation at the *CPA1*-coding sequences by the majority of 40S subunits that bypass the uORF AUG codon. For example, the peptide could arrest the movement of 80S ribosomes translating the uORF, impeding the flow of scanning 40S subunits through the uORF to the *CPA1* start codon. A *trans*-acting factor encoded by the *CPAR* gene might function in conjunction with the leader peptide to regulate translation (Fig. 8B) (Werner et al. 1987).

A closely related model could be proposed in which the mRNA sequence of the uORF adopts a conformation in the presence of arginine that arrests 80S ribosomes translating the uORF. In this scheme, the *CPA1-O* mutations would prevent formation of the inhibitory RNA structure. To help distinguish between these two possible modes of action for the uORF, it would be useful to determine the phenotype of substitution mutations in the uORF that do not change the amino acid sequence of the predicted leader peptide. These mutations should have little effect on *CPA1* expression if the leader peptide is the critical regu-

Figure 9 Regulation of *GCN4* expression by uORFs in *GCN4* mRNA. (*A*) Schematic of the *GCN4* transcription unit (*wavy arrow*) shown above the protein-coding sequences (*large closed rectangle*) and the four uORFs in the leader (*small closed boxes*, ORFs 1–4). The sequences of the uORFs are given below, along with the point mutations constructed in their ATG codons. (*B*) The effects of point mutations in the upstream ATG codons (Xs) on expression of *GCN4-lacZ* fusion enzyme activity in wild-type (wt), *gcn2*, and *gcd1* mutant cells under nonstarvation conditions (repressing, R) or under conditions of histidine starvation (derepressing, DR). (*C*) A formal model summarizing the functional interactions between uORFs in the *GCN4* mRNA leader and the way in which these interactions are modulated by positive (*GCN*) and negative (*GCD*) *trans*-acting factors in response to amino acid levels. uORFs 3 and 4 are needed for efficient repression of *GCN4* expression under nonstarvation conditions. In response to starvation, translation of uORF1 (and to a lesser extent uORF2) overcomes the barrier to initiation at the *GCN4* start codon exerted by uORFs 3 and 4, leading to increased synthesis of Gcn4p. The *GCD* factors prevent this interaction between the uORFs under nonstarvation conditions; they are, in turn, antagonized or repressed under amino acid starvation conditions by the products of *GCN1*, *GCN2*, and *GCN3*.

latory element but should have large effects if RNA structure is important.

Posttranscriptional control has been postulated for a separate arginine-specific repression mechanism that regulates several other arginine biosynthetic genes, including *ARG3* and *ARG1* (Messenguy and Dubois 1983; Crabeel et al. 1988); however, more recent measurements of the rates of mRNA synthesis and turnover indicate that arginine repression of these genes operates primarily at the transcriptional level (Crabeel et al. 1990; see chapter by Hinnebusch, other volumes). In the same experiments, *CPA1* mRNA synthesis was found to be repressed by arginine approximately twofold, indicating that arginine-specific regulation of *CPA1* expression may have both transcriptional and translational components.

2. Translational Control of GCN4 Expression by Multiple uORFs

a. Regulation of GCN4 *Expression.* Gcn4p is a transcriptional activator of a large number of genes encoding enzymes in several amino acid biosynthetic pathways. Gcn4p stimulates transcription of these genes in response to starvation for any amino acid (the general amino acid control). Analysis of *GCN4-lacZ* fusions suggests that the rate of Gcn4p synthesis itself is regulated by amino acid availability, occurring at low levels under nonstarvation conditions and increasing 10- to 50-fold upon starvation for a single amino acid. This derepression requires the products of the positive regulatory genes, *GCN1*, *GCN2*, and *GCN3* (Thireos et al. 1984; Hinnebusch 1985), that were first identified by mutations that impair the derepression of structural genes under *GCN4* control (Hinnebusch 1988). Negative regulatory factors in this system, encoded by *GCD1*, *GCD2*, *GCD10*, *GCD11*, and *GCD13*, function as repressors of *GCN4* expression under nonstarvation conditions. Gcn1p, Gcn2p, and Gcn3p are thought to act indirectly as positive regulators by antagonism of one or more *GCD* factors (Fig. 9C) because *gcd* mutations eliminate the requirement for these *GCN* products for derepression (Table 12) (Hinnebusch 1985, 1988; Mueller et al. 1987).

Regulation of *GCN4* expression has a large translational component. One piece of evidence leading to this conclusion is that *gcn2* and *gcn3* mutations impair the derepression of *GCN4-lacZ* enzyme activity under starvation conditions without reducing the steady-state amount of the fusion transcript. A second indication is that regulation of *GCN4-lacZ* enzyme expression by *GCN* and *GCD* factors remains intact following replacement of the *GCN4* promoter with that of the *GAL1* gene (Hinnebusch 1985; Mueller et al. 1987). (*GAL1* transcription is not subject to general amino acid control.)

Table 12 Expression of *GCN4-lacZ* fusions in general control regulatory mutants

| Strain | Units of β-galactosidase activity[a] | | | |
| | *GCN4-lacZ* | | *GCN4* (ΔuORFs)-*lacZ* | |
	R	DR	R	DR
Wild-type	10	105	550	470
gcn2-1	2	4	1000	1900
gcn3-102	5	15	420	630
gcd1-101	360	270	1600	1000
gcn2-1 gcd1-101	250	240	1100	940

[a] *GCN4* (ΔuORFs)-*lacZ* differs from the conventional *GCN4-lacZ* fusion by a deletion of about 240 bp in the leader of *GCN4* mRNA. The deleted sequences contain the four uORFs. R and DR refer to repressing (nonstarvation) or derepressing (starvation) conditions, respectively. (Data from Hinnebusch 1985.)

Insight into the mechanism of *GCN4* translational control came with the finding that *GCN4* mRNA has a long leader containing four AUG codons, each of which initiates a uORF of two or three codons (Fig. 9A) (Hinnebusch 1984; Thireos et al. 1984). Internal deletions that remove all four uORFs in the *GCN4* mRNA leader result in constitutive derepression of *GCN4-lacZ* enzyme expression (Table 12). As expected, amino acid biosynthetic enzymes under *GCN4* control were derepressed when the same deletions were introduced upstream of the authentic Gcn4p-coding sequences. These mutations have little or no effect on the steady-state levels of *GCN4* or *GCN4-lacZ* mRNAs, showing that the deleted leader sequences affect *GCN4* expression at the translational level (Hinnebusch 1984, 1985; Thireos et al. 1984).

Mutations in *GCD* genes cause little additional increase in *GCN4-lacZ* enzyme expression once the uORFs are removed, suggesting that the *GCD* factors mediate the inhibitory effects of the leader sequences under nonstarvation conditions. In *gcn1*, *gcn2*, and *gcn3* mutants, deletion of the uORFs produces the same high levels of *GCN4* expression as it does in wild-type cells, implying that Gcn1p, Gcn2p, and Gcn3p function to overcome the inhibitory effects of the leader sequences under amino acid starvation conditions (Table 12) (Hinnebusch 1984, 1985; Thireos et al. 1984; Mueller et al. 1987). The target of the *GCN* and *GCD* regulatory factors was narrowed down to the small sequence interval in *GCN4* mRNA containing the four uORFs by showing that translational control by these factors could be conferred upon a heterologous yeast mRNA by insertion of only this small segment of the *GCN4* mRNA leader into the 5′ end of the heterologous transcript (Mueller et al. 1987).

b. mRNA Sequence Requirements for Translational Control of GCN4.
Direct proof that the upstream AUG codons are responsible for translational control of *GCN4* expression came from the fact that removal of
these sequences by point mutations leads to the same high constitutive
GCN4-lacZ enzyme activity as a deletion of the four uORFs (Fig. 9B). In
addition, by constructing alleles containing different combinations of upstream ATG codons, it became clear that the various uORFs play different roles in the control mechanism. The third or fourth ATG codon
(counting from the 5′ end) is necessary for efficient repression of *GCN4*
expression under nonstarvation conditions. In contrast, the first and second ATG codons are relatively weak negative elements when present
alone in the mRNA leader. In fact, when ATG codon 3 or 4 is present,
the first ATG codon acts as a positive control element, being required for
efficient *GCN4* expression under starvation conditions. (The second
ATG codon also acts as a positive element in this situation, but to a much
lesser degree.)   Only when the first ATG codon is present upstream of
ATG codon 3 or 4 is there a strong requirement for *GCD* gene products
to efficiently repress *GCN4* expression. These findings led to the suggestion, summarized in Figure 9C, that (1) recognition of the first (and second) AUG codons overcomes the inhibitory effects of the third and
fourth AUG codons and (2) this interaction is prevented from occurring
under nonstarvation conditions by the *GCD* regulatory factors (Mueller
and Hinnebusch 1986; Tzamarias et al. 1986; Mueller et al. 1987).

In large part, deletions that remove a subset of the four uORFs have
the same phenotype as point mutations in the corresponding ATG
codons: (1) Deletions of uORFs 3 and 4 increase *GCN4-lacZ* expression
under repressing conditions but have much less effect under derepressing
conditions, as would be expected if they removed *cis*-acting negative elements (Fig. 10A, A+C) (Williams et al. 1988), and (2) deletions of
uORF1, or a combination of uORFs 1 and 2, reduce *GCN4-lacZ* expression only under derepressing conditions, as expected if they remove a
positive control site (Fig. 10A, construct D) (Tzamarias et al. 1986; Williams et al. 1988). Deletions of uORFs 2 and 3 have greater quantitative
effects on *GCN4* expression than removal of ATG codons 2 and 3 by
point mutations. However, even when all but 25 bp that normally separate uORFs 1 and 4 were deleted, reducing the derepression ratio to
about 5, additionally removing the uORF1 ATG codon further impaired
the ability to increase *GCN4* expression under derepressing conditions
(Fig. 10A, E+F and E′+F). Thus, uORF1 can still function as a positive
control site in the absence of most nucleotides normally found between
uORFs 1 and 4. A sizable deletion of sequences upstream of uORF1 had
little effect on *GCN4* expression (Fig. 10A, S/G), indicating that this por-

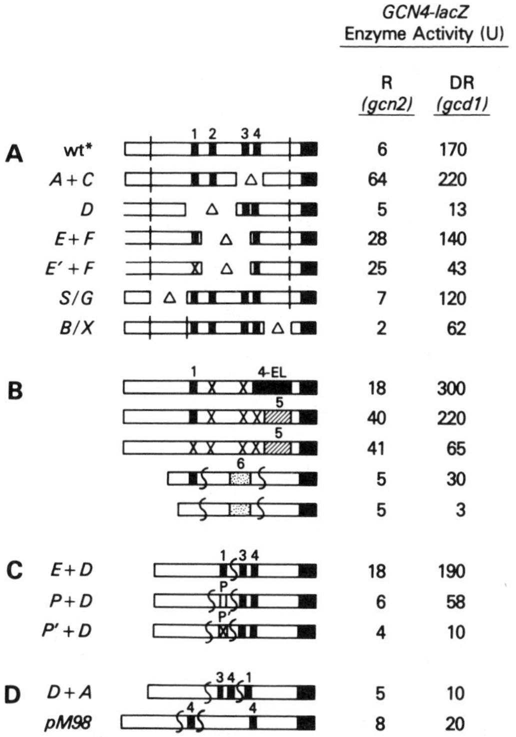

GCN4-lacZ
Enzyme Activity (U)

			R *(gcn2)*	DR *(gcd1)*
A	wt*		6	170
	A + C		64	220
	D		5	13
	E + F		28	140
	E' + F		25	43
	S/G		7	120
	B/X		2	62
B			18	300
			40	220
			41	65
			5	30
			5	3
C	E + D		18	190
	P + D		6	58
	P' + D		4	10
D	D + A		5	10
	pM98		8	20

Figure 10 Sequence and positional requirements of the uORFs for *GCN4* translational control. The uORFs (and point mutations in their ATG codons, Xs) are shown in the leader as in Fig. 9B, with deletions of leader sequences shown as gaps marked with open triangles. Vertical lines in the leader indicate restriction sites introduced to facilitate construction of deletions. *GCN4-lacZ* enzyme activity was measured in *gcn2* transformants (repressed conditions, R) or in *gcd1* transformants (derepressed conditions, DR). (*B*) 4-EL designates an elongated version of uORF4, 46 codons in length. uORF5 (*hatched*) is a 43-codon uORF constructed by insertion of a linker containing an ATG codon downstream from the normal location of uORF4. uORF6 (*stippled*) is an eight-codon uORF contained in a ~100-bp fragment of a sea urchin tubulin cDNA inserted in place of uORFs 1–4 or 2–4, the latter being removed by deletions. (*C*) P designates a three-codon uORF containing the initiation region of the *Saccharomyces PGK1* gene inserted on a 20-bp fragment in place of uORF1, the latter being removed by a deletion. P' is identical to the P sequence except that it contains a point mutation in the ATG codon of the P uORF. (*D*) D+A is a rearrangement that places a ~60-bp segment containing uORF1 downstream from uORFs 3–4; in pM98, a ~50-bp segment containing uORF1 was substituted with a fragment of the same size containing uORF4.

693

tion of the leader is not required for regulation (Williams et al. 1988). The fact that large deletions of leader sequences can be made without destroying translational control suggests that long-range mRNA structure is not an essential aspect of the *GCN4* regulatory mechanism.

Much of the leader between uORF4 and the *GCN4* start site was deleted without any effect on the derepression ratio; however, the inhibitory effect of the uORFs was increased by this deletion, resulting in lower absolute levels of *GCN4* expression under both repressing and derepressing conditions (Fig. 10A, B/X). Similarly, uORF1 was found to be more inhibitory as a solitary element when moved closer to the *GCN4* start site (Tzamarias and Thireos 1988; Williams et al. 1988), and uORF4 was less inhibitory when inserted upstream in the position normally occupied by uORF1 (see Fig. 5E) (Miller and Hinnebusch 1989). These observations recall the fact that inhibition of preproinsulin synthesis by a uORF decreased as the uORF was moved farther upstream of the preproinsulin start codon. To explain this trend, it was suggested that the probability of reinitiation following translation of the uORF is greater for larger separations between the uORF stop codon and the preproinsulin start site (Kozak 1987b). Alternatively, ribosomes translating the uORF may sterically hinder initiation at the next start site when the two AUG codons are very close together. The latter mechanism is improbable for *GCN4* mRNA because the effects of uORF proximity on *GCN4* expression have been observed for separations between the two initiation sites of 100 nucleotides or more. Thus, it seems likely that reinitiation can occur on *GCN4* mRNA, at least following translation of uORF1. As discussed below, the ability to allow reinitiation events downstream appears to be important for the novel positive regulatory function of uORF1.

c. Translational Control with Heterologous uORFs. The *GCN4* 3'-proximal uORFs were substituted with heterologous coding sequences without destroying the important qualitative features of *GCN4* translational control. The stop codons of uORF 3 or 4 were eliminated, lengthening these elements from 3 to 52 or 43 codons, respectively, with little or no effect on *GCN4* expression (see, e.g., Fig. 10B). In addition, a completely heterologous 43-codon uORF (uORF5) was inserted in place of uORFs 3 and 4, which when present alone, reduced *GCN4-lacZ* expression to a low constitutive level. Introduction of uORF1 upstream reduced the inhibitory effect of uORF5 and increased *GCN4-lacZ* expression about fivefold under derepressing conditions (Fig. 10B). Under the same circumstances, uORF1 overrides the inhibitory effect of authentic uORF3 to about the same extent (Mueller et al. 1988). Similar results were obtained with a 9-codon heterologous uORF (uORF6 in Fig. 10B) inserted downstream from uORF1 (Tzamarias and Thireos 1988).

The positive regulatory function of uORFs 1 and 2 was also reconstituted with a heterologous uORF (Williams et al. 1988). A segment containing uORFs 1 and 2 was deleted and replaced by a synthetic oligonucleotide containing a uORF with the first three codons and −1 to −7 nucleotides found at the highly expressed *PGK1* gene, which corresponds to the consensus sequence of initiation regions from highly expressed yeast genes (Hamilton et al. 1987). As expected, when present alone in the leader, this heterologous uORF greatly reduced *GCN4* expression under repressing conditions (by 95%); but more importantly, it mimicked uORF1 and stimulated *GCN4* expression under derepressing conditions when inserted upstream of uORFs 3 and 4 (Fig. 10C, P+D).

The fact that heterologous uORFs qualitatively mimic the regulatory functions of the authentic *GCN4* uORFs supports the idea that these elements participate in the regulatory mechanism as translated coding sequences. It also implies that no strict requirements exist for the nucleotide sequences or secondary structures of the uORFs for regulation per se. On the other hand, the derepression ratio for *GCN4* expression varies greatly with different combinations of uORFs, showing that particular nucleotides associated with these elements are very important in determining the efficiency of translational control.

d. Different Roles for uORFs 1 and 4 in GCN4 *Translational Control.* The importance of sequence differences between the uORFs was confirmed by showing that the 5′- and 3′-proximal uORFs are not functionally interchangeable. Inserting a segment containing uORF1 downstream from uORFs 3 and 4 resulted in low constitutive *GCN4* expression, indistinguishable from a deletion of uORF1 (Fig. 10D, D+A) (Williams et al. 1988). Similarly, replacement of a segment containing uORF1 with a segment of the same length containing uORF4 (producing an allele with two uORF4 sequences) reduced *GCN4* expression to the low constitutive level observed when uORF4 is present alone in the leader (Fig. 10D, pM98) (Miller and Hinnebusch 1989). The simplest explanation for these results is that virtually no 40S subunits can scan downstream from uORFs 3 and 4 when these sequences are present as the 5′-proximal start sites. In contrast, uORF1 is a very leaky translational barrier, reducing the number of scanning 40S subunits that reach downstream start sites by only about one-half (Mueller and Hinnebusch 1986).

The particular nucleotides at uORF4 that prevent it from functioning as a positive regulatory element were identified by making hybrids be-

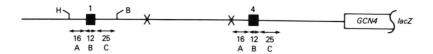

| | Origin of uORF1 Segments | | | Units of Enzyme Activity | | | |
| | | | | Constructs Containing uORF4 | | Constructs Lacking uORF4 | |
Construct	A	B	C	gcn2⁻ (R)	gcd1⁻ (DR)	gcn2⁻ (R)	gcd1⁻ (DR)
pM23	1	1	1	15	200	410	510
pM98	4	4	4	8	20	38	61
pM101	4	1	1	19	230	370	580
pM103	1	4	1	15	160	180	300
pM102	4	4	1	13	106	170	170
pM100	1	1	4	10	56	120	220
pM99	1	4	4	5	22	42	62

Figure 11 Sequences surrounding the stop codons of uORFs 1 and 4 distinguish the functions of these elements in translational control. (*Top*) Schematic of the *GCN4-lacZ* construct. (*Closed boxes*) uORFs 1 and 4; (X) mutations in the ATG codons of uORFs 2 and 3. H and B designate restriction sites employed in cassette mutagenesis of uORF1. Arrows indicate the position and length in base pairs of the A, B, and C segments that were replaced at uORF1 with the corresponding sequences from uORF4. (*Bottom*) *GCN4-lacZ* fusion enzyme activity was measured in extracts of a *gcn2-1* mutant that is constitutively repressed (R in parentheses) and in a *gcd1-101* mutant that is constitutively derepressed (DR in parentheses) for expression of wild-type *GCN4-lacZ*. Two sets of constructs were examined that differ in whether or not uORF4 is present downstream from the uORF1/uORF4 hybrids.

tween uORF1 and uORF4 (Fig. 11). Replacing the 16 bp upstream of uORF1 with the corresponding sequence from uORF4 had little effect on *GCN4* expression, suggesting that uORFs 1 and 4 have very similar initiation efficiencies. In fact, the same conclusion was reached independently by showing that *lacZ* fusions to uORFs 1 and 4 each express high constitutive levels of enzyme activity in the absence of other uORFs, comparable to that seen for the *GCN4-lacZ* fusion lacking all four uORFs (Mueller et al. 1988; Tzamarias and Thireos 1988). In contrast, replacement of the 25 nucleotides immediately following uORF1 with the corresponding sequence from uORF4 significantly lowered *GCN4* expression under derepressing conditions. Additionally replacing the coding sequence of uORF1 with that from uORF4 completely abolished the ability of uORF1 to derepress *GCN4* expression (Fig. 11). It was subsequently shown that introduction of only ten nucleotides immediately 3' to uORF4, plus the rare proline codon immediately preceding the uORF4 stop codon, is sufficient to destroy uORF1-positive function (Miller and Hinnebusch 1989). These results suggest that translation terminates differently at uORFs 1 and 4, and the way termination occurs at uORF4 is incompatible with the positive regulatory function of uORF1.

Introduction of sequences from the uORF4 termination region also makes uORF1 a stronger translational barrier when it occurs singly in the mRNA leader, whereas introduction of the nucleotides found upstream of uORF4 has no effect on the barrier function of uORF1 (Fig. 11) (Miller and Hinnebusch 1989). Thus, it appears that wild-type uORF1 blocks translation downstream less than uORF4 does, not because of different initiation efficiencies at the two start sites, but because of a greater probability for reinitiation following translation at uORF1 versus uORF4.

The fact that reducing the reinitiation potential of uORF1 destroys its ability to stimulate *GCN4* expression when uORF4 is present downstream is thought to indicate that ribosomes must first translate uORF1 and resume scanning in order to traverse uORF4 sequences. To account for this requirement, it was proposed that ribosomes (or 40S subunits) emerge from translation of uORF1 with a different complement of factors than were present during initiation at this site. For example, factors that normally associate with the 40S subunit only at the 5′ end of the mRNA might be removed from the ribosome during translation of uORF1. The absence of these factors during the reinitiation process, coupled with reduced *GCD* function under starvation conditions, would alter the behavior of 40S subunits when they reach uORF4, allowing a fraction either to scan past the uORF4 AUG codon without initiating translation or to reinitiate more efficiently following uORF4 translation (Fig. 12) (Williams et al. 1988). The important requirements for uORF1 function in this model are efficient translation plus a high probability for resumed scanning following termination. A modified version of this model was proposed in which reinitiation following uORF1 translation is restricted to starvation conditions by its dependence on positive regulators of *GCN4* expression, such as Gcn2p (Tzamarias and Thireos 1988). In this version of the model, a reinitiating ribosome generated by translation of uORF1 needs no further alteration by *trans*-acting factors to advance through uORF4.

The idea that a particular sequence context at the stop codon of uORF1 is required for efficient reinitiation downstream can also explain the fact that all mutations that move the uORF1 stop codon upstream or downstream by one or more codons impair the positive regulatory function of uORF1 (Mueller et al. 1988; Miller and Hinnebusch 1989). In addition, the negative effects on *GCN4* expression associated with a variety

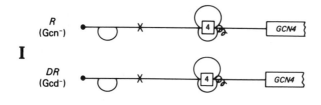

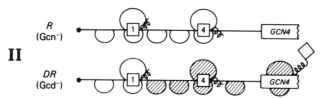

Figure 12 Hypothetical molecular mechanism for the interactions between uORFs 1 and 4 in controlling translation initiation at the *GCN4* start site. The *GCN4* mRNA is shown schematically with only uORFs 1 and 4 present upstream of the *GCN4* coding sequences. 40S subunits and 80S ribosomes are indicated. In the first two cases, uORF1 is missing due to a point mutation in its ATG codon. As a result, all 40S subunits scan to uORF4 and initiate translation. Following translation termination at uORF4, no reinitiation occurs at the *GCN4* start site under repressing (R, *gcn2*) or derepressing (DR, *gcd1*) conditions. When uORF1 is present upstream (third and fourth mRNAs), most ribosomes translate uORF1 and scanning can resume following termination at this site. Under repressing conditions, reinitiation occurs at uORF4 but no reinitiation follows termination at uORF4. Under derepressing conditions, the behavior of 40S subunits that engaged in prior translation of uORF1 is altered as the result of reduced *GCD* factor activity in the cell (*hatched* subunits). These altered initiation complexes are able to traverse uORF4 and reinitiate at the *GCN4* start site, either by skipping over the uORF4 AUG codon or by more efficient reinitiation at *GCN4* following uORF4 translation (Williams et al. 1988; Miller and Hinnebusch 1989). In a related model (Tzamarias and Thireos 1988), scanning resumes after uORF1 translation only under derepressing conditions and the reinitiation complexes thus generated traverse uORF4 by either of the two mechanisms described above.

of single-codon insertions in the middle of uORF1 (Miller and Hinnebusch 1989) could be explained by proposing that certain initiation factors remain associated with the ribosome for only the first few elongation steps and that their persistence following termination at uORF1 increases the probability of reinitiation events downstream. Of interest in this connection is the fact that certain antibiotics (e.g., verrucarin A) inhibit elongation only during formation of the first few peptide bonds, suggesting a mechanistic difference between early and advanced elongation steps and a possible overlap between initiation and elongation processes (Vazquez 1979; Warner 1983).

e. Initiation Rates at the 5′- and 3′-proximal uORFs. Expression of uORF3-*lacZ* and uORF4-*lacZ* fusions was examined in an attempt to identify the translational event that is regulated at these 3′-proximal uORFs by prior translation of uORFs 1 and 2. For example, translation of uORFs 1 and 2 could reduce the translational block at uORFs 3 and 4 by suppressing initiation at these sites without equally reducing initiation at the *GCN4* start site. Such a differential effect on initiation rates could be related to the closer proximity of uORF1 to uORFs 3 and 4 versus the *GCN4* start codon (Kozak 1987b). This would explain the fact that greatly elongating uORFs 3 and 4 has almost no effect on *GCN4* expression (Mueller et al. 1988). According to this model, translation of uORFs 3 and 4 would be reduced under derepressing conditions when translation of the *GCN4*-coding sequences increases. At odds with this expectation are the results of two different studies indicating that translation of uORF3-*lacZ* and uORF4-*lacZ* fusions is higher under derepressing conditions than under repressing conditions when uORFs 1 and 2 are present upstream (Hinnebusch et al. 1988; Tzamarias and Threos 1988). On the basis of this result, it was proposed that an event associated with termination at uORFs 3 and 4 is regulated to allow greater numbers of ribosomes to reinitiate at the *GCN4* start site under derepressing conditions. For example, 40S subunits might be more likely to remain attached to the mRNA and resume scanning following translation termination at uORFs 3 and 4 (Hinnebusch et al. 1988). However, an important shortcoming in these experiments is that insertion of *lacZ* sequences into uORFs 3 and 4 could have disrupted an important structure required to regulate initiation at these sites. One way to eliminate this possibility would be to measure translation from the authentic 3′-proximal uORFs or from modified versions of the uORFs that are still competent as translational control elements.

f. Mutations Affecting eIF-2 Resemble gcd *Mutations in Derepressing* GCN4 *Expression.* Deletions of the positive regulatory genes *GCN2* and *GCN3* have no effects on cell growth under nonstarvation conditions (Hannig and Hinnebusch 1988; Roussou et al. 1988; Wek et al. 1989), suggesting that these factors are specifically involved in general amino acid control. In contrast, nearly all *gcd* mutations are pleiotropic and lead to conditional lethality at 36°C or unconditional slow growth, irrespective of amino acid levels (Wolfner et al. 1975; Harashima and Hinnebusch 1986; Niederberger et al. 1986). The rates of incorporation of radioactive precursors into DNA, RNA, and protein are reduced, and the cells arrest early in the G_1 phase of the cell cycle when *gcd1* and *gcd2* mutants are shifted to the restrictive temperature (Wolfner et al. 1975; Harashima and Hinnebusch 1986; Hill and Struhl 1988). Moreover, deletions of *GCD1* and *GCD2* are unconditionally lethal (Hannig and Hinnebusch 1988; Hill and Struhl 1988; Paddon and Hinnebusch 1989). These results suggest that *GCD1* and *GCD2* contribute to an essential function required for entry into the cell cycle. Given that mutations in *GCD* genes impair *GCN4* translational control, this essential function could be involved with protein synthesis. If so, the products of *GCN1*, *GCN2*, and *GCN3* would function under starvation conditions to partially inactivate components of the translational machinery and thereby increase the frequency of reinitiation at the *GCN4* AUG codon.

In accord with this view, recent studies (Williams et al. 1989) indicate that mutations in *SUI2* and *SUI3*, the structural genes for the α and β subunits of eIF-2 in *S. cerevisiae*, impair translational control of *GCN4*. In addition to allowing increased utilization of a UUG codon as a translational start site in the absence of the normal *HIS4* AUG start codon, the *sui2-1* and *sui3-2* mutations lead to constitutive derepression of *HIS4* mRNA levels. This increased transcription of the *HIS4* gene was shown to occur because *GCN4* expression is translationally derepressed in these strains. This derepression is dependent on the 5′-proximal uORFs in *GCN4* mRNA and occurs in the absence of the positive regulators Gcn2p and Gcn3p. In all these respects, the mutations in the *SUI* genes resemble mutations in *GCD* genes. Another similarity noted between *SUI2*, *GCD1*, and *GCD2* is the fact that deletion of *GCN3* is lethal in the presence of certain mutations in each of these genes (see below). Because the uORFs carry out the same functions in regulating *GCN4* expression in strains containing mutations in the *SUI* genes as in wild-type strains, the derepression of *GCN4* seen in these mutants is not thought to involve increased utilization of non-AUG codons. Rather, it appears that the mutations in the *SUI* genes duplicate the effects of amino acid starvation of wild-type cells in stimulating the efficiency of reinitiation at *GCN4* following translation of uORF1.

In a related development, Tzamarias et al. (1989) reported that a temperature-sensitive lethal *gcd1* mutation leads to reduced rates of protein synthesis as the result of inefficient binding of the eIF-2/ GTP/Met-tRNA$_i^{Met}$ ternary complex to the 40S ribosomal subunit, at least after 2 hours of incubation at the nonpermissive temperature. The same defect was observed transiently in wild-type cells when the culture was shifted from medium containing all amino acids (except methionine) to minimal medium. Interestingly, the reduced translation of the majority of mRNAs that occurs after this nutritional shift-down is accompanied by a dramatic increase in the rate of Gcn4p synthesis. This increased translation of *GCN4* mRNA (as revealed by larger *GCN4* mRNA polysomes) is dependent on the presence of uORF1, suggesting that the same translational control mechanism that operates under steady-state starvation conditions is being affected transiently during a shift-down from amino-acid-rich medium to minimal medium.

The recent findings that correlate reduced or altered eIF-2 function with increased rates of *GCN4* translation raise the interesting possibility that lowering the function of one or more general initiation factors under starvation conditions is responsible for the increased ability of ribosomes to traverse the 3'-proximal uORFs and reach *GCN4*. This mechanism seems most consistent with the possibility that under starvation conditions, a certain fraction of 40S subunits scanning downstream from uORF1 ignore uORFs 3 and 4 as start sites and reinitiate at the *GCN4* AUG codon instead (one of the two possibilities depicted in Fig. 12).

The aforementioned derepression of *GCN4* translation observed in a nutritional shift-down was reported to be independent of *GCN2* function, suggesting that a different set of *trans*-acting factors is involved in this transient phenomenon than that required for steady-state derepression of *GCN4* under starvation conditions (Tzamarias et al. 1989). On the basis of previous observations suggesting that transcription of *GCN2* is stimulated by Gcn4p (Driscoll-Penn et al. 1984; Roussou et al. 1988), it was proposed that steady-state derepression of *GCN4* translation is initiated by the *GCN2*-independent mechanism evident during a nutritional shift-down. The Gcn4p thus produced would then elevate *GCN2* transcription, and the resulting increase in Gcn2p levels would permit sustained derepression of *GCN4* (Tzamarias et al. 1989). However, as mentioned below, there is now evidence that Gcn2p levels do not increase significantly under starvation conditions (Wek et al. 1990), raising some doubts about this proposal.

g. Genetic Evidence for Direct Interactions between Gcn3p and Certain GCD *Gene Products.* The idea that *GCD* factors function in general protein synthesis is consistent with the relatively large number of *GCD* genes (11) identified thus far (Hinnebusch 1988; see Hinnebusch in the accompanying volumes): Many components of the translational machinery may be required for repression of *GCN4* translation under nonstarvation conditions. However, all such factors identified genetically are not necessarily subject to regulation in wild-type cells in response to amino acid starvation. For this reason, genetic interactions detected between the dedicated positive regulators of *GCN4* expression and particular *GCD* genes should help to identify *GCD* factors that participate directly in *GCN4* translational control. Such interactions have been reported involving *GCN3* with *GCD1* and *GCD2*.

Certain *gcd1* mutations and alleles of *GCD2* (originally designated *gcd12*) were isolated as suppressors of the nonderepressible phenotype of the *gcn3-101* mutation, leading to constitutively derepressed *GCN4* expression and temperature sensitivity for growth. These *gcd* mutations also suppress the nonderepressible phenotype of a *gcn3::LEU2* deletion. Interestingly, wild-type *GCN3* partially overcomes both phenotypes associated with these *gcd1* mutations and completely overcomes the same defects in the *gcd2* mutants (Harashima et al. 1987). *GCN3* also partially suppresses the lethality associated with the *gcd1-101* mutation, isolated by a different approach, as shown by the fact that a *gcd1-101 gcn3::LEU2* double mutation is lethal (Hannig and Hinnebusch 1988). In addition, the *gcn3-102* allele was isolated by its ability to suppress the temperature-sensitive phenotype of *gcd1-101* more efficiently than *GCN3* does (Hinnebusch and Fink 1983). *GCN3* does not overcome the lethal effects of a *gcd1* or *gcd2* deletion or the constitutive derepression associated with the *gcd2-1* mutation (Hannig and Hinnebusch 1988; Hill and Struhl 1988; Paddon and Hinnebusch 1989), indicating that suppression of *gcd* mutations by *GCN3* is allele-specific.

These results suggest that Gcn3p can promote the essential and negative regulatory functions of Gcd1p and Gcd2p in certain *gcd* mutants under nonstarvation conditions, even though it acts to antagonize the corresponding wild-type *GCD* products when cells are starved for an amino acid. One hypothesis to explain this dual function is that Gcn3p exists in a complex with Gcd1p and Gcd2p and can stabilize thermolabile products encoded by mutant *gcd* alleles. Gcn3-102p would stabilize the *gcd1-101* product even more effectively than wild-type Gcn3p does. In a *gcn3* deletion strain, Gcn3p would be absent from the complex and the thermolability of the *gcd* products would be fully expressed. In wild-type cells, the presence of Gcn3p in a complex with Gcd1p and Gcd2p would enable it to antagonize these *GCD* factors in response to amino acid starvation (Harashima et al. 1987). Interestingly, sequence analysis has shown that Gcn3p has significant similarity with the carboxy-terminal segment of Gcd2p, in accord with the idea that Gcn3p and Gcd2p are subunits of the same heteromeric complex (Paddon et al. 1989).

On the basis of the lethality associated with deleting *GCN3* in the *sui2-1* mutant (Williams et al. 1989), the putative Gcd1p/Gcd2p/Gcn3p complex might be expected to interact with eIF-2 during the initiation process. Unfortunately, the deduced amino acid sequences of Gcn3p, Gcd1p, and Gcd2p provide no clues about their exact functions (Hannig and Hinnebusch 1988; Hill and Struhl 1988; Paddon et al. 1989); consequently, biochemical studies will be needed to gain further insight into the precise roles these factors play in translation initiation and in modulating ribosomal recognition of the *GCN4* upstream AUG codons in response to amino acid availability.

h. GCN2 *Functions as a Protein Kinase in Stimulating* GCN4 *Expression.* DNA sequence analysis reveals a strong similarity between a portion of the Gcn2p-coding sequences and the catalytic domain of eukaryotic protein kinases: all 33 residues highly conserved in 65 different protein kinases are present in *GCN2* (Hanks et al. 1988; Roussou et al. 1988; Wek et al. 1989). One of the best characterized of the conserved subdomains in protein kinases includes the sequence GlyX-GlyXXGlyXVal followed 13–18 residues downstream by AlaXLys. The lysine residue is believed to function in the phosphotransfer reaction, and amino acid substitutions at this position invariably abolish kinase activity (Edelman et al. 1987; Hanks et al. 1988). Substitution of the corresponding lysine in *GCN2* (Lys-559) with arginine or valine abolished *GCN2*-positive regulatory function, whereas substitution of an adjacent lysine with valine had no detectable effect (Wek et al. 1989). It was also shown that Gcn2p can autophosphorylate in vitro and that this kinase activity is destroyed by substitution of Lys-559 (Wek et al. 1990). These results strongly suggest that *GCN2* stimulates *GCN4* expression by functioning as a protein kinase.

The deduced amino acid sequence of Gcn2p may provide an important clue about how its protein kinase activity is regulated by amino acid availability: 530 residues in the carboxy-terminal domain are homologous to histidyl-tRNA synthetases (His-RS) from *S. cerevisiae*, humans, and *E. coli*. Several two-codon insertions and deletions in the His-RS-related coding sequences inactivate *GCN2*-positive regulatory function in vivo (Wek et al. 1989), and the product of one such large deletion was shown to retain autophosphorylation activity in vitro (Wek et al. 1990). These results suggest that the His-RS-related sequences are needed in vivo to stimulate *GCN2* kinase function in response to starvation, rather than being required for kinase activity per se. Given that aminoacyl-tRNA synthetases bind uncharged tRNA as a substrate and that accumulation of uncharged tRNA triggers derepression of the general control response (Messenguy and Delforge 1976), it was proposed that the

His-RS-related domain of Gcn2p monitors the concentration of uncharged tRNA and activates the adjacent protein kinase moiety under starvation conditions when uncharged tRNA accumulates. Because genes under the general control derepress in response to starvation for any of ten amino acids, Gcn2p may have diverged sufficiently from His-RS that it now lacks the ability to discriminate between different uncharged tRNA species (Wek et al. 1989).

Previous observations indicating that the level of *GCN2* mRNA increases under starvation conditions, that *GCN4* is required for normal expression of *GCN2* mRNA, and that a Gcn4p-binding site is present upstream of *GCN2* (Driscoll-Penn et al. 1984; Roussou et al. 1988) led to the idea that Gcn4p stimulates *GCN2* transcription under starvation conditions. In this way, Gcn4p would elevate its own synthesis by increasing the expression of its translational activator Gcn2p. It was also shown that extensive overexpression of wild-type Gcn2p is sufficient for partial derepression of *GCN4* and structural genes under its control (Roussou et al. 1988; Tzamarias and Thireos 1988; Wek et al. 1990), as expected if the amount of Gcn2p is the major factor limiting translation of *GCN4* mRNA. However, in a recent study, little or no increase in the levels of *GCN2* mRNA or protein was detected under starvation conditions in which *HIS4* transcription was fully derepressed; moreover, *GCN2* mRNA levels were not reduced by a *gcn4* mutation that impaired *HIS4* derepression (Wek et al. 1990). The latter results suggest that the level of *GCN2* regulatory function can be modulated by amino acid levels without changing the abundance of Gcn2p.

In support of this idea, dominant constitutive mutations were isolated (*GCN2*[c]) that derepress *GCN4* expression without increasing Gcn2p levels. Two of these mutations map in the protein kinase domain; the third maps in the carboxyl terminus, just downstream from sequences related to His-RS. It was proposed that these mutations increase *GCN2* kinase function and thereby eliminate the need for stimulation by uncharged tRNA; however, it was not possible to demonstrate increased autophosphorylation activity in vitro for the *GCN2*[c] gene products. Perhaps the stimulatory effects of the *GCN2*[c] mutations pertain only to physiological substrates; alternatively, they may influence localization of Gcn2p rather than its catalytic activity. Small deletions of the sequences containing the *GCN2*[c] mutation at the carboxyl terminus abolish *GCN2* regulatory function in vivo without lowering the level of Gcn2p or its ability to autophosphorylate in vitro, suggesting that this region is a positive regulatory domain that can be mutationally activated to circumvent the requirement for uncharged tRNA for increased *GCN2* kinase function in vivo (Wek et al. 1990).

No substrates for Gcn2p kinase activity have been identified; however, the fact that mutations in *GCD* and *SUI* genes suppress the non-derepressible phenotype of a *gcn2* deletion (Hinnebusch 1988; Williams et al. 1989) makes the products of these genes good candidates. Phosphorylation by Gcn2p would be expected to partially inactivate or modify these factors in a way that increases the probability of reinitiation at the *GCN4* AUG codon. This hypothesis is attractive given that eIF-2 activity in mammalian cells is inhibited by phosphorylation of the α subunit under certain stress conditions, which include amino acid starvation (Safer 1983; Clemens et al. 1987; Schneider and Shenk 1987). Alternatively, Gcn1p or Gcn3p may be phosphorylated and thereby converted to antagonists of *GCD* or *SUI* factors. Pursuing the substrates of Gcn2p kinase activity by genetic and biochemical means should help greatly to identify those initiation factors that are modified under starvation conditions in order to stimulate translation of *GCN4* mRNA.

B. Regulation of Translation Initiation in Yeast Mitochondria

1. Nucleus-encoded Activators of Specific Mitochondrial Genes

A number of gene products encoded in the nucleus of *S. cerevisiae* function as gene-specific translational activators in the mitochondria (Table 13). These regulatory genes were identified in genetic screens for nuclear mutations that impair respiratory function in the mitochondria and consequently prevent growth on nonfermentable carbon sources (Pet⁻). Mutations in three nuclear genes specifically impair expression of *oxi2* (encodes cytochrome *c* oxidase subunit III or COX III) without altering the steady-state level or 5′ ends of *oxi2* mRNA: *PET494* (Mueller et al. 1984), *PET54* (Costanzo et al. 1986a), and *PET122* (Costanzo and Fox 1988; Kloeckener-Gruissem et al. 1988). Similar observations were made for mutations in three nuclear genes required for expression of cytochrome *b* (COB) from the *cob* transcript: *CBP6*, *CBS1*, and *CBS2*. The defect in *cob* expression is not absolute in the *cbp6-1* mutant; however, no change whatsoever was detected in the *cob* mRNA level or its 5′ end in these cells (Dieckmann and Tzagoloff 1985). The Cob protein is undetectable in *cbs1* and *cbs2* mutants; however, unspliced *cob* precursor transcripts accumulate. In fact, the latter phenotype might be expected from a complete block to *cob* mRNA translation, since expression of an intron-encoded RNA maturase is thought to be required for *cob* mRNA processing. In addition, *cbs1* and *cbs2* mutants differ from other splicing-defective mutants in their failure to accumulate an aberrant

Table 13 Mitochondrial genes subject to translational activation by nucleus-encoded gene products

Mitochondrial gene	Nucleus-encoded activator	Mitochondrial mRNA leader is target sequence?	Activator protein found in mitochondria?
oxi2 (COX III)	Pet494	yes	yes
	Pet54	yes	yes
	Pet122	yes	not known
cob (COB)	Cbp6	not known	not known
	Cbs1	yes	yes
	Cbs2	yes	not known
oxi1 (COX II)	Pet111	probably[a]	probably[b]

[a] It was not yet shown that the *oxi1* mRNA leader is sufficient to confer *PET111*-activated translation of a heterologous transcript.

[b] A *PET111-lacZ* fusion protein localizes in the mitochondria; however, the authentic protein has not been examined.

protein translated from fused exon and intron ORFs (Roedel et al. 1985, 1986; Roedel 1986).

Cytochrome *c* oxidase subunit II (COX II) is a third mitochondrial protein whose expression is regulated posttranscriptionally by nucleus-encoded gene products. Mutations in the *PET111* gene essentially eliminate expression of COX II but reduce the amount of its transcript (*oxi1* mRNA) by only about 67% (Poutre and Fox 1987). The *sco1-1* mutation also eliminates COX II expression at a posttranscriptional step; however, this defect may involve a lesion in protein assembly rather than reduced translation of *oxi1* mRNA (Schulze and Roedel 1988).

2. *mRNA Leaders of Mitochondrial Genes Are the Regulatory Targets of Nuclear Activator Proteins*

Strong evidence that many of these nucleus-encoded activators function by stimulating translation of specific mitochondrial transcripts was provided by isolating mutations in the mRNA leaders of the mitochondrial genes that override the effects of mutations in the corresponding nuclear activator genes. The first such suppressors were obtained as Pet+ revertants of a *pet494* mutant that restored COX III expression. These revertants were mitotically unstable and gave rise to Pet⁻ segregants that carried either of two different mitochondrial genomes, wild-type or a deleted derivative. The latter were defective for respiratory function but could still restore a Pet+ phenotype when introduced back into *pet494* strains carrying wild-type mitochondrial DNA; thus, these deleted genomes carried the suppressors of *pet494*. The most important suppressor genomes that were analyzed all contained one deletion endpoint in the mRNA leader sequences at *oxi2* and the other endpoint in the leader sequences of a different mitochondrial gene. mRNA analysis confirmed the existence of fusion transcripts containing heterologous leader sequences fused to *oxi2* mRNA upstream of the *oxi2* ATG start codon (Fig. 13A) (Mueller et al. 1984; Costanzo and Fox 1986). Subsequently, it was demonstrated that the same mitochondrial rearrangements suppress the failure to express COX III in *pet54* (Costanzo et al. 1986a) and *pet122* mutants (Costanzo et al. 1988). The fact that COX III is expressed efficiently in these mutants from the mitochondrial suppressor genomes strongly suggests that Pet494p, Pet54p and Pet122p are not required for the stability of COX III protein; rather, they appear to act on 5'-untranslated sequences in *oxi2* mRNA to stimulate COX III synthesis.

A similar type of mitochondrial suppressor mutation was isolated by reversion of the nuclear *cbs1* mutation that impairs expression of the *cob* gene product. In this instance, the leader sequences of the *oli1* transcript were fused to *cob*-coding sequences (Roedel et al. 1985). The same rearrangement suppresses the *cbs2-1* mutation (Roedel 1986), suggesting that Cbs1p and Cbs2p both act upon 5'-untranslated sequences in the *cob* transcript. This conclusion was confirmed by the elegant demonstration that *CBS1* is required for expression of COX III protein from one of the aforementioned hybrid transcripts containing the COX III-coding sequences fused to the *cob* leader (Roedel and Fox 1987). Likewise, *PET494*, *PET54*, and *PET122* were all found to be required for Cob protein expression from a chimeric transcript containing the 5' two thirds of the *oxi2* mRNA leader fused to *cob* mRNA in its leader region (Costanzo and Fox 1988), thus narrowing the target site for these three activators to a portion of the *oxi2* mRNA leader.

The *CBP6*, *CBS2*, *PET494*, *PET54*, and *PET111* genes have been isolated and characterized (Dieckmann and Tzagoloff 1985; Costanzo et al. 1986b; Strick and Fox 1987; Michaelis et al. 1988). Interestingly, there is 23% identity between the carboxy-terminal approximate 100 amino acids of Pet494p and Pet111p. In addition, both proteins have very basic amino-terminal segments, a feature common among imported mitochondrial proteins. In fact, it was demonstrated that *PET494-lacZ* and *PET111-lacZ* fusion proteins are specifically associated with mitochondria in subcellular fractionation experiments. The same conclusion was reached for authentic Pet494p, Pet54p, and Cbs1p (Costanzo and Fox 1986; Strick and Fox 1987; Costanzo et al. 1988, 1989; Korte et al. 1989), suggesting that these gene products function as translational ac-

A.

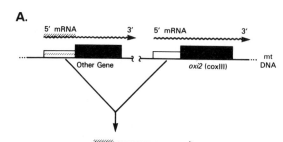

B.

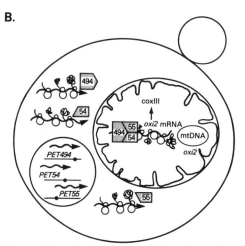

Figure 13 Translational activation of mitochondrial gene expression by nucleus-encoded proteins. (*A*) Schematic representation of deletion mutations leading to novel *oxi2* mRNAs for COX III protein synthesis that can be translated in strains lacking the activator proteins produced by *PET494*, *PET54*, and *PET122* that are required for translation of wild-type *oxi2* mRNA. The novel transcripts contain mRNA leader sequences from a heterologous mitochondrial transcript. (*B*) A hypothetical model for the functions of the *PET494*, *PET54*, and *PET122* (*PET55*) gene products in stimulating COX III protein synthesis. The three nucleus-encoded activators are depicted functioning together in the mitochondrion as a protein complex. (Adapted from Costanzo et al. 1988.)

tivators in a direct fashion in the mitochondrion itself. *pet122* mutants apparently express normal amounts of Pet494p and Pet54p, and Pet54p expression is normal in a *pet494* mutant (Costanzo et al. 1989), consistent with the idea that these three proteins function together in the mitochondria to stimulate COX III protein synthesis (Fig. 13B).

Synthesis of many mitochondrial respiratory components is subject to glucose repression. For COX III, it appears that increased protein synthesis in response to glucose depletion arises from increased translation of *oxi2* mRNA, since the steady-state amount of the transcript is unaffected by a shift from medium containing glucose to one containing lactate (Zennaro et al. 1985). Thus, the *PET494-PET54-PET122* group of translational activators could be responsible for derepression of *oxi2* expression in the absence of glucose. Supporting this idea, expression of a *PET494-lacZ* fusion protein was found to be repressed by glucose (Marykwas and Fox 1989). Similarly, a *CBS1-lacZ* fusion was repressed both by glucose and by growth under anaerobic conditions, in an additive fashion. At least the latter response appeared to involve a reduction in the steady-state level of *CBS1* mRNA (Forsbach et al. 1989).

Recently, it was shown that a *pet54* disruption impairs splicing of a particular intron from the mRNA encoding cytochrome *c* oxidase subunit I (COX I) in addition to reducing translation of the *oxi2* transcript. The splicing defect could arise from a failure to translate an intron-encoded maturase or from a more direct involvement of Pet54p in splicing. Two 30-nucleotide stretches of sequence similarity between the *oxi2* mRNA leader and the intron of the COX I mRNA that is unspliced in the *pet54* mutant were noted as potential binding sites for Pet54p (Valencik et al. 1989). Interestingly, there are several other reports suggesting dual functions of certain mitochondrial proteins in RNA processing and protein synthesis, e.g., the mitochondrial leucyl-tRNA synthetase (Labouesse et al. 1987; Herbert et al. 1988; Ben Asher et al. 1989).

C. Control of Gene Expression by Translational Frameshifting

1. Ty Elements

The yeast retrotransposon Ty encodes a single transcript containing two protein-coding sequences in different translational reading frames. The 5′-proximal ATG codon initiates an ORF of 440 codons called *TYA*; a second ORF (*TYB*) begins 38 nucleotides upstream of the *TYA* stop codon and extends for approximately 1300 codons to a position close to the 3′ end of the transcript (Fig. 14); however, no ATG codon is present near the beginning of *TYB* (Clare and Farabaugh 1985; Mellor et al. 1985). This sequence organization is similar to that described for the *gag* and *pol* genes of vertebrate retroviruses. In fact, the Tyb protein sequence shows similarity to retroviral reverse transcriptase, and Ty elements replicate by an RNA intermediate during transposition (Boeke et al. 1985). The *pol*-coding sequences of certain retroviruses are translated as a *gag-pol* fusion protein as the result of translational frameshifting in the region where the two ORFs overlap (see, e.g., Jacks and Varmus

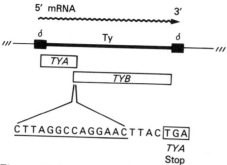

Figure 14 Expression of *TYB*-coding sequences by translational frameshifting. (*Open boxes*) Overlapping *TYA*- and *TYB*-coding sequences. A 14-nucleotide sequence (*underlined*) found just upstream of the *TYA* stop codon in the overlap region of the Ty912 element is sufficient to direct translational frameshifting when inserted into heterologous protein-coding sequences.

1985). Recent studies on the expression of *TYB* sequences suggest that a similar mechanism operates for Ty elements.

The first in-frame ATG codon in *TYB* is located approximately 900 nucleotides downstream from the beginning of the ORF. This ATG codon is not thought to initiate translation of *TYB* sequences because interferon or *lacZ*-coding sequences fused to *TYB* at a site upstream of this ATG codon, but 3′ to *TYA*, are efficiently expressed in vivo. The sizes of the resulting fusion proteins suggest that translation of *TYB* sequences actually initiates at the *TYA* start codon. Supporting this conclusion is the fact that a frameshift mutation early in *TYA* abolishes expression of a *TYB-lacZ* fusion (Clare and Farabaugh 1985; Mellor et al. 1985). These findings suggest that a +1 frameshift occurs upstream of the *TYA* stop codon to allow translation to proceed into *TYB* sequences in the correct reading frame. By comparing expression of *TYB-lacZ* sequences from the wild-type construct with that from a mutant allele in which *TYA* and *TYB-lacZ* are fused in the same reading frame with no intervening stop codon, it was concluded that a +1 frameshift occurs on about 20% of all Ty transcripts (Clare et al. 1988).

Analysis of deletions in *TYA* showed that only 31 nucleotides upstream of the *TYA* stop codon are required for *TYB* expression, indicating that the frameshift occurs somewhere within these 31 nucleotides (Wilson et al. 1986; Clare et al. 1988). This region includes a 14-nucleotide stretch, highly conserved (13/14 residues) between Ty1 and Ty2 elements, that is sufficient to direct a +1 frameshift with high efficiency when inserted into completely heterologous coding sequences (Fig. 14) (Clare et al. 1988). In fact, recent results indicate that only the first 7 of these 14 nucleotides (CTT AGG C, written as triplets in the *TYA* reading frame) are both necessary and sufficient for efficient +1 frameshifting. It now seems clear that frameshifting occurs when peptidyl-tRNA^{Leu} is in the ribosomal P site interacting with the CUU leucine codon in the beginning of the seven-nucleotide stretch. The adjacent AGG codon is decoded slowly because of the low abundance of the tRNA^{Arg} species containing the CCU anticodon. During the time when the A site is empty, the peptidyl-tRNA slips forward one nucleotide onto the UUA codon, which also encodes leucine. The tRNA^{Leu} species containing the UAG anticodon is thought to be responsible for this frameshift event (P. Farabaugh, pers. comm.). This mechanism differs importantly from that elucidated for Rous sarcoma virus involving a "simultaneous slippage" of two tRNAs occupying the P and A sites. This latter event is stimulated by a pause in translation elongation imposed by a stem-loop structure located in the mRNA downstream from the frameshift site (Jacks et al. 1988).

2. Double-stranded RNA Viruses

Recent analysis of the genome of the double-stranded RNA virus L-A of *S. cerevisiae* (Fujimura and Wickner 1988; Icho and Wickner 1989) suggests an organization very similar to that described for Ty elements and retroviruses. The L-A genome encodes a major coat protein (80 kD) and a minor protein (180 kD) thought to be the viral RNA polymerase, but it has a coding capacity for only approximately 180 kD. Two overlapping ORFs are present in the L-A genome and, on the basis of the sequences of coat protein cyanogen-bromide peptides, the 80-kD coat protein is encoded by the 5'-proximal ORF of 680 codons (ORF1). Antibody directed against a carboxy-terminal peptide encoded by ORF2 reacts with the 180-kD protein. In addition, the 180-kD protein cross-reacts with antibodies made against the 80-kD coat protein. These results suggest that the 180-kD protein is encoded by a fusion of ORF1 and ORF2. Sequences present in the overlap region suggest the occurrence of a –1 frameshift to fuse ORF1 and ORF2 during translation by the simultaneous slippage mechanism involving a UUA-decoding tRNA^{Leu} demonstrated for Rous sarcoma virus (Jacks et al. 1988).

IV. CONCLUSION

As mentioned in the Introduction, there are many similarities between the process of translation in *Saccharomyces* and that described for other organisms. In particular, yeast appears to use the same scanning mechanism for initiation described for higher eukaryotes, involving the mRNA cap structure and its associated protein factors. Recent studies suggest that eIF-2 and tRNA$_i$^{Met} play central roles in selecting the first AUG codon encountered during the scanning process as the start site for translation. The features of the scanning mechanism dictate the monocistronic nature of yeast mRNAs and provide the basis for translational control mechanisms involving uORFs as negative regulatory elements. Given the strong inhibitory effects on initiation exerted by secondary structures in mRNA leaders in yeast, it seems likely that translational control mechanisms involving such structural elements will also be discovered in *Saccharomyces*, perhaps similar to one recently described for ferritin mRNA in animal cells (Casey et al. 1988).

The study of translation in yeast is entering an exciting phase. For the first time, it should be possible to determine whether reaction mechanisms for initiation and elongation developed from in vitro studies on other eukaryotic systems are valid in vivo. In addition, genetic analysis of factors involved in translation has led to new insights about the functions of well-studied proteins, such as eIF-2 and EF-1, as well as recognition of previously undetected components of the translational machinery. One of the most interesting prospects for future research will be the identification of regulatory roles for general translation factors in altering the efficiency or fidelity of protein synthesis in response to changes in the environment.

Note added in proof

Recent findings on translational control of *GDN4* (Abastado et al. 1991) provide strong support for the idea that under amino acid starvation conditions, ribosomes translate uORF1, resume scanning, and ignore the start codons at uORFs 2–4, reinitiating translation at *GCN4* instead. This is thought to occur because the efficiency of reassembling initiation factors following termination at uORF1 is reduced under these circumstances such that ribosomes scanning downstream from uORF1 are not competent to reinitiate when they reach uORFs 2–4, but are able to do so after scanning the additional sequences between uORF4 and *GCN4*. Under nonstarvation conditions, reinitiation would be more efficient and occur exclusively at uORFs 2–4, preventing translation of *GCN4*. It was recently shown that mutations in the *PET123* gene encoding a mitochondrial small subunit ribosomal protein suppress mutations in *PET122*, one of three nucleus-encoded translational activators of the mitochondrial mRNA for cytochrome oxidase subunit III. These findings support the notion that Pet122p functions directly as a translational activator by interacting with the mitochondrial small ribosomal subunit

(McMullin et al. 1990). The unpublished observations from P. Farabaugh on the sequence requirements and mechanism of ribosomal frameshifting in translation of Ty mRNA were reported recently (Belcourt and Farabaugh 1990). Detailed information on the mechanism of frameshifting for L-A mRNA has also been published (Dinman et al. 1991). Finally, X-ray analysis of *E. coli* seryl-tRNA synthetase has revealed a three-dimensional structure that does not involve a classical nucleotide-binding fold and is thus very different from the synthetase structures described here. The seryl-tRNA synthetase appears to define a second class of synthetases that includes the enzymes for lysine, aspartic acid, asparagine, phenylalanine, glycine, alanine, and histidine (Cusack et al. 1990; Eriani et al. 1990).

ACKNOWLEDGMENTS

We thank members of our laboratories, Michael Culbertson and Philip Farabaugh for helpful comments on the contents of this paper, and Kathy Shoobridge for help in its preparation. The writing of this paper was supported in part by U.S. Public Health Service research grant GM-24189.

REFERENCES

Abastado, J.-P., P.F. Miller, B.M. Jackson, and A.G. Hinnebusch. 1991. Suppression of ribosomal reinitiation at upstream open reading frames in amino acid-starved cells forms the basis for *GCN4* translational control. *Mol. Cell. Biol.* **11:** 486.

Ahmad, M.F., N. Nasrin, A.C. Banerjee, and N.K. Gupta. 1985a. Purification and properties of eukaryotic initiation factor 2 and ancillary protein factor (Co-eIF-2A) from yeast *Saccharomyces cerevisiae. J. Biol. Chem.* **260:** 6955.

Ahmad, M.F., N. Nasrin, M.K. Bagchi, I. Chakravarty, and N.K. Gupta. 1985b. A comparative study of the characteristics of eIF-2 and eIF-2-ancillary factor activities from yeast *Saccharomyces cerevisiae* and rabbit reticulocytes. *J. Biol. Chem.* **260:** 6960.

All-Robyn, J.A., N. Brown, E. Otaka, and S.W. Liebman. 1990a. Sequence and functional similarity between a yeast ribosomal protein and the *Escherichia coli* S5 *ram* protein. *Mol. Cell. Biol.* **10:** 6544.

All-Robyn, J.A., D. Kelley-Geraghty, E. Griffin, N. Brown, and S.W. Liebman. 1990b. Isolation of omnipotent suppressors in an [eta+] yeast strain. *Genetics* **124:** 505.

Altmann, M. and H. Trachsel. 1989. Altered mRNA cap recognition activity of initiation factor 4E in the yeast cell cycle division mutant *cdc33. Nucleic Acids Res.* **17:** 5923.

Altmann, M., C. Handschin, and H. Trachsel. 1987. mRNA cap-binding protein: Cloning of the gene encoding protein synthesis initiation factor eIF-4E from *Saccharomyces cerevisiae. Mol. Cell. Biol.* **7:** 998.

Altmann, M., N. Sonenberg, and H. Trachsel. 1989a. Translation in *Saccharomyces cerevisiae*: Initiation factor 4E-dependent cell-free system. *Mol. Cell. Biol.* **9:** 4467.

Altmann, M., I. Edery, N. Sonenberg, and H. Trachsel. 1985. Purification and characterization of protein synthesis initiation factor eIF-4E from the yeast *Saccharomyces cerevisiae. Biochemistry* **24**: 6085.

Altmann, M., I. Edery, H. Trachsel, and N. Sonenberg. 1988. Site-directed mutagenesis of the tryptophan residues in yeast eukaryotic initiation factor 4E. *J. Biol. Chem.* **263**: 17229.

Altmann, M., P.P. Mueller, J. Pelletier, N. Sonenberg, and H. Trachsel. 1989b. A mammalian translation initiation factor can substitute for its yeast homologue *in vivo. J. Biol. Chem.* **264**: 12145.

An, G. and J. D. Friesen. 1980. The nucleotide sequence of *tufB* and four nearby tRNA structural genes of *Escherichia coli. Gene* **12**: 33.

Baan, R.A., P.B. Keller, and A.E. Dahlberg. 1981. Isolation of eukaryotic initiation factor 2 from yeast *Saccharomyces cerevisiae. J. Biol. Chem.* **256**: 1063.

Baim, S.B. and F. Sherman. 1988. mRNA structures influencing translation in the yeast *Saccharomyces cerevisiae. Mol. Cell. Biol.* **8**: 1591.

Ball, C.B., M.D. Mendenhall, M.G. Sandbaken, and M.R. Culbertson. 1988. The yeast *SUF5* frameshift suppressor encodes a mutant glycine tRNA$_{CCC}$. *Nucleic Acids Res.* **16**: 8712.

Bare, L. and O.C. Uhlenbeck. 1985. Aminoacylation of anticodon loop substituted yeast tyrosine transfer RNA. *Biochemistry* **24**: 2354.

―――. 1986. Specific substitution into the anticodon loop of yeast tyrosine transfer RNA. *Biochemistry* **25**: 5825.

Bedard, D.P., G.C. Johnston, and R.A. Singer. 1981. New mutations in the yeast *Saccharomyces cerevisiae* affecting completion of "start." *Curr. Genet.* **4**: 205.

Belcourt, M.F. and P.J. Farabaugh. 1990. Ribosomal frameshifting in the yeast retrotransposon Ty: tRNAs induce slippage on a 7 nucleotide minimal site. *Cell* **62**: 339.

Beltzer, J.P., S.R. Morris, and G.B. Kohlhaw. 1988. Yeast *LEU4* encodes mitochondrial and nonmitochondrial forms of α-isopropylmalate synthase. *J. Biol. Chem.* **263**: 368.

Beltzer, J.P., L.L. Chang, A.E. Hinkkanen, and G.B. Kohlhaw. 1986. Structure of yeast *LEU4*, the 5′ flanking region contains features that predict two modes of control and two productive translation starts. *J. Biol. Chem.* **261**: 5160.

Ben Asher, E., O. Groudinsky, G. Dujardin, N. Altamura, M. Kermorgant, and P.P. Slonimski. 1989. Novel class of nuclear genes involved in both mRNA splicing and protein synthesis in *Saccharomyces cerevisiae. Mol. Gen. Genet.* **215**: 517.

Bennetzen, J.L. and B.D. Hall. 1982. Codon selection in yeast. *J. Biol. Chem.* **257**: 3026.

Bhanot, O.S., Z. Kucan, S. Aoyagi, F.C. Lee, and R.W. Chambers. 1974. Purification of tyrosine:tRNA ligase, valine:tRNA ligase, alanine:tRNA ligase, isoleucine:tRNA ligase from *Saccharomcyes cerevisiae* αS288C. *Methods Enzymol.* **29**: 547.

Blum, S., M. Mueller, S.R. Schmid, P. Linder, and H. Trachsel. 1989. Translation in *Saccharomyces cerevisiae*: Initiation factor 4A-dependent cell-free system. *Proc. Natl. Acad. Sci.* **86**: 6043.

Bodley, J.W., P.C. Dunlop, and B.G. van Ness. 1984. Diphthamide in elongation factor 2: ADP-ribosylation, purification. *Methods Enzymol.* **106**: 378.

Boeke, J.D., D.J. Garfinkel, C.A. Styles, and G.R. Fink. 1985. Ty elements transpose through an RNA intermediate. *Cell* **40**: 491.

Boguta, M., M. Mieszczak, and W. Zagorski. 1988. Nuclear omnipotent suppressors of premature termination codons in mitochondrial genes affect the 37S mitoribosomal subunit. *Curr. Genet.* **13**: 129.

Brand, A.H., G. Micklem, and K. Nasmyth. 1987. A yeast silencer contains sequences that can promote autonomous plasmid replication and transcriptional activation. *Cell* **51**: 709.

Brandriss, M.C., J.W. Stewart, F. Sherman, and D. Botstein. 1976. Substitution of serine caused by a recessive lethal suppressor in yeast. *J. Mol. Biol.* **102**: 467.

Brands, J.H., J.A. Maassen, F.J. Van Hemert, R. Amons, and W. Moller. 1986. The primary structure of the alpha subunit of human elongation factor 1. Structural aspects of guanine-nucleotide-binding sites. *Eur. J. Biochem.* **155**: 167.

Breining, P. and W. Piepersberg. 1986. Yeast omnipotent suppressor *SUP1* (*sup45*): Nucleotide sequence of the wild type and a mutant gene. *Nucleic Acids Res.* **14**: 5187.

Breining, P., A.P. Surguchov, and W. Piepersberg. 1984. Cloning and identification of a DNA fragment coding for the *sup1* gene of *Saccharomyces cerevisiae*. *Curr. Genet.* **8**: 467.

Brenner, C., N. Nakayama, M. Goebl, K. Tanaka, A. Toh-e, and K. Matsumoto. 1988. *CDC33* encodes mRNA cap-binding protein eIF-4E of *Saccharomyces cerevisiae*. *Mol. Cell. Biol.* **8**: 3556.

Broach, J.R., L. Friedman, and F. Sherman. 1981. Correspondence of yeast UAA suppressors to cloned tRNA$^{Ser}_{UCA}$ genes. *J. Mol. Biol.* **150**: 375.

Brown, B.A. and J.W. Bodley. 1979. Primary structure at the site in beef and wheat elongation factor 2 of ADP-ribosylation by diphtheria toxin. *FEBS Lett.* **103**: 253.

Buchman, A.R., W.J. Kimmerly, J. Rine, and R.D. Kornberg. 1988. Two DNA-binding factors recognize specific sequences at silencers, upstream activating sequences, autonomously replicating sequences, and telomeres in *Saccharomyces cerevisiae*. *Mol. Cell. Biol.* **8**: 210.

Calderon, I.L., C.R. Contopoulou, and R.K. Mortimer. 1984. Isolation of a DNA fragment that is expressed as an amber suppressor when present in high copy number in yeast. *Gene* **29**: 69.

Carlson, M. and D. Botstein. 1982. Two differentially regulated mRNAs with different 5′ ends encode secreted and intracellular forms of yeast invertase. *Cell* **28**: 145.

Casey, J.L., M.W. Hentze, D.M. Koller, S.W. Caughman, T.A. Rouault, R.D. Klausner, and J.B. Harford. 1988. Iron-responsive elements: Regulatory RNA sequences that control mRNA levels and translation. *Science* **240**: 924.

Castilho-Valavicius, B., H. Yoon, and T.F. Donahue. 1990. Genetic characterization of the *Saccharomyces cerevisiae* translational initiation suppressors *sui1*, *sui2* and *SUI3* and their effects on *HIS4* expression. *Genetics* **124**: 483.

Chakraburtty, K. and A. Kamath. 1988. Protein synthesis in yeast. *Int. J. Biochem.* **20**: 581.

Chang, T., J. Arenas, and J. Abelson. 1990. Identification of five putative yeast RNA helicase genes. *Proc. Natl. Acad. Sci.* **87**: 1571.

Chatton, B., B. Winsor, Y. Boulanger, and F. Fasiolo. 1987. Cloning and characterization of the yeast methionyl-tRNA synthetase mutation *mes1*. *J. Biol. Chem.* **262**: 15094.

Chatton, B., P. Walter, J. Ebel, F. Lacroute, and F. Fasiolo. 1988. The yeast *VAS1* gene encodes both mitochondrial and cytoplasmic valyl-tRNA synthetases. *J. Biol. Chem.* **263**: 52.

Chattoo, B.B., E. Palmer, B. Ono, and F. Sherman. 1979. Patterns of genetic and phenotypic suppression of *lys2* mutations in the yeast *Saccharomyces cerevisiae*. *Genetics* **93**: 67.

Chen, J.Y., J.W. Bodley, and D.M. Livingston. 1985. Diphtheria toxin-resistant mutants of *Saccharomyces cerevisiae*. *Mol. Cell. Biol.* **5**: 3357.

Cherest, H., Y. Surdin-Kerjan, and H. De Robichon-Szulmajster. 1971. Methionine-mediated repression in *Saccharomyces cerevisiae*: A pleiotropic regulatory system involving methionyl transfer ribonucleic acid and the product of gene *eth2*. *J. Bacteriol.* **106**: 758.

————. 1975. Methionine- and S-adenosyl methionine-mediated repression in a methionyl-transfer ribonucleic acid synthetase mutant of *Saccharomyces cerevisiae. J. Bacteriol.* **123:** 428.

Cigan, A.M. and T.F. Donahue. 1986. The methionine initiator tRNA genes of yeast. *Gene* **41:** 343.

————. 1987. Sequence and structural features associated with translational initiator regions in yeast—A review. *Gene* **59:** 1.

Cigan, A.M., L. Feng, and T.F. Donahue. 1988a. tRNA$_i^{Met}$ functions in directing the scanning ribosome to the start site of translation. *Science* **242:** 93.

Cigan, A.M., E.K. Pabich, and T.F. Donahue. 1988b. Mutational analysis of the *HIS4* translational initiator region in *Saccharomyces cerevisiae. Mol. Cell. Biol.* **8:** 2964.

Cigan, A.M., E.K. Pabich, L. Feng, and T.F. Donahue. 1989. Yeast translation initiation suppressor *sui2* encodes the alpha subunit of eukaryotic initiation factor 2 and shares identity with the human alpha subunit. *Proc. Natl. Acad. Sci.* **86:** 2784.

Cirakoglu, B. and J.-P. Waller. 1985. Do yeast aminoacyl-tRNA synthetases exist as "soluble" enzymes within the cytoplasm? *Eur. J. Biochem.* **149:** 353.

Clare, J. and P. Farabaugh. 1985. Nucleotide sequence of a yeast Ty element: Evidence for an unusual mechanism of gene expression. *Proc. Natl. Acad. Sci.* **82:** 2829.

Clare, J. J., M. Belcourt, and P. J. Farabaugh. 1988. Efficient translational frameshifting occurs within a conserved sequence of the overlap between the two genes of a yeast Ty*1* transposon. *Proc. Natl. Acad. Sci.* **85:** 6816.

Clarke, N.D., D.C. Lien, and P. Schimmel. 1988. Evidence from cassette mutagenesis for a structure-function motif in a protein of unknown structure. *Science* **240:** 521.

Clemens, M.J., A. Galpine, S.A. Austin, R. Panniers, E.C. Henshaw, R. Duncan, J.W. Hershey, and J.W. Pollard. 1987. Regulation of polypeptide chain initiation in Chinese hamster ovary cells with a temperature-sensitive leucyl-tRNA synthetase. Changes in phosphorylation of initiation factor eIF-2 and in the activity of the guanine nucleotide exchange factor GEF. *J. Biol. Chem.* **262:** 767.

Clements, J.M., T.M. Laz, and F. Sherman. 1988. Efficiency of translation initiation by non-AUG codons in *Saccharomyces cerevisiae. Mol. Cell. Biol.* **8:** 4533.

Collier, R.J. 1975. Diphtheria toxin: Mode of action and structure. *Bacteriol. Rev.* **39:** 54.

Costanzo, M.C. and T.D. Fox. 1986. Product of *Saccharomyces cerevisiae* nuclear gene *PET494* activates translation of a specific mitochondrial mRNA. *Mol. Cell. Biol.* **6:** 3694.

————. 1988. Specific translational activation by nuclear gene products occurs in the 5′ untranslated leader of a yeast mitochondrial mRNA. *Proc. Natl. Acad. Sci.* **85:** 2677.

Costanzo, M.C., E.C. Seaver, and T.D. Fox. 1986a. At least two nuclear gene products are specifically required for translation of a single yeast mitochondrial mRNA. *EMBO J.* **5:** 3637.

————. 1989. The *PET54* gene of *Saccharomyces cerevisiae*: Characterization of a nuclear gene encoding a mitochondrial translational activator and subcellular localization of its product. *Genetics* **122:** 297.

Costanzo, M.C., P.P. Mueller, C.A. Strick, and T.D. Fox. 1986b. Primary structure of wild-type and mutant alleles of the *PET494* gene of *Saccharomyces cerevisiae. Mol. Gen. Genet.* **202:** 294.

Costanzo, M. C., E.C. Seaver, D.L. Marykwas, and T.D. Fox. 1988. Multiple nuclear gene products are specifically required to activate translation of a single yeast mitochondrial mRNA. In *Genetics of translation* (ed. M.F. Tuite), p. 373. Springer-Verlag, Berlin.

Cottrelle, P., M. Cool, P. Thuriaux, V.L. Price, D. Thiele, J.M. Buhler, and P. Fromageot.

1985a. Either one of the two yeast EF-1α genes is required for cell viability. *Curr. Genet.* **9**: 693.

Cottrelle, P., D. Thiele, V.L. Price, S. Memet, J.Y. Micouin, C. Marck, J.M. Buhler, A. Sentenac, and P. Fromageot. 1985b. Cloning, nucleotide sequence, and expression of one of two genes coding for yeast elongation factor 1-α. *J. Biol. Chem.* **260**: 3090.

Cox, B.S. 1965. A cytoplasmic suppressor of super-suppressors in yeast. *Heredity* **20**: 505.

————. 1971. A recessive lethal super-suppressor mutation in yeast and other Psi phenomena. *Heredity* **26**: 211.

————. 1977. Allosuppressors in yeast. *Genet. Res.* **30**: 187.

Cox, B.S., M.F. Tuite, and C.S. McLaughlin. 1988. The [psi] factor of yeast: A problem in inheritance. *Yeast* **4**: 159.

Crabeel, M., R. Lavalle, and N. Glansdorff. 1990. Arginine-specific repression in *Saccharomyces cerevisiae*: Kinetic data on *ARG1* and *ARG3* mRNA transcription and stability support a transcriptional control mechanism. *Mol. Cell. Biol.* **10**: 1226.

Crabeel, M., S. Seneca, K. Devos, and N. Glansdorff. 1988. Arginine repression of the *Saccharomyces cerevisiae ARG1* gene. *Curr. Genet.* **13**: 113.

Crick, F. 1966. Codon-anticodon pairing: The wobble hypothesis. *J. Mol. Biol.* **19**: 548.

Crouzet, M. and M.F. Tuite. 1987. Genetic control of translational fidelity in yeast: Molecular cloning and analysis of the allosuppressor gene *sal3*. *Mol. Gen. Genet.* **210**: 581.

Crouzet, M.F., C.M. Izgu, C.M. Grant, and M.F. Tuite. 1988. The allosuppressor gene *sal4* encodes a protein important for maintaining translational fidelity in *Saccharomyces cerevisiae*. *Curr. Genet.* **14**: 537.

Culbertson, M.R., R.F. Gaber, and C.C. Cummins. 1982. Frameshift suppression in *Saccharomyces cerevisiae*. V. Isolation and genetic properties of nongroup specific suppressors. *Genetics* **102**: 363.

Culbertson, M.R., K.M. Underbrink, and G.R. Fink. 1980. Frameshift suppression in *Saccharomyces cerevisiae*. II. Genetic properties of group II suppressors. *Genetics* **95**: 833.

Culbertson, M.R., L. Charmus, M.T. Johnson, and G.R. Fink. 1977. Frameshifts and frameshift suppressors in *Saccharomyces cerevisiae*. *Genetics* **86**: 745.

Cummins, C.M. and M.R. Culbertson. 1981. Molecular cloning of the *SUF2* frameshift suppressor gene from *Saccharomyces cerevisiae*. *Gene* **14**: 263.

Cummins, C.M., M.R. Culbertson, and G. Knapp. 1985. Frameshift suppressor mutations outside the anticodon in yeast proline tRNAs containing an intervening sequence. *Mol. Cell. Biol.* **5**: 1760.

Cummins, C.M., T.F. Donahue, and M.R. Culbertson. 1982. Nucleotide sequence of the *SUF2* frameshift suppressor gene of *Saccharomyces cerevisiae*. *Proc. Natl. Acad. Sci.* **79**: 3565.

Cummins, C.M., R.F. Gaber, M.R. Culbertson, R. Mann, and G.R. Fink. 1980. Frameshift suppression in *Saccharomyces cerevisiae*. III. Isolation and genetic properties of group III suppressors. *Genetics* **95**: 855.

Curran, J. and M. Yarus. 1987. Reading frame selection and transfer RNA anticodon loop stacking. *Science* **238**: 1545.

Cusack, S., C. Berthet-Colominas, M. Härtlein, N. Nassar, and R. Leberman. 1990. A second class of synthetase structure revealed by X-ray analysis of *Escherichia coli* seryl-tRNA synthetase at 2.5 Å. *Nature* **347**: 249.

Dardel, F., G. Fayat, and S. Blanquet. 1984. Molecular cloning and primary structure of the *Escherichia coli* methionyl-tRNA synthetase gene. *J. Bacteriol.* **160**: 1115.

Dasmahapatra, B. and K. Chakraburtty. 1981. Purification and properties of elongation factor 3 from *Saccharomyces cerevisiae. J. Biol. Chem.* **256:** 9999.

Dasmahapatra, B., L. Skogerson, and K. Chakraburtty. 1981. Purification and properties of the elongation factor 1 from *Saccharomyces cerevisiae. J. Biol. Chem.* **256:** 10005.

del Rey, F., T.F. Donahue, and G.R. Fink. 1983. The histidine tRNA genes of yeast. *J. Biol. Chem.* **258:** 8175.

Dieckmann, C.L. and A. Tzagoloff. 1985. Assembly of the mitochondrial membrane system. *J. Biol. Chem.* **260:** 1513.

Dihanich, M.E., D. Najarian, R. Clark, E.C. Gillman, N.C. Martin, and A.K. Hopper. 1987. Isolation and characterization of *MOD5*, a gene required for isopentenylation of cytoplasmic and mitochondrial tRNAs of *Saccharomyces cerevisiae. Mol. Cell. Biol.* **7:** 177.

Dinman, J.D., T. Icho, and R.B. Wickner. 1991. A −1 ribosomal frameshift in a double-stranded RNA virus of yeast forms a gag-pol fusion protein. *Proc. Natl. Acad. Sci.* **88:** 174.

Donahue, T.F. and A.M. Cigan. 1988. Genetic selection for mutations that reduce or abolish ribosomal recognition of the *HIS4* translational initiator region. *Mol. Cell. Biol.* **8:** 2955.

Donahue, T.F., P.J. Farabaugh, and G.R. Fink. 1982. The nucleotide sequence of the *HIS4* region of yeast. *Gene* **18:** 47.

Donahue, T.F., A.M. Cigan, E.K. Pabich, and B. Castilho-Valavicius. 1988. Mutations at a Zn(ll) finger motif in the yeast eIF-2β gene alter ribosomal start-site selection during the scanning process. *Cell* **54:** 621.

Driscoll-Penn, M., G. Thireos, and H. Greer. 1984. Temporal analysis of general control of amino acid biosynthesis in *Saccharomyces cerevisiae*: Role of positive regulatory genes in initiation and maintenance of mRNA derepression. *Mol. Cell. Biol.* **4:** 520.

Dujardin, G., P. Lund, and P.P. Slonimski. 1984. The effect of paromomycin and [psi] on the suppression of mitochondrial mutations in *Saccharomyces cerevisiae. Curr. Genet.* **9:** 21.

Dujardin, G., P. Pajot, O. Groudinsky, and P.P. Slonimski. 1980. Long range control circuits within mitochondria and between nucleus and mitochondria. *Mol. Gen. Genet.* **179:** 469.

Dunlop, P.C. and J.M. Bodley. 1983. Biosynthetic labeling of diphthamide in *Saccharomyces cerevisiae. J. Biol. Chem.* **258:** 4754.

Edelman, A.M., D.K. Blumenthal, and E.G. Krebs. 1987. Protein serine/threonine kinases. *Annu. Rev. Biochem.* **56:** 567.

Ellis, S.R., A.K. Hopper, and N.C. Martin. 1987. Amino-terminal extension generated from an upstream AUG codon is not required for mitochondrial import of yeast N2, N2-dimethylguanosine-specific tRNA methyltransferase. *Proc. Natl. Acad. Sci.* **84:** 5172.

Englisch, U., S. Englisch, P. Markmeyer, J. Schischkoff, and H. Sternbach. 1987. Structure of the yeast isoleucyl-tRNA synthetase gene (*ILS1*), DNA-sequence, amino-acid sequence of proteolytic peptides of the enzyme and comparison of the structure to those of other known aminoacyl-tRNA synthetases. *Biol. Chem. Hoppe-Seyler* **368:** 971.

Eriani, G., M. Delarue, O. Poch, J. Gangloff, and D. Moras. 1990. Partition of tRNA synthetases into two classes based on mutually exclusive sets of sequence motifs. *Nature* **347:** 203.

Ernst, H., R.F. Duncan, and J.W.B. Hershey. 1987. Cloning and sequencing of complementary DNAs encoding the α-subunit of translational initiation factor eIF-2. *J. Biol. Chem.* **262:** 1206.

Etcheverry, T., D. Colby, and C. Guthrie. 1979. A precursor to a minor species of yeast tRNA^{Ser} contains an intervening sequence. *Cell* **18**: 11.

Etcheverry, T., M. Salvato, and C. Guthrie. 1982. Recessive lethality of yeast strains carrying the *SUP61* suppressor results from loss of a transfer RNA with a unique decoding function. *J. Mol. Biol.* **158**: 599.

Eustice, D.C., L.P. Wakem, J.M. Wilhelm, and F. Sherman. 1986. Altered 40S ribosomal subunits in omnipotent suppressors of yeast. *J. Mol. Biol.* **188**: 207.

Fasiolo, F., J. Bonnet, and F. Lacroute. 1981. Cloning of the yeast methionyl-tRNA synthetase gene. *J. Biol. Chem.* **256**: 2324.

Feinberg, B., C.S. McLaughlin, and K. Moldave. 1982. Analysis of temperature-sensitive mutant *ts*187 of *Saccharomyces cerevisiae* altered in a component required for the initiation of protein synthesis. *J. Biol. Chem.* **257**: 10846.

Fischhoff, D. A., R. H. Waterston, and M. V. Olson. 1984. The yeast cloning vector YEp13 contains a tRNA$^{Leu}_3$ gene that can mutate to an amber suppressor. *Gene* **27**: 239.

Forchhammer, K., W. Leinfelder, and A. Bock. 1989. Identification of a novel translation factor necessary for the incorporation of selenocysteine into protein. *Nature* **342**: 453.

Forsbach, V., T. Pillar, T. Gottenof, and G. Roedel. 1989. Chromosomal localization and expression of *CBS1*, a translational activator of cytochrome *b* in yeast. *Mol. Gen. Genet.* **218**: 57.

Fox, T.D. and S. Staempfli. 1982. Suppressor of yeast mitochondrial ochre mutations that maps in or near 15S ribosomal gene of mtDNA. *Proc. Natl. Acad. Sci.* **79**: 1583.

Fox, T.D. and B. Weiss-Brummer. 1980. Leaky +1 and -1 frameshift mutations at the same site in a yeast mitochondrial gene. *Nature* **289**: 60.

Francklyn, C. and P. Schimmel. 1989. Aminoacylation of RNA minihelices with alanine. *Nature* **337**: 478.

Fujimura, T. and R.B. Wickner. 1988. Gene overlap results in a viral protein having an RNA binding domain and a coat protein domain. *Cell* **55**: 663.

Gaber, R.F. and M.R. Culbertson. 1982a. The yeast frameshift suppressor gene *SUF16-1* encodes an altered glycine tRNA containing the four-base anticodon 3'-CCCG-5'. *Gene* **19**: 163.

———. 1982b. Frameshift suppression in *Saccharomyces cerevisiae*. IV. New suppressors among spontaneous co-revertants of the group II *HIS4-206* and *LEU2-3* frameshift mutations. *Genetics* **101**: 345.

———. 1984. Codon recognition during frameshift suppression in *Saccharomyces cerevisiae*. *Mol. Cell. Biol.* **4**: 2052.

Gaber, R.F., L. Mathison, I. Edelman, and M.R. Culbertson. 1983. Frameshift suppression in *Saccharomyces cerevisiae*. VI. Complete genetic map of twenty-five suppressor genes. *Genetics* **103**: 389.

Gafner, J., E.M. DeRobertis, and P. Phillippsen. 1983. Delta sequences in the 5' noncoding region of yeast tRNA genes. *EMBO J.* **2**: 583.

Gampel, A. and A. Tzagoloff. 1989. Homology of aspartyl- and lysyl-tRNA synthetases. *Proc. Natl. Acad. Sci.* **86**: 6023.

Gasior, E., F. Herrera, C.S. McLaughlin, and K. Moldave. 1979a. The analysis of intermediary reactions involved in protein synthesis, in a cell-free extract of *Saccharomyces cerevisiae* that translates natural messenger ribonucleic acid. *J. Biol. Chem.* **254**: 3970.

Gasior, E., F. Herrera, I. Sadnik, C.S. McLaughlin, and K. Moldave. 1979b. The preparation and characterization of a cell-free system from *Saccharomyces cerevisiae* that translates natural messenger ribonucleic acid. *J. Biol. Chem.* **254**: 3965.

Gelugne, J. and J.B. Bell. 1988. Modifiers of ochre suppressors in *Saccharomyces cerevisiae* that exhibit ochre suppressor-dependent amber suppression. *Curr. Genet.* **14**:

345.

Gerlach, W.L. 1975. Mutational properties of *supP* amber-ochre supersuppressors in *Saccharomyces cerevisiae. Mol. Gen. Genet.* **144:** 213.

Gilmore, R.A., J.W. Stewart, and F. Sherman. 1971. Amino acid replacements resulting from super-suppression of nonsense mutants of iso-1-cytochrome *c* from yeast. *J. Mol. Biol.* **61:** 157.

Goodman, H.M., M.V. Olson, and B.D. Hall. 1977. Nucleotide sequence of a mutant eukaryotic gene: The yeast tyrosine-inserting ochre suppressor *SUP4-o. Proc. Natl. Acad. Sci.* **74:** 5453.

Goyer, C., M. Altmann, H. Trachsel, and N. Sonenberg. 1989. Identification and characterization of cap binding proteins from yeast. *J. Biol. Chem.* **264:** 7603.

Grossenbacher, A., B. Stadelmann, W. Heyer, P. Thuriaux, J. Kohli, C. Smith, P.F. Agris, K.C. Kuo, and C. Gehrke. 1986. Antisuppressor mutations and sulfur-carrying nucleosides in transfer RNAs of *Schizosaccharomyces pombe. J. Biol. Chem.* **261:** 16351.

Groudinsky, O., G. Dujardin, and P.P. Slonimski. 1981. Long range control circuits within mitochondria and between nucleus and mitochondria. *Mol. Gen. Genet.* **184:** 493.

Guthrie, C. and J. Abelson. 1982. Organization and expression of tRNA genes in *Saccharomyces cerevisiae.* In *The molecular biology of the yeast* Saccharomyces: *Metabolism and gene expression* (ed. J.N. Strathern et al.), p. 487. Cold Spring Harbor Laboratory, Cold Spring Harbor, New York.

Hamilton, R., C.K. Watanabe, and H.A. DeBoer. 1987. Compilation and comparison of the sequence context around the AUG startcodons in *Saccharomyces cerevisiae* mRNAs. *Nucleic Acids Res.* **15:** 3581.

Hanic-Joyce, P.J. 1985. Mapping *CDC* mutations in the yeast *S. cerevisiae* by *rad52*-mediated chromosome loss. *Genetics* **110:** 591.

Hanic-Joyce, P.J., R.A. Singer, and G.C. Johnston. 1987. Molecular characterization of the yeast *PRT1* gene in which mutations affect translation initiation and regulation of cell proliferation. *J. Biol. Chem.* **262:** 2845.

Hanks, S.K., A.M. Quinn, and T. Hunter. 1988. The protein kinase family: Conserved features and deduced phylogeny of the catalytic domains. *Science* **241:** 41.

Hannig, E.M. and A.G. Hinnebusch. 1988. Molecular analysis of *GCN3*, a translational activator of *GCN4*: Evidence for posttranslational control of *GCN3* regulatory function. *Mol. Cell. Biol.* **8:** 4808.

Hannig, E.M., D.J. Thiele, and M.J. Leibowitz. 1984. *Saccharomyces cerevisiae* killer virus transcripts contain template-coded polyadenylate tracts. *Mol. Cell. Biol.* **4:** 101.

Harashima, S. and A.G. Hinnebusch. 1986. Multiple *GCD* genes required for repression of *GCN4*, a transcriptional activator of amino acid biosynthetic genes in *Saccharomyces cerevisiae. Mol. Cell. Biol.* **6:** 3990.

Harashima, S., E.M. Hannig, and A.G. Hinnebusch. 1987. Interactions between positive and negative regulators of *GCN4* controlling gene expression and entry into the yeast cell cycle. *Genetics* **117:** 409.

Hartwell, L.H. and C.S. McLaughlin. 1968a. Temperature-sensitive mutants of yeast exhibiting a rapid inhibition of protein synthesis. *J. Bacteriol.* **96:** 1664.

———. 1968b. Mutants of yeast with temperature-sensitive isoleucyl-tRNA synthetases. *Proc. Natl. Acad. Sci.* **59:** 422.

———. 1969. A mutant of yeast apparently defective in the initiation of protein synthesis. *Proc. Natl. Acad. Sci.* **62:** 468.

Hawthorne, D.C. 1976. UGA mutations and UGA suppressors in yeast. *Biochimie* **58:**

179.

———. 1981. UGA suppressors in yeast. *Proc. Alfred Benzon Symp.* **16:** 291.

Hawthorne, D.C. and U. Leupold. 1974. Suppressor mutations in yeast. *Curr. Genet.* **64:** 1.

Hawthorne, D.C. and R.K. Mortimer. 1968. Genetic mapping of nonsense suppressors in yeast. *Genetics* **60:** 735.

Heck, J.D. and G.W. Hatfield. 1988. Valyl-tRNA synthetase gene of *Escherichia coli* K12. *J. Biol. Chem.* **263:** 868.

Heider, H., E. Gottschalk, and F. Cramer. 1971. Isolation and characterization of seryl-tRNA synthetase from yeast. *Eur. J. Biochem.* **20:** 144.

Heller, D.L., K.M. Gianda, and L. Leinwand. 1988. A highly conserved mouse gene with a propensity to form pseudogenes in mammals. *Mol. Cell. Biol.* **8:** 2797.

Herbert, C.J., M. Labouesse, G. Dujardin, and P.P. Slonimski. 1988. The NAM2 proteins from *S. cerevisiae* and *S. douglasii* are mitochondrial leucyl-tRNA synthetases, and are involved in mRNA spicing. *EMBO J.* **7:** 473.

Herrera, F., J.A. Martinez, H. Mornet, I. Sadnik, C.S. McLaughlin, B. Feinberg, and K. Moldave. 1984. Identification of an altered elongation factor in temperature-sensitive mutant ts 7' -14 of *Saccharomyces cerevisiae. J. Biol. Chem.* **259:** 14347.

Heyer, W., P. Thuriaux, J. Kohli, P. Ebert, H. Kersten, C. Gehrke, K.C. Kuo, and P.F. Agris. 1984. An antisuppressor mutation of *Schizosaccharomyces pombe* affects the post-transcriptional modification of the "wobble" base in the anticodon of tRNAs. *J. Biol. Chem.* **259:** 2856.

Higgins, C.F., I.D. Hiles, P.C. Salmond, D.R. Gill, J.A. Downie, I.J. Evans, I.B. Holland, L. Gray, S.D. Buckel, A.W. Bell, and M.A. Hermodson. 1986. A family of related ATP-binding subunits coupled to many distinct biological processes in bacteria. *Nature* **323:** 448.

Hilger, F., M. Prevot, and R. Mortimer. 1982. Genetic mapping of *arg*, *cpa*, *car* and *tsm* genes in *Saccharomyces cerevisiae* by trisomic analysis. *Curr. Genet.* **6:** 93.

Hill, D.E. and K. Struhl. 1988. Molecular characterization of *GCD1*, a yeast gene required for general control of amino acid biosynthesis and cell-cycle initiation. *Nucleic Acids. Res.* **16:** 9253.

Himmelfarb, H.J., E. Maicas, and J.D. Friesen. 1985. Isolation of the *sup45* omnipotent suppressor gene of *Saccharomyces cerevisiae* and characterization of its gene product. *Mol. Cell. Biol.* **5:** 816.

Hinnebusch, A.G. 1984. Evidence for translational regulation of the activator of general amino acid control in yeast. *Proc. Natl. Acad. Sci.* **81:** 6442.

———. 1985. A hierarchy of *trans*-acting factors modulate translation of an activator of amino acid biosynthetic genes in yeast. *Mol. Cell. Biol.* **5:** 2349.

———. 1988. Mechanisms of gene regulation in the general control of amino acid biosynthesis in *Saccharomyces cerevisiae. Microbiol. Rev.* **52:** 248.

Hinnebusch, A.G. and G.R. Fink. 1983. Positive regulation in the general amino acid control of *Saccharomyces cerevisiae. Proc. Natl. Acad. Sci.* **80:** 5374.

Hinnebusch, A.G., B.M. Jackson, and P.P. Mueller. 1988. Evidence for regulation of reinitiation in translational control of *GCN4* mRNA. *Proc. Natl. Acad. Sci.* **85:** 7279.

Hoekema, A., R.A. Kastelein, M. Vasser, and H.A. DeBoer. 1987. Codon replacement in the *PGK1* gene of *Saccharomyces cerevisiae*: Experimental approach to study the role of biased codon usage in gene expression. *Mol. Cell. Biol.* **7:** 2914.

Hopper, A.K., S.L. Nolan, J. Kurjan, and A. Hama-Furukawa. 1981. Genetic and biochemical approaches to studying *in vivo* intermediates in tRNA biosynthesis. *Proc. Alfred Benzon Symp.* **16:** 302.

Hou, Y.-M. and P. Schimmel. 1988. A simple structural feature is a major determinant of the identity of a transfer RNA. *Nature* **333**: 140.

Hountondji, C. and S. Blanquet. 1985. Methionyl-tRNA synthetase from *Escherichia coli*: Primary structure at the binding site for the 3 ′ -end of tRNA$^{Met}_f$. *Biochemistry* **24**: 1175.

Hountondji, C., P. Dessen, and S. Blanquet. 1986a. Sequence similarities among the family of aminoacyl-tRNA synthetases. *Biochimie* **68**: 1071.

Hountondji, C., F. Lederer, P. Dessen, and S. Blanquet. 1986b. *Escherichia coli* tyrosyl- and methionyl-tRNA synthetases display sequence similarity at the binding site for the 3 ′ -end of tRNA. *Biochemistry* **25**: 16.

Huang, S., R.C. Elliot, P.-S. Liu, R.K. Koduri, J.L. Weickmann, J.-H. Lee, L.C. Blair, P. Ghosh-Dastidar, R.A. Bradshaw, K.M. Bryan, B. Einarson, R.L. Kendall, K.H. Kolacz, and K. Saito. 1987. Specificity of cotranslational amino-terminal processing of proteins in yeast. *Biochemistry* **26**: 8242.

Hughes, D., J.F. Atkins, and S. Thompson. 1987. Mutants of elongation factor Tu promote ribosomal frameshifting and nonsense readthrough. *EMBO J.* **6**: 4235.

Hussain, I. and M.J. Leibowitz. 1986. Translation of homologous and heterologous messenger RNAs in a yeast cell-free system. *Gene* **46**: 13.

Hutchison, J.S., B. Feinberg, T.C. Rothwell, and K. Moldave. 1984. Monoclonal antibody specific for yeast elongation factor 3. *Biochemistry* **23**: 3055.

Icho, T. and R.B. Wickner. 1989. The double-stranded RNA genome of yeast virus L-A encodes its own putative RNA polymerase by fusing two open reading frames. *J. Biol. Chem.* **264**:6716.

Ikemura, T. 1982. Correlation between the abundance of yeast transfer RNAs and the occurrence of the respective codons in protein genes differences in synonymous codon choice patterns of yeast and *Escherichia coli* with reference to the abundance of isoaccepting transfer RNAs. *J. Mol. Biol.* **158**: 573.

Inge-Vechtomov, S.G. and V.M. Andrianova. 1970. Recessive super-suppressors in yeast. *Genetika* **6**: 103.

Ishiguro, J. 1981. Genetic and biochemical characterization of antisuppressor mutants in the yeast *Saccharomyces cerevisiae*. *Curr. Genet.* **4**: 197.

Iwasaki, K. and Y. Kaziro. 1979. Polypeptide chain elongation factors from pig liver. *Methods Enzymol.* **60**: 657.

Jacks, T. and H.E. Varmus. 1985. Expression of the rous sarcoma virus *pol* gene by ribosomal frameshifting. *Science* **230**: 1237.

Jacks, T., H.D. Madhani, F.R. Masiarz, and H.E. Varmus. 1988. Signals for ribosomal frameshifting in the rous sarcoma virus *gag-pol* region. *Cell* **55**: 447.

Janner, F., G. Vogeli, and R. Fluri. 1980. The antisuppressor strain *sin1* of *Schizosaccharomyces pombe* lacks the modification isopentenyladenosine in transfer RNA. *J. Mol. Biol.* **139**: 207.

Johansen, H., D. Schumperli, and M. Rosenberg. 1984. Affecting gene expression by altering the length and sequence of the 5 ′ leader. *Proc. Natl. Acad. Sci.* **81**: 7698.

Jordana, X., B. Chatton, M. Paz-Weisshaar, J.-M. Buhler, F. Cramer, J.P. Ebel, and F. Fasiolo. 1987. Structure of the yeast valyl-tRNA synthetase gene (*VAS1*) and the homology of its translated amino acid sequence with *Escherichia coli* isoleucyl-tRNA synthetase. *J. Biol. Chem.* **262**: 7189.

Kamath, A. and K. Chakraburtty. 1986a. Identification of an altered elongation factor in thermolabile mutants of the yeast *Saccharomyces cerevisiae*. *J. Biol. Chem.* **261**: 12593.

———. 1986b. Purification of elongation factor 3 from temperature-sensitive mutant 13-

06 of the yeast *Saccharomyces cerevisiae. J. Biol. Chem.* **261:** 12596.

———. 1989. Role of yeast elongation factor 3 in the elongation cycle. *J. Biol. Chem.* **264:** 15423.

Kikuchi, Y., H. Shimatake, and A. Kikuchi. 1988. A yeast gene required for the G_1-to-S transition encodes a protein containing an A-kinase target site and GTPase domain. *EMBO J.* **7:** 1175.

Kim, D. and J. Johnson. 1988. Construction, expression, and function of a new yeast amber suppressor, tRNA^{TrpA}. *J. Biol. Chem.* **263:** 7316.

Kimmerly, W., A. Buchman, R. Kornberg, and J. Rine. 1988. Roles of two DNA-binding factors in replication, segregation and transcriptional repression mediated by a yeast silencer. *EMBO J.* **7:** 2241.

Kimura, M. 1984. Proteins of the *Bacillus stearothermophilus* ribosome: The amino acid sequences of proteins S5 and L30. *J. Biol. Chem.* **259:** 1051.

Kloeckener-Gruissem, B., J.E. McEwen, and R.O. Poyton. 1988. Identification of a third nuclear protein-coding gene required specifically for posttranscriptional expression of the mitochondrial *COX3* gene in *Saccharomyces cerevisiae. J. Bacteriol.* **170:** 1399.

Koerner, T.J., A.M. Myers, S. Lee, and A. Tzagoloff. 1987. Isolation and characterization of the yeast gene coding for the alpha subunit of mitochondrial phenylalanyl-tRNA synthetase. *J. Biol. Chem.* **262:** 3690.

Kolman, C.J., M. Snyder, and D. Soll. 1988. Genomic organization of tRNA and aminoacyl-tRNA synthetase genes for two amino acids in *Saccharomyces cerevisiae. Genomics* **3:** 201.

Korte, A., V. Forsbach, T. Gottenof, and G. Roedel. 1989. In vitro and in vivo studies on the mitochondrial import of CBS1, a translational activator of cytochrome *b* in yeast. *Mol. Gen. Genet.* **217:** 162.

Kozak, M. 1984. Selection of initiation sites by eucaryotic ribosomes: Effect of inserting AUG triplets upstream from the coding sequence for preproinsulin. *Nucleic Acids Res.* **12:** 3873.

———. 1986a. Point mutations define a sequence flanking the AUG initiator codon that modulates translation by eukaryotic ribosomes. *Cell* **44:** 283.

———. 1986b. Influences of mRNA secondary structure on initiation by eukaryotic ribosomes. *Proc. Natl. Acad. Sci.* **83:** 2850.

———. 1987a. An analysis of 5′-noncoding sequences from 699 vertebrate messenger RNAs. *Nucleic Acids Res.* **15:** 8125.

———. 1987b. Effects of intercistronic length on the efficiency of reinitiation by eucaryotic ribosomes. *Mol. Cell. Biol.* **7:** 3438.

———. 1989. The scanning model for translation: An update. *J. Cell Biol.* **108:** 229.

Kruszewska, A. and P.P. Slonimski. 1984a. Mitochondrial and nuclear mitoribosomal suppressors that enable misreading of ochre codons in yeast mitochondria. I. Isolation, localization and allelism of suppressors. *Curr. Genet.* **9:** 1.

———. 1984b. Mitochondrial and nuclear mitoribosomal suppressors that enable misreading of ochre codons in yeast mitochondria. II. Specificity and extent of suppressor action. *Curr. Genet.* **9:** 11.

Kushnirov, V.V., M.D. Ter-Avanesyan, A.P. Surguchov, V.N. Smirnov, and S.G. Inge-Vechtomov. 1987. Localization of possible functional domains in *sup2* gene product of the yeast *Saccharomyces cerevisiae. FEBS Lett.* **215:** 257.

Kushnirov, V.V., M.D. Ter-Avanesyan, M.V. Telckov, A.P. Surguchov, V.N. Smirnov, and S.G. Inge-Vechtomov. 1988. Nucleotide sequence of the *sup2* (*sup35* gene of *Saccharomyces cerevisiae*). *Gene* **66:** 45.

Labouesse, M., C.J. Herbert, G. Dujardin, and P.P. Slonimski. 1987. Three suppressor mutations which cure a mitochondrial RNA maturase deficiency occur at the same codon in the open reading frame of the nuclear *NAM2* gene. *EMBO J.* **6:** 713.

Laten, H., J. Cramer, and R. Rownd. 1983. Thiolated nucleotides in yeast transfer RNA. *Biochim. Biophys. Acta* **741:** 1.

Laten, H., J. Gorman, and R.M. Bock. 1978. Isopentenyladenosine deficient tRNA from an antisuppressor mutant of *Saccharomyces cerevisiae*. *Nucleic Acids Res.* **5:** 4329.

Lawrence, C.W. and R. Christensen. 1976. UV mutagenesis in radiation-sensitive strains of yeast. *Genetics* **82:** 207.

Laz, T., J. Clements, and F. Sherman. 1987. The role of messenger RNA sequences and structures in eukaryotic translation. In *Translational regulation of gene expression* (ed. J. Ilan), p. 413. Plenum Press, New York.

———. 1988. Structural features of mRNA that affect translation in yeast. In *Genetics of translation: New approaches* (ed. M.F. Tuite et al.), p. 353. Springer-Verlag, Berlin.

Leer, R.J., M.M.C. Van Raamsdonk-Duin, W.H. Mager, and R. Planta. 1985. Conserved sequences upstream of yeast ribosomal protein genes. *Curr. Genet.* **9:** 273.

Liebman, S.W. and J.A. All-Robyn. 1984. A non-Mendelian factor, [*eta*+], causes lethality of yeast omnipotent suppressor strains. *Curr. Genet.* **8:** 567.

Liebman, S.W. and M.M. Cavenagh. 1980. An antisuppressor that acts on omnipotent suppressors in yeast. *Genetics* **95:** 49.

Liebman, S.W. and F. Sherman. 1979. The extrachromosomal determinant, [*psi*], suppresses nonsense mutations in yeast. *J. Bacteriol.* **139:** 1068.

Liebman, S.W., M.M. Cavenagh, and L.N. Bennett. 1980. Isolation and properties of an antisuppressor in *Saccharomyces cerevisiae* specific for an omnipotent suppressor. *J. Bacteriol.* **143:** 1527.

Liebman, S.W., F. Sherman, and J.W. Stewart. 1976. Isolation and characterization of amber suppressors in yeast. *Genetics* **82:** 251.

Liebman, S.W., J.W. Stewart, and F. Sherman. 1975. Serine substitutions caused by an ochre suppressor in yeast. *J. Mol. Biol.* **94:** 595.

Liebman, S.W., J.W. Stewart, J.H. Parker, and F. Sherman. 1977. Leucine insertion caused by a yeast amber suppressor. *J. Mol. Biol.* **109:** 13.

Liebman, S.W., J.M. Song, J. All-Robyn, E. Griffin, and D. Kelley-Geraghty. 1988. Omnipotent suppressors, allosuppressors and antisuppressors of yeast. In *Genetics of translation* (ed. M.F. Tuite et al.), p. 403. Springer-Verlag, Berlin.

Liebman, S.W., Z. Srodulski, C.R. Reed, J.W. Stewart, F. Sherman, and G. Brennan. 1984. Yeast amber suppressors corresponding to tRNA$^{Leu}_3$ genes. *J. Mol. Biol.* **178:** 209.

Leibowitz, M.J., I. Hussain, and T.L. Williams. 1988. Transcription and translation of the yeast killer virus genome. In *Viruses of fungi and simple eukaryotes* (ed. Y. Koltin), p. 133. Marcel Dekker, New York.

Leibowitz, M.J., D.J. Thiele, and E.M. Hannig. 1983. Structure of separated strands of double-stranded RNA from killer virus of yeast; a translational model. In *Double-stranded RNA viruses* (ed. D.H.L. Bishop and R.W. Compans), p. 457. Elsevier, New York.

Linder, P. and P.P. Slonimski. 1988. Sequence of the genes *TIF1* and *TIF2* from *Saccharomyces cerevisiae* coding for a translation initiation factor. *Nucleic Acids Res.* **16:** 10359.

———. 1989. An essential yeast protein, coded by duplicated genes *TIF1* and *TIF2*, and homologues to the mammalian translation initiation factor eIF-4E can suppress a mitochondrial missense mutation. *Proc. Natl. Acad. Sci.* **86:** 2286.

Linder, P., P.F. Lasko, M. Ashburner, P. Leroy, P. Nielsen, K. Nishi, J. Schnier, and P.P.

Slonimski. 1989. Birth of the D-E-A-D box. *Nature* **337**: 121.

Liu, C., C.C. Simonsen, and A.D. Levinson. 1984. Initiation of translation at internal AUG codons in mammalian cells. *Nature* **309**: 82.

Lorber, B., H. Mejdoub, J. Reinbolt, Y. Boulanger, and R. Giege. 1988. Properties of *N*-terminal truncated yeast aspartyl-tRNA synthetase and structural characteristics of the cleaved domain. *Eur. J. Biochem.* **174**: 155.

Ludmerer, S.W. and P. Schimmel. 1987a. Gene for yeast glutamine tRNA synthetase encodes a large amino-terminal extension and provides a strong confirmation of the signature sequence for a group of the aminoacyl-tRNA synthetases. *J. Biol. Chem.* **262**: 10801.

Ludmerer, S.W. and P. Schimmel. 1987b. Construction and analysis of deletions in the amino-terminal extension of glutamine tRNA synthetase of *Saccharomyces cerevisiae*. *J. Biol. Chem.* **262**: 10807.

Lund, P.M. and B.S. Cox. 1981. Reversion analysis of [*psi*⁻] mutations in *Saccharomyces cerevisiae*. *Genet. Res.* **37**: 173.

Martindale, D.W., Z. Ming, and C. Csank. 1989. Isolation and complete sequence of the yeast isoleucyl-tRNA synthetase gene (*ILS1*). *Curr. Genet.* **15**: 99.

Marykwas, D.L. and T.D. Fox. 1989. Control of the *Saccharomyces cerevisiae* regulatory gene *PET494*: Transcriptional repression by glucose and translational induction by oxygen. *Mol. Cell. Biol.* **9**: 484.

Masson, J., P. Meuris, M. Grunstein, J. Abelson, and J.H. Miller. 1987. Expression of a set of synthetic suppressor tRNA[Phe] genes in *Saccharomyces cerevisiae*. *Proc. Natl. Acad. Sci.* **84**: 6815.

Masurekar, M., E. Palmer, B. Ono, J.M. Wilhelm, and F. Sherman. 1981. Misreading of the ribosomal suppressor *SUP46* due to an altered 40S subunit in yeast. *J. Mol. Biol.* **147**: 381.

McClain, W.H. and K. Foss. 1988a. Nucleotides that contribute to the identity of *Escherichia coli* tRNA[Phe]. *J. Mol. Biol.* **202**: 697.

―――. 1988b. Changing the identity of a tRNA by introducing a G-U wobble pair near the 3′ acceptor end. *Science* **240**: 793.

McCready, S.J. and B. Cox. 1973. Antisuppressors in yeast. *Mol. Gen. Genet.* **124**: 305.

McCusker, J.H. and J.E. Haber. 1988a. Cycloheximide-resistant temperature-sensitive lethal mutations of *Saccharomyces cerevisiae*. *Genetics* **119**: 303.

―――. 1988b. *crl* mutants of *Saccharomyces cerevisiae* resemble both mutants affecting general control of amino acid biosynthesis and omnipotent translational suppressor mutants. *Genetics* **119**: 317.

McLaughlin, C.S. and L.H. Hartwell. 1969. A mutant of yeast with a defective methionyl-tRNA synthetase. *Genetics* **61**: 557.

McLaughlin, C.S., P.T. Magee, and L.H. Hartwell. 1969. Role of isoleucyl-tRNA ribonucleic acid synthetase in ribonucleic acid synthesis and enzyme repression in yeast. *J. Bacteriol.* **100**: 579.

McMullin, T.W., P. Haffner, and T.D. Fox. 1990. A novel small subunit ribosomal protein of yeast mitochondria that interacts functionally with an mRNA-specific translational activator. *Mol. Cell. Biol.* **10**: 4590.

Mellor, J., S.M. Fulton, M.J. Dobson, W. Wilson, S.M. Kingsman, and A.J. Kingsman. 1985. A retrovirus-like strategy for expression of a fusion protein encoded by yeast transposon Ty*1*. *Nature* **313**: 243.

Mendenhall, M.D. and M.R. Culbertson. 1988. The yeast *SUF3* frameshift suppressor encodes a mutant glycine tRNA$_{CCC}$. *Nucleic Acids Res.* **16**: 8713.

Mendenhall, M.D., P. Leeds, H. Fen, L. Mathison, M. Zwick, C. Sleiziz, and M.R. Culbertson. 1987. Frameshift suppressor mutations affecting the major glycine transfer RNAs of *Saccharomyces cerevisiae*. *J. Mol. Biol.* **194:** 41.

Merrick, W.C., J.F. Carvalho, and G. Carvalho. 1980. Purification and characterization of various forms of elongation factor 1. *Fed. Proc.* **39:** 2029.

Messenguy, F. and J. Delforge. 1976. Role of transfer ribonucleic acids in the regulation of several biosyntheses in *Saccharomyces cerevisiae*. *Eur. J. Biochem.* **67:** 335.

Messenguy, F. and E. Dubois. 1983. Participation of transcriptional and post-transcriptional regulatory mechanisms in the control of arginine metabolism in yeast. *Mol. Gen. Genet.* **189:** 148.

Messenguy, F., A. Feller, M. Crabeel, and A. Pierard. 1983. Control mechanisms acting at the transcriptional and post-transcriptional levels are involved in the synthesis of the arginine carbamoylphosphate synthase of yeast. *EMBO J.* **2:** 1249.

Michaelis, U., T. Schlapp, and G. Roedel. 1988. Yeast nuclear gene *CBS2*, required for translational activation of cytochrome *b*, encodes a basic protein of 45 kDa. *Mol. Gen. Genet.* **214:** 263.

Miller, D.L. and H. Weissbach. 1977. *Molecular mechanisms of protein biosynthesis*. Academic Press, New York.

Miller, P.F. and A.G. Hinnebusch. 1989. Sequences that surround the stop codons of upstream open reading frames in *GCN4* mRNA determine their distinct functions in translational control. *Genes Dev.* **3:** 1217.

Mirande, M. and J.-P. Waller. 1988. The yeast lysyl-tRNA synthetase gene: Evidence for general amino acid control of its expression and domain structure of the encoded protein. *J. Biol. Chem.* **263:** 18443.

Mitchell, A.P. and S.W. Ludmerer. 1984. Identification of a glutaminyl-tRNA synthetase mutation in *Saccharomyces cerevisiae*. *J. Bacteriol.* **158:** 530.

Miyazaki, M., M. Uritani, and H. Kagiyama. 1988. The yeast peptide elongation factor 3 (EF-3) carries an active site for ATP hydrolysis which can interact with various nucleoside triphosphates in the absence of ribosomes. *J. Biochem.* **104:** 445.

Moehring, J.M., T.J. Moehring, and D.E. Danley. 1980. Posttranslational modification of elongation factor 2 in diphtheria toxin-resistant mutants of CHO-K1 cells. *Proc. Natl. Acad. Sci.* **77:** 1010.

Moldave, K. 1985. Eukaryotic protein synthesis. *Annu. Rev. Biochem.* **54:** 1109.

Moldave, K. and C.S. McLaughlin. 1988. The analysis of temperature-sensitive mutants of *Saccharomyces cerevisiae* altered in components required for protein synthesis. In *Genetics of translation: New approaches* (ed. M.F. Tuite et al.), p. 271. Springer-Verlag, Berlin.

Mortimer, R.K. and D.C. Hawthorne. 1973. Genetic mapping in *Saccharomyces*. IV. Mapping of temperature-sensitive genes and use of disomic strains in localizing genes. *Genetics* **74:** 33.

Mueller, P.P. and A.G. Hinnebusch. 1986. Multiple upstream AUG codons mediate translational control of GCN4. *Cell* **45:** 201.

Mueller, P.P., S. Harashima, and A.G. Hinnebusch. 1987. A segment of *GCN4* mRNA containing the upstream AUG codons confers translational control upon a heterologous yeast transcript. *Proc. Natl. Acad. Sci.* **84:** 2863.

Mueller, P.P., B.M. Jackson, P.F. Miller, and A.G. Hinnebusch. 1988. The first and fourth upstream open reading frames in *GCN4* mRNA have similar initiation efficiencies but respond differently in translational control to changes in length and sequence. *Mol. Cell. Biol.* **8:** 5439.

Mueller, P.P., M.K. Reif, S. Zonghou, C. Sengstag, T.L. Mason, and T.D. Fox. 1984. A nuclear mutation that post-transcriptionally blocks accumulation of a yeast mitochondrial gene product can be suppressed by a mitochondrial gene rearrangement. *J. Mol. Biol.* **175**: 431.

Myers, A.M. and A. Tzagoloff. 1985. *MSW*, a yeast gene coding for mitochondrial tryptophanyl-tRNA synthetase. *J. Biol. Chem.* **260**: 15371.

Nagashima, K., M. Kasai, S. Nagata, and Y. Kaziro. 1986. Structure of the two genes coding for polypeptide chain elongation factor 1α (EF-1α) from *Saccharomyces cerevisiae. Gene* **45**: 265.

Nagata, S., Y. Tsunetsugu-Yokota, A. Naito, and Y. Kaziro. 1983. Molecular cloning and sequence determination of the nuclear gene coding for mitochondrial elongation factor Tu of *Saccharomyces cerevisiae. Proc. Natl. Acad. Sci.* **80**: 6192.

Nagata, S., K. Nagashima, Y. Tsunetsugu-Yokota, K. Fujimura, M. Miyazaki, and Y. Kaziro. 1984. Polypeptide chain elongation factor 1α (EF-1α) from yeast: Nucleotide sequence of one of the two genes for EF-1α from *Saccharomyces cerevisiae. EMBO J.* **3**: 1825.

Najarian, D., M.E. Dihanich, N.C. Martin, and A.K. Hopper. 1987. DNA sequence and transcript mapping of *MOD5*: features of the 5′ region which suggest two translational starts. *Mol. Cell. Biol.* **7**: 185.

Nasrin, N., M.F. Ahmad, M.K. Nag, P. Tarburton, and N.K. Gupta. 1986. Protein synthesis in yeast *Saccharomyces cerevisiae*: Purification of Co-eIF-2A and "mRNA-binding factor(s)" and studies of their roles in Met-tRNA$_f$·40S·mRNA complex formation. *Eur. J. Biochem.* **161**: 1.

Nass, G. and K. Poralla. 1976. Genetics of borrelidin resistant mutants of *Saccharomyces cerevisiae* and properties of their threonyl-tRNA-synthetase. *Mol. Gen. Genet.* **147**: 39.

Natsoulis, G., F. Hilger, and G.R. Fink. 1986. The *HTS1* gene encodes both the cytoplasmic and mitochondrial histidine tRNA synthetases of *S. cerevisiae. Cell* **46**: 235.

Niederberger, P., M. Aebi, and R. Huetter. 1983. Influence of the general control of amino acid biosynthesis on cell growth and cell viability in *Saccharomyces cerevisiae. J. Gen. Microbiol.* **129**: 2571.

―――. 1986. Identification and characterization of four new *GCD* genes in *Saccharomyces cerevisiae. Curr. Genet.* **10**: 657.

Nolan, R.D., H. Grasmuk, and J. Drews. 1979. Preparation of elongation factors from ascites cells. *Methods Enzymol.* **60**: 649.

Normanly, J., R.C. Ogden, S.J. Horvath, and J. Abelson. 1986. Changing the identity of a transfer RNA. *Nature* **321**: 213.

Nyunoya, H. and C.J. Lusty. 1984. Sequence of the small subunit of yeast carbamoyl phosphate synthetase and identification of its catalytic domain. *J. Biol. Chem.* **259**: 9790.

Ohyama, K., H. Fukuzawa, T. Kohchi, H. Shirai, T. Sano, S. Sano, K. Umesono, Y. Shiki, M. Takeuchi, Z. Chang, S. Aota, H. Inokuchi, and H. Ozeki. 1986. Chloroplast gene organization deduced from complete sequence of liverwort *Marchantia polymorpha* chloroplast DNA. *Nature* **322**: 572.

Olson, M.V., G.S. Page, A. Sentenac, P.W. Piper, M. Worthington, R.B. Weiss, and B.D. Hall. 1981. Only one of two closely related yeast suppressor tRNA genes contains an intervening sequence. *Nature* **291**: 464.

Ono, B., J.W. Stewart, and F. Sherman. 1979a. Yeast UAA suppressors effective in [psi⁺] strains: Serine-inserting suppressors. *J. Mol. Biol.* **128**: 81.

―――. 1979b. Yeast UAA suppressors effective in [psi⁺] strains: Leucine-inserting suppressors. *J. Mol. Biol.* **132**: 507.

―――. 1981. Serine insertion caused by the ribosomal suppressor *SUP46* in yeast. *J. Mol. Biol.* **147**: 373.

Ono, B., Y. Ishino-Arao, T. Shirai, N. Maeda, and S. Shinoda. 1985. Genetic mapping of leucine-inserting UAA suppressors in *Saccharomyces cerevisiae*. *Curr. Genet.* **9**: 197.

Ono, B., Y. Ishino-Arao, M. Tanaka, I. Awano, and S. Shinoda. 1986. Recessive nonsense suppressors in *Saccharomyces cerevisiae*: Action spectra, complementation groups and map positions. *Genetics* **114**: 363.

Ono, B., M. Tanaka, M. Kominami, Y. Ishino, and S. Shinoda. 1982. Recessive UAA suppressors of the yeast *Saccharomyces cerevisiae*. *Genetics* **102**: 653.

Ono, B., N. Moriga, K. Ishihara, J. Ishiguro, Y. Ishino, and S. Shinoda. 1984. Omnipotent suppressors effective in [*psi*⁺] strains of *Saccharomyces cerevisiae* recessiveness and dominance. *Genetics* **107**: 219.

Ono, B., R. Fujimoto, Y. Ohno, N. Maeda, Y. Tsuchiya, T. Usui, and Y. Ishino-Arao. 1988. UGA suppressors in *Saccharomyces cerevisiae*: Allelism, action spectra and map positions. *Genetics* **118**: 41.

Ono, B., M. Tanaka, I. Awano, F. Okamoto, R. Satoh, N. Yamagishi, and Y. Ishino-Arao. 1989. Two new loci that give rise to dominant omnipotent suppressors in *Saccharomyces cerevisiae*. *Curr. Genet.* **16**: 323.

Paddon, C.J. and A.G. Hinnebusch. 1989. *gcd12* mutations are *gcn3*-dependent alleles of *GCD2*, a negative regulator of *GCN4* in the general amino acid control of *Saccharomyces cerevisiae*. *Genetics* **122**: 543.

Paddon, C.J., E.M. Hannig, and A.G. Hinnebusch. 1989. Amino acid sequence similarity between GCN3 and GCD2, positive and negative translational regulators of *GCN4*: Evidence for antagonism by competition. *Genetics* **122**: 551.

Palmer, E., J. Wilhelm, and F. Sherman. 1979a. Phenotypic suppression of nonsense mutants in yeast by aminoglycoside antibiotics. *Nature* **277**: 148.

――――. 1979b. Variation of phenotypic suppression due to the [psi⁺] and [psi⁻] extrachromosomal determinants in yeast. *J. Mol. Biol.* **128**: 107.

Pape, L.K. and A. Tzagoloff. 1985. Cloning and characterization of the gene for the yeast cytoplasmic threonyl-tRNA synthetase. *Nucleic Acids Res.* **13**: 6171.

Pape, L.K., T.J. Koerner, and A. Tzagoloff. 1985. Characterization of a yeast nuclear gene (*MST1*) coding for the mitochondrial threonyl-tRNA₁ synthetase. *J. Biol. Chem.* **260**: 15362.

Pappenheimer, A.M., Jr. 1977. Diphtheria toxin. *Annu. Rev. Biochem.* **46**: 69.

Pathak, V.K., P.J. Nielsen, H. Trachsel, and J.W.B. Hershey. 1988. Structure of the β subunit of translational initiation factor eIF-2. *Cell* **54**: 633.

Peabody, D.S. and P. Berg. 1986. Termination-reinitiation occurs in the translation of mammalian cell mRNAs. *Mol. Cell. Biol.* **6**: 2695.

Pelletier, J. and N. Sonenberg. 1985. Insertion mutagenesis to increase secondary structure within the 5' noncoding region of a eukaryotic mRNA reduces translational efficiency. *Cell* **40**: 515.

Perez-Gosalbez, M., G.L. Rivera, and J.P.G. Ballesta. 1983. Affinity labelling of peptidyl transferase center using the 3' terminal pentanucleotide from amino-acyl-tRNA. *Biochem. Biophys. Res. Commun.* **113**: 941.

Perez-Gosalbez, M., D. Vazquez, and J.P.G. Ballesta. 1978. Affinity labelling of yeast ribosomal peptidyl transferase. *Mol. Gen. Genet.* **163**: 29.

Perlman, D., H.O. Halvorson, and L.E. Cannon. 1982. Presecretory and cytoplasmic invertase polypeptides encoded by distinct mRNAs derived from the same structural gene differ by a signal sequence. *Proc. Natl. Acad. Sci.* **79**: 781.

Perona, J.J., R.N. Swanson, M.A. Rould, T.A. Steitz, and D. Soll. 1989. Structural basis for misaminoacylation by mutant *E. coli* glutaminyl-tRNA synthetase enzymes. *Science* **246**: 1152.

Piepersberg, W., D. Geyl, H. Hummel, and A. Bock. 1980. Physiology and *biochemistry* of bacterial ribosomal mutants. In *Genetics and evolution of RNA polymerases, tRNA and ribosomes* (ed. S. Osawa et al.), p. 359. Tokyo University Press, Tokyo.

Piper, P.W. 1978. A correlation between a recessive lethal amber suppressor mutation in *S. cerevisiae* and an anticodon change in minor serine tRNA. *J. Mol. Biol.* **122:** 217.

Piper, P.W., M. Wasserstein, F. Engbaek, K. Kaltoft, J.E. Celis, J. Zeuthen, S. Liebman, and F. Sherman. 1976. Nonsense suppressors of *Saccharomyces cerevisiae* can be generated by mutation of the tyrosine tRNA anticodon. *Nature* **262:** 757.

Poutre, C.G. and T.D. Fox. 1987. *PET111*, a *Saccharomyces cerevisiae* nuclear gene required for translation of the mitochondrial mRNA encoding cytochrome *c* oxidase subunit II. *Genetics* **115:** 637.

Prather, N.E., E.J. Murgola, and B.H. Mims. 1984. Nucleotide substitution in the amino acid acceptor stem of lysine transfer RNA causes missense suppression. *J. Mol. Biol.* **172:** 177.

Pringle, J.R. and L.H. Hartwell. 1982. The *Saccharomyces cerevisiae* cell cycle. In *The molecular biology of the yeast* Saccharomyces: *Life cycle and inheritance* (ed. J.N. Strathern et al.), p. 97. Cold Spring Harbor Laboratory, Cold Spring Harbor, New York.

Pure, G.A., G.W. Robinson, L. Naumovski, and E.C. Friedberg. 1985. Partial suppression of an ochre mutation in *Saccharomyces cerevisiae* by multicopy plasmids containing a normal yeast tRNA^{Gln} gene. *J. Mol. Biol.* **183:** 31.

Qin, S., K. Moldave, and C.S. McLaughlin. 1987. Isolation of the yeast gene encoding elongation factor 3 for protein synthesis. *J. Biol. Chem.* **262:** 7802.

Qin, S., A. Xie, M.C.M. Bonato, and C.S. McLaughlin. 1990. Sequence analysis of the translational elongation factor 3 from *Saccharomyces cerevisiae*. *J. Biol. Chem.* **265:** 1903.

Ramos, F. and J.-M. Wiame. 1979. Synthesis and activation of asparagine in asparagine auxotrophs of *Saccharomyces cerevisiae*. *Eur. J. Biochem.* **94:** 409.

Reynolds, P., S. Weber, and L. Prakash. 1985. RAD6 gene of *Saccharomyces cerevisiae* encodes a protein containing a tract of 13 consecutive aspartates. *Proc. Natl. Acad. Sci.* **82:** 168.

Richter, D. and F. Flink. 1971. Isolation of peptide chain elongation factors from the yeast *Saccharomyces cerevisiae* (strain kaneka). *Methods Enzymol.* **20:** 349.

Richter, D. and F. Lipmann. 1970. Separation of mitochondrial and cytoplasmic peptide chain elongation factors from yeast. *Biochemistry* **9:** 5065.

Robertson, J.M. and W. Wintermeyer. 1987. Mechanism of ribosomal translocation: tRNA binds transiently to an exit site before leaving the ribosome during translocation. *J. Mol. Biol.* **196:** 525.

Roedel, G. 1986. Two yeast nuclear genes, *CBS1* and *CBS2*, are required for translation of mitochondrial transcripts bearing the 5'-untranslated *COB* leader. *Curr. Genet.* **11:** 41.

Roedel, G. and T.D. Fox. 1987. The yeast nuclear gene *CBS1* is required for translation of mitochondrial mRNAs bearing the *cob* 5' untranslated leader. *Mol. Gen. Genet.* **206:** 45.

Roedel, G., A. Korte, and F. Kaudewitz. 1985. Mitochondrial suppression of a yeast nuclear mutation which affects the translation of the mitochondrial apocytochrome b transcript. *Curr. Genet.* **9:** 641.

Roedel, G., U. Michaelis, V. Forsbach, J. Kreike, and F. Kaudewitz. 1986. Molecular cloning of the yeast nuclear genes *CBS1* and *CBS2*. *Curr. Genet.* **11:** 47.

Rogers, M.J. and D. Soll. 1988. Discrimination between glutaminyl-tRNA synthetase and seryl-tRNA synthetase involves nucleotides in the acceptor helix of tRNA. *Proc. Natl. Acad. Sci.* **85:** 6627.

Romby, P., D. Moras, M. Bergdoll, P. Dumas, V.V. Vlassov, E. Westhof, J.P. Ebel, and R. Giege. 1985. Yeast tRNA[Asp] tertiary structure in solution and areas of interaction of the tRNA with aspartyl-tRNA synthetase. *J. Mol. Biol.* **184:** 455.

Romero, D.P. and A.E. Dahlberg. 1986. The alpha subunit of initiation factor 2 is phosphorylated in vivo in the yeast *Saccharomyces cerevisiae. Mol. Cell. Biol.* **6:** 1044.

Rosenthal, L.P. and J.W. Bodley. 1987. Purification and characterization of *Saccharomyces cerevisiae* mitochondrial elongation factor Tu. *J. Biol. Chem.* **262:** 10955.

Rotenberg, M.O. and J.L. Woolford. 1986. Tripartite upstream promoter element essential for expression of *Saccharomyces cerevisiae* ribosomal protein genes. *Mol. Cell. Biol.* **6:** 674.

Rould, M.A., J.J. Perona, D. Soll, and T.A. Steitz. 1989. Structure of *E. coli* glutaminyl-tRNA synthetase complexed with tRNA(Gln) and ATP at 2.8 angstrom resolution. *Science* **246:** 1135.

Roussou, I., G. Thireos, and B.M. Hauge. 1988. Transcriptional-translational regulatory circuit in *Saccharomyces cerevisiae* which involves the *GCN4* transcriptional activator and the GCN2 protein kinase. *Mol. Cell. Biol.* **8:** 2132.

Rozen, F., I. Edery, K. Meerovitch, T.E. Dever, W.C. Merrick, and N. Sonenberg. 1990. Bidirectional RNA helicase activity of eucaryotic translation initiation factors 4A and 4F. *Mol. Cell. Biol.* **10:** 1134.

Safer, B. 1983. 2B or not 2B: regulation of the catalytic utilization of eIF-2. *Cell* **33:** 7.

Saha, S.K. and K. Chakraburtty. 1986. Isolation of variant forms of elongation factor 1 from the yeast *Saccharomyces cerevisiae. J. Biol. Chem.* **261:** 12599.

Sampson, J.R. and O.C. Uhlenbeck. 1988. Biochemical and physical characterization of an unmodified yeast phenylalanine transfer RNA transcribed *in vitro. Proc. Natl. Acad. Sci.* **85:** 1033.

Sampson, J.R., A.B. DiRenzo, L.S. Behlen, and O.C. Uhlenbeck. 1989. Nucleotides in yeast tRNA[Phe] required for the specific recognition by its cognate synthetase. *Science* **243:** 1363.

Sandbaken, M.G. and M.R. Culbertson. 1988. Mutations in elongation factor EF1-α affect the frequency of frameshifting and amino acid misincorporation in *Saccharomyces cerevisiae. Genetics* **120:** 923.

Sandmeyer, S.B., V.W. Bilanchone, D.J. Clark, P. Morcos, G.F. Carle, and G.M. Brodeur. 1988. Sigma elements are position-specific for many different yeast tRNA genes. *Nucleic Acids Res.* **16:** 1499.

Sanni, A., M. Mirande, J. Ebel, Y. Boulanger, J. Waller, and F. Fasiolo. 1988. Structure and expression of the genes encoding the α and β subunits of yeast phenylalanyl-tRNA synthetase. *J. Biol. Chem.* **263:** 15407.

Schiltz, E. and J. Reinbolt. 1975. Determination of the complete amino-acid sequence of protein S4 from *Escherichia coli* ribosomes. *Eur. J. Biochem.* **56:** 467.

Schimmel, P. 1987. Aminoacyl tRNA synthetases: General scheme of structure-function relationships in the polypeptides and recognition of transfer RNAs. *Annu. Rev. Biochem.* **56:** 125.

Schimmel, P.R. and D. Soll. 1979. Aminoacyl-tRNA synthetases: General features and recognition of transfer RNAs. *Annu. Rev. Biochem.* **48:** 601.

Schirmaier, F. and P. Philippsen. 1984. Identification of two genes coding for the translation elongation factor EF-1α of *S. cerevisiae. EMBO J.* **3:** 3311.

Schneider, R.J. and T. Shenk. 1987. Translational regulation by adenovirus virus-associated I RNA. In *Translational regulation of gene expression* (ed. J. Ilan), p. 431.

Plenum Press, New York.

Schulman, L.-D.H. and J. Abelson. 1988. Recent excitement in understanding transfer RNA identity. *Science* **240**: 1591.

Schulze, M. and G. Roedel. 1988. *SCO1*, a yeast nuclear gene essential for accumulation of mitochondrial cytochrome *c* oxidase subunit II. *Mol. Gen. Genet.* **211**: 492.

Sedman, S.A., P.J. Good, and J.E. Mertz. 1989. Leader-encoded open reading frames modulate both the absolute and relative rates of synthesis of the virion proteins of simian virus 40. *J. Virol.* **63**: 3884.

Sellami, M., F. Fasiolo, G. Dirheimer, J.-P. Ebel, and J. Gangloff. 1986. Nucleotide sequence of the gene coding for yeast cytoplasmic aspartyl-tRNA synthetase (APS); mapping of the 5' and 3' termini of AspRS mRNA. *Nucleic Acids Res.* **14**: 1657.

Sharp, P.M. and W.-H. Li. 1987. The codon adaptation index—A measure of directional synonomous codon usage, and its potential applications. *Nucleic Acids Res.* **15**: 1281.

Sharp, P.M., T.M.F. Tuohy, and K.R. Mosurski. 1986. Codon usage in yeast: Cluster analysis clearly differentiates highly and lowly expressed genes. *Nucleic Acids Res.* **14**: 5125.

Sharp, P.M., E. Cowe, D.G. Higgins, D.C. Shields, K.H. Wolfe, and F. Wright. 1988. Codon usage patterns in *Escherichia coli, Bacillus subtilis, Saccharomyces cerevisiae, Schizosaccharomyces pombe, Drosophila melanogaster* and *Homo sapiens*; a review of the considerable within-species diversity. *Nucleic Acids Res.* **16**: 8207.

Shen, Z. and T.D. Fox. 1989. Substitution of an invariant nucleotide at the base of the highly conserved "530 loop" of 15S rRNA causes suppression of yeast mitochondrial ochre mutations. *Nucleic Acids Res.* **17**: 4535.

Sherman, F. 1982. Suppression in the yeast *Saccharomyces cerevisiae*. In *The molecular biology of the yeast* Saccharomyces: *Metabolism and gene expression* (ed. J.N. Strathern et al.), p. 463. Cold Spring Harbor Laboratory, Cold Spring Harbor, New York.

Sherman, F. and J.W. Stewart. 1982. Mutations altering initiation of translation of yeast iso-1-cytochrome *c*; contrasts between the eukaryotic and prokaryotic initiation process. In *The molecular biology of the yeast* Saccharomyces: *Metabolism and gene expression* (ed. J.N. Strathern et al.), p. 301. Cold Spring Harbor Laboratory, Cold Spring Harbor, New York.

Sherman, F., J.W. Stewart, and S. Tsunasawa. 1985. Methionine or not methionine at the beginning of a protein. *Bioessays* **3**: 27.

Sherman, F., S.W. Liebman, J.W. Stewart, and M. Jackson. 1973. Tyrosine substitution resulting from suppression of amber mutants of iso-1-cytochrome *c* in yeast. *J. Mol. Biol.* **78**: 157.

Shinozaki, K., M. Ohme, M. Tanaka, T. Wakasugi, N. Hayashida, T. Matsubayashi, N. Zaita, J. Chunwongse, J. Obokata, K. Yamaguchi-Shinozaki, C. Ohto, K. Torazawa, B. Y. Meng, M. Sugita, H. Deno, T. Kamogashira, K. Yamada, J. Kusuda, F. Takaiwa, A. Kata, N. Tohdoh, H. Shimada, and M. Sugiura. 1980. The complete nucleotide sequence of the tobacco chloroplast genome: Its gene organization and expression. *EMBO J.* **5**: 2043.

Shore, D., D.J. Stillman, A.H. Brand, and K.A. Nasmyth. 1987. Identification of silencer binding proteins from yeast: Possible roles in *SIR* control and DNA replication. *EMBO J.* **6**: 461.

Singh, A., D. Ursic, and J. Davies. 1979. Phenotypic suppression and misreading in *Saccharomyces cerevisiae*. *Nature* **277**: 146.

Skogerson, L. 1979. Separation and characterization of yeast elongation factors. *Methods Enzymol.* **60**: 676.

Skogerson, L. and D. Engelhardt. 1977. Dissimilarity in protein chain elongation factor requirements between yeast and rat liver ribosomes. *J. Biol. Chem.* **252:** 1471.

Skogerson, L. and E. Wakatama. 1976. A ribosome-dependent GTPase from yeast distinct from elongation factor 2. *Proc. Natl. Acad. Sci.* **73:** 73.

Smirnov, V.N., V.G. Kreier, L.V. Lizlova, and S.G. Inge-Vechtomov. 1973. Recessive suppression and protein synthesis in yeast. *FEBS Lett.* **38:** 96.

Smirnov, V.N., A.P. Surguchov, V.V. Smirnov, and Y.V. Berestetskaya. 1978. Recessive nonsense suppression in yeast: Involvement of 60S ribosomal subunit. *Mol. Gen. Genet.* **163:** 87.

Smirnov, V.N., V.G. Kreier, L.V. Lizlova, V.M. Andrianova, and S.G. Inge-Vechtomov. 1974. Recessive super-suppression in yeast. *Mol. Gen. Genet.* **129:** 105.

Smirnov, V.N., A.P. Surgochov, E.S. Fominyksh, L.V. Lizlova, T.V. Saprygina, and S.H. Inge-Vechtomov. 1976. Recessive nonsense-suppression in yeast: Further characterization of a defect in translation. *FEBS Lett.* **66:** 12.

Sonenberg, N. 1988. Cap-binding proteins of eukaryotic messenger RNA: Functions in initiation and control of translation. *Prog. Nucleic Acid Res. Mol. Biol.* **35:** 173.

Song, J.M. 1987. "Genetic and molecular studies of genes that affect translational fidelity in *Saccharomyces cerevisiae*." Ph.D. thesis, University of Illinois, Chicago.

Song, J.M. and S.W. Liebman. 1985. Interaction of UAG suppressors and omnipotent suppressors in *Saccharomyces cerevisiae*. *J. Bacteriol.* **161:** 778.

———. 1987. Allosuppressors that enhance the efficiency of omnipotent suppressors in *Saccharomyces cerevisiae*. *Genetics* **115:** 451.

———. 1989. Mutations in *ADE3* reduce the efficiency of the omnipotent suppressor *sup45-2*. *Curr. Genet.* **16:** 315.

Song, J.M., S. Picologlou, C.G. Grant, M. Firoozan, M.F. Tuite, and S. Liebman. 1989. Elongation factor EF-1α gene dosage alters translational fidelity in *Saccharomyces cerevisiae*. *Mol. Cell. Biol.* **9:** 4571.

Spremulli, L. and J.M. Ravel. 1976. Characterization of elongation factor 1 from yeast. *Arch. Biochem. Biophys.* **172:** 261.

Sprinzl, M., T. Hartmann, J. Weber, J. Blank, and R. Zeidler. 1989. Compilation of tRNA sequences of tRNA genes. *Nucleic Acids Res.* **17:** 1.

Sripati, C.E., Y. Groner, and J.R. Warner. 1976. Methylated, blocked 5 ′ termini of yeast mRNA. *J. Biol. Chem.* **251:** 2898.

Starzyk, R.M., T.A. Webster, and P. Schimmel. 1987. Evidence for dispensable sequences inserted into a nucleotide fold. *Science* **237:** 1614.

Strick, C.A. and T.D. Fox. 1987. *Saccharomyces cerevisiae* positive regulatory gene *PET111* encodes a mitochondrial protein that is translated from an mRNA with a long 5 ′ leader. *Mol. Cell. Biol.* **7:** 2728.

Stucka, R., J. Hauber, and H. Feldmann. 1987. One member of the tRNA(Glu) gene family in yeast codes for a minor GAGtRNA(Glu) species and is associated with several short transposable elements. *Curr. Genet.* **12:** 323.

Subramanian, A.R., A. Steinmetz, and L. Bogorad. 1983. Maize chloroplast DNA encodes a protein sequence homologous to the bacterial ribosome assembly. *Nucleic Acids Res.* **11:** 5277.

Surguchov, A.P., M.V. Telkov, and V.N. Smirnov. 1986. Absence of structural homology between *sup1* and *sup2* genes of yeast *Saccharomyces cerevisiae* and identification of their transcripts. *FEBS Lett.* **206:** 147.

Surguchov, A.P., V.N. Smirnov, M.D. Ter-Avanesyan, and S.G. Inge-Vechtomov. 1984. Ribosomal suppression in eukaryotes. *Phys. Chem. Biol.* **4:** 147.

Surguchov, A.P., Y.V. Berestetskaya, E.S. Fominykch, E.M. Pospelova, V.N. Smirnov,

M.D. Ter-Avanesyan, and S.G. Inge-Vechtomov. 1980. Recessive suppression in yeast *Saccharomyces cerevisiae* is mediated by a ribosomal mutation. *FEBS Lett.* **111:** 175.

Szczesna, E. and W. Filipowicz. 1980. Faithful and efficient translation of viral and cellular eukaryotic mRNAs in a cell-free S-27 extract of *Saccharomyces cerevisiae*. *Biochem. Biophys. Res. Commun.* **92:** 563.

Ter-Avanesyan, M.D. and S.G. Inge-Vechtomov. 1980. Interaction of dominant and recessive suppressors in *Saccharomyces cerevisiae*. *Genetika* **16:** 86.

Thiele, D., P. Cottrelle, F. Iborra, J.M. Buhler, A. Sentenac, and P. Fromageot. 1985. Elongation factor 1 α from *Saccharomyces cerevisiae*. Rapid large-scale purification and molecular characterization. *J. Biol. Chem.* **260:** 3084.

Thireos, G., M. Driscoll-Penn, and H. Greer. 1984. 5′ untranslated sequences are required for the translational control of a yeast regulatory gene. *Proc. Natl. Acad. Sci.* **81:** 5096.

Thomas, K.R. and M.R. Capecchi. 1986. Introduction of homologous DNA sequences into mammalian cells induces mutations in the cognate gene. *Nature* **324:** 34.

Thompson, R.C. 1988. EF-Tu provides an internal kinetic standard for translational accuracy. *Trends Biochem. Sci.* **13:** 91.

Thonart, P., J. Bechet, F. Hilger, and A. Burny. 1976. Thermosensitive mutations affecting ribonucleic acid polymerases in *Saccharomyces cerevisiae*. *J. Bacteriol.* **125:** 25.

Tuite, M.F. 1989. Protein synthesis. In *Yeasts* (ed. A.H. Rose and J.S. Harrison), p. 161. Academic Press, New York.

Tuite, M.F., B.S. Cox, and C.S. McLaughlin. 1981. An homologous *in vitro* assay for yeast nonsense suppressors. *J. Biol. Chem.* **256:** 7298.

―――. 1983. *In vitro* nonsense suppression in [psi⁺] and [psi⁻] cell-free lysates of *Saccharomyces cerevisiae*. *Proc. Natl. Acad. Sci.* **80:** 2824.

Tuite, M.F., J. Plesset, K. Moldave, and C.S. McLaughlin. 1980. Faithful and efficient translation of homologous and heterologous mRNAs in an mRNA-dependent cell-free system from *Saccharomyces cerevisiae*. *J. Biol. Chem.* **255:** 8761.

Tzagoloff, A., A. Vambutas, and A. Akai. 1989. Characterization of *MSM1*, the structural gene for yeast mitochondrial methionyl-tRNA synthetase. *Eur. J. Biochem.* **179:** 365.

Tzagoloff, A., A. Akai, M. Kurkulos, and B. Repetto. 1988. Homology of yeast mitochondrial leucyl-tRNA synthetase and isoleucyl- and methionyl-tRNA synthetases of *Escherichia coli*. *J. Biol. Chem.* **263:** 850.

Tzamarias, D. and G. Thireos. 1988. Evidence that the *GCN2* protein kinase regulates reinitiation by yeast ribosomes. *EMBO J.* **7:** 3547.

Tzamarias, D., D. Alexandraki, and G. Thireos. 1986. Multiple *cis*-acting elements modulate the translational efficiency of GCN4 mRNA in yeast. *Proc. Natl. Acad. Sci.* **83:** 4849.

Tzamarias, D., I. Roussou, and G. Thireos. 1989. Coupling of GCN4 mRNA translational activation with decreased rates of polypeptide chain initiation. *Cell* **57:** 947.

Unger, M.W. 1977. Methionyl-transfer ribonucleic acid deficiency during G1 arrest of *Saccharomyces cerevisiae*. *J. Bacteriol.* **130:** 11.

Unger, M.W. and L.H. Hartwell. 1976. Control of cell division in *Saccharomyces cerevisiae* by methionyl-tRNA. *Proc. Natl. Acad. Sci.* **73:** 1664.

Uritani, M. and M. Miyazaki. 1988a. Characterization of the ATPase and GTPase activities of elongation factor 3 (EF-3) purified from yeasts. *J. Biochem.* **103:** 522.

―――. 1988b. Role of yeast peptide elongation factor 3 [EF-3] at the AA-tRNA binding step. *J. Biochem.* **104:** 118.

Valencik, M.L., B. Kloeckener-Gruissem, R.O. Poyton, and J.E. McEwen. 1989. Disruption of the yeast nuclear *PET54* gene blocks excision of mitochondrial intron a15β from pre-mRNA for cytochrome *c* oxidase subunit 1. *EMBO J.* **8:** 3899.

van den Heuvel, J.J., R.J.M. Bergkamp, R.J. Planta, and H.A. Raue. 1989. Effect of deletions in the 5´-noncoding region on the translational efficiency of phosphoglycerate kinase mRNA in yeast. *Gene* **79:** 83.

van Hemert, F.J., R. Amons, W.J. Pluijms, H. Van Ormondt, and W. Moller. 1984. The primary structure of elongation factor EF-1α from the brine shrimp *Artemia*. *EMBO J.* **3:** 1109.

van Ness, B.G., J.B. Howard, and J.W. Bodley. 1980. ADP-ribosylation of elongation factor 2 by diphtheria toxin. Isolation and properties of the novel ribosyl-amino acid and its hydrolysis products. *J. Biol. Chem.* **255:** 10710.

van Noort, J.M., B. Kraal, and L. Bosch. 1985. A second tRNA binding site on elongation factor Tu is induced while the factor is bound to the ribosome. *Proc. Natl. Acad. Sci.* **82:** 3212.

Vazquez, D. 1979. *Inhibitors of protein synthesis.* Springer-Verlag, New York.

Waekem, L.P. and F. Sherman. 1990. Isolation and characterization of omnipotent suppressors in the yeast *Saccharomyces cerevisiae*. *Genetics* **124:** 515.

Waldron, C., B. Cox, N. Willis, R. Gesteland, P. Piper, D. Colby, and C. Guthrie. 1981. Yeast ochre suppressor *SUP5-o1* is an altered tRNA$_{UCA}^{Ser}$ gene. *Nucleic Acids Res.* **9:** 3077.

Walter, P., I. Weygand-Durasevic, A. Sanni, J. Ebel, and F. Fasiolo. 1989. Deletion analysis in the amino-terminal extension of methionyl-tRNA synthetase from *Saccharomyces cerevisiae* shows that a small region is important for the activity and stability of the enzyme. *J. Biol. Chem.* **264:** 17126.

Walter, P., J. Gangloff, J. Bonnet, Y. Boulanger, J.-P. Ebel, and F. Fasiolo. 1983. Primary structure of the *Saccharomyces cerevisiae* gene for methionyl-tRNA synthetase. *Proc. Natl. Acad. Sci.* **80:** 2437.

Wang, S.S. and A.K. Hopper. 1988. Isolation of a yeast gene involved in species-specific pre-tRNA processing. *Mol. Cell. Biol.* **8:** 5140.

Warner, J.R. 1983. The yeast ribosome: Structure, function, and synthesis. In *The molecular biology of the yeast* Saccharomyces: *Metabolism and gene expression* (ed. J.N. Strathern et al.), p. 529. Cold Spring Harbor Laboratory, Cold Spring Harbor, New York.

Weiss, W.A. and E.C. Friedberg. 1986. Normal yeast tRNA$^{Gln}_{CAG}$ can suppress amber codons and is encoded by an essential gene. *J. Mol. Biol.* **192:** 725.

Weiss, W.A., I. Edelman, M.R. Culbertson, and E.C. Friedberg. 1987. Physiological levels of normal tRNA$^{Gln}_{CAG}$ can effect partial suppression of amber mutations in the yeast *Saccharomyces cerevisiae*. *Proc. Natl. Acad. Sci.* **84:** 8031.

Weiss-Brummer, B., H. Sakai, and M. Magerl-Brenner. 1987. At least two nuclear-encoded factors are involved together with a mitochondrial factor (MF1) in spontaneous mitochondrial frameshift-suppression of the yeast *S. cerevisiae*. *Curr. Genet.* **12:** 387.

Wek, R.C., B.M. Jackson, and A.G. Hinnebusch. 1989. Juxtaposition of domains homologous to protein kinases and histidyl-tRNA synthetases in GCN2 protein suggests a mechanism for coupling *GCN4* expression to amino acid availability. *Proc. Natl. Acad. Sci.* **86:** 4579.

Wek, R.C., M. Ramirez, B.M. Jackson, and A.G. Hinnebusch. 1990. Identification of positive-acting domains in GCN2 protein kinase required for translational activation of *GCN4* expression. *Mol. Cell. Biol.* **10:** 2820.

Werner, M., A. Feller, and A. Pierard. 1985. Nucleotide sequence of yeast gene *CPA1* encoding the small subunit of arginine-pathway carbamoylphosphate synthetase. *Eur. J.*

Biochem. **146**: 371.

Werner, M., A. Feller, F. Messenguy, and A. Pierard. 1987. The leader peptide of yeast gene *CPA1* is essential for the translational repression of its expression. *Cell* **49**: 805.

Weygand-Durasevic, I., D. Johnson-Burke, and D. Soll. 1987. Cloning and characterization of the gene coding for cytoplasmic seryl-tRNA synthetase from *Saccharomyces cerevisiae. Nucleic Acids Res.* **15**: 1887.

Williams, N.P., A.G. Hinnebusch, and T.F. Donahue. 1989. Mutations in the structural genes for eukaryotic initiation factors 2α and 2β of *Saccharomyces cerevisiae* disrupt translational control of *GCN4* mRNA. *Proc. Natl. Acad. Sci.* **86**: 7515.

Williams, N.P., P.P. Mueller, and A.G. Hinnebusch. 1988. The positive regulatory function of the 5′-proximal open reading frames in *GCN4* mRNA can be mimicked by heterologous, short coding sequences. *Mol. Cell. Biol.* **8**: 3827.

Wilson, P.G. and M.R. Culbertson. 1988. *SUF12* suppressor protein of yeast a fusion protein related to the EF-1 family of elongation factors. *J. Mol. Biol.* **199**: 559.

Wilson, W., M.H. Malim, J. Mellor, A.J. Kingsman, and S.M. Kingsman. 1986. Expression strategies of the yeast retrotransposon Ty: A short sequence directs ribosomal frameshifting. *Nucleic Acids Res.* **14**: 7001.

Winey, M., I. Edelman, and M.R. Culbertson. 1989a. A synthetic intron in a naturally intronless yeast pre-tRNA is spliced efficiently *in vivo. Mol. Cell. Biol.* **9**: 329.

Winey, M., L. Mathison, C.M. Soref, and M.R. Culbertson. 1989b. Distribution of introns in frameshift-suppressor proline-tRNA genes of *Saccharomyces cerevisiae. Gene* **76**: 89.

Winey, M., M.D. Mendenhall, C.M. Cummins, M.R. Culbertson, and G. Knapp. 1986. Splicing of a yeast proline tRNA containing a novel suppressor mutation in the anticodon stem. *J. Mol. Biol.* **192**: 49.

Wittmann-Liebold, B. and B. Greuer. 1978. The primary structure of the protein S5 from the small subunit of the *Escherichia coli* ribosome. *FEBS Letts.* **95**: 91.

Wold, F. 1981. *In vivo* chemical modification of proteins (post-translational modification). *Annu. Rev. Biochem.* **50**: 783.

Wolfner, M., D. Yep, F. Messenguy, and G.R. Fink. 1975. Integration of amino acid biosynthesis into the cell cycle of *Saccharomyces cerevisiae. J. Mol. Biol.* **96**: 273.

Woudt, L.P., A.B. Smit, W.H. Mager, and R.J. Planta. 1986. Conserved sequence elements upstream of the gene encoding yeast ribosomal protein L25 are involved in transcription activation. *EMBO J.* **5**: 1037.

Yarus, M. 1988. tRNA identity: A hair of the dogma that bit us. *Cell* **55**: 739.

Yokota, T., H. Sugisaki, M. Takanami, and Y. Kaziro. 1980. The nucleotide sequence of the cloned *tufA* gene of *Escherichia coli. Gene* **12**: 25.

Zennaro, E., L. Grimaldi, G. Baldacci, and L. Frontali. 1985. Mitochondrial transcription and processing of transcripts during release from glucose repression in "resting cells" of *Saccharomyces cerevisiae. Eur. J. Biochem.* **147**: 191.

Zitomer, R.S., D.A. Walthall, B.C. Rymond, and C.P. Hollenberg. 1984. *Saccharomyces cerevisiae* ribosomes recognize non-AUG initiation codons. *Mol. Cell. Biol.* **4**: 1191.

Zoladek, T., M. Boguta, and A. Putrament. 1985. Nuclear suppressors of the mitochondrial mutation *oxil-V25* in *Saccharomyces cerevisiae. Curr. Genet.* **9**: 427.

APPENDIX

Genetic and Physical Maps of
Saccharomyces cerevisiae

Robert K. Mortimer,[1,3] C. Rebecca Contopoulou,[1] and Jeff S. King[2]

[1]Department of Molecular and Cell Biology, Division of Genetics, and
[2]Graduate Group in Biophysics, and [3]Lawrence Berkeley Laboratory,
University of California, Berkeley, California 94720

Part A: Genetic and Physical Maps

Part B: Table I Glossary of Gene Symbols

Part C: Table II List of Mapped Genes

This draft of Edition 11 of the genetic maps of the chromosomes of *Saccharomyces cerevisiae* is based on the data presented in Genetic Map of *Saccharomyces cerevisiae*, Edition 10 (Mortimer et al., *Yeast 5:* 321 [1989]). Chromosomes II, XI, and XVI have undergone significant revision since the tenth edition of the map. The incorporation of the physical mapping data for most of the chromosomes into the map is under revision and thus is not included in this version of the map. New genes and new information are shown in boldface type throughout the map and tables. This draft represents work in progress toward Edition 11, and as such does not present the new mapping data that has been incorporated into the drawings and tables. Only references since Edition 10 are provided. Although efforts to assure accuracy have been made, this map has not undergone the scrutiny of a final edition. Genetic maps are composed of solid vertical lines (representing map distances determined by tetrad analysis) or dashed lines (indicating linkages established by mitotic recombination analysis), with centromeres represented as circles and with the left arm above and the right arm below the centromere. Short horizontal lines indicate the genetic position of a gene on a chromosome. In crowded regions, gene names are frequently not lined up with their corresponding horizontal lines. Vertical bars indicate a region within which a gene maps. Synonyms (alleles) are separated by commas.

Volume I. *The Molecular and Cellular Biology of the Yeast* Saccharomyces: *Genome Dynamics, Protein Synthesis, and Energetics.* Copyright 1991 Cold Spring Harbor Laboratory Press 0-87969-355-X/91 $3 + 00   **737**

Genes listed below the chromosome have been assigned to that chromosome but not to a specific location. In this group of genes, capitalized gene symbols do not necessarily indicate dominance; instead they may signify that the cloned gene was used to assign it to that chromosome. Physical maps are drawn as vertical helical lines and the gene symbols are in capital letters. The telomeres, when their positions relative to genes are known, are drawn as open arrows. Vertical bars next to the helical lines indicate the size of the restriction fragment on which a gene is located or region of uncertainty about the physical location of the gene. Parentheses are no longer used to indicate uncertain gene order; however, the order of many genes remains ambiguous. Gene symbols are defined in Table I and mapped genes are listed in Table II. In Table II, the synonyms for each gene are indented and listed below the appropriate gene.

Part A: Genetic and Physical Maps

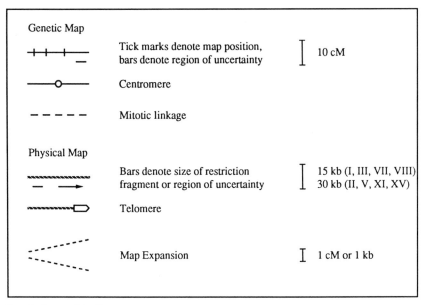

Symbols used in the maps.

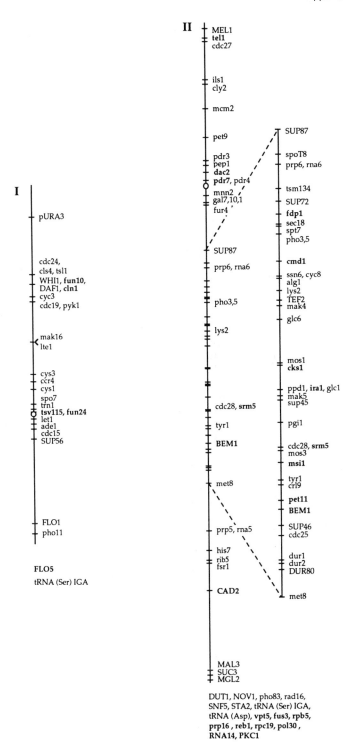

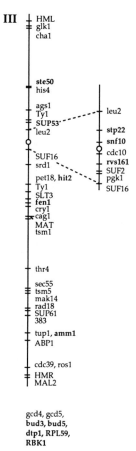

III

HML
glk1
cha1

ste50
his4

ags1
Ty1
SUP53
leu2

SUF16
srd1

pet18, **hit2**
Ty1
SLT3
fen1
cry1
cag1
MAT
tsm1

thr4

sec55
tsm5
mak14
rad18
SUP61
383

tup1, **amm1**
ABP1

cdc39, ros1
HMR
MAL2

leu2

stp22
snf10
cdc10
rvs161
SUF2
pgk1
SUF16

gcd4, gcd5,
bud3, bud5,
dtp1, RPL59,
RBK1

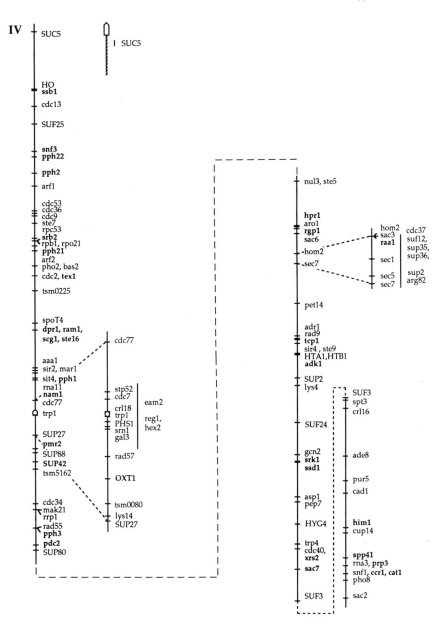

IV

SUC5

| SUC5

HO
ssb1
cdc13

SUF25

snf3
pph22

pph2

arf1

cdc53
cdc36
cdc9
ste7
rpc53
srb2
rpb1, rpo21
pph21
arf2
pho2, bas2
cdc2, **tex1**

tsm0225

spoT4
dpr1, ram1,
scg1, ste16   cdc77

aaa1
sir2, mar1
sit4, **pph1**
rna11   stp52
nam1   cdc7
cdc77   crl18   eam2
trp1   trp1   reg1,
                    PHS1   hex2
SUP27   srn1
·pmr2   gal3
SUP88   rad57
SUP42
tsm5162   **OXT1**

cdc34   tsm0080
mak21   lys14
rrp1   SUP27
rad55
pph3
pdc2
SUP80

nul3, ste5

hpr1
aro1
rgp1   hom2   cdc37
sac6   sac3   suf12,
                    raa1   sup35,
hom2   sup36,
sec7   sec1
                    sec5   sup2
                    sec7   arg82

pet14

adr1
rad9
tcp1
sir4, ste9
HTA1,HTB1
adk1

SUP2
lys4   SUF3
                    spt3
                    crl16
SUF24

gcn2
srk1   ade8
ssd1
                    pur5
asp1   cad1
pep7
HYG4   **him1**
                    cup14
trp4
cdc40,   **spp41**
xrs2,   rna3, prp3
sac7   snf1, **ccr1, cat1**
                    pho8
SUF3   sac2

ACT2, amc3, CHL4,
DBF4, kin28, mss2,
NAM1, pro1, rev1,
RPK1, ses1, SFA1,
SNF3, ssb20, ssn2,
STA1,
san1, mrp1, ctf2 = ctf12,
MEL5, OBF1, RPL44',
RPL45, **sam2,** rib3

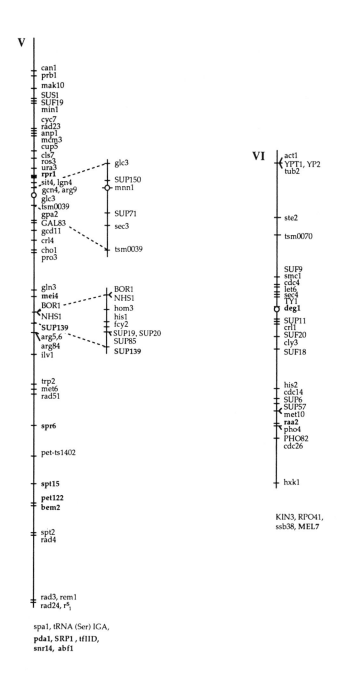

V

can1
prb1
mak10
SUS1
SUF19
min1
cyc7
rad23
anp1
mcm3
cup5
cls7
ros3
ura3
rpr1
sit4, lgn4
gcn4, arg9
glc3
tsm0039
gpa2
GAL83
gcd11
crl4
cho1
pro3

glc3

SUP150
mnn1

SUP71

sec3

tsm0039

gln3
mei4
BOR1
NHS1
SUP139
arg5,6
arg84
ilv1

BOR1
NHS1
hom3
his1
fcy2
SUP19, SUP20
SUP85
SUP139

trp2
met6
rad51

spr6

pet-ts1402

spt15

pet122
bem2

spt2
rad4

rad3, rem1
rad24, r^s_1

spa1, tRNA (Ser) IGA,
pda1, SRP1 , tfIID,
snr14, abf1

VI

act1
YPT1, YP2
tub2

ste2

tsm0070

SUF9
smc1
cdc4
let6
sec4
TY1
deg1
SUP11
crl1
SUF20
cly3
SUF18

his2
cdc14
SUP6
SUP57
met10
raa2
pho4
PHO82
cdc26

hxk1

KIN3, RPO41,
ssb38, **MEL7**

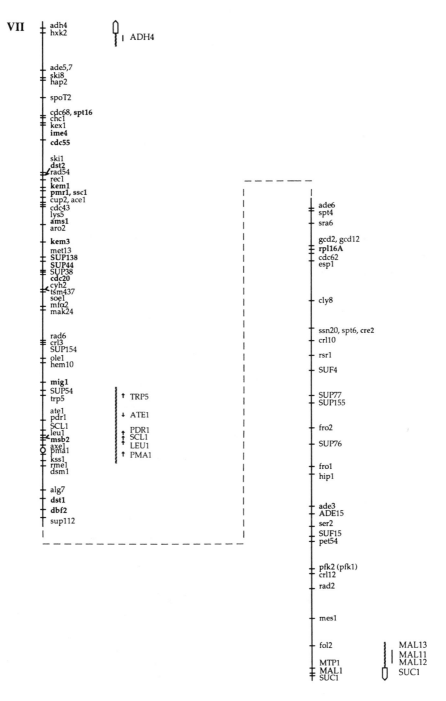

VII

adh4
hxk2

ADH4

ade5,7
ski8
hap2

spoT2

cdc68, **spt16**
chc1
kex1
ime4

cdc55

ski1
dst2
rad54
rec1
kem1
pmr1, ssc1
cup2, ace1
cdc43
lys5
ams1
aro2

kem3
met13
SUP138
SUP44
SUP38
cdc20
cyh2
tsm437
soe1
mfα2
mak24

rad6
crl3
SUP154
ole1
hem10

mig1
SUP54
trp5

ate1
pdr1
SCL1
leu1
msb2
axe1
pma1
kss1
rme1
dsm1

alg7

dst1

dbf2

sup112

ade6
spt4
sra6

gcd2, gcd12
rpl16A
cdc62
esp1

cly8

ssn20, spt6, **cre2**
crl10

rsr1

SUF4

SUP77
SUP155

fro2

SUP76

fro1
hip1

ade3
ADE15
ser2
SUF15
pet54

pfk2 (pfk1)
crl12
rad2

mes1

fol2

MTP1
MAL1
SUC1

↑ TRP5

↓ ATE1

↑ PDR1
↑ SCL1
↓ LEU1
↑ PMA1

MAL13
MAL11
MAL12
SUC1

amc4 (**chl8**), pho81, rec3, spo17,
SUP44, **hsp104, rpb9, ctf14**
ATP1, MEL2, RPL29

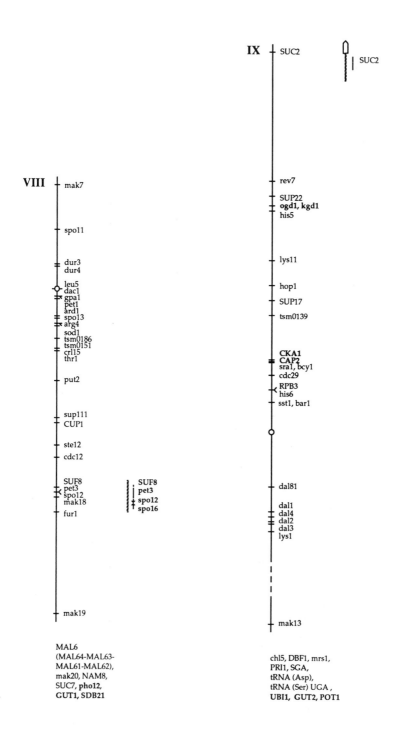

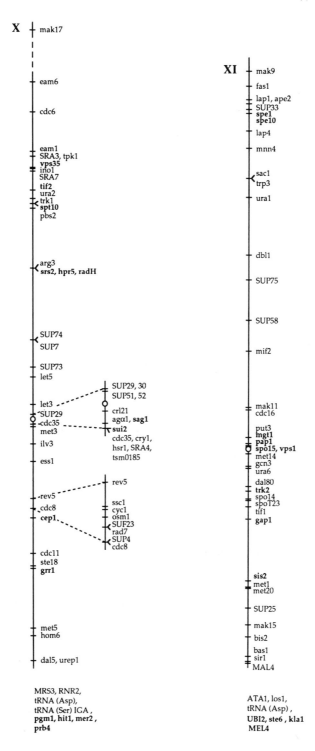

X ┼ mak17

eam6

cdc6

eam1
SRA3, tpk1
vps35
ino1
SRA7
tif2
ura2
trk1
spt10
pbs2

arg3
srs2, hpr5, radH

SUP74
SUP7

SUP73
let5

SUP29, 30
SUP51, 52

let3
crl21
agα1, **sag1**
SUP29
cdc35
sui2
met3
cdc35, cry1,
ilv3
hsr1, SRA4,
tsm0185
ess1

rev5
rev5
ssc1
cdc8
cyc1
cep1
osm1
SUF23
rad7
SUP4
cdc8

cdc11
ste18
grr1

met5
hom6

dal5, urep1

MRS3, RNR2,
tRNA (Asp),
tRNA (Ser) IGA ,
pgm1, hit1, mer2 ,
prb4

XI ┼ mak9
fas1
lap1, ape2
SUP33
spe1
spe10

lap4

mnn4

sac1
trp3

ura1

dbl1

SUP75

SUP58

mif2

mak11
cdc16

put3
mgt1
pap1
spo15, vps1
met14
gcn3
ura6
dal80
trk2
spo14
spoT23
tif1
gap1

sis2
met1
met20

SUP25

mak15

bis2

bas1
sir1
MAL4

ATA1, los1,
tRNA (Asp) ,
UBI2, ste6 , kla1
MEL4

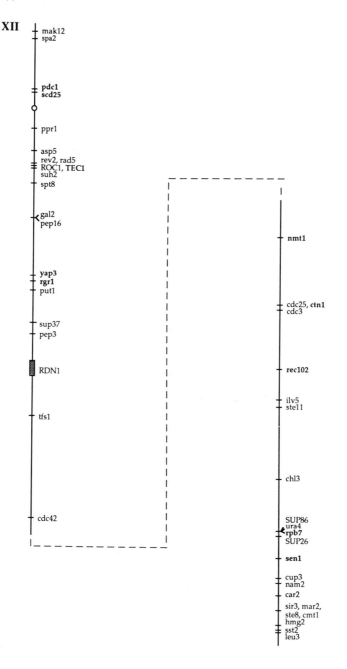

XII

mak12
spa2

pdc1
scd25

ppr1

asp5
rev2, rad5
ROC1, TEC1
suh2
spt8

gal2
pep16

yap3
rgr1
put1

sup37
pep3

RDN1

tfs1

cdc42

nmt1

cdc25, **ctn1**
cdc3

rec102

ilv5
ste11

chl3

SUP86
ura4
rpb7
SUP26

sen1

cup3
nam2
car2
sir3, mar2,
ste8, cmt1
hmg2
sst2
leu3

ass1, chl2, edr1 **(top3)**,
mak22, tRNA (Asp),
tRNA (Ser) IGA,
UBI3, UBI4, yef3,
aco1 , **rvs167** ,
sam1, tal1

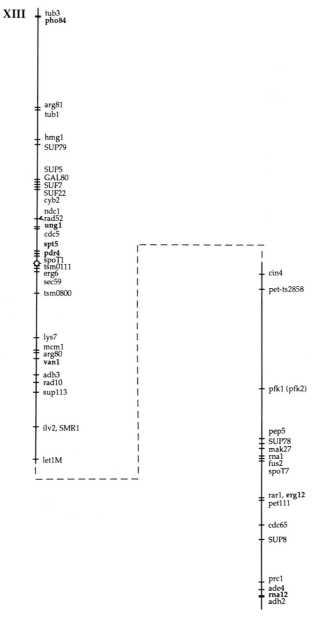

XIII
tub3
pho84

arg81
tub1

hmg1
SUP79

SUP5
GAL80
SUF7
SUF22
cyb2
ndc1
rad52
ung1
cdc5
spt5
pdr4
spoT1
tsm0111
erg6
sec59
tsm0800

lys7
mcm1
arg80
van1
adh3
rad10
sup113

ilv2, SMR1

let1M

cin4

pet-ts2858

pfk1 (pfk2)

pep5
SUP78
mak27
rna1
fus2
spoT7

rar1, **erg12**
pet111

cdc65

SUP8

prc1
ade4
rna12
adh2

crl22, gal5 (pgm2),
NAM7, pdx2,
SUP50, thi1,
tRNA (Ser) IGA,
tRNA (Asp) ,
**fun80, fun81, ctf13,
GPH1, spt21, snq1
(atr1), MEL6, ppa2,
DBP2, HSC82**

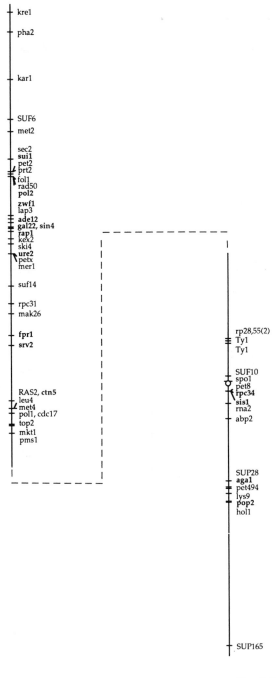

XIV

kre1

pha2

kar1

SUF6
met2

sec2
sui1
pet2
prt2
fol1
rad50
pol2
zwf1
lap3
ade12
gal22, sin4
rap1
kex2
ski4
ure2
petx
mer1

suf14

rpc31
mak26

fpr1
srv2

RAS2, **ctn5**
leu4
met4
pol1, cdc17
top2
mkt1
pms1

rp28,55(2)
Ty1
Ty1

SUF10
spo1
pet8
rpc34
sis1
rna2
abp2

SUP28
aga1
pet494
lys9
pop2
hol1

SUP165

mif1, rpd3, STA3,
tRNA (Asp) ,
pdb1 , **cms1** ,
rpd3, **RBP1**

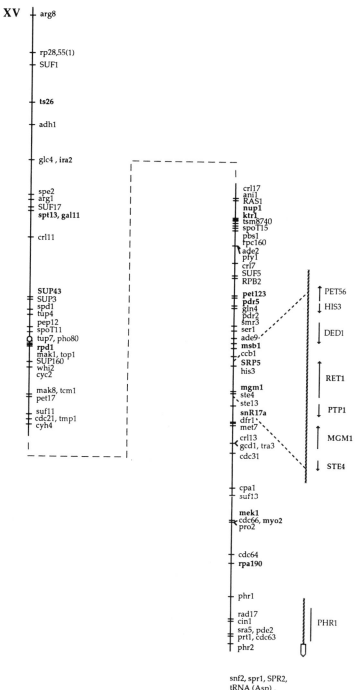

XV — arg8

rp28,55(1)
SUF1

ts26

adh1

glc4 , ira2

spe2
arg1
SUF17
spt13, gal11

crl11

SUP43
SUP3
spd1
tup4
pep12
spoT11
tup7, pho80
rpd1
mak1, top1
SUP160
whi2
cyc2

mak8, tcm1
pet17

suf11
cdc21, tmp1
cyh4

crl17
ani1
RAS1
nup1
ktr1
tsm8740
spoT15
pbs1
rpc160
ade2
pfy1
crl7
SUF5
RPB2

pet123
pdr5
gln4
pdr2
smr3
ser1
ade9
msb1
ccb1
SRP5
his3

mgm1
ste4
ste13
snR17a
dfr1
met7
crl13
gcd1, tra3
cdc31

cpa1
suf13

mek1
cdc66, myo2
pro2

cdc64
rpa190

phr1

rad17
cin1
sra5, pde2
prt1, cdc63
phr2

↑ PET56
↓ HIS3

↓ DED1

↑ RET1

↓ PTP1

↑ MGM1

↓ STE4

PHR1

snf2, spr1, SPR2,
tRNA (Asp) ,
vpt3, rpb8, rpb10 ,
CKA2, RPL3,
RPL44

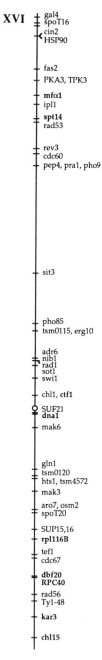

XVI
gal4
spoT16
cin2
HSP90

fas2
PKA3, TPK3
mfα1
ipl1
spt14
rad53

rev3
cdc60
pep4, pra1, pho9

sit3

pho85
tsm0115, erg10

adr6
nib1
rad1
sot1
swi1

chl1, **ctf1**
SUF21
dna1

mak6

gln1
tsm0120
hts1, tsm4572
mak3

aro7, osm2
spoT20

SUP15,16
rpl116B

tef1
cdc67

dbf20
RPC40
rad56
Ty1-48

kar3

chl15

pho9, pma2,
rna4 (prp4), rpc40,
tRNA (Ser) UGA ,
**mrp2, GPH1 , rpb6,
MEL3, RPO26,
SRP3, RNA15**

Part B: Table I

Glossary of Gene Symbols

Table I. Glossary of gene symbols

Symbol	Definition*	Symbol	Definition
aaa	Amino terminal, amino acetyl transferase	ars	Autonomously replicating sequence
aac	ATP/ADP carrier, mitochondrial	asp	Aspartic acid requiring
aar	a1-α2 repression	asr	Antisense RNA
aar	Amino acid analog resistance	asu	Antisuppressor
aas	Amino acid analog sensitive (see also gcn)	ata	Sporulation-specific gene characterized by ATA sequences
aat	Amino acid toxicity	ate	Arginyl-tRNA-protein transferase deficient
abc	Assembly of BC1 complex	atp	ATPase, mitochondrial
abf	Ars binding factor	atr	Aminotriazole resistance
abp	Actin binding protein	axe	Axenomycin resistance
ace	Activation of CUP1 expression	bap	Branched chain amino acid permease
aco	Aconitase, mitochondrial	bar	a cells lack barrier effect on α factor
acp	Acidic protein 2	bas	Basal level control
act	Actin	bcs	Branched chain amino acid sensitivity
acu	Aculeacin A resistant	bcy	Adenylate-cyclase and cAMP-dependent protein kinase deficient
ade	Adenine requiring	bem	Bud emergence
adh	Alcohol dehydrogenase defective	bet	Blocked early in the secretory pathway
adk	Adenylate kinase	bik	Bilateral defect in karyogamy
adr	Alcohol dehydrogenase regulation defective		
aep	Required for translation of *ole1* mRNA and subunit 9 assembly		
afi	Alpha factor internalization		
aga	a cell-specific sexual agglutination		

agα	α-cell specific sexual agglutination
ags	Aminoglycoside sensitive
aky	adenylate kinase
alg	Asparagine-linked glycosylation deficient
amc	Artificial minichromosome maintenance
amm	Altered minichromosome maintenance
ams	α–mannosidase
amy	Alpha amylase
amy	Antimycin resistance
anb	anaerobically induced
ani	Anisomycin resistance
anp	ANP and osmotic sensitive
ant	Antibiotic resistance
ape	Aminopeptidase
apf	Amino acid transport pleiotropic factor
aph	3'-aminoglycoside phosphotransferase
apn	Apurinic site endonuclease
aps	Aspartyl-tRNA synthetase
apt	Aminoglycoside phosphotransferase
ard	Arrest at start of cell cycle defective
arf	ADP-ribosylation factor
arg	Arginine requiring
aro	Aromatic amino acid requiring

blm	Bleomycin sensitivie
bls	Blasticidin-S resistance
bor	Borrelidin resistance
bos	Bet one suppressor
bot	Biosynthesis of turpenes
btf	Basic transcription facor
bud	Controls sites of bud emergence
byp	
cad	Cadmium resistance
cag	Constituvely agglutinable
cal	Calcium sensitive
cal	Resistant to calcofluor
can	Canavanine resistance
cap	Capping - addition of actin subunits
car	Catabolism of arginine defective
cas	Cyclic AMP suppression
cat	Catabolite repression
cbp	Centromere binding protein
cbp	Cytochrome b pre-mRNA stability; coenzyme QH2-cytochrome c reductase assembly
cbs	Cytochrome b pre-mRNA and mRNA translation factor
cca	ATP (CTP) tRNA specific nucleotidyl transferase

Table I. Glossary of gene symbols - continued

Symbol	Definition	Symbol	Definition
ccb	Cross-complementation of budding defect (see msb)	crl	Cycloheximide-resistant temperature-sensitive lethal
ccp	**Cytochrome c peroxidase, mitochondrial**	cry	Cryptopleurine resistance
		csd	Chitosan synthesis
ccr	Carbon catabolite repression	cse	Chromosome segregation
ccs	**Triethyl-tin resistance, controls oxidative phosphorylation**	csr	cyclosporin resistance
		cta	Catalase A
cdc	Cell division cycle blocked at 36°C	ctd	Phsophorylates C-terminal domain of largest RNA polII subunit
cen	Centromere		
cep	**Centromere protein**	ctf	Chromosome transmission fidelity
cha	Catabolism of hydroxy amino acids	ctn	Catalase T nutrition
chc	Clathrin heavy chain	cts	Endochitinase
chl	Chromosome loss	ctt	Catalase T
cho	Choline requiring	ctt	Cholinephosphate cytidylyltransferase
chs	**Chitin synthetase**		
cid	**Cell identity**	cup	Copper resistance
cin	Chromosome instability	cyb	Cytochrome b₂ deficiency
cip	**Complementer of ipl2**	cyc	Cytochrome c deficiency
cit	Citrate synthase, mitochondrial	cyh	Cycloheximide resistance
cka	Casein kinaseII, α subunit	cyp	Modulates expression of cyc1 and cyc7
ckb	Casein kinaseII, β subunit		
cki	**Choline kinase**	cyr	Adenylate cyclase deficient

cks	Cdc28 kinase subunit
clc	Clathrin light chain
cln	Cyclin
cls	Calcium sensitive
cly	Cell lysis at 36°C
cmd	Calmodulin gene
cmk	Calmodulin dependent protein kinase
cmp	Calmodulin binding protein
cms	Centromere mutation supprossor
cmt	Control of mating type
cna	Calcineurin subunit A
cnb	Calcineurin subunit B
cnd	Chromosome non disjunction
coq	Coenzyme Q deficient
cor	Coenzyme QH2-cytochrome c reductase
cox	Cytochrome oxidase, nuclear or mitochondrial
cpa	Carbamoyl phosphate synthetase, arginine specific
cpe	Constitutive phospholipid expression
cpf	Centromere and promoter factor
cpr	Cyclophilin
cps	Carboxypeptidase S
cre	Carbon catabolite repression effector

cys	Cysteine requiring
cyt	Cytochrome c1 heme lyase; cytochrome c (see cyc)
dac	Division arrest control for mating pheromones
daf	Dominant α factor resistance
dal	Allantoin degradation deficient
dap	Dipeptidyl aminopeptidase B
dbf	Dumbell formation; kinase required for late nuclear division
dbl	Alcian blue dye binding deficient
dbp	Dead box protein
dbr	Debranching of excized introns
dcd	dCMP deaminase
dea	Deamidation of amino terminal Asn and Gln residues
dcd	Defines essential domain, lethal
deg	Depressed growth rate
dex	Dextran utilization
dfr	Dihydrofolate reductase
dis	Phosphoserune- and phosphothreonine-specific phosphatase
dit	D,L-dityrosine in spore wall
dit	Derepressed INO1 transcription
diu	Diuron resistance
dka	Downstream of protein kinase A
dna	DNA synthesis defective

Table I. Glossary of gene symbols - continued

Symbol	Definition	Symbol	Definition
doa	Degradation of alpha factor defective	fsr	Fluphenazine resistance
dpm	Dolichol phosphate mannose synthase	fum	Fumarase
		fun	Function unknown now
dpr	Processing of ras protein	fur	Resistant to 5-FU, encodes UPRTase
dsc	Dominant suppressor of *cdc4*	fus	Fusion defective; MAP kinase involved in pheromone signal transduction, G1 arrest
dsd	Dominant suppressor of *sec14-1*		
dsm	Premeiotic DNA synthesis deficient	fus	Nuclear fusion defective
dst	DNA strand transfer	gac	Glycogen accumulation
dtp	Diadenosine 5',5''', P4-tetraphosphate phosphorylase	gal	Galactose non-utilizer
		gam	Glucoamylase
dur	Urea degradation deficient	gap	General amino acid permease
dut	dUTPase	gap	Glyceraldehyde-3-phosphate dehydrogenase, see tdh
eam	Endogenous ethanolamine biosynthesis		
edr	Enhanced delta recombination	gas	Glycophospholipid-anchored surface protein
eif	Translation initiation factor		
elm	Elongated bud morphology mutant suppressor	gcd	General control of amino acid synthesis derepressed
end	Endocytosis defective	gcn	General control of amino acid synthesis non-derepressible (see also aas)
eno	Enolase		
env	Envelope proteins	gcn	Glucosamine auxotroph
erd	Endoplasmic reticulum defective	gcr	Glycolysis regulatory protein

erg	Ergosterol biosynthesis defective; many also nystatin resistant
esp	Extra spindle pole bodies
ess	Essential
est	
eth	Ethionine resistance
exa	Extragenic suppressor subfamily A
exg	Exo-1,3-β-glucanase activity
far	Factor arrest
fas	Fatty acid synthetase deficient
fbp	Fructose-1,6-bisphosphatase
fcy	Purine-cytosine permease
fdp	Unable to grow on glucose, fructose, sucrose or mannose
fen	Fenpropimorph resistant
fer	Ferric resistance
fes	Ferric sensitivity
fkh	
fkr	
flk	Flaky
flo	Flocculation
fol	Folinic acid requiring
fox	Fatty acid oxidation
fpr	FKBP proline rotamase (isomerase)
fre	Ferric reductase
fro	Frothing

gcy	Galactose regulated, homology to aldo/keto-reductases
gdh	Glutamate dehydrogenase, NADP dependent
gef	Glycerol ethanol, ferric requiring
ger	Spore germination
ggp	GPI-glycosylated protein
glc	Glycogen storage
glk	Glucokinase deficient
gln	Glutamine synthetase non-derepressible
gpα	G protein alpha homologous gene
gpa	GTP binding protein
gph	Glycogen phosphorylase
grc	Valyl-tRNA-synthetase (see vas)
grf	General regulatory factor
grr	Glucose repression resistant
gsh	Glutathione synthetase
gst	G -1 to S transition
gsy	Glycogen synthase
gut	Glycerol utilization
ham	Supermutable
hap	Global regulator of respiratory genes; nuclear transcription factor
hem	Heme synthesis deficient
het	Hexose transport
hex	Hexose metabolism regulation
hht	Histone
him	High induced mutagenesis

Table I. Glossary of gene symbols - continued

Symbol	Definition	Symbol	Definition
hip	Histidine specific permease	ivs	Intervening sequence
hir	Histone cell cycle regulation defective	jun	Karyogamy defective
		kar	kar enhancing mutation
his	Histidine requiring	kem	Killer expression defective
hit	High temperature growth	kex	α-ketoglutarate dehydrogenase, E1
hmg	HMG-CoA reductase	kgd	component; Dihydrolipoyl
hml	Mating type cassette - left		tanssuccinylase
hmr	Mating type cassette - right		Protein kinase,
ho	Homothallic switching	kin	Potassium (k) low affinity
hol	Histidinol uptake proficient	kla	transporter
hom	Homoserine requiring		cdc28 look-alike
hop	Homologue pairing	kns	Suppression of some mak mutations
hot	Cis-acting recombination stimulating sequence fro ribosomal DNA	krb	Killer resistance
		kre	Lysyl tRNA Synthetase
hpl	Haploid lethal	krs	MAP protein kinase homolog involved in
hpr	Hyperresistance to oxidative damage	kss	pheromone signal transduction
		kti	Kluveromyces lactis Toxin Insensitive
hpr	Increased intrachromosomal recombination	ktr	Kre two related
hps	Hexaprenyl pyrophosphate synthetase	lap	Leucine aminopeptidase deficient
		lcb	Long chain base synthesis

hrr	**Involved in regulation of repair of DNA damage**
hsc	
hsf	**Heat shock transcription factor**
hsp	Heat shock protein
hsr	Heat shock resistance
hta	Histone 2A genes
htb	Histone 2B genes
hts	Histidinyl-tRNA synthetase
hxk	Hexokinase deficient
hxt	**Hexose transport**
hyg	Hygromycin resistance
idh	**isocitrate dehydrogenase**
ils	Isoleucyl-tRNA synthetase deficient; no growth at 36°C
ilv	Isoleucine-plus-valine requiring
ime	**Inducer of meiosis**
imp	**Independent of mitochondrial particle**
ino	Inositol deficient
ins	initiation of S phase = srk1
ipa	Increased accumulation of porphyrins
ipl	Increase in ploidy
ipp	Inorganic pyrophosphatase
ira	Inhibitory reulator of the RAS-cAMP pathway
ise	Inhibitor sensitive
ldd	**defective, an intermediate in sphingolipid biosynthesis**
let	Lysozyme degradation
leu	Lethal
lgn	Leucine requiring
lig	Sporulation-induced transcripts (see also sit)
lip	**Lipoic acid deficient**
lmd	**Lanosterol 14 alpha demethylase**
los	Loss of suppression and defective in tRNA processing
lpd	**Lipoamide dehydrogenase**
lte	Low temperature essential
lts	Low temperature sensitive
lys	Lysine requiring
lyt	**Lysis at 37 (see cly)**
mak	Maintenance of killer deficient
mal	Maltose fermentation
mar	Mating type cassette expression
mas	Mitochondrial assembly
mas	**temperature sensitive for import of mitochondrial proteins**
mat	Mating type locus
mbr	**Mitochondrial biogenesis of ribosomes**
mck	**Meiotic and centromere regulatory ser, tyr-kinase**

Table I. Glossary of gene symbols - continued

Symbol	Definition	Symbol	Definition
mcm	Minichromosome maintenance deficient	msi	Multicopy suppressor of *ira1*
mdh	Malate dehydrogenase	msk	Mitochondrial lysyl-tRNA synthetase
mec	Mitosis entry defective (Weinert pc)	msl	Mitochondrial leucyl-tRNA synthetase (= nam2)
mef	Mitochondrial elongation factor	msm	Mitochondrial methionyl tRNA synthetase
mei	Meiotic gene conversion defective	mss	Suppression of a mitochondrial RNA splice defect; COX1 pre-mRNA processing factor
mek	Kinase involved in meiotic chromosome pairing and recombination	mst	Mitochondrial threonyl-tRNA synthetase
mel	Melibiose fermentation	msw	Mitochondrial tryptophanyl-tRNA synthetase
mer	Meiotic recombination	msy	Mitochondrial tyrosyl-tRNA synthetase
mes	Methionyl-tRNA synthetase deficient; no growth at 36°C	mtf	Mitochondrial transcription factor
met	Methionine requiring	mtf	Mitochondrial translation initiation factor
mfα	α-mating factor	mtp	Melezitose fermentation
mfa	Mating facor a	myo	Myosin
mgl	α-Methylglucoside fermentation	nam	Nuclear accomodation of mitochondria
mgm	Mitochondrial genome maintenance		
mgt	Mitochondrial genome transmission		
mif	Mitotic frequency of chromosome transmission		

mig	Multicopy inhibitor of GAL gene expression
mih	Protein phosphatase
min	Methionine inhibited
mip	Mitochondrial DNA polymerase
mkt	Maintenance of K_2 killer factor
mnn	Mannan synthesis defective
mnt	Mannosyltransferase
mod	Modification of tRNA, IPP transferase
mom	Matrix organization mutant
mos	Modifier of ochre suppressors
mot	Modifier of transcription
mox	
mpk	Multicopy suppressor of kinase
mrf	Mitochondrial release factor
mrp	Mitochondrial ribosomal protein
mrs	Mitochondrial RNA splicing
msb	Multicopy suppressor of a budding defect (replaces ccb)
msd	Mitochondrial aspartyl-tRNA synthetase
mse	Mitochondrial glutamyl-tRNA synthetase
msf	Mitochondrial phenylalanine-tRNA synthetase
msg	Multicopy suppressor of GPA1

nat	N-terminal acetyltransferase
ndc	Nuclear division cycle
net	Negative effect on transcription
nhp	Nonhistone protein
nhs	Hydrogen sulfide production inhibitor
nib	Nibbled colony phenotype due to 2μ DNA
nmt	N-myristoyltransferase
nop	Nucleolar protein
nov	Novobiocin resistance
npi	Nitrogen permease inhibitor
npl	Nuclear protein localization
npr	Nitrogen permease reactivator
nra	Neutral red accumulation
nsp	Nuclear membrane SPB protein
nsp	Nucleoskeletal protein
nuc	DNA double-strand break repair
nul	Non-mater
nup	Nuclear pore
nur	Nuclease, recombination (tentative)
obf	Origin binding factor
odp	Oil drop protein
ogd	2-oxoglutarate dehydrogenase
ole	Oleic acid requiring
oli	Oligomycin resistance
opi	Phopholipid synthesis regulation
osm	Low osmotic pressure sensitive

Table I. Glossary of gene symbols - continued

Symbol	Definition	Symbol	Definition
oxt	Resistance to oxythiamin	ppz	Phosphoserine and phosphothreonine-specific protein phosphatase
pab	Poly A-binding protein		
pai	Plasminogen activator inhibitor	pra	Proteinase A deficient
pal	Phenylalanine Ammonia-Lyase	prb	Proteinase B deficient
pan	Poly A nuclease	prc	Proteinase C deficient
pap	Poly A polymerase	pre	Proteinase YSCE deficient
pas	Peroxisomal assembly	pri	DNA primase
pbs	Polymyxin B resistance	pro	Proline requiring
pda	Pyruvate dehydrogenase E1α	prp	Pre mRNA processing = rna
pdb	Pyruvate dehydrogenase E1β	prp	pre-mRNA splicing protein
pdc	Pyruvate decarboxylase	prt	Protein synthesis defective at 36°C
pde	Phosphodiesterase (cAMP)	pso	8-methoxypsoralen sensitive
pdh	Pyruvate dehydrogenase deficient	pss	phosphatidyl serine synthase
pdi	Protein disulfide isomerase	pta	
pdr	Pleiotropic drug resistance	ptp	Phosphotyrosine-specific protein phosphatase
pdx	Pyridoxin requiring	ptx	Preprotoxin
pem	Phospholipid methyltransferase	pur	Purine excretion
pep	Proteinase deficient	put	Proline nonutilizer
per	Suppressor of tex1 rad52 lethality	pyc	Pyruvate carboxylase
pet	Petite; unable to grow on non-fermentable carbon sources	pyk	Pyruvate kinase deficient
pfk	Phosphofructokinase		

pfy   Profilin of yeast
pgi   Phosphoglucose isomerase deficient
pgk   3-Phosphoglycerate kinase deficient
pgm   **Phosphoglucomutase**
pgm   **Phosphoglyceromutase**
pha   Phenylalanine requiring
pho   Phosphatase deficient
phr   Photoreactivation repair deficient
phs   Hydrogen sulfide production deficient
pif   **Mitochondrial repair and recombination**
pim   **PI and PIP kinase**
pka   Protein kinase catalytic subunit
pkc   **Protein kinase C homolog**
pma   Plasma membrane ATPase mutations
pmr   **Plasma membrane ATPase related**
pms   Postmeiotic segregation increased
pol   DNA polymerase
pox   **Acyl-CoA oxidase**
ppa   **Inorganic pyrophosphatase**
ppc   **Phosphoenolpyruvate carboxykinase**
ppd   Phosphoprotein phosphatase-deficient
pph   **Protein phosphatase (pph1 = sit4)**
ppr   Defective in pyrimidine biosynthetic pathway regulation

raa   **Resistant to amino acid analogs**
rad   Radiation (ultraviolet or ionizing) sensitive
rag   **Resistance to antimycin on glucose**
ram   E1α subunit of pyruvate dehydrogenase complex
rap   **Repressor activator protein**
rap   Regulation of autonomous replication
rar   Homologous to RAS proto-oncogene
ras
rbk   Ribokinase
rbp   **Rapamycin binding protein**
rca   **Rescue by cAMP of cyr1 mutants**
rcc   **Regulation of carbohydrate catabolism**
rcs   **Regulation of Cell Size**
rdn   Ribosomal RNA structural genes
reb   **DNA binding protein**
reb   **RNA polymerase I enhancer binding protein**
rec   Recombination deficient
red   **defective in the reductional division of meiosis**
ref   **Respiration negative on YPE**
reg   Regulation of galactose pathway enzymes
reo

Table I. Glossary of gene symbols - continued

Symbol	Definition	Symbol	Definition
rep	Replication and equipartion of 2 micron DNA to daughter cells	sal	Allosuppression
res	Regulation of sporulation	sam	S-adenosyl methionine synthesis
ret	Increased transcriptional read-through	san	Sir antagonist
		sar	Suppressor of sec12
ret	Reduced efficiency of termination	sca	Sporulation capable
rev	Revertibly decreased	scc	Suppressor of cdc65
rgp	Reduced growth phenotype	scd	Suppressor of cdc25 (see sdc)
rgr	Required for glucose repression	scd	Suppressor of clathrin deficiency
rgt	Restores glucose transport	scg	Constitutive arrest of cell division, pheromone-independent mating = gpa
rhm	5-aminolevulinic acid and heme biosynthesis		
		sch	cAMP-dependent protein kinase homolog; suppresses cdc25ts
rho	Mitochondrial genome		
rho	Ras homologous	sck	Suppressor of c kinase
rib	Riboflavin biosynthesis	scl	Dominant suppression of t.s. lethality of *crl3*
rip	Iron-sulfur protein of ubiquinol cytochrome c reductase; Rieske protein of coenzyme QH2-cytochrome c reductase		
		scm	Suppressor of chromosome segregation mutation
rkt	Restores pottasium transport	sco	Cytochrome oxidase deficient
rme	Repressor of meiosis	sdb	suppressor of DBF2
rna	RNA synthesis defective; unable to grow at 36°C = prp	sdc	Suppressor of cdc25 (see scd)
		sdh	Succinate dehydrogenase deficient
		sds	Suppressor of deletion of sit4

rnr	Ribonucleotide reductase
roc	Roccal resistance
ros	Relaxation of sterility
rox	**Regulation of oxygen**
rpa	**Ribosomal protein, acidic**
rpa	RNA polymerase I
rpb	RNA polymerase II
rpc	RNA polymerase III
rpd	Reduced potassium dependency
rpk	Regulatory protein kinase
rpl	**Ribosomal protein, large subunit**
rpo	**RNA polymerase II, III, IV**
rpr	RNase P
rps	**Ribosomal protein, small subunit**
rrm	**rDNA recombination mutation**
rp	rRNA processing
r1	Radiation sensitive
rsd	**Recessive suppressor of *sec14-1***
rsf	**Regulates septum formation**
rsr	Ras-related
rvs	**Reduced viability upon starvation, replaces spe**
sac	Suppressor of actin mutations
saf	**Suppressor of *adr* function**
sag	Sexual agglutination
sak	**Suppressor of cAMP-dependent kinase**

sec	Secretion deficient
sen	Splicing endonuclease
ser	Serine requiring
ses	Seryl-tRNA synthetase
sfa	Sensitive to formaldehyde
sga	**Glucoamylase, intracellular**
sga	Suppression of growth arrest of *cdc25*
sin	**Switch independent**
sir	Silent mating type information regulation
sis	**sit4 suppressor**
sit	Sporulation-induced transcripts (see also lgn, pph)
sit	Suppression of initiation of transcription
skd	**Suppressor of *kar2* defect**
ski	Superkiller
slk	**pkc-related kinase**
slo	**Small lysine pool (vacuole defective) slp1=vps33**
slp	**Sigma like sequence**
sls	Suppression at low temperature
slt	**Suppressor of loss of YPT1**
sly	Stability of minichromosomes
smc	**Suppressor of mating defect of ste7**
smd	**Regulation of meiosis (ime)**
smr	Sulfometuron methyl resistance

Table I. Glossary of gene symbols - continued

Symbol	Definition	Symbol	Definition
snf	Sucrose nonfermenting	sta	Starch hydrolysis
snq	Sensitivity to 4-nitroquinoline-N-oxide	stc	Suppressor of tsm409
snr	Small nuclear RNA	ste	Sterile (ste7, ste11-ser, thr-kinase)
snt	Stringency activity; inhibition of leaky mit- mutants on nonferm substrates	sti	Stress inducible
		stp	Species-specific pre-tRNA processing
soc	Suppressor of cdc8	stp	Ste pseudorevertants
sod	Superoxide dismutase	str	Sulfur transferase
soe	Suppression of cdc8	suc	Sucrose fermentation
sot	Suppression of deoxythymidine monophosphate uptake	suf	Suppression of frameshift mutation
		suh	Suppression of his2-1
spa	Spindle pole antigen	sui	Suppressor of initiator codon mutations
spb	Ribosomal biogenesis	sum	Suppressor of mar2
spd	Sporulation not repressed on rich media	sup	Suppression of nonsense mutations
spe	Spermidine resistance	sur	Suppressor of radH
spg	Suppressor of GPA1	sus	Suppression of ser1
spk	ser, thr, (tyr)-kinase	sut	Suppressor of TATA
spo	Sporulation deficient	swi	Homothallic switching deficient
spp	Suppressor of prp	tal	Transaldolase
spr	Sporulation regulated genes	tar	3-amino-1,2,4-triazole resistance, gcn4 suppression
sps	Sporulation specific trancript	tcm	Tricodermin resistance
spt	Suppressors of Ty transcription		

spx	Suppressor of x-ray sensitivity of rad55
sra	Suppressors of the ras mutation
srb	Osmotically fragile
srd	Suppressor of rpb1
srd	Suppressor of *rrp1*
srk	Suppressor of regulatory subunit of protein kinase
srm	Suppressor of receptor mutations
srm	Suppressor of rho mutability
sm	Suppressor of yeast *rna1-1*
srp	Serine rich protein
srp	Suppressor of rpa190
srp	Suppressor of rpb1, cold-sensitive
srs	Suppressor of rad six
srv	Suppressor of RASval119
ssa	Stress-seventy subfamily A
ssb	Single strand binding protein
ssb	Stress-seventy subfamily B
ssc	Secretion of heterologous proteins
ssc	Stress-seventy subfamily C
ssd	Stress-seventy subfamily D
ssl	Super secretion of lysozyme
ssn	Suppressor of *snf1*
ssp	Suppressor of rna8, cold sensitive.
sst	Supersensitive to α factor

tcp	t complex polypeptide
tdh	Glyceraldehyde 3-phosphate dehydrogenase, see gap
tec	Ty element enhancement control
tef	Translational elongation factor
tel	Contraction of telomere size Telomere
tex	Tn5 excision frequencies increased
tfIId	Transcription factor gene
tfp	Trifluoperazine resistance, Ca sensitive
tfs	Cdc twenty-five suppressor.
thi	Thiamine requiring
thr	Threonine requiring
ths	Threonyl tRNA synthetase
tif	Translation initiation factor
til	Thioisoleucine resistance
tmp	Thymidine monophosphate requiring
top	Topoisomerase deficient
tos	Topoisomerase II suppression,
tpd	Phosphoserine and phosphothreonine-secific protein phosphatase
tpd	tRNA production deficient
tpi	Triosephosphate isomerase
tpk	cAMP-dependent protein kinase catalytic subunit

Table I. Glossary of gene symbols - continued

Symbol	Definition	Symbol	Definition
tpk	Threonine/serine protein kinase	yak	Yet another kinase
tra	Triazylalanine resistant	yal	ORF on chromosome 1 L
trk	Transport of potassium	yap	Yeast aspartyl protease 3
trl	tRNA ligase, (trnl)	yap	Yeast aspartyl protease 3
trm	tRNA(guanine-N2, N2-)-dimethyltransferase	yar	ORF on chromosome 1 R
		ybl	ORF on chromosome 2 L
trn	Proline-tRNA gene	ybr	ORF on chromosome 2 R
trp	Tryptophan requiring	yck	Yeast casein kinase I
trx	Thioredoxin	ycl	ORF on chromosome 3 L
tsl	Temperature sensitive lethal	ycr	ORF on chromosome 3 R
tsm	Temperature sensitive lethal mutations	ydj	Yeast dnaJ
tsv	Temperature-sensitive lethal, UV-induced	ydl	ORF on chromosome 4 L
		ydr	ORF on chromosome 4 R
tub	Tubulin; MBC resistance	yef	Yeast elongation factor
tuf	Mitochondrial elongation factor	yel	ORF on chromosome 5 L
tup	Deoxythymidine monophosphate uptake	yer	ORF on chromosome 5 R
	positive	yfl	ORF on chromosome 6 L
TY	Transposable element	yfr	ORF on chromosome 6 R
tye	Ty-mediated expression	YG100	Heat shock gene
tyr	Tyrosine requiring	ygl	ORF on chromosome 7 L
ubc	Ubiquitin conjugating	ygr	ORF on chromosome 7 R
ubi	Ubiquitin	yhl	ORF on chromosome 8 L

uga	Utilization of GABA (4-aminobutyrate) as a nitrogen source
ulc	Zn resistant, cell elongation
ume	Unscheduled meiotic gene expression
umr	Ultraviolet mutability reduced
ung	Uracil DNA glycosylase
ura	Uracil requiring
ure	Ureidosuccinate transport-resistance to nitrogen repression
urep	Ureidosuccinate permease
urr	URS repression
utr	Unknown transcript
uts	Uptstream of thymidilate synthetase
vac	Vacuolar segregation
van	Vanadate resistant
vas	Valyl-tRNA-synthetase
vat	Vacuol;ar protein ATPase
vma	Vacuolar H ATPase
vph	Vacuolar pH control
vpl	Vacuolar protein localization
vps	vacuolar protein sorting
vpt	Vacuolar protein targetting
whi	Small cell size
xrs	X-ray sensitive

yhr	ORF on chromosome 8 R
yil	ORF on chromosome 9 L
yir	ORF on chromosome 9 R
yjl	ORF on chromosome 10 L
yjr	ORF on chromosome 10 R
ykl	ORF on chromosome 11 L
ykr	Kinase homolog identified with mammalian cDNA probes
ykr	ORF on chromosome 11 R
yll	ORF on chromosome 12 L
ylr	ORF on chromosome 12 R
yml	ORF on chromosome 13 L
ymr	ORF on chromosome 13 R
ynl	ORF on crhomosome 14 L
ynr	ORF on chromosome 14 R
yol	ORF on chromosome 15 L
yor	ORF on chromosome 15 R
ypk	Yeast protein kinase
ypl	ORF on chromosome 16 L
ypr	ORF on chromosome 16 R
ypt	GTP-binding protein
yro	
zfh	Zinc finger motif protein
zwf	Glucose-6-phosphate dehydrogenase

*twenty four gene symbols have two definitions: aar, amy, cal, cbp, cut, dit, gap, gcn, hpr, mtf, nsp, pgm, rap,rho, rpa, scd, sga, sit, srb, srm, ssb, ssc, stp, tel.

gap and tdh have same meaning
sga , scd, and tfs have same meaning

The gene symbols presented in this glossary were derived from the following sources:
  Genetic Map of *Saccharomyces cerevisiae*, Edition 10, Yeast **5**, 321- 401 (1989)
  XIVth International Conference on Yeast Genetics and Molecular Biology, Helsinki, 1988
  Genetics Society of America and Yeast Genetics meeting, Atlanta, 1989
  Cold Spring Harbor meeting on Yeast Cell Biology, 1989
  XVth International Conference on Yeast Genetics and Molecular Biology, The Hague, 1990
  Published literature, 1988 to present
  Personal communications

We make no claim that this list is either complete or accurate. Many investigators assign a gene symbol without giving a specifc definition, such as *clc*-clathrin light_chain. So, in these cases, we try to deduce the meaning of the symbol from the article, probably often inaccurately. We have not made an exhaustive literature search for gene symbols since our concern, until recently, was limited to genes that were on the genetic map. Nevetheless, we believe this list of symbols will serve a useful purpose in limiting the number of times a symbol has been used more than once or in which more than one symbol has been used to define a unique mutant phenotype. We invite comments, corrections and additions to this list.

Part C: Table II
List of Mapped Genes

Table II. List of mapped genes

Gene / Synonym	Map Position	References
aaal	4L	
aar1	4R	
raa1		McCusker and Haber (1990)
aar2	6R	
raa2		McCusker and Haber (1990)
aas1	4R	
gcn2		
aas2	11R	
gcn3		
aas3	5L	
arg9		
gcn4		
ABP1	3R	
abf1	**5**	Rhode *et al.* (1989)
abp2	14R	
ace1	7L	
aco1	**1 2**	Gangloff and Lauquin (1989)
act1	6L	
ACT2	4	
ade1	1R	

Gene / Synonym	Map Position	References
amm1	**3 R**	Thrash-Bingham and Fangman (1989)
tup1		
cyc9		
flk1		
umr7		
ams1	7L	Yoshihisa and Anraku (1989)
AMY1	7L	
pdr1		
AMY2	2L	
ani1	15R	
anp1	5L	
ant1	7L	
pdr1		
ape2	11L	
lap1		
ard1	8R	
arf1	4L	
arf2	4L	
arg1	15L	
arg3	10L	

Gene	Location	Reference
ade2	15R	
ade3	7R	
ade4	13R	
ade5,7	7L	
ade6	7R	
ade8	4R	
ade9	15R	
ade12	14L	D. Schild, p.c.
ADE15	7R	
adh1	15L	
adh2	13R	
adh3	13R	
adh4	7L	Walton *et al.* (1986)
adk1	4R	Konrad (1988)
adr1	4R	
adr6	16L	
agal	14R	J. Kurjan, p.c.
agαl	10R	
sagl		Lipke *et al.* (1989); Doi *et al.* (1989)
agsl	3L	
algl	2R	
alg7	7R	
amc3	4	
chl8		Larionov *et al.* (1989)
amc4	7	
chl9		Larionov *et al.* (1989)

Gene	Location	Reference
arg4	8R	
arg5,6	5R	
arg8	15L	
arg9	5L	
aas3		
gcn4		
arg80	13R	
arg81	13L	
arg82	4R	
arg84	5R	
arol	4R	
aro2	7L	
aro7	16R	
osm2		
apal	3	Plateau *et al.* (1989); Kaushal *et al.* (1990)
dtpl		
aspl	4R	
asp5	12R	
assl	12	
ATA1	11	
atel	7L	
ATP1	7	Remacha *et al.* (1990)
atrl		Gompel-Klein and Brendel (1990)
snql	13	
AXE1	7L	
barl	9L	
sstl		
basl	11R	

Table II. List of mapped genes - continued

Gene / Synonym	Map Position	References	Gene / Synonym	Map Position	References
bas2	4L		cdc21	15R	
pho2			tmp1		
bcy1	9L		cdc24	1L	
sra1			cls4		
bem1	**2R**	Bender and Pringle (1991)	ts11		
bem2	**5R**	Bender and Pringle (1991)	cdc25	12R	
BIK1	3L		**ctn1**		Bissinger et al. (1989)
bls2	11R		cdc25	2R	
BOR1	5R		cdc26	6R	
BOR2	7L		cdc27	2L	
pdr1			cdc28	2R	
bud3	**3L**	Chant and Herskowitz (1991)	**srm5**		Devin et al. (1990)
bud5	**3R**	Chant et al. (1991)	cdc29	9L	
cad1	4R		cdc31	15R	
CAD2	**2R**	Tohoyama et al. (1990)	cdc34	4R	
cag1	3R		cdc35	10R	
can1	5L		cyr1		
cap2	**9L**	Amatruda et al. (1990)	hsr1		
car2	12R		SRA4		

Gene	Location	Reference
cat1	4R	Celenza and Carlson (1989)
ccr1		
snf1		
ccb1	15R	
ccr1	4R	Celenza and Carlson (1989)
cat1		
snf1		
tex1		D. Gordinin, p.c.
cdc3	12R	
cdc4	6L	
cdc5	13L	
cdc6	10L	
cdc7	4L	
cdc8	10R	
cdc9	4L	
cdc10	3R	
cdc11	10R	
cdc12	8R	
cdc13	4L	
cdc14	6R	
cdc15	1R	
cdc16	11L	
cdc17	14L	
pol1		
cdc19	1L	
pyk1		

Gene	Location	Reference
tsm0185	4L	V. Korolev, p.c.
cdc36	4R	
cdc37	3R	
cdc39		
ros1	4R	
cdc40		
xrs2	12R	Healy *et al.* (1991)
cdc42	7L	
cdc43	4L	
cdc53		
cdc55	7L	
cdc60	16L	
cdc62	7R	
cdc63	15R	
prt1		
cdc64	15R	
cdc65	13R	
cdc66	15R	
myo2		R. Singer, p.c.
cdc67	16R	
cdc68	7L	
cdc77	4L	
ndc2		
cep1	10R	Baker and Masison (1990)
CHA1	3L	
CHC1	7L	
chl1	16L	
ctf1		Spencer *et al.* (1990)
chl2	12R	

Table II. List of mapped genes - continued

Gene / Synonym	Map Position	References
chl3	12R	
chl4	4R	
chl5	9L	
chl18	4	Larlonov et al. (1989)
amc3		
chl9	7	Larlonov et al. (1989)
amc4		
chl15	16R	
chol	5R	
cin1	15R	Stearns et al. (1990)
cin2	16L	Stearns et al. (1990)
cin4	13R	Stearns et al. (1990)
CKA1	9L	Padmanabha et al. (1990)
CKA2	15	Padmanabha et al. (1990)
cks1	2R	Hadwiger et al. (1989)
cln1	1L	Tokiwa et al. (1989)
whi1		
DAF1		
fun10		Diehl and Pringle (1991)
cls4	1L	
cdc24		
tsl1		
crl16	4R	
crl17	15R	
crl18	4L	
crl21	10R	
crl22	13R	
cry1	3R	
ctf1	16L	Spencer et al. (1990)
chl1		
ctf2	4	Spencer et al. (1988)
ctf12		
ctf13	13	Spencer et al. (1988)
ctf14	7	Spencer et al. (1988)
ctn1	12R	Bissinger et al. (1989)
cdc25		
ctn5	14L	Bissinger et al. (1989)
ras2		
CUP1	8R	
cup2	7L	
cup3	12R	
cup5	5L	
cup14	4R	
cyb2	13L	

Gene	Location		Gene	Location	
cls7	5L		cyc1	10R	
cly2	2L		cyc2	15R	
cly3	6R		cyc3	1L	
cly7	11L		cyc7	5L	
cly8	7R		cyp3		
cmd1	2R		cyc8	2R	
			ssn6		
cms1	**14L**		cyc9	3R	
cmt1	12R		flk1		
mar2			tup1		
sir3			umr7		
ste18			**amm1**		
cpal	15R				
cre2	**7R**		cyh1	2L	
spt6			cyh2	7L	
ssn20			cyh3	7L	
			pdr1		
crl1	6R		cyh4	15R	
crl3	7L		cyh10	2L	
crl4	5R		cyr1	10R	
crl7	15R		cdc35		
crl9	2R		hsr1		
crl10	7R		SRA4		
crl11	15L		tsm0185		
crl12	7R		cys1	1L	
crl13	15R		cys3	1L	
crl15	8R				

T. N. Davis and J. Thorner p.c.

J. Schero, p.c.

Denis and Malvar (1990)

Thrash-Bingham and Fangman (1991)

Table II. List of mapped genes - continued

Gene / Synonym	Map Position	References
dacl	8R	
gpal		
dac2	**2L**	**Fujimura (1990)**
dafl	*1L*	
clnl		
fun10		
whil		Tokiwa *et al.* (1989)
dall	9R	
dal2	9R	
dal3	9R	
dal4	9R	
dal5	10R	
urepl		
dal80	11R	Johnston *et al.* (1990)
dal81	9R	
DBF1	9	
dbf2	**7R**	J. Toyn and L. Johnston, p.c.
DBF4	4	
dbf20	**16R**	
dbll	11L	
DBP2	**13**	Iggo *et al.* (1991)

Gene / Synonym	Map Position	References
erg12	**13 R**	Oulmouden and Karst (1991);
rarl		Oulmouden and Karst (1990)
espl	7R	
essl	10R	Hanes *et al.* (1989)
eth2	5R	
sam2		
fasl	11L	
fas2	16L	
fcy2	5R	
fdpl	**2 R**	K. Luyten and S. Hohmann, p.c.; van de Poll and Schamhart (1977)
fenl	**3 R**	C. Marcireau, p.c.
flkl	3R	
cyc9		
tupl		
umr7		
amml		
FLO1	1R	
FLO5	**1**	F. Vezinhet and P. Barre, p.c.

Gene	Location	Reference
ded1	15R	
deg1	6R	Carbone *et al.* (1991)
dfr1	15R	
dna1	16R	Eberly *et al.* (1989)
dpr1	4L	Schafer *et al.* (1990)
ram1		
scg2		
ste16		
dsm1	7R	
dst1	7R	Clark *et al.* (1991)
dst2	7L	Dykstra *et al.* (1991)
dtp1	3	Kaushal *et al.* (1990);
apa1		Plateau *et al.* (1989)
dur1	2R	
dur2	2R	
dur3	8L	
dur4	8L	
DUR80	2R	
DUT1	2	
eam1	10L	
eam2	4	
eam6	10L	
edr1	12R	Rolfe *et al.* (1989)
top3		
erg6	13R	Gaber *et al.* (1989)
erg10	16L	
tsm0115		
fol1	14L	
fol2	7R	
fpr1	14L	Heitman *et al.* (1991)
fro1	7R	
fro2	7R	
fsr1	2R	
FUN9	1L	Diehl and Pringle (1991)
FUN10	1L	Diehl and Pringle (1991)
daf1		
cln1		Tokiwa *et al.* (1989)
whi1		
FUN11	1L	Diehl and Pringle (1991)
FUN12	1L	Diehl and Pringle (1991)
FUN19	1L	Diehl and Pringle (1991)
FUN20	1L	Diehl and Pringle (1991)
FUN21	1L	Diehl and Pringle (1991)
FUN22	1L	Diehl and Pringle (1991)
fun80	13R	Dubois *et al.* (1987)
fun81	13L	Dubois *et al.* (1987)
fur1	8R	
fur4	2R	
FUS1	3L	
fus2	13R	
fus3	2	Eliot and Fink (1989)
gac1	15R	K. Tatchell and J. M. Francois, p.c.
gal1	2R	

Table II. List of mapped genes - continued

Gene / Synonym	Map Position	References	Gene / Synonym	Map Position	References
gal2	12R		gst1	4R	
gal3	4R		suf12		
gal4	16L		sup2		
gal5	13R		sup35		
gal7	2R		sup36		
gal10	2R		GUT1	8	B. Ronnow and M. Kielland-Brandt, p.c.
gal11 spt13	15L	Fassler and Winston (1989)	GUT2	9	B. Ronnow and M. Kielland-Brandt, p.c.
gal22 sin4	14L	S. Chen, R. W. West, S. L. Johnson and H. Gans, p.c.	hap2	7L	
gal80	13L		hem1	linked to lys15	
GAL83	5R		hem10	7L	
gap1	11R	McCusker and Haber (1990)	hex2	4R	
gcd1 tra3	15R		reg1		
gcd2 gcd12	7R	Paddon and Hinnebusch (1989)	him1	4R	S.V. Kovaltzova, L.M. Gratcheva, R. A. Erstukhina and V.G. Korolev, p.c.
gcd4	3		hip1	7R	
gcd5	3		his1	5R	
			his2	6R	

Gene	Location	Reference
gcd11	5R	
gcd12	7R	
gcd2		Paddon and Hinnebusch (1989)
gcn2	4R	
aas1		
gcn3	11R	
aas2		
gcn4	5L	
aas3		
arg9		
glc1	2R	
ppd1		
ira1		Tanaka et al. (1989)
glc3	5L	
glc4	15L	
ira2		Tanaka et al. (1990)
glc6	2R	
glk1	3L	
gln1	16R	
gln3	5R	
gln4	15R	
gpa1	8R	
dac1		
gpa2	5R	
GPH1	13 or 16	Hwang et al. (1989)
grr1	10R	J. Flick and M Johnston, p.c.
his3	15R	
his8		
his4	3L	
his5	9L	
his6	9L	
his7	2R	
hit1	10	Kawakami et al. (1989)
hit2	3R	Kawakami et al. (1989)
pet18		
hmg1	13L	
hmg2	12R	
HML	3L	
HMR	3R	
HO	4L	
ho11	14R	Gaber et al. (1990)
hom2	4R	
hom3	5R	
hom6	10R	
hop1	9L	
hpr1	4R	Aguilera and Klein (1990)
hpr5	10L	Rong et al. (1991); H. Klein, p.c.
radH		
srs2		
HSC82	13	C. C. Adams and D. S. Gross, p.c.
HSP90	16L	
hsp104	7	Sanchez and Lindquist (1989)

Table II. List of mapped genes - continued

Gene / Synonym	Map Position	References	Gene / Synonym	Map Position	References
hsr1	10R		let3	10L	
cdc35			let5	10L	
cyr1			let6	6L	
SRA4			leu1	7L	
tsm0185			leu2	3L	
HTA2,B2	2R		leu3	12R	
HTA1,B1	4R		leu4	14L	
hts1	16R		leu5	8C	
tsm4572			lgn4	5L	
hxk1	6R		sit4		
hxk2	7L		los1	11	
HYG4	4R		lte1	1L	
ils1	2L		lts1	7L	
ilv1	5R		lts3	7L	
ilv2	13R		lts4	4R	
SMR1			lts10	4R	
ilv3	10R		lys1	9R	
ilv5	12R		lys2	2R	
ilp1	16L		lys4	4R	
ime4	7L	J. Shah and M. Clancy, p.c.	lys5	7L	
ino1	10L		lys7	13R	

Gene	Location	Reference
ira1	**2 R**	**Tanaka** *et al.* **(1989)**
ppd1		
glc1		
ira2	**15 L**	**Tanaka** *et al.* **(1990)**
glc4		
kar1	14L	
kar3	**16 R**	**P. Meluh and M. Rose, p.c.**
kem1	7L	**Kim** *et al.* **(1990)**
kem3	7L	**Kim** *et al.* **(1990)**
kex1	7L	
kex2	14L	
kgd1	**9 L**	**Ruttkay-Nedecky and Subik (1990)**
ogd1		
KIN3	6	
kin28	4	
kla1	**11**	**Ko (1989)**
KRB1	F12	
kre1	14L	
kre2	**15 R**	**H. Bussey, p.c.**
kss1	7R	
ktr1	**15 R**	**H. Bussey, p.c.**
lap1	11L	
ape2		
lap3	14L	
lap4	11L	
let1	1R	
let1M	13R	

Gene	Location	Reference
lys9	14R	
lys11	9L	
lys13	14R	
lys9		
lys14	4R	
lys15	linked to *hem1*	
mak1	15R	
top1		
mak3	16R	
mak4	2R	
mak5	2R	
mak6	16R	
mak7	8L	
mak8	15R	
tcm1		
mak9	11L	
mak10	5L	
mak11	11L	
mak12	12L	
mak13	9R	
mak14	3R	
mak15	11R	
mak16	1L	
mak17	10L	
mak18	8R	
mak19	8R	

Table II. List of mapped genes - continued

Gene / Synonym	Map Position	References	Gene / Synonym	Map Position	References
mak20	8		mer2	1 0	Engebrecht and Roeder (1989)
mak21	4R		mes1	7R	
mak22	12		met1	11R	
mak24	7L		met2	14L	
mak26	14L		met3	10R	
mak27	13R		met4	14L	
MAL1	7R		met5	10R	
MAL11			met6	5R	
MAL12			met7	15R	
MAL13			met8	2R	
MAL2	3R		met10	6R	
MAL21			met13	7L	
MAL22			met14	11R	
MAL23			met20	11R	
MAL3	2R		mfα1	16L	
MAL31			mfα2	7L	
MAL32			MGL2	2R	
MAL33			mgm1	15 R	B. Jones and W. Fangman, p.c.; James et al. (1990)
MAL34			mgt1	11L	Zweifel and Fangman (1991)
MAL4	11R				
MAL41					

Gene	Chromosome	Reference
MAL42		
MAL43		
MAL6	8	
MAL61		
MAL62		
MAL63		
MAL64		
mar1	4L	
sir2		
mar2	12R	
cmt1		
ste8		
sir3		
MAT	3R	Menees and Roeder (1989)
mcm1	13R	B. Rockmill, p.c.
mcm2	2L	
mcm3	5L	
mei4	5R	
mek1	15R	
MEL1	2L	
MEL2	7	Naumov et al. (1990)
MEL3	16	Naumov et al. (1990)
MEL4	11	Naumov et al. (1990)
MEL5	4	Naumov et al. (1990)
MEL6	13	Naumov et al. (1990)
MEL7	6	Naumov et al. (1990)
merl	14L	
mif1	14	
mif2	11L	M. Brown, p.c.
mig1	7L	Nehlin and Ronne (1990) and p.c.
min1	5L	
mkt1	14L	
mnn1	5C	
mnn2	2R	
mnn4	11L	
mos1	2R	
mos3	2R	
mrp1	4	Myers et al. (1987)
mrp2	16	Myers et al. (1987)
mrsl	9R	
MRS3	10	
msb1	15R	Bender and Pringle (1989)
msb2	7L	Bender and Pringle (1991)
msil	2R	Ruggieri et al. (1989)
mss2	4L	
MTP1	7R	
mut1	--	
mut2	--	
myo2 cdc66	15R	R. Singer, p.c.
nam1	4	Ben-Asher et al. (1989)
nam2	12R	
NAM7	13	

Table II. List of mapped genes - continued

Gene / Synonym	Map Position	References	Gene / Synonym	Map Position	References
NAM8	8		pdr5	15R	Leppert et al. (1990)
ndc1	13L		pdx2	13R	
ndc2	4L		pep1	2L	
cdc77			pep3	12R	
NHS1	5R		pep4	16L	
nib1	16L		pra1		
nmt1	12R	Duronio et al. (1991)	pho9		Kaneko et al. (1982)
NOV1	2		pep5	13R	
NRA2	7L		pep7	4R	
pdr1			pep12	15L	
nul3	4R		pep16	12R	
ste5			pet1	8R	
nup1	15R	Davis and Fink (1990)	pet2	14L	
obf1	4	Biswas et al. (1990)	pet3	8R	
ogd1	9L	Ruttkay-Nedecky and Subik (1990)	pet8	14R	
kgd1			pet9	2L	
ole1	7L		pet11	2R	
oli1	7L		pet14	4R	
pdr1			pet17	15R	
osm1	10R		pet18	3R	
osm2	16R		hit2		Kawakami et al. (1989)
aro7			pet54	7R	Costanzo et al. (1989)

Gene	Location	Reference
OXT1	4R	Ruml and Silhankova (1990)
pap1	11L	J. Lingner, J. Kellermann and W. Keller, p.c.
pbs1	15R	
pbs2	10L	
pda1	5	Y. de Steensma, p.c.
pdb1	14	Y. de Steensma, p.c.
pdc1	12L	Z. Lobo and P. Maitra, p.c.
pdc2	4R	Z. Lobo and P. Maitra, p.c.
pde2	15R	
sra5		
pdr1	7L	
AMY1		
ant1		
BOR2		
cyh3		
NRA2		
oli1		
smr2		
til1		
pdr2	15R	
pdr4	2R	R. A. Preston, J. D. Garman, L. B. Daniels and E. W. Jones, p.c. ; Balzi and Goffeau (1991)
pdr7		
pdr4	13R	Leppert *et al.* (1990)

Gene	Location	Reference
pet56	15R	
pet11	13R	
pet122	5R	T.W. McMullin and R.D. Fox, p.c.
pet123	15R	Haffter *et al.* (1990)
pet494	14R	Gaber *et al.* (1990)
pet-ts1402	5R	
pet-ts2858	13R	
petx	14L	
pfk1	13R	
pfk2	7R	
pjk1		
pfy1	15R	
pgi1	2R	
pgk1	3R	
pgm1	10	Rodicio and Heinisch (1987)
pha2	14L	
pho2	4L	
bas2		
pho3,5	2R	
pho4	6R	
pho8	4R	
pho9	16	Kaneko *et al.* (1982)
pep4		
pra1		

Table II. List of mapped genes - continued

Gene / Synonym	Map Position	References	Gene / Synonym	Map Position	References
phol1	1R	de Steensma et al. (1989)	pro2	15R	
pho12	8R	de Steensma et al. (1989)	pro3	5R	
pho80	15R		prp3	4R	J Maddock and J. Woolford, p.c.
tup7			ran3		
pho81	7		prp4	16	Bjorn et al. (1989)
PHO82	6R		ran4		
pho83	2		prp16	2	Burgess et al. (1990)
pho84	13L	Bun-Ya et al. (1991)	prt1	15R	
pho85	16L		cdc63		
phr1	15R		prt2	14L	
phr2	15R		prt3	---	
PHS1	4R		ptp1	15R	James et al. (1990)
PKA3	16L		pur5	4R	
TPK2			put1	12R	
PKC1	2	Levin et al. (1990)	put2	8R	
pma1	7R		put3	11L	Marczak and Brandriss (1989)
pma2	16L		pyk1	1L	
pmr1	7L	Rudolph et al. (1989); Moir et al. (1989)	cdc19		
pmr2	4R	Rudolph et al. (1989)	raa1	4R	McCusker and Haber (1990)
pms1	14L		aar1		
			raa2	6R	McCusker and Haber (1990)

Gene	Location	Synonym/Reference
pol1	14L	
pol2	14L	(cdc17) A. Morrison and A. Sugino, p.c.
pol30	2L	Bauer and Burgers (1990)
POT1	9	J. C. Perez-Ortin, p.c.
ppa2	13	M. Landin, p.c.
ppd1	2R	
ira1		(glc1) Tanaka et al. (1989)
pph1	4L	Arndt et al. (1989)
pph2	4L	(sit4) A. Sutton and K. Arndt, p.c.
pph3	4R	H. Ronne, p.c.
pph21	4L	Sneddon et al. (1990); H. Ronne, p.c.
pph22	4L	Sneddon et al. (1990); H. Ronne, p.c.
ppr1	12R	
pra1	16L	
pho9		(pep4) Kaneko et al. (1982)
prb1	5L	
prc1	13R	
PRI1	9	
pro1	4	

Gene	Location	Synonym
rad1	16L	(aar2)
rad2	7R	
rad3	5R	
rad4	5R	(rem1)
rad5	12R	
rad6	7L	(rev2)
rad7	10R	
rad9	4R	
rad10	13R	
rad16	2R	
rad17	15R	
rad18	3R	
rad23	5L	
rad24	5R	
rad50	14L	(r^s1)
rad51	5R	
rad52	13L	
rad53	16L	
rad54	7L	
rad55	4R	
rad56	16R	
rad57	4R	

Table II. List of mapped genes - continued

Gene / Synonym	Map Position	References	Gene / Synonym	Map Position	References
radH	10L	Rong *et al.* (1991);	*rna4*	16	Bjorn *et al.* (1989)
hpr5		H. Klein, p.c.	*rpr4*		
srs	5R		*rna5*	2R	
r^s_1			*rna6*	2R	
rad24			*tsm7269*		
ram1	4L	Schafer *et al.* (1990)	*rna11*	4L	
dpr11			*rna12*	13R	S. Liang and F. Lacroute, p.c.
ste16			*RNA14*	2	Minvielle-Sebastia *et al.* (1991)
scg2					
rap1	14L	de Steensma *et al.* (1990)	*RNA15*	16	Minvielle-Sebastia *et al.* (1991)
rar1	13R				
erg12		Oulmouden and Karst (1991)	*rnr2*	10C	
RAS1	15R		ROC1	12R	
RAS2	14L		*TEC1*		I. Laloux and E. Jacobs, p.c.
ctn5		Bissinger *et al.* (1989)	*ros1*	3R	
RBP1	14	Koltin *et al.* (1991)	*cdc39*		
RBK1	3	Thierry *et al.* (1990)	*ros3*	5L	
rebl	2	Ju *et al.* (1990)	*rpa190*	15R	McCusker *et al.* (1991)
RDN1	12R		*rpb1*	4L	
rec1	7L		RPB2	15R	Scafe *et al.* (1990)
rec3	7L		RPB3	9L	Kolodziej and Young (1989)

Gene	Position	Reference
rec102	12R	R. Malone and M. Cool, p.c.
reg1	4	
hex2		
rem1	5R	
rad3		
ret1	15R	James *et al.* (1990)
rev1	4	
rev2	12R	
rad5		
rev3	16L	
rev5	10R	
rev7	9L	
rgp1	4R	A. Aguilera and H. Klein, p.c.
rgr1	12R	Sakai *et al.* (1990)
rib1	2	J. L. Revuelta, M. A. Santos, J. J. Garcia-Ramirez, M.A. Plaza and M.J. Buitrago, p.c.
rib3	4	"
rib5	2R	"
rib7	2	
rme1	7R	
rna1	13R	
rna2	14R	
rna3	4R	
prp3		J. Maddock and J. Woolford, p.c.
rpb4	10	Woychik and Young (1989)
rpb5	2	Woychik and Young (1990)
rpb6	16	Woychik and Young (1990)
rpb7	12	Woychik and Young (1990)
rpb8	15	Woychik and Young (1990)
rpb9	7	Woychik and Young (1990)
rpb10	15	Woychik and Young (1990)
rpc19	2	Woychik and Young (1990)
		M. Dequardt-Chablat, M. Riva, C. Carles and A. Sentenac, p.c.
rpc31	14L	Mosrin *et al.* (1990)
rpc34	14R	S. Stettler, S. Mariotte and P. Thuriaux, p.c.
rpc40	16	
RPC40	16R	L. Johnston, p.c.
rpc53	4L	
rpc160	15R	
rpd1	15R	M. Vidal and R.F. Gaber, p.c.
rpd3	14L	Vidal *et al.* (1990)
rpo21		
RPK1	4	
RPK1	3R	Thierry *et al* (1990)
RPL3	15	
rpl16A	7R	M. Moritz and J. Woolford, p.c.
rpl16B	16R	M. Moritz and J. Woolford, p.c.

Table II. List of mapped genes - continued

Gene / Synonym	Map Position	References	Gene / Synonym	Map Position	References
RPL29	7	Remacha *et al.* (1990)	*ser1*	15R	
RPL44	1 5	Remacha *et al.* (1990)	*ser2*	7R	
RPL44'	4	Remacha *et al.* (1990)	*sesl*	4	
RPL45	4	Remacha *et al.* (1990)	*sin4*	14L	S. Chen, p.c.
RPL59	3		*SFA1*	4	
rpo21	4L		*SGA*	9	
rpb1			*sir1*	11R	
RPO41	6	Archambault *et al.* (1990)	*sir2*	4L	
RPO26	16	Lee *et al.* (1991)	*mar1*		
rpr1	5L		*sir3*	12R	
rp28,55,1	15L		*mar2*		
rp28,55,2	14L		*ste8*		
rrp1	4R		*cmt1*		
rsr1	7R	Bender and Pringle (1989)	*sir4*	4R	
rvs161	3R	Urdaci *et al.* (1990)	*ste9*		
spel61			*sis1*	14R	A. Sutton and K. Arndt, p.c.
rvs167	1 2	M. Crouzet, J. P. Bravet and M. Aigle, p.c.	*sis2*	11R	A. Sutton and K. Arndt, p.c.
sac1	11L	Novick *et al.* (1989)	*sit3*	16L	
sac2	4R	Novick *et al.* (1989)	*sit4*	4L	
sac3	4R	Novick *et al.* (1989)	*pph1*		Arndt *et al.* (1989)
			sit4	5L	

Gene	Map	Reference
sac6	4R	Adams and Botstein (1989)
sac7	4R	Dunn and Shortle (1990)
sag1	10R	Dol *et al.* (1989)
agaα		
sam1	1 2	N. Rosenberg and R. Rothstein, p.c.
sam2	4	N. Rosenberg and R. Rothstein, p.c.
san1	4	Schnell *et al.* (1989)
scg2	4L	Schafer *et al.* (1990)
dpr11		
ram1		
stel6		
SCL1	7L	
SDB21	8	V. Parker and L. H. Johnston, p.c.
sdc25	12L	Damak *et al.* (1991)
sec1	4R	
sec2	14L	
sec3	5R	
sec4	6L	
sec5	4R	
sec7	4R	
sec18	2R	
sec55	3R	
sec59	13R	
sen1	12R	

Gene	Map	Reference
lgn4		
ski1	7L	
ski4	14L	
ski8	7L	
smc1	6L	
SMR1	13R	
ilv2		
smr2	7L	
pdr1		
smr3	15R	
snf1	4R	
ccr1		
cat1		
snf2	15	Celenza and Carlson (1989)
swi3		
tye3		
snf3	4L	Marshall-Carlson *et al.* (1990)
SNF5	2	
snf10	3	
snq1	1 3	Gompel-Klein *et al.* (1989); Gompel-Klein and Brendel (1990)
atr1		
snr14	5	D. Frank and C. Guthrie, p.c.
sod1	8R	
soe1	7L	Su *et al.* (1990)
sot1	16L	

Table II. List of mapped genes - continued

Gene / Synonym	Map Position	References
spa1	5	
spa2	12L	
spd1	15L	
spe2	15L	
spe161	3R	
rvs161		Urdaci *et al.* (1990)
spe1	11L	Xie *et al.* (1990)
spe10	11L	Xie *et al.* (1990)
spo1	14L	
spo7	1L	
spo11	8L	
spo12	8R	
spo13	8R	
spo14	11R	
spo15	11L	
vps1	**8R**	Rothman *et al.* (1990)
spo16	7L	Malavasic and Elder (1990)
spo17	13C	
spoT1	7L	
spoT2	4L	

Gene / Synonym	Map Position	References
spt15	**5 R**	Eisenmann *et al.* (1989)
spt16	**7L**	C. D. Clark-Adams and F. Winston, p.c.; R. Singer, p.c.
cdc68		
spt21	**13 R**	J. D. Boeke and G. Natsoulis, p.c.
sra1	9L	
bcy1		
SRA3	10L	
tpk1		
SRA4	10R	
cyr1		
cdc35		
hsr1		
tsm0185		
sra5	15R	
pde2		
sra6	7R	
SRA7	10L	
srb2	**4 R**	Nonet and Young (1989)
srd1	3R	Fabian *et al.* (1990)
srk1	**4 R**	Wilson *et al.* (1991)

Gene	Location	Reference
spoT7	13R	
spoT8	2R	
spoT11	15L	
spoT15	15R	
spoT16	16L	
spoT20	16R	
spoT23	11R	
spp41	4R	J. Maddock and J. Woolford, p.c.
spr1	15	
SPR2	15	
spr6	5R	Kallal *et al.* (1990)
spt2	5R	
spt3	4R	
spt4	7R	
spt5	13L	M. S. Swanson and F. Winston, p.c.
spt6	7R	
ssn20		
cre2		Denis and Malvar (1990)
spt7	2R	
spt8	12R	
spt10	10L	J. D. Boeke and G. Natsoulis, p.c.
spt13	15L	Fassler and Winston (1989)
gal11		
spt14	16L	J. Fassler, p.c.
srm5	2R	Devin *et al.* (1990)
cdc28	4R	
srn1		
srp1	5	Marguet *et al.* (1988)
srp3	16	Yano and Nomura (1991)
srp5	15R	McCusker *et al.* (1991)
srs2	10L	Rong *et al.* (1991).
hpr5		
radH		H. Klein, p.c.
srv2	14L	M. Fedor-Chaiken, p.c.
ssb1	4L	Slater and Craig (1989); M. Hall, p.c.
ssb20	4	
ssb38	6	
ssc1	10R	Antebi *et al.* (1989)
ssc1	7	Moir *et al.* (1989)
pmr1		
ssd1	4R	A. Sutton and K. Arndt, p.c.
ssn2	4R	
ssn6	2R	
cyc8		
ssn20	7R	
spt6		
sst1	9L	Denis and Malvar (1990)
cre2		
bar1		
sst2	12R	

Table II. List of mapped genes - continued

Gene / Synonym	Map Position	References	Gene / Synonym	Map Position	References
STA1	4		sufl2	4R	
DEX2			gst1		
MAL5			sup2		
STA2	2		sup35		
DEX1			sup36		
STA3	14		sufl3	15R	
DEX3			sufl4	14L	
ste2	6L		SUF15	7R	
ste4	15R		SUF16	3R	
ste5	4R		SUF17	15L	
nul3			SUF18	6R	
ste6	11L	McGrath and Varshavsky (1989)	SUF19	5L	
ste7	4L		SUF20	6R	
ste8	12R		SUF21	16R	
cmt1			SUF22	13L	
mar2			SUF23	10R	
sir3			SUF24	4R	
ste9	4R		SUF25	4L	
sir4			suh2	12R	
ste11	12R		suil	14L	Castilho-Valavicius et al. (1990)

ste12	8R	
ste13	15R	
ste18	10R	
ste16	4L	Schafer et al. (1990)
dpr11		
ram1		
scg2		
ste50	3L	M. R. Rad, p.c.
stp22	3L	S. Spatrick, T. Kane and D. Jenness, p.c.
stp52	4L	
SUC1	7R	
SUC2	9L	
SUC3	2R	
SUC5	4L	
SUC7	8	
SUF1	15L	
SUF2	3R	
SUF3	4R	
SUF4	7R	
SUF5	15R	
SUF6	14L	
SUF7	13L	
SUF8	8R	
SUF9	6L	
SUF10	14L	
suf11	15R	

sui2	10R	Castilho-Valaviclus et al. (1990)
SUP-1A	---	
SUP2	4R	
sup2	4R	
gst1		
suf12		
sup35		
sup36		
SUP3	15L	
SUP4	10R	
SUP5	13L	
SUP6	6R	
SUP7	10L	
SUP8	13R	
SUP11	6R	
SUP15,16	16R	
SUP17	9L	
SUP19	5R	
SUP20		
SUP20	5R	
SUP19		
SUP22	9L	
SUP25	11R	
SUP26	12R	
SUP27	4R	
SUP28	14R	

Table II. List of mapped genes - continued

Gene / Synonym	Map Position	References	Gene / Synonym	Map Position	References
SUP29	10C		*SUP77*	7R	
SUP30			*SUP78*	13R	
SUP30	10C		*SUP79*	13L	
SUP29			*SUP80*	4R	
SUP33	11L		*SUP85*	5R	
sup35	4R		*SUP86*	12R	
gst1			*SUP87*	2R	
suf12			*SUP88*	4R	
sup2			*sup111*	8R	
sup36			*sup112*	7R	
sup36	4R		*sup113*	13R	
gst1			*SUP138*	7L	Ono *et al.* (1989)
suf12			*SUP139*	5R	Ono *et al.* (1989)
sup2			*SUP150*	5L	
sup35			*SUP154*	7L	
SUP37	12R		*SUP155*	7	
SUP38	7L	Wakem and Sherman (1990)	*SUP160*	15R	
SUP40	2		*SUP165*	14R	
SUP42	**4R**	Wakem and Sherman (1990)	*SUS1*	5L	
SUP43	**15L**	Wakem and Sherman (1990)	*swi1*	16L	
SUP44	**7L**	All-Robyn *et al.* (1990); Wakem and Sherman (1990)	*tal1*	12R	Schaff *et al.* (1990)
			tcm1	15R	

Gene	Location		Gene	Location	Reference
sup45	2R		mak8		
sup1			tcp1	4R	Ursic and Culbertson (1991)
supQ			tef1	16R	
sup47			TEF2	2R	
SUP46	2R		tell	2L	S. Kronmal and T. Petes, p.c.
sup47	2R		tfIID	5	Schmidt et al. (1989)
sup1			tfs1	12R	
sup45			thi1	13R	
supQ			thr1	8R	
SUP50	13R		thr4	3R	
SUP51	10C		tif1	11R	Muller et al. (1989)
SUP52			tif2	10L	Muller et al. (1989)
SUP52	10C		til1	7L	
SUP51			pdr1		
SUP53	3L		tmp1	15R	
SUP54	7L		cdc21		
SUP56	1R		top1	15R	
SUP57	6R		mak1		
SUP58	11L		top2	14L	
SUP61	3R		top3	12R	Rolfe et al. (1989)
SUP71	5R		edr1		
SUP72	2R		tpk1	10L	
SUP73	10L		SRA3		
SUP74	10L		TPK2	16L	
SUP75	11L		PKA3		
SUP76	7R				

Table II. List of mapped genes - continued

Gene / Synonym	Map Position	References	Gene / Synonym	Map Position	References
tra3	15R		*tup1*	3R	
gcd1			*cyc9*		
trk1	10R*t*		*flk1*		
trk2	**11R**	Ko *et al.* (1990)	*umr7*		
trn1	1R		***amm1***		Thrash-Bingham and Fangman (1989)
trp1	4R				
trp2	5R		*tup4*	15L	
trp3	11L		*tup7*	15R	
trp4	4R		*pho80*		
trp5	7L		*tyr1*	2R	
ts26	**15L**	Hasegawa *et al.* (1989)	*UBI1*	9	Monia *et al.* (1989)
tsl1	1L		*UBI2*	11	Monia *et al.* (1989)
cdc24			*UBI3*	12	Monia *et al.* (1989)
cls4			*UBI4*	12	Monia *et al.* (1989)
tsm1	3R		*umr7*	3R	
tsm5	3R		*cyc9*		
tsm0039	5R		*flk1*		
tsm0070	6L		*tup1*		
tsm0080	4R		***ung1***	**13L**	Percival *et al.* (1990)
tsm0111	13R		*ural*	11L	

Gene	Position	Reference
tsm0115	16L	
erg10	7L	
tsm0119	16R	
tsm0120	2R	
tsm134	9L	
tsm0139	8R	
tsm0151	10R	
tsm0185		
cdc35		
cyr1		
hrs1		
SRA4		
tsm0186	8R	
tsm0225	4L	
tsm437	7L	
tsm0800	13R	
tsm4572	16R	
hts1		
tsm5162	4R	
tsm7269	2R	
rna6		
tsm8740	15R	
tsv115	1L	Harris and Pringle (1991)
tub1	13L	
tub2	6L	
tub3	13L	

Gene	Position	Reference
ura2	10L	
ura3	5L	
ura4	12R	
ura6	11R	
ure2	14L	Coschigano and Magasanik (1991)
urep1	10R	
dal5		
van1	13	Kanik-Ennulat and Neff (1990)
vps1	11L	Rothman *et al.* (1990)
spol5		
vps35	10L	G. Paravicini and S. Emr, p.c.
vpt3	15R	Robinson *et al.* (1988)
vpt5	2R	Robinson *et al.* (1988)
whil	1L	
cln1		
fun10		
daf1		
whi2	15R	Tokiwa *et al.* (1989)
xrs2	4R	
yap3	12R	V. Korolev, p.c. M. Egel-Mitani and J. G Litske Petersen p.c.; Egel-Mitany *et al.* (1990).
yef3	12	Qin *et al.* (1989)
YG100	1L	

Table II. List of mapped genes - continued

Gene / Synonym	Map Position	References	Gene / Synonym	Map Position	References
YP2 YPT1	6L		383	3R	
YPT1 YP2	6L		zwf1	14L	

References

Adams, A.E.M. and D. Botstein. 1989. Dominant suppressors of yeast actin mutations that are reciprocally suppresssed. *Genetics* **121:** 675.

Aguilera, A. and H.L. Klein. 1990. *HPR1*, a novel yeast gene that prevents intrachromosomal excision recombination, shows carboxy-terminal homology to the *Saccharomyces cerevisiae TOP1* gene. *Mol. Cell. Biol.* **10:** 1439.

All-Robyn, J., D. Kelley-Geraghty, E. Griffin, N. Brown, and S. Liebman. 1990. Isolation of omnipotent suppressors in an [*eta*⁺] yeast strain. *Genetics* **124:** 505.

Amatruda, P., J.F. Cannon, K. Tatchell, C. Hug, and J.A. Cooper. 1990. Disruption of the yeast cytoskeleton in yeast capping protein mutants. *Nature* **344:** 352.

Antchi, A., H.K. Rudolph, and G.R. Fink. 1989. *PMR1*, a P-type ATPase related to Ca²⁺ pumps that affect secretion. *Yeast Genetics and Molecular Biology Meeting*, Atlanta, 1989. (Abstr. 63A.)

Archambault, J., K.T. Schappert, and J.D. Friesen. 1990. A suppressor of RNA polymerase II mutation of *Saccharomyces cerevisiae* encodes a subunit common to RNA polymerases I, II, and III. *Mol. Cell. Biol.* **10:** 6123.

Arndt, K.T., C.A. Styles, and G.R. Fink. 1989. A suppressor of a *HIS4* transcriptional defect encodes a protein with homology to the catalytic subunit of protein phosphatases. *Cell* **56:** 527.

Baker, R.E. and D.C. Masison. 1990. Isolation of the gene encoding the *Saccharomyces cerevisiae* centromere-binding protein *CEP1*. *Mol. Cell. Biol.* **10:** 2458.

Balzi, E. and A. Goffeau. 1991. Multiple or pleiotropic drug resistance in yeast. *Biochim. Biophys. Acta* **1073:** 241.

Bauer, G.A. and P.M.H. Burgers. 1990. Molecular cloning, structure and expression of the yeast proliferating cell nuclear antigen gene. *Nucleic Acids Res.* **18:** 261.

Ben-Asher, E., O. Groudinsky, G. Dujardin, N. Altamura, M. Kermorgant, and P.P. Slonimsky. 1989. Novel class of nuclear genes involved in both mRNA splicing and protein synthesis in *Saccharomyces cerevisiae*. *Mol. Gen. Genet.* **215:** 517.

Bender, A. and J. Pringle. 1989. Multicopy suppression of the *cdc24* budding defect in yeast by *CDC42* and three newly identified genes including the *ras*-related gene *RSR1*. *Proc. Natl. Acad. Sci.* **86:** 9976.

———. 1991. Use of a screen for synthetic lethal and multicopy suppressee mutants to identify two new genes involved in morphogenesis in *Saccharomyces cerevisiae. Mol. Cell. Biol.* **11:** 1295.

Bissinger, P.H., R. Wieser, B. Hamilton, and H. Ruis. 1989. Control of *Saccharomyces cerevisiae* catalase T gene (*CTT1*) expression by nutrient supply via the RAS-cyclic AMP pathway. *Mol. Cell. Biol.* **9:** 1309.

Biswas, E.E., M.J. Stefanec, and S.B. Biswas. 1990. Molecular cloning of a gene encoding an *ARS* binding factor from the yeast *Saccharomyces cerevisiae. Proc. Natl. Acad. Sci.* **87:** 6689.

Bjorn, S.P., A. Soltyk, J. Beggs, and J.D. Friesen. 1989. *PRP4* (*RNA4*) from *Saccharomyces cerevisiae:* Its gene product is associated with the U4/U6 small nuclear ribonucleoprotein particle. *Mol. Cell. Biol.* **9:** 3698.

Bun-Ya, M., M. Nishimura, S. Harashima, and Y. Oshima. 1991. The *PHO84* gene of *Saccharomyces cerevisiae* encodes an inorganic phosphate transporter. *Mol. Cell. Biol.* **11:** 3229.

Burgess, S., J. Couto, and C. Guthrie. 1990. A putative ATP binding protein influences the fidelity of break point recognition in yeast splicing. *Cell* **60:** 705.

Carbone, M.L.A., M. Solinas, S. Sora, and L. Panzeri. 1991. A gene tightly linked to *CEN6* is important for growth of *Saccharomyces cerevisiae. Curr. Genet.* **19:** 1.

Castilho-Valavicius, B., H. Yoon, and T. Donahue. 1990. Genetic characterization of the *Saccharomyces cerevisiae* translational initiation suppressors *sui1, sui2* and *SUI3* and their effects on *HIS4* expression. *Genetics* **124:** 483.

Celenza, J.L. and M. Carlson. 1989. Mutational analysis of the *Saccharomyces cerevisiae SNF1* protein kinase and evidence for functional interaction with the SNF4 protein. *Mol. Cell. Biol.* **9:** 5034.

Chant, J. and I. Herskowitz. 1991. Genetic control of bud site selection in yeast by a set of gene products that constitute a morphogenetic pathway. *Cell* **65:** 1203.

Chant, J., K. Corrado, J.R. Pringle, and I. Herskowitz. 1991. Yeast *BUD5*, encoding a putative GDP-GTP exchange factor, is necessary for bud site selection and interacts with bud formation gene *BEM1. Cell* **65:** 1213.

Clark, A.B., C.C. Dykstra, and A. Sugino. 1991. Isolation, DNA sequence and regulation of a *Saccharomyces cerevisiae* gene that encodes DNA strand transfer protein α. *Mol. Cell. Biol.* **11:** 2576.

Coschigano, P.W. and B. Magasanik. 1991. The *URE2* gene product of *Saccharomyces cerevisiae* plays an important role in the cellular response to the nitrogen source and has homology to glutathione S-transferases. *Mol. Cell. Biol.* **11:** 822.

Costanzo, M.C., E.C. Seaver, and T.D. Fox. 1989. The *PET54* gene of *Saccharomyces cerevisiae:* Characterization of a nuclear gene encoding a mitochondrial translation activator and subcellular localization of its product. *Genetics* **122:** 297.

Damak, F., E. Boy-Marcotte, D. Le-Rouscouet, R. Guilbaud, and M. Jacquet. 1991. *SCD25*, a *CDC25*-like gene which contains a RAS-activating domain and is a dispensable gene of *Saccharomyces cerevisiae. Mol. Cell. Biol.* **11:** 202.

Davis, L. and G. Fink. 1990. The *NUP1* gene encodes an essential component of the yeast nuclear pore complex. *Cell* **61:** 965.

Denis, C.L. and T. Malvar. 1990. The *CCR4* gene from *Saccharomyces cerevisiae* is required for both nonfermentative and spt-mediated gene expression. *Genetics* **124:** 283.

de Steensma, H.Y., P. de Jonge, A. Kaptein, and D.B. Kaback. 1989. Molecular cloning of chromosome I DNA from *Saccharomyces cerevisiae:* Localization of a repeated sequence containing an acid phosphatase gene near a telomere of chromosome I and chromosome VIII. *Curr. Genet.* **16:** 131.

de Steensma, H.Y., L. Holterman, I. Dekker, C.A. van Sluis, and T.J. Wenzel. 1990. Molecular cloning of the gene for the E1 α subunit of the pyruvate dehydrogenase complex from *Saccharomyces cerevisiae. Eur. J. Biochem.* **191:** 769.

Devin, A.B., T.Y. Prosvirova, V.T. Peshekonov, O.V. Chepurnaya, M.E. Smirnova, N.A. Koltovaya, E.N. Troitskaya, and I.P. Arman. 1990. The start gene *CDC28* and the genetic stability of yeast. *Yeast* **6:** 231.

Diehl, B.E. and J.R. Pringle. 1991. Molecular analysis of *Saccharomyces cerevisiae* chromosome I. Identification of additional transcribed regions and demonstration that some encode essential functions. *Genetics* **127:** 287.

Doi, S., K. Tanabe, M. Watanabe, M. Yamaguchi, and M. Yoshimura. 1989. An α-specific gene, *SAG1* is required for sexual agglutination in *Saccharomyces cerevisiae. Curr. Genet.* **15:** 393.

Dubois, E., J. Bercy, F. Descamps, and F. Messenguy. 1987. Characterization of two new genes essential for vegetative growth in *Saccharomyces cerevisiae*: Nucleotide sequence determination and chromosome mapping. *Gene* **55:** 265.

Dunn, T.M. and D. Shortle. 1990. Null alleles of SAC7 suppress temperature-sensitive actin mutations in *Saccharomyces cerevisiae. Mol. Cell. Biol.* **10:** 2308.

Duronio, R.J., D.A. Rudnick, R.L. Johnson, D.R. Johnson, and J.I. Gordon. 1991. Myristic acid auxotrophy caused by mutation of *S. cerevisiae* myristoyl-CoA:protein N-myristoyltransferase. *J. Cell Biol.* **113:** 1313.

Dykstra, C.C., K. Kitada, A.B. Clark, R.K. Hamatake, and A. Sugino. 1991. Cloning and characterization of *DST2*, the gene for DNA strand transfer protein b from *Saccharomyces cerevisiae. Mol. Cell. Biol.* **11:** 2583.

Eberly, S.L., A. Sakai, and A. Sugino. 1989. Mapping and characterizing a new DNA replication mutant in *Saccharomyces cerevisiae. Yeast* **5:** 117.

Egel-Mitani, M., H.-P. Flygenring, and M.T. Hansen. 1990. A novel yeast aspartyl protease allowing *KEX2*-independent MFα prophepheromone processing. *Yeast* **6:** 127.

Eisenmann, D.M., C. Dollard, and F. Winston. 1989. *SPT15*, the gene encoding the yeast TATA binding factor TFIID, is required for normal transcription initiation in vivo. *Cell* **58:** 1183.

Eliot, E.A. and G.R. Fink. 1989. Cell fusion mutants (Fus-) have defects in budding arrest and mating-specific structural components. *Yeast Genetics and Molecular Biology Meeting*, Atlanta, 1989. (Abstr. 12.)

Engebrecht, J. and S. Roeder. 1989. Genes that affect meiotic chromosome segregation. In *Abstracts from the Yeast Cell Biology Meeting* (ed. S. Emr et al.), p. 192. Cold Spring Harbor Laboratory, Cold Spring Harbor, New York.

Fabian, G., S.M. Hess, and A.K. Hopper. 1990. *srd1*, a *Saccharomyces cerevisiae* suppressor of the temperature-sensitive pre-rRNA processing defect in *rrp1. Genetics* **124:** 497.

Fassler, J.S. and F. Winston. 1989. The *Saccharomyces cerevisiae SPT13/GAL11* gene has both positive and negative regulatory roles in transcription. *Mol. Cell. Biol.* **9:** 5602.

Fujimura, H.-A. 1990. Identification and characterization of a mutation affecting the division arrest signaling of the pheromone response pathway in *Saccharomyces cerevisiae. Genetics* **124:** 275.

Gaber, R.F., M.C. Kielland-Brandt, and G.R. Fink. 1990. HOL1 mutations confer novel ion transport in *Saccharomyces cerevisiae. Mol. Cell. Biol.* **10:** 643.

Gaber, R.F., D.M. Copple, B.K. Kennedy, M. Vidal, and M. Bard. 1989. The yeast gene *ERG6* is required for normal membrane function but is not essential for biosynthesis of the cell-cycle-sparking sterol. *Mol. Cell. Biol.* **9:** 3447.

Gangloff, S.P. and G.J.-M. Lauquin. 1989. Aconitase: Isolation of the original *glu1-1* mutation, sequence comparison and putative existence of a second gene encoding the cytoplasmic isoform. In *Abstracts from the Yeast Cell Biology Meeting* (ed. S. Emr et al.), p. 169. Cold Spring Harbor Laboratory, Cold Spring Harbor, New York.

Gompel-Klein, P. and M. Brendel. 1990. Allelism of *SNQ1* and *ATR1*, genes of the yeast *Saccharomyces cerevisiae* required for controlling sensitivity to 4-nitroquinoline-N-oxide and aminotriazole. *Curr. Genet.* **18**: 93.

Gompel-Klein, P., M. Mack, and M. Brendel. 1989. Molecular characterization of the two genes *SNQ* and *SFA* that confer hyperresistance to 4-nitroquinoline-N-oxide and formaldehyde in *Saccharomyces cerevisiae*. *Curr. Genet.* **16**: 65.

Hadwiger, J.A., C. Wittenberg, M.D. Mendenhall, and S.I. Reed. 1989. The *Saccharomyces cerevisiae* *CKS1* gene, a homolog of the *Schizosaccharomyces pombe suc1⁺* gene, encodes a subunit of the Cdc28 protein kinase complex. *Mol. Cell. Biol.* **9**: 2034.

Haffter, P., T.W. McMullin, and T.D. Fox. 1990. A genetic link between an mRNA-specific translational activator and the translation system in yeast mitochondria. *Genetics* **125**: 495.

Hanes, S.D., P.R. Shank, and K.A. Bostian. 1989. Sequence and mutational analysis of *ESS1*, a gene essential for growth in *Saccharomyces cerevisiae*. *Yeast* **5**: 55.

Harris, S.D. and J.R. Pringle. 1991. Genetic analysis of *Saccharomyces cerevisiae* chromosome I. On the role of mutagen specificity in delimiting the set of genes identifiable using temperature-sensitive-lethal mutations. *Genetics* **127**: 279.

Hasegawa, H., A. Sakai, and A. Sugino. 1989. Isolation, DNA sequence and regulation of a new cell division cycle gene from the yeast *Saccharomyces cerevisiae*. *Yeast* **5**: 509.

Healy, A.M., S. Zolnierowicz, A.E. Stapleton, M. Goebl, A.A. Depaoli-Roach, and J.R. Pringle. 1991. *CDC55*, a *Saccharomyces cerevisiae* gene involved in cellular morphogenesis: Identification, characterization, and homology to the B subunit of mammalian type 2A protein phosphatase. *Mol. Cell. Biol.* (in press).

Heitman, J., N.R. Movva, P.C. Hiestand, and M.N. Hall. 1991. FK-506-binding protein proline rotamase is a target for the immunosuppressive agent FK506 in *Saccharomyces cerevisiae*. *Proc. Natl. Acad. Sci.* **88**: 1948.

Hwang, P.K., S. Tugendreich, and R.J. Fletterick. 1989. Molecular analysis of *GPH1*, the gene encoding glycogen phosphorylase in *Saccharomyces cerevisiae*. *Mol. Cell. Biol.* **9**: 1659.

Iggo, R.D., D.J. Jamieson, S.A. MacNeill, J. Southgate, J. McPheat, and D.P. Lane. 1991. p68 RNA helicase: Identification of a nucleolar form and cloning of related genes containing a conserved intron in yeasts. *Mol. Cell. Biol.* **11**: 1326.

James, P., S. Whelen, and B.D. Hall. 1990. The *RET1* gene of yeast encodes the second largest subunit of RNA polymerase III: Structural analysis of the wild-type and *ret1-1* mutant alleles. *J. Biol. Chem.* **266**: 5616.

Johnston, L.H., S.L. Eberly, J.W. Chapman, H. Araki, and A. Sugino. 1990. The product of the *Saccharomyces cerevisiae* cell cycle gene *DBF2* has homology with protein kinases and is periodically expressed in the cell cycle. *Mol. Cell. Biol.* **10**: 1359.

Ju, Q., B.E. Morrow, and J.R. Warner. 1990. REB1, a yeast DNA-binding protein with many targets, is essential for growth and bears some resemblance to the oncogene *myb*. *Mol. Cell. Biol.* **10**: 5226.

Kallal, L.A., M. Bhattacharyya, S.N. Grove, R.F. Iannacone, T.A. Pugh, D.A. Primerano, and M.J. Clancy. 1990. Functional analysis of the sporulation-specific *SPR6* gene of *Saccharomyces cerevisiae*. *Curr. Genet.* **18**: 293.

Kaneko, Y., A. Toh-e, and Y. Oshima. 1982. Identification of the genetic locus for the structural gene and a new regulatory gene for the synthesis of repressible alkaline phosphatase in *Saccharomyces cerevisiae*. *Mol. Cell. Biol.* **2**: 127.

Kanik-Ennulat, C. and N. Neff. 1990. Vanadate-resistant mutants of *Saccharomyces cerevisiae* show alterations in protein phosphorylation and growth control. *Mol. Cell. Biol.* **10:** 898.

Kaushal, V., D.M. Avila, S.C. Hardies, and L.D. Barnes. 1990. Sequencing and enhanced expression of the gene encoding diadenosine 5',5'''-P^1,P^4-tetraphosphate (Ap_4A) phosphorylase in *Saccharomyces cerevisiae*. *Gene* **95:** 79.

Kawakami, K., B.K. Shafer, J.N. Strathern, and Y. Nakamura. 1989. Ty insertion mutation of *S. cerevisiae* which confers temperature-sensitive growth and abolishes the synthesis of Hsp74. *Yeast Genetics and Molecular Biology Meeting*, Atlanta, 1989. (Abstr. 71B.)

Kim, J., P.O. Ljungdahl, and G.R. Fink. 1990. *kem* mutations affect nuclear fusion in *Saccharomyces cerevisiae*. *Genetics* **126:** 799.

Ko, C.H. 1989. *KLA1* encodes the putative low-affinity K^+ transporter in *Saccharomyces cerevisiae*. *Yeast Genetics and Molecular Biology Meeting*, Atlanta, 1989. (Abstr. 7.)

Ko, C.H., A.M. Buckley, and R.F. Gaber. 1990. *TRK2* is required for low affinity K^+ transport in *Saccharomyces cerevisiae*. *Genetics* **125:** 305.

Kolodziej, P. and R.A. Young. 1989. RNA polymerase II subunit RPB3 is an essential component of the mRNA transcription apparatus. *Mol. Cell. Biol.* **9:** 5387.

Koltin, Y., L. Faucette, D.J. Bergsma, M.A. Levy, R. Cafferkey, P.L. Koser, R.K. Johnson, and G.P. Livi. 1991. Rapamycin sensitivity in *Saccharomyces cerevisiae* is mediated by a peptidyl-prolyl isomerase related to human FK-506-binding protein. *Mol. Cell. Biol.* **11:** 1718.

Konrad, M. 1988. Analysis and *in vivo* disruption of the gene coding for adenylate kinase (*ADK1*) in the yeast *Saccharomyces cerevisiae*. *J. Biol. Chem.* 263: 19468.

Larionov, V.L., N.Y. Kouprina, A.V. Strunnikov, and A.V. Vlasov. 1989. A direct selection procedure for isolating yeast mutants with impaired segregation of artificial minichromosomes. *Curr. Genet.* **15:** 17.

Lee, J.Y., C.E. Rohlman, L.A. Molony, and D.R. Engelke. 1991. Characterization of *RPR1*, an essential gene encoding the RNA component of *Saccharomyces cerevisiae* nuclear RNase P. *Mol. Cell. Biol.* **11:** 721.

Leppert, G., R. McDevitt, S.C. Falco, T.K. Van Dyk, M.B. Ficke, and J. Golin. 1990. Cloning by amplification of two loci conferring multiple drug resistance in *Saccharomyces*. *Genetics* **125:** 13.

Levin, D.E., F.O. Fields, R. Kunisawa, J.M. Bishop, and J. Thorner. 1990. A candidate protein kinase C gene, PKC1, is required for the *S. cerevisiae* cell cycle. *Cell* **62:** 213.

Lipke, P.N., D. Wojciechowicz, and J. Kurjan. 1989. AGα1 is the structural gene for the *Saccharomyces cerevisiae* α-agglutinin, a cell surface glycoprotein involved in cell-cell interactions during mating. *Mol. Cell. Biol.* **9:** 3155.

Malavasic, M.J. and R.T. Elder. 1990. Complementary transcripts from two genes necessary for normal meiosis in the yeast *Saccharomyces cerevisiae*. *Mol. Cell. Biol.* **10:** 2809.

Marguet, D., X.J. Guo, and G.J.-M. Lauquin. 1988. Yeast gene *SRP1* (serine-rich protein). Intragenic repeat structure and identification of a family of SRP1-related DNA sequences. *J. Mol. Biol.* **202:** 455.

Marczak, J.E. and M.C. Brandriss. 1989. Isolation of constitutive mutations affecting the proline utilization pathway in *Saccharomyces cerevisiae* and molecular analysis of the *PUT3* transcriptional activator. *Mol. Cell. Biol.* **9:** 4696.

Marshall-Carlson, M., J.L. Celenza, B.C. Laurent, and M. Carlson. 1990. Mutational analysis of the SNF3 glucose transporter of *Saccharomyces cerevisiae*. *Mol. Cell. Biol.* **10:** 1105.

McCusker, J.H. and J.E. Haber. 1990. Mutations in *Saccharomyces cerevisiae* which confer resistance to several amino acid analogs. *Mol. Cell. Biol.* **10:** 2941.

McCusker, J.H., M. Yamagishi, J.M. Kolb, and M. Nomura. 1991. Suppressor analysis of temperature-sensitive RNA polymerase I mutations in *Sac-*

charomyces cerevisiae: Suppression of mutations in a zinc-binding motif by transposed mutant genes. Mol. Cell. Biol. 11: 746.

McGrath, J.P. and A. Varshavsky. 1989. The STE6 gene of S. cerevisiae encodes a homolog of the mammalian multidrug resistance P-glycoprotein. In Abstracts from the Yeast Cell Biology Meeting (ed. S. Emr et al.), p. 202. Cold Spring Harbor Laboratory, Cold Spring Harbor, New York.

Menees, T.M. and G.S. Roeder. 1989. MEI4, a yeast gene required for meiotic recombination. Genetics 123: 675.

Minvielle-Sebastia, L., B. Winsor, N. Bonneaud, and F. Lacroute. 1991. Mutations in the yeast RNA14 and RNA15 genes result in an abnormal mRNA decay rate: Sequence analysis reveals an RNA-binding domain in the RNA15 protein. Mol. Cell. Biol. 11: 3075.

Moir, C.M., D.T. Buckley, T.E. Dorman, J. Le Vitre, L.S. Davidow, and J. Mao. 1989. Disruption of the yeast SSC1 gene affects glycosylation and secretion. Yeast Genetics and Molecular Biology Meeting, Atlanta, 1989. (Abstr. 10.)

Monia, B.P., K.M. Haskell, J.R. Ecker, D.J. Ecker, and S.T. Crooke. 1989. Chromosomal mapping of the ubiquitin gene family in Saccharomyces cerevisiae by pulsed field gel electrophoresis. Nucleic Acids Res. 17: 3611.

Mosrin, C., M. Riva, M. Beltrame, E. Cassar, A. Sentenac, and P. Thuriaux. 1990. The RPC31 gene of Saccharomyces cerevisiae encodes a subunit of RNA polymerase C (III) with an acidic tail. Mol. Cell. Biol. 10: 4737.

Muller, P.P., H. Traschel, and P. Linder. 1989. Genetic localization of the Saccharomyces cerevisiae genes tif1 and tif2. Curr. Genet. 16: 127.

Myers, A.M., M.D. Crivellone, and A. Tzagoloff. 1987. Assembly of the mitochondrial membrane system MRP1 and MRP2, two nuclear genes coding for mitochondrial ribosomal proteins. J. Biol. Chem. 262: 3388.

Naumov, G., H. Turakainen, E. Naumova, S. Aho, and M. Korhola. 1990. A new family of polymorphic genes in Saccharomyces cerevisiae: α-galactosidase genes MEL1 - MEL7. Mol. Gen. Genet. 224: 119.

Nehlin, J.O. and H. Ronne. 1990. Yeast MIG1 repressor is related to the mammalian early growth response and Wilm's tumour finger proteins. EMBO J. 9: 2891.

Nonet, M.L. and R.A. Young. 1989. Intragenic and extragenic suppressors of mutations in the heptapeptide repeat domain of Saccharomyces cerevisiae RNA polymerase II. Genetics 123: 715.

Novick, P., B. Osmond, and D. Botstein. 1989. Suppressors of yeast actin mutants. Genetics 121: 659.

Ono, B., M. Tanaka, I. Awano, F. Okamoto, R. Satoh, N. Yamagishi, and Y. Ishino-Arao. 1989. Two new loci that give rise to dominant omnipotent suppressors in Saccharomyces cerevisiae. Curr. Genet. 16: 323.

Oulmouden, A. and F. Karst. 1990. Isolation of the ERG12 gene of Saccharomyces cerevisiae encoding mevalonate kinase. Gene 88: 253.

———. 1991. Nucleotide sequence of the ERG12 gene of Saccharomyces cerevisiae encoding mevalonate kinase. Curr. Genet. 19: 9.

Paddon, C.J. and A.G. Hinnebusch. 1989. gcd12 mutations are gcn3-dependent alleles of GCD2, a negative regulator of GCN4 in the general amino acid control of Saccharomyces cerevisiae. Genetics 122: 543.

Padmanabha, R., L.P. Chen-Wu, D.E. Hanna, and C.V.C. Glover. 1990. Isolation, sequencing, and disruption of the yeast CKA2 gene: Casein kinase II is essential for viability in Saccharomyces cerevisiae. Mol. Cell. Biol. 10: 4089.

Percival, K.J., M.B. Klein, and P.M. Burger. 1990. Molecular cloning and primary structure of the uracil-DNA-glycosylase gene from Saccharomyces cerevisiae. J. Biol. Chem. 264: 2593.

Plateau, P., M. Fromant, J-M. Schmitter, J-M. Buhler, and S. Blanquet. 1989. Islolation, characterization, and inactivation of the APA1 gene encoding yeast diadenosine 5',5'''-p1,p4-tetraphosphate phosphorylase. J. Bacteriol. 171: 6437.

Qin, S.L., A.Q. Xie, M.C.M. Bonato, and C.S. McLaughlin. 1989. Nucleotide sequence of the gene encoding elongation factor 3 for protein biosynthesis from Saccharomyces cerevisiae. In Abstracts from the Yeast Cell Biology Meeting (ed. S. Emr et al.), p. 48. Cold Spring Harbor

Laboratory, Cold Spring Harbor, New York.

Remacha, M., L. Ramirez, I. Marin, and J.P. Ballesta. 1990. Chromosome location of a family of genes encoding different acidic ribosomal proteins in *Saccharomyces cerevisiae. Curr. Genet.* **17**: 535.

Rhode, P.R., K.S. Sweder, K.F. Oegema, and J.L. Campbell. 1989. The gene encoding the *ARS*-binding factor I is essential for the viability of yeast. *Genes Dev.* **3**: 1926.

Robinson, J.S., D.J. Klionsky, L.M. Banta, and S. Emr. 1988. Protein sorting in *Saccharomyces cerevisiae:* Isolation of mutants defective in the delivery and processing of multiple vacuolar hydrolases. *Mol. Cell. Biol.* **8**: 4936.

Rodicio, R. and J. Heinisch. 1987. Isolation of the yeast phosphoglyceromutase gene and construction of deletion mutants. *Mol. Gen. Genet.* **206**: 133.

Rolfe, M., J. Wallis, G. Chebret, G. Brodsky, and R. Rothstein. 1989. Mutation of a novel eukaryotic topoisomerase in *Saccharomyces cerevisiae* causes hyperrecombination. *Yeast Genetics and Molecular Biology Meeting*, Atlanta, 1989. (Abstr. 18A.)

Rong, L., F. Palladino, A. Aguilera, and H.L. Klein. 1991. The hyper-gene conversion *hpr5-1* mutation of *Saccharomyces cerevisiae* is an allele of the *SRS2/RADH* gene. *Genetics* **127**: 75.

Rothman, J.H., C.K. Raymond, T. Gilbert, P.J. O'Hara, and T.H. Stevens. 1990. A putative GTP binding protein homologous to interferon-inducible Mx proteins performs an essential function in yeast protein sorting. *Cell* **61**: 1063.

Rudolph, H., A. Antebi, G.R. Fink, C.M. Buckley, T.E. Dorman, J. Le Vitre, L.S. Davidow, L. Mao, and D.T. Moir. 1989. The yeast secretory pathway is perturbed by mutations in *pmr1*, a member of Ca^{2+} ATPase family. *Cell* **58**: 133.

Ruggieri, R., K. Tanaka, M. Nakafuku, Y. Kaziro, Y. Toh-e, and K. Matsumoto. 1989. *MSI1*, a negative regulator of the RAS-cAMP pathway in *Saccharomyces cerevisiae. Proc. Natl. Acad. Sci.* **86**: 8778.

Ruml, T. and L. Silhankova. 1990. Dominant resistance to oxythiamin in *Saccharomyces cerevisiae* and its mapping. *Folia Microbiol.* **35**: 168.

Ruttkay-Nedecky, B. and J. Subik. 1990. The *OGD1* gene, affecting 2-oxoglutarate dehydrogenase in *S. cerevisiae*, is closely linked to *HIS5* on chromosome IX. *Curr. Genet.* **17**: 85.

Sakai, A., Y. Shimizu, S. Kondou, T. Chibazakura, and F. Hishinuma. 1990. Structure and molecular analysis of *RGR1*, a gene required for glucose repression of *Saccharomyces cerevisiae. Mol. Cell. Biol.* **10**: 4130.

Sanchez, Y. and S. Lindquist. 1989. Cloning and characterization of HSP104 gene of *Saccharomyces cerevisiae*. In *Abstracts from the Yeast Cell Biology Meeting* (ed. S. Emr et al.), p. 28. Cold Spring Harbor Laboratory, Cold Spring Harbor, New York.

Scafe, C., M. Nonet, and R.A. Young. 1990. RNA polymerase II mutants defective in transcription of a subset of genes. *Mol. Cell. Biol.* **10**: 1010.

Schaff, I., S. Hohmann, and F.K. Zimmermann. 1990. Molecular analysis of the structural gene for yeast transaldolase. *Eur. J. Biochem.* **188**: 597.

Schafer, W.R., C.E. Trueblood, C.-C. Yand, M.P. Mayer, S. Rosenberg, C.D. Poulter, S.-H. Kim, and J. Rine. 1990. Enzymatic coupling of cholesterol intermediates to a mating pheromone precursor and to the Ras protein. *Science* **249**: 1133.

Schmidt, M.C., C.C. Kao, R. Pei, and A.J. Berk. 1989. Yeast TATA-box transcription factor gene. *Proc. Natl. Acad. Sci.* **86**: 7785.

Schnell, R., L. D'Ari, M. Fass, D. Goodman, and J. Rine. 1989. Genetic and molecular characterization of suppressors of *SIR4* mutations in *Saccharomyces cerevisiae. Genetics* **122**: 29.

Slater, M.R. and E.A. Craig. 1989. The *SSB1* heat shock cognate gene of the yeast *Saccharomyces cerevisiae. Nucleic Acids Res.* **17**: 4891.

Sneddon, A. A., P.T.W. Cohen, and M.J.R. Stark. 1990. *Saccharomyces cerevisiae* protein phosphatase 2A performs an essential cellular function and is encoded by two genes. *EMBO J.* **9**: 4339.

Spencer, F., C. Connelly, S. Lee, and P. Hieter. 1988. Isolation and cloning of conditionally lethal chromosome transmission fidelity genes in *Sac-*

charomyces cerevisiae. *Cancer Cells* **6:** 441.

Spencer, F., S.L. Gerring, C. Connelly, and P. Hieter. 1990. Mitotic chromosome transmission fidelity mutants in *Saccharomyces cerevisiae. Genetics* **124:** 237.

Stearns, T., M.A. Hoyt, and D. Botstein. 1990. Yeast mutants sensitive to antimicrotubule drugs define three genes that affect microtubule function. *Genetics* **124:** 251.

Su, J.-Y., L. Belmont, and R.A. Sclafani. 1990. Genetic and molecular analysis of the *SOE* gene: A tRNA^{GLU} missense suppressor of yeast *cdc8* mutations. *Genetics* **124:** 523.

Tanaka, K., K. Matsumoto, and A. Toh-e. 1989. *IRA1*, an inhibitory regulator of the RAS-cyclic AMP pathway in *Saccharomyces cerevisiae. Mol. Cell. Biol.* **9:** 757.

Tanaka, K., M. Nakafuku, F. Tamanoi, Y. Kaziro, K. Matsumoto, and A. Toh-e. 1990. *IRA2*, a second gene of *Saccharomyces cerevisiae* that encodes a protein with a domain homologous to mammalian *ras* GTPase-activating protein. *Mol. Cell. Biol.* **10:** 4303.

Thierry, A., C. Fairhead, and B. Dujon. 1990. The complete sequence of the 8,2 kb segment left of MAT on chromosome III reveals five ORFs, including a gene for a ribokinase. *Yeast* **6:** 521.

Thrash-Bingham, C. and W. Fangman. 1989. A yeast mutation that stabilizes a plasmid bearing a mutated *ARS1* element. *Mol. Cell. Biol.* **9:** 809.

Tohoyama, H., M. Inouye, M. Joho, and T. Murayama. 1990. Resistance to cadmium is under the control of the *CAD2* gene in the yeast *Saccharomyces cerevisiae. Curr. Genet.* **18:** 181.

Tokiwa, G., R. Nash, K. Erickson, S. Anand, M. Tyers, and B. Futcher. 1989. The *whi1* gene tethers cell division to cell size, and is a cyclin homolog. *Yeast Genetics and Molecular Biology Meeting*, Atlanta, 1989. (Abstr. 23.)

Urdaci, M., L. Dulau, M. Aigle, and M. Crouzet. 1990. Sequence of the yeast gene *RVS161* located on chromosome III. *Yeast* **6:** 173.

Ursic, D. and M.R. Culbertson. 1991. The yeast homolog to mouse *tcp-1* affects microtubule-mediated processes. *Mol. Cell. Biol.* **11:** 2629.

van de Poll, K.W. and D.H.J. Schamhart. 1977. Characterization of a regulatory mutant of fructose 1,6-bisphosphatase in *Saccharomyces carlsbergensis. Mol Gen. Genet.* **154:** 61.

Vidal, M., A.M. Buckley, F. Hilger, and R.F. Gaber. 1990. Direct selection for mutants with increased K$^+$ transport in *Saccharomyces cerevisiae. Genetics* **125:** 313.

Wakem, L.P. and F. Sherman. 1990. Isolation and characterization of omnipotent suppressors in the yeast *Saccharomyces cerevisiae. Genetics* **124:** 515.

Walton, J.D., C.E. Paquin, K. Kaneko, and V.M. Williamson. 1986. Resistance to antimycin A in yeast by amplification of *ADH4* on a linear, 42 kb palindromic plasmid. *Cell* **46:** 857.

Wilson, R.B., A.A. Brenner, T.B. White, M.J. Engler, J.P. Gaughran, and K. Tatchell. 1991. The *Saccharomyces cerevisiae SRK1* gene, a suppressor of *bcy1* and *ins1*, may be involved in protein phosphatase function. *Mol. Cell. Biol.* **11:** 3369.

Woychik, N.A. and R.A. Young. 1989. RNA polymerase II subunit RPB4 is essential for high- and low-temperature yeast cell growth. *Mol. Cell. Biol.* **9:** 2854.

―――. 1990. RNA polymerase II subunit structure and function. *Trends Biochem. Sci.* **15:** 347.

Xie, Q.-W., C.W. Tabor, and H. Tabor. 1990. Ornithine decarboxylase in *Saccharomyces ccerevisiae*: Chromosomal assignment and genetic mapping of the *SPE1* gene. *Yeast* **6:** 455.

Yano, R. and M. Nomura. 1991. Suppressor analysis of temperature-sensitive mutations of the largest subunit of RNA polymerase I in *Saccharomyces cerevisiae*: A suppressor gene encodes the second-largest subunit of RNA polymerase I. *Mol. Cell. Biol.* **11**: 754.

Yoshihisa, T. and Y. Anraku. 1989. Nucleotide sequence of *AMS1*, the structural gene of vacuolar α-mannosidase of *Saccharomyces cerevisiae*. *Biochem. Biophys. Res. Commun.* **163**: 908.

Zweifel, S.G. and W.L. Fangman. 1991. A nuclear mutation reversing a biased transmission of yeast mitochondrial DNA. *Genetics* **128**: 241.

Subject Index

Aberrant segregation. *See* Postmeiotic segregation
ABF1, 66–70, 607–608
ABF1. *See* Abf1p
Abf1p, 66–70, 607–608, 612
ABFII, 69
ABFIII, 69
Acetate, 568
Acidic phospholipids, 372
Active transport system, 527, 547
ade3, 684
Adenylate cyclase, 550
Aerobic growth, 341–342
α-factor, 284
α1 protein, 106
α2 protein, 106
Amber suppressor, 668. *See also* Suppressors, informational
Aminoacyl tRNA. *See* tRNA
Aminoacyl tRNA synthetase. *See* tRNA synthetase
δ-Aminolevulinic acid synthase, 359
amm1, 109
Anaerobic growth, 341–342, 344
ANB1, 364
ANP1, 27
ant mutants, 172–173
Antibiotic resistance/sensitivity, 599, 603
AP endonuclease, 160
APN1, 160
ARC region, 27
ARD1, 603
ARG1, 690

ARG3, 690
ARS1, 56–58, 62, 70
  Abf1p-binding site, 64, 67
ARS121, 62, 66
ARS301, 50–51
ARS302, 50–51
ARS303, 50–51
ARS304, 50, 52
ARS305, 48–52, 57
ARS306, 48–50, 52
ARS307, 48–52, 57, 63–64
ARS308, 50, 52
ARS309, 48–51, 57
ARS310, 57
ARS501, 57–58
ARS-binding factors. *See* Abf1p; ABFII; ABFIII
ARS(s) (autonomous replication sequences), 9–10. *See also* Replication origin
  ARS assays, 58–60
  ARS consensus sequence, 60–62
  ARS-specific genes, 68, 105–109
  in chromosome III, 48–52
  flanking sequences, 62–66
  identification of, 45–47
  interaction with nuclear scaffold, 70
  in rDNA, 47–48
  in 2-micron-like plasmids, 320
  in 2-micron plasmid, 300–301
Artificial chromosomes. *See* Yeast artificial chromosome
asu9, 684